Studies in Systems, Decision and Control

Volume 523

The series "Studies in Systems, Decision and Control" (SSDC) covers both new developments and advances, as well as the state of the art, in the various areas of broadly perceived systems, decision making and control–quickly, up to date and with a high quality. The intent is to cover the theory, applications, and perspectives on the state of the art and future developments relevant to systems, decision making, control, complex processes and related areas, as embedded in the fields of engineering, computer science, physics, economics, social and life sciences, as well as the paradigms and methodologies behind them. The series contains monographs, textbooks, lecture notes and edited volumes in systems, decision making and control spanning the areas of Cyber-Physical Systems, Autonomous Systems, Sensor Networks, Control Systems, Energy Systems, Automotive Systems, Biological Systems, Vehicular Networking and Connected Vehicles, Aerospace Systems, Automation, Manufacturing, Smart Grids, Nonlinear Systems, Power Systems, Robotics, Social Systems, Economic Systems and other. Of particular value to both the contributors and the readership are the short publication timeframe and the world-wide distribution and exposure which enable both a wide and rapid dissemination of research output.

Indexed by SCOPUS, DBLP, WTI Frankfurt eG, zbMATH, SCImago.

All books published in the series are submitted for consideration in Web of Science.

Safoura Rezaei Aderyani · Reza Saadati ·
Chenkuan Li · Tofigh Allahviranloo

Towards Ulam Type Multi Stability Analysis

A Novel Approach for Fuzzy Dynamical Systems

Springer

Safoura Rezaei Aderyani
Department of Mathematics
Iran University of Science and Technology
Tehran, Iran

Reza Saadati
Department of Mathematics
Iran University of Science and Technology
Tehran, Iran

Chenkuan Li
Department of Mathematics and Computer Science
Brandon University
Brandon, MB, Canada

Tofigh Allahviranloo
Research Center of Performance and Productivity Analysis
Istinye University
Istanbul, Türkiye

ISSN 2198-4182 ISSN 2198-4190 (electronic)
Studies in Systems, Decision and Control
ISBN 978-3-031-55563-3 ISBN 978-3-031-55564-0 (eBook)
https://doi.org/10.1007/978-3-031-55564-0

This Springer imprint is published by the registered company Springer Nature Switzerland AG
The registered company address is: Gewerbestrasse 11, 6330 Cham, Switzerland

Paper in this product is recyclable.

To our family

Preface

The main target of this monograph is to present a new concept of Ulam type stability, i.e., Multi Stability, through the classical, well-known special functions and to obtain the best approximation error estimates by a different concept of perturbation stability including the fuzzy approach for uncertainty considerations. This stability allows us to obtain diverse approximations depending on various special functions that are initially chosen and to evaluate maximal stability and minimal error which enable us to obtain a unique optimal solution of functional equations, inequalities, and fractional equations. Stability analysis in the sense of the Ulam and its different kinds has received considerable attention from the researchers. However, how to effectively generalize the Ulam stability problems and to evaluate optimized controllability and stability are new issues. The multi stability not only covers the previous concepts but also considers the optimization of the problem and provides a comprehensive discussion of optimizing the different types of the Ulam stabilities of mathematical models used in the natural sciences and engineering disciplines with the fuzzy attitude.

Besides, this book also deals with nonlinear differential equations with various boundary conditions or initial value problems, based on the matrix Mittag–Leffler function, fixed point theory, as well as Babenko's approach to study uniqueness and existence of solutions.

In general, the benefits for the readers can be concluded as follows:

1. Evaluates maximal stability with minimal error to get a unique optimal solution.
2. Discusses an optimal method of the alternative to study existence, uniqueness, and different types of Ulam stabilities under special consideration of the fuzzy approach.

3. Delves into the new study of boundary value problems of fractional integro-differential equations with integral boundary conditions and variable coefficients.

Tehran, Iran — Safoura Rezaei Aderyani
Tehran, Iran — Reza Saadati
Brandon, Canada — Chenkuan Li
Istanbul, Türkiye — Tofigh Allahviranloo
October 2023

Acknowledgements

We would like to express our thanks to the referees for reading the manuscript and providing valuable suggestions and comments which have helped to improve the presentation of the book.

Last but not least, it is our pleasure to acknowledge the superb assistance provided by the staff of Springer for the publication of the book.

Contents

Acronyms

$\mathbb{R}$	The real numbers
$\mathbb{C}$	The complex numbers
$\mathbb{N}$	Natural numbers
$\mathbb{Z}$	Integrals
$\mathbb{R}_+$	Positive real numbers
$\mathbb{R}_-$	Negative real numbers
$\mathbb{Z}^+$	Positive integer numbers
$\mathbb{Z}^-$	Negative integer numbers
$\mathbb{N}_0$	$\mathbb{N} \cup \{0\}$
$\mathbb{Z}_0$	$\mathbb{Z} \cup \{0\}$
$\Gamma(.)$	Gamma function
$\underbrace{\varphi_i}_{1\leq i\leq n}$	Special control function
$\underbrace{\mathrm{AG}_i}_{1\leq i\leq n}$	Aggregation maps
$\bigotimes_{\mathrm{TN}}$	Triangular norm
$\bigotimes_{\mathrm{GTN}}$	Generalized triangular norm
t-norm	Triangular norm
MVFN space	Matrix valued fuzzy normed space
MVFB space	Matrix valued fuzzy Banach space
MVFN-algebra	Matrix valued fuzzy normed algebra
MVFB-algebra	Matrix valued fuzzy Banach algebra
MVFC–$\diamond$–algebra	Matrix valued fuzzy C–$\diamond$–algebra
MVFB–$\diamond$–algebra	Matrix valued fuzzy Banach–$\diamond$–algebra

Chapter 1
Introduction

The study of functional equations has a long history. In 1791 and 1809, Legendre [1] and Gauss [2] attempted to provide a solution of the following functional equation:

$$f(x+y) = f(x) + f(y),$$

for all $x, y \in \mathbb{R}$, which is called the Cauchy functional equation. A function $f : \mathbb{R} \longrightarrow \mathbb{R}$ is called an additive function if it satisfies the Cauchy functional equation. In 1821, Cauchy [3] first found the general solution of the Cauchy functional equation, that is, if $f : \mathbb{R} \longrightarrow \mathbb{R}$ is a continuous additive function, then f is linear, that is, $f(x) = mx$, where m is a constant. Further, we can consider the biadditive function on $\mathbb{R}^2$ as follows:

A function $f : \mathbb{R}^2 \longrightarrow \mathbb{R}$ is called an biadditive function if it is additive in each variable, that is,

$$f(x+y, z) = f(x, z) + f(y, z),$$

and

$$f(x, y+z) = f(x, y) + f(x, z),$$

for all $x, y, z \in \mathbb{R}$. It is well known that every continuous biadditive function $f : \mathbb{R}^2 \longrightarrow \mathbb{R}$ is of the form

$$f(x, y) = mxy,$$

for all $x, y \in \mathbb{R}$, where m is a constant.

Since the time of Legendre and Gauss, several mathematicians had dealt with additive functional equations in their books [4–8] and a number of them have studied Lagrange's mean value theorem and related functional equations, Pompeiu's mean value theorem and associated functional equations, two-dimensional mean value theorem and functional equations as well as several kinds of functional equations.

In 1940, S.M. Ulam [9] proposed the following stability problem of functional equations:

S. Rezaei Aderyani et al., *Towards Ulam Type Multi Stability Analysis*, Studies in Systems, Decision and Control 523, https://doi.org/10.1007/978-3-031-55564-0_1

Given a group G_1, a metric group G_2 with the metric $d(.,.)$ and a positive number ε, does there exist $\delta > 0$ such that, if a mapping $f : G_1 \longrightarrow G_2$ satisfies

$$d\Big(f(xy), f(x)f(y)\Big) \leq \delta,$$

for all $x, y \in G_1$, then a homomorphism $h : G_1 \longrightarrow G_2$ exists with

$$d(f(x), h(x)) \leq \varepsilon,$$

for all $x \in G_1$?

Since then, several mathematicians have dealt with special cases as well as generalizations of Ulam's problem. In fact, in 1941, D.H. Hyers [10] provided a partial solution to Ulam's problem for the case of approximately additive mappings in which G_1 and G_2 are Banach spaces with $\delta = \varepsilon$ as follows:

Let X and Y be Banach spaces and let $\varepsilon > 0$. Then, for all $g : X \longrightarrow Y$ with

$$\sup_{x,y \in X} \Big\| g(x+y) - g(x) - g(y) \Big\| \leq \varepsilon,$$

there exists a unique mapping $f : X \longrightarrow Y$ such that

$$\sup_{x \in X} \|g(x) - f(x)\| \leq \varepsilon,$$
$$f(x+y) = f(x) + f(y),$$

for all $x, y \in X$.

This proof remains unchanged if G_1 is an Abelian semigroup. Particularly, in 1968, the following theorem was proved by Forti (Proposition 1, [11]):

Theorem 1.1 *Let $(S, +)$ be an arbitrary semigroup and E be a Banach space. Assume that $f : S \longrightarrow E$ satisfies*

$$\Big\| f(x, y) - f(x) - f(y) \Big\| \leq \varepsilon. \tag{1.1}$$

Then, the limit

$$g(x) = \lim_{n \longrightarrow \infty} \frac{f(2^n x)}{2^n}, \tag{1.2}$$

exists for all $x \in S$ and $g : S \longrightarrow E$ is the unique function satisfying

$$\|f(x) - g(x)\| \leq \varepsilon, \quad g(2x) = 2g(x).$$

Finally, if the semigroup S is Abelian, then G is additive.

Note that the proof method generating the solution g by the formula like (1.2) is called a direct method.

If f is a mapping of a group or a semigroup $(S, .)$ into a vector space E, then we call the following expression:

$$Cf(x, y) = f(x.y) - f(x) - f(y),$$

the Cauchy difference of f on $S \times S$. In the case that E is a topological vector space, we call the equation of homomorphism stable if, whenever the Cauchy difference Cf is bounded on $S \times S$, there exists a homomorphism $g : S \longrightarrow E$ such that $f - g$ is bounded on S.

In 1980, Rätz [12] generalized Theorem 1.1 as follows: Let $(X, *)$ be a power associative groupoid, that is, X is a nonempty set with a binary relation $x_1 * x_2 \in X$ such that the left powers satisfy $x^{m+n} = x^m * x^n$ for all $m, n \geq 1$ and $x \in X$. Let $(Y, |\,.\,|)$ be a topological vector space over the field $\mathbb{Q}$ of rational numbers with $\mathbb{Q}$ topologized by its usual absolute value $|\,.\,|$.

Theorem 1.2 *Let V be a nonempty bounded Q-convex subset of Y containing the origin and assume that Y is sequentially complete. Let $f : X \longrightarrow Y$ satisfy the following conditions: for all $x_1, x_2 \in X$, there exist $k \geq 2$ such that*

$$f\left((x_1 * x_2)^{k^n}\right) = f\left(x_1^{k^n} * x_2^{k^n}\right), \tag{1.3}$$

for all $n \geq 1$ and

$$f(x_1) + f(x_2) - f(x_1 * x_2) \in V. \tag{1.4}$$

Then there exists a function $g : X \to Y$ such that $g(x_1) * g(x_2)$ and $f(x) - g(x) \in \overline{V}$, where $\overline{V}$ is the sequential closure of V for all $x \in X$. When Y is a Hausdorff space, then g is uniquely determined.

Note that the condition (1.3) is satisfied when X is commutative and it takes the place of the commutativity in proving the additivity of g. However, as Rätz pointed out in his paper, the condition

$$(x_1 * x_2)^{k^n} = x_1^{k^n} * x_2^{k^n},$$

for all $x_1, x_2 \in X$, where X is a semigroup, and, for all $k \geq 1$, does not imply the commutativity.

In the proofs of Theorems 1.1 and 1.2, the completeness of the image space E and the sequential completeness of Y, respectively, were essential in proving the existence of the limit which defined the additive function g. The question arises whether the completeness is necessary for the existence of an odd additive function g such that $f - g$ is uniformly bounded, given that the Cauchy difference is bounded.

For this problem, in 1988, Schwaiger [13] proved the following:

Theorem 1.3 *Let E be a normed space with the property that, for each function* $f : \mathbb{Z} \to E$, *whose Cauchy difference* $Cf = f(x+y) - f(x) - f(y)$ *is bounded for all* $x, y \in \mathbb{Z}$ *and there exists an additive mapping* $g : \mathbb{Z} \to E$, *such that* $f(x) - g(x)$ *is bounded for all* $x \in \mathbb{Z}$. *Then E is complete.*

Corollary 1.1 *The statement of Theorem 1.3 remains true if* $\mathbb{Z}$ *is replaced by any vector space over* $\mathbb{Q}$.

In 1950, Aoki [14] generalized Hyers' theorem as follows:

Theorem 1.4 *Let* E_1 *and* E_2 *be two Banach spaces. If there exist* $K > 0$ *and* $0 \leq p < 1$ *such that*

$$\left\| f(x+y) - f(x) - f(y) \right\| \leq K\left(\|x\|^p + \|y\|^p \right),$$

for all $x, y \in E_1$, *then there exists a unique additive mapping* $g : E_1 \longrightarrow E_2$ *such that*

$$\|f(x) - g(x)\| \leq \frac{2K}{2 - 2^p} \|X\|^p,$$

for all $x \in E_1$.

In 1978, Rassias [15] formulated and proved the stability theorem for the linear mapping between Banach spaces E_1 and E_2 subject to the continuity of $f(tx)$ with respect to $t \in \mathbb{R}$ for each fixed $x \in E_1$. Thus, Rassias' theorem implies Aoki's theorem as a special case. Later, in 1990, Rassias [16] observed that the proof of his stability theorem also holds true for $p < 0$. In 1991, Gajda [17] showed that the proof of Rassias' theorem can be proved also for the case $p > 1$ by just replacing n by $-n$ in (1.2). These results are stated in a generalized form as follows (see Rassias and Šemrl [18]):

Theorem 1.5 *Let* $\beta(s, t)$ *be nonnegative for all nonnegative real numbers* s, t *and positive homogeneous of degree p, where p is real and* $p \neq 1$, *that is,* $\beta(\lambda s, \lambda t) = \lambda^p \beta(s, t)$, *for all nonnegative* λ, s, t. *Given a normed space* E_1 *and a Banach space* E_2, *assume that* $f : E_1 \longrightarrow E_2$ *satisfies the inequality*

$$\left\| f(x+y) - f(x) - f(y) \right\| \leq \beta(\|x\|, \|y\|),$$

for all $x, y \in E_1$. *Then there exists a unique additive mapping* $g : E_1 \longrightarrow E_2$ *such that*

$$\|f(x) - g(x)\| \leq \delta \|x\|^p,$$

for all $x \in E_1$, *where*

$$\delta := \begin{cases} \dfrac{\beta(1,1)}{2-2^p}, & p < 1, \\ \dfrac{\beta(1,1)}{2-2^p}, & p > 1. \end{cases} \tag{1.5}$$

The proofs for the cases $p < 1$ and $p > 1$ were provided by applying the direct methods. For $p < 1$, the additive mapping g is given by (1.2), while in case $p > 1$ the formula is

$$g(x) = \lim_{n\to\infty} 2^n f\left(\frac{x}{2^n}\right).$$

Corollary 1.2 *Let $f : E_1 \longrightarrow E_2$ be a mapping satisfying the hypotheses of Theorem 1.5 and suppose that f is continuous at a single point $y \in E_1$, then the additive mapping g is continuous.*

Corollary 1.3 *If, under the hypotheses of Theorem 1.5, we assume that, for each fixed $x \in E_1$, the mapping $t \longrightarrow f(tx)$ from $\mathbb{R}$ to E_2 is continuous, then the additive mapping g is linear.*

Remark 1.1 (1) For $p = 0$, Theorem 1.5, Corollaries 1.2 and 1.3 reduce to the results of Hyers in 1941. If we put $\beta(s,t) = \varepsilon(sp + tp)$, then we obtain the results of Rassias [15] in 1978 and Gajda [17] in 1991.

(2) The case $p = 1$ was excluded in Theorem 1.5. Simple counterexamples prove that one can not extend Rassias' Theorem when p takes the value one (see Z. Gajda [17], Rassias and Šemrl [18] and Hyers and Rassias [19] in 1992).

A further generalization of the Hyers-Ulam stability for a large class of mappings was obtained by Isac and Rassias [20] by introducing the following:

Definition 1.1 A mapping $f : E_1 \longrightarrow E_2$ is said to be ϕ-additive if there exist $\Phi \geq 0$ and a function $\phi : \mathbb{R}^+ \longrightarrow \mathbb{R}^+$ satisfying

$$\lim_{t\longrightarrow\infty} \frac{\phi(t)}{t} = 0,$$

such that

$$\Big\| f(x+y) - f(x) - f(y) \Big\| \leq \Phi[\phi(\|x\|) + \phi(\|y\|)],$$

for all $x, y \in E_1$.

In [20], Isac and Rassias proved the following:

Theorem 1.6 *Let E_1 be a real normed vector space and E_2 be a real Banach space. Let $f : E_1 \longrightarrow E_2$ be a mapping such that $f(tx)$ is continuous in t for each fixed $x \in E_1$. If f is ϕ-additive and ϕ satisfies the following conditions:*

(a) $\phi(ts) \leq \phi(t)\phi(s)$ *for all* $s, t \in \mathbb{R}$;
(b) $\phi(t) < t$ *for all* $t > 1$,

then there exists a unique linear mapping $T : E_1 \longrightarrow E_2$ *such that*

$$\|f(x) - T(x)\| \leq \frac{2\Phi}{2 - \phi(2)}\phi(\|x\|),$$

for all $x \in E_1$.

Remark 1.2 (1) If $\phi(t) = t^p$ with $p < 1$, then, from Theorem 1.6, we obtain Rassias' theorem [15].

(2) If $p < 0$ and $\phi(t) = t^p$ with $t > 0$, then Theorem 1.6 is implied by the result of Gajda in 1991.

In [21], Diaz and Margolis proved a "theorem of the alternative" for any "contraction mappin" T on a "generalized complete metric space" X. The conclusion of the theorem, speaking in general terms, asserts that: either all consecutive pairs of the sequence of successive approximations (starting from an element x_0 of X) are infinitely far apart, or the sequence of successive approximations, with initial element x_0 converges to a fixed point of T (what particular fixed point depends, in general, on the initial element x_0). The present theorem contains as special cases both Banach's contraction mapping theorem [22] for complete metric spaces, and Luxemburg's contraction mapping theorem [23] for generalized metric spaces.

Following Luxemburg [23], the concept of a "generalized complete metric spac" may be introduced as in this quotation: "Let X be an abstract (nonempty) set, the elements of which are denoted by $x, y, \ldots$ and assume that on the Cartesian product $X \times X$ a distance function $d(x, y)(0 \leq d(x, y) \leq \infty)$ is defined, satisfying the following conditions:

(D1) $d(x, y) = 0$ if and only if $x = y$,

(D2) d(x,y)=d(y,x) (symmetry),

(D3) $d(x, y) \leq d(x, z) + d(z, y)$, (triangle inequality),

(D4) every d-Cauchy sequence in X is d-convergent, i.e. $\lim_{n,m\longrightarrow\infty} d(x_n, x_m) = 0$, for a sequence $x_n \in X (n = 1, 2, \ldots)$ implies the existence of an element $x \in X$ with $\lim_{n\longrightarrow\infty} d(x, x_n) = 0$, ($x$ is unique by (D1) and (D3)).

This concept differs from the usual concept of a complete metric space by the fact that not every two points in X have necessarily a finite distance. One might call such a space a generalized complete metric space".

Using this notion, one has the following:

Theorem 1.7 ([21]) *Suppose that* (X, d) *is a generalized complete metric space, and that the function* $T : X \longrightarrow X$ *is a "contraction," that is,* T *satisfies the condition: There exists a constant* q, *with* $0 < q < 1$, *such that whenever* $d(x, y) < \infty$ *one has*

$$d(Tx, Ty) \leq qd(x, y).$$

Let $x_0 \in X$, *and consider the "sequence of successive approximations with initial element* x_0*":* $x_0, Tx_0, T^2x_0, \ldots, T^lx_0, \ldots$. *Then, the following alternative holds: either*

(A) for every integer $l = 0, 1, 2, \ldots,$ *one has*

$$d(T^l x_0, T^{l+1} x_0) = \infty, \quad or$$

(B) the sequence of successive approximations $x_0, Tx_0, T^2x_0, \ldots, T^l x_0, \ldots,$ *is* d*-convergent to a fixed point of* T.

In [24, 25], Cadariu and Radu and then Radu and Mihet presented the Cadariu-Radu theory (for classical spaces) and the Radu-Mihet theory (for fuzzy spaces) derived from the Diaz-Margolis theorem, respectively, as follows:

Theorem 1.8 *Let* $x, y \in X$. *Assume the complete* $[0, \infty]$*-valued metric* d *on* X *and strictly contractive function* T *on* X *with* $d(Tx, Ty) \leq qd(x, y)$, *where* $q < 1$. *If we obtain a* $l_0 \in \mathbb{N}$ *s.t.* $d(T^l x, T^{l+1} x) < \infty$, *for any* $l \geq l_0$, *therefore we get the following:*

- *the fixed point* y^* *of* T *is the convergence point of* $\{T^l x\}$;
- *in* $\{y \in X \mid d(T^{p_0} x, y) < \infty\}$, y^* *is the unique fixed point of* T;
- $(1 - q)d(y, y^*) \leq d(y, Ty)$ *for every* $y \in X$.

Since the time the above stated results were proven, several mathematicians have extensively studied stability theorems for several kinds of functional equations in various spaces, for example, Banach spaces, 2-Banach spaces, Banach n-Lie algebras, quasi-Banach spaces, Banach ternary algebras, non-Archimedean normed and Banach spaces, metric and ultra metric spaces, Menger probabilistic normed spaces, probabilistic normed space, p-2-normed spaces, C-*-algebras, C-*-ternary algebras, Banach ternary algebras, Banach modules, inner product spaces, Heisenberg groups, random normed spaces, fuzzy normed space and others. Further, researchers focused on the applications of the Hyers-Ulam-Rassias stability problems, for example, (partial) differential equations, fractional differential and integral equations, Volterra integral equations, group and ring theory, mathematical biology modeling, bending beam problems of mechanical engineering also, some kind of models in population dynamics, and some kinds of equations [26–31].

As mentioned at the beginning, the primary target of this monograph is to provide a new interpretation of the Ulam type stability, i.e., multi stability, with the application of classical, well–known special functions. This stability facilitates us to obtain diverse estimations based on the various special functions that are initially selected and to estimate optimal stability with minimal error which provides a unique optimized solution (see [32–52]).

The monograph is divided into 21 chapters:

Chapters 2–8 present a background to the classical well-known special functions which play an important role in mathematical physics, especially in boundary value and initial condition problems of differential equations. Generally speaking, we call a function "special" when the function, just as logarithmic, exponential, and trigonometric functions (the elementary transcendental functions), belongs to the toolbox of applied mathematicians, physicists or engineers. Usually there are a particular standardized notation, and a number of known properties of the function. This branch of mathematics has a respectable history with great names such as Gauss, Euler, Fourier, Legendre, Mittag, Leffler, Bessel, and Riemann. They all made good contributions to the area. A great part of their work was inspired by physics and driven by differential equations. About 70 years ago, these activities were summarized in the standard work "A Course of Modern Analysis" by Whittaker and Watson, which has had great influence and is still important nowadays. Many special functions appear as solutions of differential equations or integrals of elementary functions. Therefore, tables of integrals usually include descriptions of special functions, and tables of special functions include most important integrals and the integral representation of special functions. The main target of these chapters are to provide the detailed investigations to several newly established special functions involving the Euler gamma function, Pochhammer symbols, Gaussian hypergeometric series, Clausen hypergeometric series, supertrigonometric and superhyperbolic functions via the hypergeometric function, the Wright function, Wright's generalized hypergeometric function, supertrigonometric and superhyperbolic functions via the Wright function, Wright's generalized hypergeometric function, Mittag-Leffler function, supertrigonometric functions and superhyperbolic functions via the Mittag-Leffler function, the truncated the Mittag-Leffler function, Wiman function, supertrigonometric functions and superhyperbolic functions via the Wiman function, the truncated Wiman functions, Prabhakar function, the supertrigonometric and superhyperbolic functions via the Prabhakar function, the truncated Prabhakar functions, and so on.

In Chap. 9, the material can be formally divided into two main parts, which are discussed as follows. At first, we recall some definitions and results which will be used later on in the book. Then, starting from a novel view on the stability problem in the sense of the Ulam, we define a new concept of the multi stability to provide a comprehensive discussion of optimizing the different types of the Ulam stabilities.

In Chaps. 10–14, we use Radu's approach derived from the theorem of Diaz and Margolis to study existence, uniqueness and the multi stability results of mathematical equations in classical spaces.

In Chap. 15, we introduce basic and standard properties often required for fuzzy spaces.

In Chaps. 16–21, we consider both functional and fractional equations containing fuzzy uncertainties and prove their multi stability via the fixed point theory in diverse fuzzy normed spaces.

Chapter 22 is comprised of seven independent and self–contained sections which deal with nonlinear differential equations with various boundary conditions or initial value problems, based on the matrix Mittag–Leffler function, fixed point theory, as well as Babenko's approach.

Throughout the book, we let $\mathbb{C}$, $\mathbb{R}$, $\mathbb{Z}$ and $\mathbb{N}$ be the sets of the complex numbers, real numbers, integrals, and natural numbers, respectively. Let $\mathbb{Z}^+$, $\mathbb{R}_+$, $\mathbb{Z}^-$ and $\mathbb{R}_-$, be the sets of the positive integers, positive real numbers, negative integer numbers, and negative real numbers, respectively. Let $\mathbb{N}_0 = \mathbb{N} \cup 0$ and $\mathbb{Z}_0^- = \mathbb{Z}^- \cup 0$. Finally, $\Re(\nu)$ denotes the real part of ν if $\nu \in \mathbb{C}$.

References

1. Legendre AM (1791) Elements de geometrie, vol II. Didot, Paris
2. Gauss CF (1809) Theoria moyus corporum caelestium. Perthes-Besser, Hamburg
3. Cauchy AL (1821) Analyse algebrique, cours d'analyse de l'ecole polytechnique, vol 1. Debure, Paris
4. Aczel J (1966) Lectures on functional equations and their applications. Academic Press, New York
5. Aczél J (1987) A short course on functional equations. Reidel, Dordrecht
6. Aczél J, Dhombres J (1989) Functional equations in several variables. Cambridge University Press, Cambridge
7. Jung SM, Min S (2009) On approximate Euler differential equations. Abstr Appl Anal 537963:8 pp
8. Sahoo PK, Riedel T (1998) Mean value theorem and functional equations. World Scientific, Singapore
9. Ulam SM (1960) A collection of mathematical problems. Interscience, New York
10. Hyers DH (1941) On the stability of the linear functional equation. Proc Natl Acad Sci USA 27:222–224
11. Forti GL (1980) An existence and stability theorem for a class of functional equations. Stochastica 4:22–30
12. Rätz J (1980) On approximately additive mappings. In: Beckenbach EF (ed) International series of numerical mathematics, vol 47, (eds) General Inequalities 2. Birkhäuser, Basel, pp 233–251
13. Schwaiger J (1988) Remark 12, in Report on the 25th international symposium on functional equations. Aequat Math 35:120–121
14. Aoki T (1950) On the stability of the linear transformation in Banach spaces. J Math Soc Jpn 2:64–66
15. Rassias ThM (1978) On the stability of the linear mapping in Banach spaces. Proc Am Math Soc 72:297–300
16. Rassias ThM (1990) The stability of mappings and related topics, in Report on the 27th international symposium on functional equations. Aequat Math 39:292–293
17. Gajda Z (1991) On stability of additive mappings. Int J Math Math Sci 149:431–434
18. Rassias ThM, Šemrl P (1993) On the Hyers-Ulam stability of linear mappings. J Math Anal Appl 173:325–338
19. Hyers DH, Rassias ThM (1992) Approximate homomorphism. Aequ Math 44:125–153
20. Isac G, Rassias ThM (1996) Stability of Ψ-additive mappings: applications to nonlinear analysis. Int J Math Math Sci 19:219–228
21. Diaz JB, Margolis B (1968). A fixed point theorem of the alternative, for contractions on a generalized complete metric space
22. Banach S (1922) Sur les opérations dans les ensembles abstraits et leur application aux équations intégrales. Fundam Math 3(1):133–181
23. Luxemburg WAJ (1958) On the convergence of successive approximations in the theory of ordinary differential equations II. Indag Math 20(5):540–546
24. Cadariu L, Radu V (2008) Fixed point methods for the generalized stability of functional equations in a single variable. Fixed Point Theory Appl 749392
25. Mihet D, Radu V (2008) On the stability of the additive Cauchy functional equation in random normed spaces. J Math Anl Appl 343(1):567–572
26. Brzdek J (2023) On Ulam stability with respect to 2-norm. Symmetry 15(9):1664
27. Benzarouala C, Brzdek J, El-hady ES, Oubbi L (2023) On ulam stability of the inhomogeneous version of the general linear functional equation. Results Math 78(3):76
28. Benzarouala C, Brzdek J, Oubbi L (2023) A fixed point theorem and Ulam stability of a general linear functional equation in random normed spaces. J Fixed Point Theory Appl 25(1):33
29. Novac A, Otrocol D, Popa D (2023) On ulam stability of a partial differential operator in Banach spaces. Mathematics 11(11):2488

30. Tamilvanan K, Mohiuddine SA, Revathi N (2023) 12 ulam stability of mixed type functional equation. In: Recent Advances and Applications in Engineering and Mathematical Sciences, Soft Computing, p. 167
31. Jiddah JA, Shehu Shagari M, Imam AT (2023) On fixed points of a general class of hybrid contractions with ulam-type stability. Sahand Commun Math Anal 20(2):39–64
32. Aderyani SR, Saadati R (2021) Best approximations of the ϕ-Hadamard fractional Volterra integro-differential equation by matrix valued fuzzy control functions. Adv Differ Equ 154
33. Aderyani SR, Saadati R, Feckan M (2021) The Cădariu-Radu method for existence, uniqueness and gauss hypergeometric stability of Ω-Hilfer fractional differential equations. Mathematics 9(12):1408
34. Aderyani SR, Saadati R, Yang XJ (2021) Radu-Mihet method for UHML stability for a class of ξ-Hilfer fractional differential equations in matrix valued fuzzy Banach spaces. Math Methods Appl Sci 44(18):14619–14631
35. Aderyani SR, Saadati R, Mesiar R (2022) Estimation of permuting tri-homomorphisms and permuting tri-derivations associated with the tri-additive Υ-random operator inequality in matrix MB-algebra. Int J Gen Syst 51(6):547–569
36. Aderyani SR, Saadati R, Abdeljawad T, Mlaiki N (2022) Multi-stability of non homogenous vector-valued fractional differential equations in matrix-valued Menger spaces. Alexandria Eng J 61(12):10913–10923
37. Aderyani SR, Saadati R, O'Regan D, Abdeljawad T UHML stability of a class of Δ-Hilfer FDEs via CRM
38. Aderyani SR, Saadati R, Rassias TM, Park C (2022) Best approximation of (G1, G2)-random operator inequality in matrix Menger Banach algebras with application of stochastic Mittag-Leffler and H-Fox control functions
39. Aderyani SR, Saadati R, ORegan D (2022) The Cadariu–Radu method for existence, uniqueness and Gauss Hypergeometric stability of a class of Ξ-Hilfer fractional differential equations. Int J Nonlinear Sci Numer Simul
40. Aderyani SR, Saadati R, ORegan D, Alshammari FS (2022) Multi-super-stability of antiderivations in Banach algebras. AIMS Math 7(11):20143–20163
41. Aderyani SR, Saadati R, Allahviranloo T (2022) Existence, uniqueness and matrix-valued fuzzy Mittag-Leffler-Hypergeometric-Wright stability for P-Hilfer fractional differential equations in matrix-valued fuzzy Banach space. Comput Appl Math 41(6):234
42. Aderyani SR, Saadati R, ORegan D, Alshammari FS (2022) Existence, uniqueness and stability analysis with the multiple exp function method for NPDEs. Mathematics 10(21):4151
43. Aderyani SR, Saadati R, Vahidi J (2023) Multiple exp-function method to solve the nonlinear space-time fractional partial differential symmetric regularized long wave (SRLW) equation and the (1+ 1)-dimensional Benjamin-Ono equation. Int J Modern Phys B 37(22):2350213
44. Rezaei Aderyani S, Saadati R, Li C, Rassias TM (2023) Park C (2023) Special functions and multi-stability of the Jensen type random operator equation in C^*-algebras via fixed point. J Inequal Appl 1:1–24
45. Aderyani SR, Saadati R, ORegan D, Alshammari FS (2023) Fuzzy approximate solutions of matrix-valued fractional differential equations by fuzzy control functions. Mathematics 11(6):1386
46. Aderyani SR, Saadati R, Allahviranloo T, Abbasbandy S, Catak M (2023) Fuzzy approximation of a fractional Lorenz system and a fractional financial crisis. Iran J Fuzzy Syst 20(7):27–36
47. Aderyani SR, Saadati R (2023) Stability and controllability results by n–ary aggregation functions in matrix valued fuzzy n–normed spaces. Inf Sci 119265
48. Aderyani SR, Saadati R, Rassias TM, Srivastava HM (2023) Existence, uniqueness and the multi-stability results for a W-Hilfer fractional differential equation. Axioms 12(7):681
49. Aderyani SR, Saadati R, ORegan D, Li C (2023) On a new approach for stability and controllability analysis of functional equations. Mathematics 11(16):3458
50. ORegan D, Aderyani SR, Saadati R, Li C (2023) Stability results and parametric delayed Mittag-Leffler matrices in symmetric fuzzy-random spaces with application. Symmetry 15(10):1880

51. Aderyani SR, Saadati R, O'Regan D, Alshammari FS (2023) Application of aggregated control functions for approximating $\mathscr{C}$-Hilfer fractional differential equations. AIMS Math 8(11):28010–28032
52. Aderyani SR, Saadati R (2022) Approximation of derivation–homomorphism fuzzy functional inequalities in matrix valued FC-◇-algebras. In: 2022 9th Iranian joint congress on fuzzy and intelligent systems (CFIS). IEEE, pp 1–6

Chapter 2
The Hypergeometric, Supertrigonometric, and Superhyperbolic Functions

In this chapter, we introduce the Euler gamma function, the Pochhammer symbols, the Gaussian hypergeometric series, the Clausen hypergeometric series and the Supertrigonometric, and Superhyperbolic functions via Gaussian hypergeometric series and Clausen hypergeometric series.

2.1 The Euler Gamma Function and the Pochhammer Symbols

In this section, we present the Euler gamma function and the Pochhammer symbols. We begin with the definition of the gamma function.

Definition 2.1 ([1, 2]) The gamma function first defined by Euler is given by

$$\Gamma(X) = \int_0^\infty e^{-Y} Y^{X-1} dY,$$

where $\Re(X) > 0$ and $X \in \mathbb{C}$.

Definition 2.2 ([1]) Let $\Re(X) > 0$ and $X \in \mathbb{C}$. Then the Euler gamma function satisfies

$$\Gamma(X+1) = X\Gamma(X).$$

The result was discovered by Euler in 1729 [3] and reported by Weierstrass [4], Brunel [5], Gronwall [6], and Olver [7].

S. Rezaei Aderyani et al., *Towards Ulam Type Multi Stability Analysis*, Studies in Systems, Decision and Control 523, https://doi.org/10.1007/978-3-031-55564-0_2

Theorem 2.1 ([1]) *If* $X \in \mathbb{N}_0$, *then we have*

$$\Gamma(X+1) = X!.$$

This work was presented by Euler in 1729 [1] and also studied by Weierstrass [8], Brunel [5], and Gronwall [6].

Theorem 2.2 ([1]) $\Gamma(0.5) = \pi^{0.5}$.

Indeed,

$$\Gamma(0.5) = \int_0^\infty e^{-Y} Y^{-0.5} dY = 2 \int_0^\infty e^{-x^2} dx = \sqrt{\pi},$$

by making the variable change $Y = x^2$.

This work was initially discovered by Euler in 1729 [6] and then re-discussed in Bell [9], Luke [10], and Bendersky [11].

We now introduce the Pochhammer symbols.

Definition 2.3 ([12]) The Pochhammer symbol is given by

$$\begin{aligned}(\mathsf{C})_k &= \prod_{i=1}^{k} (\mathsf{C} + i - 1) \\ &= \frac{\Gamma(\mathsf{C}+k)}{\Gamma(\mathsf{C})} \\ &= \begin{cases} 1 & k = 0 \\ \mathsf{C}(\mathsf{C}+1)\cdots(\mathsf{C}+k-1) & k \in \mathbb{N}, \end{cases}\end{aligned}$$

where $\mathsf{C} \in \mathbb{C}$ and $k \in \mathbb{N}_0$.

The Pochhammer symbol was first suggested and used by Pochhammer in 1870 [12]. In addition, Weierstrass noticed in 1856 [13] that

$$\Gamma(\mathsf{C}+k) = \mathsf{C}(\mathsf{C}+1)\cdots(\mathsf{C}+k-1)\Gamma(\mathsf{C}),$$

where $k \in \mathbb{N}_0$.

2.2 The Gaussian Hypergeometric Series, Supertrigonometric and Superhyperbolic Functions

In this section, we present the definition, theorems, and properties for the Gaussian hypergeometric series, and the relationships among the hypergeometric supertrigonometric and superhyperbolic functions via Gauss's hypergeometric series.

2.2.1 The Gaussian Hypergeometric Series

Let us start with the definition of the Gaussian hypergeometric series.

Definition 2.4 ([14]) The Gaussian hypergeometric series of form is given by

$$\begin{aligned}
&{}_2\mathbb{F}_1(\mathsf{A}, \mathsf{B}; \mathsf{C}; X) \\
&= 1 + \frac{\mathsf{AB}}{\mathsf{C}} X + \frac{\mathsf{A}(\mathsf{A}+1)\mathsf{B}(\mathsf{B}+1)}{\mathsf{C}(\mathsf{C}+1)} \frac{X^2}{2} + \cdots \\
&= \sum_{n=0}^{\infty} \frac{(\mathsf{A})_n (\mathsf{B})_n}{(\mathsf{C})_n} \frac{X^n}{n!},
\end{aligned}$$

where $\mathsf{A}, \mathsf{B}, \mathsf{C}, X \in \mathbb{C}$, and $|X| < 1$.

The hypergeometric series is a solution of the Gaussian differential equation of the following form

$$X(1-X)\frac{d^2\omega}{dX^2} + (\mathsf{C} - (\mathsf{A}+\mathsf{B}+1)X)\frac{d\omega}{dX} - \mathsf{AB}\omega = 0,$$

where $\mathsf{A}, \mathsf{B}, X \in \mathbb{C}$, $\mathsf{C} \in \mathbb{C}\backslash\mathbb{Z}_0^-$, and $|X| < 1$.

The above result was proved by Whittaker and Watson in 1927 [15], and by Bailey in 1935 [18].

Theorem 2.3 ([19]) *If* $\mathsf{A}, \mathsf{B}, \mathsf{C}, X \in \mathbb{C}$ *and* $|X| < 1$, *then*

$$\begin{aligned}
&{}_2\mathbb{F}_1(\mathsf{A}, \mathsf{B}; \mathsf{C}; X) \\
&= \frac{\Gamma(\mathsf{C})}{\Gamma(\mathsf{B})\Gamma(\mathsf{C}-\mathsf{B})} \int_0^1 Y^{\mathsf{B}-1}(1-Y)^{\mathsf{C}-\mathsf{B}-1}(1-XY)^{-\mathsf{A}} dY,
\end{aligned}$$

where $\Re(\mathsf{C}) > \Re(\mathsf{B}) > 0$.

Theorem 2.4 ([15]) *If* $\Re(\mathsf{C}) > 0$, $|X| < 1$, *and* $|\arg(-X)| < \pi$, *then*

$$\begin{aligned}
&{}_2\mathbb{F}_1(\mathsf{A}, \mathsf{B}; \mathsf{C}; X) \\
&= \frac{\Gamma(\mathsf{C})}{\Gamma(\mathsf{A})\Gamma(\mathsf{B})} \frac{1}{2\pi i} \int_{-i\infty}^{+i\infty} \frac{\Gamma(\mathsf{A}+Y)\Gamma(\mathsf{B}+Y)\Gamma(-Y)}{\Gamma(\mathsf{C}+\mathsf{Y})} (-X)^Y dY.
\end{aligned}$$

2.2.2 The Hypergeometric Supertrigonometric Functions via the Gaussian Superhyperbolic Series

Let $\mathsf{A}, \mathsf{B}; \mathsf{C}, X \in \mathbb{C}$. We now define the hypergeometric supertrigonometric functions [18] via the Gaussian superhyperbolic series as follows:

▷ The hypergeometric supersine through the Gaussian superhyperbolic series:

$$ {}_2Supersin_1(\mathsf{A}, \mathsf{B}; \mathsf{C}; X) = \sum_{k=0}^{\infty} \frac{(\mathsf{A})_{2k+1}(\mathsf{B})_{2k+1}}{(\mathsf{C})_{2k+1}} \frac{(-1)^k X^{2k+1}}{(2k+1)!}. $$

▷ The hypergeometric supercosine through the Gaussian superhyperbolic series:

$$ {}_2Supercos_1(\mathsf{A}, \mathsf{B}; \mathsf{C}; X) = \sum_{k=0}^{\infty} \frac{(\mathsf{A})_{2k}(\mathsf{B})_{2k}}{(\mathsf{C})_{2k}} \frac{(-1)^k X^{2k}}{(2k)!}. $$

▷ The hypergeometric supertangent through the Gaussian superhyperbolic series:

$$ {}_2Supertan_1(\mathsf{A}, \mathsf{B}; \mathsf{C}; X) = \frac{{}_2Supersin_1(\mathsf{A}, \mathsf{B}; \mathsf{C}; X)}{{}_2Supercos_1(\mathsf{A}, \mathsf{B}; \mathsf{C}; X)}. $$

▷ The hypergeometric supercotangent through the Gaussian superhyperbolic series:

$$ {}_2Supercot_1(\mathsf{A}, \mathsf{B}; \mathsf{C}; X) = \frac{{}_2Supercos_1(\mathsf{A}, \mathsf{B}; \mathsf{C}; X)}{{}_2\mathrm{Supersin}_1(\mathsf{A}, \mathsf{B}; \mathsf{C}; X)}. $$

▷ The hypergeometric superscent through the Gaussian superhyperbolic series:

$$ {}_2Supersec_1(\mathsf{A}, \mathsf{B}; \mathsf{C}; X) = \frac{1}{{}_2Supercos_1(\mathsf{A}, \mathsf{B}; \mathsf{C}; X)}. $$

▷ The hypergeometric supercosecant through the Gaussian superhyperbolic series:

$$ {}_2Supercosec_1(\mathsf{A}, \mathsf{B}; \mathsf{C}; X) = \frac{1}{{}_2Supersin_1(\mathsf{A}, \mathsf{B}; \mathsf{C}; X)}. $$

Theorem 2.5 ([18]) *Let* $\mathsf{A}, \mathsf{B}; \mathsf{C}, X \in \mathbb{C}$, *and* $i^2 = -1$. *Then,*

$$ {}_2\mathbb{F}_1(\mathsf{A}, \mathsf{B}; \mathsf{C}; iX) = {}_2Supercos_1(\mathsf{A}, \mathsf{B}; \mathsf{C}; X) + i\,{}_2Supersin_1(\mathsf{A}, \mathsf{B}; \mathsf{C}; X). $$

This result was presented by Yang in 2020 [18].

2.2.3 *The Hypergeometric Superhyperbolic Functions via the Gaussian Hypergeometric Series*

Let $\mathsf{A}, \mathsf{B}; \mathsf{C}, X \in \mathbb{C}$. We now have the hypergeometric superhyperbolic functions [18] via the Gaussian hypergeometric series, as follows:

- The hypergeometric superhyperbolic sine by the Gaussian hypergeometric series:

$$ {}_2Supersinh_1(\mathsf{A}, \mathsf{B}; \mathsf{C}; X) = \sum_{k=0}^{\infty} \frac{(\mathsf{A})_{2k+1}(\mathsf{B})_{2k+1}}{(\mathsf{C})_{2k+1}} \frac{X^{2k+1}}{(2k+1)!}. $$

- The hypergeometric superhyperbolic cosine by the Gaussian hypergeometric series:

$$ {}_2Supercosh_1(\mathsf{A}, \mathsf{B}; \mathsf{C}; X) = \sum_{k=0}^{\infty} \frac{(\mathsf{A})_{2k}(\mathsf{B})_{2k}}{(\mathsf{C})_{2k}} \frac{X^{2k}}{(2k)!}. $$

- The hypergeometric superhyperbolic tangent by the Gaussian hypergeometric series:

$$ {}_2Supertanh_1(\mathsf{A}, \mathsf{B}; \mathsf{C}; X) = \frac{{}_2Supersinh_1(\mathsf{A}, \mathsf{B}; \mathsf{C}; X)}{{}_2\mathrm{Supercosh}_1(\mathsf{A}, \mathsf{B}; \mathsf{C}; X)}. $$

- The hypergeometric superhyperbolic cotangent by the Gaussian hypergeometric series:

$$ {}_2Supercoth_1(\mathsf{A}, \mathsf{B}; \mathsf{C}; X) = \frac{{}_2Supercosh_1(\mathsf{A}, \mathsf{B}; \mathsf{C}; X)}{{}_2\mathrm{Supersinh}_1(\mathsf{A}, \mathsf{B}; \mathsf{C}; X)}. $$

- The hypergeometric superhyperbolic secant by the Gaussian hypergeometric series:

$$ {}_2Supersech_1(\mathsf{A}, \mathsf{B}; \mathsf{C}; X) = \frac{1}{{}_2Supercosh_1(\mathsf{A}, \mathsf{B}; \mathsf{C}; X)}. $$

- The hypergeometric superhyperbolic cosecant by the Gaussian hypergeometric series:

$$ {}_2Supercosech_1(\mathsf{A}, \mathsf{B}; \mathsf{C}; X) = \frac{1}{{}_2Supersinh_1(\mathsf{A}, \mathsf{B}; \mathsf{C}; X)}. $$

Theorem 2.6 ([18]) *Let* $\mathsf{A}, \mathsf{B}; \mathsf{C}, X \in \mathbb{C}$. *Then,*

$$ {}_2\mathbb{F}_1(\mathsf{A}, \mathsf{B}; \mathsf{C}; X) = {}_2Supercosh_1(\mathsf{A}, \mathsf{B}; \mathsf{C}; X) + {}_2Supersinh_1(\mathsf{A}, \mathsf{B}; \mathsf{C}; X). $$

2.3 The Clausen Hypergeometric Series, Supertrigonometric and Superhyperbolic Functions

In this section, we introduce the definition, theorems, and properties for the Clausen hypergeometric series.

2.3.1 *The Clausen Hypergeometric Series*

We now consider the definition and theorems for the Clausen hypergeometric series.

Definition 2.5 ([20]) The Clausen hypergeometric series of form is given by

$$\begin{aligned}
&{}_p\mathbb{F}_q((\mathsf{A});(\mathsf{C});X)\\
&= {}_p\mathbb{F}_q(\mathsf{A}_1,\ldots,\mathsf{A}_p;\mathsf{C}_1,\ldots,\mathsf{C}_q;X)\\
&= {}_p\mathbb{F}_q\left(\begin{matrix}\mathsf{A}_1,\ldots,\mathsf{A}_p\\ \mathsf{C}_1,\ldots,\mathsf{C}_q\end{matrix};X\right)\\
&= \sum_{k=0}^{\infty}\frac{(\mathsf{A}_1)_k\cdots(\mathsf{A}_p)_k}{(\mathsf{C}_1)_k\cdots(\mathsf{C}_q)_k}\frac{X^k}{k!},
\end{aligned}$$

where $\mathsf{A}_p, \mathsf{C}_q, X \in \mathbb{C}$ and $p, q \in \mathbb{N}_0$.

This work, as an extended version of the Gaussian hypergeometric series, was presented by Clausen in 1828 [20].

Let $X \in \mathbb{C}$, $p, q \in \mathbb{N}_0$, and $|X| < 1$. Then the Clausen hypergeometric series is a solution of the following differential equation

$$\Big(M(q,\eta,\mathsf{C}_q)\omega\Big)(X) - \Big(N(p,\eta,\mathsf{A}_p)\omega\Big)(X) = 0,$$

where

$$\begin{aligned}
&\Big(M(q,\eta,\mathsf{C}_q)\omega\Big)(X)\\
&= \left(X\frac{d}{dX}\right)\prod_{n=1}^{q}\left(\left(X\frac{d}{dX}\right)\omega(X) + (\mathsf{C}_n - 1)\omega(X)\right)\\
&= X\frac{d}{dX}\left(\prod_{n=1}^{q}\left(\left(X\frac{d}{dX} + (\mathsf{C}_n - 1)\right)\omega\right)(X)\right),
\end{aligned}$$

and

$$\begin{aligned}&\Big(N(p,\eta,\mathsf{A}_p)\omega\Big)(X)\\&=X\prod_{n=1}^{p}\left(X\frac{d\omega(X)}{dX}+\mathsf{A}_n\omega(X)\right)\\&=X\prod_{n=1}^{p}\left(\left(X\frac{d}{dX}+\mathsf{A}_n\right)\omega\right)(X).\end{aligned}$$

The above result was presented by Srivastava and Kashyap [21], Rainville [22], Andrews et al. [23], and Luke [24].

We now present the convergence for the Clausen Hypergeometric Series.

Theorem 2.7 *The cases of the convergence of the Clausen hypergeometric series hold for* $\mathsf{A}_n \in \mathbb{C}\backslash\mathbb{Z}_0^-$ *and* $n \in \mathbb{N}_0$*:*

- *if* $p < q$*, then the series converges absolutely for* $X \in \mathbb{C}$*,*
- *if* $p = q + 1$*, then the series converges absolutely for* $|X| < 1$ *and diverges for* $|X| > 1$ *and for* $|X| = 1$ *it converges absolutely for* $\Re\left(\sum_{k=1}^{q}\mathsf{C}_k - \sum_{k=1}^{p}\mathsf{A}_k\right) > 0$,
- *if* $p > q + 1$*, then the series converges only for* $X = 0$.

These conditions were reported on by Bailey [25], Srivastava and Kashyap [21], Slater [26], Andrews et al. [23], and Rainville [22].

2.3.2 The Hypergeometric Supertrigonometric Functions via the Clausen Superhyperbolic Series

Let $\mathsf{A}_k, \mathsf{C}_k \in \mathbb{C}$, and $k, p, q \in \mathbb{N}_0$. We now present the hypergeometric supertrigonometric functions [18] via the Clausen superhyperbolic series as follows:

▷ The hypergeometric supersine through the Clausen hypergeometric series:

$${}_pSupersin_q(\mathsf{A}_1,\ldots,\mathsf{A}_p;\mathsf{C}_1,\ldots,\mathsf{C}_q;X)=\sum_{k=0}^{\infty}\frac{(\mathsf{A}_1)_{2k+1}\cdots(\mathsf{A}_p)_{2k+1}}{(\mathsf{C}_1)_{2k+1}\cdots(\mathsf{C}_q)_{2k+1}}\frac{(-1)^kX^{2k+1}}{(2k+1)!}.$$

▷ The hypergeometric supercosine through the Clausen hypergeometric series:

$${}_pSupercos_q(\mathsf{A}_1,\ldots,\mathsf{A}_p;\mathsf{C}_1,\ldots,\mathsf{C}_q;X)=\sum_{k=0}^{\infty}\frac{(\mathsf{A}_1)_{2k}\cdots(\mathsf{A}_p)_{2k}}{(\mathsf{C}_1)_{2k}\cdots(\mathsf{C}_q)_{2k}}\frac{(-1)^kX^{2k}}{(2k)!}.$$

▷ The hypergeometric supertangent through the Clausen hypergeometric series:

$$ {}_pSupertan_q(\mathsf{A}_1,\ldots,\mathsf{A}_p;\mathsf{C}_1,\ldots,\mathsf{C}_q;X)=\frac{{}_pSupersin_q(\mathsf{A}_1,\ldots,\mathsf{A}_p;\mathsf{C}_1,\ldots,\mathsf{C}_q;X)}{{}_pSupercos_q(\mathsf{A}_1,\ldots,\mathsf{A}_p;\mathsf{C}_1,\ldots,\mathsf{C}_q;X)}. $$

▷ The hypergeometric supercotangent through the Clausen hypergeometric series:

$$ {}_pSupercot_q(\mathsf{A}_1,\ldots,\mathsf{A}_p;\mathsf{C}_1,\ldots,\mathsf{C}_q;X)=\frac{{}_pSupercos_q(\mathsf{A}_1,\ldots,\mathsf{A}_p;\mathsf{C}_1,\ldots,\mathsf{C}_q;X)}{{}_pSupersin_q(\mathsf{A}_1,\ldots,\mathsf{A}_p;\mathsf{C}_1,\ldots,\mathsf{C}_q;X)}. $$

▷ The hypergeometric supersecant through the Clausen hypergeometric series:

$$ {}_pSupersec_q(\mathsf{A}_1,\ldots,\mathsf{A}_p;\mathsf{C}_1,\ldots,\mathsf{C}_q;X)=\frac{1}{{}_pSupercos_q(\mathsf{A}_1,\ldots,\mathsf{A}_p;\mathsf{C}_1,\ldots,\mathsf{C}_q;X)}. $$

▷ The hypergeometric supercosecant through the Clausen hypergeometric series:

$$ {}_pSupercosec_q(\mathsf{A}_1,\ldots,\mathsf{A}_p;\mathsf{C}_1,\ldots,\mathsf{C}_q;X)=\frac{1}{{}_pSupersin_q(\mathsf{A}_1,\ldots,\mathsf{A}_p;\mathsf{C}_1,\ldots,\mathsf{C}_q;X)}. $$

2.3.3 The Hypergeometric Superhyperbolic Functions via the Clausen Superhyperbolic Series

Let $\mathsf{A}_k, \mathsf{C}_k \in \mathbb{C}$, and $p, q \in \mathbb{N}_0$. We now have the hypergeometric superhyperbolic functions [18] via the Clausen superhyperbolic series as follows:

▷ The hypergeometric supersine through the Clausen hypergeometric series:

$$ {}_pSupersinh_q(\mathsf{A}_1,\ldots,\mathsf{A}_p;\mathsf{C}_1,\ldots,\mathsf{C}_q;X)=\sum_{k=0}^{\infty}\frac{(\mathsf{A}_1)_{2k+1}\ldots(\mathsf{A}_p)_{2k+1}}{(\mathsf{C}_1)_{2k+1}\ldots(\mathsf{C}_q)_{2k+1}}\frac{X^{2k+1}}{(2k+1)!}. $$

▷ The hypergeometric supercosine through the Clausen hypergeometric series:

$$ {}_pSupercosh_q(\mathsf{A}_1,\ldots,\mathsf{A}_p;\mathsf{C}_1,\ldots,\mathsf{C}_q;X)=\sum_{k=0}^{\infty}\frac{(\mathsf{A}_1)_{2k}\cdots(\mathsf{A}_p)_{2k}}{(\mathsf{C}_1)_{2k}\cdots(\mathsf{C}_q)_{2k}}\frac{X^{2k}}{(2k)!}. $$

▷ The hypergeometric supertangent through the Clausen hypergeometric series:

$$ {}_pSupertanh_q(\mathsf{A}_1,\ldots,\mathsf{A}_p;\mathsf{C}_1,\ldots,\mathsf{C}_q;X)=\frac{{}_pSupersinh_q(\mathsf{A}_1,\ldots,\mathsf{A}_p;\mathsf{C}_1,\ldots,\mathsf{C}_q;X)}{{}_pSupercosh_q(\mathsf{A}_1,\ldots,\mathsf{A}_p;\mathsf{C}_1,\cdots,\mathsf{C}_q;X)}. $$

▷ The hypergeometric supercotangent through the Clausen hypergeometric series:

$$ {}_pSupercoth_q(\mathsf{A}_1,\ldots,\mathsf{A}_p;\mathsf{C}_1,\ldots,\mathsf{C}_q;X)=\frac{{}_pSupercosh_q(\mathsf{A}_1,\ldots,\mathsf{A}_p;\mathsf{C}_1,\ldots,\mathsf{C}_q;X)}{{}_pSupersinh_q(\mathsf{A}_1,\ldots,\mathsf{A}_p;\mathsf{C}_1,\ldots,\mathsf{C}_q;X)}. $$

▷ The hypergeometric supersecant through the Clausen hypergeometric series:

$$ {}_pSupersech_q(\mathsf{A}_1,\ldots,\mathsf{A}_p;\mathsf{C}_1,\ldots,\mathsf{C}_q;X)=\frac{1}{{}_pSupercosh_q(\mathsf{A}_1,\ldots,\mathsf{A}_p;\mathsf{C}_1,\ldots,\mathsf{C}_q;X)}. $$

▷ The hypergeometric supercosecant through the Clausen hypergeometric series:

$$ {}_pSupercosech_q(\mathsf{A}_1,\ldots,\mathsf{A}_p;\mathsf{C}_1,\ldots,\mathsf{C}_q;X)=\frac{1}{{}_pSupersinh_q(\mathsf{A}_1,\ldots,\mathsf{A}_p;\mathsf{C}_1,\ldots,\mathsf{C}_q;X)}. $$

The above results are based on the idea of separation of a power series into its even and odd terms, which were presented by Euler in 1748 [27]. Yang further derived the following new functions through the Clausen hypergeometric series in 2020 [18].

Theorem 2.8 ([18]) *Let* $\mathsf{A}_k, \mathsf{C}_k \in \mathbb{C}$, *and* $k, p, q \in \mathbb{N}_0$. *Then,*

- $ {}_pSupersinh_q(\mathsf{A}_1,\ldots,\mathsf{A}_p;\mathsf{C}_1,\ldots,\mathsf{C}_q;X)=-i\,{}_pSupersin_q(\mathsf{A}_1,\ldots,\mathsf{A}_p;\mathsf{C}_1,\ldots,\mathsf{C}_q;iX).$
- $ {}_pSupercosh_q(\mathsf{A}_1,\ldots,\mathsf{A}_p;\mathsf{C}_1,\ldots,\mathsf{C}_q;-X)={}_pSupercos_q(\mathsf{A}_1,\ldots,\mathsf{A}_p;\mathsf{C}_1,\ldots,\mathsf{C}_q;iX).$
- $ {}_pSupertanh_q(\mathsf{A}_1,\ldots,\mathsf{A}_p;\mathsf{C}_1,\ldots,\mathsf{C}_q;-X)=-i\,{}_pSupertan_q(\mathsf{A}_1,\ldots,\mathsf{A}_p;\mathsf{C}_1,\ldots,\mathsf{C}_q;iX).$
- $ {}_pSupercoth_q(\mathsf{A}_1,\ldots,\mathsf{A}_p;\mathsf{C}_1,\ldots,\mathsf{C}_q;-X)=i\,{}_pSupercot_q(\mathsf{A}_1,\ldots,\mathsf{A}_p;\mathsf{C}_1,\ldots,\mathsf{C}_q;iX).$
- $ {}_pSupersech_q(\mathsf{A}_1,\ldots,\mathsf{A}_p;\mathsf{C}_1,\ldots,\mathsf{C}_q;-X)={}_pSupersec_q(\mathsf{A}_1,\ldots,\mathsf{A}_p;\mathsf{C}_1,\ldots,\mathsf{C}_q;iX).$
- $ {}_pSupercosech_q(\mathsf{A}_1,\ldots,\mathsf{A}_p;\mathsf{C}_1,\ldots,\mathsf{C}_q;-X)=i\,{}_pSupercsec_q(\mathsf{A}_1,\ldots,\mathsf{A}_p;\mathsf{C}_1,\ldots,\mathsf{C}_q;iX).$

References

1. Euler L (1843) Letter to Goldbach, 13 Oct 1729. Correspondence math. et phys. de quelques celebres geometres du 18e siecle, publiee par Fuss, vol. 1, St. Petersburg
2. Legendre AM (1814) Exercises de calcul intégral, vol 2. Courcier, Paris
3. Euler L (1738) De progressionibus transcendentibus sen quaroum termini generales algebrare dari nequeunt. Commentarii academiae scientiarum Petropolitanae 5:36–57, presented to the St. Petersburg Academy on 28 Nov 1729
4. Weierstrass K (1856) Über die Theorie der analytischen Facultäten. J Reine Angew Math 51:1–60
5. Brunel GEA (1886) Monographie de la fonction gamma. Memoires de la societe des sciences physiques et naturelles de Bordeaux, ser 3, vol 3, pp 1–184
6. Gronwall TH (1916) An elementary exposition of the theory of the gamma function. Ann Math 17(3):124–166

7. Olver F (1997) Asymptotics and special functions. CRC Press, Routledge
8. Srivastava HM, Choi J (2012) Zeta and q-zeta functions and associated series and integrals. Elsevier, Amsterdam
9. Bell WW (1968) Special functions for scientists and engineers. Courier Corporation, D. Van Nostrand Company, London
10. Luke YL (ed) (1969) Special functions and their approximations, vol 1. Academic, New York
11. Bendersky L (1933) Sur la fonction gamma généralisée. Acta Math 61:263–322
12. Pochhammer L (1870) Ueber Hypergeometrischen Funktionen Hoheren Ordungen. J Reine Angew Math 71(216):316–352
13. Srivastava HM, Choi J (2012) Zeta and q-zeta functions and associated series and integrals. Elsevier, Amsterdam
14. Gauss CF (1876) Disquisitiones generales circa seriem infinitam $1+\frac{\alpha\beta}{1.\gamma}x+\frac{\alpha(\alpha+1)\beta(\beta+1)}{1.2.\gamma(\gamma+1)}xx+ect\cdots$ Commentationes Societiones Regiae Scientiarum Gottingensis Recentiores, vol 2, pp 1–46 (1812). Reprinted in Werke 3(2):122–162
15. Whittaker ET, Watson GN (1927) A course of modern analysis. Cambridge University Press, Cambridge (1927); Reprinted by Cambridge University Press in 1927
16. Bailey WN (1935) Generalized hypergeometric series. Camrbridge University Press, Camrbridge
17. Srivastava HM, Choi J (2012) Zeta and q-zeta functions and associated series and integrals. Elsevier, Amsterdam
18. Yang X-J (2021) An introduction to hypergeometric supertrigonometric and superhyperbolic functions. Academic, New York
19. Barnes EW (1908) A new development of the theory of the hypergeometric functions. Proc Lond Math Soc 2(1):141–177
20. Clausen T (1828) Ueber die Fälle, wenn die Reihe von der Form y= 1+. . . etc. ein Quadrat von der Form z= 1. . . etc. hat. J Reine Angew Math 3:89–95
21. Srivastava HM, Kashyap BRK (1982) Special functions in queuing theory and related stochastic processes. Academic, New York
22. Rainville ED (1960) Special functions. Macmillan, New York
23. Andrews GE, Askey R, Roy R (1999) Special functions, vol 71. Cambridge University Press, Cambridge
24. Luke YL (ed) (1969) Special functions and their approximations, vol 1. Academic, New York
25. Bailey WN (1935) Generalized hypergeometric series. Camrbridge University Press, Camrbridge
26. Slater LJ (1966) Generalized hypergeometric functions. Cambridge University Press, Cambridge
27. Luke YL (ed) (1969) Special functions and their approximations, vol 1. Academic, New York

Chapter 3
The Wright, Supertrigonometric and Superhyperbolic Functions

In this chapter, we study the concepts and theorems for Wright's generalized hypergeometric function and the Wright function, and introduce the related supertrigonometric and superhyperbolic functions.

3.1 Wright's Generalized Hypergeometric Function

We now begin with the definitions and theorems for Wright's generalized hypergeometric function. To formulate the results we use the following notation:

$$\Xi := \sum_{j=1}^{p} a_j - \sum_{k=1}^{q} b_k, \tag{3.1}$$

$$\sigma := \prod_{j=1}^{p} a_j^{-a_j} - \prod_{k=1}^{q} b_k^{-b_k}, \tag{3.2}$$

and

$$\chi := \sum_{k=1}^{q} \vartheta_k - \sum_{j=1}^{p} \kappa_j + \frac{p-q}{2}, \tag{3.3}$$

where $\kappa_j, \vartheta_k \in \mathbb{C}$, and $a_j, b_k \in \mathbb{R}_+$.

S. Rezaei Aderyani et al., *Towards Ulam Type Multi Stability Analysis*, Studies in Systems, Decision and Control 523, https://doi.org/10.1007/978-3-031-55564-0_3

Definition 3.1 ([1, 2]) Wright's generalized hypergeometric series is given by

$$
\begin{aligned}
&{}_p\mathbb{W}_q(X) \\
&= {}_p\mathbb{W}_q\left(\begin{smallmatrix}(\kappa_1,a_1),\cdots,(\kappa_p,a_p)\\(\vartheta_1,b_1),\cdots,(\vartheta_q,b_q)\end{smallmatrix}; X\right) \\
&= {}_p\mathbb{W}_q\left(\begin{smallmatrix}(\kappa_p,a_p)_{1,p}\\(\vartheta_q,b_q)_{1,q}\end{smallmatrix}; X\right) \\
&= \sum_{s=0}^{\infty} \frac{\left\{\prod_{j=1}^{p} \Gamma(\kappa_j + a_j s)\right\}}{\left\{\prod_{k=1}^{q} \Gamma(\upsilon_k + b_k s)\right\}} \frac{X^s}{s!},
\end{aligned}
\tag{3.4}
$$

where $X \in \mathbb{C}$, $\kappa_j, \upsilon_k \in \mathbb{C}$, $a_j, b_k \in \mathbb{R}_+$, and $\Xi > -1$.

This definition was presented by Wright in 1935 [1] and further investigated by Wright in 1940 [2].

Theorem 3.1 *Suppose* $X \in \mathbb{C}$, $\kappa_j, \upsilon_k \in \mathbb{C}$, $a_j, b_k \in \mathbb{R}_+$. *Then*

- *If* $\Xi + 1 > 0$, *then the series in* (3.4) *is absolutely convergent for* $X \in \mathbb{C}$.
- *If* $\Xi + 1 = 0$, *then the series in* (3.4) *is absolutely convergent for each values of* $|X| < \sigma$, *and of* $|X| = \sigma$ *and* $\Re(\chi) > 0.5$.

This theorem was first presented by Kibas in 2005 [3] and further reported on in [4].

Definition 3.2 ([5]) The Wright function is given by the following series representation:

$$
\mathbb{K}(\vartheta, b, X) = {}_0\mathbb{W}_1\left(\begin{smallmatrix}-\\(b,\vartheta)\end{smallmatrix}; X\right) = \sum_{k=0}^{\infty} \frac{1}{\Gamma(\vartheta + bk)} \frac{X^k}{k!}, \tag{3.5}
$$

where $X, \vartheta \in \mathbb{C}$, and $b \in \mathbb{R}$.

This definition was introduced by Wright in 1933 [5].

Theorem 3.2 *The Wright function for* $\vartheta > -1$ *and* $b \in \mathbb{C}$ ($b \in \mathbb{Z}_0^-$ *if* $\vartheta = 0$) *is an entire function of finite order* $\frac{1}{1+\vartheta}$ *and type* $(1 + \vartheta)|\vartheta|^{-\frac{\vartheta}{1+\vartheta}}$.

This result was presented by Luchko in 2019 [6].

Theorem 3.3 *The Wright function is an entire function of X for each values of the parameters* $\vartheta > -1$, *and* $b \in \mathbb{C}$.

This theorem was proven by Wright in 1935 [1, 7], reclaimed by Kilbas, Saigo, and Trujillo in 2002 [8], and further reported on by Luchko in 2019 [6].

Definition 3.3 ([3]) The Bessel-Maitland function (also called the Wright generalized Bessel function) is given by

$$\mathbb{J}(\kappa, a, X) = {}_0\mathbb{W}_1\left({-\atop (\kappa+1,b)}; -X\right)$$
$$= \sum_{k=0}^{\infty} \frac{1}{\Gamma(\kappa + 1 + ak)} \frac{(-X)^k}{k!},$$

where $X, \kappa \in \mathbb{C}$, and $a \in \mathbb{R}$.

Theorem 3.4 *Suppose $X \in \mathbb{C}$, $\kappa_j, \upsilon_k \in \mathbb{C}$, and $a_j, b_k \in \mathbb{R}_+$. Then, Wright's generalized hypergeometric function is an entire function of X.*

This theorem was presented by Kibas in 2005 [3].

Theorem 3.5 *Suppose $b \in \mathbb{R}$ and $\upsilon_k \in \mathbb{C}$.*

- *If $b + 1 > 0$, then the series in (3.5) is absolutely convergent for $X \in \mathbb{C}$.*
- *If $b + 1 = 0$, then the series in (3.5) is absolutely convergent for all values of $|X| < 1$ and of $|X| = 1$, and $\Re(\chi) > 0.5$.*

This result was presented by Kibas in 2005 [3].

Theorem 3.6 *Suppose $b > -1$, $\vartheta \in \mathbb{C}$. Then, the Wright function $\mathbb{K}(\vartheta, b, X)$ is an entire function of X.*

This work was presented by Kibas in 2005 [3].

Theorem 3.7 *Suppose $X \in \mathbb{C}$, $\kappa_j, \vartheta_k \in \mathbb{C}$, and $a_j, b_k \in \mathbb{R}_+$. Then, we get*

$${}_p\mathbb{W}_q\left({(\kappa_1,1),\cdots,(\kappa_p,1) \atop (\vartheta_1,1),\cdots,(\vartheta_q,1)}; X\right) = \frac{\prod_{j=1}^{p} \Gamma(\kappa_j)}{\prod_{k=1}^{p} \Gamma(\vartheta_k)} {}_p\mathbb{F}_q\left({\kappa_1,\cdots,\kappa_p \atop \vartheta_1,\cdots,\vartheta_q}; X\right),$$

where $\Xi + 1 \geq 0$.

This work was presented by Srivastava and Karlsson in 1985 [9].

Definition 3.4 ([10]) The shifted Wright generalized hypergeometric series is given by

$$\begin{aligned}&{}_p\mathbb{B}_q(X)\\ &= {}_p\mathbb{B}_q\left({(\kappa_1,a_1;\vartheta_1,b_1),\cdots,(\kappa_p,a_p;\vartheta_p,b_p) \atop (\widehat{\kappa}_1,c_1;\widehat{\vartheta}_1,d_1),\cdots,(\widehat{\kappa}_p,c_p;\widehat{\vartheta}_p,d_p)}; X\right)\\ &= {}_p\mathbb{B}_q\left({(\kappa_p,a_p;\vartheta_p,b_p)_{1,p} \atop (\widehat{\kappa}_p,c_p;\widehat{\vartheta}_p,d_p)_{1,q}}; X\right)\end{aligned}$$

$$= \sum_{k=0}^{\infty} \frac{\left\{\prod_{m=1}^{p} b(\kappa_m + a_m k; \vartheta_m + b_m k)\right\}}{\left\{\prod_{n=1}^{q} b(\widehat{\kappa}_n + c_n k; \widehat{\vartheta}_n + d_n k)\right\}} \frac{X^k}{k!}$$

$$= \sum_{k=0}^{\infty} \frac{\prod_{m=1}^{p}\Big(\Gamma(\kappa_m + a_m k)\Gamma(\vartheta_m + b_m k)\Big)\prod_{n=1}^{q}\Gamma\Big((\widehat{\kappa}_n + \widehat{\vartheta}_n) + (c_n + d_n)k\Big)}{\prod_{m=1}^{p}\Gamma\Big((\vartheta_m + \kappa_m) + (b_m + a_m)k\Big)\prod_{n=1}^{q}\Big(\Gamma(\widehat{\kappa}_n + c_n k)\Gamma(\widehat{\vartheta}_n + d_n k)\Big)} \frac{X^k}{k!},$$

where $\kappa_m, \vartheta_m, \widehat{\kappa}_n, \widehat{\vartheta}_n, X \in \mathbb{C}$, and $a_m, b_m, c_n, d_n \in \mathbb{R}_+$.

There are some special cases as follows [10]:

$${}_0\mathbb{B}_0 = e^X,$$

$$\begin{aligned}
&{}_1\mathbb{B}_0(X)\\
&= {}_1\mathbb{B}_0\left(\begin{smallmatrix}(\kappa,a;\vartheta,b)\\ -\end{smallmatrix}; X\right)\\
&= \sum_{k=0}^{\infty} b(\kappa + ak; \vartheta + bk)\frac{X^k}{k!}\\
&= \sum_{k=0}^{\infty} \frac{\Gamma(\kappa + ak)\Gamma(\vartheta + bk)}{\Gamma[(\vartheta + \kappa) + (b + b)k]} \frac{X^k}{k!}\\
&= {}_2\mathbb{W}_1\left(\begin{smallmatrix}(\kappa,b),(\vartheta,b)\\ (\vartheta+\kappa,b+b)\end{smallmatrix}; X\right),
\end{aligned}$$

$$\begin{aligned}
&{}_0\mathbb{B}_1(X)\\
&= {}_0\mathbb{B}_1\left(\begin{smallmatrix}-\\ (\kappa,a;\vartheta,b)\end{smallmatrix}; X\right)\\
&= \sum_{k=0}^{\infty} \frac{1}{b(\kappa + ak; \vartheta + bk)} \frac{X^k}{k!}\\
&= \sum_{k=0}^{\infty} \frac{\Gamma[(\vartheta + \kappa) + (b + b)k]}{\Gamma(\kappa + ak)\Gamma(\vartheta + bk)} \frac{X^k}{k!}\\
&= {}_1\mathbb{W}_2\left(\begin{smallmatrix}(\vartheta+\kappa,b+b)\\ (\kappa,b),(\vartheta,b)\end{smallmatrix}; X\right),
\end{aligned}$$

$$
\begin{aligned}
&{}_1\mathbb{B}_1(X) \\
&= {}_1\mathbb{B}_1\left({}^{(\widehat{\kappa},c;\widehat{\vartheta},d)}_{(\kappa,a;\vartheta,b)};X\right) \\
&= \sum_{k=0}^{\infty} \frac{b(\kappa+ak;\vartheta+bk)}{b(\widehat{\kappa}+ck;\widehat{\vartheta}+dk)} \frac{X^k}{k!} \\
&= \sum_{k=0}^{\infty} \frac{\Gamma(\kappa+ak)\Gamma(\vartheta+bk)\Gamma[(\widehat{\kappa}+\widehat{\vartheta})+(c+d)k]}{\Gamma[(\kappa+\vartheta)+(b+b)k]\Gamma(\widehat{\kappa}+ck)\Gamma(\widehat{\vartheta}+dk)} \frac{X^k}{k!} \\
&= {}_3\mathbb{W}_3\left({}^{(\kappa,b),(\vartheta,b),(\widehat{\kappa}+\widehat{\vartheta},c+d)}_{(\widehat{\kappa},c),(\widehat{\vartheta},d),(\vartheta+\kappa,b+b)};X\right),
\end{aligned}
$$

where $\kappa, \vartheta, \widehat{\kappa}, \widehat{\vartheta}, X \in \mathbb{C}$, and $a, b, c, d \in \mathbb{R}_+$.

3.2 The Supertrigonometric Functions via Wright's Generalized Hypergeometric Function

Let $i^2 = -1$, $X \in \mathbb{C}$, $\kappa_j, \vartheta_k \in \mathbb{C}$, and $a_j, b_k \in \mathbb{R}_+$. In this part, based on the above formulas, we propose the definitions for the supertrigonometric functions [10] via Wright's generalized hypergeometric function as follows:

▹ The supersine through Wright's generalized hypergeometric function:

$$
\begin{aligned}
{}_p\sin_q(X) &= \frac{1}{2i}\left({}_p\mathbb{W}_q(iX) - {}_p\mathbb{W}_q(-iX)\right) \\
&= \sum_{k=0}^{\infty}(-1)^k \frac{\left\{\prod_{j=1}^{p}\Gamma(\kappa_j+a_j(2k+1)\right\}}{\left\{\prod_{j=1}^{p}\Gamma(\vartheta_j+b_j(2k+1)\right\}} \frac{X^{2k+1}}{(2k+1)!}.
\end{aligned}
\tag{3.6}
$$

▹ The supercosine through Wright's generalized hypergeometric function:

$$
\begin{aligned}
{}_p\cos_q(X) &= \frac{1}{2}\left({}_p\mathbb{W}_q(iX) + {}_p\mathbb{W}_q(-iX)\right) \\
&= \sum_{k=0}^{\infty}(-1)^k \frac{\left\{\prod_{j=1}^{p}\Gamma(\kappa_j+a_j(2k)\right\}}{\left\{\prod_{j=1}^{p}\Gamma(\vartheta_j+b_j(2k)\right\}} \frac{X^{2k}}{(2k)!}.
\end{aligned}
\tag{3.7}
$$

▹ The supertangent through Wright's generalized hypergeometric function:

$$
\begin{aligned}
{}_p\tan_q(X) &= \frac{{}_p\sin_q(X)}{{}_p\cos_q(X)} \\
&= \frac{1}{i}\frac{{}_p\mathbb{W}_q(iX) - {}_p\mathbb{W}_q(-iX)}{{}_p\mathbb{W}_q(iX) + {}_p\mathbb{W}_q(-iX)} \\
&= \frac{\sum_{k=0}^{\infty}(-1)^k \frac{\left\{\prod_{j=1}^{p}\Gamma(\kappa_j + a_j(2k+1)\right\}}{\left\{\prod_{j=1}^{p}\Gamma(\vartheta_j + b_j(2k+1)\right\}}\frac{X^{2k+1}}{(2k+1)!}}{\sum_{k=0}^{\infty}(-1)^k \frac{\left\{\prod_{j=1}^{p}\Gamma(\kappa_j + a_j(2k)\right\}}{\left\{\prod_{j=1}^{p}\Gamma(\vartheta_j + b_j(2k)\right\}}\frac{X^{2k}}{(2k)!}}.
\end{aligned}
\tag{3.8}
$$

▷ The supercotangent through Wright's generalized hypergeometric function:

$$
\begin{aligned}
{}_p\cot_q(X) &= \frac{{}_p\cos_q(X)}{{}_p\sin_q(X)} \\
&= i\frac{{}_p\mathbb{W}_q(iX) + {}_p\mathbb{W}_q(-iX)}{{}_p\mathbb{W}_q(iX) - {}_p\mathbb{W}_q(-iX)} \\
&= \frac{\sum_{k=0}^{\infty}(-1)^k \frac{\left\{\prod_{j=1}^{p}\Gamma(\kappa_j + a_j(2k)\right\}}{\left\{\prod_{j=1}^{p}\Gamma(\vartheta_j + b_j(2k)\right\}}\frac{X^{2k}}{(2k)!}}{\sum_{k=0}^{\infty}(-1)^k \frac{\left\{\prod_{j=1}^{p}\Gamma(\kappa_j + a_j(2k+1)\right\}}{\left\{\prod_{j=1}^{p}\Gamma(\vartheta_j + b_j(2k+1)\right\}}\frac{X^{2k+1}}{(2k+1)!}}.
\end{aligned}
\tag{3.9}
$$

▷ The supersecant through Wright's generalized hypergeometric function:

$$
\begin{aligned}
{}_p\sec_q(X) &= \frac{1}{{}_p\cos_q(X)} \\
&= \frac{2}{{}_p\mathbb{W}_q(iX) + {}_p\mathbb{W}_q(-iX)} \\
&= \frac{1}{\sum_{k=0}^{\infty}(-1)^k \frac{\left\{\prod_{j=1}^{p}\Gamma(\kappa_j + a_j(2k)\right\}}{\left\{\prod_{j=1}^{p}\Gamma(\vartheta_j + b_j(2k)\right\}}\frac{X^{2k}}{(2k)!}}.
\end{aligned}
\tag{3.10}
$$

▷ The supercosecant through Wright's generalized hypergeometric function:

$$
\begin{aligned}
{}_p\mathrm{cosec}_q(X) &= \frac{1}{{}_p\sin_q(X)} \\
&= \frac{2i}{{}_p\mathbb{W}_q(iX) - {}_p\mathbb{W}_q(-iX)} \\
&= \frac{1}{\sum_{k=0}^{\infty}(-1)^k \frac{\left\{\prod_{j=1}^{p}\Gamma(\kappa_j + a_j(2k+1)\right\}}{\left\{\prod_{j=1}^{p}\Gamma(\vartheta_j + b_j(2k+1)\right\}}\frac{X^{2k+1}}{(2k+1)!}}.
\end{aligned}
\tag{3.11}
$$

The supercosine ${}_p\cos_q(X)$, supersine ${}_p\sin_q(X)$, supercotangent ${}_p\cot_q(X)$, supertangent ${}_p\tan_q(X)$, supercosecant ${}_p\mathrm{cosec}_q(X)$, and supersecant ${}_p\sec_q(X)$ are called the supertrigonometric functions through Wright's generalized hypergeometric function.

Theorem 3.8 ([10]) *Suppose $X \in \mathbb{C}$, $\kappa_j, \vartheta_k \in \mathbb{C}$, and $a_j, b_k \in \mathbb{R}_+$. Then*

- *If $\Xi + 1 > 0$, then the series in (3.6) is absolutely convergent for $X \in \mathbb{C}$;*
- *If $\Xi + 1 = 0$, then the series in (3.6) is absolutely convergent for each value of $|X| < \sigma$, and of $|X| = \sigma$, and $\Re(\chi) > 0.5$.*

Theorem 3.9 ([10]) *Suppose $X \in \mathbb{C}$, $\kappa_j, \vartheta_k \in \mathbb{C}$, and $a_j, b_k \in \mathbb{R}_+$. Then*

- *If $\Xi + 1 > 0$, then the series in (3.7) is absolutely convergent for $X \in \mathbb{C}$;*
- *If $\Xi + 1 = 0$, then the series in (3.7) is absolutely convergent for each values of $|X| < \sigma$, and of $|X| = \sigma$, and $\Re(\chi) > 0.5$.*

Theorem 3.10 ([10]) *Suppose $i^2 = -1$, $X, \kappa_j, \vartheta_k \in \mathbb{C}$, and $a_j, b_k \in \mathbb{R}_+$, and $\sum_{j=1}^{p} a_j - \sum_{j=1}^{q} b_j = \Xi > -1$. Then*

$$
{}_p\mathbb{W}_q(iX) = {}_p\cos_q(X) + i\,{}_p\sin_q(X),
$$

and

$$
{}_p\mathbb{W}_q(-iX) = {}_p\cos_q(X) - i\,{}_p\sin_q(X).
$$

Theorem 3.11 ([10]) *Suppose $X \in \mathbb{C}$, $\Xi + 1 \geq 0$, $\kappa_j \in \mathbb{C}$, $\vartheta_k \in \mathbb{C} \setminus \{0, -1, -2, \ldots\}$ and $a_j, b_k \in \mathbb{R}_+$. Then*

$$
{}_p\sin_q\left({}^{(\kappa_1,1),\cdots,(\kappa_p,1)}_{(\vartheta_1,1),\cdots,(\vartheta_q,1)}; X\right) = \frac{\prod_{j=1}^{p}\Gamma(\kappa_j)}{\prod_{k=1}^{p}\Gamma(\vartheta_k)}\,{}_p supersine_q\left({}^{\kappa_1,\cdots,\kappa_p}_{\vartheta_1\cdots,\vartheta_q}; X\right).
$$

3.3 The Superhyperbolic Functions via Wright's Generalized Hypergeometric Function

Let $i^2 = -1$, $X, \kappa_j, \vartheta_k \in \mathbb{C}$, and $a_j, b_k \in \mathbb{R}_+$. In this part, we introduce the definitions for the superhyperbolic functions [10] via Wright's generalized hypergeometric function as follows:

▷ The superhyperbolic sine by Wright's generalized hypergeometric function:

$$\begin{aligned} {}_p\sinh_q(X) &= \frac{1}{2}\Big({}_p\mathbb{W}_q(X) - {}_p\mathbb{W}_q(-X)\Big) \\ &= \sum_{k=0}^{\infty} \frac{\left\{\prod_{j=1}^{p} \Gamma(\kappa_j + a_j(2k+1)\right\}}{\left\{\prod_{j=1}^{p} \Gamma(\vartheta_j + b_j(2k+1)\right\}} \frac{X^{2k+1}}{(2k+1)!}. \end{aligned} \tag{3.12}$$

▷ The superhyperbolic cosine by Wright's generalized hypergeometric function:

$$\begin{aligned} {}_p\cosh_q(X) &= \frac{1}{2}\Big({}_p\mathbb{W}_q(X) + {}_p\mathbb{W}_q(-X)\Big) \\ &= \sum_{k=0}^{\infty} \frac{\left\{\prod_{j=1}^{p} \Gamma(\kappa_j + a_j(2k)\right\}}{\left\{\prod_{j=1}^{p} \Gamma(\vartheta_j + b_j(2k)\right\}} \frac{X^{2k}}{(2k)!}. \end{aligned} \tag{3.13}$$

▷ The superhyperbolic tangent by Wright's generalized hypergeometric function:

$$\begin{aligned} {}_p\tanh_q(X) &= \frac{{}_p\sinh_q(X)}{{}_p\cosh_q(X)} \\ &= \frac{{}_p\mathbb{W}_q(X) - {}_p\mathbb{W}_q(-X)}{{}_p\mathbb{W}_q(X) + {}_p\mathbb{W}_q(-X)} \\ &= \frac{\sum_{k=0}^{\infty} \frac{\left\{\prod_{j=1}^{p} \Gamma(\kappa_j + a_j(2k+1)\right\}}{\left\{\prod_{j=1}^{p} \Gamma(\vartheta_j + b_j(2k+1)\right\}} \frac{X^{2k+1}}{(2k+1)!}}{\sum_{k=0}^{\infty} \frac{\left\{\prod_{j=1}^{p} \Gamma(\kappa_j + a_j(2k)\right\}}{\left\{\prod_{j=1}^{p} \Gamma(\vartheta_j + b_j(2k)\right\}} \frac{X^{2k}}{(2k)!}}. \end{aligned} \tag{3.14}$$

▷ The superhyperbolic cotangent by Wright's generalized hypergeometric function:

$$
\begin{aligned}
{}_p\coth_q(X) &= \frac{{}_p\cosh_q(X)}{{}_p\sinh_q(X)} \\
&= \frac{{}_p\mathbb{W}_q(X) + {}_p\mathbb{W}_q(-X)}{{}_p\mathbb{W}_q(X) - {}_p\mathbb{W}_q(-X)} \\
&= \frac{\sum_{k=0}^{\infty} \dfrac{\left\{\prod_{j=1}^{p} \Gamma(\kappa_j + a_j(2k)\right\}}{\left\{\prod_{j=1}^{p} \Gamma(\vartheta_j + b_j(2k)\right\}} \dfrac{X^{2k}}{(2k)!}}{\sum_{k=0}^{\infty} \dfrac{\left\{\prod_{j=1}^{p} \Gamma(\kappa_j + a_j(2k+1)\right\}}{\left\{\prod_{j=1}^{p} \Gamma(\vartheta_j + b_j(2k+1)\right\}} \dfrac{X^{2k+1}}{(2k+1)!}}.
\end{aligned}
\tag{3.15}
$$

▷ The superhyperbolic secant by Wright's generalized hypergeometric function:

$$
\begin{aligned}
{}_p\mathrm{sech}_q(X) &= \frac{1}{{}_p\cosh_q(X)} \\
&= \frac{2}{{}_p\mathbb{W}_q(X) + {}_p\mathbb{W}_q(-X)} \\
&= \frac{1}{\sum_{k=0}^{\infty} \dfrac{\left\{\prod_{j=1}^{p} \Gamma(\kappa_j + a_j(2k)\right\}}{\left\{\prod_{j=1}^{p} \Gamma(\vartheta_j + b_j(2k)\right\}} \dfrac{X^{2k}}{(2k)!}}.
\end{aligned}
\tag{3.16}
$$

▷ The superhyperbolic cosecant by Wright's generalized hypergeometric function:

$$
\begin{aligned}
{}_p\mathrm{cosech}_q(X) &= \frac{1}{{}_p\sinh_q(X)} \\
&= \frac{2}{{}_p\mathbb{W}_q(X) - {}_p\mathbb{W}_q(-X)} \\
&= \frac{1}{\sum_{k=0}^{\infty} \dfrac{\left\{\prod_{j=1}^{p} \Gamma(\kappa_j + a_j(2k+1)\right\}}{\left\{\prod_{j=1}^{p} \Gamma(\vartheta_j + b_j(2k+1)\right\}} \dfrac{X^{2k+1}}{(2k+1)!}}.
\end{aligned}
\tag{3.17}
$$

Notice that the superhyperbolic cosine ${}_p\cosh_q(X)$, superhyperbolic sine ${}_p\sinh_q(X)$, superhyperbolic cotangent ${}_p\coth_q(X)$, superhyperbolic tangent ${}_p\tanh_q(X)$, superhyperbolic cosecant ${}_p\mathrm{cosech}_q(X)$, and superhyperbolic secant

${}_p\mathrm{sech}_q(X)$ are called the superhyperbolic functions through Wright's generalized hypergeometric function.

Theorem 3.12 ([10]) *Suppose $X, \kappa_j, \vartheta_k \in \mathbb{C}$, and $a_j, b_k \in \mathbb{R}_+$. Then*

- *If $\Xi + 1 > 0$, then the series in* (3.12) *is absolutely convergent for $X \in \mathbb{C}$;*
- *If $\Xi + 1 = 0$, then the series in* (3.12) *is absolutely convergent for each value of $|X| < \sigma$, and of $|X| = \sigma$, and $\Re(\chi) > 0.5$.*

Theorem 3.13 ([10]) *Suppose $X, \kappa_j, \vartheta_k \in \mathbb{C}$, and $a_j, b_k \in \mathbb{R}_+$. Then*

- *If $\Xi + 1 > 0$, then the series in* (3.13) *is absolutely convergent for $X \in \mathbb{C}$;*
- *If $\Xi + 1 = 0$, then the series in* (3.13) *is absolutely convergent for each value of $|X| < \sigma$, and of $|X| = \sigma$, and $\Re(\chi) > 0.5$.*

Theorem 3.14 ([10]) *Suppose $i^2 = -1$, $X, \kappa_j, \vartheta_k \in \mathbb{C}$, and $a_j, b_k \in \mathbb{R}_+$, and $\sum_{j=1}^{p} a_j - \sum_{j=1}^{q} b_j = \Xi > -1$. Then*

$$ {}_p\mathbb{W}_q(X) = {}_p\cosh_q(X) + {}_p\sinh_q(X), $$

and

$$ {}_p\mathbb{W}_q(-X) = {}_p\cosh_q(X) - {}_p\sinh_q(X). $$

Let $X, \kappa_j, \vartheta_k \in \mathbb{C}$, and $a_j, b_k \in \mathbb{R}_+$. Then we have the following properties [10]:

$$ {}_p\sinh_q\left({}^{(\kappa_1,a_1),\cdots,(\kappa_p,a_p)}_{(\vartheta_1,b_1),\ldots,(\vartheta_q,b_q)}; X\right) = -i\,{}_p\sin_q\left({}^{(\kappa_1,a_1),\ldots,(\kappa_p,a_p)}_{(\vartheta_1,b_1),\ldots,(\vartheta_q,b_q)}; iX\right), $$

$$ {}_p\cosh_q\left({}^{(\kappa_1,a_1),\ldots,(\kappa_p,a_p)}_{(\vartheta_1,b_1),\ldots,(\vartheta_q,b_q)}; X\right) = {}_p\cos_q\left({}^{(\kappa_1,a_1),\ldots,(\kappa_p,a_p)}_{(\vartheta_1,b_1),\ldots,(\vartheta_q,b_q)}; iX\right), $$

$$ {}_p\tanh_q\left({}^{(\kappa_1,a_1),\ldots,(\kappa_p,a_p)}_{(\vartheta_1,b_1),\ldots,(\vartheta_q,b_q)}; X\right) = -i\,{}_p\tan_q\left({}^{(\kappa_1,a_1),\ldots,(\kappa_p,a_p)}_{(\vartheta_1,b_1),\ldots,(\vartheta_q,b_q)}; iX\right), $$

$$ {}_p\coth_q\left({}^{(\kappa_1,a_1),\ldots,(\kappa_p,a_p)}_{(\vartheta_1,b_1),\ldots,(\vartheta_q,b_q)}; X\right) = i\,{}_p\cot_q\left({}^{(\kappa_1,a_1),\ldots,(\kappa_p,a_p)}_{(\vartheta_1,b_1),\ldots,(\vartheta_q,b_q)}; iX\right), $$

$$ {}_p\mathrm{sech}_q\left({}^{(\kappa_1,a_1),\ldots,(\kappa_p,a_p)}_{(\vartheta_1,b_1),\ldots,(\vartheta_q,b_q)}; X\right) = {}_p\sec_q\left({}^{(\kappa_1,a_1),\ldots,(\kappa_p,a_p)}_{(\vartheta_1,b_1),\ldots,(\vartheta_q,b_q)}; iX\right), $$

$$ {}_p\mathrm{cosech}_q\left({}^{(\kappa_1,a_1),\ldots,(\kappa_p,a_p)}_{(\vartheta_1,b_1),\ldots,(\vartheta_q,b_q)};X\right)=i\,{}_p\mathrm{cosec}_q\left({}^{(\kappa_1,a_1),\ldots,(\kappa_p,a_p)}_{(\vartheta_1,b_1),\ldots,(\vartheta_q,b_q)};iX\right). $$

Also, we have [10],

$$ {}_2\mathbb{W}_2\left({}^{(\kappa_1,a_1),(\kappa_2,a_2)}_{(\vartheta_1,b_1),(\vartheta_2,b_2)};X\right)=\sum_{k=0}^{\infty}\frac{\Gamma(\kappa_1+a_1k)\Gamma(\kappa_2+a_2k)}{\Gamma(\vartheta_1+b_1k)\Gamma(\vartheta_2+b_2k)}\frac{X^k}{k!}, $$

$$ {}_2\sin_2\left({}^{(\kappa_1,a_1),(\kappa_2,a_2)}_{(\vartheta_1,b_1),(\vartheta_2,b_2)};X^\kappa\right) $$
$$ =\frac{1}{2i}\left[{}_2\mathbb{W}_2\left({}^{(\kappa_1,a_1),(\kappa_2,a_2)}_{(\vartheta_1,b_1),(\vartheta_2,b_2)};iX^\kappa\right)-{}_2\mathbb{W}_2\left({}^{(\kappa_1,a_1),(\kappa_2,a_2)}_{(\vartheta_1,b_1),(\vartheta_2,b_2)};-iX^\kappa\right)\right], $$

$$ {}_2\cos_2\left({}^{(\kappa_1,a_1),(\kappa_2,a_2)}_{(\vartheta_1,b_1),(\vartheta_2,b_2)};X^\kappa\right) $$
$$ =\frac{1}{2}\left[{}_2\mathbb{W}_2\left({}^{(\kappa_1,a_1),(\kappa_2,a_2)}_{(\vartheta_1,b_1),(\vartheta_2,b_2)};iX^\kappa\right)+{}_2\mathbb{W}_2\left({}^{(\kappa_1,a_1),(\kappa_2,a_2)}_{(\vartheta_1,b_1),(\vartheta_2,b_2)};-iX^\kappa\right)\right], $$

$$ {}_2\sinh_2\left({}^{(\kappa_1,a_1),(\kappa_2,a_2)}_{(\vartheta_1,b_1),(\vartheta_2,b_2)};X^\kappa\right) $$
$$ =\frac{1}{2}\left[{}_2\mathbb{W}_2\left({}^{(\kappa_1,a_1),(\kappa_2,a_2)}_{(\vartheta_1,b_1),(\vartheta_2,b_2)};X^\kappa\right)-{}_2\mathbb{W}_2\left({}^{(\kappa_1,a_1),(\kappa_2,a_2)}_{(\vartheta_1,b_1),(\vartheta_2,b_2)};-X^\kappa\right)\right], $$

and

$$ {}_2\cosh_2\left({}^{(\kappa_1,a_1),(\kappa_2,a_2)}_{(\vartheta_1,b_1),(\vartheta_2,b_2)};X^\kappa\right) $$
$$ =\frac{1}{2}\left[{}_2\mathbb{W}_2\left({}^{(\kappa_1,a_1),(\kappa_2,a_2)}_{(\vartheta_1,b_1),(\vartheta_2,b_2)};X^\kappa\right)+{}_2\mathbb{W}_2\left({}^{(\kappa_1,a_1),(\kappa_2,a_2)}_{(\vartheta_1,b_1),(\vartheta_2,b_2)};-X^\kappa\right)\right], $$

where $\kappa_1,\kappa_2,\vartheta_1,\vartheta_2,X\in\mathbb{C}$ and $a_1,a_2,b_1,b_2\in\mathbb{R}_+$.

3.4 The Supertrigonometric Functions via Wright's Function

Let $X,\vartheta\in\mathbb{C}$, and $b\in\mathbb{R}$. In this part, we introduce the definitions for the supertrigonometric functions [10] via Wright's function as follows:

▷ The supersine through Wright's function:

$$\begin{aligned}
&{}_0\sin_1\left(\substack{-\\(b,\vartheta)};X\right)\\
&=\frac{1}{2i}\Big(\mathbb{K}(\vartheta,b,iX)-\mathbb{K}(\vartheta,b,-iX)\Big)\\
&=\frac{1}{2i}\left({}_0\mathbb{W}_1\left(\substack{-\\(b,\vartheta)};iX\right)-{}_0\mathbb{W}_1\left(\substack{-\\(b,\vartheta)};-iX\right)\right)\\
&=\sum_{k=0}^{\infty}\frac{(-1)^k}{\Gamma(\vartheta+b(2k+1))}\frac{X^{2k+1}}{(2k+1)!}.
\end{aligned}$$

▷ The supercosine through Wright's function:

$$\begin{aligned}
&{}_0\cos_1\left(\substack{-\\(b,\vartheta)};X\right)\\
&=\frac{1}{2}\Big(\mathbb{K}(\vartheta,b,iX)+\mathbb{K}(\vartheta,b,-iX)\Big)\\
&=\frac{1}{2}\left({}_0\mathbb{W}_1\left(\substack{-\\(b,\vartheta)};iX\right)-{}_0\mathbb{W}_1\left(\substack{-\\(b,\vartheta)};-iX\right)\right)\\
&=\sum_{k=0}^{\infty}\frac{(-1)^k}{\Gamma(\vartheta+b(2k))}\frac{X^{2k}}{(2k)!}.
\end{aligned}$$

▷ The supertangent through Wright's function:

$$\begin{aligned}
&{}_0\tan_1\left(\substack{-\\(b,\vartheta)};X\right)\\
&=\frac{{}_0\sin_1\left(\substack{-\\(b,\vartheta)};X\right)}{{}_0\cos_1\left(\substack{-\\(b,\vartheta)};X\right)}\\
&=\frac{1}{i}\frac{\mathbb{K}(\vartheta,b,iX)-\mathbb{K}(\vartheta,b,-iX)}{\mathbb{K}(\vartheta,b,iX)+\mathbb{K}(\vartheta,b,-iX)}\\
&=\frac{\sum_{k=0}^{\infty}\frac{(-1)^k}{\Gamma(\vartheta+b(2k+1))}\frac{X^{2k+1}}{(2k+1)!}}{\sum_{k=0}^{\infty}\frac{(-1)^k}{\Gamma(\vartheta+b(2k))}\frac{X^{2k}}{(2k)!}}.
\end{aligned}$$

▷ The supercotangent through Wright's function:

$$
\begin{aligned}
&{}_0\cot_1\left(\overline{(b,\vartheta)}; X\right)\\
&= \frac{{}_0\cos_1\left(\overline{(b,\vartheta)}; X\right)}{{}_0\sin_1\left(\overline{(b,\vartheta)}; X\right)}\\
&= i\frac{\mathbb{K}(\vartheta, b, iX) + \mathbb{K}(\vartheta, b, -iX)}{\mathbb{K}(\vartheta, b, iX) - \mathbb{K}(\vartheta, b, -iX)}\\
&= \frac{\sum_{k=0}^{\infty} \dfrac{(-1)^k}{\Gamma(\vartheta + b(2k))} \dfrac{X^{2k}}{(2k)!}}{\sum_{k=0}^{\infty} \dfrac{(-1)^k}{\Gamma(\vartheta + b(2k+1))} \dfrac{X^{2k+1}}{(2k+1)!}}.
\end{aligned}
$$

▷ The supersecant through Wright's function:

$$
\begin{aligned}
&{}_0\sec_1\left(\overline{(b,\vartheta)}; X\right)\\
&= \frac{1}{{}_0\cos_1\left(\overline{(b,\vartheta)}; X\right)}\\
&= \frac{2}{\mathbb{K}(\vartheta, b, iX) + \mathbb{K}(\vartheta, b, -iX)}.
\end{aligned}
$$

▷ The supercosecant through Wright's function:

$$
\begin{aligned}
&{}_0\mathrm{cosec}_1\left(\overline{(b,\vartheta)}; X\right)\\
&= \frac{1}{{}_0\sin_1\left(\overline{(b,\vartheta)}; X\right)}\\
&= \frac{2i}{\mathbb{K}(\vartheta, b, iX) - \mathbb{K}(\vartheta, b, -iX)}.
\end{aligned}
$$

Theorem 3.15 ([10]) *Suppose $i^2 = -1$, $X, \vartheta \in \mathbb{C}$, and $b \in \mathbb{R}$. Then*

$$
\mathbb{K}(\vartheta, b, iX) = {}_0\cos_1\left(\overline{(b,\vartheta)}; X\right) + i\,{}_0\sin_1\left(\overline{(b,\vartheta)}; X\right),
$$

and

$$
\mathbb{K}(\vartheta, b, -iX) = {}_0\cos_1\left(\overline{(b,\vartheta)}; X\right) - i\,{}_0\sin_1\left(\overline{(b,\vartheta)}; X\right).
$$

3.5 The Superhyperbolic Functions via Wright's Function

Let $X, \vartheta \in \mathbb{C}$, and $b \in \mathbb{R}$. In this part, we introduce the definitions for the superhyperbolic functions [10] via Wright's function as follows:

▷ The superhyperbolic sine through Wright's function:

$$\begin{aligned}
&{}_0\sinh_1\left(\overline{(b,\vartheta)}; X\right)\\
&= \frac{1}{2}\Big(\mathbb{K}(\vartheta, b, X) - \mathbb{K}(\vartheta, b, -X)\Big)\\
&= \frac{1}{2}\left({}_0\mathbb{W}_1\left(\overline{(b,\vartheta)}; X\right) - {}_0\mathbb{W}_1\left(\overline{(b,\vartheta)}; -X\right)\right)\\
&= \sum_{k=0}^{\infty} \frac{1}{\Gamma(\vartheta + b(2k+1))} \frac{X^{2k+1}}{(2k+1)!}.
\end{aligned}$$

▷ The superhyperbolic cosine through Wright's function:

$$\begin{aligned}
&{}_0\cosh_1\left(\overline{(b,\vartheta)}; X\right)\\
&= \frac{1}{2}\Big(\mathbb{K}(\vartheta, b, X) + \mathbb{K}(\vartheta, b, -X)\Big)\\
&= \frac{1}{2}\left({}_0\mathbb{W}_1\left(\overline{(b,\vartheta)}; X\right) - {}_0\mathbb{W}_1\left(\overline{(b,\vartheta)}; -X\right)\right)\\
&= \sum_{k=0}^{\infty} \frac{1}{\Gamma(\vartheta + b(2k))} \frac{X^{2k}}{(2k)!}.
\end{aligned}$$

▷ The superhyperbolic tangent through Wright's function:

$$\begin{aligned}
&{}_0\tanh_1\left(\overline{(b,\vartheta)}; X\right)\\
&= \frac{{}_0\sinh_1\left(\overline{(b,\vartheta)}; X\right)}{{}_0\cosh_1\left(\overline{(b,\vartheta)}; X\right)}\\
&= \frac{\mathbb{K}(\vartheta, b, X) - \mathbb{K}(\vartheta, b, -X)}{\mathbb{K}(\vartheta, b, X) + \mathbb{K}(\vartheta, b, -X)}\\
&= \frac{\sum_{k=0}^{\infty} \frac{1}{\Gamma(\vartheta + b(2k+1))} \frac{X^{2k+1}}{(2k+1)!}}{\sum_{k=0}^{\infty} \frac{1}{\Gamma(\vartheta + b(2k))} \frac{X^{2k}}{(2k)!}}.
\end{aligned}$$

▷ The superhyperbolic cotangent through Wright's function:

$$
\begin{aligned}
&{}_0\coth_1\left(\overline{}_{(b,\vartheta)}; X\right) \\
&= \frac{{}_0\cosh_1\left(\overline{}_{(b,\vartheta)}; X\right)}{{}_0\sinh_1\left(\overline{}_{(b,\vartheta)}; X\right)} \\
&= \frac{\mathbb{K}(\vartheta, b, X) + \mathbb{K}(\vartheta, b, -X)}{\mathbb{K}(\vartheta, b, X) - \mathbb{K}(\vartheta, b, -X)} \\
&= \frac{\sum_{k=0}^{\infty} \dfrac{1}{\Gamma(\vartheta + b(2k))} \dfrac{X^{2k}}{(2k)!}}{\sum_{k=0}^{\infty} \dfrac{1}{\Gamma(\vartheta + b(2k+1))} \dfrac{X^{2k+1}}{(2k+1)!}}.
\end{aligned}
$$

▷ The superhyperbolic secant through Wright's function:

$$
\begin{aligned}
&{}_0\mathrm{sech}_1\left(\overline{}_{(b,\vartheta)}; X\right) \\
&= \frac{1}{{}_0\cosh_1\left(\overline{}_{(b,\vartheta)}; X\right)} \\
&= \frac{2}{\mathbb{K}(\vartheta, b, iX) + \mathbb{K}(\vartheta, b, -iX)}.
\end{aligned}
$$

▷ The superhyperbolic cosecant through Wright's function:

$$
\begin{aligned}
&{}_0\mathrm{cosech}_1\left(\overline{}_{(b,\vartheta)}; X\right) \\
&= \frac{1}{{}_0\sinh_1\left(\overline{}_{(b,\vartheta)}; X\right)} \\
&= \frac{2i}{\mathbb{K}(\vartheta, b, iX) - \mathbb{K}(\vartheta, b, -iX)}.
\end{aligned}
$$

Theorem 3.16 ([10]) *Suppose $X, \vartheta \in \mathbb{C}$, and $b \in \mathbb{R}$. Then,*

$$
\mathbb{K}(\vartheta, b, X) = {}_0\cosh_1\left(\overline{}_{(b,\vartheta)}; X\right) + {}_0\sinh_1\left(\overline{}_{(b,\vartheta)}; X\right),
$$

and

$$\mathbb{K}(\vartheta, b, -X) = {}_0\cosh_1\left({}_{(b,\vartheta)}^{-}; X\right) - {}_0\sinh_1\left({}_{(b,\vartheta)}^{-}; X\right).$$

3.6 The Truncated Wright Generalized Hypergeometric Function

Definition 3.5 ([10]) The generalized Wright hypergeometric series is given by

$$[{}_p\mathbb{W}_q]^n(X) = \sum_{s=0}^{n} \frac{\left\{\prod_{j=1}^{p} \Gamma(\kappa_j + a_j s)\right\}}{\left\{\prod_{k=1}^{q} \Gamma(\vartheta_k + b_k s)\right\}} \frac{X^s}{s!},$$

where $X, \kappa_j, \vartheta_j \in \mathbb{C}$, $q, p \in \mathbb{N}$, and $a_j, b_k \in \mathbb{R}_+$.

We let $X, \kappa_j, \vartheta_j \in \mathbb{C}$, $q, p \in \mathbb{N}$, and $a_j, b_k \in \mathbb{R}_+$. Now, we present the truncated Wright Generalized Hypergeometric Functions [10] as follows:

▷ The truncated supersine through Wright's generalized hypergeometric function:

$$\begin{aligned}
&[{}_p\sin_q]^n(X) \\
&= \frac{1}{2i}\Big([{}_p\mathbb{W}_q]^n(iX) - [{}_p\mathbb{W}_q]^n(-iX)\Big) \\
&= \sum_{k=0}^{n}(-1)^k \frac{\left\{\prod_{j=1}^{p} \Gamma(\kappa_j + a_j(2k+1))\right\}}{\left\{\prod_{j=1}^{q} \Gamma(\vartheta_j + b_j(2k+1))\right\}} \frac{X^{2k+1}}{(2k+1)!}, \quad n = 0, 1, \cdots .
\end{aligned}$$

▷ The truncated supercosine through Wright's generalized hypergeometric function:

$$\begin{aligned}
&[{}_p\cos_q]^n(X) \\
&= \frac{1}{2}\Big([{}_p\mathbb{W}_q]^n(iX) + [{}_p\mathbb{W}_q]^n(-iX)\Big) \\
&= \sum_{k=0}^{n}(-1)^k \frac{\left\{\prod_{j=1}^{p} \Gamma(\kappa_j + a_j(2k))\right\}}{\left\{\prod_{j=1}^{q} \Gamma(\vartheta_j + b_j(2k))\right\}} \frac{X^{2k}}{(2k)!}, \quad n = 0, 1, \cdots .
\end{aligned}$$

▷ The truncated superhyperbolic sine through Wright's generalized hypergeometric function:

$$
\begin{aligned}
&[{}_p\sinh_q]^n(X)\\
&= \frac{1}{2}\Big([{}_p\mathbb{W}_q]^n(X) - [{}_p\mathbb{W}_q]^n(-X)\Big)\\
&= \sum_{k=0}^{n} \frac{\left\{\prod_{j=1}^{p} \Gamma(\kappa_j + a_j(2k+1))\right\}}{\left\{\prod_{j=1}^{q} \Gamma(\vartheta_j + b_j(2k+1))\right\}} \frac{X^{2k+1}}{(2k+1)!}, \quad n = 0, 1, \cdots .
\end{aligned}
$$

▷ The truncated superhyperbolic cosine through Wright's generalized hypergeometric function:

$$
\begin{aligned}
&[{}_p\cosh_q]^n(X)\\
&= \frac{1}{2}\Big([{}_p\mathbb{W}_q]^n(X) + [{}_p\mathbb{W}_q]^n(-X)\Big)\\
&= \sum_{k=0}^{n} \frac{\left\{\prod_{j=1}^{p} \Gamma(\kappa_j + a_j(2k))\right\}}{\left\{\prod_{j=1}^{q} \Gamma(\vartheta_j + b_j(2k))\right\}} \frac{X^{2k}}{(2k)!}, \quad n = 0, 1, \cdots .
\end{aligned}
$$

References

1. Wright EM (1935) The asymptotic expansion of the generalized hypergeometric function. J Lond Math Soc 1(4):286–293
2. Wright EM (1940) The asymptotic expansion of integral functions defined by Taylor series. Philos Trans R Soc Lond Ser A Math Phys Sci 238(795):423–451
3. Kilbas A (2005) Fractional calculus of the generalized Wright function. Fract Calc Appl Anal 8(2):113–126
4. Kilbas AA, Srivastava HM, Trujillo JJ (2006) Theory and applications of fractional differential equations. Elsevier, Amsterdam
5. Wright EM (1933) On the coefficients of power series having exponential singularities. J Lond Math Soc 1(1):71–79
6. Luchko Y (2019) The Wright function and its applications. In: Kochubei A, Luchko Yu (eds) Handbook of fractional calculus with applications, vol 1. De Gruyter, Berlin, pp 241–268
7. Wright EM (1935) The asymptotic expansion of the generalized Bessel function. Proc Lond Math Soc 2(1):257–270

8. Kilbas AA, Saigo M, Trujillo JJ (2002) On the generalized Wright function. Fract Calc Appl Anal 5(4):437–460
9. Srivastava HM, Karlsson PW (1985) Multiple Gaussian hypergeometric series. Halsted Press (Ellis Horwood Limited), Chichester
10. Yang X-J (2021) Theory and applications of special functions for scientists and engineers. Springer

Chapter 4
The Mittag–Leffler, Supertrigonometric, and Superhyperbolic Functions

In this chapter, we introduce the theory of the Mittag–Leffler function, supertrigonometric functions, superhyperbolic functions and the truncated Mittag–Leffler functions. Then, we present the concepts of the generalized Hyperbolic function and the matrix Mittag–Leffler function.

4.1 The Mittag–Leffler Function

In this part, we present the results related to the Mittag-Leffler function. It is well-known that Swedish mathematician Gösta Magnus Mittag-Leffler defined the Mittag-Leffler function [1] in 1903 as the following:

Definition 4.1 The Mittag–Leffler function is given by

$$\mathbb{E}_{\kappa}(X) = \sum_{n=0}^{\infty} \frac{X^n}{\Gamma(n\kappa + 1)},$$

where $\kappa, X \in \mathbb{C}$, $\Re(\kappa) > 0$.

The special form of Mittag-Leffler function can be written as [2]:

$$\mathbb{E}_1(-X) = \exp(-X),$$

$$\mathbb{E}_0(-X) = \frac{1}{1+X} \ (|X| < 1),$$

$$\mathbb{E}_0(X) = \frac{1}{1-X} \ (|X| < 1),$$

S. Rezaei Aderyani et al., *Towards Ulam Type Multi Stability Analysis*, Studies in Systems, Decision and Control 523, https://doi.org/10.1007/978-3-031-55564-0_4

$$\mathbb{E}_2(X) = \cosh(X^{0.5}),$$

$$\mathbb{E}_2(-X^2) = \cos(X),$$

where $X \in \mathbb{C}$.

The Pollard function, denoted as $\mathbb{L}_\vartheta(s)$, was given by Pollard in 1948 [3]

$$e^{-X^\vartheta} = \int_0^\infty e^{sX}\mathbb{L}_\vartheta(s)ds,$$

where e^{-X^ϑ} is the Kohlrausch-Williams-Watts function, which leads to the following integral

$$\mathbb{E}_\kappa(-X) = \int_0^\infty e^{-s^\vartheta X}\mathbb{L}_\vartheta(s)ds,$$

where $\vartheta \in (0, 1)$.

The above result was discovered by Pollard [3] and reported by Berberan-Santos [4].

This implies

$$\mathbb{E}_\kappa(-X) = \int_0^\infty e^{-sX}\frac{s^{-(1+\frac{1}{\kappa})}}{\vartheta}\mathbb{L}_\vartheta(s^{-\frac{1}{\vartheta}})ds,$$

where $\vartheta \in (0, 1)$.

Definition 4.2 ([5]) The Mittag–Leffler function of the second type is given by

$$\mathbb{E}_\kappa(X^\kappa) = \sum_{n=0}^{\infty}\frac{X^{n\kappa}}{\Gamma(n\kappa+1)},$$

where $\kappa, X \in \mathbb{C}$, and $\Re(\kappa) > 0$.

Figures 4.1, 4.2, 4.3, and 4.4 display the plots of $\mathbb{E}_\kappa(X^\kappa)$, $\mathbb{E}_\kappa(X)$ $\mathbb{E}_\kappa(-X^\kappa)$, and $\mathbb{E}_\kappa(-X)$ with different values of κ.

The above graphs were presented by Hille and Tamarkin due to expression [5]

$$M(X, \varpi) = \frac{d}{dX}\mathbb{E}_{1-\kappa}\left(\varpi\Gamma(1-\kappa)X^{1-\kappa}\right)$$

with the following Hille-Tamarkin integral equation

$$\begin{aligned}
&\nu(X)\\
&= h(X) - \varpi\int_0^0 M(X-Y)h(Y)dY\\
&= h(X) - \varpi\int_0^0 \left[\frac{d}{dY}\mathbb{E}_{1-\kappa}\left(\varpi\Gamma(1-\kappa)(X-Y)^{1-\kappa}\right)\right]h(Y)dY,
\end{aligned}$$

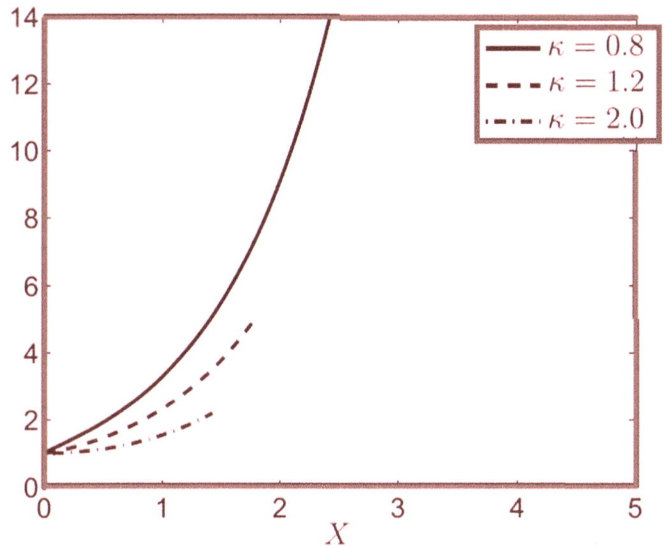

Fig. 4.1 The plots of $\mathbb{E}_\kappa(X^\kappa)$ with different values of κ

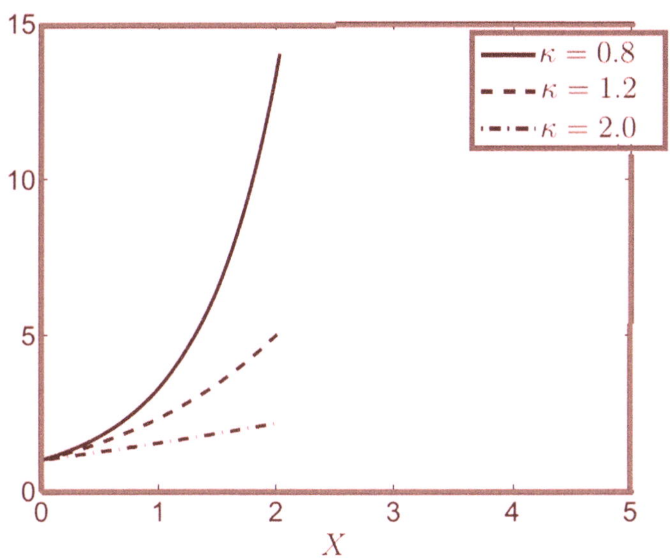

Fig. 4.2 The plots of $\mathbb{E}_\kappa(X)$ with different values $of\,\kappa$

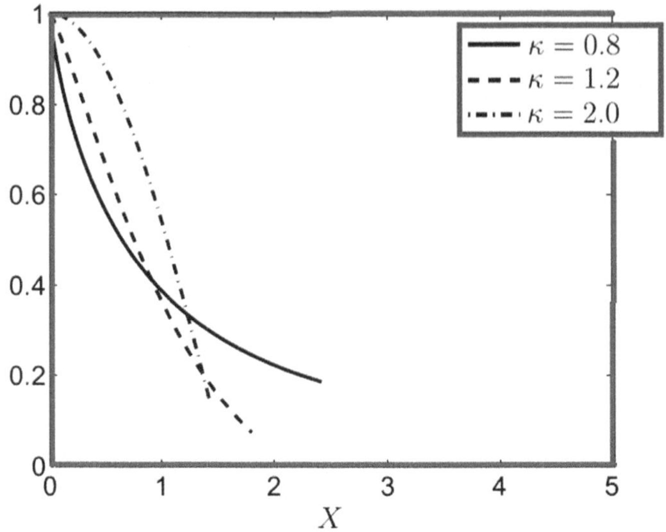

Fig. 4.3 The plots of $\mathbb{E}_\kappa(-X^\kappa)$ with different values of κ

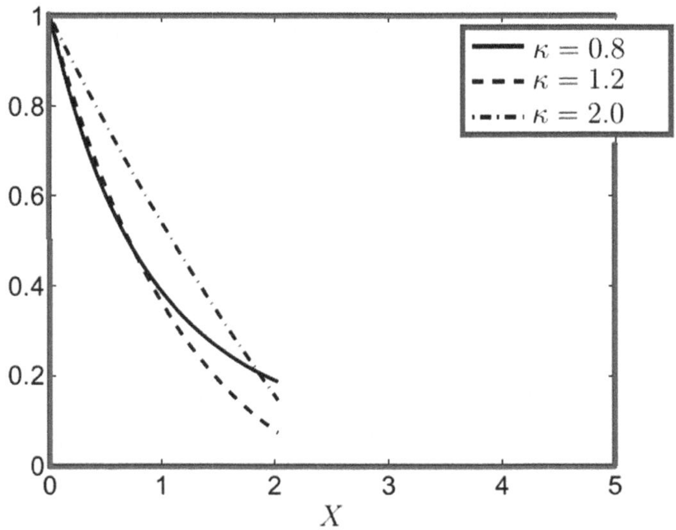

Fig. 4.4 The plots of $\mathbb{E}_\kappa(-X)$ with different values of κ

where $\varpi \in \mathbb{R}$.

Using the Pollard function, one can show the following integral representation of the form [4]

$$\mathbb{E}_{\kappa}(-X^{\kappa}) = \int_0^{\infty} e^{-s^{\kappa}X^{\kappa}} \mathbb{L}_{\kappa}(s)ds,$$

where $\kappa \in (0, 1)$.

Definition 4.3 ([3, 4]) The Mittag–Leffler function of the third type is given by

$$\mathbb{E}_{\kappa}(X^{\vartheta}) = \sum_{n=0}^{\infty} \frac{X^{n\vartheta}}{\Gamma(n\kappa + 1)},$$

where $\kappa, \vartheta, X \in \mathbb{C}, \Re(\kappa), \Re(\vartheta) > 0$.

This definition was introduced by Pollard [3] and reported by Berberan-Santos [4].

With the aid of the Pollard function, one can obtain the following integral representation of the form

$$\mathbb{E}_{\kappa}(-X^{\vartheta}) = \int_0^{\infty} e^{-s^{\vartheta}X^{\vartheta}} \mathbb{L}_{\vartheta}(s)ds,$$

where $\vartheta \in (0, 1)$.

Definition 4.4 ([2]) The Mittag–Leffler function of the Gaussian type is given by

$$\mathbb{E}_{\kappa}(X^{2\vartheta}) = \sum_{n=0}^{\infty} \frac{X^{2n\vartheta}}{\Gamma(n\kappa + 1)},$$

where $\kappa, \vartheta, X \in \mathbb{C}$, and $\Re(\kappa), \Re(\vartheta) > 0$.

Theorem 4.1 ([6]) *Suppose $X \geq 0$, and $\kappa \in [0, 1]$. The function $\mathbb{E}_{\kappa}(X)$ is an entire function of order $\frac{1}{\kappa}$ and type* 1.

The above theorem was presented by Phragmén [6].

Theorem 4.2 ([3]) *The function $\mathbb{E}_{\kappa}(X)$ is completely monotonic for $X \geq 0$, and $\kappa \in [0, 1]$.*

Or alternatively

$$(-1)^n \frac{d^n}{dX^n} \mathbb{E}_{\kappa}(-X) > 0, \quad n \in \mathbb{N}.$$

Also, for $X \in \mathbb{R}$, and $\kappa \in [0, 1]$,

$$\frac{d^n}{dX^n} \mathbb{E}_{\kappa}(X) > 0, \quad n \in \mathbb{N}.$$

Theorem 4.3 ([3]) *The function* $\mathbb{E}_\kappa(X)$ *has no real zeros when* $\kappa \in [0, 1]$ *and* $X \in \mathbb{R}$.

This result was conjectured by Feller [3] and proved by Pollard [3].

4.2 The Supertrigonometric Functions via the Mittag–Leffler Function

In 1988, Bagley introduced the supersine and supercosine functions based on the Mittag-Leffler function [7]. Zhang presented supertangent and supercotangent in 2011 [8] and Yang presented the supertrigonometric functions according to the Mittag-Leffler function [2].

Let κ, $X \in \mathbb{C}$, $\Re(\kappa) > 0$, and $i^2 = -1$. We now introduce the definitions of the supertrigonometric functions based on the Mittag-Leffler function [2, 7, 8], as follows:

▷ The Bagley supersine function is based on the Mittag-Leffler function:

$$supersin_\kappa(X^\kappa) = \frac{\mathbb{E}_\kappa(iX^\kappa) - \mathbb{E}_\kappa(-iX^\kappa)}{2i} = \sum_{n=0}^{\infty} \frac{(-1)^n X^{(2n+1)\kappa}}{\Gamma((2n+1)\kappa + 1)}.$$

▷ The Bagley supercosine function is based on the Mittag-Leffler function:

$$supercos_\kappa(X^\kappa) = \frac{\mathbb{E}_\kappa(iX^\kappa) + \mathbb{E}_\kappa(-iX^\kappa)}{2} = \sum_{n=0}^{\infty} \frac{(-1)^n X^{(2n)\kappa}}{\Gamma((2n)\kappa + 1)}.$$

These results were presented by Bagley [7].

▷ The Zhang supertangent function is based on the Mittag-Leffler function:

$$supertan_\kappa(X^\kappa) = \frac{supersin_\kappa(X^\kappa)}{supercos_\kappa(X^\kappa)}.$$

▷ The Zhang supercotangent function is based on the Mittag-Leffler function:

$$supercotan_\kappa(X^\kappa) = \frac{supercos_\kappa(X^\kappa)}{supersin_\kappa(X^\kappa)}.$$

These results were presented by Zhang [8].

▷ The Yang supercosecant function is based on the Mittag-Leffler function:

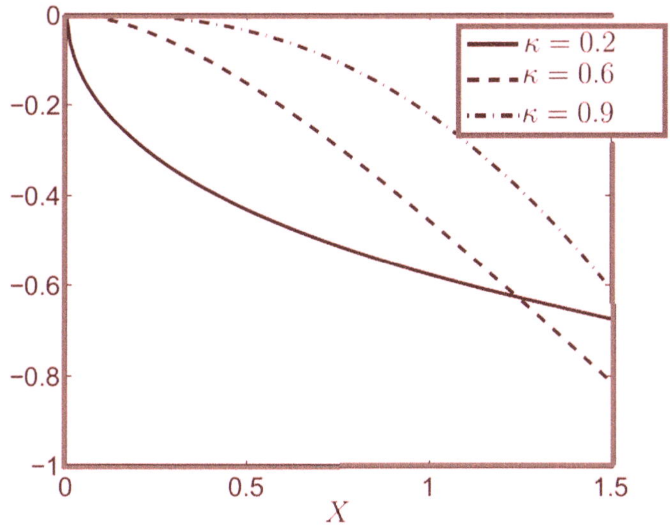

Fig. 4.5 The plots of $supersin_\kappa(X^\kappa)$ with different values of κ

$$supercsc_\kappa(X^\kappa) = \frac{1}{supersin_\kappa(X^\kappa)}, \quad \left(supersin_\kappa(X^\kappa) \neq 0\right).$$

▷ The Yang supersecant function is based on the Mittag-Leffler function:

$$supersec_\kappa(X^\kappa) = \frac{1}{supercos_\kappa(X^\kappa)}, \quad \left(supercos_\kappa(X^\kappa) \neq 0\right).$$

These results were presented by Yang [2].

Figures 4.5 and 4.6 display the plots of $supersin_\kappa(X^\kappa)$ and $supercos_\kappa(X^\kappa)$ with different values of κ.

Theorem 4.4 ([2])

Assume κ, $X \in \mathbb{C}$, $\Re(\kappa) > 0$, *and* $i^2 = -1$. *Then,*

$$\mathbb{E}_\kappa(iX^\kappa) = supercos_\kappa(X^\kappa) + i\ supersin_\kappa(X^\kappa),$$

$$\mathbb{E}_\kappa(-iX^\kappa) = supercos_\kappa(X^\kappa) - i\ supersin_\kappa(X^\kappa),$$

and

$$\mathbb{E}_\kappa(iX^\kappa)\mathbb{E}_\kappa(-iX^\kappa) = supercos^2_\kappa(X^\kappa) + \ supersin^2_\kappa(X^\kappa),$$

where

$$supercos^2_\kappa(X^\kappa) = supercos_\kappa(X^\kappa)supercos_\kappa(X^\kappa),$$

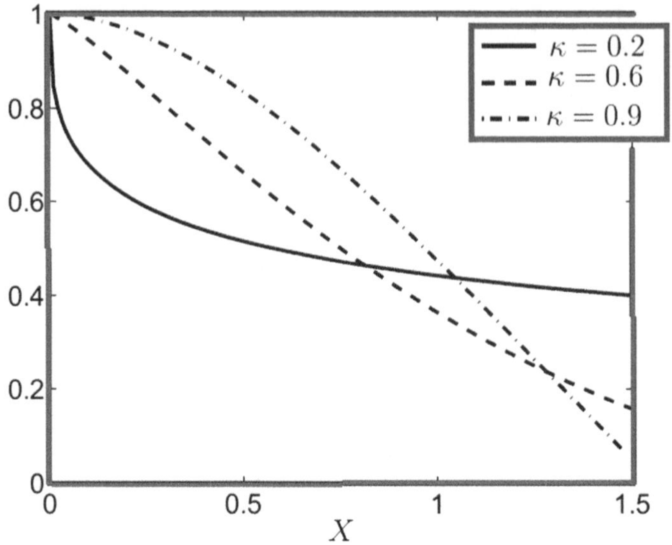

Fig. 4.6 The plots of $supercos_\kappa(X^\kappa)$ with different values of κ

and

$$supersin_\kappa^2(X^\kappa) = supersin_\kappa(X^\kappa)supersin_\kappa(X^\kappa).$$

4.3 The Superhyperbolic Functions via the Mittag–Leffler Function

Let X, $\kappa \in \mathbb{C}$, and $\Re(\kappa) > 0$. In the section, we present the superhyperbolic functions [2, 7, 8] based on the Mittag-Leffler function as follows:

▷ The Bagley superhyperbolic supersine function is based on the Mittag-Leffler function:

$$\begin{aligned}
&supersinh_\kappa(X^\kappa) \\
&= \frac{\mathbb{E}_\kappa(X^\kappa) - \mathbb{E}_\kappa(-X^\kappa)}{2} \\
&= \sum_{n=0}^{\infty} \frac{X^{(2n+1)\kappa}}{\Gamma((2n+1)\kappa + 1)}.
\end{aligned}$$

▷ The Bagley superhyperbolic supercosine function is based on the Mittag-Leffler function:

$$
\begin{aligned}
& supercosh_{\kappa}(X^{\kappa}) \\
&= \frac{\mathbb{E}_{\kappa}(X^{\kappa}) + \mathbb{E}_{\kappa}(-X^{\kappa})}{2} \\
&= \sum_{n=0}^{\infty} \frac{X^{(2n)\kappa}}{\Gamma((2n)\kappa + 1)}.
\end{aligned}
$$

These results were presented by Bagley [7].

- ▹ The Zhang superhyperbolic supertangent function is based on the Mittag-Leffler function:

$$
supertanh_{\kappa}(X^{\kappa}) = \frac{supersinh_{\kappa}(X^{\kappa})}{supercosh_{\kappa}(X^{\kappa})}.
$$

- ▹ The Zhang superhyperbolic supercotangent function based on the Mittag-Leffler function:

$$
supercotanh_{\kappa}(X^{\kappa}) = \frac{supercosh_{\kappa}(X^{\kappa})}{supersinh_{\kappa}(X^{\kappa})}.
$$

These results were presented by Zhang [8].

- ▹ The Yang superhyperbolic supercosecant function is based on the Mittag-Leffler function:

$$
\begin{aligned}
& supercsch_{\kappa}(X^{\kappa}) \\
&= \frac{1}{supersinh_{\kappa}(X^{\kappa})}, \left(supersinh_{\kappa}(X^{\kappa}) \neq 0 \right).
\end{aligned}
$$

- ▹ The Yang superhyperbolic supersecant function is based on the Mittag-Leffler function:

$$
\begin{aligned}
& supersech_{\kappa}(X^{\kappa}) \\
&= \frac{1}{supercosh_{\kappa}(X^{\kappa})}, \left(supercosh_{\kappa}(X^{\kappa}) \neq 0 \right).
\end{aligned}
$$

The results were presented by Yang [2].

Figures 4.7 and 4.8 display the plots of $supersinh_{\kappa}(X^{\kappa})$ and $supercosh_{\kappa}(X^{\kappa})$ with different values of κ.

Theorem 4.5 ([2]) *Assume* $\kappa, X \in \mathbb{C}$, *and* $\Re(\kappa) > 0$. *Then,*

$$
\mathbb{E}_{\kappa}(X^{\kappa}) = supercosh_{\kappa}(X^{\kappa}) + supersinh_{\kappa}(X^{\kappa}),
$$

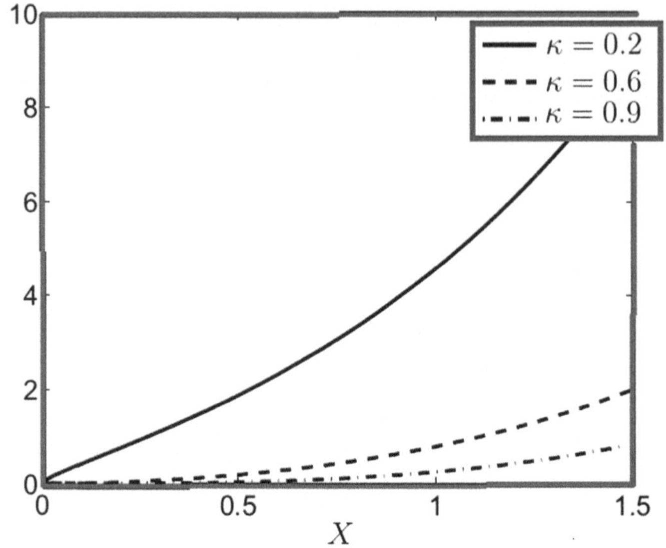

Fig. 4.7 The plots of $supersinh_\kappa(X^\kappa)$ with different values of κ

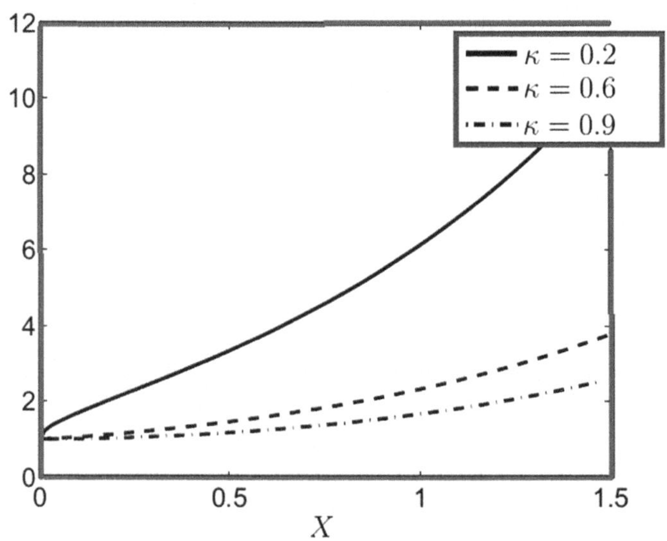

Fig. 4.8 The plots of $supercosh_\kappa(X^\kappa)$ with different values of κ

$$\mathbb{E}_{\kappa}(-X^{\kappa}) = supercosh_{\kappa}(X^{\kappa}) - supersinh_{\kappa}(X^{\kappa}),$$

and

$$\mathbb{E}_{\kappa}(X^{\kappa})\mathbb{E}_{\kappa}(-X^{\kappa}) = supercosh^2_{\kappa}(X^{\kappa}) - supersinh^2_{\kappa}(X^{\kappa}),$$

where

$$supercosh^2_{\kappa}(X^{\kappa}) = supercosh_{\kappa}(X^{\kappa})supercosh_{\kappa}(X^{\kappa}),$$

and

$$supersinh^2_{\kappa}(X^{\kappa}) = supersinh_{\kappa}(X^{\kappa})supersinh_{\kappa}(X^{\kappa}).$$

4.4 The Pre-supertrigonometric Functions via the Mittag–Leffler Function

Here, let $X, \kappa \in \mathbb{C}$, and $\Re(\kappa) > 0$. We now present the concepts of the pre-supercosine, pre-supersine, pre-supercotangent, pre-supertangent, pre-supersecant, and pre-supercosecant [2, 9] through the Mittag-Leffler function, as follows:

▷ The pre-supersine function through the Mittag-Leffler function:

$$\begin{aligned} &presin_{\kappa}(X) \\ &= \frac{1}{2i}\Big(\mathbb{E}_{\kappa}(iX) - \mathbb{E}_{\kappa}(-iX)\Big) \\ &= \sum_{n=0}^{\infty} \frac{(-1)^n X^{2n+1}}{\Gamma((2n+1)\kappa + 1)}. \end{aligned}$$

▷ The pre-supercosine function through the Mittag-Leffler function:

$$\begin{aligned} &precos_{\kappa}(X) \\ &= \frac{1}{2}\Big(\mathbb{E}_{\kappa}(iX) + \mathbb{E}_{\kappa}(-iX)\Big) \\ &= \sum_{n=0}^{\infty} \frac{(-1)^n X^{2n}}{\Gamma((2n)\kappa + 1)}. \end{aligned}$$

▷ The pre-supertangent through the Mittag-Leffler function:

$$pretan_{\kappa}(X) = \frac{presin_{\kappa}(X)}{precos_{\kappa}(X)}, \quad \Big(precos_{\kappa}(X) \neq 0\Big).$$

▷ The pre-supercotangent through the Mittag-Leffler function:

$$precotan_\kappa(X) = \frac{precos_\kappa(X)}{presin_\kappa(X)}, \quad \left(presin_\kappa(X) \neq 0\right).$$

▷ The pre-supercosecant through the Mittag-Leffler function:

$$precsc_\kappa(X) = \frac{1}{presin_\kappa(X)}, \quad \left(presin_\kappa(X) \neq 0\right).$$

▷ The pre-supersecant through the Mittag-Leffler function:

$$presec_\kappa(X) = \frac{1}{precos_\kappa(X)}, \quad \left(precos_\kappa(X) \neq 0\right).$$

Figures 4.9 and 4.10 display the plots of $presin_\kappa(X)$ and $precos_\kappa(X)$, with different values of κ.

Also, we have

$$\mathbb{E}_\kappa(iX) = precos_\kappa(X) + ipresin_\kappa(X),$$

$$\mathbb{E}_\kappa(-iX) = precos_\kappa(X) - ipresin_\kappa(X),$$

and

$$\mathbb{E}_\kappa(iX)\mathbb{E}_\kappa(-iX) = precos_\kappa^2(X) + presin_\kappa^2(X),$$

where

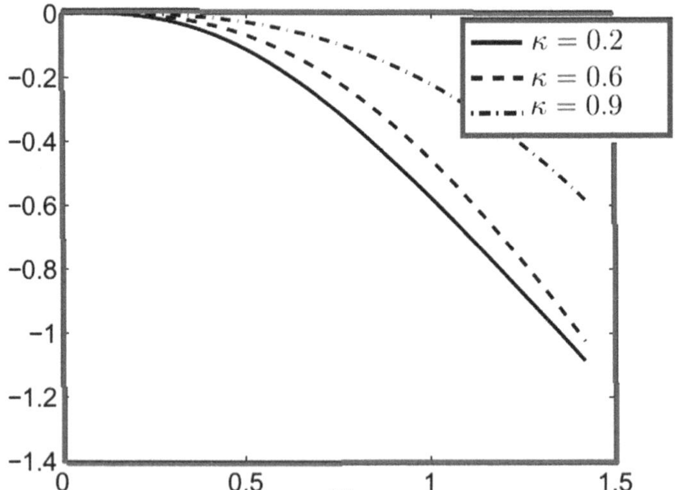

Fig. 4.9 The plots of $presin_\kappa(X)$ with different values of κ

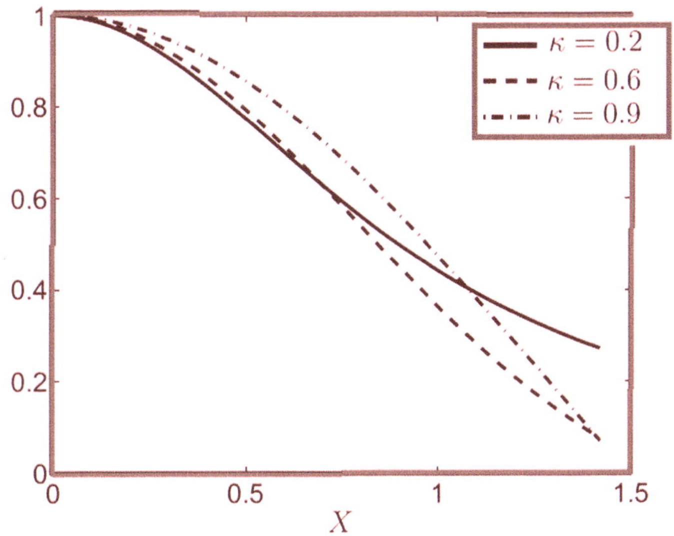

Fig. 4.10 The plots of $precos_{\kappa}(X)$ with different values of κ

$$precos_{\kappa}^{2}(X) = precos_{\kappa}(X)precos_{\kappa}(X),$$

and

$$presin_{\kappa}^{2}(X) = presin_{\kappa}(X)presin_{\kappa}(X).$$

The results were presented by Dattoli et al. [9], and Yang [2].

Theorem 4.6 ([2]) *Let $X, \kappa \in \mathbb{C}$, and $\Re(\kappa) > 0$. Then,*

$$supersin_{\kappa}(X^{\kappa}) = presin_{\kappa}(X^{\kappa}),$$

$$supercos_{\kappa}(X^{\kappa}) = precos_{\kappa}(X^{\kappa}),$$

$$supertan_{\kappa}(X^{\kappa}) = pretan_{\kappa}(X^{\kappa}),$$

$$supercotan_{\kappa}(X^{\kappa}) = precotan_{\kappa}(X^{\kappa}),$$

$$supercsc_{\kappa}(X^{\kappa}) = precsc_{\kappa}(X^{\kappa}),$$

and

$$supersec_{\kappa}(X^{\kappa}) = presec_{\kappa}(X^{\kappa}).$$

Let $X, \kappa, \vartheta \in \mathbb{C}$, and $\Re(\kappa), \Re(\vartheta) > 0$. Now, we have the following [2]:

▷ The pre-supersine through the Mittag-Leffler function of the third type:

$$
\begin{aligned}
&presin_{\kappa}(X^{\vartheta})\\
&= \frac{1}{2i}\Big(\mathbb{E}_{\kappa}(iX^{\vartheta}) - \mathbb{E}_{\kappa}(-iX^{\vartheta})\Big)\\
&= \sum_{n=0}^{\infty} \frac{(-1)^n X^{(2n+1)\vartheta}}{\Gamma((2n+1)\kappa+1)}.
\end{aligned}
$$

▷ The pre-supercosine through the Mittag-Leffler function of the third type:

$$
\begin{aligned}
&precos_{\kappa}(X^{\vartheta})\\
&= \frac{1}{2}\Big(\mathbb{E}_{\kappa}(iX^{\vartheta}) + \mathbb{E}_{\kappa}(-iX^{\vartheta})\Big)\\
&= \sum_{n=0}^{\infty} \frac{(-1)^n X^{(2n)\vartheta}}{\Gamma((2n)\kappa+1)}.
\end{aligned}
$$

▷ The pre-presupertangent through the Mittag-Leffler function of the third type:

$$
pretan_{\kappa}(X^{\vartheta}) = \frac{presin_{\kappa}(X^{\vartheta})}{precos_{\kappa}(X^{\vartheta})}, \quad \Big(precos_{\kappa}(X^{\vartheta}) \neq 0\Big).
$$

▷ The pre-supercotangent through the Mittag-Leffler of the third type:

$$
precotan_{\kappa}(X^{\vartheta}) = \frac{precos_{\kappa}(X^{\vartheta})}{presin_{\kappa}(X^{\vartheta})}, \quad \Big(presin_{\kappa}(X^{\vartheta}) \neq 0\Big).
$$

▷ The pre-supercosecant through the Mittag-Leffler function of the third type:

$$
precsc_{\kappa}(X^{\vartheta}) = \frac{1}{presin_{\kappa}(X^{\vartheta})}, \quad \Big(presin_{\kappa}(X^{\vartheta}) \neq 0\Big).
$$

▷ The pre-supersecant through the Mittag-Leffler function of the third type:

$$
presec_{\kappa}(X^{\vartheta}) = \frac{1}{precos_{\kappa}(X^{\vartheta})}, \quad \Big(precos_{\kappa}(X^{\vartheta}) \neq 0\Big).
$$

Theorem 4.7 ([2]) *Let $X, \kappa, \vartheta \in \mathbb{C}$, and $\Re(\kappa), \Re(\vartheta) > 0$. Then,*

$$
\mathbb{E}_{\kappa}(iX^{\vartheta}) = precos_{\kappa}(X^{\vartheta}) + ipresin_{\kappa}(X^{\vartheta}),
$$

$$
\mathbb{E}_{\kappa}(-iX^{\vartheta}) = precos_{\kappa}(X^{\vartheta}) - ipresin_{\kappa}(X^{\vartheta}),
$$

and

$$
\mathbb{E}_{\kappa}(iX^{\vartheta})\mathbb{E}_{\kappa}(-iX^{\vartheta}) = precos_{\kappa}^2(X^{\vartheta}) + presin_{\kappa}^2(X^{\vartheta}),
$$

where

$$precos_{\kappa}^{2}(X^{\vartheta}) = precos_{\kappa}(X^{\vartheta})precos_{\kappa}(X^{\vartheta}),$$

and

$$presin_{\kappa}^{2}(X^{\vartheta}) = presin_{\kappa}(X^{\vartheta})presin_{\kappa}(X^{\vartheta}).$$

4.5 The Pre-superhyperbolic Functions via the Mittag–Leffler Function

In this section, we now study the concepts of the pre-superhyperbolic supersine, pre-superhyperbolic supercosine, pre-seuperhyperbolic supertangent, pre-superhyperbolic supercotangent, pre-superhyperbolic supercosecant, and pre-superhyperbolic supersecant based on the Mittag-Leffler function.

Let $X, \kappa \in \mathbb{C}$, and $\Re(\kappa) > 0$. We now have the following [2]:

▷ The pre-superhyperbolic supersine through the Mittag-Leffler function:

$$\begin{aligned} & presinh_{\kappa}(X) \\ &= \frac{1}{2}\Big(\mathbb{E}_{\kappa}(X) - \mathbb{E}_{\kappa}(-X)\Big) \\ &= \sum_{n=0}^{\infty} \frac{X^{2n+1}}{\Gamma((2n+1)\kappa + 1)}. \end{aligned}$$

▷ The pre-superhyperbolic supercosine through the Mittag-Leffler function:

$$\begin{aligned} & precosh_{\kappa}(X) \\ &= \frac{1}{2}\Big(\mathbb{E}_{\kappa}(X) + \mathbb{E}_{\kappa}(-X)\Big) \\ &= \sum_{n=0}^{\infty} \frac{X^{2n}}{\Gamma((2n)\kappa + 1)}. \end{aligned}$$

▷ The pre-superhyperbolic supertangent through the Mittag-Leffler function:

$$pretanh_{\kappa}(X) = \frac{presinh_{\kappa}(X)}{precosh_{\kappa}(X)}, \quad \Big(precosh_{\kappa}(X) \neq 0\Big).$$

▷ The pre-superhyperbolic supercotangent through the Mittag-Leffler function:

$$precotanh_{\kappa}(X) = \frac{precosh_{\kappa}(X)}{presinh_{\kappa}(X)}, \quad \Big(presinh_{\kappa}(X) \neq 0\Big).$$

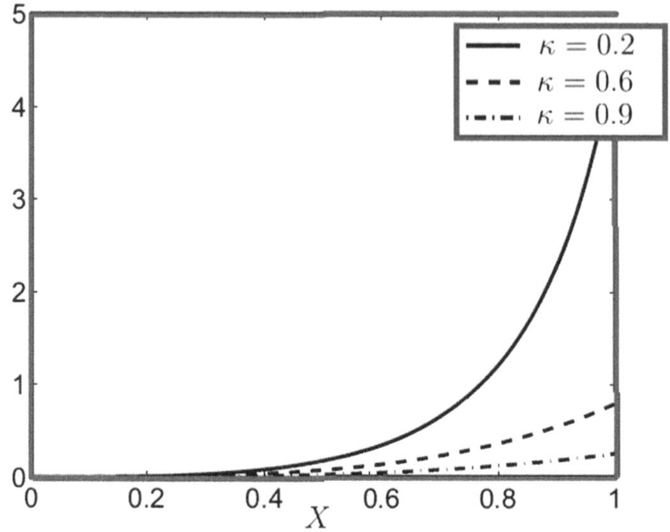

Fig. 4.11 The plots of $presinh_\kappa(X)$ with different values of κ

▷ The pre-superhyperbolic supercosecant through the Mittag-Leffler function:

$$precsch_\kappa(X) = \frac{1}{presinh_\kappa(X)}, \quad \Big(presinh_\kappa(X) \neq 0\Big).$$

▷ The pre-superhyperbolic supersecant through the Mittag-Leffler function:

$$presech_\kappa(X) = \frac{1}{precosh_\kappa(X)}, \quad \Big(precosh_\kappa(X) \neq 0\Big).$$

Figures 4.11 and 4.12 display the plots of $presinh_\kappa(X)$ and $precosh_\kappa(X)$ with different values of κ.

Also, we have [2],

$$\mathbb{E}_\kappa(X) = precosh_\kappa(X) + presinh_\kappa(X),$$

$$\mathbb{E}_\kappa(-X) = precosh_\kappa(X) - presinh_\kappa(X),$$

and

$$\mathbb{E}_\kappa(X)\mathbb{E}_\kappa(-X) = precosh^2_\kappa(X) - presinh^2_\kappa(X),$$

where

$$precosh^2_\kappa(X) = precosh_\kappa(X)precosh_\kappa(X),$$

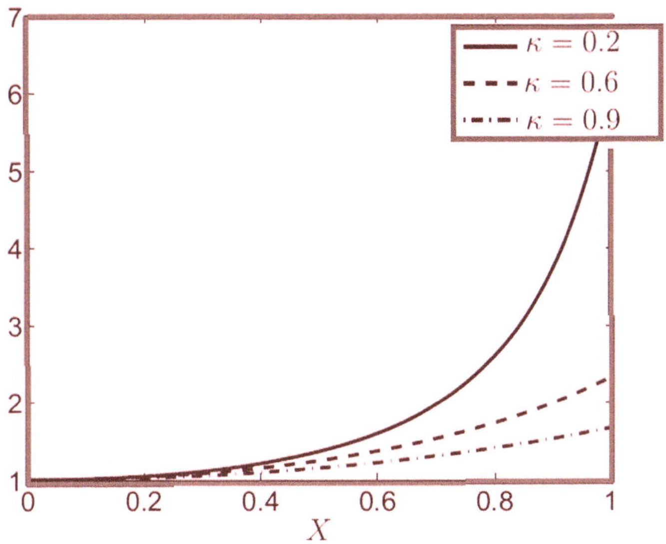

Fig. 4.12 The plots of $precosh_{\kappa}(X)$ with different values of κ

and

$$presinh_{\kappa}^{2}(X) = presinh_{\kappa}(X)presinh_{\kappa}(X).$$

Theorem 4.8 ([2]) *Let* $X, \kappa \in \mathbb{C}$, *and* $\Re(\kappa) > 0$. *Then,*

$$supersinh_{\kappa}(X^{\kappa}) = presinh_{\kappa}(X^{\kappa}),$$

$$supercosh_{\kappa}(X^{\kappa}) = precosh_{\kappa}(X^{\kappa}),$$

$$supertanh_{\kappa}(X^{\kappa}) = pretanh_{\kappa}(X^{\kappa}),$$

$$supercotanh_{\kappa}(X^{\kappa}) = precotanh_{\kappa}(X^{\kappa}),$$

$$supercsch_{\kappa}(X^{\kappa}) = precsch_{\kappa}(X^{\kappa}),$$

and

$$supersech_{\kappa}(X^{\kappa}) = presech_{\kappa}(X^{\kappa}).$$

Let $X, \kappa, \vartheta \in \mathbb{C}$, and $\Re(\kappa), \Re(\vartheta) > 0$. Now, we have the following [2]:

▷ The pre-superhyperbolic supersine through the Mittag-Leffler function of the third type:

$$\begin{aligned} &presinh_{\kappa}(X^{\vartheta}) \\ &= \frac{1}{2}\Big(\mathbb{E}_{\kappa}(X^{\vartheta}) - \mathbb{E}_{\kappa}(-X^{\vartheta})\Big) \\ &= \sum_{n=0}^{\infty} \frac{X^{(2n+1)\vartheta}}{\Gamma((2n+1)\kappa+1)}. \end{aligned}$$

▷ The pre-superhyperbolic supercosine through the Mittag-Leffler function of the third type:

$$\begin{aligned} &precosh_{\kappa}(X^{\vartheta}) \\ &= \frac{1}{2}\Big(\mathbb{E}_{\kappa}(X^{\vartheta}) + \mathbb{E}_{\kappa}(-X^{\vartheta})\Big) \\ &= \sum_{n=0}^{\infty} \frac{X^{(2n)\vartheta}}{\Gamma((2n)\kappa+1)}. \end{aligned}$$

▷ The pre-superhyperbolic presupertangent through the Mittag-Leffler function of the third type:

$$pretanh_{\kappa}(X^{\vartheta}) = \frac{presinh_{\kappa}(X^{\vartheta})}{precosh_{\kappa}(X^{\vartheta})}, \quad \Big(precosh_{\kappa}(X^{\vartheta}) \neq 0\Big).$$

▷ The pre-superhyperbolic supercotangent through the Mittag-Leffler function of the third type:

$$precotanh_{\kappa}(X^{\vartheta}) = \frac{precosh_{\kappa}(X^{\vartheta})}{presinh_{\kappa}(X^{\vartheta})}, \quad \Big(presinh_{\kappa}(X^{\vartheta}) \neq 0\Big).$$

▷ The pre-superhyperbolic supercosecant through the Mittag-Leffler function of the third type:

$$precsch_{\kappa}(X^{\vartheta}) = \frac{1}{presinh_{\kappa}(X^{\vartheta})}, \quad \Big(presinh_{\kappa}(X^{\vartheta}) \neq 0\Big).$$

▷ The pre-superhyperbolic supersecant through the Mittag-Leffler function of the third type:

$$presech_{\kappa}(X^{\vartheta}) = \frac{1}{precosh_{\kappa}(X^{\vartheta})}, \quad \Big(precosh_{\kappa}(X^{\vartheta}) \neq 0\Big).$$

Theorem 4.9 ([2]) *Let* $X, \kappa, \vartheta \in \mathbb{C}$, *and* $\Re(\kappa), \Re(\vartheta) > 0$. *Then,*

$$\mathbb{E}_{\kappa}(X^{\vartheta}) = precosh_{\kappa}(X^{\vartheta}) + presinh_{\kappa}(X^{\vartheta}),$$

$$\mathbb{E}_{\kappa}(-X^{\vartheta}) = precosh_{\kappa}(X^{\vartheta}) - presinh_{\kappa}(X^{\vartheta}),$$

and

$$\mathbb{E}_{\kappa}(X^{\vartheta})\mathbb{E}_{\kappa}(-X^{\vartheta}) = precosh_{\kappa}^{2}(X^{\vartheta}) - presinh_{\kappa}^{2}(X^{\vartheta}),$$

where

$$precosh_{\kappa}^{2}(X^{\vartheta}) = precosh_{\kappa}(X^{\vartheta})precosh_{\kappa}(X^{\vartheta}),$$

and

$$presinh_{\kappa}^{2}(X^{\vartheta}) = presinh_{\kappa}(X^{\vartheta})presinh_{\kappa}(X^{\vartheta}).$$

Now, we list the relationships between the Mittag-Leffler function and Wright's generalized hypergeometric function [2].

We have

$$\mathbb{E}_{\kappa}(X^{\kappa}) = {}_1\mathbb{W}_1\left(\begin{matrix}(1,1)\\(1,\kappa)\end{matrix}; X^{\kappa}\right),$$

$$\mathbb{E}_{\kappa}(X) = {}_1\mathbb{W}_1\left(\begin{matrix}(1,1)\\(1,\kappa)\end{matrix}; X\right),$$

$$\mathbb{E}_{\kappa}(X^{\vartheta}) = {}_1\mathbb{W}_1\left(\begin{matrix}(1,1)\\(1,\kappa)\end{matrix}; X^{\vartheta}\right),$$

$$supersin_{\kappa}(X^{\kappa}) = {}_1\sin_1\left(\begin{matrix}(1,1)\\(1,\kappa)\end{matrix}; X^{\kappa}\right),$$

$$supercos_{\kappa}(X^{\kappa}) = {}_1\cos_1\left(\begin{matrix}(1,1)\\(1,\kappa)\end{matrix}; X^{\kappa}\right),$$

$$supersinh_{\kappa}(X^{\kappa}) = {}_1\sinh_1\left(\begin{matrix}(1,1)\\(1,\kappa)\end{matrix}; X^{\kappa}\right),$$

$$supercosh_{\kappa}(X^{\kappa}) = {}_1\cosh_1\left(\begin{matrix}(1,1)\\(1,\kappa)\end{matrix}; X^{\kappa}\right),$$

$$presin_{\kappa}(X) = {}_1\sin_1\left(\begin{matrix}(1,1)\\(1,\kappa)\end{matrix}; X\right),$$

$$precos_{\kappa}(X) = {}_1\cos_1\left(\begin{matrix}(1,1)\\(1,\kappa)\end{matrix}; X\right),$$

$$presinh_{\kappa}(X) = {}_1\sinh_1\left(\begin{matrix}(1,1)\\(1,\kappa)\end{matrix}; X\right),$$

$$precosh_{\kappa}(X) = {}_1\cosh_1\left(\substack{(1,1)\\(1,\kappa)}; X\right),$$

$$presin_{\kappa}(X^{\vartheta}) = {}_1\sin_1\left(\substack{(1,1)\\(1,\kappa)}; X^{\vartheta}\right),$$

$$precos_{\kappa}(X^{\vartheta}) = {}_1\cos_1\left(\substack{(1,1)\\(1,\kappa)}; X^{\vartheta}\right),$$

$$presinh_{\kappa}(X^{\vartheta}) = {}_1\sinh_1\left(\substack{(1,1)\\(1,\kappa)}; X^{\vartheta}\right),$$

$$precosh_{\kappa}(X^{\vartheta}) = {}_1\cosh_1\left(\substack{(1,1)\\(1,\kappa)}; X^{\vartheta}\right).$$

where $\kappa, \vartheta, X \in \mathbb{C}$, and $\Re(\kappa), \Re(\vartheta) > 0$.

4.6 The Truncated Mittag–Leffler Functions and Related Functions

In this section, we detail the truncated Mittag-Leffler functions and related functions.

Definition 4.5 ([2]) The truncated Mittag–Leffler function is given by

$$\{\mathbb{E}_{\kappa}\}^{n}(X) = \sum_{m=0}^{n} \frac{X^m}{\Gamma(m\kappa+1)},$$

where $X, \kappa \in \mathbb{C}$, $\Re(\kappa) > 0$, and $n \in \mathbb{N}_0$.

In particular,

$$\{\mathbb{E}_{\kappa}\}^{0}(X) = 1,$$

$$\{\mathbb{E}_{\kappa}\}^{1}(X) = 1 + \frac{X}{\Gamma(m\kappa+1)},$$

and

$$\{\mathbb{E}_{\kappa}\}^{2}(X) = 1 + \frac{X}{\Gamma(\kappa+1)} + \frac{X^2}{\Gamma(2\kappa+1)}.$$

Let $X, \kappa \in \mathbb{C}$, $\Re(\kappa) > 0$, and $n \in \mathbb{N}_0$. We now have the following [2]:

▷ The truncated pre-supersine function is based on the truncated Mittag-Leffler function:

$$\begin{aligned}
&\{presin_\kappa\}^n(X) \\
&= \frac{1}{2i}\Big(\{\mathbb{E}_\kappa\}^n(iX) - \{\mathbb{E}_\kappa\}^n(-iX)\Big) \\
&= \sum_{m=0}^{n} \frac{(-1)^m X^{2m+1}}{\Gamma((2m+1)\kappa+1)}.
\end{aligned}$$

In particular,

$$\begin{aligned}
&\{presin_\kappa\}^0(X) \\
&= \frac{1}{2i}\Big(\{\mathbb{E}_\kappa\}^0(iX) - \{\mathbb{E}_\kappa\}^0(-iX)\Big) \\
&= \frac{X}{\Gamma(\kappa+1)},
\end{aligned}$$

$$\begin{aligned}
&\{presin_\kappa\}^1(X) \\
&= \frac{1}{2i}\Big(\{\mathbb{E}_\kappa\}^1(iX) - \{\mathbb{E}_\kappa\}^1(-iX)\Big) \\
&= \frac{X}{\Gamma(\kappa+1)} - \frac{X^3}{\Gamma(3\kappa+1)},
\end{aligned}$$

$$\begin{aligned}
&\{presin_\kappa\}^2(X) \\
&= \frac{1}{2i}\Big(\{\mathbb{E}_\kappa\}^2(iX) - \{\mathbb{E}_\kappa\}^2(-iX)\Big) \\
&= \frac{X}{\Gamma(\kappa+1)} - \frac{X^3}{\Gamma(3\kappa+1)} + \frac{X^5}{\Gamma(5\kappa+1)}.
\end{aligned}$$

▷ The truncated pre-supercosine function is based on the truncated Mittag-Leffler function:

$$\begin{aligned}
&\{precos_\kappa\}^n(X) \\
&= \frac{1}{2}\Big(\{\mathbb{E}_\kappa\}^n(iX) + \{\mathbb{E}_\kappa\}^n(-iX)\Big) \\
&= \sum_{m=0}^{n} \frac{(-1)^m X^{2m}}{\Gamma((2m)\kappa+1)}.
\end{aligned}$$

For $n = 0, 1, 2$,

$$\begin{aligned}
&\{precos_\kappa\}^0(X) \\
&= \frac{1}{2}\Big(\{\mathbb{E}_\kappa\}^0(iX) + \{\mathbb{E}_\kappa\}^0(-iX)\Big) \\
&= 1,
\end{aligned}$$

$$\begin{aligned}&\{precos_{\kappa}\}^{1}(X)\\&=\frac{1}{2}\Big(\{\mathbb{E}_{\kappa}\}^{1}(iX)+\{\mathbb{E}_{\kappa}\}^{1}(-iX)\Big)\\&=1-\frac{X^{2}}{\Gamma(2\kappa+1)},\end{aligned}$$

$$\begin{aligned}&\{precos_{\kappa}\}^{2}(X)\\&=\frac{1}{2}\Big(\{\mathbb{E}_{\kappa}\}^{2}(iX)+\{\mathbb{E}_{\kappa}\}^{2}(-iX)\Big)\\&=1-\frac{X^{2}}{\Gamma(2\kappa+1)}+\frac{X^{4}}{\Gamma(4\kappa+1)}.\end{aligned}$$

▷ The truncated superhyperbolic pre-supersine function is based on the truncated Mittag-Leffler function:

$$\begin{aligned}&\{presinh_{\kappa}\}^{n}(X)\\&=\frac{1}{2}\Big(\{\mathbb{E}_{\kappa}\}^{n}(X)-\{\mathbb{E}_{\kappa}\}^{n}(-X)\Big)\\&=\sum_{m=0}^{n}\frac{X^{2m+1}}{\Gamma((2m+1)\kappa+1)}.\end{aligned}$$

For $n=0,1,2,$

$$\begin{aligned}&\{presinh_{\kappa}\}^{0}(X)\\&=\frac{1}{2}\Big(\{\mathbb{E}_{\kappa}\}^{0}(X)-\{\mathbb{E}_{\kappa}\}^{0}(-X)\Big)\\&=\frac{X}{\Gamma(\kappa+1)},\end{aligned}$$

$$\begin{aligned}&\{presinh_{\kappa}\}^{1}(X)\\&=\frac{1}{2}\Big(\{\mathbb{E}_{\kappa}\}^{1}(X)-\{\mathbb{E}_{\kappa}\}^{1}(-X)\Big)\\&=\frac{X}{\Gamma(\kappa+1)}+\frac{X^{3}}{\Gamma(3\kappa+1)},\end{aligned}$$

$$\begin{aligned}&\{presinh_{\kappa}\}^{2}(X)\\&=\frac{1}{2}\Big(\{\mathbb{E}_{\kappa}\}^{2}(X)-\{\mathbb{E}_{\kappa}\}^{2}(-X)\Big)\\&=\frac{X}{\Gamma(\kappa+1)}+\frac{X^{3}}{\Gamma(3\kappa+1)}+\frac{X^{5}}{\Gamma(5\kappa+1)}.\end{aligned}$$

▷ The truncated superhyperbolic pre-supercosine function is based on the truncated Mittag-Leffler function:

$$\begin{aligned}
&\{precosh_\kappa\}^n(X)\\
&= \frac{1}{2}\Big(\{\mathbb{E}_\kappa\}^n(X) + \{\mathbb{E}_\kappa\}^n(-X)\Big)\\
&= \sum_{m=0}^{n} \frac{X^{2m}}{\Gamma((2m)\kappa+1)}.
\end{aligned}$$

For $n = 0, 1, 2,$

$$\begin{aligned}
&\{precosh_\kappa\}^0(X)\\
&= \frac{1}{2}\Big(\{\mathbb{E}_\kappa\}^0(X) + \{\mathbb{E}_\kappa\}^0(-X)\Big)\\
&= 1,
\end{aligned}$$

$$\begin{aligned}
&\{precosh_\kappa\}^1(X)\\
&= \frac{1}{2}\Big(\{\mathbb{E}_\kappa\}^1(X) + \{\mathbb{E}_\kappa\}^1(-X)\Big)\\
&= 1 + \frac{X^2}{\Gamma(2\kappa+1)},
\end{aligned}$$

$$\begin{aligned}
&\{precosh_\kappa\}^2(X)\\
&= \frac{1}{2}\Big(\{\mathbb{E}_\kappa\}^2(X) + \{\mathbb{E}_\kappa\}^2(-X)\Big)\\
&= 1 + \frac{X^2}{\Gamma(2\kappa+1)} + \frac{X^4}{\Gamma(4\kappa+1)}.
\end{aligned}$$

▷ The truncated pre-supersine function is based on the truncated Mittag-Leffler function:

$$\begin{aligned}
&\{presin_\kappa\}^n(X^\vartheta)\\
&= \frac{1}{2i}\Big(\{\mathbb{E}_\kappa\}^n(iX^\vartheta) - \{\mathbb{E}_\kappa\}^n(-iX^\vartheta)\Big)\\
&= \sum_{m=0}^{n} \frac{(-1)^m X^{(2m+1)\vartheta}}{\Gamma((2m+1)\kappa+1)}.
\end{aligned}$$

For $n = 0, 1, 2,$

$$
\begin{aligned}
&\{presin_\kappa\}^0(X^\vartheta)\\
&= \frac{1}{2i}\Big(\{\mathbb{E}_\kappa\}^0(iX^\vartheta) - \{\mathbb{E}_\kappa\}^0(-iX^\vartheta)\Big)\\
&= \frac{X^\vartheta}{\Gamma(\kappa+1)},
\end{aligned}
$$

$$
\begin{aligned}
&\{presin_\kappa\}^1(X^\vartheta)\\
&= \frac{1}{2i}\Big(\{\mathbb{E}_\kappa\}^1(iX^\vartheta) - \{\mathbb{E}_\kappa\}^1(-iX^\vartheta)\Big)\\
&= \frac{X^\vartheta}{\Gamma(\kappa+1)} - \frac{X^{3\vartheta}}{\Gamma(3\kappa+1)},
\end{aligned}
$$

$$
\begin{aligned}
&\{presin_\kappa\}^2(X^\vartheta)\\
&= \frac{1}{2i}\Big(\{\mathbb{E}_\kappa\}^2(iX^\vartheta) - \{\mathbb{E}_\kappa\}^2(-iX^\vartheta)\Big)\\
&= \frac{X^\vartheta}{\Gamma(\kappa+1)} - \frac{X^{3\vartheta}}{\Gamma(3\kappa+1)} + \frac{X^{5\vartheta}}{\Gamma(5\kappa+1)}.
\end{aligned}
$$

▷ The truncated pre-supercosine function is based on the truncated Mittag-Leffler function:

$$
\begin{aligned}
&\{precos_\kappa\}^n(X^\vartheta)\\
&= \frac{1}{2}\Big(\{\mathbb{E}_\kappa\}^n(iX^\vartheta) + \{\mathbb{E}_\kappa\}^n(-iX^\vartheta)\Big)\\
&= \sum_{m=0}^{n} \frac{(-1)^m X^{2m\vartheta}}{\Gamma((2m)\kappa+1)}.
\end{aligned}
$$

For $n = 0, 1, 2,$

$$
\begin{aligned}
&\{precos_\kappa\}^0(X^\vartheta)\\
&= \frac{1}{2}\Big(\{\mathbb{E}_\kappa\}^0(iX^\vartheta) + \{\mathbb{E}_\kappa\}^0(-iX^\vartheta)\Big)\\
&= 1,
\end{aligned}
$$

$$\begin{aligned}
&\{precos_\kappa\}^1(X^\vartheta)\\
&=\frac{1}{2}\Big(\{\mathbb{E}_\kappa\}^1(iX^\vartheta)+\{\mathbb{E}_\kappa\}^1(-iX^\vartheta)\Big)\\
&=1-\frac{X^{2\vartheta}}{\Gamma(2\kappa+1)},
\end{aligned}$$

$$\begin{aligned}
&\{precos_\kappa\}^2(X^\vartheta)\\
&=\frac{1}{2}\Big(\{\mathbb{E}_\kappa\}^2(iX^\vartheta)+\{\mathbb{E}_\kappa\}^2(-iX^\vartheta)\Big)\\
&=1-\frac{X^{2\vartheta}}{\Gamma(2\kappa+1)}+\frac{X^{4\vartheta}}{\Gamma(4\kappa+1)}.
\end{aligned}$$

▷ The truncated superhyperbolic pre-supersine function is based on the truncated Mittag-Leffler function:

$$\begin{aligned}
&\{presinh_\kappa\}^n(X^\vartheta)\\
&=\frac{1}{2}\Big(\{\mathbb{E}_\kappa\}^n(X^\vartheta)-\{\mathbb{E}_\kappa\}^n(-X^\vartheta)\Big)\\
&=\sum_{m=0}^{n}\frac{X^{(2m+1)\vartheta}}{\Gamma((2m+1)\kappa+1)}.
\end{aligned}$$

For $n=0,1,2,$

$$\begin{aligned}
&\{presinh_\kappa\}^0(X^\vartheta)\\
&=\frac{1}{2}\Big(\{\mathbb{E}_\kappa\}^0(X^\vartheta)-\{\mathbb{E}_\kappa\}^0(-X^\vartheta)\Big)\\
&=\frac{X^{\vartheta}}{\Gamma(\kappa+1)},
\end{aligned}$$

$$\begin{aligned}
&\{presinh_\kappa\}^1(X^\vartheta)\\
&=\frac{1}{2}\Big(\{\mathbb{E}_\kappa\}^1(X^\vartheta)-\{\mathbb{E}_\kappa\}^1(-X^\vartheta)\Big)\\
&=\frac{X^{\vartheta}}{\Gamma(\kappa+1)}+\frac{X^{3\vartheta}}{\Gamma(3\kappa+1)},
\end{aligned}$$

$$\begin{aligned}
&\{presinh_\kappa\}^2(X^\vartheta)\\
&=\frac{1}{2}\Big(\{\mathbb{E}_\kappa\}^2(X^\vartheta)-\{\mathbb{E}_\kappa\}^2(-X^\vartheta)\Big)\\
&=\frac{X^{\vartheta}}{\Gamma(\kappa+1)}+\frac{X^{3\vartheta}}{\Gamma(3\kappa+1)}+\frac{X^{5\vartheta}}{\Gamma(5\kappa+1)}.
\end{aligned}$$

▷ The truncated superhyperbolic pre-supercosine function is based on the truncated Mittag-Leffler function:

$$\begin{aligned}&\{precosh_\kappa\}^n(X^\vartheta)\\&=\frac{1}{2}\Big(\{\mathbb{E}_\kappa\}^n(X^\vartheta)+\{\mathbb{E}_\kappa\}^n(-X^\vartheta)\Big)\\&=\sum_{m=0}^{n}\frac{X^{2m\vartheta}}{\Gamma((2m)\kappa+1)}.\end{aligned}$$

For $n = 0, 1, 2,$

$$\begin{aligned}&\{precosh_\kappa\}^0(X^\vartheta)\\&=\frac{1}{2}\Big(\{\mathbb{E}_\kappa\}^0(X^\vartheta)+\{\mathbb{E}_\kappa\}^0(-X^\vartheta)\Big)\\&=1,\end{aligned}$$

$$\begin{aligned}&\{precosh_\kappa\}^1(X^\vartheta)\\&=\frac{1}{2}\Big(\{\mathbb{E}_\kappa\}^1(X^\vartheta)+\{\mathbb{E}_\kappa\}^1(-X^\vartheta)\Big)\\&=1+\frac{X^{2\vartheta}}{\Gamma(2\kappa+1)},\end{aligned}$$

$$\begin{aligned}&\{precosh_\kappa\}^2(X^\vartheta)\\&=\frac{1}{2}\Big(\{\mathbb{E}_\kappa\}^2(X^\vartheta)+\{\mathbb{E}_\kappa\}^2(-X^\vartheta)\Big)\\&=1+\frac{X^{2\vartheta}}{\Gamma(2\kappa+1)}+\frac{X^{4\vartheta}}{\Gamma(4\kappa+1)}.\end{aligned}$$

▷ The truncated supersine function is based on the truncated Mittag-Leffler function:

$$\begin{aligned}&\{presin_\kappa\}^n(X^\kappa)\\&=\frac{1}{2i}\Big(\{\mathbb{E}_\kappa\}^n(iX^\kappa)-\{\mathbb{E}_\kappa\}^n(-iX^\kappa)\Big)\\&=\sum_{m=0}^{n}\frac{(-1)^m X^{(2m+1)\kappa}}{\Gamma((2m+1)\kappa+1)}.\end{aligned}$$

For $n = 0, 1, 2,$

$$\begin{aligned}&\{presin_\kappa\}^0(X^\kappa)\\&= \frac{1}{2i}\Big(\{\mathbb{E}_\kappa\}^0(iX^\kappa) - \{\mathbb{E}_\kappa\}^0(-iX^\kappa)\Big)\\&= \frac{X^\kappa}{\Gamma(\kappa+1)},\end{aligned}$$

$$\begin{aligned}&\{presin_\kappa\}^1(X^\kappa)\\&= \frac{1}{2i}\Big(\{\mathbb{E}_\kappa\}^1(iX^\kappa) - \{\mathbb{E}_\kappa\}^1(-iX^\kappa)\Big)\\&= \frac{X^\kappa}{\Gamma(\kappa+1)} - \frac{X^{3\kappa}}{\Gamma(3\kappa+1)},\end{aligned}$$

$$\begin{aligned}&\{presin_\kappa\}^2(X^\kappa)\\&= \frac{1}{2i}\Big(\{\mathbb{E}_\kappa\}^2(iX^\kappa) - \{\mathbb{E}_\kappa\}^2(-iX^\kappa)\Big)\\&= \frac{X^\kappa}{\Gamma(\kappa+1)} - \frac{X^{3\kappa}}{\Gamma(3\kappa+1)} + \frac{X^{5\kappa}}{\Gamma(5\kappa+1)}.\end{aligned}$$

▹ The truncated supercosine function is based on the truncated Mittag- Leffler function:

$$\begin{aligned}&\{precos_\kappa\}^n(X^\kappa)\\&= \frac{1}{2}\Big(\{\mathbb{E}_\kappa\}^n(iX^\kappa) + \{\mathbb{E}_\kappa\}^n(-iX^\kappa)\Big)\\&= \sum_{m=0}^{n} \frac{(-1)^m X^{2m\kappa}}{\Gamma((2m)\kappa+1)}.\end{aligned}$$

For $n = 0, 1, 2,$

$$\begin{aligned}&\{precos_\kappa\}^0(X^\kappa)\\&= \frac{1}{2}\Big(\{\mathbb{E}_\kappa\}^0(iX^\kappa) + \{\mathbb{E}_\kappa\}^0(-iX^\kappa)\Big)\\&= 1,\end{aligned}$$

$$\begin{aligned}&\{precos_\kappa\}^1(X^\kappa)\\&= \frac{1}{2}\Big(\{\mathbb{E}_\kappa\}^1(iX^\kappa) + \{\mathbb{E}_\kappa\}^1(-iX^\kappa)\Big)\\&= 1 - \frac{X^{2\kappa}}{\Gamma(2\kappa+1)},\end{aligned}$$

$$\begin{aligned}&\{precos_\kappa\}^2(X^\kappa)\\&=\frac{1}{2}\Big(\{\mathbb{E}_\kappa\}^2(iX^\kappa)+\{\mathbb{E}_\kappa\}^2(-iX^\kappa)\Big)\\&=1-\frac{X^{2\kappa}}{\Gamma(2\kappa+1)}+\frac{X^{4\kappa}}{\Gamma(4\kappa+1)}.\end{aligned}$$

▷ The truncated superhyperbolic supersine function is based on the truncated Mittag-Leffler function:

$$\begin{aligned}&\{presinh_\kappa\}^n(X^\kappa)\\&=\frac{1}{2}\Big(\{\mathbb{E}_\kappa\}^n(X^\kappa)-\{\mathbb{E}_\kappa\}^n(-X^\kappa)\Big)\\&=\sum_{m=0}^{n}\frac{X^{(2m+1)\kappa}}{\Gamma((2m+1)\kappa+1)}.\end{aligned}$$

For $n = 0, 1, 2,$

$$\begin{aligned}&\{presinh_\kappa\}^0(X^\kappa)\\&=\frac{1}{2}\Big(\{\mathbb{E}_\kappa\}^0(X^\kappa)-\{\mathbb{E}_\kappa\}^0(-X^\kappa)\Big)\\&=\frac{X^\kappa}{\Gamma(\kappa+1)},\end{aligned}$$

$$\begin{aligned}&\{presinh_\kappa\}^1(X^\kappa)\\&=\frac{1}{2}\Big(\{\mathbb{E}_\kappa\}^1(X^\kappa)-\{\mathbb{E}_\kappa\}^1(-X^\vartheta)\Big)\\&=\frac{X^\kappa}{\Gamma(\kappa+1)}+\frac{X^{3\kappa}}{\Gamma(3\kappa+1)},\end{aligned}$$

$$\begin{aligned}&\{presinh_\kappa\}^2(X^\kappa)\\&=\frac{1}{2}\Big(\{\mathbb{E}_\kappa\}^2(X^\kappa)-\{\mathbb{E}_\kappa\}^2(-X^\kappa)\Big)\\&=\frac{X^\kappa}{\Gamma(\kappa+1)}+\frac{X^{3\kappa}}{\Gamma(3\kappa+1)}+\frac{X^{5\kappa}}{\Gamma(5\kappa+1)}.\end{aligned}$$

▷ The truncated superhyperbolic supercosine function is based on the truncated Mittag-Leffler function:

$$\begin{aligned}
&\{precosh_\kappa\}^n(X^\kappa)\\
&= \frac{1}{2}\Big(\{\mathbb{E}_\kappa\}^n(X^\kappa) + \{\mathbb{E}_\kappa\}^n(-X^\kappa)\Big)\\
&= \sum_{m=0}^{n} \frac{X^{2m\kappa}}{\Gamma((2m)\kappa + 1)}.
\end{aligned}$$

For $n = 0, 1, 2,$

$$\begin{aligned}
&\{precosh_\kappa\}^0(X^\kappa)\\
&= \frac{1}{2}\Big(\{\mathbb{E}_\kappa\}^0(X^\kappa) + \{\mathbb{E}_\kappa\}^0(-X^\kappa)\Big)\\
&= 1,
\end{aligned}$$

$$\begin{aligned}
&\{precosh_\kappa\}^1(X^\kappa)\\
&= \frac{1}{2}\Big(\{\mathbb{E}_\kappa\}^1(X^\kappa) + \{\mathbb{E}_\kappa\}^1(-X^\kappa)\Big)\\
&= 1 + \frac{X^{2\kappa}}{\Gamma(2\kappa + 1)},
\end{aligned}$$

$$\begin{aligned}
&\{precosh_\kappa\}^2(X^\kappa)\\
&= \frac{1}{2}\Big(\{\mathbb{E}_\kappa\}^2(X^\kappa) + \{\mathbb{E}_\kappa\}^2(-X^\kappa)\Big)\\
&= 1 + \frac{X^{2\kappa}}{\Gamma(2\kappa + 1)} + \frac{X^{4\kappa}}{\Gamma(4\kappa + 1)}.
\end{aligned}$$

4.7 The Generalized Hyperbolic Function

The generalized hyperbolic function, as an extended version of the Mittag-Leffler function, is given by

$$\mathbb{M}^{\kappa}_{m,r}(X) = \sum_{n=0}^{\infty} \frac{\kappa^n}{\Gamma(mn + r)} X^{nm+r},$$

where $\kappa \in \mathbb{C}\backslash\{0\}$, $r = 0, 1, \cdots, m - 1$, and $m \in \mathbb{N}$.

This work was presented by Riccati [10].

The special cases of the generalized hyperbolic function are presented as follows [2]:

$$\mathbb{M}^{\kappa}_{m,0}(0) = 1,$$

$$\mathbb{M}^{\kappa}_{1,0}(X) = \sum_{n=0}^{\infty} \frac{(\kappa X)^n}{\Gamma(n)} = exp(\kappa X),$$

$$\mathbb{M}^{\kappa}_{2,0}(X) = \cosh(\kappa^{0.5} X),$$

$$\mathbb{M}^{\kappa}_{2,1}(X) = \kappa^{-0.5} \sinh(\kappa^{0.5} X),$$

and

$$\mathbb{M}^{1}_{m,0}(X) = \sum_{n=0}^{\infty} \frac{X^{nm}}{\Gamma(nm)} = X\mathbb{E}_m(X).$$

These results were discovered by Muldoon and Ungar [11]. For more details of the generalized hyperbolic function, see [12].

4.8 The Matrix Mittag–Leffler Function

To end off this Chapter we are going to introduce our newly established matrix Mittag-Leffler function, which is a generalization of the famous multivariate Mittag-Leffler function and is something that we will make use of later in Chap. 22 to study uniqueness of several nonlinear fractional integro-differential equations with initial or boundary condition problems by fixed point theory and Babenko's approach [13].

Let $\alpha_{ij} \geq 0, \gamma_i > 0$ for all $i = 0, \cdots, n,\ j = 1, \cdots, m$, and

$$M = \begin{bmatrix} \alpha_{01} \cdots \alpha_{0m}\ \gamma_0 \\ \alpha_{11} \cdots \alpha_{1m}\ \gamma_1 \\ \cdots \\ \alpha_{n1} \cdots \alpha_{nm}\ \gamma_n \end{bmatrix}.$$

Definition 4.6 A matrix Mittag–Leffler function is defined by the following series

$$\mathbb{E}_M(X_1, \cdots, X_m) = \sum_{k=0}^{\infty} \sum_{\substack{k_1+\cdots+k_m=k \\ k_1\geq 0,\cdots,k_m\geq 0}} \binom{k}{k_1, \cdots, k_m} \frac{X_1^{k_1} \cdots X_m^{k_m}}{\Gamma(\alpha_{01}k_1 + \cdots + \alpha_{0m}k_m + \gamma_0)}$$
$$\cdot \frac{1}{\Gamma(\alpha_{11}k_1 + \cdots + \alpha_{1m}k_m + \gamma_1) \cdots \Gamma(\alpha_{n1}k_1 + \cdots + \alpha_{nm}k_m + \gamma_n)}, \tag{4.1}$$

where $X_i \in \mathbb{C}$ for $i = 1, 2, \cdots, m$ and

$$\binom{k}{k_1, \cdots, k_m} = \frac{k!}{k_1! \cdots k_m!}.$$

Clearly, the above series converges since there is a positive number δ such that

$$\begin{aligned} &\Gamma(\alpha_{11}k_1 + \cdots + \alpha_{1m}k_m + \gamma_1) \geq \delta, \\ &\cdots, \\ &\Gamma(\alpha_{n1}k_1 + \cdots + \alpha_{nm}k_m + \gamma_n) \geq \delta. \end{aligned}$$

In particular,

$$\begin{aligned} \mathbb{E}_P(X_1, \cdots, X_m) &= \mathbb{E}_{(\alpha_{01}, \cdots, \alpha_{0m}), \gamma_0}(X_1, \cdots, X_m) \\ &= \sum_{k=0}^{\infty} \sum_{\substack{k_1 + \cdots + k_m = k \\ k_1 \geq 0, \cdots, k_m \geq 0}} \binom{k}{k_1, \cdots, k_m} \frac{X_1^{k_1} \cdots X_m^{k_m}}{\Gamma(\alpha_{01}k_1 + \cdots + \alpha_{0m}k_m + \gamma_0)}, \end{aligned}$$

which is the multivariate Mittag–Leffler function given in [14], with

$$P = \begin{bmatrix} \alpha_{01} \cdots \alpha_{0m} & \gamma_0 \\ 0 \cdots \quad 0 & 1 \\ \cdots & \\ 0 \cdots \quad 0 & 1 \end{bmatrix}.$$

Moreover,

$$\mathbb{E}_{P_0}(X) = \mathbb{E}_{\alpha_{01}, \gamma_0}(X) = \sum_{k=0}^{\infty} \frac{X^k}{\Gamma(\alpha_{01}k + \gamma_0)}, \quad X \in \mathbb{C},$$

which is the well known two-parameter Mittag-Leffler function, with

$$P_0 = \begin{bmatrix} \alpha_{01} & 0 \cdots 0 & \gamma_0 \\ 0 & 0 \cdots 0 & 1 \\ \cdots & & \\ 0 & 0 \cdots 0 & 1 \end{bmatrix}.$$

References

1. Mittag-Leffler GM (1903) Sur la nouvelle fonction $E_a(x)$. C R Acad Sci 137:554–558
2. Xiao-Jun Y (2021) Theory and applications of special functions for scientists and engineers. Springer
3. Pollard H (1948) The completely monotonic character of the Mittag-Leffler function $E_a\{(-x)\}$. Bull Am Math Soc 54(12):1115–1117

4. Berberan-Santos MN (2005) Properties of the Mittag-Leffler relaxation function. J Math Chem 38(4):629–635
5. Hille E, Tamarkin JD (1933) On moment functions. Proc Natl Acad Sci USA 19(10):902
6. Phragmén E (1904) Sur une extension d'un théorème classique de la théorie des fonctions. Acta Math 28:351–368
7. Bagley RL (1988) The initial value problem for fractional order differential equations with constant coefficients. Air Force Institute of Technology Report, AFIT-TR-EN-88-1
8. Zhang S, Zhang HQ (2011) Fractional sub-equation method and its applications to nonlinear fractional PDEs. Phys Lett A 375(7):1069–1073
9. Dattoli G, Gorska K, Horzela A, Licciardi S, Pidatella RM (2017) Comments on the properties of Mittag-Leffler function. Eur Phys J Spec Top 226(16–18):3427–3443
10. Riccati V (1757) Vincentii Riccati soc. Jesu Opusculorum ad res physicas and mathematicas pertinentium Tomus primus. [-secundus]
11. Muldoon ME, Ungar AA (1996) Beyond sin and cos. Math Mag 69(1):3–14
12. Ungar A (1982) Generalized hyperbolic functions. Am Math Mon 89(9):688–691
13. Babenko YI (1986) Heat and mass transfer. Leningrad, Russian, Khimiya (In Russian)
14. Hadid SB, Luchko YF (1996) An operational method for solving fractional differential equations of an arbitrary real order. Panam Math J 6:57–73

Chapter 5
The Wiman, Supertrigonometric, and Superhyperbolic Functions

In this chapter, we introduce the Wiman function and present the supertrigonometric and superhyperbolic functions via the Wiman function.

5.1 The Wiman Function

In 1905, Anders Wiman (a Swedish mathematician) worked on algebraic geometry, applications of group theory and generalized the Mittag-Leffler function [1], which is called the two-parametric Mittag-Leffler function, the Mittag-Leffler type function, or the Wiman function.

We now present the following definition of the Wiman function.

Definition 5.1 ([1]) The Wiman function is given by

$$\mathbb{E}_{\kappa,\vartheta}(X) = \sum_{n=0}^{\infty} \frac{X^n}{\Gamma(n\kappa + \vartheta)},$$

where $X, \kappa, \vartheta \in \mathbb{C}$, and $\Re(\kappa), \Re(\vartheta) > 0$.

The relationship between the generalized Wright function and the Wiman function is given by [2]

$$\mathbb{E}_{\kappa,\vartheta}(X) = {}_1\mathbb{W}_1\left({}^{(1,1)}_{(\vartheta,\kappa)}; X\right),$$

where $X, \kappa, \vartheta \in \mathbb{C}$, and $\Re(\kappa), \Re(\vartheta) > 0$.

S. Rezaei Aderyani et al., *Towards Ulam Type Multi Stability Analysis*, Studies in Systems, Decision and Control 523, https://doi.org/10.1007/978-3-031-55564-0_5

Theorem 5.1 *Let* $\kappa, \vartheta > 0$. *Then,*

$$X\mathbb{E}_{\kappa,\kappa+\vartheta}(X) = \mathbb{E}_{\kappa,\vartheta}(X) - \frac{1}{\Gamma(\vartheta)},$$

where $X \in \mathbb{C}$.

This work was discovered by Erdelyi et al. [3] and reported on by Gorenflo et al. [4].

Corollary 5.1 *For any* $X \in \mathbb{C}$, *we have*

$$X\mathbb{E}_{\kappa,\kappa+1}(X) = \mathbb{E}_{\kappa}(X) - 1.$$

The result was presented by Gorenflo et al. [4].

We now introduce the function of Wiman type as follows:

Definition 5.2 The function of Wiman type is defined as

$$X^{\vartheta-1}\mathbb{E}_{\kappa,\vartheta}(X^{\kappa}) = \sum_{n=0}^{\infty} \frac{X^{n\kappa+\vartheta-1}}{\Gamma(n\kappa+\vartheta)}, \tag{5.1}$$

where $X, \kappa, \vartheta \in \mathbb{C}$, and $\Re(\kappa), \Re(\vartheta) > 0$.

The definition was introduced by Erdelyi et al. [3] and mentioned by Srivastava [5].

Figures 5.1, 5.2, 5.3, 5.4, 5.5, 5.6, 5.7, and 5.8 display the plots of $\mathbb{E}_{\kappa,\vartheta}(X)$, $\mathbb{E}_{\kappa,\vartheta}(-X)$, $\mathbb{E}_{\kappa,\vartheta}(X^{\kappa})$, $\mathbb{E}_{\kappa,\vartheta}(-X^{\kappa})$, $X^{\vartheta-1}\mathbb{E}_{\kappa,\vartheta}(X)$, $X^{\vartheta-1}\mathbb{E}_{\kappa,\vartheta}(-X)$, $X^{\vartheta-1}\mathbb{E}_{\kappa,\vartheta}(X^{\kappa})$, and $X^{\vartheta-1}\mathbb{E}_{\kappa,\vartheta}(-X^{\kappa})$ with different values of κ and ϑ.

Notice that we have

$$\mathbb{E}_{\kappa,1}(X) = \mathbb{E}_{\kappa}(X),$$

where $X, \kappa \in \mathbb{C}$, and $\Re(\kappa) > 0$.

Theorem 5.2 ([6]) *Suppose* $X, \kappa, \vartheta \in \mathbb{C}$, *and* $\Re(\kappa), \Re(\vartheta) > 0$, *then*

$$\mathbb{E}_{1,1}(X) = \exp(X),$$

$$\mathbb{E}_{1,2}(X) = \frac{\exp(X) - 1}{X},$$

$$\mathbb{E}_{2,1}(X) = \cosh\sqrt{X},$$

$$\mathbb{E}_{2,2}(X) = \frac{\sinh\sqrt{X}}{\sqrt{X}}.$$

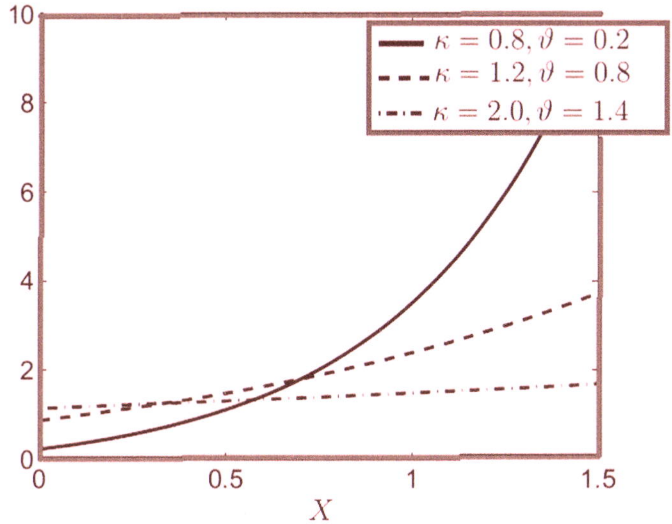

Fig. 5.1 The plots of $\mathbb{E}_{\kappa,\vartheta}(X)$ with different values of κ and ϑ

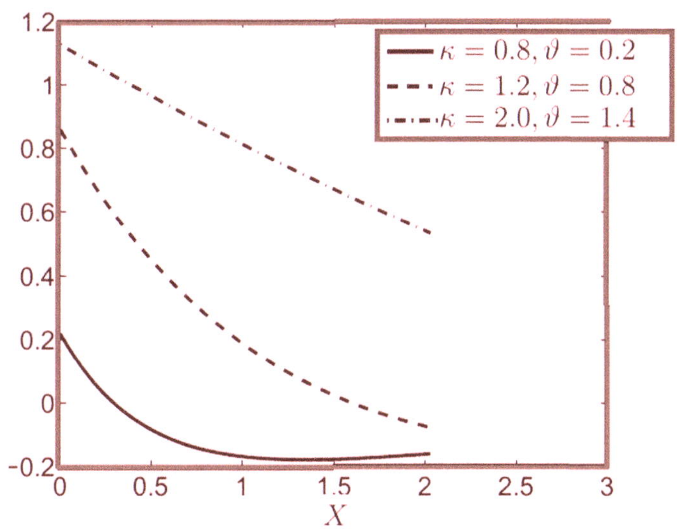

Fig. 5.2 The plots of $\mathbb{E}_{\kappa,\vartheta}(-X)$ with different values of κ and ϑ

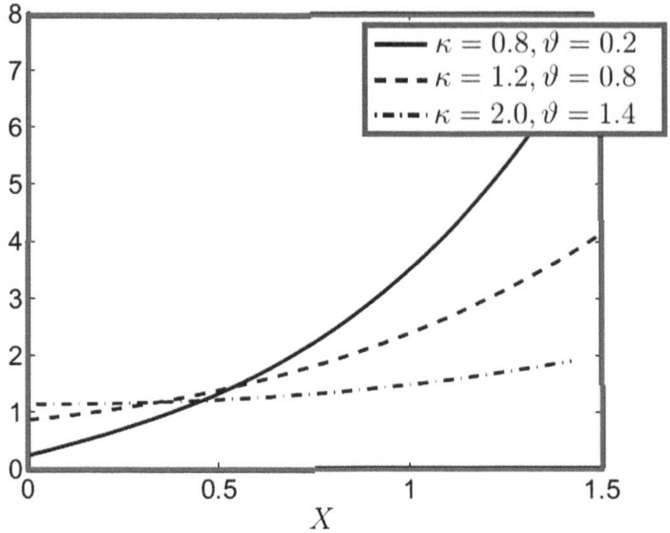

Fig. 5.3 The plots of $\mathbb{E}_{\kappa,\vartheta}(X^{\kappa})$ with different values of κ and ϑ

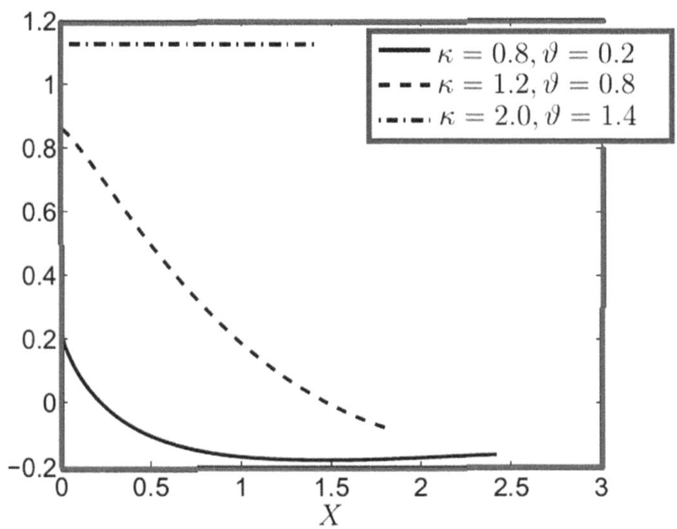

Fig. 5.4 The plots of $\mathbb{E}_{\kappa,\vartheta}(-X^{\kappa})$ with different values of κ and ϑ

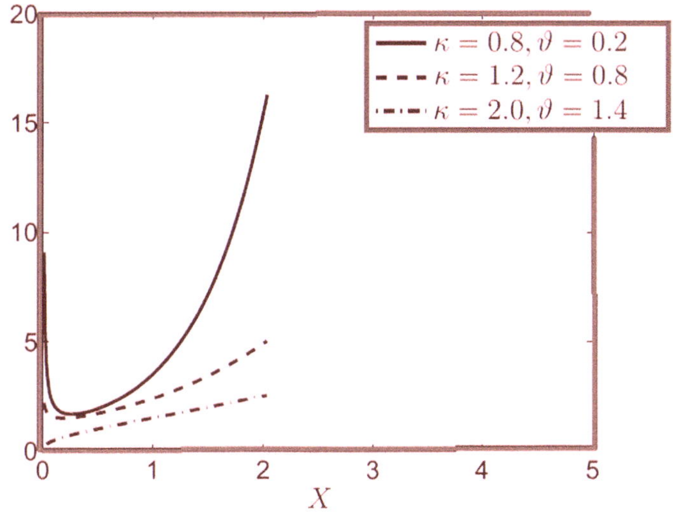

Fig. 5.5 The plots of $X^{\vartheta-1}\mathbb{E}_{\kappa,\vartheta}(X)$ with different values of κ and ϑ

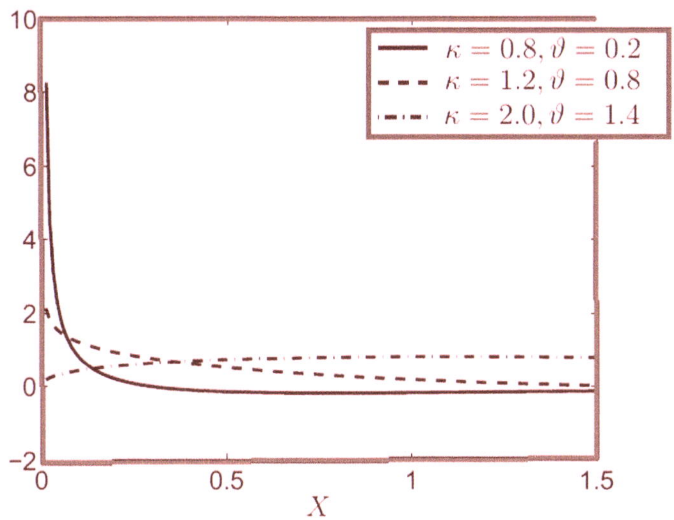

Fig. 5.6 The plots of $X^{\vartheta-1}\mathbb{E}_{\kappa,\vartheta}(-X)$ with different values of κ and ϑ

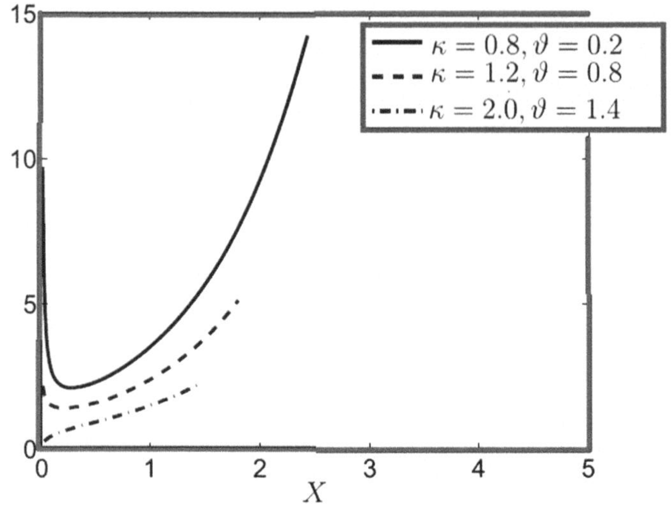

Fig. 5.7 The plots of $X^{\vartheta-1}\mathbb{E}_{\kappa,\vartheta}(X^{\kappa})$ with different values of κ and ϑ

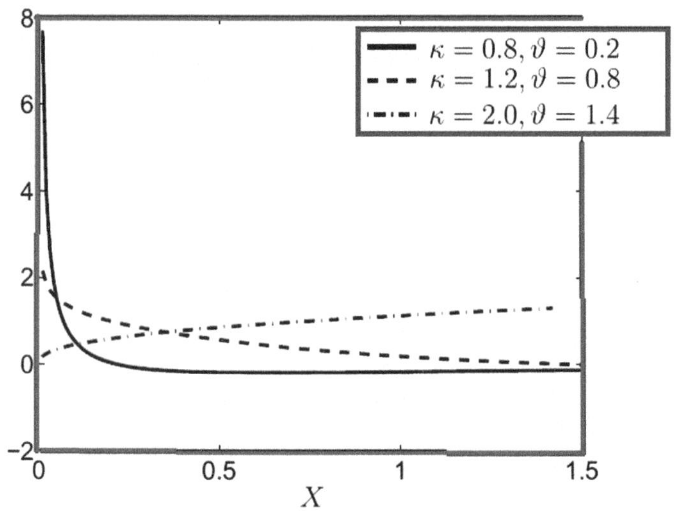

Fig. 5.8 The plots of $X^{\vartheta-1}\mathbb{E}_{\kappa,\vartheta}(-X^{\kappa})$ with different values of κ and ϑ

The investigation history of the above functions is as follows:

↪ The function $\mathbb{E}_{\kappa,\vartheta}(X^{\kappa})$ was presented by Erdélyi et al. [7].

↪ The function $\mathbb{E}_{\kappa,\vartheta}(-X)$ was presented by Schneider [8] and discussed by Miller and Samko [9].

↪ The function $X^{\vartheta-1}\mathbb{E}_{\kappa,\vartheta}(-X^{\kappa})$ was first reported by Erdélyi et al. [7], and further reported on by Dzherbashian [10].

5.2 The Supertrigonometric Functions via the Wiman Function

Let X, κ, $\vartheta \in \mathbb{C}$, and $\Re(\kappa)$, $\Re(\vartheta) > 0$. We now define the Supertrigonometric Functions through the Wiman Function [11], as follows:

▷ The supersine defined through the Wiman function is given by

$$\begin{aligned}&supersin_{\kappa,\vartheta}(X^{\kappa})\\&=\frac{X^{\vartheta-1}}{2i}\left(\mathbb{E}_{\kappa,\vartheta}(iX^{\kappa})-\mathbb{E}_{\kappa,\vartheta}(-iX^{\kappa})\right)\\&=\sum_{n=0}^{\infty}\frac{(-1)^{n}X^{(2n+1)\kappa+\vartheta-1}}{\Gamma((2n+1)\kappa+\vartheta)}.\end{aligned}$$

▷ The supercosine defined through the Wiman function is given by

$$\begin{aligned}&supercos_{\kappa,\vartheta}(X^{\kappa})\\&=\frac{X^{\vartheta-1}}{2}\left(\mathbb{E}_{\kappa,\vartheta}(iX^{\kappa})+\mathbb{E}_{\kappa,\vartheta}(-iX^{\kappa})\right)\\&=\sum_{n=0}^{\infty}\frac{(-1)^{n}X^{(2n)\kappa+\vartheta-1}}{\Gamma((2n)\kappa+\vartheta)}.\end{aligned}$$

▷ The supertangent defined through the Wiman function is given by

$$supertan_{\kappa,\vartheta}(X^{\kappa})=\frac{supersin_{\kappa,\vartheta}(X^{\kappa})}{supercos_{\kappa,\vartheta}(X^{\kappa})},\quad (supercos_{\kappa,\vartheta}(X^{\kappa}\neq 0).$$

▷ The supercotangent defined through the Wiman function is given by

$$supercotan_{\kappa,\vartheta}(X^{\kappa})=\frac{supercos_{\kappa,\vartheta}(X^{\kappa})}{supersin_{\kappa,\vartheta}(X^{\kappa})},\quad (supersin_{\kappa,\vartheta}(X^{\kappa})\neq 0).$$

▷ The supercosecant defined through the Wiman function is given by

$$supercsc_{\kappa,\vartheta}(X^{\kappa})=\frac{1}{supersin_{\kappa,\vartheta}(X^{\kappa})},\quad (supersin_{\kappa,\vartheta}(X^{\kappa})\neq 0).$$

▷ The supersecant defined through the Wiman function is given by

$$supersec_{\kappa,\vartheta}(X^{\kappa})=\frac{1}{supercos_{\kappa,\vartheta}(X^{\kappa})},\quad (supercos_{\kappa,\vartheta}(X^{\kappa})\neq 0).$$

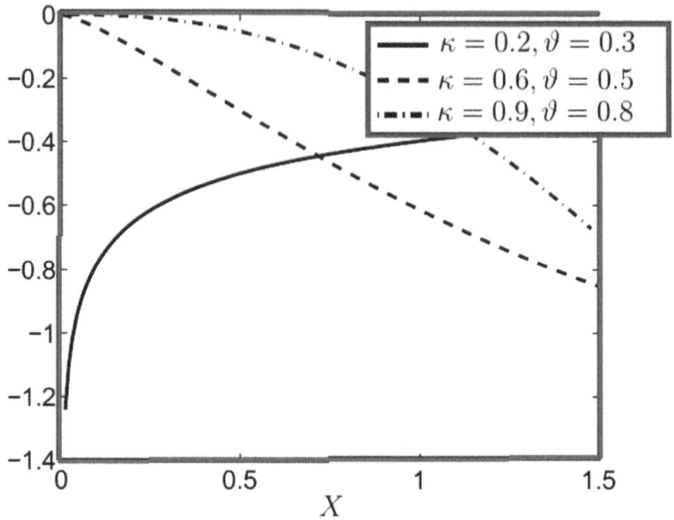

Fig. 5.9 The plots of $supersin_{\kappa,\vartheta}(X^{\kappa})$ with different values of κ and ϑ

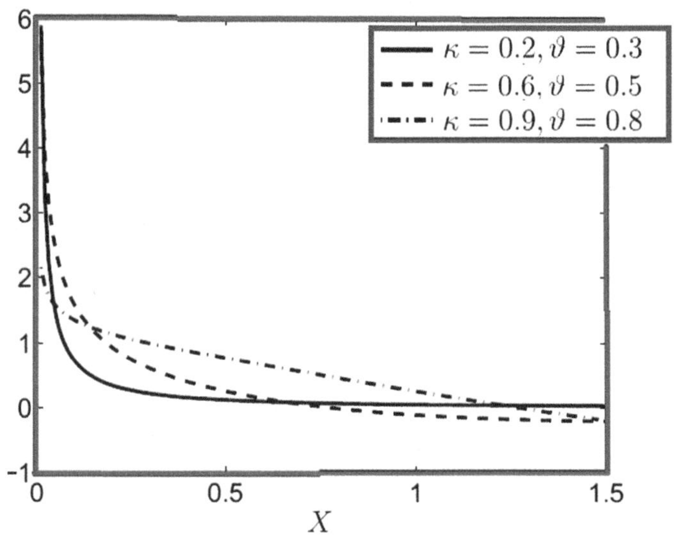

Fig. 5.10 The plots of $supercos_{\kappa,\vartheta}(X^{\kappa})$ with different values of κ and ϑ

These results were presented by Yang [11].

Figures 5.9 and 5.10 display the plots of $supersin_{\kappa,\vartheta}(X^{\kappa})$ and $supercos_{\kappa,\vartheta}(X^{\kappa})$ with different values of κ and ϑ.

Theorem 5.3 ([11]) *Let* $X, \kappa, \vartheta \in \mathbb{C}$, *and* $\Re(\kappa), \Re(\vartheta) > 0$. *Then,*

$$X^{\vartheta-1}\mathbb{E}_{\kappa,\vartheta}(iX^{\kappa}) = supercos_{\kappa,\vartheta}(X^{\vartheta}) + i\, supersin_{\kappa,\vartheta}(X^{\vartheta}),$$

$$X^{\vartheta-1}\mathbb{E}_{\kappa,\vartheta}(-iX^{\kappa}) = supercos_{\kappa,\vartheta}(X^{\vartheta}) - i\, supersin_{\kappa,\vartheta}(X^{\vartheta}),$$

and

$$[X^{\vartheta-1}\mathbb{E}_{\kappa,\vartheta}(iX^{\kappa})][X^{\vartheta-1}\mathbb{E}_{\kappa,\vartheta}(-iX^{\kappa})] = supercos^2_{\kappa,\vartheta}(X^{\vartheta}) + supersin^2_{\kappa,\vartheta}(X^{\vartheta}).$$

The above theorem was also presented by Yang [11].

5.3 The Superhyperbolic Functions via the Wiman Function

Let $X, \kappa, \vartheta \in \mathbb{C}$, and $\Re(\kappa), \Re(\vartheta) > 0$. We now present the superhyperbolic functions [11] via the Wiman function as follows:

▷ The superhyperbolic supersine through the Wiman function is given by

$$\begin{aligned} &supersinh_{\kappa,\vartheta}(X^{\kappa}) \\ &= \frac{X^{\vartheta-1}}{2}\Big(\mathbb{E}_{\kappa,\vartheta}(X^{\kappa}) - \mathbb{E}_{\kappa,\vartheta}(-X^{\kappa})\Big) \\ &= \sum_{n=0}^{\infty} \frac{X^{(2n+1)\kappa+\vartheta-1}}{\Gamma((2n+1)\kappa+\vartheta)}. \end{aligned}$$

▷ The superhyperbolic supercosine through the Wiman function is given by

$$\begin{aligned} &supercosh_{\kappa,\vartheta}(X^{\kappa}) \\ &= \frac{X^{\vartheta-1}}{2}\Big(\mathbb{E}_{\kappa,\vartheta}(X^{\kappa}) + \mathbb{E}_{\kappa,\vartheta}(-X^{\kappa})\Big) \\ &= \sum_{n=0}^{\infty} \frac{X^{(2n)\kappa+\vartheta-1}}{\Gamma((2n)\kappa+\vartheta)}. \end{aligned}$$

▷ The superhyperbolic supertangent through the Wiman function is given by

$$supertanh_{\kappa,\vartheta}(X^{\kappa}) = \frac{supersinh_{\kappa,\vartheta}(X^{\kappa})}{supercosh_{\kappa,\vartheta}(X^{\kappa})}, \quad (supercosh_{\kappa,\vartheta}(X^{\kappa}) \neq 0).$$

▷ The superhyperbolic supercotangent through the Wiman function is given by

$$supercotanh_{\kappa,\vartheta}(X^{\kappa}) = \frac{supercosh_{\kappa,\vartheta}(X^{\kappa})}{supersinh_{\kappa,\vartheta}(X^{\kappa})}, \quad (supersinh_{\kappa,\vartheta}(X^{\kappa}) \neq 0).$$

▷ The superhyperbolic supercosecant through the Wiman function is given by

$$supercsch_{\kappa,\vartheta}(X^{\kappa}) = \frac{1}{supersinh_{\kappa,\vartheta}(X^{\kappa})}, \quad (supersinh_{\kappa,\vartheta}(X^{\kappa}) \neq 0).$$

▷ The superhyperbolic supersecant through the Wiman function is given by

$$supersech_{\kappa,\vartheta}(X^{\kappa}) = \frac{1}{supercosh_{\kappa,\vartheta}(X^{\kappa})}, \quad (supercosh_{\kappa,\vartheta}(X^{\kappa}) \neq 0).$$

The above results were also due to Yang [11].

Figures 5.11 and 5.12 display the plots of $supersinh_{\kappa,\vartheta}(X^{\kappa})$ and $supercosh_{\kappa,\vartheta}(X^{\kappa})$ with different values of κ and ϑ.

Theorem 5.4 ([11]) *Let* $X, \kappa, \vartheta \in \mathbb{C}$, *and* $\Re(\kappa), \Re(\vartheta) > 0$. *Then,*

$$X^{\vartheta-1}\mathbb{E}_{\kappa,\vartheta}(X^{\kappa}) = supercos_{\kappa,\vartheta}(X^{\vartheta}) + supersin_{\kappa,\vartheta}(X^{\vartheta}),$$

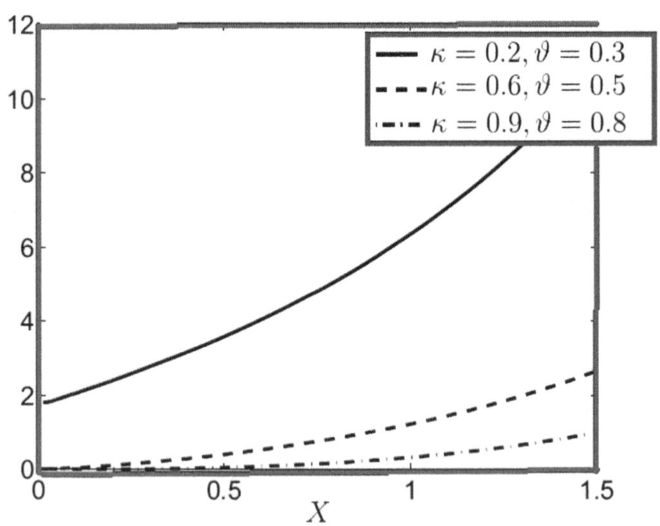

Fig. 5.11 The plots of $supersinh_{\kappa,\vartheta}(X^{\kappa})$ with different values of κ and ϑ

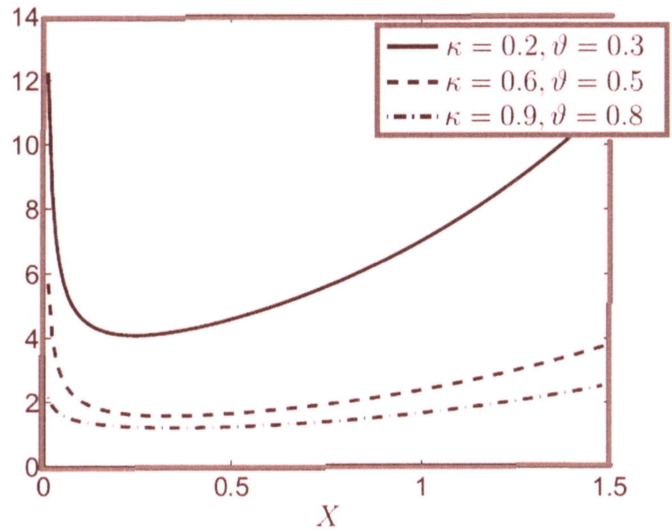

Fig. 5.12 The plots of $supercosh_{\kappa,\vartheta}(X^{\kappa})$ with different values of κ and ϑ

$$X^{\vartheta-1}\mathbb{E}_{\kappa,\vartheta}(-X^{\kappa}) = supercos_{\kappa,\vartheta}(X^{\vartheta}) - i\, supersin_{\kappa,\vartheta}(X^{\vartheta}),$$

and

$$[X^{\vartheta-1}\mathbb{E}_{\kappa,\vartheta}(X^{\kappa})][X^{\vartheta-1}\mathbb{E}_{\kappa,\vartheta}(-X^{\kappa})] = supercos^2_{\kappa,\vartheta}(X^{\vartheta}) - supersin^2_{\kappa,\vartheta}(X^{\vartheta}).$$

5.4 The Pre-supertrigonometric Functions via the Wiman Function

Let $X, \kappa, \vartheta \in \mathbb{C}$, and $\Re(\kappa), \Re(\vartheta) > 0$. We now introduce the pre-supertrigonometric functions [11] through the Wiman function as follows:

▹ The presupersine through the Wiman function is given by

$$\begin{aligned} & presin_{\kappa,\vartheta}(X) \\ &= \frac{1}{2i}\left(\mathbb{E}_{\kappa,\vartheta}(iX) - \mathbb{E}_{\kappa,\vartheta}(-iX)\right) \\ &= \sum_{n=0}^{\infty} \frac{(-1)^n X^{(2n+1)}}{\Gamma((2n+1)\kappa + \vartheta)}. \end{aligned}$$

▷ The precosine through the Wiman function is given by

$$\begin{aligned} & precos_{\kappa,\vartheta}(X) \\ &= \frac{1}{2}\Big(\mathbb{E}_{\kappa,\vartheta}(iX) + \mathbb{E}_{\kappa,\vartheta}(-iX)\Big) \\ &= \sum_{n=0}^{\infty} \frac{(-1)^n X^{(2n)}}{\Gamma((2n)\kappa + \vartheta)}. \end{aligned}$$

▷ The pretangent through the Wiman function is given by

$$pretan_{\kappa,\vartheta}(X) = \frac{presin_{\kappa,\vartheta}(X)}{precos_{\kappa,\vartheta}(X)}, \quad (precos_{\kappa,\vartheta}(X) \neq 0).$$

▷ The precotangent through the Wiman function is given by

$$precotan_{\kappa,\vartheta}(X) = \frac{precos_{\kappa,\vartheta}(X)}{presin_{\kappa,\vartheta}(X)}, \quad (presin_{\kappa,\vartheta}(X) \neq 0).$$

▷ The precosecant through the Wiman function is given by

$$precsc_{\kappa,\vartheta}(X) = \frac{1}{presin_{\kappa,\vartheta}(X)}, \quad (presin_{\kappa,\vartheta}(X) \neq 0).$$

▷ The presecant through the Wiman function is given by

$$presec_{\kappa,\vartheta}(X) = \frac{1}{precos_{\kappa,\vartheta}(X)}, \quad (precos_{\kappa,\vartheta}(X) \neq 0).$$

These results were presented by Yang in 2019 [11].

Theorem 5.5 ([11]) *Let* $X, \kappa, \vartheta \in \mathbb{C}$, *and* $\Re(\kappa), \Re(\vartheta) > 0$. *Then,*

$$\mathbb{E}_{\kappa,\vartheta}(iX) = precos_{\kappa,\vartheta}(X) + i\, presin_{\kappa,\vartheta}(X),$$

$$\mathbb{E}_{\kappa,\vartheta}(-iX) = precos_{\kappa,\vartheta}(X) - i\, presin_{\kappa,\vartheta}(X),$$

and

$$[\mathbb{E}_{\kappa,\vartheta}(iX)][\mathbb{E}_{\kappa,\vartheta}(-iX)] = precos^2_{\kappa,\vartheta}(X) + presin^2_{\kappa,\vartheta}(X).$$

This result was presented by Yang [11].

5.5 The Pre-superhyperbolic Functions via the Wiman Function

Let X, κ, $\vartheta \in \mathbb{C}$, and $\Re(\kappa)$, $\Re(\vartheta) > 0$. We now present the pre-superhyperbolic functions [11] via the Wiman function as follows:

▷ The hyperbolic presupersine through the Wiman function is given by

$$\begin{aligned} & presinh_{\kappa,\vartheta}(X) \\ &= \frac{1}{2}\Big(\mathbb{E}_{\kappa,\vartheta}(X) - \mathbb{E}_{\kappa,\vartheta}(-X)\Big) \\ &= \sum_{n=0}^{\infty} \frac{X^{(2n+1)}}{\Gamma((2n+1)\kappa + \vartheta)}. \end{aligned}$$

▷ The hyperbolic precosine through the Wiman function is given by

$$\begin{aligned} & precosh_{\kappa,\vartheta}(X) \\ &= \frac{1}{2}\Big(\mathbb{E}_{\kappa,\vartheta}(X) + \mathbb{E}_{\kappa,\vartheta}(-X)\Big) \\ &= \sum_{n=0}^{\infty} \frac{X^{(2n)}}{\Gamma((2n)\kappa + \vartheta)}. \end{aligned}$$

▷ The hyperbolic pretangent through the Wiman function is given by

$$pretanh_{\kappa,\vartheta}(X) = \frac{presinh_{\kappa,\vartheta}(X)}{precosh_{\kappa,\vartheta}(X)}, \quad (precosh_{\kappa,\vartheta}(X) \neq 0).$$

▷ The hyperbolic precotangent through the Wiman function is given by

$$precotanh_{\kappa,\vartheta}(X) = \frac{precosh_{\kappa,\vartheta}(X)}{presinh_{\kappa,\vartheta}(X)}, \quad (presin_{\kappa,\vartheta}(X) \neq 0).$$

▷ The hyperbolic precosecant through the Wiman function is given by

$$precsch_{\kappa,\vartheta}(X) = \frac{1}{presinh_{\kappa,\vartheta}(X)}, \quad (presinh_{\kappa,\vartheta}(X) \neq 0).$$

▷ The hyperbolic presecant through the Wiman function is given by

$$presech_{\kappa,\vartheta}(X) = \frac{1}{precosh_{\kappa,\vartheta}(X)}, \quad (precosh_{\kappa,\vartheta}(X) \neq 0).$$

These results were presented by Yang [11].

Theorem 5.6 ([11]) *Let X, κ, $\vartheta \in \mathbb{C}$, and $\Re(\kappa)$, $\Re(\vartheta) > 0$. Then,*

$$\mathbb{E}_{\kappa,\vartheta}(X) = precosh_{\kappa,\vartheta}(X) + presinh_{\kappa,\vartheta}(X),$$

$$\mathbb{E}_{\kappa,\vartheta}(-X) = precosh_{\kappa,\vartheta}(X) - presinh_{\kappa,\vartheta}(X),$$

and

$$[\mathbb{E}_{\kappa,\vartheta}(iX)][\mathbb{E}_{\kappa,\vartheta}(-X)] = precosh^2_{\kappa,\vartheta}(X) + presinh^2_{\kappa,\vartheta}(X).$$

5.6 The Supertrigonometric and Superhyperbolic Functions with the Power Law via the Wiman Function

Let X, κ, ϑ, $\gamma \in \mathbb{C}$, and $\Re(\kappa)$, $\Re(\vartheta) > 0$. Now, we present the supertrigonometric and superhyperbolic functions with the power law [6] as follows:

▷ The presupersine with power law through the Wiman function is given by

$$\begin{aligned} &presin_{\kappa,\vartheta}(X^{\gamma}) \\ &= \frac{1}{2i}\Big(\mathbb{E}_{\kappa,\vartheta}(iX^{\gamma}) - \mathbb{E}_{\kappa,\vartheta}(-iX^{\gamma})\Big) \\ &= \sum_{n=0}^{\infty} \frac{(-1)^n X^{(2n+1)\gamma}}{\Gamma((2n+1)\kappa+\vartheta)}. \end{aligned}$$

▷ The precosine with power law through the Wiman function is given by

$$\begin{aligned} &precos_{\kappa,\vartheta}(X^{\gamma}) \\ &= \frac{1}{2}\Big(\mathbb{E}_{\kappa,\vartheta}(iX^{\gamma}) + \mathbb{E}_{\kappa,\vartheta}(-iX^{\gamma})\Big) \\ &= \sum_{n=0}^{\infty} \frac{(-1)^n X^{(2n)\gamma}}{\Gamma((2n)\kappa+\vartheta)}. \end{aligned}$$

▷ The pretangent with power law through the Wiman function is given by

$$pretan_{\kappa,\vartheta}(X^{\gamma}) = \frac{presin_{\kappa,\vartheta}(X^{\gamma})}{precos_{\kappa,\vartheta}(X^{\gamma})}, \quad (precos_{\kappa,\vartheta}(X^{\gamma}) \neq 0).$$

▷ The precotangent with power law through the Wiman function is given by

$$precotan_{\kappa,\vartheta}(X^{\gamma}) = \frac{precos_{\kappa,\vartheta}(X^{\gamma})}{presin_{\kappa,\vartheta}(X^{\gamma})}, \quad (presin_{\kappa,\vartheta}(X^{\gamma}) \neq 0).$$

▷ The precosecant with power law through the Wiman function is given by

$$precsc_{\kappa,\vartheta}(X^{\gamma}) = \frac{1}{presin_{\kappa,\vartheta}(X^{\gamma})}, \quad (presin_{\kappa,\vartheta}(X^{\gamma}) \neq 0).$$

▷ The presecant with power law through the Wiman function is given by

$$presec_{\kappa,\vartheta}(X^{\gamma}) = \frac{1}{precos_{\kappa,\vartheta}(X^{\gamma})}, \quad (precos_{\kappa,\vartheta}(X^{\gamma}) \neq 0).$$

Theorem 5.7 ([6]) *Let* $X, \kappa, \vartheta, \gamma \in \mathbb{C}$, *and* $\Re(\kappa), \Re(\vartheta) > 0$. *Then,*

$$\mathbb{E}_{\kappa,\vartheta}(iX^{\gamma}) = precos_{\kappa,\vartheta}(X^{\gamma}) + i\, presin_{\kappa,\vartheta}(X^{\gamma}),$$

$$\mathbb{E}_{\kappa,\vartheta}(-iX^{\gamma}) = precos_{\kappa,\vartheta}(X^{\gamma}) - i\, presin_{\kappa,\vartheta}(X^{\gamma}),$$

and

$$[\mathbb{E}_{\kappa,\vartheta}(iX^{\gamma})][\mathbb{E}_{\kappa,\vartheta}(-iX^{\gamma})] = precos^{2}_{\kappa,\vartheta}(X^{\gamma}) + presin^{2}_{\kappa,\vartheta}(X^{\gamma}).$$

Let $X, \kappa, \vartheta, \gamma \in \mathbb{C}$, and $\Re(\kappa), \Re(\vartheta) > 0$. We now have the following [6]:

▷ The hyperbolic presupersine with power law through the Wiman function is given by

$$\begin{aligned} &presinh_{\kappa,\vartheta}(X^{\gamma}) \\ &= \frac{1}{2}\Big(\mathbb{E}_{\kappa,\vartheta}(X^{\gamma}) - \mathbb{E}_{\kappa,\vartheta}(-X^{\gamma})\Big) \\ &= \sum_{n=0}^{\infty} \frac{X^{(2n+1)\gamma}}{\Gamma((2n+1)\kappa + \vartheta)}. \end{aligned}$$

▷ The hyperbolic precosine with power law through the Wiman function is given by

$$\begin{aligned} &precosh_{\kappa,\vartheta}(X^{\gamma}) \\ &= \frac{1}{2}\Big(\mathbb{E}_{\kappa,\vartheta}(X^{\gamma}) + \mathbb{E}_{\kappa,\vartheta}(-X^{\gamma})\Big) \\ &= \sum_{n=0}^{\infty} \frac{X^{(2n)\gamma}}{\Gamma((2n)\kappa + \vartheta)}. \end{aligned}$$

▷ The hyperbolic pretangent with power law through the Wiman function is given by

$$pretanh_{\kappa,\vartheta}(X^{\gamma}) = \frac{presinh_{\kappa,\vartheta}(X^{\gamma})}{precosh_{\kappa,\vartheta}(X^{\gamma})}, \quad (precosh_{\kappa,\vartheta}(X^{\gamma}) \neq 0).$$

▷ The hyperbolic precotangent with power law through the Wiman function is given by

$$precotanh_{\kappa,\vartheta}(X^{\gamma}) = \frac{precosh_{\kappa,\vartheta}(X^{\gamma})}{presinh_{\kappa,\vartheta}(X^{\gamma})}, \quad (presin_{\kappa,\vartheta}(X^{\gamma}) \neq 0).$$

▷ The hyperbolic precosecant with power law through the Wiman function is given by

$$precsch_{\kappa,\vartheta}(X^{\gamma}) = \frac{1}{presinh_{\kappa,\vartheta}(X^{\gamma})}, \quad (presinh_{\kappa,\vartheta}(X^{\gamma}) \neq 0).$$

▷ The hyperbolic presecant with power law through the Wiman function is given by

$$presech_{\kappa,\vartheta}(X^{\gamma}) = \frac{1}{precosh_{\kappa,\vartheta}(X^{\gamma})}, \quad (precosh_{\kappa,\vartheta}(X^{\gamma}) \neq 0).$$

Theorem 5.8 ([6]) *Suppose* $X, \kappa, \vartheta, \gamma \in \mathbb{C}$, *and* $\Re(\kappa), \Re(\vartheta) > 0$. *Then*

$$\mathbb{E}_{\kappa,\vartheta}(X^{\gamma}) = precosh_{\kappa,\vartheta}(X^{\gamma}) + presinh_{\kappa,\vartheta}(X^{\gamma}),$$

$$\mathbb{E}_{\kappa,\vartheta}(-X^{\gamma}) = precosh_{\kappa,\vartheta}(X^{\gamma}) - presinh_{\kappa,\vartheta}(X^{\gamma}),$$

and

$$[\mathbb{E}_{\kappa,\vartheta}(iX^{\gamma})][\mathbb{E}_{\kappa,\vartheta}(-X^{\gamma})] = precosh^2_{\kappa,\vartheta}(X^{\gamma}) + presinh^2_{\kappa,\vartheta}(X^{\gamma}).$$

5.7 The Supertrigonometric and Superhyperbolic Functions with the Parameter via the Wiman Function

Let $X, \kappa, \vartheta \in \mathbb{C}$, and $\Re(\kappa), \Re(\vartheta) > 0$. Now, we present the supertrigonometric and superhyperbolic functions with parameters [6] as follows:

▷ The presupersine through the Wiman function is given by

$$
\begin{aligned}
& Tresin_{\kappa,\vartheta}(X) \\
&= \frac{X^{\vartheta-1}}{2i}\Big(\mathbb{E}_{\kappa,\vartheta}(iX) - \mathbb{E}_{\kappa,\vartheta}(-iX)\Big) \\
&= X^{\vartheta-1}\sum_{n=0}^{\infty}\frac{(-1)^n X^{(2n+1)}}{\Gamma((2n+1)\kappa+\vartheta)}.
\end{aligned}
$$

▷ The precosine through the Wiman function is given by

$$
\begin{aligned}
& Trecos_{\kappa,\vartheta}(X) \\
&= \frac{X^{\vartheta-1}}{2}\Big(\mathbb{E}_{\kappa,\vartheta}(iX) + \mathbb{E}_{\kappa,\vartheta}(-iX)\Big) \\
&= X^{\vartheta-1}\sum_{n=0}^{\infty}\frac{(-1)^n X^{(2n)}}{\Gamma((2n)\kappa+\vartheta)}.
\end{aligned}
$$

▷ The pretangent through the Wiman function is given by

$$
Tretan_{\kappa,\vartheta}(X) = \frac{Tresin_{\kappa,\vartheta}(X)}{Trecos_{\kappa,\vartheta}(X)}, \quad (Trecos_{\kappa,\vartheta}(X) \neq 0).
$$

▷ The precotangent through the Wiman function is given by

$$
Trecotan_{\kappa,\vartheta}(X) = \frac{Trecos_{\kappa,\vartheta}(X)}{Tresin_{\kappa,\vartheta}(X)}, \quad (Tresin_{\kappa,\vartheta}(X) \neq 0).
$$

▷ The precosecant through the Wiman function is given by

$$
Trecsc_{\kappa,\vartheta}(X) = \frac{1}{Tresin_{\kappa,\vartheta}(X)}, \quad (Tresin_{\kappa,\vartheta}(X) \neq 0).
$$

▷ The presecant through the Wiman function is given by

$$
Tresec_{\kappa,\vartheta}(X) = \frac{1}{Trecos_{\kappa,\vartheta}(X)}, \quad (Trecos_{\kappa,\vartheta}(X) \neq 0).
$$

Theorem 5.9 ([6]) *Let* X, κ, $\vartheta \in \mathbb{C}$, *and* $\Re(\kappa)$, $\Re(\vartheta) > 0$. *Then,*

$$
Tresin_{\kappa,\vartheta}(X) = X^{\vartheta-1} presin_{\kappa,\vartheta}(X),
$$

$$
Trecos_{\kappa,\vartheta}(X) = X^{\vartheta-1} precos_{\kappa,\vartheta}(X).
$$

Theorem 5.10 ([6]) *Let* $X, \kappa, \vartheta \in \mathbb{C}$, *and* $\Re(\kappa), \Re(\vartheta) > 0$. *Then,*

$$X^{\vartheta-1}\mathbb{E}_{\kappa,\vartheta}(iX) = Trecos_{\kappa,\vartheta}(X) + i\, Tresin_{\kappa,\vartheta}(X),$$

$$X^{\vartheta-1}\mathbb{E}_{\kappa,\vartheta}(-iX) = Trecos_{\kappa,\vartheta}(X) - i\, Tresin_{\kappa,\vartheta}(X),$$

and

$$[X^{\vartheta-1}\mathbb{E}_{\kappa,\vartheta}(iX)][X^{\vartheta-1}\mathbb{E}_{\kappa,\vartheta}(-iX)] = Trecos^2_{\kappa,\vartheta}(X) + Tresin^2_{\kappa,\vartheta}(X).$$

Let $X, \kappa, \vartheta \in \mathbb{C}$, and $\Re(\kappa), \Re(\vartheta) > 0$. Now, we have the following [6]:

- ▷ The hyperbolic presupersine through the Wiman function is given by

$$\begin{aligned} &Tresinh_{\kappa,\vartheta}(X) \\ &= \frac{X^{\vartheta-1}}{2}\Big(\mathbb{E}_{\kappa,\vartheta}(X) - \mathbb{E}_{\kappa,\vartheta}(-X)\Big) \\ &= X^{\vartheta-1}\sum_{n=0}^{\infty}\frac{X^{(2n+1)}}{\Gamma((2n+1)\kappa+\vartheta)}. \end{aligned}$$

- ▷ The hyperbolic precosine through the Wiman function is given by

$$\begin{aligned} &Trecosh_{\kappa,\vartheta}(X) \\ &= \frac{X^{\vartheta-1}}{2}\Big(\mathbb{E}_{\kappa,\vartheta}(X) + \mathbb{E}_{\kappa,\vartheta}(-X)\Big) \\ &= X^{\vartheta-1}\sum_{n=0}^{\infty}\frac{X^{(2n)}}{\Gamma((2n)\kappa+\vartheta)}. \end{aligned}$$

- ▷ The hyperbolic pretangentthrough the Wiman function is given by

$$Tretanh_{\kappa,\vartheta}(X) = \frac{Tresinh_{\kappa,\vartheta}(X)}{Trecosh_{\kappa,\vartheta}(X)}, \quad (Trecosh_{\kappa,\vartheta}(X) \neq 0).$$

- ▷ The hyperbolic precotangent through the Wiman function is given by

$$Trecotanh_{\kappa,\vartheta}(X) = \frac{Trecosh_{\kappa,\vartheta}(X)}{Tresinh_{\kappa,\vartheta}(X)}, \quad (Tresinh_{\kappa,\vartheta}(X) \neq 0).$$

- ▷ The hyperbolic precosecant through the Wiman function is given by

$$Trecsch_{\kappa,\vartheta}(X) = \frac{1}{Tresinh_{\kappa,\vartheta}(X)}, \quad (Tresinh_{\kappa,\vartheta}(X) \neq 0).$$

▷ The hyperbolic presecant through the Wiman function is given by

$$Tresech_{\kappa,\vartheta}(X) = \frac{1}{Trecosh_{\kappa,\vartheta}(X)}, \quad (Trecosh_{\kappa,\vartheta}(X) \neq 0).$$

Theorem 5.11 ([6]) *Let* $X, \kappa, \vartheta \in \mathbb{C}$, *and* $\Re(\kappa), \Re(\vartheta) > 0$. *Then,*

$$X^{\vartheta-1}\mathbb{E}_{\kappa,\vartheta}(X) = Trecosh_{\kappa,\vartheta}(X) + Tresinh_{\kappa,\vartheta}(X),$$

$$X^{\vartheta-1}\mathbb{E}_{\kappa,\vartheta}(-X) = Trecosh_{\kappa,\vartheta}(X) - Tresinh_{\kappa,\vartheta}(X),$$

and

$$[X^{\vartheta-1}\mathbb{E}_{\kappa,\vartheta}(X)][X^{\vartheta-1}\mathbb{E}_{\kappa,\vartheta}(-X)] = Trecosh^2_{\kappa,\vartheta}(X) + Tresinh^2_{\kappa,\vartheta}(X).$$

5.8 The Supertrigonometric and Superhyperbolic Functions with the Power Law and Parameter via the Wiman Function

Let $X, \kappa, \vartheta, \gamma \in \mathbb{C}$, and $\Re(\kappa), \Re(\vartheta) > 0$. We now present the supertrigonometric and superhyperbolic functions with the power law and Parameter [6] as follows:

▷ The presupersine with the power law X^{γ} and parameter $X^{\vartheta-1}$ through the Wiman function is given by

$$\begin{aligned} &Gresin_{\kappa,\vartheta}(X^{\gamma}) \\ &= \frac{X^{\vartheta-1}}{2i}\Big(\mathbb{E}_{\kappa,\vartheta}(iX^{\gamma}) - \mathbb{E}_{\kappa,\vartheta}(-iX^{\gamma})\Big) \\ &= X^{\vartheta-1}\sum_{n=0}^{\infty}\frac{(-1)^n X^{(2n+1)\gamma}}{\Gamma((2n+1)\kappa+\vartheta)}. \end{aligned}$$

▷ The precosine with the power law X^{γ} and parameter $X^{\vartheta-1}$ through the Wiman function is given by

$$\begin{aligned} &Grecos_{\kappa,\vartheta}(X^{\gamma}) \\ &= \frac{X^{\vartheta-1}}{2}\Big(\mathbb{E}_{\kappa,\vartheta}(iX^{\gamma}) + \mathbb{E}_{\kappa,\vartheta}(-iX^{\gamma})\Big) \\ &= X^{\vartheta-1}\sum_{n=0}^{\infty}\frac{(-1)^n X^{(2n)\gamma}}{\Gamma((2n)\kappa+\vartheta)}. \end{aligned}$$

▷ The pretangent with the power law X^{γ} and parameter $X^{\vartheta-1}$ through the Wiman function is given by

$$Gretan_{\kappa,\vartheta}(X^{\gamma}) = \frac{Gresin_{\kappa,\vartheta}(X^{\gamma})}{Grecos_{\kappa,\vartheta}(X^{\gamma})}, \quad (Grecos_{\kappa,\vartheta}(X^{\gamma}) \neq 0).$$

▷ The precotangent with the power law X^{γ} and parameter $X^{\vartheta-1}$ through the Wiman function is given by

$$Grecotan_{\kappa,\vartheta}(X^{\gamma}) = \frac{Grecos_{\kappa,\vartheta}(X^{\gamma})}{Gresin_{\kappa,\vartheta}(X^{\gamma})}, \quad (Gresin_{\kappa,\vartheta}(X^{\gamma}) \neq 0).$$

▷ The precosecant with the power law X^{γ} and parameter $X^{\vartheta-1}$ through the Wiman function is given by

$$Grecsc_{\kappa,\vartheta}(X^{\gamma}) = \frac{1}{Gresin_{\kappa,\vartheta}(X^{\gamma})}, \quad (Gresin_{\kappa,\vartheta}(X^{\gamma}) \neq 0).$$

▷ The presecant with the power law X^{γ} and parameter $X^{\vartheta-1}$ through the Wiman function is given by

$$Gresec_{\kappa,\vartheta}(X^{\gamma}) = \frac{1}{Grecos_{\kappa,\vartheta}(X^{\gamma})}, \quad (Grecos_{\kappa,\vartheta}(X^{\gamma}) \neq 0).$$

Theorem 5.12 ([6]) *Let $X, \kappa, \vartheta, \gamma \in \mathbb{C}$, and $\Re(\kappa), \Re(\vartheta) > 0$. Then,*

$$Gresin_{\kappa,\vartheta}(X^{\gamma}) = X^{\vartheta-1} presin_{\kappa,\vartheta}(X^{\gamma}),$$

$$Grecos_{\kappa,\vartheta}(X^{\gamma}) = X^{\vartheta-1} precos_{\kappa,\vartheta}(X^{\gamma}).$$

Theorem 5.13 ([6]) *Let $X, \kappa, \vartheta, \gamma \in \mathbb{C}$, and $\Re(\kappa), \Re(\vartheta) > 0$. Then,*

$$X^{\vartheta-1}\mathbb{E}_{\kappa,\vartheta}(iX^{\gamma}) = Grecos_{\kappa,\vartheta}(X^{\gamma}) + i\, Gresin_{\kappa,\vartheta}(X^{\gamma}),$$

$$X^{\vartheta-1}\mathbb{E}_{\kappa,\vartheta}(-iX^{\gamma}) = Grecos_{\kappa,\vartheta}(X^{\gamma}) - i\, Gresin_{\kappa,\vartheta}(X^{\gamma}),$$

and

$$[X^{\vartheta-1}\mathbb{E}_{\kappa,\vartheta}(iX^{\gamma})][X^{\vartheta-1}\mathbb{E}_{\kappa,\vartheta}(-iX^{\gamma})] = Grecos^2_{\kappa,\vartheta}(X^{\gamma}) + Gresin^2_{\kappa,\vartheta}(X^{\gamma}).$$

Let $X, \kappa, \vartheta \in \mathbb{C}$, and $\Re(\kappa), \Re(\vartheta) > 0$. Now, we have the following [6]:

▷ The hyperbolic presupersine with the power law X^{γ} and parameter $X^{\vartheta-1}$ through the Wiman function is given by

$$\begin{aligned} &Gresinh_{\kappa,\vartheta}(X^{\gamma})\\ &= \frac{X^{\vartheta-1}}{2}\left(\mathbb{E}_{\kappa,\vartheta}(X^{\gamma}) - \mathbb{E}_{\kappa,\vartheta}(-X^{\gamma})\right)\\ &= X^{\vartheta-1}\sum_{n=0}^{\infty}\frac{X^{(2n+1)\gamma}}{\Gamma((2n+1)\kappa+\vartheta)}. \end{aligned}$$

▷ The hyperbolic precosine with the power law X^{γ} and parameter $X^{\vartheta-1}$ through the Wiman function is given by

$$\begin{aligned} &Grecosh_{\kappa,\vartheta}(X^{\gamma})\\ &= \frac{X^{\vartheta-1}}{2}\left(\mathbb{E}_{\kappa,\vartheta}(X^{\gamma}) + \mathbb{E}_{\kappa,\vartheta}(-X^{\gamma})\right)\\ &= X^{\vartheta-1}\sum_{n=0}^{\infty}\frac{X^{(2n)\gamma}}{\Gamma((2n)\kappa+\vartheta)}. \end{aligned}$$

▷ The hyperbolic pretangent with the power law X^{γ} and parameter $X^{\vartheta-1}$ through the Wiman function is given by

$$Gretanh_{\kappa,\vartheta}(X^{\gamma}) = \frac{Gresinh_{\kappa,\vartheta}(X^{\gamma})}{Grecosh_{\kappa,\vartheta}(X^{\gamma})}, \quad (Grecosh_{\kappa,\vartheta}(X^{\gamma}) \neq 0).$$

▷ The hyperbolic precotangent with the power law X^{γ} and parameter $X^{\vartheta-1}$ through the Wiman function is given by

$$Grecotanh_{\kappa,\vartheta}(X^{\gamma}) = \frac{Grecosh_{\kappa,\vartheta}(X)}{Gresinh_{\kappa,\vartheta}(X^{\gamma})}, \quad (Gresinh_{\kappa,\vartheta}(X^{\gamma}) \neq 0).$$

▷ The hyperbolic precosecant with the power law X^{γ} and parameter $X^{\vartheta-1}$ through the Wiman function is given by

$$Grecsch_{\kappa,\vartheta}(X^{\gamma}) = \frac{1}{Gresinh_{\kappa,\vartheta}(X^{\gamma})}, \quad (Gresinh_{\kappa,\vartheta}(X^{\gamma}) \neq 0).$$

▷ The hyperbolic presecant with the power law X^{γ} and parameter $X^{\vartheta-1}$ through the Wiman function is given by

$$Gresech_{\kappa,\vartheta}(X^{\gamma}) = \frac{1}{Grecosh_{\kappa,\vartheta}(X^{\gamma})}, \quad (Grecosh_{\kappa,\vartheta}(X^{\gamma}) \neq 0).$$

Theorem 5.14 ([6]) *Let* $X, \kappa, \vartheta, \gamma \in \mathbb{C}$, *and* $\Re(\kappa), \Re(\vartheta) > 0$. *Then,*

$$X^{\vartheta-1}\mathbb{E}_{\kappa,\vartheta}(X^{\gamma}) = Grecosh_{\kappa,\vartheta}(X^{\gamma}) + Gresinh_{\kappa,\vartheta}(X^{\gamma}),$$

$$X^{\vartheta-1}\mathbb{E}_{\kappa,\vartheta}(-X^{\gamma}) = Grecosh_{\kappa,\vartheta}(X^{\gamma}) - Gresinh_{\kappa,\vartheta}(X^{\gamma}),$$

and

$$[X^{\vartheta-1}\mathbb{E}_{\kappa,\vartheta}(X^{\gamma})][X^{\vartheta-1}\mathbb{E}_{\kappa,\vartheta}(-X^{\gamma})] = Grecosh^2_{\kappa,\vartheta}(X^{\gamma}) + Gresinh^2_{\kappa,\vartheta}(X^{\gamma}).$$

5.9 Special Cases of the Wiman Function

Let X, ϑ, $\kappa \in \mathbb{C}$, and $\Re(\kappa)$, $\Re(\vartheta) > 0$. We now introduce special cases according to the Wiman function [6] as follows:

▹ The supersine given by the Wiman function

$$\begin{aligned} & supersin_{1,\vartheta}(X^{\kappa}) \\ &= \frac{X^{\vartheta-1}}{2i}\Big(\mathbb{E}_{1,\vartheta}(iX^{\kappa}) - \mathbb{E}_{1,\vartheta}(-iX^{\kappa})\Big) \\ &= \sum_{n=0}^{\infty} \frac{(-1)^n X^{(2n+1)\kappa}}{\Gamma(2n+1+\vartheta)}. \end{aligned}$$

▹ The supercosine given by the Wiman function

$$\begin{aligned} & supercos_{1,\vartheta}(X^{\kappa}) \\ &= \frac{X^{\vartheta-1}}{2}\Big(\mathbb{E}_{1,\vartheta}(iX^{\kappa}) + \mathbb{E}_{1,\vartheta}(-iX^{\kappa})\Big) \\ &= \sum_{n=0}^{\infty} \frac{(-1)^n X^{(2n)\kappa}}{\Gamma(2n+\vartheta)}. \end{aligned}$$

▹ The supertangent given by the Wiman function

$$supertan_{1,\vartheta}(X^{\kappa}) = \frac{supersin_{1,\vartheta}(X^{\kappa})}{supercos_{1,\vartheta}(X^{\kappa})}, \quad (supercos_{1,\vartheta}(X^{\kappa}) \neq 0).$$

▹ The supercotangent given by the Wiman function

$$supercotan_{1,\vartheta}(X^{\kappa}) = \frac{supercos_{1,\vartheta}(X^{\kappa})}{supersin_{1,\vartheta}(X^{\kappa})}, \quad (supersin_{1,\vartheta}(X^{\kappa}) \neq 0).$$

▷ The supercosecant given by the Wiman function

$$supercsc_{1,\vartheta}(X^{\kappa}) = \frac{1}{supersin_{1,\vartheta}(X^{\kappa})}, \quad (supersin_{1,\vartheta}(X^{\kappa}) \neq 0).$$

▷ The supersecant given by the Wiman function

$$supersec_{1,\vartheta}(X^{\kappa}) = \frac{1}{supercos_{1,\vartheta}(X^{\kappa})}, \quad (supercos_{1,\vartheta}(X^{\kappa}) \neq 0).$$

▷ The hyperbolic supersine given by the Wiman function

$$\begin{aligned} & supersinh_{1,\vartheta}(X^{\kappa}) \\ &= \frac{X^{\vartheta-1}}{2}\Big(\mathbb{E}_{1,\vartheta}(X^{\kappa}) - \mathbb{E}_{1,\vartheta}(-X^{\kappa})\Big) \\ &= \sum_{n=0}^{\infty} \frac{X^{(2n+1)\kappa}}{\Gamma(2n+1+\vartheta)}. \end{aligned}$$

▷ The hyperbolic supercosine given by the Wiman function

$$\begin{aligned} & supercosh_{1,\vartheta}(X^{\kappa}) \\ &= \frac{X^{\vartheta-1}}{2}\Big(\mathbb{E}_{1,\vartheta}(X^{\kappa}) + \mathbb{E}_{1,\vartheta}(-X^{\kappa})\Big) \\ &= \sum_{n=0}^{\infty} \frac{X^{(2n)\kappa}}{\Gamma(2n+\vartheta)}. \end{aligned}$$

▷ The hyperbolic supertangent given by the Wiman function

$$supertanh_{1,\vartheta}(X^{\kappa}) = \frac{supersinh_{1,\vartheta}(X^{\kappa})}{supercosh_{1,\vartheta}(X^{\kappa})}, \quad (supercosh_{1,\vartheta}(X^{\kappa}) \neq 0).$$

▷ The hyperbolic supercotangent given by the Wiman function

$$supercotanh_{1,\vartheta}(X^{\kappa}) = \frac{supercosh_{1,\vartheta}(X^{\kappa})}{supersinh_{1,\vartheta}(X^{\kappa})}, \quad (supersinh_{1,\vartheta}(X^{\kappa}) \neq 0).$$

▷ The hyperbolic supercosecant given by the Wiman function

$$supercsch_{1,\vartheta}(X^{\kappa}) = \frac{1}{supersinh_{1,\vartheta}(X^{\kappa})}, \quad (supersinh_{1,\vartheta}(X^{\kappa}) \neq 0).$$

▷ The hyperbolic supersecant given by the Wiman function

$$supersech_{1,\vartheta}(X^{\kappa}) = \frac{1}{supercosh_{1,\vartheta}(X^{\kappa})}, \quad (supercosh_{1,\vartheta}(X^{\kappa}) \neq 0).$$

▷ The presupersine given by the Wiman function

$$\begin{aligned} & presin_{1,\vartheta}(X) \\ &= \frac{1}{2i}\Big(\mathbb{E}_{1,\vartheta}(iX) - \mathbb{E}_{1,\vartheta}(-iX)\Big) \\ &= \sum_{n=0}^{\infty} \frac{(-1)^n X^{(2n+1)}}{\Gamma(2n+1+\vartheta)}. \end{aligned}$$

▷ The precosine given by the Wiman function

$$\begin{aligned} & precos_{1,\vartheta}(X) \\ &= \frac{1}{2}\Big(\mathbb{E}_{1,\vartheta}(iX) + \mathbb{E}_{1,\vartheta}(-iX)\Big) \\ &= \sum_{n=0}^{\infty} \frac{(-1)^n X^{(2n)}}{\Gamma(2n+\vartheta)}. \end{aligned}$$

▷ The pretangent given by the Wiman function

$$pretan_{1,\vartheta}(X) = \frac{presin_{1,\vartheta}(X)}{precos_{1,\vartheta}(X)}, \quad (precos_{1,\vartheta}(X) \neq 0).$$

▷ The precotangent given by the Wiman function

$$precotan_{1,\vartheta}(X) = \frac{precos_{1,\vartheta}(X)}{presin_{1,\vartheta}(X)}, \quad (presin_{1,\vartheta}(X) \neq 0).$$

▷ The precosecant given by the Wiman function

$$precsc_{1,\vartheta}(X) = \frac{1}{presin_{1,\vartheta}(X)}, \quad (presin_{1,\vartheta}(X) \neq 0).$$

▷ The presecant given by the Wiman function

$$presec_{1,\vartheta}(X) = \frac{1}{precos_{1,\vartheta}(X)}, \quad (precos_{1,\vartheta}(X) \neq 0).$$

▷ The hyperbolic presupersine given by the Wiman function

$$\begin{aligned} & presinh_{1,\vartheta}(X) \\ & = \frac{1}{2}\Big(\mathbb{E}_{1,\vartheta}(X) - \mathbb{E}_{1,\vartheta}(-X)\Big) \\ & = \sum_{n=0}^{\infty} \frac{X^{(2n+1)}}{\Gamma(2n+1+\vartheta)}. \end{aligned}$$

▷ The hyperbolic precosine given by the Wiman function

$$\begin{aligned} & precosh_{1,\vartheta}(X) \\ & = \frac{1}{2}\Big(\mathbb{E}_{1,\vartheta}(X) + \mathbb{E}_{1,\vartheta}(-X)\Big) \\ & = \sum_{n=0}^{\infty} \frac{X^{(2n)}}{\Gamma(2n+\vartheta)}. \end{aligned}$$

▷ The hyperbolic pretangent given by the Wiman function

$$pretanh_{1,\vartheta}(X) = \frac{presinh_{1,\vartheta}(X)}{precosh_{1,\vartheta}(X)}, \ (precosh_{1,\vartheta}(X) \neq 0).$$

▷ The hyperbolic precotangent given by the Wiman function

$$precotanh_{1,\vartheta}(X) = \frac{precosh_{1,\vartheta}(X)}{presinh_{1,\vartheta}(X)}, \ (presinh_{1,\vartheta}(X) \neq 0).$$

▷ The hyperbolic precosecant given by the Wiman function

$$precsch_{1,\vartheta}(X) = \frac{1}{supersinh_{1,\vartheta}(X)}, \ (supersinh_{1,\vartheta}(X) \neq 0).$$

▷ The hyperbolic presecant given by the Wiman function

$$presech_{1,\vartheta}(X) = \frac{1}{precosh_{1,\vartheta}(X)}, \ (precosh_{1,\vartheta}(X) \neq 0).$$

▷ The presupersine with the power law X^{γ} and parameter $X^{\vartheta-1}$ by the Wiman function is given by

$$\begin{aligned}
&Gresin_{1,\vartheta}(X^{\gamma}) \\
&= \frac{X^{\vartheta-1}}{2i}\Big(\mathbb{E}_{1,\vartheta}(iX^{\gamma}) - \mathbb{E}_{1,\vartheta}(-iX^{\gamma})\Big) \\
&= \sum_{n=0}^{\infty} \frac{(-1)^n X^{(2n+1)\gamma}}{\Gamma(2n+1+\vartheta)}.
\end{aligned}$$

▷ The precosine with the power law X^{γ} and parameter $X^{\vartheta-1}$ by the Wiman function is given by

$$\begin{aligned}
&Grecos_{1,\vartheta}(X^{\gamma}) \\
&= \frac{X^{\vartheta-1}}{2}\Big(\mathbb{E}_{1,\vartheta}(iX^{\gamma}) + \mathbb{E}_{1,\vartheta}(-iX^{\gamma})\Big) \\
&= \sum_{n=0}^{\infty} \frac{(-1)^n X^{(2n)\gamma}}{\Gamma(2n+\vartheta)}.
\end{aligned}$$

▷ The pretangent with the power law X^{γ} and parameter $X^{\vartheta-1}$ given by the Wiman function

$$Gretan_{1,\vartheta}(X^{\gamma}) = \frac{Gresin_{1,\vartheta}(X^{\gamma})}{Grecos_{1,\vartheta}(X^{\gamma})}, \quad (Grecos_{1,\vartheta}(X^{\gamma}) \neq 0).$$

▷ The precotangent with the power law X^{γ} and parameter $X^{\vartheta-1}$ given by the Wiman function

$$Grecotan_{1,\vartheta}(X^{\gamma}) = \frac{Grecos_{1,\vartheta}(X^{\gamma})}{presin_{1,\vartheta}(X^{\gamma})}, \quad (Gresin_{1,\vartheta}(X^{\gamma}) \neq 0).$$

▷ The precosecant with the power law X^{γ} and parameter $X^{\vartheta-1}$ given by the Wiman function

$$Grecsc_{1,\vartheta}(X^{\gamma}) = \frac{1}{Gresin_{1,\vartheta}(X^{\gamma})}, \quad (Gresin_{1,\vartheta}(X^{\gamma}) \neq 0).$$

▷ The presecant with the power law X^{γ} and parameter $X^{\vartheta-1}$ given by the Wiman function

$$Gresec_{1,\vartheta}(X^{\gamma}) = \frac{1}{precos_{1,\vartheta}(X^{\gamma})}, \quad (Grecos_{1,\vartheta}(X^{\gamma}) \neq 0).$$

▷ The hyperbolic presupersine with the power law X^{γ} and parameter $X^{\vartheta-1}$ given by the Wiman function

$$\begin{aligned}&Gresinh_{1,\vartheta}(X^{\gamma})\\&=\frac{X^{\vartheta-1}}{2}\Big(\mathbb{E}_{1,\vartheta}(X^{\gamma})-\mathbb{E}_{1,\vartheta}(-X^{\gamma})\Big)\\&=X^{\vartheta-1}\sum_{n=0}^{\infty}\frac{X^{(2n+1)\gamma}}{\Gamma(2n+1+\vartheta)}.\end{aligned}$$

▷ The hyperbolic precosine with the power law X^{γ} and parameter $X^{\vartheta-1}$ given by the Wiman function

$$\begin{aligned}&Grecosh_{1,\vartheta}(X^{\gamma})\\&=\frac{X^{\vartheta-1}}{2}\Big(\mathbb{E}_{1,\vartheta}(X^{\gamma})+\mathbb{E}_{1,\vartheta}(-X^{\gamma})\Big)\\&=X^{\vartheta-1}\sum_{n=0}^{\infty}\frac{X^{(2n)\gamma}}{\Gamma(2n+\vartheta)}.\end{aligned}$$

▷ The hyperbolic pretangent with the power law X^{γ} and parameter $X^{\vartheta-1}$ given by the Wiman function

$$Gretanh_{1,\vartheta}(X^{\gamma})=\frac{Gresinh_{1,\vartheta}(X^{\gamma})}{Grecosh_{1,\vartheta}(X^{\gamma})},\quad (Grecosh_{1,\vartheta}(X^{\gamma})\neq 0).$$

▷ The hyperbolic precotangent with the power law X^{γ} and parameter $X^{\vartheta-1}$ given by the Wiman function

$$Grecotanh_{1,\vartheta}(X^{\gamma})=\frac{Grecosh_{1,\vartheta}(X^{\gamma})}{Gresinh_{1,\vartheta}(X^{\gamma})},\quad (Gresinh_{1,\vartheta}(X^{\gamma})\neq 0).$$

▷ The hyperbolic precosecant with the power law X^{γ} and parameter $X^{\vartheta-1}$ given by the Wiman function

$$Grecsch_{1,\vartheta}(X^{\gamma})=\frac{1}{Gresinh_{1,\vartheta}(X^{\gamma})},\quad (Gresinh_{1,\vartheta}(X^{\gamma})\neq 0).$$

▷ The hyperbolic presecant with the power law X^{γ} and parameter $X^{\vartheta-1}$ by the Wiman function

$$Gresech_{1,\vartheta}(X^{\gamma})=\frac{1}{Grecosh_{1,\vartheta}(X^{\gamma})},\ (Grecosh_{1,\vartheta}(X^{\gamma})\neq 0).$$

Now, we have the following results [6]:

$$\mathbb{E}_{\kappa,\vartheta}(X^{\kappa}) = {}_1\mathbb{W}_1\left({(1,1) \atop (\vartheta,\kappa)}; X^{\kappa}\right),$$

$$\mathbb{E}_{\kappa,\vartheta}(X) = {}_1\mathbb{W}_1\left({(1,1) \atop (\vartheta,\kappa)}; X\right),$$

$$\mathbb{E}_{\kappa,\vartheta}(X^{\widehat{\vartheta}}) = {}_1\mathbb{W}_1\left({(1,1) \atop (\vartheta,\kappa)}; X^{\widehat{\vartheta}}\right),$$

$$supersin_{\kappa,\vartheta}(X^{\kappa}) = X^{\vartheta-1}\,{}_1\sin_1\left({(1,1) \atop (\vartheta,\kappa)}; X^{\kappa}\right),$$

$$supercos_{\kappa,\vartheta}(X^{\kappa}) = X^{\vartheta-1}\,{}_1\cos_1\left({(1,1) \atop (\vartheta,\kappa)}; X^{\kappa}\right),$$

$$supersinh_{\kappa,\vartheta}(X^{\kappa}) = X^{\vartheta-1}\,{}_1\sinh_1\left({(1,1) \atop (\vartheta,\kappa)}; X^{\kappa}\right),$$

$$supercosh_{\kappa,\vartheta}(X^{\kappa}) = X^{\vartheta-1}\,{}_1\cosh_1\left({(1,1) \atop (\vartheta,\kappa)}; X^{\kappa}\right),$$

$$presin_{\kappa,\vartheta}(X) = {}_1\sin_1\left({(1,1) \atop (\vartheta,\kappa)}; X\right),$$

$$precos_{\kappa,\vartheta}(X) = {}_1\cos_1\left({(1,1) \atop (\vartheta,\kappa)}; X\right),$$

$$presinh_{\kappa,\vartheta}(X) = {}_1\sinh_1\left({(1,1) \atop (\vartheta,\kappa)}; X\right),$$

$$precosh_{\kappa,\vartheta}(X) = {}_1\cosh_1\left({(1,1) \atop (\vartheta,\kappa)}; X\right),$$

$$presin_{\kappa,\vartheta}(X^{\widehat{\vartheta}}) = {}_1\sin_1\left({(1,1) \atop (\vartheta,\kappa)}; X^{\widehat{\vartheta}}\right),$$

$$precos_{\kappa,\vartheta}(X^{\widehat{\vartheta}}) = {}_1\cos_1\left({(1,1) \atop (\vartheta,\kappa)}; X^{\widehat{\vartheta}}\right),$$

$$presinh_{\kappa,\vartheta}(X^{\widehat{\vartheta}}) = {}_1\sinh_1\left({}^{(1,1)}_{(\vartheta,\kappa)}; X^{\widehat{\vartheta}}\right),$$

$$precosh_{\kappa,\vartheta}(X^{\widehat{\vartheta}}) = {}_1\cosh_1\left({}^{(1,1)}_{(\vartheta,\kappa)}; X^{\widehat{\vartheta}}\right),$$

$$Tresin_{\kappa,\vartheta}(X) = X^{\vartheta-1}{}_1\sin_1\left({}^{(1,1)}_{(\vartheta,\kappa)}; X\right),$$

$$Trecos_{\kappa,\vartheta}(X) = X^{\vartheta-1}{}_1\cos_1\left({}^{(1,1)}_{(\vartheta,\kappa)}; X\right),$$

$$Tresinh_{\kappa,\vartheta}(X) = X^{\vartheta-1}{}_1\sinh_1\left({}^{(1,1)}_{(\vartheta,\kappa)}; X\right),$$

$$Trecosh_{\kappa,\vartheta}(X) = X^{\vartheta-1}{}_1\cosh_1\left({}^{(1,1)}_{(\vartheta,\kappa)}; X\right),$$

$$Gresin_{\kappa,\vartheta}(X^{\widehat{\vartheta}}) = X^{\vartheta-1}{}_1\sin_1\left({}^{(1,1)}_{(\vartheta,\kappa)}; X^{\widehat{\vartheta}}\right),$$

$$Grecos_{\kappa,\vartheta}(X^{\widehat{\vartheta}}) = X^{\vartheta-1}{}_1\cos_1\left({}^{(1,1)}_{(\vartheta,\kappa)}; X^{\widehat{\vartheta}}\right),$$

$$Gresinh_{\kappa,\vartheta}(X^{\widehat{\vartheta}}) = X^{\vartheta-1}{}_1\sinh_1\left({}^{(1,1)}_{(\vartheta,\kappa)}; X^{\widehat{\vartheta}}\right),$$

and

$$Grecosh_{\kappa,\vartheta}(X^{\widehat{\vartheta}}) = X^{\vartheta-1}{}_1\cosh_1\left({}^{(1,1)}_{(\vartheta,\kappa)}; X^{\widehat{\vartheta}}\right),$$

where

$$\begin{aligned}
&{}_1\sin_1\left({}^{(1,1)}_{(\vartheta,\kappa)}; X^{\kappa}\right)\\
&= \frac{1}{2i}\left({}_1\mathbb{W}_1\left({}^{(1,1)}_{(\vartheta,\kappa)}; iX^{\kappa}\right) - {}_1\mathbb{W}_1\left({}^{(1,1)}_{(\vartheta,\kappa)}; -iX^{\kappa}\right)\right)\\
&= \sum_{n=0}^{\infty} \frac{(-1)^n X^{(2n+1)\kappa}}{\Gamma((2n+1)\kappa+\vartheta)},
\end{aligned}$$

$$
\begin{aligned}
&{}_1\cos_1\left({}^{(1,1)}_{(\vartheta,\kappa)}; X^\kappa\right)\\
&= \frac{1}{2}\left({}_1\mathbb{W}_1\left({}^{(1,1)}_{(\vartheta,\kappa)}; iX^\kappa\right) + {}_1\mathbb{W}_1\left({}^{(1,1)}_{(\vartheta,\kappa)}; -iX^\kappa\right)\right)\\
&= \sum_{n=0}^{\infty} \frac{(-1)^n X^{(2n)\kappa}}{\Gamma((2n+1)\kappa+\vartheta)},
\end{aligned}
$$

$$
\begin{aligned}
&{}_1\sinh_1\left({}^{(1,1)}_{(\vartheta,\kappa)}; X^\kappa\right)\\
&= \frac{1}{2}\left({}_1\mathbb{W}_1\left({}^{(1,1)}_{(\vartheta,\kappa)}; X^\kappa\right) - {}_1\mathbb{W}_1\left({}^{(1,1)}_{(\vartheta,\kappa)}; -X^\kappa\right)\right)\\
&= \sum_{n=0}^{\infty} \frac{X^{(2n+1)\kappa}}{\Gamma((2n+1)\kappa+\vartheta)},
\end{aligned}
$$

$$
\begin{aligned}
&{}_1\cosh_1\left({}^{(1,1)}_{(\vartheta,\kappa)}; X^\kappa\right)\\
&= \frac{1}{2}\left({}_1\mathbb{W}_1\left({}^{(1,1)}_{(\vartheta,\kappa)}; X^\kappa\right) + {}_1\mathbb{W}_1\left({}^{(1,1)}_{(\vartheta,\kappa)}; -X^\kappa\right)\right)\\
&= \sum_{n=0}^{\infty} \frac{X^{(2n)\kappa}}{\Gamma((2n+1)\kappa+\vartheta)},
\end{aligned}
$$

$$
\begin{aligned}
&{}_1\sin_1\left({}^{(1,1)}_{(\vartheta,\kappa)}; X\right)\\
&= \frac{1}{2i}\left({}_1\mathbb{W}_1\left({}^{(1,1)}_{(\vartheta,\kappa)}; iX\right) - {}_1\mathbb{W}_1\left({}^{(1,1)}_{(\vartheta,\kappa)}; -iX\right)\right)\\
&= \sum_{n=0}^{\infty} \frac{(-1)^n X^{(2n+1)}}{\Gamma((2n+1)\kappa+\vartheta)},
\end{aligned}
$$

$$
\begin{aligned}
&{}_1\cos_1\left({}^{(1,1)}_{(\vartheta,\kappa)}; X\right)\\
&= \frac{1}{2}\left({}_1\mathbb{W}_1\left({}^{(1,1)}_{(\vartheta,\kappa)}; iX\right) + {}_1\mathbb{W}_1\left({}^{(1,1)}_{(\vartheta,\kappa)}; -iX\right)\right)\\
&= \sum_{n=0}^{\infty} \frac{(-1)^n X^{(2n)}}{\Gamma((2n+1)+\vartheta)},
\end{aligned}
$$

$$
\begin{aligned}
&{}_1\sinh_1\left(\begin{smallmatrix}(1,1)\\(\vartheta,\kappa)\end{smallmatrix};X\right)\\
&=\frac{1}{2}\left({}_1\mathbb{W}_1\left(\begin{smallmatrix}(1,1)\\(\vartheta,\kappa)\end{smallmatrix};X\right)-{}_1\mathbb{W}_1\left(\begin{smallmatrix}(1,1)\\(\vartheta,\kappa)\end{smallmatrix};-X\right)\right)\\
&=\sum_{n=0}^{\infty}\frac{X^{(2n+1)}}{\Gamma((2n+1)\kappa+\vartheta)},
\end{aligned}
$$

$$
\begin{aligned}
&{}_1\cosh_1\left(\begin{smallmatrix}(1,1)\\(\vartheta,\kappa)\end{smallmatrix};X\right)\\
&=\frac{1}{2}\left({}_1\mathbb{W}_1\left(\begin{smallmatrix}(1,1)\\(\vartheta,\kappa)\end{smallmatrix};X\right)+{}_1\mathbb{W}_1\left(\begin{smallmatrix}(1,1)\\(\vartheta,\kappa)\end{smallmatrix};-X\right)\right)\\
&=\sum_{n=0}^{\infty}\frac{X^{(2n)}}{\Gamma((2n+1)\kappa+\vartheta)},
\end{aligned}
$$

$$
\begin{aligned}
&{}_1\sin_1\left(\begin{smallmatrix}(1,1)\\(\vartheta,\kappa)\end{smallmatrix};X^{\widehat{\vartheta}}\right)\\
&=\frac{1}{2i}\left({}_1\mathbb{W}_1\left(\begin{smallmatrix}(1,1)\\(\vartheta,\kappa)\end{smallmatrix};iX^{\widehat{\vartheta}}\right)-{}_1\mathbb{W}_1\left(\begin{smallmatrix}(1,1)\\(\vartheta,\kappa)\end{smallmatrix};-iX^{\widehat{\vartheta}}\right)\right)\\
&=\sum_{n=0}^{\infty}\frac{(-1)^nX^{(2n+1)\widehat{\vartheta}}}{\Gamma((2n+1)\kappa+\vartheta)},
\end{aligned}
$$

$$
\begin{aligned}
&{}_1\cos_1\left(\begin{smallmatrix}(1,1)\\(\vartheta,\kappa)\end{smallmatrix};X^{\widehat{\vartheta}}\right)\\
&=\frac{1}{2}\left({}_1\mathbb{W}_1\left(\begin{smallmatrix}(1,1)\\(\vartheta,\kappa)\end{smallmatrix};iX^{\widehat{\vartheta}}\right)+{}_1\mathbb{W}_1\left(\begin{smallmatrix}(1,1)\\(\vartheta,\kappa)\end{smallmatrix};-iX^{\widehat{\vartheta}}\right)\right)\\
&=\sum_{n=0}^{\infty}\frac{(-1)^nX^{(2n)\widehat{\vartheta}}}{\Gamma((2n+1)\kappa+\vartheta)},
\end{aligned}
$$

$$
\begin{aligned}
&{}_1\sinh_1\left(\begin{smallmatrix}(1,1)\\(\vartheta,\kappa)\end{smallmatrix};X^{\widehat{\vartheta}}\right)\\
&=\frac{1}{2}\left({}_1\mathbb{W}_1\left(\begin{smallmatrix}(1,1)\\(\vartheta,\kappa)\end{smallmatrix};X^{\widehat{\vartheta}}\right)-{}_1\mathbb{W}_1\left(\begin{smallmatrix}(1,1)\\(\vartheta,\kappa)\end{smallmatrix};-X^{\widehat{\vartheta}}\right)\right)\\
&=\sum_{n=0}^{\infty}\frac{X^{(2n+1)\widehat{\vartheta}}}{\Gamma((2n+1)\kappa+\vartheta)},
\end{aligned}
$$

and

$$\begin{aligned}
&{}_1\cosh_1\left({}^{(1,1)}_{(\vartheta,\kappa)}; X^{\widehat{\vartheta}}\right)\\
&= \frac{1}{2}\left({}_1\mathbb{W}_1\left({}^{(1,1)}_{(\vartheta,\kappa)}; X^{\widehat{\vartheta}}\right) + {}_1\mathbb{W}_1\left({}^{(1,1)}_{(\vartheta,\kappa)}; -X^{\widehat{\vartheta}}\right)\right)\\
&= \sum_{n=0}^{\infty} \frac{X^{(2n)\widehat{\vartheta}}}{\Gamma((2n+1)\kappa+\vartheta)},
\end{aligned}$$

where $\kappa, \vartheta, \widehat{\vartheta}, X \in \mathbb{C}$, and $\Re(\kappa), \Re(\vartheta) > 0$.

Also, we have that [6]

$${}_2\mathbb{W}_2\left({}^{(\kappa_1,a_1),(\kappa_2,a_2)}_{(\vartheta_1,b_1),(\vartheta_2,b_2)}; X\right) = \sum_{k=0}^{\infty} \frac{\Gamma(\kappa_1+a_1k)\Gamma(\kappa_2+a_2k)}{\Gamma(\vartheta_1+b_1k)\Gamma(\vartheta_2+b_2k)} \frac{X^k}{k!},$$

$${}_2\sin_2\left({}^{(\kappa_1,a_1),(\kappa_2,a_2)}_{(\vartheta_1,b_1),(\vartheta_2,b_2)}; X^{\kappa}\right) = \sum_{k=0}^{\infty}(-1)^k \frac{\Gamma(\kappa_1+a_1(2k+1))\Gamma(\kappa_2+a_2(2k+1))}{\Gamma(\vartheta_1+b_1(2k+1))\Gamma(\vartheta_2+b_2(2k+1))} \frac{X^{(2k+1)\kappa}}{(2k+1)!},$$

$${}_2\cos_2\left({}^{(\kappa_1,a_1),(\kappa_2,a_2)}_{(\vartheta_1,b_1),(\vartheta_2,b_2)}; X^{\kappa}\right) = \sum_{k=0}^{\infty}(-1)^k \frac{\Gamma(\kappa_1+a_1(2k))\Gamma(\kappa_2+a_2(2k))}{\Gamma(\vartheta_1+b_1(2k))\Gamma(\vartheta_2+b_2(2k))} \frac{X^{(2k)\kappa}}{(2k)!},$$

$${}_2\sinh_2\left({}^{(\kappa_1,a_1),(\kappa_2,a_2)}_{(\vartheta_1,b_1),(\vartheta_2,b_2)}; X^{\kappa}\right) = \sum_{k=0}^{\infty} \frac{\Gamma(\kappa_1+a_1(2k+1))\Gamma(\kappa_2+a_2(2k+1))}{\Gamma(\vartheta_1+b_1(2k+1))\Gamma(\vartheta_2+b_2(2k+1))} \frac{X^{(2k+1)\kappa}}{(2k+1)!},$$

$${}_2\cosh_2\left({}^{(\kappa_1,a_1),(\kappa_2,a_2)}_{(\vartheta_1,b_1),(\vartheta_2,b_2)}; X^{\kappa}\right) = \sum_{k=0}^{\infty} \frac{\Gamma(\kappa_1+a_1(2k))\Gamma(\kappa_2+a_2(2k))}{\Gamma(\vartheta_1+b_1(2k))\Gamma(\vartheta_2+b_2(2k))} \frac{X^{(2k)\kappa}}{(2k)!},$$

$${}_2\sin_2\left({}^{(\kappa_1,a_1),(\kappa_2,a_2)}_{(\vartheta_1,b_1),(\vartheta_2,b_2)}; X^{\widehat{\vartheta}}\right) = \sum_{k=0}^{\infty}(-1)^k \frac{\Gamma(\kappa_1+a_1(2k+1))\Gamma(\kappa_2+a_2(2k+1))}{\Gamma(\vartheta_1+b_1(2k+1))\Gamma(\vartheta_2+b_2(2k+1))} \frac{X^{(2k+1)\widehat{\vartheta}}}{(2k+1)!},$$

$${}_2\cos_2\left({}^{(\kappa_1,a_1),(\kappa_2,a_2)}_{(\vartheta_1,b_1),(\vartheta_2,b_2)}; X^{\widehat{\vartheta}}\right) = \sum_{k=0}^{\infty}(-1)^k \frac{\Gamma(\kappa_1+a_1(2k))\Gamma(\kappa_2+a_2(2k))}{\Gamma(\vartheta_1+b_1(2k))\Gamma(\vartheta_2+b_2(2k))} \frac{X^{(2k)\widehat{\vartheta}}}{(2k)!},$$

$${}_2\sinh_2\left({}^{(\kappa_1,a_1),(\kappa_2,a_2)}_{(\vartheta_1,b_1),(\vartheta_2,b_2)}; X^{\widehat{\vartheta}}\right) = \sum_{k=0}^{\infty} \frac{\Gamma(\kappa_1+a_1(2k+1))\Gamma(\kappa_2+a_2(2k+1))}{\Gamma(\vartheta_1+b_1(2k+1))\Gamma(\vartheta_2+b_2(2k+1))} \frac{X^{(2k+1)\widehat{\vartheta}}}{(2k+1)!},$$

$$
{}_2\cosh_2\left({(\kappa_1,a_1),(\kappa_2,a_2) \atop (\vartheta_1,b_1),(\vartheta_2,b_2)}; X^{\widehat{\vartheta}} \right) = \sum_{k=0}^{\infty} \frac{\Gamma(\kappa_1 + a_1(2k))\Gamma(\kappa_2 + a_2(2k))}{\Gamma(\vartheta_1 + b_1(2k))\Gamma(\vartheta_2 + b_2(2k))} \frac{X^{(2k)\widehat{\vartheta}}}{(2k)!},
$$

$$
{}_2\sin_2\left({(\kappa_1,a_1),(\kappa_2,a_2) \atop (\vartheta_1,b_1),(\vartheta_2,b_2)}; X \right) = \sum_{k=0}^{\infty} (-1)^k \frac{\Gamma(\kappa_1 + a_1(2k+1))\Gamma(\kappa_2 + a_2(2k+1))}{\Gamma(\vartheta_1 + b_1(2k+1))\Gamma(\vartheta_2 + b_2(2k+1))} \frac{X^{(2k+1)}}{(2k+1)!},
$$

$$
{}_2\cos_2\left({(\kappa_1,a_1),(\kappa_2,a_2) \atop (\vartheta_1,b_1),(\vartheta_2,b_2)}; X \right) = \sum_{k=0}^{\infty} (-1)^k \frac{\Gamma(\kappa_1 + a_1(2k))\Gamma(\kappa_2 + a_2(2k))}{\Gamma(\vartheta_1 + b_1(2k))\Gamma(\vartheta_2 + b_2(2k))} \frac{X^{(2k)}}{(2k)!},
$$

$$
{}_2\sinh_2\left({(\kappa_1,a_1),(\kappa_2,a_2) \atop (\vartheta_1,b_1),(\vartheta_2,b_2)}; X \right) = \sum_{k=0}^{\infty} \frac{\Gamma(\kappa_1 + a_1(2k+1))\Gamma(\kappa_2 + a_2(2k+1))}{\Gamma(\vartheta_1 + b_1(2k+1))\Gamma(\vartheta_2 + b_2(2k+1))} \frac{X^{(2k+1)}}{(2k+1)!},
$$

and

$$
{}_2\cosh_2\left({(\kappa_1,a_1),(\kappa_2,a_2) \atop (\vartheta_1,b_1),(\vartheta_2,b_2)}; X \right) = \sum_{k=0}^{\infty} \frac{\Gamma(\kappa_1 + a_1(2k))\Gamma(\kappa_2 + a_2(2k))}{\Gamma(\vartheta_1 + b_1(2k))\Gamma(\vartheta_2 + b_2(2k))} \frac{X^{(2k)}}{(2k)!},
$$

which lead to

$$
{}_2\sin_2\left({(\kappa_1,a_1),(\vartheta_1,b_1) \atop (\vartheta_1,b_1),(\vartheta_2,b_2)}; X \right) = {}_1\sin_1\left({(\kappa_1,a_1) \atop (\vartheta_2,b_2)}; X \right),
$$

$$
{}_2\cos_2\left({(\kappa_1,a_1),(\vartheta_1,b_1) \atop (\vartheta_1,b_1),(\vartheta_2,b_2)}; X \right) = {}_1\cos_1\left({(\kappa_1,a_1) \atop (\vartheta_2,b_2)}; X \right),
$$

$$
{}_2\sinh_2\left({(\kappa_1,a_1),(\vartheta_1,b_1) \atop (\vartheta_1,b_1),(\vartheta_2,b_2)}; X \right) = {}_1\sinh_1\left({(\kappa_1,a_1) \atop (\vartheta_2,b_2)}; X \right),
$$

$$
{}_2\cosh_2\left({(\kappa_1,a_1),(\vartheta_1,b_1) \atop (\vartheta_1,b_1),(\vartheta_2,b_2)}; X \right) = {}_1\cosh_1\left({(\kappa_1,a_1) \atop (\vartheta_2,b_2)}; X \right),
$$

where $\kappa_1, \vartheta_1, \kappa_2, \vartheta_2 \in \mathbb{C}$, and $a_1, b_1, a_2, b_2 \in \mathbb{R}_+$.

5.10 The Truncated Wiman Functions and Related Functions

In this section, we define the truncated Wiman functions and some truncated functions related to the Wiman function.

Definition 5.3 ([6]) The truncated Wiman function is given by

$$\{\mathbb{E}_{\kappa,\vartheta}\}^n(X) = \sum_{k=0}^{n} \frac{X^k}{\Gamma(k\kappa+\vartheta)},$$

where $X, \kappa, \vartheta \in \mathbb{C}$, $n \in \mathbb{N}_0$, and $\Re(\kappa), \Re(\vartheta) > 0$.

The special cases for the truncated Wiman function are given as follows [6]:

$$\{\mathbb{E}_{\kappa,\vartheta}\}^n(X^\kappa) = \sum_{k=0}^{n} \frac{X^{k\kappa}}{\Gamma(k\kappa+\vartheta)},$$

and

$$\{\mathbb{E}_{\kappa,\vartheta}\}^n(X^\gamma) = \sum_{k=0}^{n} \frac{X^{k\gamma}}{\Gamma(k\kappa+\vartheta)},$$

where $X, \kappa, \vartheta, \gamma \in \mathbb{C}$, $n \in \mathbb{N}_0$, and $\Re(\kappa), \Re(\vartheta) > 0$.

We have the following results [6]:

The truncated supersine via the Wiman function is given by

$$\begin{aligned}
&\left\{supersin_{\kappa,\vartheta}\right\}^n(X^\kappa) \\
&= \frac{X^{\vartheta-1}}{2i}\left(\left\{\mathbb{E}_{\kappa,\vartheta}\right\}^n(iX^\kappa) - \left\{\mathbb{E}_{\kappa,\vartheta}\right\}^n(-iX^\kappa)\right) \\
&= \sum_{k=0}^{n} \frac{(-1)^n X^{(2k+1)\kappa}}{\Gamma((2k+1)\kappa+\vartheta},
\end{aligned}$$

where $X, \kappa, \vartheta \in \mathbb{C}$, $n \in \mathbb{N}_0$, and $\Re(\kappa), \Re(\vartheta) > 0$.

So, we have that

$$\begin{aligned}
&\left\{supersin_{\kappa,\vartheta}\right\}^0(X^\kappa) \\
&= \frac{X^{\vartheta-1}}{2}\left(\left\{\mathbb{E}_{\kappa,\vartheta}\right\}^0(iX^\kappa) - \left\{\mathbb{E}_{\kappa,\vartheta}\right\}^0(-iX^\kappa)\right) \\
&= \frac{X^{\kappa+\vartheta-1}}{\Gamma(\kappa+\vartheta)},
\end{aligned}$$

$$\left\{supersin_{\kappa,\vartheta}\right\}^{1}(X^{\kappa})$$
$$=\frac{X^{\vartheta-1}}{2}\left(\left\{\mathbb{E}_{\kappa,\vartheta}\right\}^{1}(iX^{\kappa})-\left\{\mathbb{E}_{\kappa,\vartheta}\right\}^{1}(-iX^{\kappa})\right)$$
$$=X^{\vartheta-1}\sum_{k=0}^{1}\frac{(-1)^{k}X^{(2k+1)\kappa}}{\Gamma((2k+1)\kappa+\vartheta)},$$

and

$$\left\{supersin_{\kappa,\vartheta}\right\}^{2}(X^{\kappa})$$
$$=\frac{X^{\vartheta-1}}{2}\left(\left\{\mathbb{E}_{\kappa,\vartheta}\right\}^{2}(iX^{\kappa})-\left\{\mathbb{E}_{\kappa,\vartheta}\right\}^{2}(-iX^{\kappa})\right)$$
$$=X^{\vartheta-1}\sum_{k=0}^{2}\frac{(-1)^{k}X^{(2k+1)\kappa}}{\Gamma((2k+1)\kappa+\vartheta)},$$

where $X, \kappa, \vartheta \in \mathbb{C}$, $n \in \mathbb{N}_0$, and $\Re(\kappa), \Re(\vartheta) > 0$.

The truncated supercosine via the Wiman function is given by:

$$\left\{supercos_{\kappa,\vartheta}\right\}^{n}(X^{\kappa})$$
$$=\frac{X^{\vartheta-1}}{2}\left(\left\{\mathbb{E}_{\kappa,\vartheta}\right\}^{n}(iX^{\kappa})+\left\{\mathbb{E}_{\kappa,\vartheta}\right\}^{n}(-iX^{\kappa})\right)$$
$$=\sum_{k=0}^{n}\frac{(-1)^{n}X^{(2k)\kappa}}{\Gamma((2k)\kappa+\vartheta},$$

where $X, \kappa, \vartheta \in \mathbb{C}$, $n \in \mathbb{N}_0$, and $\Re(\kappa), \Re(\vartheta) > 0$.

Furthermore, we give

$$\left\{supercos_{\kappa,\vartheta}\right\}^{0}(X^{\kappa})$$
$$=\frac{X^{\vartheta-1}}{2}\left(\left\{\mathbb{E}_{\kappa,\vartheta}\right\}^{0}(iX^{\kappa})+\left\{\mathbb{E}_{\kappa,\vartheta}\right\}^{0}(-iX^{\kappa})\right)$$
$$=\frac{X^{\vartheta-1}}{\Gamma(\vartheta)},$$

$$\begin{aligned}
&\left\{supercos_{\kappa,\vartheta}\right\}^{1}(X^{\kappa})\\
&=\frac{X^{\vartheta-1}}{2}\left(\left\{\mathbb{E}_{\kappa,\vartheta}\right\}^{1}(iX^{\kappa})+\left\{\mathbb{E}_{\kappa,\vartheta}\right\}^{1}(-iX^{\kappa})\right)\\
&=X^{\vartheta-1}\sum_{k=0}^{1}\frac{(-1)^{k}X^{(2k)\kappa}}{\Gamma((2k)\kappa+\vartheta)},
\end{aligned}$$

and

$$\begin{aligned}
&\left\{supercos_{\kappa,\vartheta}\right\}^{2}(X^{\kappa})\\
&=\frac{X^{\vartheta-1}}{2}\left(\left\{\mathbb{E}_{\kappa,\vartheta}\right\}^{2}(iX^{\kappa})+\left\{\mathbb{E}_{\kappa,\vartheta}\right\}^{2}(-iX^{\kappa})\right)\\
&=X^{\vartheta-1}\sum_{k=0}^{2}\frac{(-1)^{k}X^{(2k)\kappa}}{\Gamma((2k)\kappa+\vartheta)},
\end{aligned}$$

where $X, \kappa, \vartheta \in \mathbb{C}$, $n \in \mathbb{N}_0$, and $\Re(\kappa), \Re(\vartheta) > 0$.

The truncated hyperbolic supersine via the Wiman function is given by

$$\begin{aligned}
&\left\{supersinh_{\kappa,\vartheta}\right\}^{n}(X^{\kappa})\\
&=\frac{X^{\vartheta-1}}{2}\left(\left\{\mathbb{E}_{\kappa,\vartheta}\right\}^{n}(X^{\kappa})-\left\{\mathbb{E}_{\kappa,\vartheta}\right\}^{n}(-X^{\kappa})\right)\\
&=X^{\vartheta-1}\sum_{k=0}^{n}\frac{X^{(2k+1)\kappa}}{\Gamma((2k+1)\kappa+\vartheta},
\end{aligned}$$

where $X, \kappa, \vartheta \in \mathbb{C}$, $n \in \mathbb{N}_0$, and $\Re(\kappa), \Re(\vartheta) > 0$.

So, we have that

$$\begin{aligned}
&\left\{supersinh_{\kappa,\vartheta}\right\}^{0}(X^{\kappa})\\
&=\frac{X^{\vartheta-1}}{2}\left(\left\{\mathbb{E}_{\kappa,\vartheta}\right\}^{0}(X^{\kappa})-\left\{\mathbb{E}_{\kappa,\vartheta}\right\}^{0}(-X^{\kappa})\right)\\
&=\frac{X^{\kappa+\vartheta-1}}{\Gamma(\kappa+\vartheta)},
\end{aligned}$$

$$
\begin{aligned}
&\left\{ supersinh_{\kappa,\vartheta} \right\}^{1} (X^{\kappa}) \\
&= \frac{X^{\vartheta-1}}{2} \left(\left\{ \mathbb{E}_{\kappa,\vartheta} \right\}^{1} (X^{\kappa}) - \left\{ \mathbb{E}_{\kappa,\vartheta} \right\}^{1} (-X^{\kappa}) \right) \\
&= X^{\vartheta-1} \sum_{k=0}^{1} \frac{X^{(2k+1)\kappa}}{\Gamma((2k+1)\kappa+\vartheta)},
\end{aligned}
$$

and

$$
\begin{aligned}
&\left\{ supersinh_{\kappa,\vartheta} \right\}^{2} (X^{\kappa}) \\
&= \frac{X^{\vartheta-1}}{2} \left(\left\{ \mathbb{E}_{\kappa,\vartheta} \right\}^{2} (X^{\kappa}) - \left\{ \mathbb{E}_{\kappa,\vartheta} \right\}^{2} (-X^{\kappa}) \right) \\
&= X^{\vartheta-1} \sum_{k=0}^{2} \frac{X^{(2k+1)\kappa}}{\Gamma((2k+1)\kappa+\vartheta)},
\end{aligned}
$$

where $X, \kappa, \vartheta \in \mathbb{C}$, $n \in \mathbb{N}_0$, and $\Re(\kappa), \Re(\vartheta) > 0$.

The truncated hyperbolic supercosine via the Wiman function is given by:

$$
\begin{aligned}
&\left\{ supercosh_{\kappa,\vartheta} \right\}^{n} (X^{\kappa}) \\
&= \frac{X^{\vartheta-1}}{2} \left(\left\{ \mathbb{E}_{\kappa,\vartheta} \right\}^{n} (X^{\kappa}) + \left\{ \mathbb{E}_{\kappa,\vartheta} \right\}^{n} (-X^{\kappa}) \right) \\
&= X^{\vartheta-1} \sum_{k=0}^{n} \frac{X^{(2k)\kappa}}{\Gamma((2k)\kappa+\vartheta},
\end{aligned}
$$

where $X, \kappa, \vartheta \in \mathbb{C}$, $n \in \mathbb{N}_0$, and $\Re(\kappa), \Re(\vartheta) > 0$.

Thus, we have that

$$
\begin{aligned}
&\left\{ supercosh_{\kappa,\vartheta} \right\}^{0} (X^{\kappa}) \\
&= \frac{X^{\vartheta-1}}{2} \left(\left\{ \mathbb{E}_{\kappa,\vartheta} \right\}^{0} (X^{\kappa}) + \left\{ \mathbb{E}_{\kappa,\vartheta} \right\}^{0} (-X^{\kappa}) \right) \\
&= \frac{X^{\vartheta-1}}{\Gamma(\vartheta)},
\end{aligned}
$$

$$\begin{aligned}
&\left\{supercosh_{\kappa,\vartheta}\right\}^{1}(X^{\kappa})\\
&=\frac{X^{\vartheta-1}}{2}\left(\left\{\mathbb{E}_{\kappa,\vartheta}\right\}^{1}(X^{\kappa})+\left\{\mathbb{E}_{\kappa,\vartheta}\right\}^{1}(-X^{\kappa})\right)\\
&=X^{\vartheta-1}\sum_{k=0}^{1}\frac{X^{(2k)\kappa}}{\Gamma((2k)\kappa+\vartheta)},
\end{aligned}$$

and

$$\begin{aligned}
&\left\{supercosh_{\kappa,\vartheta}\right\}^{2}(X^{\kappa})\\
&=\frac{X^{\vartheta-1}}{2}\left(\left\{\mathbb{E}_{\kappa,\vartheta}\right\}^{2}(X^{\kappa})+\left\{\mathbb{E}_{\kappa,\vartheta}\right\}^{2}(-X^{\kappa})\right)\\
&=X^{\vartheta-1}\sum_{k=0}^{2}\frac{X^{(2k)\kappa}}{\Gamma((2k)\kappa+\vartheta)},
\end{aligned}$$

where $X, \kappa, \vartheta \in \mathbb{C}$, $n \in \mathbb{N}_0$, and $\Re(\kappa), \Re(\vartheta) > 0$.

The truncated presine via the Wiman function is given by

$$\begin{aligned}
&\left\{presin_{\kappa,\vartheta}\right\}^{n}(X^{\gamma})\\
&=\frac{1}{2i}\left(\left\{\mathbb{E}_{\kappa,\vartheta}\right\}^{n}(iX^{\gamma})-\left\{\mathbb{E}_{\kappa,\vartheta}\right\}^{n}(-iX^{\gamma})\right)\\
&=\sum_{k=0}^{n}\frac{(-1)^{n}X^{(2k+1)\gamma}}{\Gamma((2k+1)\kappa+\vartheta},
\end{aligned}$$

where $X, \kappa, \vartheta, \gamma \in \mathbb{C}$, $n \in \mathbb{N}_0$, and $\Re(\kappa), \Re(\vartheta) > 0$.

Therefore, we have that

$$\begin{aligned}
&\left\{presin_{\kappa,\vartheta}\right\}^{0}(X^{\gamma})\\
&=\frac{1}{2}\left(\left\{\mathbb{E}_{\kappa,\vartheta}\right\}^{0}(iX^{\gamma})-\left\{\mathbb{E}_{\kappa,\vartheta}\right\}^{0}(-iX^{\gamma})\right)\\
&=\frac{X^{\gamma}}{\Gamma(1+\vartheta)},
\end{aligned}$$

$$\begin{aligned}&\Big\{presin_{\kappa,\vartheta}\Big\}^{1}(X^{\gamma})\\&=\frac{1}{2}\left(\Big\{\mathbb{E}_{\kappa,\vartheta}\Big\}^{1}(iX^{\gamma})-\Big\{\mathbb{E}_{\kappa,\vartheta}\Big\}^{1}(-iX^{\gamma})\right)\\&=\sum_{k=0}^{1}\frac{(-1)^{k}X^{(2k+1)\gamma}}{\Gamma((2k+1)\kappa+\vartheta)},\end{aligned}$$

and

$$\begin{aligned}&\Big\{presin_{\kappa,\vartheta}\Big\}^{2}(X^{\gamma})\\&=\frac{1}{2}\left(\Big\{\mathbb{E}_{\kappa,\vartheta}\Big\}^{2}(iX^{\gamma})-\Big\{\mathbb{E}_{\kappa,\vartheta}\Big\}^{2}(-iX^{\gamma})\right)\\&=\sum_{k=0}^{2}\frac{(-1)^{k}X^{(2k+1)\gamma}}{\Gamma((2k+1)\kappa+\vartheta)},\end{aligned}$$

where $X, \kappa, \vartheta, \gamma \in \mathbb{C}$, $n \in \mathbb{N}_0$, and $\Re(\kappa), \Re(\vartheta) > 0$.

The truncated precosine via the Wiman function is given by:

$$\begin{aligned}&\Big\{precos_{\kappa,\vartheta}\Big\}^{n}(X^{\gamma})\\&=\frac{1}{2}\left(\Big\{\mathbb{E}_{\kappa,\vartheta}\Big\}^{n}(iX^{\gamma})+\Big\{\mathbb{E}_{\kappa,\vartheta}\Big\}^{n}(-iX^{\gamma})\right)\\&=\sum_{k=0}^{n}\frac{(-1)^{n}X^{(2k)\gamma}}{\Gamma((2k)\kappa+\vartheta)},\end{aligned}$$

where $X, \kappa, \vartheta, \gamma \in \mathbb{C}$, $n \in \mathbb{N}_0$, and $\Re(\kappa), \Re(\vartheta) > 0$.

So, we have that

$$\begin{aligned}&\Big\{precos_{\kappa,\vartheta}\Big\}^{0}(X^{\gamma})\\&=\frac{1}{2}\left(\Big\{\mathbb{E}_{\kappa,\vartheta}\Big\}^{0}(iX^{\gamma})+\Big\{\mathbb{E}_{\kappa,\vartheta}\Big\}^{0}(-iX^{\gamma})\right)\\&=\frac{1}{\Gamma(\vartheta)},\end{aligned}$$

$$\left\{precos_{\kappa,\vartheta}\right\}^{1}(X^{\gamma})$$
$$=\frac{1}{2}\left(\left\{\mathbb{E}_{\kappa,\vartheta}\right\}^{1}(iX^{\gamma})+\left\{\mathbb{E}_{\kappa,\vartheta}\right\}^{1}(-iX^{\gamma})\right)$$
$$=\sum_{k=0}^{1}\frac{(-1)^{k}X^{(2k)\gamma}}{\Gamma((2k)\kappa+\vartheta)},$$

and

$$\left\{precos_{\kappa,\vartheta}\right\}^{2}(X^{\gamma})$$
$$=\frac{1}{2}\left(\left\{\mathbb{E}_{\kappa,\vartheta}\right\}^{2}(iX^{\gamma})+\left\{\mathbb{E}_{\kappa,\vartheta}\right\}^{2}(-iX^{\gamma})\right)$$
$$=\sum_{k=0}^{2}\frac{(-1)^{k}X^{(2k)\gamma}}{\Gamma((2k)\kappa+\vartheta)},$$

where $X, \kappa, \vartheta, \gamma \in \mathbb{C}$, $n \in \mathbb{N}_0$, and $\Re(\kappa), \Re(\vartheta) > 0$.

The truncated hyperbolic presine via the Wiman function is given by

$$\left\{presinh_{\kappa,\vartheta}\right\}^{n}(X^{\gamma})$$
$$=\frac{1}{2}\left(\left\{\mathbb{E}_{\kappa,\vartheta}\right\}^{n}(X^{\gamma})-\left\{\mathbb{E}_{\kappa,\vartheta}\right\}^{n}(-X^{\gamma})\right)$$
$$=\sum_{k=0}^{n}\frac{X^{(2k+1)\gamma}}{\Gamma((2k+1)\kappa+\vartheta},$$

where $X, \kappa, \vartheta, \gamma \in \mathbb{C}$, $n \in \mathbb{N}_0$, and $\Re(\kappa), \Re(\vartheta) > 0$.

So, we have that

$$\left\{presinh_{\kappa,\vartheta}\right\}^{0}(X^{\gamma})$$
$$=\frac{1}{2}\left(\left\{\mathbb{E}_{\kappa,\vartheta}\right\}^{0}(X^{\gamma})-\left\{\mathbb{E}_{\kappa,\vartheta}\right\}^{0}(-X^{\gamma})\right)$$
$$=\frac{X^{\gamma}}{\Gamma(1+\vartheta)},$$

$$
\begin{aligned}
&\left\{presinh_{\kappa,\vartheta}\right\}^{1}(X^{\gamma})\\
&=\frac{1}{2}\left(\left\{\mathbb{E}_{\kappa,\vartheta}\right\}^{1}(X^{\gamma})-\left\{\mathbb{E}_{\kappa,\vartheta}\right\}^{1}(-X^{\gamma})\right)\\
&=\sum_{k=0}^{1}\frac{X^{(2k+1)\gamma}}{\Gamma((2k+1)\kappa+\vartheta)},
\end{aligned}
$$

and

$$
\begin{aligned}
&\left\{presinh_{\kappa,\vartheta}\right\}^{2}(X^{\gamma})\\
&=\frac{1}{2}\left(\left\{\mathbb{E}_{\kappa,\vartheta}\right\}^{2}(X^{\gamma})-\left\{\mathbb{E}_{\kappa,\vartheta}\right\}^{2}(-X^{\gamma})\right)\\
&=\sum_{k=0}^{2}\frac{X^{(2k+1)\gamma}}{\Gamma((2k+1)\kappa+\vartheta)},
\end{aligned}
$$

where $X, \kappa, \vartheta, \gamma \in \mathbb{C}$, $n \in \mathbb{N}_0$, and $\Re(\kappa), \Re(\vartheta) > 0$.

The truncated hyperbolic precosine via the Wiman function is given by

$$
\begin{aligned}
&\left\{precosh_{\kappa,\vartheta}\right\}^{n}(X^{\gamma})\\
&=\frac{1}{2}\left(\left\{\mathbb{E}_{\kappa,\vartheta}\right\}^{n}(X^{\gamma})+\left\{\mathbb{E}_{\kappa,\vartheta}\right\}^{n}(-X^{\gamma})\right)\\
&=\sum_{k=0}^{n}\frac{X^{(2k)\gamma}}{\Gamma((2k)\kappa+\vartheta)},
\end{aligned}
$$

where $X, \kappa, \vartheta, \gamma \in \mathbb{C}$, $n \in \mathbb{N}_0$, and $\Re(\kappa), \Re(\vartheta) > 0$.

Thus, we have that

$$
\begin{aligned}
&\left\{precosh_{\kappa,\vartheta}\right\}^{0}(X^{\gamma})\\
&=\frac{1}{2}\left(\left\{\mathbb{E}_{\kappa,\vartheta}\right\}^{0}(X^{\gamma})+\left\{\mathbb{E}_{\kappa,\vartheta}\right\}^{0}(-X^{\gamma})\right)\\
&=\frac{1}{\Gamma(\vartheta)},
\end{aligned}
$$

$$\begin{aligned}
&\left\{precosh_{\kappa,\vartheta}\right\}^{1}(X^{\gamma})\\
&=\frac{1}{2}\left(\left\{\mathbb{E}_{\kappa,\vartheta}\right\}^{1}(X^{\gamma})+\left\{\mathbb{E}_{\kappa,\vartheta}\right\}^{1}(-X^{\gamma})\right)\\
&=\sum_{k=0}^{1}\frac{X^{(2k)\gamma}}{\Gamma((2k)\kappa+\vartheta)},
\end{aligned}$$

and

$$\begin{aligned}
&\left\{precosh_{\kappa,\vartheta}\right\}^{2}(X^{\gamma})\\
&=\frac{1}{2}\left(\left\{\mathbb{E}_{\kappa,\vartheta}\right\}^{2}(X^{\gamma})+\left\{\mathbb{E}_{\kappa,\vartheta}\right\}^{2}(-X^{\gamma})\right)\\
&=\sum_{k=0}^{2}\frac{X^{(2k)\gamma}}{\Gamma((2k)\kappa+\vartheta)},
\end{aligned}$$

where $X, \kappa, \vartheta, \gamma \in \mathbb{C}$, $n \in \mathbb{N}_0$, and $\Re(\kappa), \Re(\vartheta) > 0$.

The truncated presine via the Wiman function is given by

$$\begin{aligned}
&\left\{presin_{\kappa,\vartheta}\right\}^{n}(X)\\
&=\frac{1}{2i}\left(\left\{\mathbb{E}_{\kappa,\vartheta}\right\}^{n}(iX)-\left\{\mathbb{E}_{\kappa,\vartheta}\right\}^{n}(-iX)\right)\\
&=\sum_{k=0}^{n}\frac{(-1)^{n}X^{(2k+1)}}{\Gamma((2k+1)\kappa+\vartheta},
\end{aligned}$$

where $X, \kappa, \vartheta \in \mathbb{C}$, $n \in \mathbb{N}_0$, and $\Re(\kappa), \Re(\vartheta) > 0$.

Therefore, we have that

$$\begin{aligned}
&\left\{presin_{\kappa,\vartheta}\right\}^{0}(X)\\
&=\frac{1}{2}\left(\left\{\mathbb{E}_{\kappa,\vartheta}\right\}^{0}(iX)-\left\{\mathbb{E}_{\kappa,\vartheta}\right\}^{0}(-iX)\right)\\
&=\frac{X}{\Gamma(1+\vartheta)},
\end{aligned}$$

$$\begin{aligned}&\Big\{presin_{\kappa,\vartheta}\Big\}^{1}(X)\\&=\frac{1}{2}\left(\Big\{\mathbb{E}_{\kappa,\vartheta}\Big\}^{1}(iX)-\Big\{\mathbb{E}_{\kappa,\vartheta}\Big\}^{1}(-iX)\right)\\&=\sum_{k=0}^{1}\frac{(-1)^{k}X^{(2k+1)}}{\Gamma((2k+1)\kappa+\vartheta)},\end{aligned}$$

and

$$\begin{aligned}&\Big\{presin_{\kappa,\vartheta}\Big\}^{2}(X)\\&=\frac{1}{2}\left(\Big\{\mathbb{E}_{\kappa,\vartheta}\Big\}^{2}(iX)-\Big\{\mathbb{E}_{\kappa,\vartheta}\Big\}^{2}(-iX)\right)\\&=\sum_{k=0}^{2}\frac{(-1)^{k}X^{(2k+1)}}{\Gamma((2k+1)\kappa+\vartheta)},\end{aligned}$$

where $X, \kappa, \vartheta \in \mathbb{C}$, $n \in \mathbb{N}_0$, and $\Re(\kappa), \Re(\vartheta) > 0$.

The truncated precosine via the Wiman function is given by:

$$\begin{aligned}&\Big\{precos_{\kappa,\vartheta}\Big\}^{n}(X)\\&=\frac{1}{2}\left(\Big\{\mathbb{E}_{\kappa,\vartheta}\Big\}^{n}(iX)+\Big\{\mathbb{E}_{\kappa,\vartheta}\Big\}^{n}(-iX)\right)\\&=\sum_{k=0}^{n}\frac{(-1)^{n}X^{(2k)}}{\Gamma((2k)\kappa+\vartheta)},\end{aligned}$$

where $X, \kappa, \vartheta \in \mathbb{C}$, $n \in \mathbb{N}_0$, and $\Re(\kappa), \Re(\vartheta) > 0$.

So, we give

$$\begin{aligned}&\Big\{precos_{\kappa,\vartheta}\Big\}^{0}(X)\\&=\frac{1}{2}\left(\Big\{\mathbb{E}_{\kappa,\vartheta}\Big\}^{0}(iX)+\Big\{\mathbb{E}_{\kappa,\vartheta}\Big\}^{0}(-iX)\right)\\&=\frac{1}{\Gamma(\vartheta)},\end{aligned}$$

$$\begin{aligned}
&\left\{precos_{\kappa,\vartheta}\right\}^{1}(X)\\
&=\frac{1}{2}\left(\left\{\mathbb{E}_{\kappa,\vartheta}\right\}^{1}(iX)+\left\{\mathbb{E}_{\kappa,\vartheta}\right\}^{1}(-iX)\right)\\
&=\sum_{k=0}^{1}\frac{(-1)^{k}X^{(2k)}}{\Gamma((2k)\kappa+\vartheta)},
\end{aligned}$$

and

$$\begin{aligned}
&\left\{precos_{\kappa,\vartheta}\right\}^{2}(X)\\
&=\frac{1}{2}\left(\left\{\mathbb{E}_{\kappa,\vartheta}\right\}^{2}(iX)+\left\{\mathbb{E}_{\kappa,\vartheta}\right\}^{2}(-iX)\right)\\
&=\sum_{k=0}^{2}\frac{(-1)^{k}X^{(2k)}}{\Gamma((2k)\kappa+\vartheta)},
\end{aligned}$$

where $X, \kappa, \vartheta \in \mathbb{C}$, $n \in \mathbb{N}_0$, and $\Re(\kappa), \Re(\vartheta) > 0$.

The truncated hyperbolic presine via the Wiman function is given by

$$\begin{aligned}
&\left\{presinh_{\kappa,\vartheta}\right\}^{n}(X)\\
&=\frac{1}{2}\left(\left\{\mathbb{E}_{\kappa,\vartheta}\right\}^{n}(X)-\left\{\mathbb{E}_{\kappa,\vartheta}\right\}^{n}(-X)\right)\\
&=\sum_{k=0}^{n}\frac{X^{(2k+1)}}{\Gamma((2k+1)\kappa+\vartheta},
\end{aligned}$$

where $X, \kappa, \vartheta \in \mathbb{C}$, $n \in \mathbb{N}_0$, and $\Re(\kappa), \Re(\vartheta) > 0$.

So, we have that

$$\begin{aligned}
&\left\{presinh_{\kappa,\vartheta}\right\}^{0}(X)\\
&=\frac{1}{2}\left(\left\{\mathbb{E}_{\kappa,\vartheta}\right\}^{0}(X)-\left\{\mathbb{E}_{\kappa,\vartheta}\right\}^{0}(-X)\right)\\
&=\frac{X}{\Gamma(1+\vartheta)},
\end{aligned}$$

$$
\begin{aligned}
&\left\{presinh_{\kappa,\vartheta}\right\}^{1}(X)\\
&=\frac{1}{2}\left(\left\{\mathbb{E}_{\kappa,\vartheta}\right\}^{1}(X)-\left\{\mathbb{E}_{\kappa,\vartheta}\right\}^{1}(-X)\right)\\
&=\sum_{k=0}^{1}\frac{X^{(2k+1)}}{\Gamma((2k+1)\kappa+\vartheta)},
\end{aligned}
$$

and

$$
\begin{aligned}
&\left\{presinh_{\kappa,\vartheta}\right\}^{2}(X)\\
&=\frac{1}{2}\left(\left\{\mathbb{E}_{\kappa,\vartheta}\right\}^{2}(X)-\left\{\mathbb{E}_{\kappa,\vartheta}\right\}^{2}(-X)\right)\\
&=\sum_{k=0}^{2}\frac{X^{(2k+1)}}{\Gamma((2k+1)\kappa+\vartheta)},
\end{aligned}
$$

where $X, \kappa, \vartheta \in \mathbb{C}$, $n \in \mathbb{N}_0$, and $\Re(\kappa), \Re(\vartheta) > 0$.

The truncated hyperbolic precosine via the Wiman function is given by

$$
\begin{aligned}
&\left\{precosh_{\kappa,\vartheta}\right\}^{n}(X)\\
&=\frac{1}{2}\left(\left\{\mathbb{E}_{\kappa,\vartheta}\right\}^{n}(X)+\left\{\mathbb{E}_{\kappa,\vartheta}\right\}^{n}(-X)\right)\\
&=\sum_{k=0}^{n}\frac{X^{(2k)}}{\Gamma((2k)\kappa+\vartheta)},
\end{aligned}
$$

where $X, \kappa, \vartheta \in \mathbb{C}$, $n \in \mathbb{N}_0$, and $\Re(\kappa), \Re(\vartheta) > 0$.

Thus, we have that

$$
\begin{aligned}
&\left\{precosh_{\kappa,\vartheta}\right\}^{0}(X)\\
&=\frac{1}{2}\left(\left\{\mathbb{E}_{\kappa,\vartheta}\right\}^{0}(X)+\left\{\mathbb{E}_{\kappa,\vartheta}\right\}^{0}(-X)\right)\\
&=\frac{1}{\Gamma(\vartheta)},
\end{aligned}
$$

$$
\begin{aligned}
&\left\{precosh_{\kappa,\vartheta}\right\}^{1}(X)\\
&=\frac{1}{2}\left(\left\{\mathbb{E}_{\kappa,\vartheta}\right\}^{1}(X)+\left\{\mathbb{E}_{\kappa,\vartheta}\right\}^{1}(-X)\right)\\
&=\sum_{k=0}^{1}\frac{X^{(2k)}}{\Gamma((2k)\kappa+\vartheta)},
\end{aligned}
$$

and

$$
\begin{aligned}
&\left\{precosh_{\kappa,\vartheta}\right\}^{2}(X)\\
&=\frac{1}{2}\left(\left\{\mathbb{E}_{\kappa,\vartheta}\right\}^{2}(X)+\left\{\mathbb{E}_{\kappa,\vartheta}\right\}^{2}(-X)\right)\\
&=\sum_{k=0}^{2}\frac{X^{(2k)}}{\Gamma((2k)\kappa+\vartheta)},
\end{aligned}
$$

where $X, \kappa, \vartheta \in \mathbb{C}$, $n \in \mathbb{N}_0$, and $\Re(\kappa), \Re(\vartheta) > 0$.

The truncated presine via the Wiman function is given by

$$
\begin{aligned}
&\left\{Tresin_{\kappa,\vartheta}\right\}^{n}(X)\\
&=\frac{X^{\vartheta-1}}{2i}\left(\left\{\mathbb{E}_{\kappa,\vartheta}\right\}^{n}(iX)-\left\{\mathbb{E}_{\kappa,\vartheta}\right\}^{n}(-iX)\right)\\
&=X^{\vartheta-1}\sum_{k=0}^{n}\frac{(-1)^{n}X^{(2k+1)}}{\Gamma((2k+1)\kappa+\vartheta)},
\end{aligned}
$$

where $X, \kappa, \vartheta \in \mathbb{C}$, $n \in \mathbb{N}_0$, and $\Re(\kappa), \Re(\vartheta) > 0$.

Therefore, we have that

$$
\begin{aligned}
&\left\{Tresin_{\kappa,\vartheta}\right\}^{0}(X)\\
&=\frac{X^{\vartheta-1}}{2}\left(\left\{\mathbb{E}_{\kappa,\vartheta}\right\}^{0}(iX)-\left\{\mathbb{E}_{\kappa,\vartheta}\right\}^{0}(-iX)\right)\\
&=\frac{X^{\vartheta}}{\Gamma(1+\vartheta)},
\end{aligned}
$$

$$\left\{Tresin_{\kappa,\vartheta}\right\}^{1}(X)$$
$$=\frac{X^{\vartheta-1}}{2}\left(\left\{\mathbb{E}_{\kappa,\vartheta}\right\}^{1}(iX)-\left\{\mathbb{E}_{\kappa,\vartheta}\right\}^{1}(-iX)\right)$$
$$=X^{\vartheta-1}\sum_{k=0}^{1}\frac{(-1)^{k}X^{(2k+1)}}{\Gamma((2k+1)\kappa+\vartheta)},$$

and

$$\left\{Tresin_{\kappa,\vartheta}\right\}^{2}(X)$$
$$=\frac{X^{\vartheta-1}}{2}\left(\left\{\mathbb{E}_{\kappa,\vartheta}\right\}^{2}(iX)-\left\{\mathbb{E}_{\kappa,\vartheta}\right\}^{2}(-iX)\right)$$
$$=X^{\vartheta-1}\sum_{k=0}^{2}\frac{(-1)^{k}X^{(2k+1)}}{\Gamma((2k+1)\kappa+\vartheta)},$$

where $X, \kappa, \vartheta \in \mathbb{C}$, $n \in \mathbb{N}_0$, and $\Re(\kappa), \Re(\vartheta) > 0$.

The truncated precosine via the Wiman function is given by:

$$\left\{Trecos_{\kappa,\vartheta}\right\}^{n}(X)$$
$$=\frac{X^{\vartheta-1}}{2}\left(\left\{\mathbb{E}_{\kappa,\vartheta}\right\}^{n}(iX)+\left\{\mathbb{E}_{\kappa,\vartheta}\right\}^{n}(-iX)\right)$$
$$=X^{\vartheta-1}\sum_{k=0}^{n}\frac{(-1)^{n}X^{(2k)}}{\Gamma((2k)\kappa+\vartheta)},$$

where $X, \kappa, \vartheta \in \mathbb{C}$, $n \in \mathbb{N}_0$, and $\Re(\kappa), \Re(\vartheta) > 0$.

So, we give

$$\left\{Trecos_{\kappa,\vartheta}\right\}^{0}(X)$$
$$=\frac{X^{\vartheta-1}}{2}\left(\left\{\mathbb{E}_{\kappa,\vartheta}\right\}^{0}(iX)+\left\{\mathbb{E}_{\kappa,\vartheta}\right\}^{0}(-iX)\right)$$
$$=\frac{X^{\vartheta-1}}{\Gamma(\vartheta)},$$

$$
\begin{aligned}
&\left\{Trecos_{\kappa,\vartheta}\right\}^{1}(X) \\
&= \frac{X^{\vartheta-1}}{2}\left(\left\{\mathbb{E}_{\kappa,\vartheta}\right\}^{1}(iX) + \left\{\mathbb{E}_{\kappa,\vartheta}\right\}^{1}(-iX)\right) \\
&= X^{\vartheta-1}\sum_{k=0}^{1}\frac{(-1)^k X^{(2k)}}{\Gamma((2k)\kappa+\vartheta)},
\end{aligned}
$$

and

$$
\begin{aligned}
&\left\{Trecos_{\kappa,\vartheta}\right\}^{2}(X) \\
&= \frac{X^{\vartheta-1}}{2}\left(\left\{\mathbb{E}_{\kappa,\vartheta}\right\}^{2}(iX) + \left\{\mathbb{E}_{\kappa,\vartheta}\right\}^{2}(-iX)\right) \\
&= X^{\vartheta-1}\sum_{k=0}^{2}\frac{(-1)^k X^{(2k)}}{\Gamma((2k)\kappa+\vartheta)},
\end{aligned}
$$

where $X, \kappa, \vartheta \in \mathbb{C}$, $n \in \mathbb{N}_0$, and $\Re(\kappa), \Re(\vartheta) > 0$.

The truncated hyperbolic presine via the Wiman function is given by

$$
\begin{aligned}
&\left\{Tresinh_{\kappa,\vartheta}\right\}^{n}(X) \\
&= \frac{X^{\vartheta-1}}{2}\left(\left\{\mathbb{E}_{\kappa,\vartheta}\right\}^{n}(X) - \left\{\mathbb{E}_{\kappa,\vartheta}\right\}^{n}(-X)\right) \\
&= X^{\vartheta-1}\sum_{k=0}^{n}\frac{X^{(2k+1)}}{\Gamma((2k+1)\kappa+\vartheta},
\end{aligned}
$$

where $X, \kappa, \vartheta \in \mathbb{C}$, $n \in \mathbb{N}_0$, and $\Re(\kappa), \Re(\vartheta) > 0$.

So, we have that

$$
\begin{aligned}
&\left\{Tresinh_{\kappa,\vartheta}\right\}^{0}(X) \\
&= \frac{X^{\vartheta-1}}{2}\left(\left\{\mathbb{E}_{\kappa,\vartheta}\right\}^{0}(X) - \left\{\mathbb{E}_{\kappa,\vartheta}\right\}^{0}(-X)\right) \\
&= \frac{X^{\vartheta}}{\Gamma(1+\vartheta)},
\end{aligned}
$$

$$\begin{aligned}
&\left\{Tresinh_{\kappa,\vartheta}\right\}^{1}(X)\\
&=\frac{X^{\vartheta-1}}{2}\left(\left\{\mathbb{E}_{\kappa,\vartheta}\right\}^{1}(X)-\left\{\mathbb{E}_{\kappa,\vartheta}\right\}^{1}(-X)\right)\\
&=X^{\vartheta-1}\sum_{k=0}^{1}\frac{X^{(2k+1)}}{\Gamma((2k+1)\kappa+\vartheta)},
\end{aligned}$$

and

$$\begin{aligned}
&\left\{Tresinh_{\kappa,\vartheta}\right\}^{2}(X)\\
&=\frac{X^{\vartheta-1}}{2}\left(\left\{\mathbb{E}_{\kappa,\vartheta}\right\}^{2}(X)-\left\{\mathbb{E}_{\kappa,\vartheta}\right\}^{2}(-X)\right)\\
&=X^{\vartheta-1}\sum_{k=0}^{2}\frac{X^{(2k+1)}}{\Gamma((2k+1)\kappa+\vartheta)},
\end{aligned}$$

where $X, \kappa, \vartheta \in \mathbb{C}$, $n \in \mathbb{N}_0$, and $\Re(\kappa), \Re(\vartheta) > 0$.

The truncated hyperbolic precosine via the Wiman function is given by

$$\begin{aligned}
&\left\{Trecosh_{\kappa,\vartheta}\right\}^{n}(X)\\
&=\frac{X^{\vartheta-1}}{2}\left(\left\{\mathbb{E}_{\kappa,\vartheta}\right\}^{n}(X)+\left\{\mathbb{E}_{\kappa,\vartheta}\right\}^{n}(-X)\right)\\
&=X^{\vartheta-1}\sum_{k=0}^{n}\frac{X^{(2k)}}{\Gamma((2k)\kappa+\vartheta)},
\end{aligned}$$

where $X, \kappa, \vartheta \in \mathbb{C}$, $n \in \mathbb{N}_0$, and $\Re(\kappa), \Re(\vartheta) > 0$.

Thus, we have that

$$\begin{aligned}
&\left\{Trecosh_{\kappa,\vartheta}\right\}^{0}(X)\\
&=\frac{X^{\vartheta-1}}{2}\left(\left\{\mathbb{E}_{\kappa,\vartheta}\right\}^{0}(X)+\left\{\mathbb{E}_{\kappa,\vartheta}\right\}^{0}(-X)\right)\\
&=\frac{X^{\vartheta-1}}{\Gamma(\kappa+\vartheta)},
\end{aligned}$$

$$\begin{aligned}
&\left\{Trecosh_{\kappa,\vartheta}\right\}^{1}(X)\\
&=\frac{X^{\vartheta-1}}{2}\left(\left\{\mathbb{E}_{\kappa,\vartheta}\right\}^{1}(X)+\left\{\mathbb{E}_{\kappa,\vartheta}\right\}^{1}(-X)\right)\\
&=X^{\vartheta-1}\sum_{k=0}^{1}\frac{X^{(2k)}}{\Gamma((2k)\kappa+\vartheta)},
\end{aligned}$$

and

$$\begin{aligned}
&\left\{Trecosh_{\kappa,\vartheta}\right\}^{2}(X)\\
&=\frac{X^{\vartheta-1}}{2}\left(\left\{\mathbb{E}_{\kappa,\vartheta}\right\}^{2}(X)+\left\{\mathbb{E}_{\kappa,\vartheta}\right\}^{2}(-X)\right)\\
&=X^{\vartheta-1}\sum_{k=0}^{2}\frac{X^{(2k)}}{\Gamma((2k)\kappa+\vartheta)},
\end{aligned}$$

where $X, \kappa, \vartheta \in \mathbb{C}$, $n \in \mathbb{N}_0$, and $\Re(\kappa), \Re(\vartheta) > 0$.

The truncated presine through the Wiman function is given by

$$\begin{aligned}
&\left\{Gresin_{\kappa,\vartheta}\right\}^{n}(X^{\gamma})\\
&=\frac{X^{\vartheta-1}}{2i}\left(\left\{\mathbb{E}_{\kappa,\vartheta}\right\}^{n}(iX^{\gamma})-\left\{\mathbb{E}_{\kappa,\vartheta}\right\}^{n}(-iX^{\gamma})\right)\\
&=X^{\vartheta-1}\sum_{k=0}^{n}\frac{(-1)^{n}X^{(2k+1)\gamma}}{\Gamma((2k+1)\kappa+\vartheta)},
\end{aligned}$$

where $X, \kappa, \vartheta \in \mathbb{C}$, $n \in \mathbb{N}_0$, and $\Re(\kappa), \Re(\vartheta) > 0$.

Therefore, we have that

$$\begin{aligned}
&\left\{Gresin_{\kappa,\vartheta}\right\}^{0}(X^{\gamma})\\
&=\frac{X^{\vartheta-1}}{2}\left(\left\{\mathbb{E}_{\kappa,\vartheta}\right\}^{0}(iX^{\gamma})-\left\{\mathbb{E}_{\kappa,\vartheta}\right\}^{0}(-iX^{\gamma})\right)\\
&=\frac{X^{\vartheta+\gamma-1}}{\Gamma(\kappa+\vartheta)},
\end{aligned}$$

$$\begin{aligned}
&\left\{Gresin_{\kappa,\vartheta}\right\}^{1}(X^{\gamma})\\
&=\frac{X^{\vartheta-1}}{2}\left(\left\{\mathbb{E}_{\kappa,\vartheta}\right\}^{1}(iX^{\gamma})-\left\{\mathbb{E}_{\kappa,\vartheta}\right\}^{1}(-iX^{\gamma})\right)\\
&=X^{\vartheta-1}\sum_{k=0}^{1}\frac{(-1)^{k}X^{(2k+1)\gamma}}{\Gamma((2k+1)\kappa+\vartheta)},
\end{aligned}$$

and

$$\begin{aligned}
&\left\{Gresin_{\kappa,\vartheta}\right\}^{2}(X^{\gamma})\\
&=\frac{X^{\vartheta-1}}{2}\left(\left\{\mathbb{E}_{\kappa,\vartheta}\right\}^{2}(iX^{\gamma})-\left\{\mathbb{E}_{\kappa,\vartheta}\right\}^{2}(-iX^{\gamma})\right)\\
&=X^{\vartheta-1}\sum_{k=0}^{2}\frac{(-1)^{k}X^{(2k+1)\gamma}}{\Gamma((2k+1)\kappa+\vartheta)},
\end{aligned}$$

where $X, \kappa, \vartheta \in \mathbb{C}$, $n \in \mathbb{N}_0$, and $\Re(\kappa), \Re(\vartheta) > 0$.

The truncated precosine through the Wiman function is given by:

$$\begin{aligned}
&\left\{Grecos_{\kappa,\vartheta}\right\}^{n}(X^{\gamma})\\
&=\frac{X^{\vartheta-1}}{2}\left(\left\{\mathbb{E}_{\kappa,\vartheta}\right\}^{n}(iX^{\gamma})+\left\{\mathbb{E}_{\kappa,\vartheta}\right\}^{n}(-iX^{\gamma})\right)\\
&=X^{\vartheta-1}\sum_{k=0}^{n}\frac{(-1)^{n}X^{(2k)\gamma}}{\Gamma((2k)\kappa+\vartheta)},
\end{aligned}$$

where $X, \kappa, \vartheta \in \mathbb{C}$, $n \in \mathbb{N}_0$, and $\Re(\kappa), \Re(\vartheta) > 0$.

So, we give

$$\begin{aligned}
&\left\{Grecos_{\kappa,\vartheta}\right\}^{0}(X^{\gamma})\\
&=\frac{X^{\vartheta-1}}{2}\left(\left\{\mathbb{E}_{\kappa,\vartheta}\right\}^{0}(iX^{\gamma})+\left\{\mathbb{E}_{\kappa,\vartheta}\right\}^{0}(-iX^{\gamma})\right)\\
&=\frac{X^{\vartheta-1}}{\Gamma(\kappa+\vartheta)},
\end{aligned}$$

$$\begin{aligned}&\left\{Grecos_{\kappa,\vartheta}\right\}^{1}(X^{\gamma})\\&=\frac{X^{\vartheta-1}}{2}\left(\left\{\mathbb{E}_{\kappa,\vartheta}\right\}^{1}(iX^{\gamma})+\left\{\mathbb{E}_{\kappa,\vartheta}\right\}^{1}(-iX^{\gamma})\right)\\&=X^{\vartheta-1}\sum_{k=0}^{1}\frac{(-1)^{k}X^{(2k)\gamma}}{\Gamma((2k)\kappa+\vartheta)},\end{aligned}$$

and

$$\begin{aligned}&\left\{Grecos_{\kappa,\vartheta}\right\}^{2}(X^{\gamma})\\&=\frac{X^{\vartheta-1}}{2}\left(\left\{\mathbb{E}_{\kappa,\vartheta}\right\}^{2}(iX^{\gamma})+\left\{\mathbb{E}_{\kappa,\vartheta}\right\}^{2}(-iX^{\gamma})\right)\\&=X^{\vartheta-1}\sum_{k=0}^{2}\frac{(-1)^{k}X^{(2k)\gamma}}{\Gamma((2k)\kappa+\vartheta)},\end{aligned}$$

where $X, \kappa, \vartheta, \gamma \in \mathbb{C}$, $n \in \mathbb{N}_0$, and $\Re(\kappa), \Re(\vartheta) > 0$.

The truncated hyperbolic presine through the Wiman function is given by

$$\begin{aligned}&\left\{Gresinh_{\kappa,\vartheta}\right\}^{n}(X^{\gamma})\\&=\frac{X^{\vartheta-1}}{2}\left(\left\{\mathbb{E}_{\kappa,\vartheta}\right\}^{n}(X^{\gamma})-\left\{\mathbb{E}_{\kappa,\vartheta}\right\}^{n}(-X^{\gamma})\right)\\&=X^{\vartheta-1}\sum_{k=0}^{n}\frac{X^{(2k+1)\gamma}}{\Gamma((2k+1)\kappa+\vartheta},\end{aligned}$$

where $X, \kappa, \vartheta, \gamma \in \mathbb{C}$, $n \in \mathbb{N}_0$, and $\Re(\kappa), \Re(\vartheta) > 0$.

So, we have that

$$\begin{aligned}&\left\{Gresinh_{\kappa,\vartheta}\right\}^{0}(X^{\gamma})\\&=\frac{X^{\vartheta-1}}{2}\left(\left\{\mathbb{E}_{\kappa,\vartheta}\right\}^{0}(X^{\gamma})-\left\{\mathbb{E}_{\kappa,\vartheta}\right\}^{0}(-X^{\gamma})\right)\\&=\frac{X^{\vartheta+\gamma-1}}{\Gamma(\kappa+\vartheta)},\end{aligned}$$

$$
\begin{aligned}
&\left\{Gresinh_{\kappa,\vartheta}\right\}^{1}(X^{\gamma})\\
&=\frac{X^{\vartheta-1}}{2}\left(\left\{\mathbb{E}_{\kappa,\vartheta}\right\}^{1}(X^{\gamma})-\left\{\mathbb{E}_{\kappa,\vartheta}\right\}^{1}(-X^{\gamma})\right)\\
&=X^{\vartheta-1}\sum_{k=0}^{1}\frac{X^{(2k+1)\gamma}}{\Gamma((2k+1)\kappa+\vartheta)},
\end{aligned}
$$

and

$$
\begin{aligned}
&\left\{Gresinh_{\kappa,\vartheta}\right\}^{2}(X^{\gamma})\\
&=\frac{X^{\vartheta-1}}{2}\left(\left\{\mathbb{E}_{\kappa,\vartheta}\right\}^{2}(X^{\gamma})-\left\{\mathbb{E}_{\kappa,\vartheta}\right\}^{2}(-X^{\gamma})\right)\\
&=X^{\vartheta-1}\sum_{k=0}^{2}\frac{X^{(2k+1)\gamma}}{\Gamma((2k+1)\kappa+\vartheta)},
\end{aligned}
$$

where $X, \kappa, \vartheta, \gamma \in \mathbb{C}$, $n \in \mathbb{N}_0$, and $\Re(\kappa), \Re(\vartheta) > 0$.

The truncated hyperbolic precosine via the Wiman function is given by

$$
\begin{aligned}
&\left\{Grecosh_{\kappa,\vartheta}\right\}^{n}(X^{\gamma})\\
&=\frac{X^{\vartheta-1}}{2}\left(\left\{\mathbb{E}_{\kappa,\vartheta}\right\}^{n}(X^{\gamma})+\left\{\mathbb{E}_{\kappa,\vartheta}\right\}^{n}(-X^{\gamma})\right)\\
&=X^{\vartheta-1}\sum_{k=0}^{n}\frac{X^{(2k)\gamma}}{\Gamma((2k)\kappa+\vartheta)},
\end{aligned}
$$

where $X, \kappa, \vartheta, \gamma \in \mathbb{C}$, $n \in \mathbb{N}_0$, and $\Re(\kappa), \Re(\vartheta) > 0$.

Thus, we have that

$$
\begin{aligned}
&\left\{Grecosh_{\kappa,\vartheta}\right\}^{0}(X^{\gamma})\\
&=\frac{X^{\vartheta-1}}{2}\left(\left\{\mathbb{E}_{\kappa,\vartheta}\right\}^{0}(X^{\gamma})+\left\{\mathbb{E}_{\kappa,\vartheta}\right\}^{0}(-X^{\gamma})\right)\\
&=\frac{X^{\vartheta-1}}{\Gamma(\kappa+\vartheta)},
\end{aligned}
$$

$$\begin{aligned}
&\left\{Grecosh_{\kappa,\vartheta}\right\}^{1}(X^{\gamma})\\
&=\frac{X^{\vartheta-1}}{2}\left(\left\{\mathbb{E}_{\kappa,\vartheta}\right\}^{1}(X^{\gamma})+\left\{\mathbb{E}_{\kappa,\vartheta}\right\}^{1}(-X^{\gamma})\right)\\
&=X^{\vartheta-1}\sum_{k=0}^{1}\frac{X^{(2k)\gamma}}{\Gamma((2k)\kappa+\vartheta)},
\end{aligned}$$

and

$$\begin{aligned}
&\left\{Grecosh_{\kappa,\vartheta}\right\}^{2}(X^{\gamma})\\
&=\frac{X^{\vartheta-1}}{2}\left(\left\{\mathbb{E}_{\kappa,\vartheta}\right\}^{2}(X^{\gamma})+\left\{\mathbb{E}_{\kappa,\vartheta}\right\}^{2}(-X^{\gamma})\right)\\
&=X^{\vartheta-1}\sum_{k=0}^{2}\frac{X^{(2k)\gamma}}{\Gamma((2k)\kappa+\vartheta)},
\end{aligned}$$

where $X, \kappa, \vartheta, \gamma \in \mathbb{C}$, $n \in \mathbb{N}_0$, and $\Re(\kappa), \Re(\vartheta) > 0$.

References

1. Wiman A (1905) Über den Fundamentalsatz in der Teorie der Funktionen Ea (x). Acta Math 29:191–201
2. Mathai AM, Haubold HJ (2008) Special functions for applied scientists. Springer, New York
3. Bateman H (1953) Higher transcendental functions. California Institute of Technology Bateman Manuscript Project, vol 1. McGraw-Hill, New York, vol. 3 (1955)
4. Gorenflo R, Kilbas AA, Rogosin SV (1998) On the generalized Mittag-Leffler type functions. Integral Transform Spec Funct 7(3–4):215–224
5. Srivastava HM (1994) A certain family of sub-exponential series. Int J Math Educ Sci Technol 25(2):211–216
6. Yang X-J (2021) Theory and applications of special functions for scientists and engineers. Springer
7. Erdelyi A, Magnus W, Oberhettinger F, Tricomi FG (1955) Higher transcendental functions, vol 3. McGraw-Hill, New York, NY
8. Schneider WR (1996) Completely monotone generalized Mittag-Leffler functions. Expo Math 14(1):3–24
9. Miller KS, Samko SG (1999) A note on the complete monotonicity of the generalized Mittag-Leffler function. Real Anal Exch 23(2):753–756
10. Djrbashian MM (1966) Integral transforms and representations of functions in the complex domain (Russian). Nauka, Moscow
11. Yang XJ, Gao F, Ju Y (2020) General fractional derivatives with applications in viscoelasticity. Academic, New York

Chapter 6
The Prabhakar, Supertrigonometric, and Superhyperbolic Functions

In this chapter, we introduce the Prabhakar function and express the supertrigonometric and superhyperbolic functions via the Prabhakar function.

6.1 The Prabhakar Function

In 1971, Tilak Raj Prabhakar, an Indian mathematician, worked on the generalized Mittag-Leffler function, which is called the three-parametric Mittag-Leffler function, Mittag-Leffler type function, or the Prabhakar function [1].

Definition 6.1 ([1, 2]) The Prabhakar function is given by

$$\mathbb{E}^{\gamma}_{\kappa,\vartheta}(X) = \sum_{k=0}^{\infty} \frac{(\gamma)_k}{\Gamma(k\kappa + \vartheta)} \frac{X^k}{k!},$$

where $\kappa, \vartheta, \gamma, X \in \mathbb{C}$, and $\Re(\kappa), \Re(\vartheta) > 0$.

Some special cases of the Prabhakar function include:

$$\mathbb{E}^{1}_{\kappa,\vartheta}(X) = \mathbb{E}_{\kappa,\vartheta}(X) = \sum_{k=0}^{\infty} \frac{X^k}{\Gamma(k\kappa + \vartheta)},$$

$$\mathbb{E}^{1}_{\kappa,1}(X) = \mathbb{E}_{\kappa,1}(X) = \mathbb{E}_{\kappa}(X) = \sum_{k=0}^{\infty} \frac{X^k}{\Gamma(k\kappa + 1)},$$

where $\kappa, \vartheta, X \in \mathbb{C}$, and $\Re(\kappa), \Re(\vartheta) > 0$.

S. Rezaei Aderyani et al., *Towards Ulam Type Multi Stability Analysis*, Studies in Systems, Decision and Control 523, https://doi.org/10.1007/978-3-031-55564-0_6

The relationships between the Wright hypergeometric function and the Prabhakar function function are given as follows [3]:

$$\mathbb{E}^{\gamma}_{\kappa,\vartheta}(X) = \frac{1}{\Gamma(\gamma)}{}_1\mathbb{W}_1\left({}^{(\gamma,1)}_{(\vartheta,\kappa)}; X\right),$$

$$\mathbb{E}^{\gamma}_{1,\vartheta}(X) = \frac{1}{\Gamma(\gamma)}{}_1\mathbb{W}_1\left({}^{(\gamma,1)}_{(\vartheta,1)}; X\right),$$

$$\mathbb{E}^{\gamma}_{\kappa,1}(X) = \frac{1}{\Gamma(\gamma)}{}_1\mathbb{W}_1\left({}^{(\gamma,1)}_{(1,\kappa)}; X\right),$$

$$\mathbb{E}^{\gamma}_{1,1}(X) = \frac{1}{\Gamma(\gamma)}{}_1\mathbb{W}_1\left({}^{(\gamma,1)}_{(1,1)}; X\right),$$

$$\mathbb{E}^{1}_{\kappa,\vartheta}(X) = {}_1\mathbb{W}_1\left({}^{(1,1)}_{(\vartheta,\kappa)}; X\right),$$

$$\mathbb{E}^{1}_{\kappa,1}(X) = {}_1\mathbb{W}_1\left({}^{(1,1)}_{(1,\kappa)}; X\right),$$

where $\kappa, \vartheta, \gamma, X \in \mathbb{C}$, and $\Re(\kappa), \Re(\vartheta) > 0$.

In addition, the family of the Gaussian-like functions can be written as follows [4]:

$$\mathbb{E}^{\gamma}_{\kappa,\vartheta}(X^2) = \frac{1}{\Gamma(\gamma)}{}_1\mathbb{W}_1\left({}^{(\gamma,1)}_{(\vartheta,\kappa)}; X^2\right),$$

$$\mathbb{E}^{\gamma}_{1,\vartheta}(X^2) = \frac{1}{\Gamma(\gamma)}{}_1\mathbb{W}_1\left({}^{(\gamma,1)}_{(\vartheta,1)}; X^2\right),$$

$$\mathbb{E}^{\gamma}_{\kappa,1}(X^2) = \frac{1}{\Gamma(\gamma)}{}_1\mathbb{W}_1\left({}^{(\gamma,1)}_{(1,\kappa)}; X^2\right),$$

$$\mathbb{E}^{\gamma}_{1,1}(X^2) = \frac{1}{\Gamma(\gamma)}{}_1\mathbb{W}_1\left({}^{(\gamma,1)}_{(1,1)}; X^2\right),$$

$$\mathbb{E}^{1}_{\kappa,\vartheta}(X^2) = {}_1\mathbb{W}_1\left({}^{(1,1)}_{(\vartheta,\kappa)}; X^2\right),$$

$$\mathbb{E}^{1}_{\kappa,1}(X^2) = {}_1\mathbb{W}_1\left({}^{(1,1)}_{(1,\kappa)}; X^2\right),$$

where $\kappa, \vartheta, \gamma, X \in \mathbb{C}$, and $\Re(\kappa), \Re(\vartheta) > 0$.

Theorem 6.1 ([1]) *Let* $\vartheta, \gamma, X \in \mathbb{C}$, *and* $\kappa > 0$. *Then,* $\mathbb{E}^{\gamma}_{\kappa,\vartheta}(X)$ *is an entire function of order* $\frac{1}{\kappa}$.

This result was discovered by Prabhakar [1].

6.2 The Supertrigonometric Functions via the Prabhakar Function

Let $\vartheta, \gamma, X \in \mathbb{C}$, and $\kappa > 0$. In this section, we introduce the supertrigonometric functions via the Prabhakar function [5] as follows:

▷ The supersine through the Prabhakar function is given by

$$\begin{aligned}
&supersin^{\gamma}_{\kappa,\vartheta}(X^{\kappa})\\
&= \frac{X^{\vartheta-1}}{2i}\left(\mathbb{E}^{\gamma}_{\kappa,\vartheta}(iX^{\kappa}) - \mathbb{E}^{\gamma}_{\kappa,\vartheta}(-iX^{\kappa})\right)\\
&= \frac{X^{\vartheta-1}}{2i}\frac{1}{\Gamma(\gamma)}{}_1\sin_1\left(\begin{smallmatrix}(\gamma,1)\\(\vartheta,\kappa)\end{smallmatrix}; X^{\kappa}\right)\\
&= \sum_{k=0}^{\infty}\frac{(-1)^k(\gamma)_{2k+1}}{\Gamma((2k+1)\kappa+\vartheta)}\frac{X^{(2k+1)\kappa+\vartheta-1}}{(2k+1)!}.
\end{aligned}$$

▷ The supercosine through the Prabhakar function is given by

$$\begin{aligned}
&supercos^{\gamma}_{\kappa,\vartheta}(X^{\kappa})\\
&= \frac{X^{\vartheta-1}}{2}\left(\mathbb{E}^{\gamma}_{\kappa,\vartheta}(iX^{\kappa}) + \mathbb{E}^{\gamma}_{\kappa,\vartheta}(-iX^{\kappa})\right)\\
&= \frac{X^{\vartheta-1}}{2}\frac{1}{\Gamma(\gamma)}{}_1\cos_1\left(\begin{smallmatrix}(\gamma,1)\\(\vartheta,\kappa)\end{smallmatrix}; X^{\kappa}\right)\\
&= \sum_{k=0}^{\infty}\frac{(-1)^k(\gamma)_{2k}}{\Gamma((2k)\kappa+\vartheta)}\frac{X^{(2k)\kappa+\vartheta-1}}{(2k)!}.
\end{aligned}$$

▷ The supertangent through the Prabhakar function is given by

$$\begin{aligned}
&supertan^{\gamma}_{\kappa,\vartheta}(X^{\kappa})\\
&= \frac{supersin^{\gamma}_{\kappa,\vartheta}(X^{\kappa})}{supercos^{\gamma}_{\kappa,\vartheta}(X^{\kappa})}, \ (supercos^{\gamma}_{\kappa,\vartheta}(X^{\kappa}) \neq 0).
\end{aligned}$$

▷ The supercotangent through the Prabhakar function is given by

$$\begin{aligned} & supercotan^{\gamma}_{\kappa,\vartheta}(X^{\kappa}) \\ & = \frac{supercos^{\gamma}_{\kappa,\vartheta}(X^{\kappa})}{supersin^{\gamma}_{\kappa,\vartheta}(X^{\kappa})}, \ (supersin^{\gamma}_{\kappa,\vartheta}(X^{\kappa}) \neq 0). \end{aligned}$$

▷ The supercosecant through the Prabhakar function is given by

$$\begin{aligned} & supercsc^{\gamma}_{\kappa,\vartheta}(X^{\kappa}) \\ & = \frac{1}{supersin^{\gamma}_{\kappa,\vartheta}(X^{\kappa})}, \ (supersin^{\gamma}_{\kappa,\vartheta}(X^{\kappa}) \neq 0). \end{aligned}$$

▷ The supersecant through the Prabhakar function is given by

$$\begin{aligned} & supersec^{\gamma}_{\kappa,\vartheta}(X^{\kappa}) \\ & = \frac{1}{supercos^{\gamma}_{\kappa,\vartheta}(X^{\kappa})}, \ (supercos^{\gamma}_{\kappa,\vartheta}(X^{\kappa}) \neq 0). \end{aligned}$$

The above work was presented by Yang [4].

The following theorem can be shown based on these results.

Theorem 6.2 ([4]) *Let $\kappa, \vartheta, \gamma, X \in \mathbb{C}$, and $\Re(\kappa), \Re(\vartheta) > 0$. Then,*

$$X^{\vartheta-1}\mathbb{E}^{\gamma}_{\kappa,\vartheta}(iX^{\kappa}) = supercos^{\gamma}_{\kappa,\vartheta}(X^{\kappa}) + isupersin^{\gamma}_{\kappa,\vartheta}(X^{\kappa}),$$

and

$$X^{\vartheta-1}\mathbb{E}^{\gamma}_{\kappa,\vartheta}(-iX^{\kappa}) = supercos^{\gamma}_{\kappa,\vartheta}(X^{\kappa}) - isupersin^{\gamma}_{\kappa,\vartheta}(X^{\kappa}).$$

6.3 The Superhyperbolic Functions via the Prabhakar Function

Let $\vartheta, \gamma, X \in \mathbb{C}$, and $\kappa > 0$. In the section, we present the superhyperbolic functions defined via the Prabhakar function [4] as follows:

▷ The superhyperbolic supersine through the Prabhakar function is given by

$$\begin{aligned} & supersinh^{\gamma}_{\kappa,\vartheta}(X^{\kappa}) \\ & = \frac{X^{\vartheta-1}}{2}\left(\mathbb{E}^{\gamma}_{\kappa,\vartheta}(X^{\kappa}) - \mathbb{E}^{\gamma}_{\kappa,\vartheta}(-X^{\kappa})\right) \end{aligned}$$

$$= \frac{X^{\vartheta-1}}{2}\frac{1}{\Gamma(\gamma)}{}_1\sinh_1\left({(\gamma,1) \atop (\vartheta,\kappa)}; X^{\kappa}\right)$$
$$= \sum_{k=0}^{\infty}\frac{(\gamma)_{2k+1}X^{(2k+1)\kappa+\vartheta-1}}{\Gamma((2k+1)\kappa+\vartheta)}.$$

▷ The superhyperbolic supercosine through the Prabhakar function is given by

$$\begin{aligned}&supercosh^{\gamma}_{\kappa,\vartheta}(X^{\kappa})\\ &= \frac{X^{\vartheta-1}}{2}\left(\mathbb{E}^{\gamma}_{\kappa,\vartheta}(X^{\kappa}) + \mathbb{E}^{\gamma}_{\kappa,\vartheta}(-X^{\kappa})\right)\\ &= \frac{X^{\vartheta-1}}{2}\frac{1}{\Gamma(\gamma)}{}_1\cosh_1\left({(\gamma,1) \atop (\vartheta,\kappa)}; X^{\kappa}\right)\\ &= \sum_{k=0}^{\infty}\frac{(\gamma)_{2k}X^{(2k)\kappa+\vartheta-1}}{\Gamma((2k)\kappa+\vartheta)}.\end{aligned}$$

▷ The superhyperbolic supertangent through the Prabhakar function is given by

$$\begin{aligned}&supertanh^{\gamma}_{\kappa,\vartheta}(X^{\kappa})\\ &= \frac{supersinh^{\gamma}_{\kappa,\vartheta}(X^{\kappa})}{supercosh^{\gamma}_{\kappa,\vartheta}(X^{\kappa})}, \ (supercosh^{\gamma}_{\kappa,\vartheta}(X^{\kappa}) \neq 0).\end{aligned}$$

▷ The superhyperbolic supercotangent through the Prabhakar function is given by

$$\begin{aligned}&supercotanh^{\gamma}_{\kappa,\vartheta}(X^{\kappa})\\ &= \frac{supercosh^{\gamma}_{\kappa,\vartheta}(X^{\kappa})}{supersinh^{\gamma}_{\kappa,\vartheta}(X^{\kappa})}, \ (supersinh^{\gamma}_{\kappa,\vartheta}(X^{\kappa}) \neq 0).\end{aligned}$$

▷ The superhyperbolic supercosecant through the Prabhakar function is given by

$$\begin{aligned}&supercsch^{\gamma}_{\kappa,\vartheta}(X^{\kappa})\\ &= \frac{1}{supersinh^{\gamma}_{\kappa,\vartheta}(X^{\kappa})}, \ (supersinh^{\gamma}_{\kappa,\vartheta}(X^{\kappa}) \neq 0).\end{aligned}$$

▷ The superhyperbolic supersecant through the Prabhakar function is given by

$$\begin{aligned}&supersech^{\gamma}_{\kappa,\vartheta}(X^{\kappa})\\ &= \frac{1}{supercosh^{\gamma}_{\kappa,\vartheta}(X^{\kappa})}, \ (supercosh^{\gamma}_{\kappa,\vartheta}(X^{\kappa}) \neq 0).\end{aligned}$$

The above work was also presented by Yang in 2019 [4].

From these results, the following theorem can be shown.

Theorem 6.3 ([4]) *Let* $\kappa, \vartheta, \gamma, X \in \mathbb{C}$, *and* $\Re(\kappa), \Re(\vartheta) > 0$. *Then,*

$$X^{\vartheta-1}\mathbb{E}^{\gamma}_{\kappa,\vartheta}(X^{\kappa}) = supercosh^{\gamma}_{\kappa,\vartheta}(X^{\kappa}) + supersinh^{\gamma}_{\kappa,\vartheta}(X^{\kappa}),$$

$$X^{\vartheta-1}\mathbb{E}^{\gamma}_{\kappa,\vartheta}(-X^{\kappa}) = supercosh^{\gamma}_{\kappa,\vartheta}(X^{\kappa}) - supersinh^{\gamma}_{\kappa,\vartheta}(X^{\kappa}).$$

This work was presented by Yang [4].

6.4 The Pre-supertrigonometric Functions via the Prabhakar Function

Let $\vartheta, \gamma, X \in \mathbb{C}$, and $\kappa > 0$. In this section, we introduce the pre-supertrigonometric functions defined via the Prabhakar function [4], as follows:

▹ The presine through the Prabhakar function is given by

$$\begin{aligned} & presin^{\gamma}_{\kappa,\vartheta}(X) \\ &= \frac{1}{2i}\left(\mathbb{E}^{\gamma}_{\kappa,\vartheta}(iX) - \mathbb{E}^{\gamma}_{\kappa,\vartheta}(-iX)\right) \\ &= \frac{1}{2i}\frac{1}{\Gamma(\gamma)}{}_1\sin_1\left(\begin{smallmatrix}(\gamma,1) \\ (\vartheta,\kappa)\end{smallmatrix}; X\right) \\ &= \sum_{k=0}^{\infty}\frac{(-1)^k(\gamma)_{2k+1}X^{(2k+1)}}{\Gamma((2k+1)\kappa+\vartheta)}. \end{aligned}$$

▹ The precosine through the Prabhakar function is given by

$$\begin{aligned} & precos^{\gamma}_{\kappa,\vartheta}(X) \\ &= \frac{1}{2}\left(\mathbb{E}^{\gamma}_{\kappa,\vartheta}(iX) + \mathbb{E}^{\gamma}_{\kappa,\vartheta}(-iX)\right) \\ &= \frac{1}{2}\frac{1}{\Gamma(\gamma)}{}_1\cos_1\left(\begin{smallmatrix}(\gamma,1) \\ (\vartheta,\kappa)\end{smallmatrix}; X\right) \\ &= \sum_{k=0}^{\infty}\frac{(-1)^k(\gamma)_{2k}X^{(2k)}}{\Gamma((2k)\kappa+\vartheta)}. \end{aligned}$$

▷ The pretangent through the Prabhakar function is given by

$$\begin{aligned}&pretan^{\gamma}_{\kappa,\vartheta}(X)\\&=\frac{presin^{\gamma}_{\kappa,\vartheta}(X)}{precos^{\gamma}_{\kappa,\vartheta}(X)},\ (precos^{\gamma}_{\kappa,\vartheta}(X)\neq 0).\end{aligned}$$

▷ The precotangent through the Prabhakar function is given by

$$\begin{aligned}&preotan^{\gamma}_{\kappa,\vartheta}(X)\\&=\frac{precos^{\gamma}_{\kappa,\vartheta}(X)}{presin^{\gamma}_{\kappa,\vartheta}(X)},\ (presin^{\gamma}_{\kappa,\vartheta}(X)\neq 0).\end{aligned}$$

▷ The precosecant through the Prabhakar function is given by

$$\begin{aligned}&precsc^{\gamma}_{\kappa,\vartheta}(X)\\&=\frac{1}{presin^{\gamma}_{\kappa,\vartheta}(X)},\ (presin^{\gamma}_{\kappa,\vartheta}(X)\neq 0).\end{aligned}$$

▷ The presecant through the Prabhakar function is given by

$$\begin{aligned}&presec^{\gamma}_{\kappa,\vartheta}(X)\\&=\frac{1}{precos^{\gamma}_{\kappa,\vartheta}(X)},\ (precos^{\gamma}_{\kappa,\vartheta}(X)\neq 0).\end{aligned}$$

The above work was given by Yang [4].

Based on this result, we may have the following theorem:

Theorem 6.4 ([4]) *Suppose* $\kappa, \vartheta, \gamma, X \in \mathbb{C}$, *and* $\Re(\kappa), \Re(\vartheta) > 0$. *Then,*

$$\mathbb{E}^{\gamma}_{\kappa,\vartheta}(iX) = precos^{\gamma}_{\kappa,\vartheta}(X) + ipresin^{\gamma}_{\kappa,\vartheta}(X),$$

$$\mathbb{E}^{\gamma}_{\kappa,\vartheta}(-iX) = precos^{\gamma}_{\kappa,\vartheta}(X) - ipresin^{\gamma}_{\kappa,\vartheta}(X).$$

This work was presented by Yang [4].

6.5 The Pre-superhyperbolic Functions via the Prabhakar Function

Let $\vartheta, \gamma, X \in \mathbb{C}$, and $\kappa > 0$. In this section, we introduce the pre-superhyperbolic functions via the Prabhakar function [4] as follows:

▷ The hyperbolic presine through the Prabhakar function is given by

$$\begin{aligned} & presinh^{\gamma}_{\kappa,\vartheta}(X) \\ &= \frac{1}{2}\Big(\mathbb{E}^{\gamma}_{\kappa,\vartheta}(X) - \mathbb{E}^{\gamma}_{\kappa,\vartheta}(-X)\Big) \\ &= \frac{1}{2}\frac{1}{\Gamma(\gamma)}{}_1\sinh_1\left({}^{(\gamma,1)}_{(\vartheta,\kappa)}; X\right) \\ &= \sum_{k=0}^{\infty}\frac{(\gamma)_{2k+1}X^{(2k+1)}}{\Gamma((2k+1)\kappa+\vartheta)}. \end{aligned}$$

▷ The hyperbolic precosine through the Prabhakar function is given by

$$\begin{aligned} & precosh^{\gamma}_{\kappa,\vartheta}(X) \\ &= \frac{1}{2}\Big(\mathbb{E}^{\gamma}_{\kappa,\vartheta}(X) + \mathbb{E}^{\gamma}_{\kappa,\vartheta}(-X)\Big) \\ &= \frac{1}{2}\frac{1}{\Gamma(\gamma)}{}_1\cosh_1\left({}^{(\gamma,1)}_{(\vartheta,\kappa)}; X\right) \\ &= \sum_{k=0}^{\infty}\frac{(\gamma)_{2k}X^{(2k)}}{\Gamma((2k)\kappa+\vartheta)}. \end{aligned}$$

▷ The hyperbolic pretangent through the Prabhakar function is given by

$$\begin{aligned} & pretanh^{\gamma}_{\kappa,\vartheta}(X) \\ &= \frac{presinh^{\gamma}_{\kappa,\vartheta}(X)}{supercosh^{\gamma}_{\kappa,\vartheta}(X)}, \ (precosh^{\gamma}_{\kappa,\vartheta}(X \neq 0). \end{aligned}$$

▷ The hyperbolic precotangent through the Prabhakar function is given by

$$\begin{aligned} & precotanh^{\gamma}_{\kappa,\vartheta}(X) \\ &= \frac{precosh^{\gamma}_{\kappa,\vartheta}(X)}{presinh^{\gamma}_{\kappa,\vartheta}(X)}, \ (presinh^{\gamma}_{\kappa,\vartheta}(X) \neq 0). \end{aligned}$$

▷ The hyperbolic precosecant through the Prabhakar function is given by

$$\begin{aligned} & precsch^{\gamma}_{\kappa,\vartheta}(X) \\ & = \frac{1}{presinh^{\gamma}_{\kappa,\vartheta}(X)}, \ (presinh^{\gamma}_{\kappa,\vartheta}(X) \neq 0). \end{aligned}$$

▷ The hyperbolic presecant through the Prabhakar function is given by

$$\begin{aligned} & presech^{\gamma}_{\kappa,\vartheta}(X) \\ & = \frac{1}{precosh^{\gamma}_{\kappa,\vartheta}(X)}, \ (precosh^{\gamma}_{\kappa,\vartheta}(X) \neq 0). \end{aligned}$$

This work was presented by Yang [4].

Based on the result, we may deduce the following theorem.

Theorem 6.5 ([4]) *Let* $\kappa, \vartheta, \gamma, X \in \mathbb{C}$, *and* $\Re(\kappa), \Re(\vartheta) > 0$. *Then,*

$$\mathbb{E}^{\gamma}_{\kappa,\vartheta}(X) = precosh^{\gamma}_{\kappa,\vartheta}(X) + presinh^{\gamma}_{\kappa,\vartheta}(X),$$

$$\mathbb{E}^{\gamma}_{\kappa,\vartheta}(-X) = precosh^{\gamma}_{\kappa,\vartheta}(X) - presinh^{\gamma}_{\kappa,\vartheta}(X).$$

This work was presented by Yang [4].

6.6 The Pre-supertrigonometric Functions with Power Law via the Prabhakar Function

Let $\kappa, \vartheta, \gamma, \zeta, X \in \mathbb{C}$, and $\Re(\kappa), \Re(\vartheta) > 0$. In this section, we give the pre-supertrigonometric functions with the power law defined via the Prabhakar function [4], as follows:

▷ The presine with power law through the Prabhakar function is given by

$$\begin{aligned} & presin^{\gamma}_{\kappa,\vartheta}(X^{\zeta}) \\ & = \frac{1}{2i}\left(\mathbb{E}^{\gamma}_{\kappa,\vartheta}(iX^{\zeta}) - \mathbb{E}^{\gamma}_{\kappa,\vartheta}(-iX^{\zeta})\right) \\ & = \frac{1}{\Gamma(\gamma)}{}_1\sin_1\left(\substack{(\gamma,1) \\ (\vartheta,\kappa)}; X^{\zeta}\right) \\ & = \sum_{k=0}^{\infty} \frac{(-1)^k (\gamma)_{2k+1} X^{(2k+1)\zeta}}{\Gamma((2k+1)\kappa + \vartheta)}. \end{aligned}$$

▷ The precosine with power law through the Prabhakar function is given by

$$\begin{aligned} &precos^{\gamma}_{\kappa,\vartheta}(X^{\varsigma}) \\ &= \frac{1}{2}\Big(\mathbb{E}^{\gamma}_{\kappa,\vartheta}(iX^{\varsigma}) + \mathbb{E}^{\gamma}_{\kappa,\vartheta}(-iX^{\varsigma})\Big) \\ &= \frac{1}{\Gamma(\gamma)} {}_1\cos_1\left(\begin{smallmatrix}(\gamma,1)\\(\vartheta,\kappa)\end{smallmatrix}; X^{\varsigma}\right) \\ &= \sum_{k=0}^{\infty} \frac{(-1)^k (\gamma)_{2k} X^{(2k)\varsigma}}{\Gamma((2k)\kappa + \vartheta)}. \end{aligned}$$

▷ The pretangent with power law through the Prabhakar function is given by

$$\begin{aligned} &pretan^{\gamma}_{\kappa,\vartheta}(X^{\varsigma}) \\ &= \frac{presin^{\gamma}_{\kappa,\vartheta}(X^{\varsigma})}{precos^{\gamma}_{\kappa,\vartheta}(X^{\varsigma})}, \quad \Big(precos^{\gamma}_{\kappa,\vartheta}(X^{\varsigma}) \neq 0\Big). \end{aligned}$$

▷ The precotangentwith power law through the Prabhakar function is given by

$$\begin{aligned} &precotan^{\gamma}_{\kappa,\vartheta}(X^{\varsigma}) \\ &= \frac{precos^{\gamma}_{\kappa,\vartheta}(X^{\varsigma})}{presin^{\gamma}_{\kappa,\vartheta}(X^{\varsigma})}, \quad \Big(presin^{\gamma}_{\kappa,\vartheta}(X^{\varsigma}) \neq 0\Big). \end{aligned}$$

▷ The precosecant with power law through the Prabhakar function is given by

$$\begin{aligned} &precsc^{\gamma}_{\kappa,\vartheta}(X^{\varsigma}) \\ &= \frac{1}{presin^{\gamma}_{\kappa,\vartheta}(X^{\varsigma})}, \quad \Big(presin^{\gamma}_{\kappa,\vartheta}(X^{\varsigma}) \neq 0\Big). \end{aligned}$$

▷ The presecant with power law through the Prabhakar function is given by

$$\begin{aligned} &presec^{\gamma}_{\kappa,\vartheta}(X^{\varsigma}) \\ &= \frac{1}{precos^{\gamma}_{\kappa,\vartheta}(X^{\varsigma})}, \quad \Big(precos^{\gamma}_{\kappa,\vartheta}(X^{\varsigma}) \neq 0\Big). \end{aligned}$$

Theorem 6.6 ([4]) *Let* $\kappa, \vartheta, \gamma, \varsigma, X \in \mathbb{C}$, *and* $\Re(\kappa), \Re(\vartheta) > 0$. *Then,*

$$\mathbb{E}^{\gamma}_{\kappa,\vartheta}(iX^{\varsigma}) = precos^{\gamma}_{\kappa,\vartheta}(X^{\varsigma}) + i\, presin^{\gamma}_{\kappa,\vartheta}(X^{\varsigma}),$$

$$\mathbb{E}^{\gamma}_{\kappa,\vartheta}(-iX^{\varsigma}) = precos^{\gamma}_{\kappa,\vartheta}(X^{\varsigma}) - i\, presin^{\gamma}_{\kappa,\vartheta}(X^{\varsigma}).$$

6.7 The Pre-superhyperbolic Functions with the Power Law via the Prabhakar Function

Let $\kappa, \vartheta, \gamma, \zeta, X \in \mathbb{C}$, and $\Re(\kappa), \Re(\vartheta) > 0$. In this section, we give the pre-superhyperbolic functions with the power law defined via the Prabhakar function [4], as follows:

▷ The hyperbolic presine with the power law through the Prabhakar function is given by

$$\begin{aligned} & presinh^{\gamma}_{\kappa,\vartheta}(X^{\zeta}) \\ &= \frac{1}{2}\Big(\mathbb{E}^{\gamma}_{\kappa,\vartheta}(X^{\zeta}) - \mathbb{E}^{\gamma}_{\kappa,\vartheta}(-X^{\zeta})\Big) \\ &= \frac{1}{\Gamma(\gamma)} {}_1\sinh_1\left(\begin{smallmatrix}(\gamma,1)\\(\vartheta,\kappa)\end{smallmatrix}; X^{\zeta}\right) \\ &= \sum_{k=0}^{\infty} \frac{(\gamma)_{2k+1} X^{(2k+1)\zeta}}{\Gamma((2k+1)\kappa+\vartheta)}. \end{aligned}$$

▷ The hyperbolic precosine with the power law through the Prabhakar function is given by

$$\begin{aligned} & precosh^{\gamma}_{\kappa,\vartheta}(X^{\zeta}) \\ &= \frac{1}{2}\Big(\mathbb{E}^{\gamma}_{\kappa,\vartheta}(X^{\zeta}) + \mathbb{E}^{\gamma}_{\kappa,\vartheta}(-X^{\zeta})\Big) \\ &= \frac{1}{\Gamma(\gamma)} {}_1\cosh_1\left(\begin{smallmatrix}(\gamma,1)\\(\vartheta,\kappa)\end{smallmatrix}; X^{\zeta}\right) \\ &= \sum_{k=0}^{\infty} \frac{(\gamma)_{2k} X^{(2k)\zeta}}{\Gamma((2k)\kappa+\vartheta)}. \end{aligned}$$

▷ The hyperbolic pretangent with the power law through the Prabhakar function is given by

$$\begin{aligned} & pretanh^{\gamma}_{\kappa,\vartheta}(X^{\zeta}) \\ &= \frac{presinh^{\gamma}_{\kappa,\vartheta}(X^{\zeta})}{precosh^{\gamma}_{\kappa,\vartheta}(X^{\zeta})}, \quad \Big(precosh^{\gamma}_{\kappa,\vartheta}(X^{\zeta}) \neq 0\Big). \end{aligned}$$

▷ The hyperbolic precotangent with the power law through the Prabhakar function is given by

$$\begin{aligned} & precotanh^{\gamma}_{\kappa,\vartheta}(X^{\zeta}) \\ &= \frac{precosh^{\gamma}_{\kappa,\vartheta}(X^{\zeta})}{presinh^{\gamma}_{\kappa,\vartheta}(X^{\zeta})}, \quad \Big(presinh^{\gamma}_{\kappa,\vartheta}(X^{\zeta}) \neq 0\Big). \end{aligned}$$

▷ The hyperbolic precosecant with the power law through the Prabhakar function is given by

$$precsch^{\gamma}_{\kappa,\vartheta}(X^{\varsigma}) = \frac{1}{presinh^{\gamma}_{\kappa,\vartheta}(X^{\varsigma})}, \quad \Big(presinh^{\gamma}_{\kappa,\vartheta}(X^{\varsigma}) \neq 0\Big).$$

▷ The hyperbolic presecant with the power law through the Prabhakar function is given by

$$presech^{\gamma}_{\kappa,\vartheta}(X^{\varsigma}) = \frac{1}{precosh^{\gamma}_{\kappa,\vartheta}(X^{\varsigma})}, \quad \Big(precosh^{\gamma}_{\kappa,\vartheta}(X^{\varsigma}) \neq 0\Big).$$

Theorem 6.7 ([4]) *Let $\kappa, \vartheta, \gamma, \varsigma, X \in \mathbb{C}$, and $\Re(\kappa), \Re(\vartheta) > 0$. Then,*

$$\mathbb{E}^{\gamma}_{\kappa,\vartheta}(X^{\varsigma}) = precosh^{\gamma}_{\kappa,\vartheta}(X^{\varsigma}) + presinh^{\gamma}_{\kappa,\vartheta}(X^{\varsigma}),$$

and

$$\mathbb{E}^{\gamma}_{\kappa,\vartheta}(-X^{\varsigma}) = precosh^{\gamma}_{\kappa,\vartheta}(X^{\varsigma}) - presinh^{\gamma}_{\kappa,\vartheta}(X^{\varsigma}).$$

6.8 The Pre-supertrigonometric Functions with Parameter via the Prabhakar Function

Let $\kappa, \vartheta, \gamma, X \in \mathbb{C}$, and $\Re(\kappa), \Re(\vartheta) > 0$. In this section, we present the pre-supertrigonometric functions with parameter $X^{\vartheta-1}$ via the Prabhakar function [4], as follows:

▷ The presine with parameter $X^{\vartheta-1}$ through the Prabhakar function is given by

$$\begin{aligned} Tresin^{\gamma}_{\kappa,\vartheta}(X) &= \frac{X^{\vartheta-1}}{2i}\Big(\mathbb{E}^{\gamma}_{\kappa,\vartheta}(iX) - \mathbb{E}^{\gamma}_{\kappa,\vartheta}(-iX)\Big) \\ &= \frac{X^{\vartheta-1}}{\Gamma(\gamma)}\, {}_1\sin_1\left({}^{(\gamma,1)}_{(\vartheta,\kappa)}; X\right) \\ &= X^{\vartheta-1}\sum_{k=0}^{\infty}\frac{(-1)^k(\gamma)_{2k+1}X^{(2k+1)}}{\Gamma((2k+1)\kappa+\vartheta)}. \end{aligned}$$

▷ The precosine with parameter $X^{\vartheta-1}$ through the Prabhakar function is given by

$$\begin{aligned}
&Trecos^{\gamma}_{\kappa,\vartheta}(X)\\
&=\frac{X^{\vartheta-1}}{2}\Big(\mathbb{E}^{\gamma}_{\kappa,\vartheta}(iX)+\mathbb{E}^{\gamma}_{\kappa,\vartheta}(-iX)\Big)\\
&=\frac{X^{\vartheta-1}}{\Gamma(\gamma)}{}_1\cos_1\left({}^{(\gamma,1)}_{(\vartheta,\kappa)};X\right)\\
&=X^{\vartheta-1}\sum_{k=0}^{\infty}\frac{(-1)^k(\gamma)_{2k}X^{(2k)}}{\Gamma((2k)\kappa+\vartheta)}.
\end{aligned}$$

▷ The pretangent through the Prabhakar function is given by

$$\begin{aligned}
&Tretan^{\gamma}_{\kappa,\vartheta}(X)\\
&=\frac{Tresin^{\gamma}_{\kappa,\vartheta}(X)}{Trecos^{\gamma}_{\kappa,\vartheta}(X)},\quad\Big(Trecos^{\gamma}_{\kappa,\vartheta}(X)\neq 0\Big).
\end{aligned}$$

▷ The precotangent with parameter through the Prabhakar function is given by

$$\begin{aligned}
&Trecotan^{\gamma}_{\kappa,\vartheta}(X)\\
&=\frac{Trecos^{\gamma}_{\kappa,\vartheta}(X)}{Tresin^{\gamma}_{\kappa,\vartheta}(X)},\quad\Big(Tresin^{\gamma}_{\kappa,\vartheta}(X)\neq 0\Big).
\end{aligned}$$

▷ The precosecant with parameter through the Prabhakar function is given by

$$\begin{aligned}
&Trecsc^{\gamma}_{\kappa,\vartheta}(X)\\
&=\frac{1}{Tresin^{\gamma}_{\kappa,\vartheta}(X)},\quad\Big(Tresin^{\gamma}_{\kappa,\vartheta}(X)\neq 0\Big).
\end{aligned}$$

▷ The presecant with parameter through the Prabhakar function is given by

$$\begin{aligned}
&Tresec^{\gamma}_{\kappa,\vartheta}(X)\\
&=\frac{1}{Trecos^{\gamma}_{\kappa,\vartheta}(X)},\quad\Big(Trecos^{\gamma}_{\kappa,\vartheta}(X)\neq 0\Big).
\end{aligned}$$

Using the above results, we may deduce the following theorem.

Theorem 6.8 ([4]) *Let* $\kappa, \vartheta, \gamma, X \in \mathbb{C}$, *and* $\Re(\kappa), \Re(\vartheta) > 0$. *Then,*

$$X^{\vartheta-1}\mathbb{E}^{\gamma}_{\kappa,\vartheta}(iX)=Trecos^{\gamma}_{\kappa,\vartheta}(X)+i\,Tresin^{\gamma}_{\kappa,\vartheta}(X),$$

and

$$X^{\vartheta-1}\mathbb{E}^{\gamma}_{\kappa,\vartheta}(-iX)=Trecos^{\gamma}_{\kappa,\vartheta}(X)-i\,Tresin^{\gamma}_{\kappa,\vartheta}(X).$$

6.9 The Pre-superhyperbolic Functions with Parameter via the Prabhakar Function

Let $\kappa, \vartheta, \gamma, X \in \mathbb{C}$, and $\Re(\kappa), \Re(\vartheta) > 0$. In this section, we give the pre-superhyperbolic functions with parameter via the Prabhakar function [4], as follows:

▷ The hyperbolic presine with parameter through the Prabhakar function is given by

$$\begin{aligned} &Tresinh^{\gamma}_{\kappa,\vartheta}(X) \\ &= \frac{X^{\vartheta-1}}{2}\Big(\mathbb{E}^{\gamma}_{\kappa,\vartheta}(X) - \mathbb{E}^{\gamma}_{\kappa,\vartheta}(-X)\Big) \\ &= \frac{X^{\vartheta-1}}{\Gamma(\gamma)}\,{}_1\sinh_1\left(\begin{smallmatrix}(\gamma,1)\\(\vartheta,\kappa)\end{smallmatrix}; X\right) \\ &= X^{\vartheta-1}\sum_{k=0}^{\infty}\frac{(\gamma)_{2k+1}X^{(2k+1)}}{\Gamma((2k+1)\kappa+\vartheta)}. \end{aligned}$$

▷ The hyperbolic precosine with parameter through the Prabhakar function is given by

$$\begin{aligned} &Trecosh^{\gamma}_{\kappa,\vartheta}(X) \\ &= \frac{X^{\vartheta-1}}{2}\Big(\mathbb{E}^{\gamma}_{\kappa,\vartheta}(X) + \mathbb{E}^{\gamma}_{\kappa,\vartheta}(-X)\Big) \\ &= \frac{X^{\vartheta-1}}{\Gamma(\gamma)}\,{}_1\cosh_1\left(\begin{smallmatrix}(\gamma,1)\\(\vartheta,\kappa)\end{smallmatrix}; X\right) \\ &= X^{\vartheta-1}\sum_{k=0}^{\infty}\frac{(\gamma)_{2k}X^{(2k)}}{\Gamma((2k)\kappa+\vartheta)}. \end{aligned}$$

▷ The hyperbolic pretangent with parameter through the Prabhakar function is given by

$$\begin{aligned} &Tretanh^{\gamma}_{\kappa,\vartheta}(X) \\ &= \frac{Tresinh^{\gamma}_{\kappa,\vartheta}(X)}{Trecosh^{\gamma}_{\kappa,\vartheta}(X)}, \quad \Big(Trecosh^{\gamma}_{\kappa,\vartheta}(X) \neq 0\Big). \end{aligned}$$

▷ The hyperbolic precotangent with parameter through the Prabhakar function is given by

$$\begin{aligned} &Trecotanh^{\gamma}_{\kappa,\vartheta}(X) \\ &= \frac{Trecosh^{\gamma}_{\kappa,\vartheta}(X)}{Tresinh^{\gamma}_{\kappa,\vartheta}(X)}, \quad \Big(Tresinh^{\gamma}_{\kappa,\vartheta}(X) \neq 0\Big). \end{aligned}$$

▷ The hyperbolic precosecant with parameter through the Prabhakar function is given by

$$Trecsch^{\gamma}_{\kappa,\vartheta}(X) = \frac{1}{Tresinh^{\gamma}_{\kappa,\vartheta}(X)}, \quad \left(Tresinh^{\gamma}_{\kappa,\vartheta}(X) \neq 0\right).$$

▷ The hyperbolic presecant with parameter through the Prabhakar function is given by

$$Tresech^{\gamma}_{\kappa,\vartheta}(X) = \frac{1}{Trecosh^{\gamma}_{\kappa,\vartheta}(X)}, \quad \left(Trecosh^{\gamma}_{\kappa,\vartheta}(X) \neq 0\right).$$

With the aid of the above results, we may have the following theorem.

Theorem 6.9 ([4]) *Let $\kappa, \vartheta, \gamma, X \in \mathbb{C}$, and $\Re(\kappa), \Re(\vartheta) > 0$. Then,*

$$X^{\vartheta-1}\mathbb{E}^{\gamma}_{\kappa,\vartheta}(X) = Trecosh^{\gamma}_{\kappa,\vartheta}(X) + Tresinh^{\gamma}_{\kappa,\vartheta}(X),$$

and

$$X^{\vartheta-1}\mathbb{E}^{\gamma}_{\kappa,\vartheta}(-X) = Trecosh^{\gamma}_{\kappa,\vartheta}(X) - Tresinh^{\gamma}_{\kappa,\vartheta}(X).$$

6.10 The Pre-supertrigonometric Functions with the Power Law and Parameter via the Prabhakar Function

Let $\kappa, \vartheta, \gamma, \zeta, X \in \mathbb{C}$, and $\Re(\kappa), \Re(\vartheta) > 0$. In this section, we present the pre-supertrigonometric functions with the power law and parameter via the Prabhakar function [4], as follows:

▷ The presupersine with the power law and parameter through the Prabhakar function is given by

$$\begin{aligned} Gresin^{\gamma}_{\kappa,\vartheta}(X^{\zeta}) &= \frac{X^{\vartheta-1}}{2i}\left(\mathbb{E}^{\gamma}_{\kappa,\vartheta}(iX^{\zeta}) - \mathbb{E}^{\gamma}_{\kappa,\vartheta}(-iX^{\zeta})\right) \\ &= \frac{X^{\vartheta-1}}{\Gamma(\gamma)}\, {}_1\sinh_1\left(\begin{smallmatrix}(\gamma,1)\\(\vartheta,\kappa)\end{smallmatrix}; X^{\zeta}\right) \\ &= X^{\vartheta-1}\sum_{k=0}^{\infty}\frac{(-1)^k(\gamma)_{2k+1}X^{(2k+1)\zeta}}{\Gamma((2k+1)\kappa+\vartheta)}. \end{aligned}$$

▷ The precosine with the power law and parameter through the Prabhakar function is given by

$$
\begin{aligned}
& Grecos^{\gamma}_{\kappa,\vartheta}(X^{\varsigma}) \\
&= \frac{X^{\vartheta-1}}{2}\left(\mathbb{E}^{\gamma}_{\kappa,\vartheta}(iX^{\varsigma}) + \mathbb{E}^{\gamma}_{\kappa,\vartheta}(-iX^{\varsigma})\right) \\
&= \frac{X^{\vartheta-1}}{\Gamma(\gamma)}\,{}_1\cos_1\left({(\gamma,1) \atop (\vartheta,\kappa)}; X^{\varsigma}\right) \\
&= X^{\vartheta-1}\sum_{k=0}^{\infty}\frac{(-1)^k(\gamma)_{2k}X^{(2k)\varsigma}}{\Gamma((2k)\kappa+\vartheta)}.
\end{aligned}
$$

▷ The pretangent with the power law and parameter through the Prabhakar function is given by

$$
\begin{aligned}
& Gretan^{\gamma}_{\kappa,\vartheta}(X^{\varsigma}) \\
&= \frac{Gresin^{\gamma}_{\kappa,\vartheta}(X^{\varsigma})}{Grecos^{\gamma}_{\kappa,\vartheta}(X^{\varsigma})}, \quad \left(Grecos^{\gamma}_{\kappa,\vartheta}(X^{\varsigma}) \neq 0\right).
\end{aligned}
$$

▷ The pretangent with the power law and parameter through the Prabhakar function is given by

$$
\begin{aligned}
& Grecotan^{\gamma}_{\kappa,\vartheta}(X^{\varsigma}) \\
&= \frac{Grecos^{\gamma}_{\kappa,\vartheta}(X^{\varsigma})}{Gresin^{\gamma}_{\kappa,\vartheta}(X^{\varsigma})}, \quad \left(Gresin^{\gamma}_{\kappa,\vartheta}(X^{\varsigma}) \neq 0\right).
\end{aligned}
$$

▷ The precotangent with the power law and parameter through the Prabhakar function is given by

$$
\begin{aligned}
& Grecsc^{\gamma}_{\kappa,\vartheta}(X^{\varsigma}) \\
&= \frac{1}{Gresin^{\gamma}_{\kappa,\vartheta}(X^{\varsigma})}, \quad \left(Gresin^{\gamma}_{\kappa,\vartheta}(X^{\varsigma}) \neq 0\right).
\end{aligned}
$$

▷ The precosecant with the power law and parameter through the Prabhakar function is given by

$$
\begin{aligned}
& Gresec^{\gamma}_{\kappa,\vartheta}(X^{\varsigma}) \\
&= \frac{1}{Grecos^{\gamma}_{\kappa,\vartheta}(X^{\varsigma})}, \quad \left(Grecos^{\gamma}_{\kappa,\vartheta}(X^{\varsigma}) \neq 0\right).
\end{aligned}
$$

With the use of the above results, we may deduce the following theorem.

Theorem 6.10 ([4]) *Let* $\kappa, \vartheta, \gamma, \zeta, X \in \mathbb{C}$, *and* $\Re(\kappa), \Re(\vartheta) > 0$. *Then,*

$$X^{\vartheta-1}\mathbb{E}^{\gamma}_{\kappa,\vartheta}(iX^{\zeta}) = Grecos^{\gamma}_{\kappa,\vartheta}(X^{\zeta}) + i\, Gresin^{\gamma}_{\kappa,\vartheta}(X^{\zeta}),$$

and

$$X^{\vartheta-1}\mathbb{E}^{\gamma}_{\kappa,\vartheta}(-iX^{\zeta}) = Grecos^{\gamma}_{\kappa,\vartheta}(X^{\zeta}) - i\, Gresin^{\gamma}_{\kappa,\vartheta}(X^{\zeta}).$$

6.11 The Pre-superhyperbolic Functions with the Power Law and Parameter via the Prabhakar Function

Let $\kappa, \vartheta, \gamma, \zeta, X \in \mathbb{C}$, and $\Re(\kappa), \Re(\vartheta) > 0$. In this section, we define the pre-superhyperbolic functions with the power law and parameter via the Prabhakar function [4], as follows:

▷ The hyperbolic presine with the power law and parameter through the Prabhakar function is given by

$$\begin{aligned} &Gresinh^{\gamma}_{\kappa,\vartheta}(X^{\zeta}) \\ &= \frac{X^{\vartheta-1}}{2}\left(\mathbb{E}^{\gamma}_{\kappa,\vartheta}(X^{\zeta}) - \mathbb{E}^{\gamma}_{\kappa,\vartheta}(-X^{\zeta})\right) \\ &= \frac{X^{\vartheta-1}}{\Gamma(\gamma)}\, {}_1\sinh_1\left(\begin{smallmatrix}(\gamma,1)\\(\vartheta,\kappa)\end{smallmatrix}; X^{\zeta}\right) \\ &= X^{\vartheta-1}\sum_{k=0}^{\infty}\frac{(\gamma)_{2k+1}X^{(2k+1)\zeta}}{\Gamma((2k+1)\kappa+\vartheta)}. \end{aligned}$$

▷ The hyperbolic precosine with the power law and parameter through the Prabhakar function is given by

$$\begin{aligned} &Grecosh^{\gamma}_{\kappa,\vartheta}(X^{\zeta}) \\ &= \frac{X^{\vartheta-1}}{2}\left(\mathbb{E}^{\gamma}_{\kappa,\vartheta}(X^{\zeta}) + \mathbb{E}^{\gamma}_{\kappa,\vartheta}(-X^{\zeta})\right) \\ &= \frac{X^{\vartheta-1}}{\Gamma(\gamma)}\, {}_1\cosh_1\left(\begin{smallmatrix}(\gamma,1)\\(\vartheta,\kappa)\end{smallmatrix}; X^{\zeta}\right) \\ &= X^{\vartheta-1}\sum_{k=0}^{\infty}\frac{(\gamma)_{2k}X^{(2k)\zeta}}{\Gamma((2k)\kappa+\vartheta)}. \end{aligned}$$

▷ The hyperbolic pretangent with the power law and parameter through the Prabhakar function is given by

$$\begin{aligned} &Gretanh^{\gamma}_{\kappa,\vartheta}(X^{\varsigma}) \\ &= \frac{Gresinh^{\gamma}_{\kappa,\vartheta}(X^{\varsigma})}{Grecosh^{\gamma}_{\kappa,\vartheta}(X^{\varsigma})}, \quad \left(Grecosh^{\gamma}_{\kappa,\vartheta}(X^{\varsigma}) \neq 0\right). \end{aligned}$$

▷ The hyperbolic precotangentwith the power law and parameter through the Prabhakar function is given by

$$\begin{aligned} &Grecotanh^{\gamma}_{\kappa,\vartheta}(X^{\varsigma}) \\ &= \frac{Grecosh^{\gamma}_{\kappa,\vartheta}(X^{\varsigma})}{Gresinh^{\gamma}_{\kappa,\vartheta}(X^{\varsigma})}, \quad \left(Gresinh^{\gamma}_{\kappa,\vartheta}(X^{\varsigma}) \neq 0\right). \end{aligned}$$

▷ The hyperbolic precosecant with the power law and parameter through the Prabhakar function is given by

$$\begin{aligned} &Grecsch^{\gamma}_{\kappa,\vartheta}(X^{\varsigma}) \\ &= \frac{1}{Gresinh^{\gamma}_{\kappa,\vartheta}(X^{\varsigma})}, \quad \left(Gresinh^{\gamma}_{\kappa,\vartheta}(X^{\varsigma}) \neq 0\right). \end{aligned}$$

▷ The hyperbolic presecant with the power law and parameter through the Prabhakar function is given by

$$\begin{aligned} &Gresech^{\gamma}_{\kappa,\vartheta}(X^{\varsigma}) \\ &= \frac{1}{Grecosh^{\gamma}_{\kappa,\vartheta}(X^{\varsigma})}, \quad \left(Grecosh^{\gamma}_{\kappa,\vartheta}(X^{\varsigma}) \neq 0\right). \end{aligned}$$

Using these results, we may deduce the following theorem.

Theorem 6.11 ([4]) *Let* $\kappa, \vartheta, \zeta, X \in \mathbb{C}$, *and* $\Re(\kappa), \Re(\vartheta) > 0$. *Then,*

$$X^{\vartheta-1}\mathbb{E}^{\gamma}_{\kappa,\vartheta}(X^{\varsigma}) = Grecosh^{\gamma}_{\kappa,\vartheta}(X^{\varsigma}) + Gresinh^{\gamma}_{\kappa,\vartheta}(X^{\varsigma}),$$

and

$$X^{\vartheta-1}\mathbb{E}^{\gamma}_{\kappa,\vartheta}(-X^{\varsigma}) = Grecosh^{\gamma}_{\kappa,\vartheta}(X^{\varsigma}) - Gresinh^{\gamma}_{\kappa,\vartheta}(X^{\varsigma}).$$

6.12 The Truncated Prabhakar Functions and Related Functions

In this section, we present the truncated Prabhakar functions and some functions related to the Prabhakar function.

Definition 6.2 ([4]) The truncated Prabhakar function is given by

$$\left\{\mathbb{E}^{\gamma}_{\kappa,\vartheta}\right\}^{n}(X) = \sum_{k=0}^{n} \frac{(\gamma)_k}{\Gamma(k\kappa+\vartheta)} \frac{X^k}{k!},$$

where $\kappa, \vartheta, \gamma, X \in \mathbb{C}$, and $\Re(\kappa), \Re(\vartheta) > 0$.

Let $\kappa, \vartheta, \gamma, X \in \mathbb{C}$, and $\Re(\kappa), \Re(\vartheta) > 0$. We have the following results [4]:

▷ The truncated supersine through the Prabhakar function is given by

$$\begin{aligned}
&\left\{supersin^{\gamma}_{\kappa,\vartheta}\right\}^{n}(X^{\kappa}) \\
&= \frac{X^{\vartheta-1}}{2i}\left(\left\{\mathbb{E}^{\gamma}_{\kappa,\vartheta}\right\}^{n}(iX^{\kappa}) - \left\{\mathbb{E}^{\gamma}_{\kappa,\vartheta}\right\}^{n}(-iX^{\kappa})\right) \\
&= \sum_{k=0}^{n} \frac{(-1)^k(\gamma)_{2k+1}}{\Gamma((2k+1)\kappa+\vartheta)} \frac{X^{(2k+1)\kappa+\vartheta-1}}{(2k+1)!}.
\end{aligned}$$

Thus,

$$\begin{aligned}
&\left\{supersin^{\gamma}_{\kappa,\vartheta}\right\}^{0}(X^{\kappa}) \\
&= \frac{X^{\vartheta-1}}{2}\left(\left\{\mathbb{E}^{\gamma}_{\kappa,\vartheta}\right\}^{0}(iX^{\kappa}) - \left\{\mathbb{E}^{\gamma}_{\kappa,\vartheta}\right\}^{0}(-iX^{\kappa})\right) \\
&= \frac{\Gamma(\gamma+1)}{\Gamma(\kappa+\vartheta)} X^{\kappa+\vartheta-1},
\end{aligned}$$

$$\begin{aligned}
&\left\{supersin^{\gamma}_{\kappa,\vartheta}\right\}^{1}(X^{\kappa}) \\
&= \frac{X^{\vartheta-1}}{2}\left(\left\{\mathbb{E}^{\gamma}_{\kappa,\vartheta}\right\}^{1}(iX^{\kappa}) - \left\{\mathbb{E}^{\gamma}_{\kappa,\vartheta}\right\}^{1}(-iX^{\kappa})\right) \\
&= \sum_{k=0}^{1} \frac{(-1)^k(\gamma)_{2k+1}}{\Gamma((2k+1)\kappa+\vartheta)} \frac{X^{(2k+1)\kappa+\vartheta-1}}{(2k+1)!},
\end{aligned}$$

$$
\begin{aligned}
&\left\{supersin^{\gamma}_{\kappa,\vartheta}\right\}^{2}(X^{\kappa})\\
&=\frac{X^{\vartheta-1}}{2}\left(\left\{\mathbb{E}^{\gamma}_{\kappa,\vartheta}\right\}^{2}(iX^{\kappa})-\left\{\mathbb{E}^{\gamma}_{\kappa,\vartheta}\right\}^{2}(-iX^{\kappa})\right)\\
&=\sum_{k=0}^{2}\frac{(-1)^{k}(\gamma)_{2k+1}}{\Gamma((2k+1)\kappa+\vartheta)}\frac{X^{(2k+1)\kappa+\vartheta-1}}{(2k+1)!}.
\end{aligned}
$$

▷ The truncated supercosine through the Prabhakar function is given by

$$
\begin{aligned}
&\left\{supercos^{\gamma}_{\kappa,\vartheta}\right\}^{n}(X^{\kappa})\\
&=\frac{X^{\vartheta-1}}{2}\left(\left\{\mathbb{E}^{\gamma}_{\kappa,\vartheta}\right\}^{n}(iX^{\kappa})+\left\{\mathbb{E}^{\gamma}_{\kappa,\vartheta}\right\}^{n}(-iX^{\kappa})\right)\\
&=\sum_{k=0}^{n}\frac{(-1)^{k}(\gamma)_{2k}}{\Gamma((2k)\kappa+\vartheta)}\frac{X^{(2k)\kappa+\vartheta-1}}{(2k)!}.
\end{aligned}
$$

Thus,

$$
\begin{aligned}
&\left\{supercos^{\gamma}_{\kappa,\vartheta}\right\}^{0}(X^{\kappa})\\
&=\frac{X^{\vartheta-1}}{2}\left(\left\{\mathbb{E}^{\gamma}_{\kappa,\vartheta}\right\}^{0}(iX^{\kappa})+\left\{\mathbb{E}^{\gamma}_{\kappa,\vartheta}\right\}^{0}(-iX^{\kappa})\right)\\
&=\frac{X^{\vartheta-1}}{\Gamma(\vartheta)},
\end{aligned}
$$

$$
\begin{aligned}
&\left\{supercos^{\gamma}_{\kappa,\vartheta}\right\}^{1}(X^{\kappa})\\
&=\frac{X^{\vartheta-1}}{2}\left(\left\{\mathbb{E}^{\gamma}_{\kappa,\vartheta}\right\}^{1}(iX^{\kappa})+\left\{\mathbb{E}^{\gamma}_{\kappa,\vartheta}\right\}^{1}(-iX^{\kappa})\right)\\
&=\sum_{k=0}^{1}\frac{(-1)^{k}(\gamma)_{2k}}{\Gamma((2k)\kappa+\vartheta)}\frac{X^{(2k)\kappa+\vartheta-1}}{(2k)!},
\end{aligned}
$$

$$
\begin{aligned}
&\left\{supercos_{\kappa,\vartheta}^{\gamma}\right\}^{2}(X^{\kappa})\\
&=\frac{X^{\vartheta-1}}{2}\left(\left\{\mathbb{E}_{\kappa,\vartheta}^{\gamma}\right\}^{2}(iX^{\kappa})+\left\{\mathbb{E}_{\kappa,\vartheta}^{\gamma}\right\}^{2}(-iX^{\kappa})\right)\\
&=\sum_{k=0}^{2}\frac{(-1)^{k}(\gamma)_{2k}}{\Gamma((2k)\kappa+\vartheta)}\frac{X^{(2k)\kappa+\vartheta-1}}{(2k)!}.
\end{aligned}
$$

▷ The truncated hyperbolic supersine through the Prabhakar function is given by

$$
\begin{aligned}
&\left\{supersinh_{\kappa,\vartheta}^{\gamma}\right\}^{n}(X^{\kappa})\\
&=\frac{X^{\vartheta-1}}{2}\left(\left\{\mathbb{E}_{\kappa,\vartheta}^{\gamma}\right\}^{n}(X^{\kappa})-\left\{\mathbb{E}_{\kappa,\vartheta}^{\gamma}\right\}^{n}(-X^{\kappa})\right)\\
&=\sum_{k=0}^{n}\frac{(\gamma)_{2k+1}X^{(2k+1)\kappa+\vartheta-1}}{\Gamma((2k+1)\kappa+\vartheta)}.
\end{aligned}
$$

Thus,

$$
\begin{aligned}
&\left\{supersinh_{\kappa,\vartheta}^{\gamma}\right\}^{0}(X^{\kappa})\\
&=\frac{X^{\vartheta-1}}{2}\left(\left\{\mathbb{E}_{\kappa,\vartheta}^{\gamma}\right\}^{0}(X^{\kappa})-\left\{\mathbb{E}_{\kappa,\vartheta}^{\gamma}\right\}^{0}(-X^{\kappa})\right)\\
&=\frac{\Gamma(\gamma+1)}{\Gamma(\kappa+\vartheta)}X^{\kappa+\vartheta-1},
\end{aligned}
$$

$$
\begin{aligned}
&\left\{supersinh_{\kappa,\vartheta}^{\gamma}\right\}^{1}(X^{\kappa})\\
&=\frac{X^{\vartheta-1}}{2}\left(\left\{\mathbb{E}_{\kappa,\vartheta}^{\gamma}\right\}^{1}(X^{\kappa})-\left\{\mathbb{E}_{\kappa,\vartheta}^{\gamma}\right\}^{1}(-X^{\kappa})\right)\\
&=\sum_{k=0}^{1}\frac{(\gamma)_{2k+1}X^{(2k+1)\kappa+\vartheta-1}}{\Gamma((2k+1)\kappa+\vartheta)},
\end{aligned}
$$

$$\begin{aligned}
&\left\{supersinh^{\gamma}_{\kappa,\vartheta}\right\}^{2}(X^{\kappa})\\
&=\frac{X^{\vartheta-1}}{2}\left(\left\{\mathbb{E}^{\gamma}_{\kappa,\vartheta}\right\}^{2}(X^{\kappa})-\left\{\mathbb{E}^{\gamma}_{\kappa,\vartheta}\right\}^{2}(-X^{\kappa})\right)\\
&=\sum_{k=0}^{2}\frac{(\gamma)_{2k+1}X^{(2k+1)\kappa+\vartheta-1}}{\Gamma((2k+1)\kappa+\vartheta)}.
\end{aligned}$$

▷ The truncated hyperbolic supercosine through the Prabhakar function is given by

$$\begin{aligned}
&\left\{supercosh^{\gamma}_{\kappa,\vartheta}\right\}^{n}(X^{\kappa})\\
&=\frac{X^{\vartheta-1}}{2}\left(\left\{\mathbb{E}^{\gamma}_{\kappa,\vartheta}\right\}^{n}(X^{\kappa})+\left\{\mathbb{E}^{\gamma}_{\kappa,\vartheta}\right\}^{n}(-X^{\kappa})\right)\\
&=\sum_{k=0}^{n}\frac{(\gamma)_{2k}X^{(2k)\kappa+\vartheta-1}}{\Gamma((2k)\kappa+\vartheta)}.
\end{aligned}$$

Thus,

$$\begin{aligned}
&\left\{supercosh^{\gamma}_{\kappa,\vartheta}\right\}^{0}(X^{\kappa})\\
&=\frac{X^{\vartheta-1}}{2}\left(\left\{\mathbb{E}^{\gamma}_{\kappa,\vartheta}\right\}^{0}(X^{\kappa})+\left\{\mathbb{E}^{\gamma}_{\kappa,\vartheta}\right\}^{0}(-X^{\kappa})\right)\\
&=\frac{X^{\vartheta-1}}{\Gamma(\vartheta)},
\end{aligned}$$

$$\begin{aligned}
&\left\{supercosh^{\gamma}_{\kappa,\vartheta}\right\}^{1}(X^{\kappa})\\
&=\frac{X^{\vartheta-1}}{2}\left(\left\{\mathbb{E}^{\gamma}_{\kappa,\vartheta}\right\}^{1}(X^{\kappa})+\left\{\mathbb{E}^{\gamma}_{\kappa,\vartheta}\right\}^{1}(-X^{\kappa})\right)\\
&=\sum_{k=0}^{1}\frac{(\gamma)_{2k}X^{(2k)\kappa+\vartheta-1}}{\Gamma((2k)\kappa+\vartheta)},
\end{aligned}$$

$$
\begin{aligned}
&\left\{supercosh_{\kappa,\vartheta}^{\gamma}\right\}^{2}(X^{\kappa})\\
&=\frac{X^{\vartheta-1}}{2}\left(\left\{\mathbb{E}_{\kappa,\vartheta}^{\gamma}\right\}^{2}(X^{\kappa})+\left\{\mathbb{E}_{\kappa,\vartheta}^{\gamma}\right\}^{2}(-X^{\kappa})\right)\\
&=\sum_{k=0}^{2}\frac{(\gamma)_{2k}X^{(2k)\kappa+\vartheta-1}}{\Gamma((2k)\kappa+\vartheta)}.
\end{aligned}
$$

▷ The truncated presine through the Prabhakar function is given by

$$
\begin{aligned}
&\left\{presin_{\kappa,\vartheta}^{\gamma}\right\}^{n}(X)\\
&=\frac{1}{2i}\left(\left\{\mathbb{E}_{\kappa,\vartheta}^{\gamma}\right\}^{n}(iX)-\left\{\mathbb{E}_{\kappa,\vartheta}^{\gamma}\right\}^{n}(-iX)\right)\\
&=\sum_{k=0}^{n}\frac{(-1)^{k}(\gamma)_{2k+1}X^{(2k+1)}}{\Gamma((2k+1)\kappa+\vartheta)}.
\end{aligned}
$$

Thus,

$$
\begin{aligned}
&\left\{presin_{\kappa,\vartheta}^{\gamma}\right\}^{0}(X)\\
&=\frac{1}{2}\left(\left\{\mathbb{E}_{\kappa,\vartheta}^{\gamma}\right\}^{0}(iX)-\left\{\mathbb{E}_{\kappa,\vartheta}^{\gamma}\right\}^{0}(-iX)\right)\\
&=\frac{\Gamma(\gamma+1)}{\Gamma(\kappa+\vartheta)}X,
\end{aligned}
$$

$$
\begin{aligned}
&\left\{presin_{\kappa,\vartheta}^{\gamma}\right\}^{1}(X)\\
&=\frac{1}{2}\left(\left\{\mathbb{E}_{\kappa,\vartheta}^{\gamma}\right\}^{1}(iX)-\left\{\mathbb{E}_{\kappa,\vartheta}^{\gamma}\right\}^{1}(-iX)\right)\\
&=\sum_{k=0}^{1}\frac{(-1)^{k}(\gamma)_{2k+1}X^{(2k+1)}}{\Gamma((2k+1)\kappa+\vartheta)},
\end{aligned}
$$

$$
\begin{aligned}
&\left\{presin^{\gamma}_{\kappa,\vartheta}\right\}^{2}(X)\\
&=\frac{1}{2}\left(\left\{\mathbb{E}^{\gamma}_{\kappa,\vartheta}\right\}^{2}(iX)-\left\{\mathbb{E}^{\gamma}_{\kappa,\vartheta}\right\}^{2}(-iX)\right)\\
&=\sum_{k=0}^{2}\frac{(-1)^{k}(\gamma)_{2k+1}X^{(2k+1)}}{\Gamma((2k+1)\kappa+\vartheta)}.
\end{aligned}
$$

▷ The truncated precosine through the Prabhakar function is given by

$$
\begin{aligned}
&\left\{precos^{\gamma}_{\kappa,\vartheta}\right\}^{n}(X)\\
&=\frac{1}{2}\left(\left\{\mathbb{E}^{\gamma}_{\kappa,\vartheta}\right\}^{n}(iX)+\left\{\mathbb{E}^{\gamma}_{\kappa,\vartheta}\right\}^{n}(-iX)\right)\\
&=\sum_{k=0}^{n}\frac{(-1)^{k}(\gamma)_{2k}X^{(2k)}}{\Gamma((2k)\kappa+\vartheta)}.
\end{aligned}
$$

Thus,

$$
\begin{aligned}
&\left\{precos^{\gamma}_{\kappa,\vartheta}\right\}^{0}(X)\\
&=\frac{1}{2}\left(\left\{\mathbb{E}^{\gamma}_{\kappa,\vartheta}\right\}^{0}(iX)+\left\{\mathbb{E}^{\gamma}_{\kappa,\vartheta}\right\}^{0}(-iX)\right)\\
&=\frac{1}{\Gamma(\vartheta)},
\end{aligned}
$$

$$
\begin{aligned}
&\left\{precos^{\gamma}_{\kappa,\vartheta}\right\}^{1}(X)\\
&=\frac{1}{2}\left(\left\{\mathbb{E}^{\gamma}_{\kappa,\vartheta}\right\}^{1}(iX)+\left\{\mathbb{E}^{\gamma}_{\kappa,\vartheta}\right\}^{1}(-iX)\right)\\
&=\sum_{k=0}^{1}\frac{(-1)^{k}(\gamma)_{2k}X^{(2k)}}{\Gamma((2k)\kappa+\vartheta)},
\end{aligned}
$$

$$
\begin{aligned}
&\left\{precos^{\gamma}_{\kappa,\vartheta}\right\}^{2}(X)\\
&=\frac{1}{2}\left(\left\{\mathbb{E}^{\gamma}_{\kappa,\vartheta}\right\}^{2}(iX)+\left\{\mathbb{E}^{\gamma}_{\kappa,\vartheta}\right\}^{2}(-iX)\right)\\
&=\sum_{k=0}^{2}\frac{(-1)^{k}(\gamma)_{2k}X^{(2k)}}{\Gamma((2k)\kappa+\vartheta)}.
\end{aligned}
$$

▷ The truncated hyperbolic presine through the Prabhakar function is given by

$$
\begin{aligned}
&\left\{presinh^{\gamma}_{\kappa,\vartheta}\right\}^{n}(X)\\
&=\frac{1}{2}\left(\left\{\mathbb{E}^{\gamma}_{\kappa,\vartheta}\right\}^{n}(X)-\left\{\mathbb{E}^{\gamma}_{\kappa,\vartheta}\right\}^{n}(-X)\right)\\
&=\sum_{k=0}^{n}\frac{(\gamma)_{2k+1}X^{(2k+1)}}{\Gamma((2k+1)\kappa+\vartheta)}.
\end{aligned}
$$

Thus,

$$
\begin{aligned}
&\left\{presinh^{\gamma}_{\kappa,\vartheta}\right\}^{0}(X)\\
&=\frac{1}{2}\left(\left\{\mathbb{E}^{\gamma}_{\kappa,\vartheta}\right\}^{0}(X)-\left\{\mathbb{E}^{\gamma}_{\kappa,\vartheta}\right\}^{0}(-X)\right)\\
&=\frac{\Gamma(\gamma+1)}{\Gamma(\kappa+\vartheta)}X,
\end{aligned}
$$

$$
\begin{aligned}
&\left\{presinh^{\gamma}_{\kappa,\vartheta}\right\}^{1}(X)\\
&=\frac{1}{2}\left(\left\{\mathbb{E}^{\gamma}_{\kappa,\vartheta}\right\}^{1}(X)-\left\{\mathbb{E}^{\gamma}_{\kappa,\vartheta}\right\}^{1}(-X)\right)\\
&=\sum_{k=0}^{1}\frac{(\gamma)_{2k+1}X^{(2k+1)}}{\Gamma((2k+1)\kappa+\vartheta)},
\end{aligned}
$$

$$
\begin{aligned}
&\left\{presinh^{\gamma}_{\kappa,\vartheta}\right\}^{2}(X)\\
&=\frac{1}{2}\left(\left\{\mathbb{E}^{\gamma}_{\kappa,\vartheta}\right\}^{2}(X)-\left\{\mathbb{E}^{\gamma}_{\kappa,\vartheta}\right\}^{2}(-X)\right)\\
&=\sum_{k=0}^{2}\frac{(\gamma)_{2k+1}X^{(2k+1)}}{\Gamma((2k+1)\kappa+\vartheta)}.
\end{aligned}
$$

▷ The truncated hyperbolic precosine through the Prabhakar function is given by

$$
\begin{aligned}
&\left\{precosh^{\gamma}_{\kappa,\vartheta}\right\}^{n}(X)\\
&=\frac{1}{2}\left(\left\{\mathbb{E}^{\gamma}_{\kappa,\vartheta}\right\}^{n}(X)+\left\{\mathbb{E}^{\gamma}_{\kappa,\vartheta}\right\}^{n}(-X)\right)\\
&=\sum_{k=0}^{n}\frac{(\gamma)_{2k}X^{(2k)}}{\Gamma((2k)\kappa+\vartheta)}.
\end{aligned}
$$

Thus,

$$
\begin{aligned}
&\left\{precosh^{\gamma}_{\kappa,\vartheta}\right\}^{0}(X)\\
&=\frac{1}{2}\left(\left\{\mathbb{E}^{\gamma}_{\kappa,\vartheta}\right\}^{0}(X)+\left\{\mathbb{E}^{\gamma}_{\kappa,\vartheta}\right\}^{0}(-X)\right)\\
&=\frac{1}{\Gamma(\vartheta)},
\end{aligned}
$$

$$
\begin{aligned}
&\left\{precosh^{\gamma}_{\kappa,\vartheta}\right\}^{1}(X)\\
&=\frac{1}{2}\left(\left\{\mathbb{E}^{\gamma}_{\kappa,\vartheta}\right\}^{1}(X)+\left\{\mathbb{E}^{\gamma}_{\kappa,\vartheta}\right\}^{1}(-X)\right)\\
&=\sum_{k=0}^{1}\frac{(\gamma)_{2k}X^{(2k)}}{\Gamma((2k)\kappa+\vartheta)},
\end{aligned}
$$

$$
\begin{aligned}
&\left\{precosh^{\gamma}_{\kappa,\vartheta}\right\}^{2}(X)\\
&=\frac{1}{2}\left(\left\{\mathbb{E}^{\gamma}_{\kappa,\vartheta}\right\}^{2}(X)+\left\{\mathbb{E}^{\gamma}_{\kappa,\vartheta}\right\}^{2}(-X)\right)\\
&=\sum_{k=0}^{2}\frac{(\gamma)_{2k}X^{(2k)}}{\Gamma((2k)\kappa+\vartheta)}.
\end{aligned}
$$

▷ The truncated presine through the Prabhakar function is given by

$$
\begin{aligned}
&\left\{presin^{\gamma}_{\kappa,\vartheta}\right\}^{n}(X^{\varsigma})\\
&=\frac{1}{2i}\left(\left\{\mathbb{E}^{\gamma}_{\kappa,\vartheta}\right\}^{n}(iX^{\varsigma})-\left\{\mathbb{E}^{\gamma}_{\kappa,\vartheta}\right\}^{n}(-iX^{\varsigma})\right)\\
&=\sum_{k=0}^{n}\frac{(-1)^{k}(\gamma)_{2k+1}X^{(2k+1)\varsigma}}{\Gamma((2k+1)\kappa+\vartheta)}.
\end{aligned}
$$

Thus,

$$
\begin{aligned}
&\left\{presin^{\gamma}_{\kappa,\vartheta}\right\}^{0}(X^{\varsigma})\\
&=\frac{1}{2}\left(\left\{\mathbb{E}^{\gamma}_{\kappa,\vartheta}\right\}^{0}(iX^{\varsigma})-\left\{\mathbb{E}^{\gamma}_{\kappa,\vartheta}\right\}^{0}(-iX^{\varsigma})\right)\\
&=\frac{\Gamma(\gamma+1)}{\Gamma(\kappa+\vartheta)}X^{\varsigma},
\end{aligned}
$$

$$
\begin{aligned}
&\left\{presin^{\gamma}_{\kappa,\vartheta}\right\}^{1}(X^{\varsigma})\\
&=\frac{1}{2}\left(\left\{\mathbb{E}^{\gamma}_{\kappa,\vartheta}\right\}^{1}(iX^{\varsigma})-\left\{\mathbb{E}^{\gamma}_{\kappa,\vartheta}\right\}^{1}(-iX^{\varsigma})\right)\\
&=\sum_{k=0}^{1}\frac{(-1)^{k}(\gamma)_{2k+1}X^{(2k+1)\varsigma}}{\Gamma((2k+1)\kappa+\vartheta)},
\end{aligned}
$$

$$\begin{aligned}&\left\{presin^{\gamma}_{\kappa,\vartheta}\right\}^{2}(X^{\varsigma})\\&=\frac{1}{2}\left(\left\{\mathbb{E}^{\gamma}_{\kappa,\vartheta}\right\}^{2}(iX^{\varsigma})-\left\{\mathbb{E}^{\gamma}_{\kappa,\vartheta}\right\}^{2}(-iX^{\varsigma})\right)\\&=\sum_{k=0}^{2}\frac{(-1)^{k}(\gamma)_{2k+1}X^{(2k+1)\varsigma}}{\Gamma((2k+1)\kappa+\vartheta)}.\end{aligned}$$

▷ The truncated precosine through the Prabhakar function is given by

$$\begin{aligned}&\left\{precos^{\gamma}_{\kappa,\vartheta}\right\}^{n}(X^{\varsigma})\\&=\frac{1}{2}\left(\left\{\mathbb{E}^{\gamma}_{\kappa,\vartheta}\right\}^{n}(iX^{\varsigma})+\left\{\mathbb{E}^{\gamma}_{\kappa,\vartheta}\right\}^{n}(-iX^{\varsigma})\right)\\&=\sum_{k=0}^{n}\frac{(-1)^{k}(\gamma)_{2k}X^{(2k)\varsigma}}{\Gamma((2k)\kappa+\vartheta)}.\end{aligned}$$

Thus,

$$\begin{aligned}&\left\{precos^{\gamma}_{\kappa,\vartheta}\right\}^{0}(X^{\varsigma})\\&=\frac{1}{2}\left(\left\{\mathbb{E}^{\gamma}_{\kappa,\vartheta}\right\}^{0}(iX^{\varsigma})+\left\{\mathbb{E}^{\gamma}_{\kappa,\vartheta}\right\}^{0}(-iX^{\varsigma})\right)\\&=\frac{1}{\Gamma(\vartheta)},\end{aligned}$$

$$\begin{aligned}&\left\{precos^{\gamma}_{\kappa,\vartheta}\right\}^{1}(X^{\varsigma})\\&=\frac{1}{2}\left(\left\{\mathbb{E}^{\gamma}_{\kappa,\vartheta}\right\}^{1}(iX^{\varsigma})+\left\{\mathbb{E}^{\gamma}_{\kappa,\vartheta}\right\}^{1}(-iX^{\varsigma})\right)\\&=\sum_{k=0}^{1}\frac{(-1)^{k}(\gamma)_{2k}X^{(2k)\varsigma}}{\Gamma((2k)\kappa+\vartheta)},\end{aligned}$$

$$\begin{aligned}
&\left\{precos^{\gamma}_{\kappa,\vartheta}\right\}^{2}(X^{\varsigma})\\
&=\frac{1}{2}\left(\left\{\mathbb{E}^{\gamma}_{\kappa,\vartheta}\right\}^{2}(iX^{\varsigma})+\left\{\mathbb{E}^{\gamma}_{\kappa,\vartheta}\right\}^{2}(-iX^{\varsigma})\right)\\
&=\sum_{k=0}^{2}\frac{(-1)^k(\gamma)_{2k}X^{(2k)\varsigma}}{\Gamma((2k)\kappa+\vartheta)}.
\end{aligned}$$

▷ The truncated hyperbolic presine through the Prabhakar function is given by

$$\begin{aligned}
&\left\{presinh^{\gamma}_{\kappa,\vartheta}\right\}^{n}(X^{\varsigma})\\
&=\frac{1}{2}\left(\left\{\mathbb{E}^{\gamma}_{\kappa,\vartheta}\right\}^{n}(X^{\varsigma})-\left\{\mathbb{E}^{\gamma}_{\kappa,\vartheta}\right\}^{n}(-X^{\varsigma})\right)\\
&=\sum_{k=0}^{n}\frac{(\gamma)_{2k+1}X^{(2k+1)\varsigma}}{\Gamma((2k+1)\kappa+\vartheta)}.
\end{aligned}$$

Thus,

$$\begin{aligned}
&\left\{presinh^{\gamma}_{\kappa,\vartheta}\right\}^{0}(X^{\varsigma})\\
&=\frac{1}{2}\left(\left\{\mathbb{E}^{\gamma}_{\kappa,\vartheta}\right\}^{0}(X^{\varsigma})-\left\{\mathbb{E}^{\gamma}_{\kappa,\vartheta}\right\}^{0}(-X^{\varsigma})\right)\\
&=\frac{\Gamma(\gamma+1)}{\Gamma(\kappa+\vartheta)}X^{\varsigma},
\end{aligned}$$

$$\begin{aligned}
&\left\{presinh^{\gamma}_{\kappa,\vartheta}\right\}^{1}(X^{\varsigma})\\
&=\frac{1}{2}\left(\left\{\mathbb{E}^{\gamma}_{\kappa,\vartheta}\right\}^{1}(X^{\varsigma})-\left\{\mathbb{E}^{\gamma}_{\kappa,\vartheta}\right\}^{1}(-X^{\varsigma})\right)\\
&=\sum_{k=0}^{1}\frac{(\gamma)_{2k+1}X^{(2k+1)\varsigma}}{\Gamma((2k+1)\kappa+\vartheta)},
\end{aligned}$$

$$\begin{aligned}
&\left\{presinh_{\kappa,\vartheta}^{\gamma}\right\}^{2}(X^{\varsigma})\\
&=\frac{1}{2}\left(\left\{\mathbb{E}_{\kappa,\vartheta}^{\gamma}\right\}^{2}(X^{\varsigma})-\left\{\mathbb{E}_{\kappa,\vartheta}^{\gamma}\right\}^{2}(-X^{\varsigma})\right)\\
&=\sum_{k=0}^{2}\frac{(\gamma)_{2k+1}X^{(2k+1)\varsigma}}{\Gamma((2k+1)\kappa+\vartheta)}.
\end{aligned}$$

▷ The truncated hyperbolic precosine through the Prabhakar function is given by

$$\begin{aligned}
&\left\{precosh_{\kappa,\vartheta}^{\gamma}\right\}^{n}(X^{\varsigma})\\
&=\frac{1}{2}\left(\left\{\mathbb{E}_{\kappa,\vartheta}^{\gamma}\right\}^{n}(X^{\varsigma})+\left\{\mathbb{E}_{\kappa,\vartheta}^{\gamma}\right\}^{n}(-X^{\varsigma})\right)\\
&=\sum_{k=0}^{n}\frac{(\gamma)_{2k}X^{(2k)\varsigma}}{\Gamma((2k)\kappa+\vartheta)}.
\end{aligned}$$

Hence,

$$\begin{aligned}
&\left\{precosh_{\kappa,\vartheta}^{\gamma}\right\}^{0}(X^{\varsigma})\\
&=\frac{1}{2}\left(\left\{\mathbb{E}_{\kappa,\vartheta}^{\gamma}\right\}^{0}(X^{\varsigma})+\left\{\mathbb{E}_{\kappa,\vartheta}^{\gamma}\right\}^{0}(-X^{\varsigma})\right)\\
&=\frac{1}{\Gamma(\vartheta)},
\end{aligned}$$

$$\begin{aligned}
&\left\{precosh_{\kappa,\vartheta}^{\gamma}\right\}^{1}(X^{\varsigma})\\
&=\frac{1}{2}\left(\left\{\mathbb{E}_{\kappa,\vartheta}^{\gamma}\right\}^{1}(X^{\varsigma})+\left\{\mathbb{E}_{\kappa,\vartheta}^{\gamma}\right\}^{1}(-X^{\varsigma})\right)\\
&=\sum_{k=0}^{1}\frac{(\gamma)_{2k}X^{(2k)\varsigma}}{\Gamma((2k)\kappa+\vartheta)},
\end{aligned}$$

$$
\begin{aligned}
&\left\{precosh_{\kappa,\vartheta}^{\gamma}\right\}^{2}(X^{\zeta})\\
&=\frac{1}{2}\left(\left\{\mathbb{E}_{\kappa,\vartheta}^{\gamma}\right\}^{2}(X^{\zeta})+\left\{\mathbb{E}_{\kappa,\vartheta}^{\gamma}\right\}^{2}(-X^{\zeta})\right)\\
&=\sum_{k=0}^{2}\frac{(\gamma)_{2k}X^{(2k)\zeta}}{\Gamma((2k)\kappa+\vartheta)}.
\end{aligned}
$$

▷ The truncated presine through the Prabhakar function is given by

$$
\begin{aligned}
&\left\{Tresin_{\kappa,\vartheta}^{\gamma}\right\}^{n}(X)\\
&=\frac{X^{\vartheta-1}}{2i}\left(\left\{\mathbb{E}_{\kappa,\vartheta}^{\gamma}\right\}^{n}(iX)-\left\{\mathbb{E}_{\kappa,\vartheta}^{\gamma}\right\}^{n}(-iX)\right)\\
&=X^{\vartheta-1}\sum_{k=0}^{n}\frac{(-1)^{k}(\gamma)_{2k+1}X^{(2k+1)}}{\Gamma((2k+1)\kappa+\vartheta)}.
\end{aligned}
$$

Thus,

$$
\begin{aligned}
&\left\{Tresin_{\kappa,\vartheta}^{\gamma}\right\}^{0}(X)\\
&=\frac{X^{\vartheta-1}}{2}\left(\left\{\mathbb{E}_{\kappa,\vartheta}^{\gamma}\right\}^{0}(iX)-\left\{\mathbb{E}_{\kappa,\vartheta}^{\gamma}\right\}^{0}(-iX)\right)\\
&=\frac{\Gamma(\gamma+1)}{\Gamma(\kappa+\vartheta)}X^{\vartheta},
\end{aligned}
$$

$$
\begin{aligned}
&\left\{Tresin_{\kappa,\vartheta}^{\gamma}\right\}^{1}(X)\\
&=\frac{X^{\vartheta-1}}{2}\left(\left\{\mathbb{E}_{\kappa,\vartheta}^{\gamma}\right\}^{1}(iX)-\left\{\mathbb{E}_{\kappa,\vartheta}^{\gamma}\right\}^{1}(-iX)\right)\\
&=X^{\vartheta-1}\sum_{k=0}^{1}\frac{(-1)^{k}(\gamma)_{2k+1}X^{(2k+1)}}{\Gamma((2k+1)\kappa+\vartheta)},
\end{aligned}
$$

$$
\begin{aligned}
&\left\{Tresin_{\kappa,\vartheta}^{\gamma}\right\}^{2}(X)\\
&=\frac{X^{\vartheta-1}}{2}\left(\left\{\mathbb{E}_{\kappa,\vartheta}^{\gamma}\right\}^{2}(iX)-\left\{\mathbb{E}_{\kappa,\vartheta}^{\gamma}\right\}^{2}(-iX)\right)\\
&=X^{\vartheta-1}\sum_{k=0}^{2}\frac{(-1)^{k}(\gamma)_{2k+1}X^{(2k+1)}}{\Gamma((2k+1)\kappa+\vartheta)}.
\end{aligned}
$$

▷ The truncated precosine through the Prabhakar function is given by

$$
\begin{aligned}
&\left\{Trecos_{\kappa,\vartheta}^{\gamma}\right\}^{n}(X)\\
&=\frac{X^{\vartheta-1}}{2}\left(\left\{\mathbb{E}_{\kappa,\vartheta}^{\gamma}\right\}^{n}(iX)+\left\{\mathbb{E}_{\kappa,\vartheta}^{\gamma}\right\}^{n}(-iX)\right)\\
&=X^{\vartheta-1}\sum_{k=0}^{n}\frac{(-1)^{k}(\gamma)_{2k}X^{(2k)}}{\Gamma((2k)\kappa+\vartheta)}.
\end{aligned}
$$

Therefore,

$$
\begin{aligned}
&\left\{Trecos_{\kappa,\vartheta}^{\gamma}\right\}^{0}(X)\\
&=\frac{X^{\vartheta-1}}{2}\left(\left\{\mathbb{E}_{\kappa,\vartheta}^{\gamma}\right\}^{0}(iX)+\left\{\mathbb{E}_{\kappa,\vartheta}^{\gamma}\right\}^{0}(-iX)\right)\\
&=\frac{X^{\vartheta-1}}{\Gamma(\vartheta)},
\end{aligned}
$$

$$
\begin{aligned}
&\left\{Trecos_{\kappa,\vartheta}^{\gamma}\right\}^{1}(X)\\
&=\frac{X^{\vartheta-1}}{2}\left(\left\{\mathbb{E}_{\kappa,\vartheta}^{\gamma}\right\}^{1}(iX)+\left\{\mathbb{E}_{\kappa,\vartheta}^{\gamma}\right\}^{1}(-iX)\right)\\
&=X^{\vartheta-1}\sum_{k=0}^{1}\frac{(-1)^{k}(\gamma)_{2k}X^{(2k)}}{\Gamma((2k)\kappa+\vartheta)},
\end{aligned}
$$

$$\begin{aligned}
&\left\{Trecos_{\kappa,\vartheta}^{\gamma}\right\}^{2}(X)\\
&=\frac{X^{\vartheta-1}}{2}\left(\left\{\mathbb{E}_{\kappa,\vartheta}^{\gamma}\right\}^{2}(iX)+\left\{\mathbb{E}_{\kappa,\vartheta}^{\gamma}\right\}^{2}(-iX)\right)\\
&=X^{\vartheta-1}\sum_{k=0}^{2}\frac{(-1)^{k}(\gamma)_{2k}X^{(2k)}}{\Gamma((2k)\kappa+\vartheta)}.
\end{aligned}$$

▷ The truncated hyperbolic presine through the Prabhakar function is given by

$$\begin{aligned}
&\left\{Tresinh_{\kappa,\vartheta}^{\gamma}\right\}^{n}(X)\\
&=\frac{X^{\vartheta-1}}{2}\left(\left\{\mathbb{E}_{\kappa,\vartheta}^{\gamma}\right\}^{n}(X)-\left\{\mathbb{E}_{\kappa,\vartheta}^{\gamma}\right\}^{n}(-X)\right)\\
&=X^{\vartheta-1}\sum_{k=0}^{n}\frac{(\gamma)_{2k+1}X^{(2k+1)}}{\Gamma((2k+1)\kappa+\vartheta)}.
\end{aligned}$$

Thus,

$$\begin{aligned}
&\left\{Tresinh_{\kappa,\vartheta}^{\gamma}\right\}^{0}(X)\\
&=\frac{X^{\vartheta-1}}{2}\left(\left\{\mathbb{E}_{\kappa,\vartheta}^{\gamma}\right\}^{0}(X)-\left\{\mathbb{E}_{\kappa,\vartheta}^{\gamma}\right\}^{0}(-X)\right)\\
&=\frac{\Gamma(\gamma+1)}{\Gamma(\kappa+\vartheta)}X^{\vartheta},
\end{aligned}$$

$$\begin{aligned}
&\left\{Tresinh_{\kappa,\vartheta}^{\gamma}\right\}^{1}(X)\\
&=\frac{X^{\vartheta-1}}{2}\left(\left\{\mathbb{E}_{\kappa,\vartheta}^{\gamma}\right\}^{1}(X)-\left\{\mathbb{E}_{\kappa,\vartheta}^{\gamma}\right\}^{1}(-X)\right)\\
&=X^{\vartheta-1}\sum_{k=0}^{1}\frac{(\gamma)_{2k+1}X^{(2k+1)}}{\Gamma((2k+1)\kappa+\vartheta)},
\end{aligned}$$

$$\begin{aligned}
&\left\{Tresinh^{\gamma}_{\kappa,\vartheta}\right\}^{2}(X)\\
&=\frac{X^{\vartheta-1}}{2}\left(\left\{\mathbb{E}^{\gamma}_{\kappa,\vartheta}\right\}^{2}(X)-\left\{\mathbb{E}^{\gamma}_{\kappa,\vartheta}\right\}^{2}(-X)\right)\\
&=X^{\vartheta-1}\sum_{k=0}^{2}\frac{(\gamma)_{2k+1}X^{(2k+1)}}{\Gamma((2k+1)\kappa+\vartheta)}.
\end{aligned}$$

▷ The truncated hyperbolic precosine through the Prabhakar function is given by

$$\begin{aligned}
&\left\{Trecosh^{\gamma}_{\kappa,\vartheta}\right\}^{n}(X)\\
&=\frac{X^{\vartheta-1}}{2}\left(\left\{\mathbb{E}^{\gamma}_{\kappa,\vartheta}\right\}^{n}(X)+\left\{\mathbb{E}^{\gamma}_{\kappa,\vartheta}\right\}^{n}(-X)\right)\\
&=X^{\vartheta-1}\sum_{k=0}^{n}\frac{(\gamma)_{2k}X^{(2k)}}{\Gamma((2k)\kappa+\vartheta)}.
\end{aligned}$$

Thus,

$$\begin{aligned}
&\left\{Trecosh^{\gamma}_{\kappa,\vartheta}\right\}^{0}(X)\\
&=\frac{X^{\vartheta-1}}{2}\left(\left\{\mathbb{E}^{\gamma}_{\kappa,\vartheta}\right\}^{0}(X)+\left\{\mathbb{E}^{\gamma}_{\kappa,\vartheta}\right\}^{0}(-X)\right)\\
&=\frac{X^{\vartheta-1}}{\Gamma(\vartheta)},
\end{aligned}$$

$$\begin{aligned}
&\left\{Trecosh^{\gamma}_{\kappa,\vartheta}\right\}^{1}(X)\\
&=\frac{X^{\vartheta-1}}{2}\left(\left\{\mathbb{E}^{\gamma}_{\kappa,\vartheta}\right\}^{1}(X)+\left\{\mathbb{E}^{\gamma}_{\kappa,\vartheta}\right\}^{1}(-X)\right)\\
&=X^{\vartheta-1}\sum_{k=0}^{1}\frac{(\gamma)_{2k}X^{(2k)}}{\Gamma((2k)\kappa+\vartheta)},
\end{aligned}$$

$$\begin{aligned}
&\left\{Trecosh_{\kappa,\vartheta}^{\gamma}\right\}^{2}(X)\\
&=\frac{X^{\vartheta-1}}{2}\left(\left\{\mathbb{E}_{\kappa,\vartheta}^{\gamma}\right\}^{2}(X)+\left\{\mathbb{E}_{\kappa,\vartheta}^{\gamma}\right\}^{2}(-X)\right)\\
&=X^{\vartheta-1}\sum_{k=0}^{2}\frac{(\gamma)_{2k}X^{(2k)}}{\Gamma((2k)\kappa+\vartheta)}.
\end{aligned}$$

▷ The truncated presine through the Prabhakar function is given by

$$\begin{aligned}
&\left\{Gresin_{\kappa,\vartheta}^{\gamma}\right\}^{n}(X^{\zeta})\\
&=\frac{X^{\vartheta-1}}{2i}\left(\left\{\mathbb{E}_{\kappa,\vartheta}^{\gamma}\right\}^{n}(iX^{\zeta})-\left\{\mathbb{E}_{\kappa,\vartheta}^{\gamma}\right\}^{n}(-iX^{\zeta})\right)\\
&=X^{\vartheta-1}\sum_{k=0}^{n}\frac{(-1)^{k}(\gamma)_{2k+1}X^{(2k+1)\zeta}}{\Gamma((2k+1)\kappa+\vartheta)}.
\end{aligned}$$

Thus,

$$\begin{aligned}
&\left\{Gresin_{\kappa,\vartheta}^{\gamma}\right\}^{0}(X^{\zeta})\\
&=\frac{X^{\vartheta-1}}{2}\left(\left\{\mathbb{E}_{\kappa,\vartheta}^{\gamma}\right\}^{0}(iX^{\zeta})-\left\{\mathbb{E}_{\kappa,\vartheta}^{\gamma}\right\}^{0}(-iX^{\zeta})\right)\\
&=\frac{\Gamma(\gamma+1)}{\Gamma(\kappa+\vartheta)}X^{\zeta+\vartheta-1},
\end{aligned}$$

$$\begin{aligned}
&\left\{Gresin_{\kappa,\vartheta}^{\gamma}\right\}^{1}(X^{\zeta})\\
&=\frac{X^{\vartheta-1}}{2}\left(\left\{\mathbb{E}_{\kappa,\vartheta}^{\gamma}\right\}^{1}(iX^{\zeta})-\left\{\mathbb{E}_{\kappa,\vartheta}^{\gamma}\right\}^{1}(-iX^{\zeta})\right)\\
&=X^{\vartheta-1}\sum_{k=0}^{1}\frac{(-1)^{k}(\gamma)_{2k+1}X^{(2k+1)\zeta}}{\Gamma((2k+1)\kappa+\vartheta)},
\end{aligned}$$

$$\begin{aligned}
&\left\{Gresin^{\gamma}_{\kappa,\vartheta}\right\}^{2}(X^{\varsigma})\\
&=\frac{X^{\vartheta-1}}{2}\left(\left\{\mathbb{E}^{\gamma}_{\kappa,\vartheta}\right\}^{2}(iX^{\varsigma})-\left\{\mathbb{E}^{\gamma}_{\kappa,\vartheta}\right\}^{2}(-iX^{\varsigma})\right)\\
&=X^{\vartheta-1}\sum_{k=0}^{2}\frac{(-1)^{k}(\gamma)_{2k+1}X^{(2k+1)\varsigma}}{\Gamma((2k+1)\kappa+\vartheta)}.
\end{aligned}$$

▷ The truncated precosine through the Prabhakar function is given by

$$\begin{aligned}
&\left\{Grecos^{\gamma}_{\kappa,\vartheta}\right\}^{n}(X^{\varsigma})\\
&=\frac{X^{\vartheta-1}}{2}\left(\left\{\mathbb{E}^{\gamma}_{\kappa,\vartheta}\right\}^{n}(iX^{\varsigma})+\left\{\mathbb{E}^{\gamma}_{\kappa,\vartheta}\right\}^{n}(-iX^{\varsigma})\right)\\
&=X^{\vartheta-1}\sum_{k=0}^{n}\frac{(-1)^{k}(\gamma)_{2k}X^{(2k)\varsigma}}{\Gamma((2k)\kappa+\vartheta)}.
\end{aligned}$$

Thus,

$$\begin{aligned}
&\left\{Grecos^{\gamma}_{\kappa,\vartheta}\right\}^{0}(X^{\varsigma})\\
&=\frac{X^{\vartheta-1}}{2}\left(\left\{\mathbb{E}^{\gamma}_{\kappa,\vartheta}\right\}^{0}(iX^{\varsigma})+\left\{\mathbb{E}^{\gamma}_{\kappa,\vartheta}\right\}^{0}(-iX^{\varsigma})\right)\\
&=\frac{X^{\vartheta-1}}{\Gamma(\vartheta)},
\end{aligned}$$

$$\begin{aligned}
&\left\{Grecos^{\gamma}_{\kappa,\vartheta}\right\}^{1}(X^{\varsigma})\\
&=\frac{X^{\vartheta-1}}{2}\left(\left\{\mathbb{E}^{\gamma}_{\kappa,\vartheta}\right\}^{1}(iX^{\varsigma})+\left\{\mathbb{E}^{\gamma}_{\kappa,\vartheta}\right\}^{1}(-iX^{\varsigma})\right)\\
&=X^{\vartheta-1}\sum_{k=0}^{1}\frac{(-1)^{k}(\gamma)_{2k}X^{(2k)\varsigma}}{\Gamma((2k)\kappa+\vartheta)},
\end{aligned}$$

$$
\begin{aligned}
&\left\{Grecos^{\gamma}_{\kappa,\vartheta}\right\}^{2}(X^{\varsigma}) \\
&= \frac{X^{\vartheta-1}}{2}\left(\left\{\mathbb{E}^{\gamma}_{\kappa,\vartheta}\right\}^{2}(iX^{\varsigma}) + \left\{\mathbb{E}^{\gamma}_{\kappa,\vartheta}\right\}^{2}(-iX^{\varsigma})\right) \\
&= X^{\vartheta-1}\sum_{k=0}^{2}\frac{(-1)^{k}(\gamma)_{2k}X^{(2k)\varsigma}}{\Gamma((2k)\kappa+\vartheta)}.
\end{aligned}
$$

▷ The truncated hyperbolic presine through the Prabhakar function is given by

$$
\begin{aligned}
&\left\{Gresinh^{\gamma}_{\kappa,\vartheta}\right\}^{n}(X^{\varsigma}) \\
&= \frac{X^{\vartheta-1}}{2}\left(\left\{\mathbb{E}^{\gamma}_{\kappa,\vartheta}\right\}^{n}(X^{\varsigma}) - \left\{\mathbb{E}^{\gamma}_{\kappa,\vartheta}\right\}^{n}(-X^{\varsigma})\right) \\
&= X^{\vartheta-1}\sum_{k=0}^{n}\frac{(\gamma)_{2k+1}X^{(2k+1)\varsigma}}{\Gamma((2k+1)\kappa+\vartheta)}.
\end{aligned}
$$

Thus,

$$
\begin{aligned}
&\left\{Gresinh^{\gamma}_{\kappa,\vartheta}\right\}^{0}(X^{\varsigma}) \\
&= \frac{X^{\vartheta-1}}{2}\left(\left\{\mathbb{E}^{\gamma}_{\kappa,\vartheta}\right\}^{0}(X^{\varsigma}) - \left\{\mathbb{E}^{\gamma}_{\kappa,\vartheta}\right\}^{0}(-X^{\varsigma})\right) \\
&= \frac{\Gamma(\gamma+1)}{\Gamma(\kappa+\vartheta)}X^{\varsigma+\vartheta-1},
\end{aligned}
$$

$$
\begin{aligned}
&\left\{Gresinh^{\gamma}_{\kappa,\vartheta}\right\}^{1}(X^{\varsigma}) \\
&= \frac{X^{\vartheta-1}}{2}\left(\left\{\mathbb{E}^{\gamma}_{\kappa,\vartheta}\right\}^{1}(X^{\varsigma}) - \left\{\mathbb{E}^{\gamma}_{\kappa,\vartheta}\right\}^{1}(-X^{\varsigma})\right) \\
&= X^{\vartheta-1}\sum_{k=0}^{1}\frac{(\gamma)_{2k+1}X^{(2k+1)\varsigma}}{\Gamma((2k+1)\kappa+\vartheta)},
\end{aligned}
$$

$$\begin{aligned}
&\left\{Gresinh^{\gamma}_{\kappa,\vartheta}\right\}^{2}(X^{\zeta})\\
&=\frac{X^{\vartheta-1}}{2}\left(\left\{\mathbb{E}^{\gamma}_{\kappa,\vartheta}\right\}^{2}(X^{\zeta})-\left\{\mathbb{E}^{\gamma}_{\kappa,\vartheta}\right\}^{2}(-X^{\zeta})\right)\\
&=X^{\vartheta-1}\sum_{k=0}^{2}\frac{(\gamma)_{2k+1}X^{(2k+1)\zeta}}{\Gamma((2k+1)\kappa+\vartheta)}.
\end{aligned}$$

▷ The truncated hyperbolic precosine through the Prabhakar function is given by

$$\begin{aligned}
&\left\{Grecosh^{\gamma}_{\kappa,\vartheta}\right\}^{n}(X^{\zeta})\\
&=\frac{X^{\vartheta-1}}{2}\left(\left\{\mathbb{E}^{\gamma}_{\kappa,\vartheta}\right\}^{n}(X^{\zeta})+\left\{\mathbb{E}^{\gamma}_{\kappa,\vartheta}\right\}^{n}(-X^{\zeta})\right)\\
&=X^{\vartheta-1}\sum_{k=0}^{n}\frac{(\gamma)_{2k}X^{(2k)\zeta}}{\Gamma((2k)\kappa+\vartheta)}.
\end{aligned}$$

Thus,

$$\begin{aligned}
&\left\{Grecosh^{\gamma}_{\kappa,\vartheta}\right\}^{0}(X^{\zeta})\\
&=\frac{X^{\vartheta-1}}{2}\left(\left\{\mathbb{E}^{\gamma}_{\kappa,\vartheta}\right\}^{0}(X^{\zeta})+\left\{\mathbb{E}^{\gamma}_{\kappa,\vartheta}\right\}^{0}(-X^{\zeta})\right)\\
&=\frac{X^{\vartheta-1}}{\Gamma(\vartheta)},
\end{aligned}$$

$$\begin{aligned}
&\left\{Grecosh^{\gamma}_{\kappa,\vartheta}\right\}^{1}(X^{\zeta})\\
&=\frac{X^{\vartheta-1}}{2}\left(\left\{\mathbb{E}^{\gamma}_{\kappa,\vartheta}\right\}^{1}(X^{\zeta})+\left\{\mathbb{E}^{\gamma}_{\kappa,\vartheta}\right\}^{1}(-X^{\zeta})\right)\\
&=X^{\vartheta-1}\sum_{k=0}^{1}\frac{(\gamma)_{2k}X^{(2k)\zeta}}{\Gamma((2k)\kappa+\vartheta)},
\end{aligned}$$

$$\begin{aligned}
&\left\{Grecosh^{\gamma}_{\kappa,\vartheta}\right\}^{2}(X^{\zeta})\\
&=\frac{X^{\vartheta-1}}{2}\left(\left\{\mathbb{E}^{\gamma}_{\kappa,\vartheta}\right\}^{2}(X^{\zeta})+\left\{\mathbb{E}^{\gamma}_{\kappa,\vartheta}\right\}^{2}(-X^{\zeta})\right)\\
&=X^{\vartheta-1}\sum_{k=0}^{2}\frac{(\gamma)_{2k}X^{(2k)\zeta}}{\Gamma((2k)\kappa+\vartheta)}.
\end{aligned}$$

6.13 Other Special Functions Related to the Prabhakar Function

In this section, we present some other special functions related to the Prabhakar function. We have the following from [4]:

$$\begin{aligned}
&supersin^{\gamma}_{1,\vartheta}(X)\\
&=\frac{X^{\vartheta-1}}{2i}\left(\mathbb{E}^{\gamma}_{1,\vartheta}(iX)-\mathbb{E}^{\gamma}_{1,\vartheta}(-iX)\right)\\
&=\frac{X^{\vartheta-1}}{\Gamma(\gamma)}\,{}_{1}\sin_{1}\left({(\gamma,1)\atop(\vartheta,1)};X\right)\\
&=\sum_{k=0}^{\infty}\frac{(-1)^{k}(\gamma)_{2k+1}}{\Gamma(2k+1+\vartheta)}\frac{X^{2k+\vartheta}}{(2k+1)!},
\end{aligned}$$

$$\begin{aligned}
&supercos^{\gamma}_{1,\vartheta}(X)\\
&=\frac{X^{\vartheta-1}}{2}\left(\mathbb{E}^{\gamma}_{1,\vartheta}(iX)+\mathbb{E}^{\gamma}_{1,\vartheta}(-iX)\right)\\
&=\frac{X^{\vartheta-1}}{\Gamma(\gamma)}\,{}_{1}\cos_{1}\left({(\gamma,1)\atop(\vartheta,1)};X\right)\\
&=\sum_{k=0}^{\infty}\frac{(-1)^{k}(\gamma)_{2k}}{\Gamma(2k+\vartheta)}\frac{X^{2k+\vartheta-1}}{(2k)!},
\end{aligned}$$

$$
\begin{aligned}
&supersinh^{\gamma}_{1,\vartheta}(X)\\
&= \frac{X^{\vartheta-1}}{2}\left(\mathbb{E}^{\gamma}_{1,\vartheta}(X) - \mathbb{E}^{\gamma}_{1,\vartheta}(-X)\right)\\
&= \frac{X^{\vartheta-1}}{\Gamma(\gamma)}\ {}_1\sinh_1\left({(\gamma,1) \atop (\vartheta,1)}; X\right)\\
&= \sum_{k=0}^{\infty} \frac{(\gamma)_{2k+1} X^{2k+\vartheta}}{\Gamma(2k+1+\vartheta)},
\end{aligned}
$$

$$
\begin{aligned}
&supercosh^{\gamma}_{1,\vartheta}(X)\\
&= \frac{X^{\vartheta-1}}{2}\left(\mathbb{E}^{\gamma}_{1,\vartheta}(X) + \mathbb{E}^{\gamma}_{1,\vartheta}(-X)\right)\\
&= \frac{X^{\vartheta-1}}{\Gamma(\gamma)}\ {}_1\cosh_1\left({(\gamma,1) \atop (\vartheta,1)}; X\right)\\
&= \sum_{k=0}^{\infty} \frac{(\gamma)_{2k} X^{2k+\vartheta-1}}{\Gamma(2k+\vartheta)},
\end{aligned}
$$

$$
\begin{aligned}
&presin^{\gamma}_{1,\vartheta}(X)\\
&= \frac{1}{2i}\left(\mathbb{E}^{\gamma}_{1,\vartheta}(iX) - \mathbb{E}^{\gamma}_{1,\vartheta}(-iX)\right)\\
&= \frac{1}{\Gamma(\gamma)}\ {}_1\sin_1\left({(\gamma,1) \atop (\vartheta,1)}; X\right)\\
&= \sum_{k=0}^{\infty} \frac{(-1)^k(\gamma)_{2k+1} X^{2k+1}}{\Gamma(2k+1+\vartheta)},
\end{aligned}
$$

$$
\begin{aligned}
&precos^{\gamma}_{1,\vartheta}(X)\\
&= \frac{1}{2}\left(\mathbb{E}^{\gamma}_{1,\vartheta}(iX) + \mathbb{E}^{\gamma}_{1,\vartheta}(-iX)\right)\\
&= \frac{1}{\Gamma(\gamma)}\ {}_1\cos_1\left({(\gamma,1) \atop (\vartheta,1)}; X\right)\\
&= \sum_{k=0}^{\infty} \frac{(-1)^k(\gamma)_{2k} X^{2k}}{\Gamma(2k+\vartheta)},
\end{aligned}
$$

$$
\begin{aligned}
&presinh^{\gamma}_{1,\vartheta}(X)\\
&= \frac{1}{2}\Big(\mathbb{E}^{\gamma}_{1,\vartheta}(X) - \mathbb{E}^{\gamma}_{1,\vartheta}(-X)\Big)\\
&= \frac{1}{\Gamma(\gamma)}\ {}_1\sinh_1\left({}^{(\gamma,1)}_{(\vartheta,1)}; X\right)\\
&= \sum_{k=0}^{\infty} \frac{(\gamma)_{2k+1} X^{2k+1}}{\Gamma(2k+1+\vartheta)},
\end{aligned}
$$

$$
\begin{aligned}
&precosh^{\gamma}_{1,\vartheta}(X)\\
&= \frac{1}{2}\Big(\mathbb{E}^{\gamma}_{1,\vartheta}(X) + \mathbb{E}^{\gamma}_{1,\vartheta}(-X)\Big)\\
&= \frac{1}{\Gamma(\gamma)}\ {}_1\cosh_1\left({}^{(\gamma,1)}_{(\vartheta,1)}; X\right)\\
&= \sum_{k=0}^{\infty} \frac{(\gamma)_{2k} X^{2k}}{\Gamma(2k+\vartheta)},
\end{aligned}
$$

$$
\begin{aligned}
&Gresin^{\gamma}_{1,\vartheta}(X^{\zeta})\\
&= \frac{X^{\vartheta-1}}{2i}\Big(\mathbb{E}^{\gamma}_{1,\vartheta}(iX^{\zeta}) - \mathbb{E}^{\gamma}_{1,\vartheta}(-iX^{\zeta})\Big)\\
&= \frac{X^{\vartheta-1}}{\Gamma(\gamma)}\ {}_1\sin_1\left({}^{(\gamma,1)}_{(\vartheta,1)}; X^{\zeta}\right)\\
&= X^{\vartheta-1} \sum_{k=0}^{\infty} \frac{(-1)^k (\gamma)_{2k+1} X^{(2k+1)\zeta}}{\Gamma(2k+1+\vartheta)},
\end{aligned}
$$

$$
\begin{aligned}
&Grecos^{\gamma}_{1,\vartheta}(X^{\zeta})\\
&= \frac{X^{\vartheta-1}}{2}\Big(\mathbb{E}^{\gamma}_{1,\vartheta}(iX^{\zeta}) + \mathbb{E}^{\gamma}_{1,\vartheta}(-iX^{\zeta})\Big)\\
&= \frac{X^{\vartheta-1}}{\Gamma(\gamma)}\ {}_1\cos_1\left({}^{(\gamma,1)}_{(\vartheta,1)}; X^{\zeta}\right)\\
&= X^{\vartheta-1} \sum_{k=0}^{\infty} \frac{(-1)^k (\gamma)_{2k} X^{2k\zeta}}{\Gamma(2k+\vartheta)},
\end{aligned}
$$

$$
\begin{aligned}
&Gresinh^{\gamma}_{1,\vartheta}(X^{\zeta})\\
&= \frac{X^{\vartheta-1}}{2}\Big(\mathbb{E}^{\gamma}_{1,\vartheta}(X^{\zeta}) - \mathbb{E}^{\gamma}_{1,\vartheta}(-X^{\zeta})\Big)\\
&= \frac{X^{\vartheta-1}}{\Gamma(\gamma)}\ {}_1\sinh_1\left({}^{(\gamma,1)}_{(\vartheta,1)}; X^{\zeta}\right)\\
&= X^{\vartheta-1}\sum_{k=0}^{\infty}\frac{(\gamma)_{2k+1}X^{(2k+1)\zeta}}{\Gamma(2k+1+\vartheta)},
\end{aligned}
$$

$$
\begin{aligned}
&Grecosh^{\gamma}_{1,\vartheta}(X^{\zeta})\\
&= \frac{X^{\vartheta-1}}{2}\Big(\mathbb{E}^{\gamma}_{1,\vartheta}(X^{\zeta}) + \mathbb{E}^{\gamma}_{1,\vartheta}(-X^{\zeta})\Big)\\
&= \frac{X^{\vartheta-1}}{\Gamma(\gamma)}\ {}_1\cosh_1\left({}^{(\gamma,1)}_{(\vartheta,1)}; X^{\zeta}\right)\\
&= X^{\vartheta-1}\sum_{k=0}^{\infty}\frac{(\gamma)_{2k}X^{2k\zeta}}{\Gamma(2k+\vartheta)},
\end{aligned}
$$

$$
\begin{aligned}
&supersin^{\gamma}_{\kappa,1}(X^{\kappa})\\
&= \frac{1}{2i}\Big(\mathbb{E}^{\gamma}_{\kappa,1}(iX^{\kappa}) - \mathbb{E}^{\gamma}_{\kappa,1}(-iX^{\kappa})\Big)\\
&= \frac{1}{\Gamma(\gamma)}\ {}_1\sin_1\left({}^{(\gamma,1)}_{(1,\kappa)}; X^{\kappa}\right)\\
&= \sum_{k=0}^{\infty}\frac{(-1)^k(\gamma)_{2k+1}}{\Gamma((2k+1)\kappa+1)}\frac{X^{(2k+1)\kappa}}{(2k+1)!},
\end{aligned}
$$

$$
\begin{aligned}
&supercos^{\gamma}_{\kappa,1}(X^{\kappa})\\
&= \frac{1}{2}\Big(\mathbb{E}^{\gamma}_{\kappa,1}(iX^{\kappa}) + \mathbb{E}^{\gamma}_{\kappa,1}(-iX^{\kappa})\Big)\\
&= \frac{1}{\Gamma(\gamma)}\ {}_1\cos_1\left({}^{(\gamma,1)}_{(1,\kappa)}; X^{\kappa}\right)\\
&= \sum_{k=0}^{\infty}\frac{(-1)^k(\gamma)_{2k}}{\Gamma((2k)\kappa+1)}\frac{X^{(2k)\kappa}}{(2k)!},
\end{aligned}
$$

$$
\begin{aligned}
&supersinh^{\gamma}_{\kappa,1}(X^{\kappa})\\
&= \frac{1}{2}\Big(\mathbb{E}^{\gamma}_{\kappa,1}(X^{\kappa}) - \mathbb{E}^{\gamma}_{\kappa,1}(-X^{\kappa})\Big)\\
&= \frac{1}{\Gamma(\gamma)}\ {}_1\sinh_1\left({(\gamma,1) \atop (1,\kappa)}; X^{\kappa}\right)\\
&= \sum_{k=0}^{\infty} \frac{(\gamma)_{2k+1} X^{(2k+1)\kappa}}{\Gamma((2k+1)\kappa+1)},
\end{aligned}
$$

$$
\begin{aligned}
&supercosh^{\gamma}_{\kappa,1}(X^{\kappa})\\
&= \frac{1}{2}\Big(\mathbb{E}^{\gamma}_{\kappa,1}(X^{\kappa}) + \mathbb{E}^{\gamma}_{\kappa,1}(-X^{\kappa})\Big)\\
&= \frac{1}{\Gamma(\gamma)}\ {}_1\cosh_1\left({(\gamma,1) \atop (1,\kappa)}; X^{\kappa}\right)\\
&= \sum_{k=0}^{\infty} \frac{(\gamma)_{2k} X^{(2k)\kappa}}{\Gamma((2k)\kappa+1)},
\end{aligned}
$$

$$
\begin{aligned}
&presin^{\gamma}_{\kappa,1}(X)\\
&= \frac{1}{2i}\Big(\mathbb{E}^{\gamma}_{\kappa,1}(iX) - \mathbb{E}^{\gamma}_{\kappa,1}(-iX)\Big)\\
&= \frac{1}{\Gamma(\gamma)}\ {}_1\sin_1\left({(\gamma,1) \atop (1,\kappa)}; X\right)\\
&= \sum_{k=0}^{\infty} \frac{(-1)^k(\gamma)_{2k+1} X^{(2k+1)}}{\Gamma((2k+1)\kappa+1)},
\end{aligned}
$$

$$
\begin{aligned}
&precos^{\gamma}_{\kappa,1}(X)\\
&= \frac{1}{2}\Big(\mathbb{E}^{\gamma}_{\kappa,1}(iX) + \mathbb{E}^{\gamma}_{\kappa,1}(-iX)\Big)\\
&= \frac{1}{\Gamma(\gamma)}\ {}_1\cos_1\left({(\gamma,1) \atop (1,\kappa)}; X\right)\\
&= \sum_{k=0}^{\infty} \frac{(-1)^k(\gamma)_{2k} X^{(2k)}}{\Gamma((2k)\kappa+1)},
\end{aligned}
$$

$$\begin{aligned}
&presinh^{\gamma}_{\kappa,1}(X)\\
&= \frac{1}{2}\Big(\mathbb{E}^{\gamma}_{\kappa,1}(X) - \mathbb{E}^{\gamma}_{\kappa,1}(-X)\Big)\\
&= \frac{1}{\Gamma(\gamma)}\ {}_1\sinh_1\left({}^{(\gamma,1)}_{(1,\kappa)}; X\right)\\
&= \sum_{k=0}^{\infty} \frac{(\gamma)_{2k+1} X^{(2k+1)}}{\Gamma((2k+1)\kappa+1)},
\end{aligned}$$

$$\begin{aligned}
&precosh^{\gamma}_{\kappa,1}(X)\\
&= \frac{1}{2}\Big(\mathbb{E}^{\gamma}_{\kappa,1}(X) + \mathbb{E}^{\gamma}_{\kappa,1}(-X)\Big)\\
&= \frac{1}{\Gamma(\gamma)}\ {}_1\cosh_1\left({}^{(\gamma,1)}_{(1,\kappa)}; X\right)\\
&= \sum_{k=0}^{\infty} \frac{(\gamma)_{2k} X^{(2k)}}{\Gamma((2k)\kappa+1)},
\end{aligned}$$

$$\begin{aligned}
&supersin^{\gamma}_{1,1}(X)\\
&= \frac{1}{2i}\Big(\mathbb{E}^{\gamma}_{1,1}(iX) - \mathbb{E}^{\gamma}_{1,1}(-iX)\Big)\\
&= \frac{1}{\Gamma(\gamma)}\ {}_1\sin_1\left({}^{(\gamma,1)}_{(1,1)}; X\right)\\
&= \sum_{k=0}^{\infty} \frac{(-1)^k(\gamma)_{2k+1}}{\Gamma(2k+2)} \frac{X^{2k+1}}{(2k+1)!},
\end{aligned}$$

$$\begin{aligned}
&supercos^{\gamma}_{1,1}(X)\\
&= \frac{1}{2}\Big(\mathbb{E}^{\gamma}_{1,1}(iX) - \mathbb{E}^{\gamma}_{1,1}(-iX)\Big)\\
&= \frac{1}{\Gamma(\gamma)}\ {}_1\cos_1\left({}^{(\gamma,1)}_{(1,1)}; X\right)\\
&= \sum_{k=0}^{\infty} \frac{(-1)^k(\gamma)_{2k}}{\Gamma(2k+1)} \frac{X^{2k}}{(2k)!},
\end{aligned}$$

$$\begin{aligned}
&supersinh^{\gamma}_{1,1}(X)\\
&=\frac{1}{2}\left(\mathbb{E}^{\gamma}_{1,1}(X)-\mathbb{E}^{\gamma}_{1,1}(-X)\right)\\
&=\frac{1}{\Gamma(\gamma)}\;{}_1\sinh_1\left({}^{(\gamma,1)}_{(1,1)};X\right)\\
&=\sum_{k=0}^{\infty}\frac{(\gamma)_{2k+1}X^{2k+1}}{\Gamma(2k+2)},
\end{aligned}$$

$$\begin{aligned}
&supercosh^{\gamma}_{1,1}(X)\\
&=\frac{1}{2}\left(\mathbb{E}^{\gamma}_{1,1}(X)-\mathbb{E}^{\gamma}_{1,1}(-X)\right)\\
&=\frac{1}{\Gamma(\gamma)}\;{}_1\cosh_1\left({}^{(\gamma,1)}_{(1,1)};X\right)\\
&=\sum_{k=0}^{\infty}\frac{(\gamma)_{2k}X^{2k}}{\Gamma(2k+1)}.
\end{aligned}$$

References

1. Prabhakar TR (1971) A singular integral equation with a generalized Mittag Leffler function in the kernel. Yokohama Math J 19:7–15
2. Gorenflo R, Kilbas AA, Mainardi F, Rogosin SV (2014) Mittag-Leffler functions, related topics and applications, vol 2. Springer, Berlin
3. Mathai AM, Saxena RK (1973) Generalized hypergeometric functions with applications in statistics and physical sciences, vol 348. Lecture Notes in Mathematics. Springer, Berlin
4. Yang X-J (2021) Theory and applications of special functions for scientists and engineers. Springer
5. Chaudhry MA, Qadir A, Rafique M, Zubair SM (1997) Extension of Euler's beta function. J Comput Appl Math 78(1):19–32

Chapter 7
The Sonine, Supertrigonometric and Superhyperbolic Functions

In this chapter, we introduce special functions related to the Mittag- Leffler function containing the Sonine functions.

7.1 The Sonine Function of the First Type

In 1884, Nikolay Yakovlevich Sonin (Sonine), a Russian mathematician, worked on special functions and numerical integration and proposed the Sonine functions [1], which are called Sonine functions of the first type and Sonine functions of the second type.

Definition 7.1 ([1]) The Sonine function of first type is given by

$$
\begin{aligned}
&\Psi_{\psi}(X)\\
&= X^{-\psi} \sum_{k=0}^{\infty} \frac{X^k}{\Gamma(k+1)\Gamma(k-\psi+1)}\\
&= \sum_{k=0}^{\infty} \frac{X^{k-\psi}}{\Gamma(k+1)\Gamma(k-\psi+1)},
\end{aligned}
$$

where $X, \psi \in \mathbb{C}$.

This result was discovered by Sonine in 1884 [1] and was further reported on in 2020 by Yang [2], which represents the pairs of the Sonine function as:

S. Rezaei Aderyani et al., *Towards Ulam Type Multi Stability Analysis*, Studies in Systems, Decision and Control 523, https://doi.org/10.1007/978-3-031-55564-0_7

$$\begin{aligned}
&\Psi_{\psi}(\varsigma, X) \\
&= X^{-\psi} \sum_{k=0}^{\infty} \frac{(\varsigma X)^k}{\Gamma(k+1)\Gamma(k-\psi+1)} \\
&= \sum_{k=0}^{\infty} \frac{\varsigma^k X^{k-\psi}}{\Gamma(k+1)\Gamma(k-\psi+1)},
\end{aligned}$$

where $X, \psi \in \mathbb{C}$, and ς is a constant.

Furthermore from [3],

$$\Psi_{\psi}(X) = \frac{1}{2\pi i} \int_{-i\infty}^{i\infty} \frac{\Gamma(Y-\psi)}{\Gamma(1-Y)} (-X)^{Y-1} dY,$$

where $\Re(\psi) < \Re(Y) < 1$, and $\arg(X) < \pi$.

We also have [3],

$$\begin{aligned}
&\Psi_{\psi}(\varsigma, X) \\
&= X^{-\psi} \sum_{k=0}^{\infty} \frac{(\varsigma X)^k}{\Gamma(k+1)\Gamma(k-\psi+1)} \\
&= X^{-\psi} {}_0\mathbb{W}_1\left({\overline{} \atop (-\psi+1,1)} ; \varsigma X \right).
\end{aligned}$$

7.2 The Supertrigonometric Functions via the Sonine Function of the First Type

Let $X, \psi, \varsigma \in \mathbb{C}$. In this part, we give the supertrigonometric functions via the Sonine function of the first type [3], as follows:

▷ The supersine through the Sonine function of first type is given by

$$\begin{aligned}
&\Psi \sin_{\psi}(X) \\
&= \frac{1}{2i}\Big(\Psi_{\psi}(i, X) - \Psi_{\psi}(-i, X) \Big) \\
&= X^{-\psi} \sum_{k=0}^{\infty} \frac{(-1)^k X^{2k+1}}{\Gamma(2k+2)\Gamma(2k-\psi+2)}.
\end{aligned}$$

▷ The supercosine through the Sonine function of the first type is given by

$$\begin{aligned}
&\Psi \cos_{\psi}(X) \\
&= \frac{1}{2}\Big(\Psi_{\psi}(i, X) + \Psi_{\psi}(-i, X)\Big) \\
&= X^{-\psi} \sum_{k=0}^{\infty} \frac{(-1)^k X^{2k}}{\Gamma(2k+1)\Gamma(2k-\psi+1)}.
\end{aligned}$$

▷ The supertangent through the Sonine function of the first type is given by

$$\Psi \tan_{\psi}(X) = \frac{\Psi \sin_{\psi}(X)}{\Psi \cos_{\psi}(X)}, \ \Big(\Psi \cos_{\psi}(X) \neq 0\Big).$$

▷ The supercotangent through the Sonine function of the first type is given by

$$\Psi \text{cotan}_{\psi}(X) = \frac{\Psi \cos_{\psi}(X)}{\Psi \sin_{\psi}(X)}, \ \Big(\Psi \sin_{\psi}(X) \neq 0\Big).$$

▷ The supercosecant through the Sonine function of the first type is given by

$$\Psi \csc_{\psi}(X) = \frac{1}{\Psi \sin_{\psi}(X)}, \ \Big(\Psi \sin_{\psi}(X) \neq 0\Big).$$

▷ The supersecant through the Sonine function of the first type is given by

$$\Psi \sec_{\psi}(X) = \frac{1}{\Psi \cos_{\psi}(X)}, \ \Big(\Psi \cos_{\psi}(X) \neq 0\Big).$$

Thus, we get the following [3]:

$$\begin{aligned}
\Psi \sin_{\psi}(\varsigma, X) &= \frac{1}{2i}\Big(\Psi_{\psi}(i\varsigma, X) - \Psi_{\psi}(-i\varsigma, X)\Big) \\
&= X^{-\psi} \sum_{k=0}^{\infty} \frac{(-1)^k (X\varsigma)^{2k+1}}{\Gamma(2k+2)\Gamma(2k-\psi+2)},
\end{aligned}$$

$$\begin{aligned}
\Psi \cos_{\psi}(\varsigma, X) &= \frac{1}{2}\Big(\Psi_{\psi}(i\varsigma, X) + \Psi_{\psi}(-i\varsigma, X)\Big) \\
&= X^{-\psi} \sum_{k=0}^{\infty} \frac{(-1)^k (X\varsigma)^{2k}}{\Gamma(2k+1)\Gamma(2k-\psi+1)},
\end{aligned}$$

$$\Psi \tan_{\psi}(\varsigma, X) = \frac{\Psi \sin_{\psi}(\varsigma, X)}{\Psi \cos_{\psi}(\varsigma, X)}, \left(\Psi \cos_{\psi}(\varsigma, X) \neq 0\right),$$

$$\Psi \mathrm{cotan}_{\psi}(\varsigma, X) = \frac{\Psi \cos_{\psi}(\varsigma, X)}{\Psi \sin_{\psi}(\varsigma, X)}, \left(\Psi \sin_{\psi}(\varsigma, X) \neq 0\right),$$

$$\Psi \csc_{\psi}(\varsigma, X) = \frac{1}{\Psi \sin_{\psi}(\varsigma, X)}, \left(\Psi \sin_{\psi}(\varsigma, X) \neq 0\right),$$

$$\Psi \sec_{\psi}(\varsigma, X) = \frac{1}{\Psi \cos_{\psi}(\varsigma, X)}, \left(\Psi \cos_{\psi}(\varsigma, X) \neq 0\right).$$

Theorem 7.1 ([3]) *Let* $\varsigma, X, \psi \in \mathbb{C}$. *Then,*

$$\Psi_{\psi}(i\varsigma, X) = \Psi \cos_{\psi}(\varsigma, X) + i\Psi \cos_{\psi}(\varsigma, X).$$

7.3 The Superhyperbolic Functions via the Sonine Function of the First Type

Suppose $X, \psi, \varsigma \in \mathbb{C}$. In this part, we introduce the superhyperbolic functions via the Sonine functions of the first type [3], as follows:

▷ The hyperbolic supersine through the Sonine function of the first type is given by

$$\begin{aligned} &\Psi \sinh_{\psi}(X) \\ &= \frac{1}{2}\left(\Psi_{\psi}(1, X) - \Psi_{\psi}(-1, X)\right) \\ &= X^{-\psi} \sum_{k=0}^{\infty} \frac{X^{2k+1}}{\Gamma(2k+2)\Gamma(2k-\psi+2)}. \end{aligned}$$

▷ The hyperbolic supercosine through the Sonine function of the first type is given by

$$
\begin{aligned}
&\Psi \cosh_{\psi}(X) \\
&= \frac{1}{2}\Big(\Psi_{\psi}(1, X) + \Psi_{\psi}(-1, X)\Big) \\
&= X^{-\psi} \sum_{k=0}^{\infty} \frac{X^{2k}}{\Gamma(2k+1)\Gamma(2k-\psi+1)}.
\end{aligned}
$$

▷ The hyperbolic supertangent through the Sonine function of the first type is given by

$$
\Psi \tanh_{\psi}(X) = \frac{\Psi \sinh_{\psi}(X)}{\Psi \cosh_{\psi}(X)}, \; \Big(\Psi \cosh_{\psi}(X) \neq 0\Big).
$$

▷ The hyperbolic supercotangent through the Sonine function of the first type is given by

$$
\Psi \mathrm{cotanh}_{\psi}(X) = \frac{\Psi \cosh_{\psi}(X)}{\Psi \sin_{\psi}(X)}, \; \Big(\Psi \sinh_{\psi}(X) \neq 0\Big).
$$

▷ The hyperbolic supercosecant through the Sonine function of the first type is given by

$$
\Psi \mathrm{csch}_{\psi}(X) = \frac{1}{\Psi \sinh_{\psi}(X)}, \; \Big(\Psi \sinh_{\psi}(X) \neq 0\Big).
$$

▷ The hyperbolic supersecant through the Sonine function of the first type is given by

$$
\Psi \mathrm{sech}_{\psi}(X) = \frac{1}{\Psi \cosh_{\psi}(X)}, \; \Big(\Psi \cosh_{\psi}(X) \neq 0\Big).
$$

Thus, we have the following [3]:

$$
\begin{aligned}
\Psi \sinh_{\psi}(\varsigma, X) &= \frac{1}{2}\Big(\Psi_{\psi}(\varsigma, X) - \Psi_{\psi}(-\varsigma, X)\Big) \\
&= X^{-\psi} \sum_{k=0}^{\infty} \frac{(X\varsigma)^{2k+1}}{\Gamma(2k+2)\Gamma(2k-\psi+2)},
\end{aligned}
$$

$$\Psi \cosh_{\psi}(\varsigma, X) = \frac{1}{2}\Big(\Psi_{\psi}(\varsigma, X) + \Psi_{\psi}(-\varsigma, X)\Big)$$
$$= X^{-\psi} \sum_{k=0}^{\infty} \frac{(X\varsigma)^{2k}}{\Gamma(2k+1)\Gamma(2k-\psi+1)},$$

$$\Psi \tanh_{\psi}(\varsigma, X) = \frac{\Psi \sinh_{\psi}(\varsigma, X)}{\Psi \cosh_{\psi}(\varsigma, X)}, \ \Big(\Psi \cosh_{\psi}(\varsigma, X) \neq 0\Big),$$

$$\Psi \mathrm{cotanh}_{\psi}(\varsigma, X) = \frac{\Psi \cosh_{\psi}(\varsigma, X)}{\Psi \sinh_{\psi}(\varsigma, X)}, \ \Big(\Psi \sinh_{\psi}(\varsigma, X) \neq 0\Big),$$

$$\Psi \mathrm{csch}_{\psi}(\varsigma, X) = \frac{1}{\Psi \sinh_{\psi}(\varsigma, X)}, \ \Big(\Psi \sinh_{\psi}(\varsigma, X) \neq 0\Big),$$

$$\Psi \mathrm{sech}_{\psi}(\varsigma, X) = \frac{1}{\Psi \cosh_{\psi}(\varsigma, X)}, \ \Big(\Psi \cosh_{\psi}(\varsigma, X) \neq 0\Big).$$

7.4 The Sonine Function of the Second Type

Definition 7.2 ([1]) The Sonine function of the second type is given by

$$\Phi_{\psi}^{\phi}(X)$$
$$= X^{-\psi} \sum_{k=0}^{\infty} \frac{(\phi)_k X^k}{\Gamma(k+1)\Gamma(k-\psi+1)}$$
$$= \sum_{k=0}^{\infty} \frac{(\phi)_k X^{k-\psi}}{\Gamma(k+1)\Gamma(k-\psi+1)},$$

where $\psi, \phi, X \in \mathbb{C}$.

This result was discovered by Sonine in 1884.

In addition, we have that from [1],

$$
\begin{aligned}
&\Phi_{\psi}^{\phi}(\varsigma, X) \\
&= X^{-\psi} \sum_{k=0}^{\infty} \frac{(\phi)_k}{\Gamma(k-\psi+1)} \frac{(\varsigma X)^k}{k!} \\
&= \sum_{k=0}^{\infty} \frac{(\phi)_k \varsigma^k X^{k-\psi}}{\Gamma(k+1)\Gamma(k-\psi+1)} \\
&= \frac{1}{\Gamma(\phi+1)} \sum_{k=0}^{\infty} \frac{\Gamma(\phi+k)\varsigma^k X^{k-\psi}}{\Gamma(k+1)\Gamma(k-\psi+1)} \\
&= \frac{X^{-\psi}}{\Gamma(\phi+1)} {}_1\mathbb{W}_1\left({}^{(\phi,1)}_{(-\psi+1,1)}; \varsigma X \right), \qquad \phi \neq 0, -1, -2, \ldots,
\end{aligned}
$$

where $\varsigma, X, \psi, \phi \in \mathbb{C}$.

7.5 The Supertrigonometric Functions via the Sonine Function of the Second Type

Let $X, \varsigma, \psi, \phi \in \mathbb{C}$. In this part, we present the supertrigonometric functions via the Sonine function of the second type [3], as follows:

▷ The supersine through the Sonine function of the second type is given by

$$
\begin{aligned}
&\Phi \sin_{\psi}^{\phi}(X) \\
&= \frac{1}{2i}\left(\Phi_{\psi}^{\phi}(i, X) - \Phi_{\psi}^{\phi}(-i, X) \right) \\
&= X^{-\psi} \sum_{k=0}^{\infty} \frac{(\phi)_{2k+1}}{\Gamma(2k-\psi+2)} \frac{(-1)^k X^{2k+1}}{(2k+1)!}.
\end{aligned}
$$

▷ The supercosine through the Sonine function of the second type is given by

$$
\begin{aligned}
&\Phi \cos_{\psi}^{\phi}(X) \\
&= \frac{1}{2}\left(\Phi_{\psi}^{\phi}(i, X) + \Phi_{\psi}^{\phi}(-i, X) \right) \\
&= X^{-\psi} \sum_{k=0}^{\infty} \frac{(\phi)_{2k}}{\Gamma(2k-\psi+1)} \frac{(-1)^k X^{2k}}{(2k)!}.
\end{aligned}
$$

▷ The supertangent through the Sonine function of the second type is given by

$$\Phi \tan_{\psi}^{\phi}(X) = \frac{\Phi \sin_{\psi}^{\phi}(X)}{\Phi \cos_{\psi}^{\phi}(X)}, \quad \left(\Phi \cos_{\psi}^{\phi}(X) \neq 0\right).$$

▷ The supercotangent through the Sonine function of the second type is given by

$$\Phi \operatorname{cotan}_{\psi}^{\phi}(X) = \frac{\Phi \cos_{\psi}^{\phi}(X)}{\Phi \sin_{\psi}^{\phi}(X)}, \quad \left(\Phi \sin_{\psi}^{\phi}(X) \neq 0\right).$$

▷ The supercosecant through the Sonine function of the second type is given by

$$\Phi \csc_{\psi}^{\phi}(X) = \frac{1}{\Phi \sin_{\psi}^{\phi}(X)}, \quad \left(\Phi \sin_{\psi}^{\phi}(X) \neq 0\right).$$

▷ The supersecant through the Sonine function of the second type is given by

$$\Phi \sec_{\psi}^{\phi}(X) = \frac{1}{\Phi \cos_{\psi}^{\phi}(X)}, \quad \left(\Phi \cos_{\psi}^{\phi}(X) \neq 0\right).$$

Thus, we have that [3],

$$\begin{aligned}
\Phi \sin_{\psi}^{\phi}(\varsigma, X) &= \frac{1}{2i}\left(\Phi_{\psi}^{\phi}(i\varsigma, X) - \Phi_{\psi}^{\phi}(-i\varsigma, X)\right) \\
&= X^{-\psi} \sum_{k=0}^{\infty} \frac{(\phi)_{2k+1}}{\Gamma(2k-\psi+2)} \frac{(-1)^k (X\varsigma)^{2k+1}}{(2k+1)!},
\end{aligned}$$

$$\begin{aligned}
\Phi \cos_{\psi}^{\phi}(\varsigma, X) &= \frac{1}{2}\left(\Phi_{\psi}^{\phi}(i\varsigma, X) + \Phi_{\psi}^{\phi}(-i\varsigma, X)\right) \\
&= X^{-\psi} \sum_{k=0}^{\infty} \frac{(\phi)_{2k}}{\Gamma(2k-\psi+1)} \frac{(-1)^k (X\varsigma)^{2k}}{(2k)!},
\end{aligned}$$

$$\Phi \tan_{\psi}^{\phi}(\varsigma, X) = \frac{\Phi \sin_{\psi}^{\phi}(\varsigma, X)}{\Phi \cos_{\psi}^{\phi}(\varsigma, X)} \left(\Phi \cos_{\psi}^{\phi}(\varsigma, X) \neq 0\right),$$

$$\Phi\ \mathrm{cotan}_{\psi}^{\phi}(\varsigma, X) = \frac{\Phi\ \cos_{\psi}^{\phi}(\varsigma, X)}{\Phi\ \sin_{\psi}^{\phi}(\varsigma, X)}\left(\Phi\ \sin_{\psi}^{\phi}(\varsigma, X) \neq 0\right),$$

$$\Phi\ \csc_{\psi}^{\phi}(\varsigma, X) = \frac{1}{\Phi\ \sin_{\psi}^{\phi}(\varsigma, X)}\left(\Phi\ \sin_{\psi}^{\phi}(\varsigma, X) \neq 0\right),$$

$$\Phi\ \sec_{\psi}^{\phi}(\varsigma, X) = \frac{1}{\Phi\ \cos_{\psi}^{\phi}(\varsigma, X)}\left(\Phi\ \cos_{\psi}^{\phi}(\varsigma, X) \neq 0\right).$$

Also, we have

$$\Phi_{\psi}^{\phi}(i\varsigma, X) = \Phi\ \sin_{\psi}^{\phi}(\varsigma, X) + i\Phi\ \cos_{\psi}^{\phi}(\varsigma, X).$$

7.6 The Superhyperbolic Functions via the Sonine Function of The Second Type

Let $X, \varsigma, \psi, \phi \in \mathbb{C}$. In this part, we present the superhyperbolic functions via the Sonine function of the second type [3], as follows:

- ▷ The hyperbolic supersine through the Sonine function of the second type is given by

$$\begin{aligned}
&\Phi\ \sinh_{\psi}^{\phi}(X) \\
&= \frac{1}{2}\left(\Phi_{\psi}^{\phi}(1, X) - \Phi_{\psi}^{\phi}(-1, X)\right) \\
&= X^{-\psi} \sum_{k=0}^{\infty} \frac{(\phi)_{2k+1}}{\Gamma(2k-\psi+2)} \frac{X^{2k+1}}{(2k+1)!}.
\end{aligned}$$

- ▷ The hyperbolic supercosine through the Sonine function of the second type is given by

$$\begin{aligned}
&\Phi\ \cosh_{\psi}^{\phi}(X) \\
&= \frac{1}{2}\left(\Phi_{\psi}^{\phi}(1, X) + \Phi_{\psi}^{\phi}(-1, X)\right) \\
&= X^{-\psi} \sum_{k=0}^{\infty} \frac{(\phi)_{2k}}{\Gamma(2k-\psi+1)} \frac{X^{2k}}{(2k)!}.
\end{aligned}$$

▷ The hyperbolic supertangent through the Sonine function of the second type is given by

$$\Phi\ \tanh_{\psi}^{\phi}(X) = \frac{\Phi\ \sinh_{\psi}^{\phi}(X)}{\Phi\ \cosh_{\psi}^{\phi}(X)}, \quad \left(\Phi\ \cosh_{\psi}^{\phi}(X) \neq 0\right).$$

▷ The hyperbolic supercotangent through the Sonine function of the second type is given by

$$\Phi\ \mathrm{cotanh}_{\psi}^{\phi}(X) = \frac{\Phi\ \cosh_{\psi}^{\phi}(X)}{\Phi\ \sinh_{\psi}^{\phi}(X)}, \quad \left(\Phi\ \sinh_{\psi}^{\phi}(X) \neq 0\right).$$

▷ The hyperbolic supercosecant through the Sonine function of the second type is given by

$$\Phi\ \mathrm{csch}_{\psi}^{\phi}(X) = \frac{1}{\Phi\ \sinh_{\psi}^{\phi}(X)}, \quad \left(\Phi\ \sinh_{\psi}^{\phi}(X) \neq 0\right).$$

▷ The hyperbolic supersecant through the Sonine function of the second type is given by

$$\Phi\ \mathrm{sech}_{\psi}^{\phi}(X) = \frac{1}{\Phi\ \cosh_{\psi}^{\phi}(X)}, \quad \left(\Phi\ \cosh_{\psi}^{\phi}(X) \neq 0\right).$$

Thus, we have that [3],

$$\begin{aligned}\Phi\ \sinh_{\psi}^{\phi}(\varsigma, X) &= \frac{1}{2}\left(\Phi_{\psi}^{\phi}(\varsigma, X) - \Phi_{\psi}^{\phi}(-\varsigma, X)\right)\\ &= X^{-\psi}\sum_{k=0}^{\infty}\frac{(\phi)_{2k+1}}{\Gamma(2k-\psi+2)}\frac{(X\varsigma)^{2k+1}}{(2k+1)!},\end{aligned}$$

$$\begin{aligned}\Phi\ \cosh_{\psi}^{\phi}(\varsigma, X) &= \frac{1}{2}\left(\Phi_{\psi}^{\phi}(\varsigma, X) + \Phi_{\psi}^{\phi}(-\varsigma, X)\right)\\ &= X^{-\psi}\sum_{k=0}^{\infty}\frac{(\phi)_{2k}}{\Gamma(2k-\psi+1)}\frac{(X\varsigma)^{2k}}{(2k)!},\end{aligned}$$

$$\Phi \tanh_{\psi}^{\phi}(\varsigma, X) = \frac{\Phi \sinh_{\psi}^{\phi}(\varsigma, X)}{\Phi \cosh_{\psi}^{\phi}(\varsigma, X)}, \quad \left(\Phi \cosh_{\psi}^{\phi}(\varsigma, X) \neq 0\right),$$

$$\Phi \operatorname{cotanh}_{\psi}^{\phi}(\varsigma, X) = \frac{\Phi \cosh_{\psi}^{\phi}(\varsigma, X)}{\Phi \sinh_{\psi}^{\phi}(\varsigma, X)}, \quad \left(\Phi \sinh_{\psi}^{\phi}(\varsigma, X) \neq 0\right),$$

$$\Phi \operatorname{csch}_{\psi}^{\phi}(\varsigma, X) = \frac{1}{\Phi \sinh_{\psi}^{\phi}(\varsigma, X)}, \quad \left(\Phi \sinh_{\psi}^{\phi}(\varsigma, X) \neq 0\right),$$

$$\Phi \operatorname{sech}_{\psi}^{\phi}(\varsigma, X) = \frac{1}{\Phi \cosh_{\psi}^{\phi}(\varsigma, X)}, \quad \left(\Phi \cosh_{\psi}^{\phi}(\varsigma, X) \neq 0\right).$$

Also, we have

$$\Phi_{\psi}^{\phi}(\varsigma, X) = \Phi \sinh_{\psi}^{\phi}(\varsigma, X) + \Phi \cosh_{\psi}^{\phi}(\varsigma, X).$$

7.7 The Sonine Function of the Third Type

Definition 7.3 ([1]) The Sonine function of the third type is given by

$$\begin{aligned} &\Upsilon_{\kappa}(X) \\ &= X^{\kappa} \sum_{k=0}^{\infty} \frac{X^k}{\Gamma(k+\kappa)} \\ &= \sum_{k=0}^{\infty} \frac{X^{k+\kappa}}{\Gamma(k+\kappa)}, \quad X, \kappa \in \mathbb{C}. \end{aligned}$$

The Sonine function of the third type can alternatively also be given as [1],

$$\begin{aligned} &\Upsilon_{\kappa}(\varsigma, X) \\ &= X^{\kappa} \sum_{k=0}^{\infty} \frac{(X\varsigma)^k}{\Gamma(k+\kappa)} \\ &= \sum_{k=0}^{\infty} \frac{\varsigma^k (X)^{k+\kappa}}{\Gamma(k+\kappa)}, \end{aligned}$$

where $X, \kappa \in \mathbb{C}$.

Notice we have that [3],

$$\begin{aligned}
&\Upsilon_\kappa(\varsigma, X) \\
&= \sum_{k=0}^{\infty} \frac{\varsigma^k (X)^k}{\Gamma(k+\kappa)} \\
&= X^\kappa \mathbb{E}_{1,\kappa}(\varsigma X) \\
&= X^\kappa {}_1\mathbb{W}_1\left(\begin{smallmatrix}(1,1) \\ (\kappa,1)\end{smallmatrix}; \varsigma X\right),
\end{aligned}$$

which leads to

$$\Upsilon_\kappa(X) = \Upsilon_\kappa(1, X).$$

Notice that for any $\kappa, \varsigma \in \mathbb{C}$, $\Re(\kappa) > 0$, and $0 < \Re(Y) < 1$, we have the following from [3],

$$\Upsilon_\kappa(X) = \frac{1}{2\pi i} \int_{-i\infty}^{+i\infty} \frac{\Gamma(Y+\kappa)\Gamma(1-Y-\kappa)}{\Gamma(1-Y)} (-X)^{-(Y+\kappa)} dY.$$

7.8 The Supertrigonometric Functions via the Sonine Function of the Third Type

Let $X, \kappa \in \mathbb{C}$. In this part, we propose the supertrigonometric functions via the Sonine function of the third type [3], as follows:

▷ The supersine through the Sonine function of the third type is given by

$$\begin{aligned}
&\Upsilon \sin_\kappa(X) \\
&= \frac{1}{2i}\Big(\Upsilon_\kappa(i, X) - \Upsilon_\kappa(-i, X)\Big) \\
&= X^\kappa \sum_{k=0}^{\infty} \frac{(-1)^k X^{2k+1}}{\Gamma(2k+1+\kappa)}.
\end{aligned}$$

▷ The supercosine through the Sonine function of the third type is given by

$$\begin{aligned}
&\Upsilon \cos_{\kappa}(X)\\
&= \frac{1}{2}\Big(\Upsilon_{\kappa}(i, X) + \Upsilon_{\kappa}(-i, X)\Big)\\
&= X^{\kappa} \sum_{k=0}^{\infty} \frac{(-1)^k X^{2k}}{\Gamma(2k+\kappa)}.
\end{aligned}$$

▷ The supertangent through the Sonine function of the third type is given by

$$\Upsilon \tan_{\kappa}(X) = \frac{\Upsilon \sin_{\kappa}(X)}{\Upsilon \cos_{\kappa}(X)}, \quad \Big(\Upsilon \cos_{\kappa}(X) \neq 0\Big).$$

▷ The supercotangent through the Sonine function of the third type is given by

$$\Upsilon \mathrm{cotan}_{\kappa}(X) = \frac{\Upsilon \cos_{\kappa}(X)}{\Upsilon \sin_{\kappa}(X)}, \quad \Big(\Upsilon \sin_{\kappa}(X) \neq 0\Big).$$

▷ The supercosecant through the Sonine function of the third type is given by

$$\Upsilon \csc_{\kappa}(X) = \frac{1}{\Upsilon \sin_{\kappa}(X)}, \quad \Big(\Upsilon \sin_{\kappa}(X) \neq 0\Big).$$

▷ The supersecant through the Sonine function of the third type is given by

$$\Upsilon \sec_{\kappa}(X) = \frac{1}{\Upsilon \cos_{\kappa}(X)}, \quad \Big(\Upsilon \cos_{\kappa}(X) \neq 0\Big).$$

Thus, we get [3],

$$\begin{aligned}
\Upsilon \sin_{\kappa}(\varsigma, X) &= \frac{1}{2i}\Big(\Upsilon_{\kappa}(i\varsigma, X) - \Upsilon_{\kappa}(-i\varsigma, X)\Big)\\
&= X^{\kappa} \sum_{k=0}^{\infty} \frac{(-1)^k (X\varsigma)^{2k+1}}{\Gamma(2k+1+\kappa)},
\end{aligned}$$

$$\begin{aligned}
\Upsilon \cos_{\kappa}(\varsigma, X) &= \frac{1}{2}\Big(\Upsilon_{\kappa}(i\varsigma, X) + \Upsilon_{\kappa}(-i\varsigma, X)\Big)\\
&= X^{\kappa} \sum_{k=0}^{\infty} \frac{(-1)^k (X\varsigma)^{2k}}{\Gamma(2k+\kappa)},
\end{aligned}$$

$$\Upsilon \tan_{\kappa}(\varsigma, X) = \frac{\Upsilon \sin_{\kappa}(\varsigma, X)}{\Upsilon \cos_{\kappa}(\varsigma, X)}, \quad \Big(\Upsilon \cos_{\kappa}(\varsigma, X) \neq 0\Big),$$

$$\Upsilon \mathrm{cotan}_{\kappa}(\varsigma, X) = \frac{\Upsilon \cos_{\kappa}(\varsigma, X)}{\Upsilon \sin_{\kappa}(\varsigma, X)}, \quad \Big(\Upsilon \sin_{\kappa}(\varsigma, X) \neq 0\Big),$$

$$\Upsilon \csc_{\kappa}(\varsigma, X) = \frac{1}{\Upsilon \sin_{\kappa}(\varsigma, X)}, \quad \Big(\Upsilon \sin_{\kappa}(\varsigma, X) \neq 0\Big),$$

$$\Upsilon \sec_{\kappa}(\varsigma, X) = \frac{1}{\Upsilon \cos_{\kappa}(\varsigma, X)}, \quad \Big(\Upsilon \cos_{\kappa}(\varsigma, X) \neq 0\Big).$$

Theorem 7.2 ([3]) *Let $X, \varsigma, \kappa \in \mathbb{C}$. Then,*

$$\Upsilon_{\kappa}(i\varsigma, X) = \Upsilon \cos_{\kappa}(\varsigma, X) + i\Upsilon \sin_{\kappa}(\varsigma, X).$$

In particular,

$$\Upsilon_{\kappa}(X) = \Upsilon \cos_{\kappa}(X) + i\Upsilon \sin_{\kappa}(X).$$

7.9 The Superhyperbolic Functions via the Sonine Function of the Third Type

We suppose $X, \kappa \in \mathbb{C}$. In this part, we introduce the superhyperbolic functions via the Sonine function of the third type [3], as follows:

▷ The hyperbolic supersine through the Sonine function of the third type is given by

$$\begin{aligned} &\Upsilon \sinh_{\kappa}(X) \\ &= \frac{1}{2}\Big(\Upsilon_{\kappa}(1, X) - \Upsilon_{\kappa}(-1, X)\Big) \\ &= X^{\kappa} \sum_{k=0}^{\infty} \frac{X^{2k+1}}{\Gamma(2k+1+\kappa)}. \end{aligned}$$

▷ The hyperbolic supercosine through the Sonine function of the third type is given by

$$\begin{aligned}
&\Upsilon \cosh_{\kappa}(X) \\
&= \frac{1}{2}\Big(\Upsilon_{\kappa}(1, X) + \Upsilon_{\kappa}(-1, X)\Big) \\
&= X^{\kappa} \sum_{k=0}^{\infty} \frac{X^{2k}}{\Gamma(2k+\kappa)}.
\end{aligned}$$

▷ The hyperbolic supertangent through the Sonine function of the third type is given by

$$\Upsilon \tanh_{\kappa}(X) = \frac{\Upsilon \sinh_{\kappa}(X)}{\Upsilon \cosh_{\kappa}(X)}, \quad \Big(\Upsilon \cosh_{\kappa}(X) \neq 0\Big).$$

▷ The hyperbolic supercotangent through the Sonine function of the third type is given by

$$\Upsilon \mathrm{cotanh}_{\kappa}(X) = \frac{\Upsilon \cosh_{\kappa}(X)}{\Upsilon \sinh_{\kappa}(X)}, \quad \Big(\Upsilon \sinh_{\kappa}(X) \neq 0\Big).$$

▷ The hyperbolic supercosecant through the Sonine function of the third type is given by

$$\Upsilon \mathrm{csch}_{\kappa}(X) = \frac{1}{\Upsilon \sinh_{\kappa}(X)}, \quad \Big(\Upsilon \sinh_{\kappa}(X) \neq 0\Big).$$

▷ The hyperbolic supersecant through the Sonine function of the third type is given by

$$\Upsilon \mathrm{sech}_{\kappa}(X) = \frac{1}{\Upsilon \cosh_{\kappa}(X)}, \quad \Big(\Upsilon \cosh_{\kappa}(X) \neq 0\Big).$$

Thus,

$$\begin{aligned}
\Upsilon \sinh_{\kappa}(\varsigma, X) &= \frac{1}{2}\Big(\Upsilon_{\kappa}(\varsigma, X) - \Upsilon_{\kappa}(-\varsigma, X)\Big) \\
&= X^{\kappa} \sum_{k=0}^{\infty} \frac{(X\varsigma)^{2k+1}}{\Gamma(2k+1+\kappa)},
\end{aligned}$$

$$\Upsilon \cosh_\kappa(\varsigma, X) = \frac{1}{2}\Big(\Upsilon_\kappa(\varsigma, X) + \Upsilon_\kappa(-\varsigma, X)\Big) = X^\kappa \sum_{k=0}^{\infty} \frac{(X\varsigma)^{2k}}{\Gamma(2k+\kappa)},$$

$$\Upsilon \tanh_\kappa(\varsigma, X) = \frac{\Upsilon \sinh_\kappa(\varsigma, X)}{\Upsilon \cosh_\kappa(\varsigma, X)}, \quad \Big(\Upsilon \cosh_\kappa(\varsigma, X) \neq 0\Big),$$

$$\Upsilon \mathrm{cotanh}_\kappa(\varsigma, X) = \frac{\Upsilon \cosh_\kappa(\varsigma, X)}{\Upsilon \sinh_\kappa(\varsigma, X)}, \quad \Big(\Upsilon \sinh_\kappa(\varsigma, X) \neq 0\Big),$$

$$\Upsilon \mathrm{csch}_\kappa(\varsigma, X) = \frac{1}{\Upsilon \sinh_\kappa(\varsigma, X)}, \quad \Big(\Upsilon \sinh_\kappa(\varsigma, X) \neq 0\Big),$$

$$\Upsilon \mathrm{sech}_\kappa(\varsigma, X) = \frac{1}{\Upsilon \cosh_\kappa(\varsigma, X)}, \quad \Big(\Upsilon \cosh_\kappa(\varsigma, X) \neq 0\Big).$$

Theorem 7.3 ([3]) *Let* $X, \varsigma, \kappa \in \mathbb{C}$. *Then,*

$$\Upsilon_\kappa(\varsigma, X) = \Upsilon \cosh_\kappa(\varsigma, X) + \Upsilon \sinh_\kappa(\varsigma, X).$$

In particular,

$$\Upsilon_\kappa(X) = \Upsilon \cosh_\kappa(X) + \Upsilon \sinh_\kappa(X).$$

References

1. Sonine N (1884) Sur la généralisation d'une formule d'Abel. Acta Math 4(1):171–176
2. Yang XJ, Gao F, Ju Y (2020) General fractional derivatives with applications in viscoelasticity. Academic, New York
3. Yang X-J (2021) Theory and applications of special functions for scientists and engineers. Springer (2021)

Chapter 8
Other Special Functions, Supertrigonometric and Superhyperbolic Functions

In this chapter, we propose some other special functions including the Rabotnov fractional exponential function, the Miller–Ross function, the Lorenzo–Hartley function, the Kohlrausch–Williams–Watts function, the Fox's $\mathbb{H}$ function, the Meijer $\mathbb{G}$–function, and introduce the related supertrigonometric and superhyperbolic functions.

8.1 The Rabotnov Fractional Exponential Function, Supertrigonometric and Superhyperbolic Functions

In this section, we introduce some special functions related to the Mittag-Leffler function containing the Rabotnov fractional exponential function.

8.1.1 The Rabotnov Function

In 1948, Yuri Nikolayevich Rabotnov (a Russian academician and Russian mechanical engineer) worked in the theory of shells, plasticity, creep, and failure of materials, and proposed the special function [1], which is called the Rabotnov function or the Rabotnov fractional exponential function [2] applied in viscoelasticity.

Definition 8.1 ([1]) The Rabotnov function is given by

$$\begin{aligned}&\Lambda_{\kappa}(X)\\&=X^{\kappa}\sum_{k=0}^{\infty}\frac{X^{k(\kappa+1)}}{\Gamma((k+1)(\kappa+1))}\end{aligned}$$

S. Rezaei Aderyani et al., *Towards Ulam Type Multi Stability Analysis*, Studies in Systems, Decision and Control 523, https://doi.org/10.1007/978-3-031-55564-0_8

$$=\sum_{k=0}^{\infty}\frac{X^{k(\kappa+1)+\kappa}}{\Gamma((k+1)(\kappa+1))},$$

where $X, \kappa \in \mathbb{C}$.

The Rabotnov function can be written as [1]:

$$\Lambda_{\kappa}(\varsigma, X)=X^{\kappa}\sum_{k=0}^{\infty}\frac{\varsigma^{k}X^{k(\kappa+1)}}{\Gamma((k+1)(\kappa+1))},$$

where $X, \kappa, \varsigma \in \mathbb{C}$.

The relationship between the Mittag-Leffler function and the Rabotnov function can be written as [3]:

$$\begin{aligned}\Lambda_{\kappa}(\varsigma, X)&=X^{\kappa}\sum_{k=0}^{\infty}\frac{\varsigma^{k}X^{k(\kappa+1)}}{\Gamma((k+1)(\kappa+1))}\\&=X^{\kappa}\mathbb{E}_{\kappa+1,\kappa+1}(\varsigma X^{\kappa+1}),\end{aligned}$$

where $\kappa, X, \varsigma \in \mathbb{C}$.

The relationship between the generalized Wright function and the Rabotnov function can be given as [4]:

$$\begin{aligned}\Lambda_{\kappa}(\varsigma, X)&=X^{\kappa}\sum_{k=0}^{\infty}\frac{\varsigma^{k}X^{k(\kappa+1)}}{\Gamma((k+1)(\kappa+1))}\\&=X^{\kappa}{}_{1}\mathbb{W}_{1}\left(\begin{smallmatrix}(1,1)\\(\kappa+1,\kappa+1)\end{smallmatrix};\varsigma X^{\kappa+1}\right),\end{aligned}$$

where $\kappa, X, \varsigma \in \mathbb{C}$.

Now, we get the following special case:

The Gauss-Rabotnov type function can be written as a special case of the Rabotnov function [4]:

$$\begin{aligned}&\Lambda_{\kappa}(\varsigma, X^{2})\\&=X^{2\kappa}\sum_{k=0}^{\infty}\frac{\varsigma^{k}X^{2k(\kappa+1)}}{\Gamma((k+1)(\kappa+1))}\\&=X^{2\kappa}\mathbb{E}_{\kappa+1,\kappa+1}(\varsigma X^{2(\kappa+1)})\\&=X^{2\kappa}{}_{1}\mathbb{W}_{1}\left(\begin{smallmatrix}(1,1)\\(\kappa+1,\kappa+1)\end{smallmatrix};\varsigma X^{2(\kappa+1)}\right),\end{aligned}$$

where $X, \kappa, \varsigma \in \mathbb{C}$.

8.1.2 The Supertrigonometric Functions via the Rabotnov Function

Let $X, \kappa \in \mathbb{C}$. In this part, we propose the supertrigonometric functions via the Rabotnov function [4], as follows:

▷ The supersine through the Rabotnov function is given by

$$\begin{aligned}
&\Lambda \sin_\kappa(X)\\
&= \frac{1}{2i}\Big(\Lambda_\kappa(i, X) - \Lambda_\kappa(-i, X)\Big)\\
&= X^\kappa \sum_{k=0}^{\infty} \frac{(-1)^k X^{(2k+1)(\kappa+1)}}{\Gamma((2k+2)(\kappa+1))}.
\end{aligned}$$

▷ The supercosine through the Rabotnov function is given by

$$\begin{aligned}
&\Lambda \cos_\kappa(X)\\
&= \frac{1}{2}\Big(\Lambda_\kappa(i, X) + \Lambda_\kappa(-i, X)\Big)\\
&= X^\kappa \sum_{k=0}^{\infty} \frac{(-1)^k X^{(2k)(\kappa+1)}}{\Gamma((2k+1)(\kappa+1))}.
\end{aligned}$$

▷ The supertangent through the Rabotnov function is given by

$$\Lambda \tan_\kappa(X) = \frac{\Lambda \sin_\kappa(X)}{\Lambda \cos_\kappa(X)}, \quad \Big(\Lambda \cos_\kappa(X) \neq 0\Big).$$

▷ The supercotangent through the Rabotnov function is given by

$$\Lambda \text{cotan}_\kappa(X) = \frac{\Lambda \cos_\kappa(X)}{\Lambda \sin_\kappa(X)}, \quad \Big(\Lambda \sin_\kappa(X) \neq 0\Big).$$

▷ The supercosecant through the Rabotnov function is given by

$$\Lambda \csc_\kappa(X) = \frac{1}{\Lambda \sin_\kappa(X)}, \quad \Big(\Lambda \sin_\kappa(X) \neq 0\Big).$$

▷ The supersecant through Rabotnov function is given by

$$\Lambda \sec_\kappa(X) = \frac{1}{\Lambda \cos_\kappa(X)}, \quad \Big(\Lambda \cos_\kappa(X) \neq 0\Big).$$

Thus, we get [4] tat,

$$\Lambda \sin_\kappa(\varsigma, X) = \frac{1}{2i}\Big(\Lambda_\kappa(i\varsigma, X) - \Lambda_\kappa(-i\varsigma, X)\Big)$$
$$= X^\kappa \sum_{k=0}^{\infty} \frac{(-1)^k \varsigma^{2k+1} X^{(2k+1)(\kappa+1)}}{\Gamma((2k+2)(\kappa+1))},$$

$$\Lambda \cos_\kappa(\varsigma, X) = \frac{1}{2}\Big(\Lambda_\kappa(i\varsigma, X) + \Lambda_\kappa(-i\varsigma, X)\Big)$$
$$= X^\kappa \sum_{k=0}^{\infty} \frac{(-1)^k \varsigma^{2k} X^{(2k)(\kappa+1)}}{\Gamma((2k+1)(\kappa+1))},$$

$$\Lambda \tan_\kappa(\varsigma, X) = \frac{\Lambda \sin_\kappa(\varsigma, X)}{\Lambda \cos_\kappa(\varsigma, X)}, \quad \Big(\Lambda \cos_\kappa(\varsigma, X) \neq 0\Big),$$

$$\Lambda\mathrm{cotan}_\kappa(\varsigma, X) = \frac{\Lambda \cos_\kappa(\varsigma, X)}{\Lambda \sin_\kappa(\varsigma, X)}, \quad \Big(\Lambda \sin_\kappa(\varsigma, X) \neq 0\Big),$$

$$\Lambda\mathrm{csc}_\kappa(\varsigma, X) = \frac{1}{\Lambda \sin_\kappa(\varsigma, X)}, \quad \Big(\Lambda \sin_\kappa(\varsigma, X) \neq 0\Big),$$

$$\Lambda\mathrm{sec}_\kappa(\varsigma, X) = \frac{1}{\Lambda \cos_\kappa(\varsigma, X)}, \quad \Big(\Lambda \cos_\kappa(\varsigma, X) \neq 0\Big).$$

Theorem 8.1 ([4]) *Let* $\varsigma, \kappa, X \in \mathbb{C}$. *Then,*

$$\Lambda_\kappa(i\varsigma, X) = \Lambda \cos_\kappa(\varsigma, X) + i\Lambda \sin_\kappa(\varsigma, X).$$

8.1.3 The Superhyperbolic Functions via the Rabotnov Function

Let $X, \kappa \in \mathbb{C}$. In this part, we propose the superhyperbolic functions via the Rabotnov function [4], as follows:

- ▷ The hyperbolic supersine through the Rabotnov function is given by

$$
\begin{aligned}
&\Lambda \sinh_{\kappa}(X)\\
&= \frac{1}{2}\Big(\Lambda_{\kappa}(1, X) - \Lambda_{\kappa}(-1, X)\Big)\\
&= X^{\kappa} \sum_{k=0}^{\infty} \frac{X^{(2k+1)(\kappa+1)}}{\Gamma((2k+2)(\kappa+1))}.
\end{aligned}
$$

▷ The hyperbolic supercosine through the Rabotnov function is given by

$$
\begin{aligned}
&\Lambda \cosh_{\kappa}(X)\\
&= \frac{1}{2}\Big(\Lambda_{\kappa}(1, X) + \Lambda_{\kappa}(-1, X)\Big)\\
&= X^{\kappa} \sum_{k=0}^{\infty} \frac{X^{(2k)(\kappa+1)}}{\Gamma((2k+1)(\kappa+1))}.
\end{aligned}
$$

▷ The hyperbolic supertangent through the Rabotnov function is given by

$$
\Lambda \tanh_{\kappa}(X) = \frac{\Lambda \sinh_{\kappa}(X)}{\Lambda \cosh_{\kappa}(X)}, \quad \Big(\Lambda \cosh_{\kappa}(X) \neq 0\Big).
$$

▷ The hyperbolic supercotangent through the Rabotnov function is given by

$$
\Lambda \text{cotanh}_{\kappa}(X) = \frac{\Lambda \cosh_{\kappa}(X)}{\Lambda \sinh_{\kappa}(X)}, \quad \Big(\Lambda \sinh_{\kappa}(X) \neq 0\Big).
$$

▷ The hyperbolic supercosecant through the Rabotnov function is given by

$$
\Lambda \text{csch}_{\kappa}(X) = \frac{1}{\Lambda \sinh_{\kappa}(X)}, \quad \Big(\Lambda \sinh_{\kappa}(X) \neq 0\Big).
$$

▷ The hyperbolic supersecant through the Rabotnov function is given by

$$
\Lambda \text{sech}_{\kappa}(X) = \frac{1}{\Lambda \cosh_{\kappa}(X)}, \quad \Big(\Lambda \cosh_{\kappa}(X) \neq 0\Big).
$$

Thus, we get [4],

$$
\begin{aligned}
\Lambda \sinh_{\kappa}(\varsigma, X) &= \frac{1}{2}\Big(\Lambda_{\kappa}(\varsigma, X) - \Lambda_{\kappa}(-\varsigma, X)\Big)\\
&= X^{\kappa} \sum_{k=0}^{\infty} \frac{\varsigma^{2k+1} X^{(2k+1)(\kappa+1)}}{\Gamma((2k+2)(\kappa+1))},
\end{aligned}
$$

$$\Lambda \cosh_{\kappa}(\varsigma, X) = \frac{1}{2}\Big(\Lambda_{\kappa}(\varsigma, X) + \Lambda_{\kappa}(-\varsigma, X)\Big)$$
$$= X^{\kappa} \sum_{k=0}^{\infty} \frac{\varsigma^{2k} X^{(2k)(\kappa+1)}}{\Gamma((2k+1)(\kappa+1))},$$

$$\Lambda \tanh_{\kappa}(\varsigma, X) = \frac{\Lambda \sinh_{\kappa}(\varsigma, X)}{\Lambda \cosh_{\kappa}(\varsigma, X)}, \quad \Big(\Lambda \cosh_{\kappa}(\varsigma, X) \neq 0\Big),$$

$$\Lambda \mathrm{cotanh}_{\kappa}(\varsigma, X) = \frac{\Lambda \cosh_{\kappa}(\varsigma, X)}{\Lambda \sinh_{\kappa}(\varsigma, X)}, \quad \Big(\Lambda \sinh_{\kappa}(\varsigma, X) \neq 0\Big),$$

$$\Lambda \mathrm{csch}_{\kappa}(\varsigma, X) = \frac{1}{\Lambda \sinh_{\kappa}(\varsigma, X)}, \quad \Big(\Lambda \sinh_{\kappa}(\varsigma, X) \neq 0\Big),$$

$$\Lambda \mathrm{sech}_{\kappa}(\varsigma, X) = \frac{1}{\Lambda \cosh_{\kappa}(\varsigma, X)}, \quad \Big(\Lambda \cosh_{\kappa}(\varsigma, X) \neq 0\Big).$$

Theorem 8.2 ([4]) *Let* $\varsigma, \kappa, X \in \mathbb{C}$. *Then,*

$$\Lambda_{\kappa}(\varsigma, X) = \Lambda \cosh_{\kappa}(\varsigma, X) + \Lambda \sinh_{\kappa}(\varsigma, X).$$

8.2 The Miller–Ross, Supertrigonometric, and Superhyperbolic Functions

In this section, we introduce some special functions related to the Mittag- Leffler function containing the Miller-Ross function.

8.2.1 The Miller–Ross Function

In 1993, Kenneth S. Miller and Bertram Ross proposed the special function called the Miller-Ross function [2, 5].

Definition 8.2 ([2, 5]) The Miller–Ross function is given by

$$\begin{aligned}
&\Theta_{\kappa}(X)\\
&= X^{\kappa}\sum_{k=0}^{\infty}\frac{X^{k}}{\Gamma(k+1+\kappa)}\\
&= \sum_{k=0}^{\infty}\frac{X^{k+\kappa}}{\Gamma(k+1+\kappa)},
\end{aligned}$$

where $X, \kappa \in \mathbb{C}$.

The Miller-Ross function can be written as follows [2, 5]:

$$\begin{aligned}
&\Theta_{\kappa}(\varsigma, X)\\
&= X^{\kappa}\sum_{k=0}^{\infty}\frac{\varsigma^{k}X^{k}}{\Gamma(k+1+\kappa)}\\
&= \sum_{k=0}^{\infty}\frac{\varsigma^{k}X^{k+\kappa}}{\Gamma(k+1+\kappa)},
\end{aligned}$$

where $X, \kappa, \varsigma \in \mathbb{C}$.

It is straightforward to show that [3]

$$\begin{aligned}
\Theta_{\kappa}(\varsigma, X) &= \sum_{k=0}^{\infty}\frac{(\varsigma X)^{k}}{\Gamma(k+1+\kappa)}\\
&= X^{\kappa}\mathbb{E}_{1,1+\kappa}(\varsigma X)\\
&= X^{\kappa}{}_{1}\mathbb{W}_{1}\left({}^{(1,1)}_{(1+\kappa,1)}; \varsigma X\right),
\end{aligned}$$

which leads to

$$\Theta_{\kappa}(X) = \Theta_{\kappa}(1, X).$$

Notice that we have [4],

$$\Theta_{\kappa}(X) = \frac{1}{2\pi i}\int_{-i\infty}^{i\infty}\frac{\Gamma(Y+\kappa)\Gamma(1-Y-\kappa)}{\Gamma(1-Y)}(-X)^{-(Y+\kappa)}dY,$$

where $X, \kappa \in \mathbb{C}$, $\Re(\kappa) > -1$, and $0 < \Re(Y) < 1$.

8.2.2 The Supertrigonometric Functions via the Miller–Ross Function

Let $\kappa, X \in \mathbb{C}$. In this part, we propose the supertrigonometric functions via the Miller-Ross function [4], as follows:

▷ The Miller-Ross supersine function is given by

$$\begin{aligned}&\Theta \sin_{\kappa}(X)\\&=\frac{1}{2i}\Big(\Theta_{\kappa}(i, X)-\Theta_{\kappa}(-i, X)\Big)\\&=X^{\kappa}\sum_{k=0}^{\infty}\frac{(-1)^k X^{2k+1}}{\Gamma(2k+2+\kappa)},\end{aligned}$$

▷ The Miller-Ross supercosine function is given by

$$\begin{aligned}&\Theta \cos_{\kappa}(X)\\&=\frac{1}{2}\Big(\Theta_{\kappa}(i, X)+\Theta_{\kappa}(-i, X)\Big)\\&=X^{\kappa}\sum_{k=0}^{\infty}\frac{(-1)^k X^{2k}}{\Gamma(2k+1+\kappa)},\end{aligned}$$

▷ The Miller-Ross supertangent function is given by

$$\Theta \tan_{\kappa}(X)=\frac{\Theta \sin_{\kappa}(X)}{\Theta \cos_{\kappa}(X)},\quad \Big(\Theta \cos_{\kappa}(X)\neq 0\Big).$$

▷ The Miller-Ross supercotangent function is given by

$$\Theta \mathrm{cotan}_{\kappa}(X)=\frac{\Theta \cos_{\kappa}(X)}{\Theta \sin_{\kappa}(X)},\quad \Big(\Theta \sin_{\kappa}(X)\neq 0\Big).$$

▷ The Miller-Ross supercosecant function is given by

$$\Theta \csc_{\kappa}(X)=\frac{1}{\Theta \sin_{\kappa}(X)},\quad \Big(\Theta \sin_{\kappa}(X)\neq 0\Big).$$

▷ The Miller-Ross supersecant function is given by

$$\Theta \sec_{\kappa}(X)=\frac{1}{\Theta \cos_{\kappa}(X)},\quad \Big(\Theta \cos_{\kappa}(X)\neq 0\Big).$$

Thus, we get [4],

$$\Theta \sin_{\kappa}(\varsigma, X) = \frac{1}{2i}\Big(\Theta_{\kappa}(i\varsigma, X) - \Theta_{\kappa}(-i\varsigma, X)\Big)$$
$$= X^{\kappa} \sum_{k=0}^{\infty} \frac{(-1)^k (\varsigma X)^{2k+1}}{\Gamma(2k+2+\kappa)},$$

$$\Theta \cos_{\kappa}(\varsigma, X) = \frac{1}{2}\Big(\Theta_{\kappa}(i\varsigma, X) + \Theta_{\kappa}(-i\varsigma, X)\Big)$$
$$= X^{\kappa} \sum_{k=0}^{\infty} \frac{(-1)^k (\varsigma X)^{2k}}{\Gamma(2k+1+\kappa)},$$

$$\Theta \tan_{\kappa}(\varsigma, X) = \frac{\Theta \sin_{\kappa}(\varsigma, X)}{\Theta \cos_{\kappa}(\varsigma, X)}, \quad \Big(\Theta \cos_{\kappa}(\varsigma, X) \neq 0\Big),$$

$$\Theta \mathrm{cotan}_{\kappa}(\varsigma, X) = \frac{\Theta \cos_{\kappa}(\varsigma, X)}{\Theta \sin_{\kappa}(\varsigma, X)}, \quad \Big(\Theta \sin_{\kappa}(\varsigma, X) \neq 0\Big),$$

$$\Theta \mathrm{csc}_{\kappa}(\varsigma, X) = \frac{1}{\Theta \sin_{\kappa}(\varsigma, X)}, \quad \Big(\Theta \sin_{\kappa}(\varsigma, X) \neq 0\Big),$$

$$\Theta \sec_{\kappa}(\varsigma, X) = \frac{1}{\Theta \cos_{\kappa}(\varsigma, X)}, \quad \Big(\Theta \cos_{\kappa}(\varsigma, X) \neq 0\Big).$$

Theorem 8.3 ([4]) *Let* $\kappa, X, \varsigma \in \mathbb{C}$*. Then, we get*

$$\Theta_{\kappa}(i\varsigma, X) = \Theta \cos_{\kappa}(\varsigma, X) + i\Theta \sin_{\kappa}(\varsigma, X),$$

and

$$\Theta_{\kappa}(X) = \Theta \cos_{\kappa}(X) + i\Theta \sin_{\kappa}(X).$$

8.2.3 The Superhyperbolic Functions via the Miller–Ross Function

Let $\kappa, X, \varsigma \in \mathbb{C}$. In this part, we define the superhyperbolic functions via the Miller-Ross function [4], as follows:

▹ The Miller-Ross hyperbolic supersine function is given by

$$\begin{aligned}&\Theta \sinh_{\kappa}(X)\\&=\frac{1}{2}\Big(\Theta_{\kappa}(1, X)-\Theta_{\kappa}(-1, X)\Big)\\&=X^{\kappa}\sum_{k=0}^{\infty}\frac{X^{2k+1}}{\Gamma(2k+2+\kappa)}.\end{aligned}$$

▹ The Miller-Ross hyperbolic supercosine function is given by

$$\begin{aligned}&\Theta \cosh_{\kappa}(X)\\&=\frac{1}{2}\Big(\Theta_{\kappa}(1, X)+\Theta_{\kappa}(-1, X)\Big)\\&=X^{\kappa}\sum_{k=0}^{\infty}\frac{X^{2k}}{\Gamma(2k+1+\kappa)}.\end{aligned}$$

▹ The Miller-Ross hyperbolic supertangent function is given by

$$\Theta \tanh_{\kappa}(X)=\frac{\Theta \sinh_{\kappa}(X)}{\Theta \cos_{\kappa}(X)}, \quad \Big(\Theta \cosh_{\kappa}(X)\neq 0\Big).$$

▹ The Miller-Ross hyperbolic supercotangent function is given by

$$\Theta \text{cotanh}_{\kappa}(X)=\frac{\Theta \cosh_{\kappa}(X)}{\Theta \sinh_{\kappa}(X)}, \quad \Big(\Theta \sinh_{\kappa}(X)\neq 0\Big).$$

▹ The Miller-Ross hyperbolic supercosecant function is given by

$$\Theta \text{csch}_{\kappa}(X)=\frac{1}{\Theta \sinh_{\kappa}(X)}, \quad \Big(\Theta \sinh_{\kappa}(X)\neq 0\Big).$$

▹ The Miller-Ross hyperbolic supersecant function is given by

$$\Theta \text{sech}_{\kappa}(X)=\frac{1}{\Theta \cosh_{\kappa}(X)}, \quad \Big(\Theta \cosh_{\kappa}(X)\neq 0\Big).$$

Thus, we get [4] the following:

$$\begin{aligned}\Theta \sinh_{\kappa}(\varsigma, X) &= \frac{1}{2}\Big(\Theta_{\kappa}(\varsigma, X) - \Theta_{\kappa}(-\varsigma, X)\Big)\\ &= X^{\kappa}\sum_{k=0}^{\infty}\frac{(\varsigma X)^{2k+1}}{\Gamma(2k+2+\kappa)},\end{aligned}$$

$$\begin{aligned}\Theta \cosh_{\kappa}(\varsigma, X) &= \frac{1}{2}\Big(\Theta_{\kappa}(\varsigma, X) + \Theta_{\kappa}(-\varsigma, X)\Big)\\ &= X^{\kappa}\sum_{k=0}^{\infty}\frac{(\varsigma X)^{2k}}{\Gamma(2k+1+\kappa)},\end{aligned}$$

$$\Theta \tanh_{\kappa}(\varsigma, X) = \frac{\Theta \sinh_{\kappa}(\varsigma, X)}{\Theta \cosh_{\kappa}(\varsigma, X)}, \quad \Big(\Theta \cosh_{\kappa}(\varsigma, X) \neq 0\Big),$$

$$\Theta \mathrm{cotanh}_{\kappa}(\varsigma, X) = \frac{\Theta \cosh_{\kappa}(\varsigma, X)}{\Theta \sinh_{\kappa}(\varsigma, X)}, \quad \Big(\Theta \sinh_{\kappa}(\varsigma, X) \neq 0\Big),$$

$$\Theta \mathrm{csch}_{\kappa}(\varsigma, X) = \frac{1}{\Theta \sinh_{\kappa}(\varsigma, X)}, \quad \Big(\Theta \sinh_{\kappa}(\varsigma, X) \neq 0\Big),$$

$$\Theta \mathrm{sech}_{\kappa}(\varsigma, X) = \frac{1}{\Theta \cosh_{\kappa}(\varsigma, X)}, \quad \Big(\Theta \cosh_{\kappa}(\varsigma, X) \neq 0\Big).$$

Theorem 8.4 *Let $\kappa, X, \varsigma \in \mathbb{C}$. Then, we have from [4]*

$$\Theta_{\kappa}(\varsigma, X) = \Theta \cosh_{\kappa}(\varsigma, X) + \Theta \sinh_{\kappa}(\varsigma, X),$$

and

$$\Theta_{\kappa}(X) = \Theta \cosh_{\kappa}(X) + \Theta \sinh_{\kappa}(X).$$

8.3 The Lorenzo–Hartley, Supertrigonometric, and Superhyperbolic Functions

In this section, we introduce some special functions related to the Mittag-Leffler function containing the Lorenzo-Hartley functions.

In 1998 and 2008, Carl F. Lorenzo and Tom T. Hartley proposed some special functions, which are called the Lorenzo-Hartley function of the first type (one-parametric Lorenzo-Hartley function) and Lorenzo-Hartley function of the second type (two-parametric Lorenzo-Hartley function) [2, 6–8], respectively.

8.3.1 The Lorenzo–Hartley Function of the First Type

Definition 8.3 ([2, 6, 7]) The Lorenzo–Hartley function of the first type is given by

$$\Omega_{\kappa}(X) = X^{\kappa-1}\sum_{k=0}^{\infty}\frac{X^{k\kappa}}{\Gamma((k+1)\kappa)},$$

where $\kappa, X \in \mathbb{C}$.

We may consider the Lorenzo-Hartley function of the first type [2, 6, 7] as

$$\begin{aligned}
&\Omega_{\kappa}(\varsigma, X)\\
&= X^{\kappa-1}\sum_{k=0}^{\infty}\frac{\varsigma^{k}X^{k\kappa}}{\Gamma((k+1)\kappa)}\\
&= X^{\kappa-1}\mathbb{E}_{\kappa,\kappa}(\varsigma X^{\kappa})\\
&= X^{\kappa-1}{}_1\mathbb{W}_1\left(\begin{smallmatrix}(1,1)\\(\kappa,\kappa)\end{smallmatrix}; \varsigma X^{\kappa}\right).
\end{aligned}$$

This leads to

$$\Omega_{\kappa}(X) = \Omega_{\kappa}(1, X),$$

where $\kappa, X \in \mathbb{C}$.

8.3.2 The Supertrigonometric Functions via the Lorenzo–Hartley Function of the First Type

Let $X, \kappa, \varsigma \in \mathbb{C}$. In this part, we define the supertrigonometric functions via the Lorenzo-Hartley function of the first type [4], as follows:

▷ The supersine through the Lorenzo-Hartley function of the first type is given by

$$\begin{aligned}&\Omega \sin_{\kappa}(X)\\&=\frac{1}{2i}\Big(\Omega_{\kappa}(i, X)-\Omega_{\kappa}(-i, X)\Big)\\&=X^{\kappa-1}\sum_{k=0}^{\infty}\frac{(-1)^{k}X^{(2k+1)\kappa}}{\Gamma(2k\kappa+2\kappa)}.\end{aligned}$$

▷ The supercosine through the Lorenzo-Hartley function of the first type is given by

$$\begin{aligned}&\Omega \cos_{\kappa}(X)\\&=\frac{1}{2}\Big(\Omega_{\kappa}(i, X)+\Omega_{\kappa}(-i, X)\Big)\\&=X^{\kappa-1}\sum_{k=0}^{\infty}\frac{(-1)^{k}X^{(2k)\kappa}}{\Gamma(2k\kappa+\kappa)}.\end{aligned}$$

▷ The supertangent through the Lorenzo-Hartley function of the first type is given by

$$\Omega \tan_{\kappa}(X)=\frac{\Omega \sin_{\kappa}(X)}{\Omega \cos_{\kappa}(X)},\quad \Big(\Omega \cos_{\kappa}(X)\neq 0\Big).$$

▷ The supercotangent through the Lorenzo-Hartley function of the first type is given by

$$\Omega\mathrm{cotan}_{\kappa}(X)=\frac{\Omega \cos_{\kappa}(X)}{\Omega \sin_{\kappa}(X)},\quad \Big(\Omega \sin_{\kappa}(X)\neq 0\Big).$$

▷ The supercosecant through the Lorenzo-Hartley function of the first type is given by

$$\Omega\mathrm{csc}_{\kappa}(X)=\frac{1}{\Omega \sin_{\kappa}(X)},\quad \Big(\Omega \sin_{\kappa}(X)\neq 0\Big).$$

▷ The supersecant through the Lorenzo-Hartley function of the first type is given by

$$\Omega\mathrm{sec}_{\kappa}(X)=\frac{1}{\Omega \cos_{\kappa}(X)},\quad \Big(\Omega \cos_{\kappa}(X)\neq 0\Big).$$

Therefore, we get [4],

$$\Omega \sin_\kappa(\varsigma, X) = \frac{1}{2i}\Big(\Omega_\kappa(i\varsigma, X) - \Omega_\kappa(-i\varsigma, X)\Big)$$
$$= X^{\kappa-1}\sum_{k=0}^{\infty}\frac{(-1)^k\varsigma^{2k+1}X^{(2k+1)\kappa}}{\Gamma(2k\kappa+2\kappa)},$$

$$\Omega \cos_\kappa(\varsigma, X) = \frac{1}{2}\Big(\Omega_\kappa(i\varsigma, X) + \Omega_\kappa(-i\varsigma, X)\Big)$$
$$= X^{\kappa-1}\sum_{k=0}^{\infty}\frac{(-1)^k\varsigma^{2k}X^{(2k)\kappa}}{\Gamma(2k\kappa+\kappa)},$$

$$\Omega \tan_\kappa(\varsigma, X) = \frac{\Omega \sin_\kappa(\varsigma, X)}{\Omega \cos_\kappa(\varsigma, X)}, \quad \Big(\Omega \cos_\kappa(\varsigma, X) \neq 0\Big),$$

$$\Omega\mathrm{cotan}_\kappa(\varsigma, X) = \frac{\Omega \cos_\kappa(\varsigma, X)}{\Omega \sin_\kappa(\varsigma, X)}, \quad \Big(\Omega \sin_\kappa(\varsigma, X) \neq 0\Big),$$

$$\Omega\csc_\kappa(\varsigma, X) = \frac{1}{\Omega \sin_\kappa(\varsigma, X)}, \quad \Big(\Omega \sin_\kappa(\varsigma, X) \neq 0\Big),$$

$$\Omega\sec_\kappa(\varsigma, X) = \frac{1}{\Omega \cos_\kappa(\varsigma, X)}, \quad \Big(\Omega \cos_\kappa(\varsigma, X) \neq 0\Big).$$

Theorem 8.5 ([4]) *Let $X, \varsigma, \kappa \in \mathbb{C}$. Then,*

$$\Omega_\kappa(i\varsigma, X) = \Omega \cos_\kappa(\varsigma, X) + i\Omega \sin_\kappa(\varsigma, X),$$

and

$$\Omega_\kappa(X) = \Omega \cos_\kappa(X) + i\Omega \sin_\kappa(X).$$

8.3.3 The Superhyperbolic Functions via the Lorenzo–Hartley Function of the First Type

Let $X, \kappa, \varsigma \in \mathbb{C}$. In this part, we present the superhyperbolic functions via the Lorenzo-Hartley function of the first type [4], as follows:

▷ The hyperbolic supersine through the Lorenzo-Hartley function of the first type is given by

$$\begin{aligned}
&\Omega \sinh_\kappa(X) \\
&= \frac{1}{2}\Big(\Omega_\kappa(1, X) - \Omega_\kappa(-1, X)\Big) \\
&= X^{\kappa-1} \sum_{k=0}^{\infty} \frac{X^{(2k+1)\kappa}}{\Gamma(2k\kappa + 2\kappa)}.
\end{aligned}$$

▷ The hyperbolic supercosine through the Lorenzo-Hartley function of the first type is given by

$$\begin{aligned}
&\Omega \cosh_\kappa(X) \\
&= \frac{1}{2}\Big(\Omega_\kappa(1, X) + \Omega_\kappa(-1, X)\Big) \\
&= X^{\kappa-1} \sum_{k=0}^{\infty} \frac{X^{(2k)\kappa}}{\Gamma(2k\kappa + \kappa)}.
\end{aligned}$$

▷ The hyperbolic supertangent through the Lorenzo-Hartley function of the first type is given by

$$\Omega \tanh_\kappa(X) = \frac{\Omega \sinh_\kappa(X)}{\Omega \cosh_\kappa(X)}, \quad \Big(\Omega \cosh_\kappa(X) \neq 0\Big).$$

▷ The hyperbolic supercotangent through the Lorenzo-Hartley function of the first type is given by

$$\Omega \text{cotanh}_\kappa(X) = \frac{\Omega \cosh_\kappa(X)}{\Omega \sinh_\kappa(X)}, \quad \Big(\Omega \sinh_\kappa(X) \neq 0\Big).$$

▷ The hyperbolic supercosecant through the Lorenzo-Hartley function of the first type is given by

$$\Omega \text{csch}_\kappa(X) = \frac{1}{\Omega \sinh_\kappa(X)}, \quad \Big(\Omega \sinh_\kappa(X) \neq 0\Big).$$

▷ The hyperbolic supersecant through the Lorenzo-Hartley function of the first type is given by

$$\Omega \text{sech}_\kappa(X) = \frac{1}{\Omega \cosh_\kappa(X)}, \quad \Big(\Omega \cosh_\kappa(X) \neq 0\Big).$$

Therefore, we get from [4]

$$\begin{aligned}\Omega \sinh_{\kappa}(\varsigma, X) &= \frac{1}{2}\Big(\Omega_{\kappa}(\varsigma, X) - \Omega_{\kappa}(-\varsigma, X)\Big)\\ &= X^{\kappa-1}\sum_{k=0}^{\infty}\frac{\varsigma^{2k+1}X^{(2k+1)\kappa}}{\Gamma(2k\kappa+2\kappa)},\end{aligned}$$

$$\begin{aligned}\Omega \cosh_{\kappa}(\varsigma, X) &= \frac{1}{2}\Big(\Omega_{\kappa}(\varsigma, X) + \Omega_{\kappa}(-\varsigma, X)\Big)\\ &= X^{\kappa-1}\sum_{k=0}^{\infty}\frac{\varsigma^{2k}X^{(2k)\kappa}}{\Gamma(2k\kappa+\kappa)},\end{aligned}$$

$$\Omega \tanh_{\kappa}(\varsigma, X) = \frac{\Omega \sinh_{\kappa}(\varsigma, X)}{\Omega \cosh_{\kappa}(\varsigma, X)}, \quad \Big(\Omega \cosh_{\kappa}(\varsigma, X) \neq 0\Big),$$

$$\Omega \mathrm{cotanh}_{\kappa}(\varsigma, X) = \frac{\Omega \cosh_{\kappa}(\varsigma, X)}{\Omega \sinh_{\kappa}(\varsigma, X)}, \quad \Big(\Omega \sinh_{\kappa}(\varsigma, X) \neq 0\Big),$$

$$\Omega \mathrm{csch}_{\kappa}(\varsigma, X) = \frac{1}{\Omega \sinh_{\kappa}(\varsigma, X)}, \quad \Big(\Omega \sinh_{\kappa}(\varsigma, X) \neq 0\Big),$$

$$\Omega \mathrm{sech}_{\kappa}(\varsigma, X) = \frac{1}{\Omega \cosh_{\kappa}(\varsigma, X)}, \quad \Big(\Omega \cosh_{\kappa}(\varsigma, X) \neq 0\Big).$$

Theorem 8.6 ([4]) *Let $X, \varsigma, \kappa \in \mathbb{C}$. Then,*

$$\Omega_{\kappa}(\varsigma, X) = \Omega \cosh_{\kappa}(\varsigma, X) + \Omega \sinh_{\kappa}(\varsigma, X),$$

and

$$\Omega_{\kappa}(X) = \Omega \cosh_{\kappa}(X) + \Omega \sinh_{\kappa}(X).$$

8.3.4 The Lorenzo–Hartley Function of the Second Type

In this part, we investigate the special functions related to the Lorenzo-Hartley function of the second type.

Definition 8.4 ([2, 7, 8]) The Lorenzo-Hartley function of the second type is given by

$$\Omega_{\kappa,\vartheta}(X) = X^{\kappa-\vartheta-1} \sum_{k=0}^{\infty} \frac{X^{k\kappa}}{\Gamma((k+1)\kappa - \vartheta)},$$

where $\kappa, \vartheta, X \in \mathbb{C}$, and $\Re(\kappa) > \Re(\vartheta)$.

For $\kappa, \vartheta, X, \varsigma \in \mathbb{C}$, and $\Re(\kappa) > \Re(\vartheta)$, we also have the Lorenzo-Hartley function of the second type [2, 7, 8] as

$$\begin{aligned}
&\Omega_{\kappa,\vartheta}(\varsigma, X) \\
&= X^{\kappa-\vartheta-1} \sum_{k=0}^{\infty} \frac{\varsigma^k X^{k\kappa}}{\Gamma((k+1)\kappa - \vartheta)} \\
&= X^{\kappa-\vartheta-1} \mathbb{E}_{\kappa,\kappa-\vartheta}(\varsigma X^{\kappa}) \\
&= X^{\kappa-\vartheta-1} {}_1\mathbb{W}_1\left({}^{(1,1)}_{(\kappa,\kappa-\vartheta)}; \varsigma X^{\kappa} \right),
\end{aligned}$$

which leads to

$$\Omega_{\kappa,\vartheta}(1, X) = \Omega_{\kappa,\vartheta}(X).$$

8.3.5 The Supertrigonometric Functions via the Lorenzo–Hartley Function of the Second Type

Assume that $X, \varsigma, \kappa, \vartheta \in \mathbb{C}$. In this part, we give the supertrigonometric functions via the Lorenzo-Hartley function of the second type [4], as follows:

- ▷ The supersine through the Lorenzo-Hartley function of the second type is given by

$$\begin{aligned}
&\Omega \sin_{\kappa,\vartheta}(X) \\
&= \frac{1}{2i}\Big(\Omega_{\kappa,\vartheta}(i, X) - \Omega_{\kappa,\vartheta}(-i, X) \Big) \\
&= X^{\kappa-\vartheta-1} \sum_{k=0}^{\infty} \frac{(-1)^k X^{(2k+1)\kappa}}{\Gamma((2k+2)\kappa - \vartheta)}.
\end{aligned}$$

- ▷ The supercosine through the Lorenzo-Hartley function of the second type is given by

$$\begin{aligned}
&\Omega\cos_{\kappa,\vartheta}(X)\\
&=\frac{1}{2}\Big(\Omega_{\kappa,\vartheta}(i,X)+\Omega_{\kappa,\vartheta}(-i,X)\Big)\\
&=X^{\kappa-\vartheta-1}\sum_{k=0}^{\infty}\frac{(-1)^k X^{(2k)\kappa}}{\Gamma((2k+1)\kappa-\vartheta)}.
\end{aligned}$$

▷ The supertangent through the Lorenzo-Hartley function of the second type is given by

$$\Omega\tan_{\kappa,\vartheta}(X)=\frac{\Omega\sin_{\kappa,\vartheta}(X)}{\Omega\cos_{\kappa,\vartheta}(X)},\quad \Big(\Omega\cos_{\kappa,\vartheta}(X)\neq 0\Big).$$

▷ The supercotangent through the Lorenzo-Hartley function of the second type is given by

$$\Omega\mathrm{cotan}_{\kappa,\vartheta}(X)=\frac{\Omega\cos_{\kappa,\vartheta}(X)}{\Omega\sin_{\kappa,\vartheta}(X)},\quad \Big(\Omega\sin_{\kappa,\vartheta}(X)\neq 0\Big),$$

▷ The supercosecant through the Lorenzo-Hartley function of the second type is given by

$$\Omega\csc_{\kappa,\vartheta}(X)=\frac{1}{\Omega\sin_{\kappa,\vartheta}(X)},\quad \Big(\Omega\sin_{\kappa,\vartheta}(X)\neq 0\Big),$$

▷ The supersecant through the Lorenzo-Hartley function of the second type is given by

$$\Omega\sec_{\kappa,\vartheta}(X)=\frac{1}{\Omega\cos_{\kappa,\vartheta}(X)},\quad \Big(\Omega\cos_{\kappa,\vartheta}(X)\neq 0\Big).$$

Thus, we get [4]

$$\begin{aligned}
&\Omega\sin_{\kappa,\vartheta}(\varsigma,X)\\
&=\frac{1}{2i}\Big(\Omega_{\kappa,\vartheta}(i\varsigma,X)-\Omega_{\kappa,\vartheta}(-i\varsigma,X)\Big)\\
&=X^{\kappa-\vartheta-1}\sum_{k=0}^{\infty}\frac{(-1)^k\varsigma^{2k+1}X^{(2k+1)\kappa}}{\Gamma((2k+2)\kappa-\vartheta)},
\end{aligned}$$

$$
\begin{aligned}
&\Omega \cos_{\kappa,\vartheta}(\varsigma, X) \\
&= \frac{1}{2}\left(\Omega_{\kappa,\vartheta}(i\varsigma, X) + \Omega_{\kappa,\vartheta}(-i\varsigma, X)\right) \\
&= X^{\kappa-\vartheta-1}\sum_{k=0}^{\infty}\frac{(-1)^k\varsigma^{2k}X^{(2k)\kappa}}{\Gamma((2k+1)\kappa-\vartheta)},
\end{aligned}
$$

$$
\Omega \tan_{\kappa,\vartheta}(\varsigma, X) = \frac{\Omega \sin_{\kappa,\vartheta}(\varsigma, X)}{\Omega \cos_{\kappa,\vartheta}(\varsigma, X)}, \quad \left(\Omega \cos_{\kappa,\vartheta}(\varsigma, X) \neq 0\right),
$$

$$
\Omega \mathrm{cotan}_{\kappa,\vartheta}(\varsigma, X) = \frac{\Omega \cos_{\kappa,\vartheta}(\varsigma, X)}{\Omega \sin_{\kappa,\vartheta}(\varsigma, X)}, \quad \left(\Omega \sin_{\kappa,\vartheta}(\varsigma, X) \neq 0\right),
$$

$$
\Omega \csc_{\kappa,\vartheta}(\varsigma, X) = \frac{1}{\Omega \sin_{\kappa,\vartheta}(\varsigma, X)}, \quad \left(\Omega \sin_{\kappa,\vartheta}(\varsigma, X) \neq 0\right),
$$

$$
\Omega \sec_{\kappa,\vartheta}(\varsigma, X) = \frac{1}{\Omega \cos_{\kappa,\vartheta}(\varsigma, X)}, \quad \left(\Omega \cos_{\kappa,\vartheta}(\varsigma, X) \neq 0\right),
$$

where $X, \varsigma, \kappa, \vartheta \in \mathbb{C}$.

Theorem 8.7 ([4]) *Let $X, \varsigma, \kappa, \vartheta \in \mathbb{C}$. Then,*

$$
\Omega_{\kappa,\vartheta}(i\varsigma, X) = \Omega cos_{\kappa,\vartheta}(\varsigma, X) + i\Omega sin_{\kappa,\vartheta}(\varsigma, X),
$$

and

$$
\Omega_{\kappa,\vartheta}(X) = \Omega cos_{\kappa,\vartheta}(X) + i\Omega sin_{\kappa,\vartheta}(X).
$$

8.3.6 The Superhyperbolic Functions via the Lorenzo–Hartley Function of the Second Type

Let $X, \varsigma, \kappa, \vartheta \in \mathbb{C}$. In this part, we give the superhyperbolic functions via the Lorenzo-Hartley function of the second type [4], as follows:

▷ The hyperbolic supersine through the Lorenzo-Hartley function of the second type is given by

$$\begin{aligned}
&\Omega \sinh_{\kappa,\vartheta}(X) \\
&= \frac{1}{2}\Big(\Omega_{\kappa,\vartheta}(1, X) - \Omega_{\kappa,\vartheta}(-1, X)\Big) \\
&= X^{\kappa-\vartheta-1} \sum_{k=0}^{\infty} \frac{X^{(2k+1)\kappa}}{\Gamma((2k+2)\kappa - \vartheta)}.
\end{aligned}$$

▷ The hyperbolic supercosine through the Lorenzo-Hartley function of the second type is given by

$$\begin{aligned}
&\Omega \cosh_{\kappa,\vartheta}(X) \\
&= \frac{1}{2}\Big(\Omega_{\kappa,\vartheta}(1, X) + \Omega_{\kappa,\vartheta}(-1, X)\Big) \\
&= X^{\kappa-\vartheta-1} \sum_{k=0}^{\infty} \frac{X^{(2k)\kappa}}{\Gamma((2k+1)\kappa - \vartheta)}.
\end{aligned}$$

▷ The hyperbolic supertangent through the Lorenzo-Hartley function of the second type is given by

$$\Omega \tanh_{\kappa,\vartheta}(X) = \frac{\Omega \sinh_{\kappa,\vartheta}(X)}{\Omega \cosh_{\kappa,\vartheta}(X)}, \quad \Big(\Omega \cosh_{\kappa,\vartheta}(X) \neq 0\Big).$$

▷ The hyperbolic supercotangent through the Lorenzo-Hartley function of the second type is given by

$$\Omega \mathrm{cotanh}_{\kappa,\vartheta}(X) = \frac{\Omega \cosh_{\kappa,\vartheta}(X)}{\Omega \sinh_{\kappa,\vartheta}(X)}, \quad \Big(\Omega \sinh_{\kappa,\vartheta}(X) \neq 0\Big).$$

▷ The hyperbolic supercosecant through the Lorenzo-Hartley function of the second type is given by

$$\Omega \mathrm{csch}_{\kappa,\vartheta}(X) = \frac{1}{\Omega \sinh_{\kappa,\vartheta}(X)}, \quad \Big(\Omega \sinh_{\kappa,\vartheta}(X) \neq 0\Big).$$

▷ The hyperbolic supersecant through the Lorenzo-Hartley function of the second type is given by

$$\Omega \mathrm{sech}_{\kappa,\vartheta}(X) = \frac{1}{\Omega \cosh_{\kappa,\vartheta}(X)}, \quad \Big(\Omega \cosh_{\kappa,\vartheta}(X) \neq 0\Big).$$

Thus, we get [4]

$$\begin{aligned}
&\Omega \sinh_{\kappa,\vartheta}(\varsigma, X) \\
&= \frac{1}{2}\Big(\Omega_{\kappa,\vartheta}(\varsigma, X) - \Omega_{\kappa,\vartheta}(-\varsigma, X)\Big) \\
&= X^{\kappa-\vartheta-1} \sum_{k=0}^{\infty} \frac{\varsigma^{2k+1} X^{(2k+1)\kappa}}{\Gamma((2k+2)\kappa - \vartheta)},
\end{aligned}$$

$$\begin{aligned}
&\Omega \cosh_{\kappa,\vartheta}(\varsigma, X) \\
&= \frac{1}{2}\Big(\Omega_{\kappa,\vartheta}(\varsigma, X) + \Omega_{\kappa,\vartheta}(-\varsigma, X)\Big) \\
&= X^{\kappa-\vartheta-1} \sum_{k=0}^{\infty} \frac{\varsigma^{2k} X^{(2k)\kappa}}{\Gamma((2k+1)\kappa - \vartheta)},
\end{aligned}$$

$$\Omega \tanh_{\kappa,\vartheta}(\varsigma, X) = \frac{\Omega \sinh_{\kappa,\vartheta}(\varsigma, X)}{\Omega \cosh_{\kappa,\vartheta}(\varsigma, X)}, \quad \Big(\Omega \cosh_{\kappa,\vartheta}(\varsigma, X) \neq 0\Big),$$

$$\Omega \mathrm{cotanh}_{\kappa,\vartheta}(\varsigma, X) = \frac{\Omega \cosh_{\kappa,\vartheta}(\varsigma, X)}{\Omega \sinh_{\kappa,\vartheta}(\varsigma, X)}, \quad \Big(\Omega \sinh_{\kappa,\vartheta}(\varsigma, X) \neq 0\Big),$$

$$\Omega \mathrm{csch}_{\kappa,\vartheta}(\varsigma, X) = \frac{1}{\Omega \sinh_{\kappa,\vartheta}(\varsigma, X)}, \quad \Big(\Omega \sinh_{\kappa,\vartheta}(\varsigma, X) \neq 0\Big),$$

$$\Omega \mathrm{sech}_{\kappa,\vartheta}(\varsigma, X) = \frac{1}{\Omega \cosh_{\kappa,\vartheta}(\varsigma, X)}, \quad \Big(\Omega \cosh_{\kappa,\vartheta}(\varsigma, X) \neq 0\Big),$$

where $X, \varsigma, \kappa, \vartheta \in \mathbb{C}$.

Theorem 8.8 ([4]) *Let $X, \varsigma, \kappa, \vartheta \in \mathbb{C}$. Then,*

$$\Omega_{\kappa,\vartheta}(\varsigma, X) = \Omega cosh_{\kappa,\vartheta}(\varsigma, X) + \Omega sinh_{\kappa,\vartheta}(\varsigma X),$$

and

$$\Omega_{\kappa,\vartheta}(X) = \Omega cosh_{\kappa,\vartheta}(X) + \Omega sinh_{\kappa,\vartheta}(X).$$

8.4 The Kohlrausch–Williams–Watts Function and Related Topics

In this section, we discuss the history, definitions, and properties of the Kohlrausch-Williams-Watts function. The subtrigonometric and subhyperbolic functions are addressed based on the Kohlrausch-Williams-Watts function.

8.4.1 The Kohlrausch–Williams–Watts Function

In 1854, Rudolph Kohlrausch proposed the stretched exponential function of integral order and application to electric circuit [9]. In 1970, Graham Williams and David C. Watts rediscovered use of the stretched exponential function of fractional order considered in to the field of the non-symmetrical dielectric relaxation with the characteristics of the decay function [10]. The special function is called the Kohlrausch-Williams-Watts function [11, 12] or the Kohlrausch-Williams-Watts stretched exponential function [13]. The function is well-known as the "Material Function".

Definition 8.5 ([9, 10]) The Kohlrausch–Williams–Watts function is given by

$$\mho_{\kappa}(X) = e^{X^{\kappa}},$$

where $X, \kappa \in \mathbb{C}$.

The series form of the Kohlrausch–Williams–Watts function can be written as [4]:

$$\begin{aligned} &\mho_{\kappa}(\varsigma, X) \\ &= e^{\varsigma X^{\kappa}} \\ &= \sum_{k=0}^{\infty} \frac{\varsigma^k X^{k\kappa}}{k!} \\ &= \mathbb{E}_1(\varsigma X^{\kappa}) \\ &= {}_0\mathbb{W}_0(-; \varsigma X^{\kappa}), \end{aligned}$$

which leads to

$$\mho_{\kappa}(1, X) = \mho_{\kappa}(X),$$

where $X, \kappa, \varsigma \in \mathbb{C}$.

Figure 8.1 displays the plots of $e^{-\varsigma X^{\kappa}}$ with different values of κ and ς.

Definition 8.6 ([14]) The Kohlrausch–Williams–Watts type function is given by

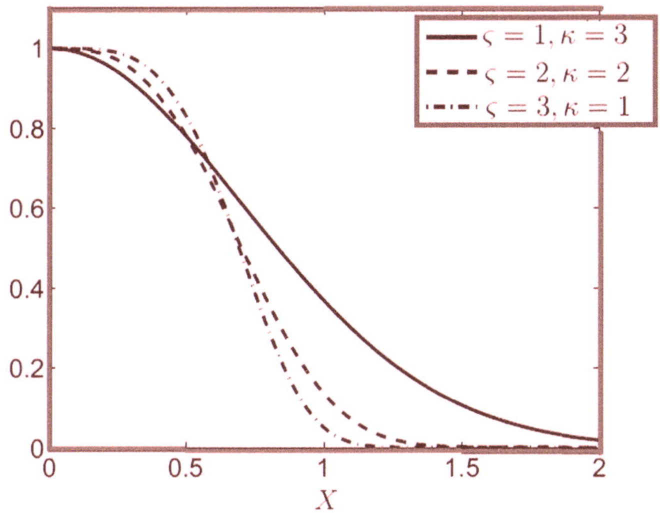

Fig. 8.1 The plots of $e^{-\varsigma X^{\kappa}}$ with different values of κ and ς

$$
\begin{aligned}
&\mho_{\kappa,\vartheta}(\varsigma, X)\\
&= X^{\vartheta-1}e^{\varsigma X^{\kappa}}\\
&= X^{\vartheta-1}\sum_{k=0}^{\infty}\frac{\varsigma^{k}X^{k\kappa}}{k!}\\
&= X^{\vartheta-1}\mathbb{E}_{1}(\varsigma X^{\kappa})\\
&= X^{\vartheta-1}{}_{0}\mathbb{W}_{0}(-;\varsigma X^{\kappa}),
\end{aligned}
$$

which implies that

$$\mho_{\kappa,\vartheta}(1, X) = \mho_{\kappa,\vartheta}(X),$$

and

$$\mho_{\kappa}(\varsigma, X) = \mho_{\kappa,1}(\varsigma, X),$$

where $X, \kappa, \varsigma, \vartheta \in \mathbb{C}$.

Figure 8.2 displays The plots of $\alpha e^{-X^{\kappa}}$ with different values of κ and α. In particular, we have [4],

$$
\begin{aligned}
&\mho_{\kappa,\kappa}(\varsigma, X)\\
&= X^{\kappa-1}e^{\varsigma X^{\kappa}}\\
&= X^{\kappa-1}\sum_{k=0}^{\infty}\frac{\varsigma^{k}X^{k\kappa}}{k!}\\
&= X^{\kappa-1}\mathbb{E}_{1}(\varsigma X^{\kappa})
\end{aligned}
$$

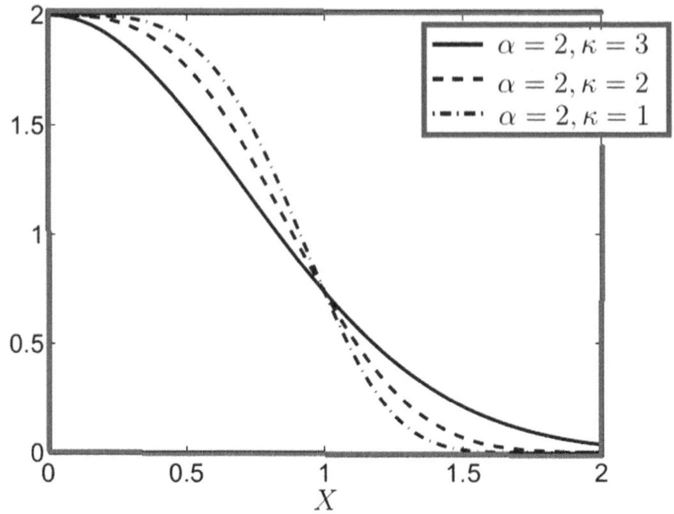

Fig. 8.2 The plots of $\alpha e^{-X^{\kappa}}$ with different values of κ and α

$$= X^{\kappa-1}{}_0\mathbb{W}_0(-;\varsigma X^{\kappa}),$$

and

$$\begin{aligned}&\mho_{0.5,0.5}(-1,X)\\&= X^{-0.5}e^{-X^{0.5}}\\&= X^{-0.5}\mathbb{E}_1(-X^{\kappa})\\&= X^{-0.5}{}_0\mathbb{W}_0(-;-X^{\kappa}).\end{aligned}$$

Figure 8.3 displays the plots of $\alpha e^{-\varsigma X^{\kappa}}$ with different values of κ, ς, and α.

8.4.2 The Subtrigonometric Functions via the Kohlrausch–Williams–Watts Function

Let $\kappa, X \in \mathbb{C}$. In this part, we propose the subtrigonometric functions via the Kohlrausch-Williams-Watts function [4], as follows:

▷ The subsine through the Kohlrausch-Williams-Watts function is given by

$$\begin{aligned}&\operatorname{subsin}_{\kappa}(X)\\&= \frac{1}{2i}\Big(\mho_{\kappa}(i,X) - \mho_{\kappa}(-i,X)\Big)\end{aligned}$$

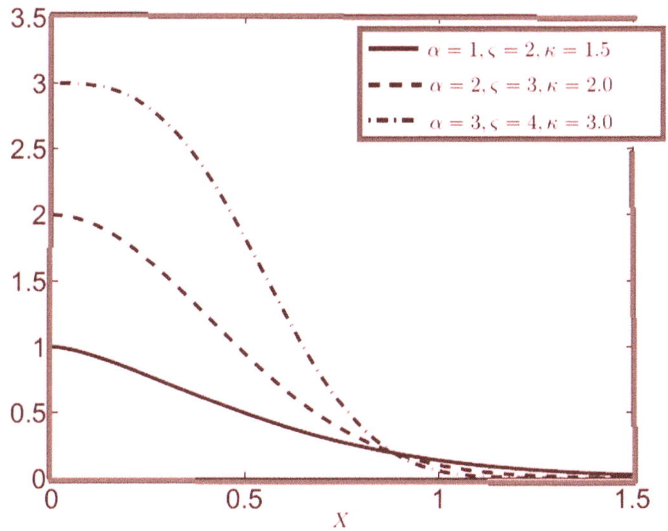

Fig. 8.3 The plots of $\alpha e^{-\varsigma X^{\kappa}}$ with different values of κ, ς, and α

$$
\begin{aligned}
&= \sum_{k=0}^{\infty} \frac{(-1)^k X^{(2k+1)\kappa}}{\Gamma(2k+2)} \\
&= \sum_{k=0}^{\infty} \frac{(-1)^k X^{(2k+1)\kappa}}{(2k+1)!}.
\end{aligned}
$$

▷ The subcosine through the Kohlrausch-Williams-Watts function is given by

$$
\begin{aligned}
&\operatorname{subcos}_{\kappa}(X) \\
&= \frac{1}{2}\Big(\mho_{\kappa}(i, X) + \mho_{\kappa}(-i, X)\Big) \\
&= \sum_{k=0}^{\infty} \frac{(-1)^k X^{(2k)\kappa}}{\Gamma(2k+1)} \\
&= \sum_{k=0}^{\infty} \frac{(-1)^k X^{(2k)\kappa}}{(2k)!}.
\end{aligned}
$$

▷ The subtangent through the Kohlrausch-Williams-Watts function is given by

$$
\operatorname{subtan}_{\kappa}(X) = \frac{\operatorname{subsin}_{\kappa}(X)}{\operatorname{subcos}_{\kappa}(X)}, \quad \Big(\operatorname{subcos}_{\kappa}(X) \neq 0\Big).
$$

▷ The subcotangent through the Kohlrausch-Williams-Watts function is given by

$$\mathrm{subcotan}_{\kappa}(X) = \frac{\mathrm{subcos}_{\kappa}(X)}{\mathrm{subsin}_{\kappa}(X)}, \quad \left(\mathrm{subsin}_{\kappa}(X) \neq 0\right).$$

▷ The subcosecant through the Kohlrausch-Williams-Watts function is given by

$$\mathrm{subcsc}_{\kappa}(X) = \frac{1}{\mathrm{subsin}_{\kappa}(X)}, \quad \left(\mathrm{subsin}_{\kappa}(X) \neq 0\right).$$

▷ The subsecant through the Kohlrausch-Williams-Watts function is given by

$$\mathrm{subsec}_{\kappa}(X) = \frac{1}{\mathrm{subcos}_{\kappa}(X)}, \quad \left(\mathrm{subcos}_{\kappa}(X) \neq 0\right).$$

Thus, we get the following from [4]

$$\begin{aligned} &\mathrm{subsin}_{\kappa}(\varsigma, X) \\ &= \frac{1}{2i}\left(\mho_{\kappa}(i\varsigma, X) - \mho_{\kappa}(-i\varsigma, X)\right) \\ &= \sum_{k=0}^{\infty} \frac{(-1)^k \varsigma^{2k+1} X^{(2k+1)\kappa}}{\Gamma(2k+2)} \\ &= \sum_{k=0}^{\infty} \frac{(-1)^k \varsigma^{2k+1} X^{(2k+1)\kappa}}{(2k+1)!}, \end{aligned}$$

$$\begin{aligned} &\mathrm{subcos}_{\kappa}(\varsigma, X) \\ &= \frac{1}{2}\left(\mho_{\kappa}(i\varsigma, X) + \mho_{\kappa}(-i\varsigma, X)\right) \\ &= \sum_{k=0}^{\infty} \frac{(-1)^k \varsigma^{2k} X^{(2k)\kappa}}{\Gamma(2k+1)} \\ &= \sum_{k=0}^{\infty} \frac{(-1)^k \varsigma^{2k} X^{(2k)\kappa}}{(2k)!}, \end{aligned}$$

$$\mathrm{subtan}_{\kappa}(\varsigma, X) = \frac{\mathrm{subsin}_{\kappa}(\varsigma, X)}{\mathrm{subcos}_{\kappa}(\varsigma, X)}, \quad \left(\mathrm{subcos}_{\kappa}(\varsigma, X) \neq 0\right),$$

$$\mathrm{subcotan}_{\kappa}(\varsigma, X) = \frac{\mathrm{subcos}_{\kappa}(\varsigma, X)}{\mathrm{subsin}_{\kappa}(\varsigma, X)}, \quad \left(\mathrm{subsin}_{\kappa}(\varsigma, X) \neq 0\right),$$

$$\mathrm{subcsc}_\kappa(\varsigma, X) = \frac{1}{\mathrm{subsin}_\kappa(\varsigma, X)}, \quad \Big(\mathrm{subsin}_\kappa(\varsigma, X) \neq 0\Big),$$

$$\mathrm{subsec}_\kappa(\varsigma, X) = \frac{1}{\mathrm{subcos}_\kappa(\varsigma, X)}, \quad \Big(\mathrm{subcos}_\kappa(\varsigma, X) \neq 0\Big),$$

where $\kappa, X \in \mathbb{C}$.

Theorem 8.9 ([4]) *Let $\kappa, X, \varsigma \in \mathbb{C}$. Then,*

$$\mho_\kappa(i\varsigma, X) = subcos_\kappa(\varsigma, X) + i\,subsin_\kappa(\varsigma, X).$$

8.4.3 The Subhyperbolic Functions via the Kohlrausch–Williams–Watts Function

Let $\kappa, X \in \mathbb{C}$. In this part, we define the subhyperbolic functions via the Kohlrausch-Williams-Watts function [4], as follows:

▷ The hyperbolic subsine through the Kohlrausch-Williams-Watts function is given by

$$\begin{aligned}
&\mathrm{subsinh}_\kappa(X)\\
&= \frac{1}{2}\Big(\mho_\kappa(1, X) - \mho_\kappa(-1, X)\Big)\\
&= \sum_{k=0}^{\infty} \frac{X^{(2k+1)\kappa}}{\Gamma(2k+2)}\\
&= \sum_{k=0}^{\infty} \frac{X^{(2k+1)\kappa}}{(2k+1)!}.
\end{aligned}$$

▷ The hyperbolic subcosine through the Kohlrausch-Williams-Watts function is given by

$$\begin{aligned}
&\mathrm{subcosh}_\kappa(X)\\
&= \frac{1}{2}\Big(\mho_\kappa(1, X) + \mho_\kappa(-1, X)\Big)\\
&= \sum_{k=0}^{\infty} \frac{X^{(2k)\kappa}}{\Gamma(2k+1)}\\
&= \sum_{k=0}^{\infty} \frac{X^{(2k)\kappa}}{(2k)!}.
\end{aligned}$$

▷ The hyperbolic subtangent through the Kohlrausch-Williams-Watts function is given by

$$\operatorname{subtanh}_{\kappa}(X) = \frac{\operatorname{subsinh}_{\kappa}(X)}{\operatorname{subcosh}_{\kappa}(X)}, \quad \Big(\operatorname{subcosh}_{\kappa}(X) \neq 0\Big).$$

▷ The hyperbolic subcotangent through the Kohlrausch-Williams-Watts function is given by

$$\operatorname{subcotanh}_{\kappa}(X) = \frac{\operatorname{subcosh}_{\kappa}(X)}{\operatorname{subsinh}_{\kappa}(X)}, \quad \Big(\operatorname{subsinh}_{\kappa}(X) \neq 0\Big).$$

▷ The hyperbolic subcosecant through the Kohlrausch-Williams-Watts function is given by

$$\operatorname{subcsch}_{\kappa}(X) = \frac{1}{\operatorname{subsinh}_{\kappa}(X)}, \quad \Big(\operatorname{subsinh}_{\kappa}(X) \neq 0\Big).$$

▷ The hyperbolic subsecant through the Kohlrausch-Williams-Watts function is given by

$$\operatorname{subsech}_{\kappa}(X) = \frac{1}{\operatorname{subcosh}_{\kappa}(X)}, \quad \Big(\operatorname{subcosh}_{\kappa}(X) \neq 0\Big).$$

Thus, we get [4]

$$\begin{aligned}
&\operatorname{subsinh}_{\kappa}(\varsigma, X) \\
&= \frac{1}{2}\Big(\mho_{\kappa}(\varsigma, X) - \mho_{\kappa}(-\varsigma, X)\Big) \\
&= \sum_{k=0}^{\infty} \frac{\varsigma^{2k+1} X^{(2k+1)\kappa}}{\Gamma(2k+2)} \\
&= \sum_{k=0}^{\infty} \frac{\varsigma^{2k+1} X^{(2k+1)\kappa}}{(2k+1)!},
\end{aligned}$$

$$
\begin{aligned}
&\operatorname{subcosh}_{\kappa}(\varsigma, X) \\
&= \frac{1}{2}\Big(\mho_{\kappa}(\varsigma, X) + \mho_{\kappa}(-\varsigma, X)\Big) \\
&= \sum_{k=0}^{\infty} \frac{\varsigma^{2k} X^{(2k)\kappa}}{\Gamma(2k+1)} \\
&= \sum_{k=0}^{\infty} \frac{\varsigma^{2k} X^{(2k)\kappa}}{(2k)!},
\end{aligned}
$$

$$
\operatorname{subtanh}_{\kappa}(\varsigma, X) = \frac{\operatorname{subsinh}_{\kappa}(\varsigma, X)}{\operatorname{subcosh}_{\kappa}(\varsigma, X)}, \quad \Big(\operatorname{subcosh}_{\kappa}(\varsigma, X) \neq 0\Big),
$$

$$
\operatorname{subcotanh}_{\kappa}(\varsigma, X) = \frac{\operatorname{subcosh}_{\kappa}(\varsigma, X)}{\operatorname{subsinh}_{\kappa}(\varsigma, X)}, \quad \Big(\operatorname{subsinh}_{\kappa}(\varsigma, X) \neq 0\Big),
$$

$$
\operatorname{subcsch}_{\kappa}(\varsigma, X) = \frac{1}{\operatorname{subsinh}_{\kappa}(\varsigma, X)}, \quad \Big(\operatorname{subsinh}_{\kappa}(\varsigma, X) \neq 0\Big),
$$

$$
\operatorname{subsech}_{\kappa}(\varsigma, X) = \frac{1}{\operatorname{subcosh}_{\kappa}(\varsigma, X)}, \quad \Big(\operatorname{subcosh}_{\kappa}(\varsigma, X) \neq 0\Big),
$$

where $\kappa, X, \varsigma \in \mathbb{C}$.

Theorem 8.10 ([4]) *Let $\kappa, X, \varsigma \in \mathbb{C}$. Then,*

$$
\mho_{\kappa}(\varsigma, X) = \mathit{subcosh}_{\kappa}(\varsigma, X) + \mathit{subsinh}_{\kappa}(\varsigma, X).
$$

8.4.4 The Subtrigonometric Functions via the Kohlrausch–Williams–Watts Type Function

Let $\kappa, \vartheta, X \in \mathbb{C}$. In this part, we define the subtrigonometric functions via the Kohlrausch-Williams-Watts type function [4], as follows:

▷ The subsine through the Kohlrausch-Williams-Watts type function is given by

$$\begin{aligned}
&\mathrm{subsin}_{\kappa,\vartheta}(X)\\
&=\frac{1}{2i}\Big(\mho_{\kappa,\vartheta}(i,X)-\mho_{\kappa,\vartheta}(-i,X)\Big)\\
&=X^{\vartheta-1}\sum_{k=0}^{\infty}\frac{(-1)^k X^{(2k+1)\kappa}}{\Gamma(2k+2)}\\
&=X^{\vartheta-1}\sum_{k=0}^{\infty}\frac{(-1)^k X^{(2k+1)\kappa}}{(2k+1)!}.
\end{aligned}$$

▷ The subcosine through the Kohlrausch-Williams-Watts type function is given by

$$\begin{aligned}
&\mathrm{subcosin}_{\kappa,\vartheta}(X)\\
&=\frac{1}{2}\Big(\mho_{\kappa,\vartheta}(i,X)+\mho_{\kappa,\vartheta}(-i,X)\Big)\\
&=X^{\vartheta-1}\sum_{k=0}^{\infty}\frac{(-1)^k X^{(2k)\kappa}}{\Gamma(2k+1)}\\
&=X^{\vartheta-1}\sum_{k=0}^{\infty}\frac{(-1)^k X^{(2k)\kappa}}{(2k)!}.
\end{aligned}$$

▷ The subtangent through the Kohlrausch-Williams-Watts type function is given by

$$\mathrm{subtan}_{\kappa,\vartheta}(X)=\frac{\mathrm{subsin}_{\kappa,\vartheta}(X)}{\mathrm{subcos}_{\kappa,\vartheta}(X)},\quad \Big(\mathrm{subcos}_{\kappa,\vartheta}(X)\neq 0\Big).$$

▷ The subcotangent through the Kohlrausch-Williams-Watts type function is given by

$$\mathrm{subcotan}_{\kappa,\vartheta}(X)=\frac{\mathrm{subcos}_{\kappa,\vartheta}(X)}{\mathrm{subsin}_{\kappa,\vartheta}(X)},\quad \Big(\mathrm{subsin}_{\kappa,\vartheta}(X)\neq 0\Big).$$

▷ The subcosecant through the Kohlrausch-Williams-Watts type function is given by

$$\mathrm{subcsc}_{\kappa,\vartheta}(X)=\frac{1}{\mathrm{subsin}_{\kappa,\vartheta}(X)},\quad \Big(\mathrm{subsin}_{\kappa,\vartheta}(X)\neq 0\Big).$$

▷ The subsecant through the Kohlrausch-Williams-Watts type function is given by

$$\text{subsec}_{\kappa,\vartheta}(X) = \frac{1}{\text{subcos}_{\kappa,\vartheta}(X)}, \quad \Big(\text{subcos}_{\kappa,\vartheta}(X) \neq 0\Big).$$

Thus, we get [4],

$$\begin{aligned}
&\text{subsin}_{\kappa,\vartheta}(\varsigma, X)\\
&= \frac{1}{2i}\Big(\mho_{\kappa,\vartheta}(i\varsigma, X) - \mho_{\kappa,\vartheta}(-i\varsigma, X)\Big)\\
&= X^{\vartheta-1}\sum_{k=0}^{\infty}\frac{(-1)^k\varsigma^{2k+1}X^{(2k+1)\kappa}}{\Gamma(2k+2)}\\
&= X^{\vartheta-1}\sum_{k=0}^{\infty}\frac{(-1)^k\varsigma^{2k+1}X^{(2k+1)\kappa}}{(2k+1)!},
\end{aligned}$$

$$\begin{aligned}
&\text{subcosin}_{\kappa,\vartheta}(\varsigma, X)\\
&= \frac{1}{2}\Big(\mho_{\kappa,\vartheta}(i\varsigma, X) + \mho_{\kappa,\vartheta}(-i\varsigma, X)\Big)\\
&= X^{\vartheta-1}\sum_{k=0}^{\infty}\frac{(-1)^k\varsigma^{2k}X^{(2k)\kappa}}{\Gamma(2k+1)}\\
&= X^{\vartheta-1}\sum_{k=0}^{\infty}\frac{(-1)^k\varsigma^{2k}X^{(2k)\kappa}}{(2k)!},
\end{aligned}$$

$$\text{subtan}_{\kappa,\vartheta}(\varsigma, X) = \frac{\text{subsin}_{\kappa,\vartheta}(\varsigma, X)}{\text{subcos}_{\kappa,\vartheta}(\varsigma, X)}, \quad \Big(\text{subcos}_{\kappa,\vartheta}(\varsigma, X) \neq 0\Big),$$

$$\text{subcotan}_{\kappa,\vartheta}(\varsigma, X) = \frac{\text{subcos}_{\kappa,\vartheta}(\varsigma, X)}{\text{subsin}_{\kappa,\vartheta}(X)}, \quad \Big(\text{subsin}_{\kappa,\vartheta}(\varsigma, X) \neq 0\Big),$$

$$\text{subcsc}_{\kappa,\vartheta}(\varsigma, X) = \frac{1}{\text{subsin}_{\kappa,\vartheta}(\varsigma, X)}, \quad \Big(\text{subsin}_{\kappa,\vartheta}(\varsigma, X) \neq 0\Big),$$

$$\text{subsec}_{\kappa,\vartheta}(\varsigma, X) = \frac{1}{\text{subcos}_{\kappa,\vartheta}(\varsigma, X)}, \quad \Big(\text{subcos}_{\kappa,\vartheta}(\varsigma, X) \neq 0\Big),$$

where $\kappa, \vartheta, X \in \mathbb{C}$.

Theorem 8.11 ([4]) *Let $\kappa, \vartheta, X \in \mathbb{C}$. Then,*

$$\mho_{\kappa,\vartheta}(i\varsigma, X) = subcos_{\kappa,\vartheta}(\varsigma, X) + i subsin_{\kappa,\vartheta}(\varsigma, X).$$

8.4.5 The Subhyperbolic Functions via the Kohlrausch–Williams–Watts Type Function

Let $\kappa, \vartheta, X \in \mathbb{C}$. In this part, we define the subhyperbolic functions via Kohlrausch-Williams-Watts type function [4], as follows:

▷ The hyperbolic subsine through the Kohlrausch-Williams-Watts type function is given by

$$\begin{aligned}
&\mathrm{subsinh}_{\kappa,\vartheta}(X)\\
&= \frac{1}{2}\Big(\mho_{\kappa,\vartheta}(1, X) - \mho_{\kappa,\vartheta}(-1, X)\Big)\\
&= X^{\vartheta-1}\sum_{k=0}^{\infty}\frac{X^{(2k+1)\kappa}}{\Gamma(2k+2)}\\
&= X^{\vartheta-1}\sum_{k=0}^{\infty}\frac{X^{(2k+1)\kappa}}{(2k+1)!},
\end{aligned}$$

▷ The hyperbolic subcosine through the Kohlrausch-Williams-Watts type function is given by

$$\begin{aligned}
&\mathrm{subcosh}_{\kappa}(X)\\
&= \frac{1}{2}\Big(\mho_{\kappa}(1, X) + \mho_{\kappa}(-1, X)\Big)\\
&= \sum_{k=0}^{\infty}\frac{X^{(2k)\kappa}}{\Gamma(2k+1)}\\
&= \sum_{k=0}^{\infty}\frac{X^{(2k)\kappa}}{(2k)!}.
\end{aligned}$$

▷ The hyperbolic subtangent through the Kohlrausch-Williams-Watts type function is given by

$$\mathrm{subtanh}_{\kappa,\vartheta}(X) = \frac{\mathrm{subsinh}_{\kappa,\vartheta}(X)}{\mathrm{subcosh}_{\kappa,\vartheta}(X)}, \quad \Big(\mathrm{subcosh}_{\kappa,\vartheta}(X) \neq 0\Big),$$

▷ The hyperbolic subcotangent through the Kohlrausch-Williams-Watts type function is given by

$$\operatorname{subcotanh}_{\kappa,\vartheta}(X) = \frac{\operatorname{subcosh}_{\kappa,\vartheta}(X)}{\operatorname{subsinh}_{\kappa,\vartheta}(X)}, \quad \Big(\operatorname{subsinh}_{\kappa,\vartheta}(X) \neq 0\Big),$$

▷ The hyperbolic subcosecant through the Kohlrausch-Williams-Watts type function is given by

$$\operatorname{subcsch}_{\kappa,\vartheta}(X) = \frac{1}{\operatorname{subsinh}_{\kappa,\vartheta}(X)}, \quad \Big(\operatorname{subsinh}_{\kappa,\vartheta}(X) \neq 0\Big),$$

▷ The hyperbolic subsecant through the Kohlrausch-Williams-Watts type function is given by

$$\operatorname{subsech}_{\kappa,\vartheta}(X) = \frac{1}{\operatorname{subcosh}_{\kappa,\vartheta}(X)}, \quad \Big(\operatorname{subcosh}_{\kappa,\vartheta}(X) \neq 0\Big).$$

Thus, we get [4],

$$\begin{aligned}
&\operatorname{subsinh}_{\kappa,\vartheta}(\varsigma, X) \\
&= \frac{1}{2}\Big(\mho_{\kappa,\vartheta}(\varsigma, X) - \mho_{\kappa,\vartheta}(-\varsigma, X)\Big) \\
&= X^{\vartheta-1} \sum_{k=0}^{\infty} \frac{\varsigma^{2k+1} X^{(2k+1)\kappa}}{\Gamma(2k+2)} \\
&= X^{\vartheta-1} \sum_{k=0}^{\infty} \frac{\varsigma^{2k+1} X^{(2k+1)\kappa}}{(2k+1)!},
\end{aligned}$$

$$\begin{aligned}
&\operatorname{subcosinh}_{\kappa,\vartheta}(\varsigma, X) \\
&= \frac{1}{2}\Big(\mho_{\kappa,\vartheta}(\varsigma, X) + \mho_{\kappa,\vartheta}(-\varsigma, X)\Big) \\
&= X^{\vartheta-1} \sum_{k=0}^{\infty} \frac{\varsigma^{2k} X^{(2k)\kappa}}{\Gamma(2k+1)} \\
&= X^{\vartheta-1} \sum_{k=0}^{\infty} \frac{\varsigma^{2k} X^{(2k)\kappa}}{(2k)!},
\end{aligned}$$

$$\mathrm{subtanh}_{\kappa,\vartheta}(\varsigma, X) = \frac{\mathrm{subsinh}_{\kappa,\vartheta}(\varsigma, X)}{\mathrm{subcosh}_{\kappa,\vartheta}(\varsigma, X)}, \quad \Big(\mathrm{subcosh}_{\kappa,\vartheta}(\varsigma, X) \neq 0\Big),$$

$$\mathrm{subcotanh}_{\kappa,\vartheta}(\varsigma, X) = \frac{\mathrm{subcosh}_{\kappa,\vartheta}(\varsigma, X)}{\mathrm{subsinh}_{\kappa,\vartheta}(X)}, \quad \Big(\mathrm{subsinh}_{\kappa,\vartheta}(\varsigma, X) \neq 0\Big),$$

$$\mathrm{subcsch}_{\kappa,\vartheta}(\varsigma, X) = \frac{1}{\mathrm{subsinh}_{\kappa,\vartheta}(\varsigma, X)}, \quad \Big(\mathrm{subsinh}_{\kappa,\vartheta}(\varsigma, X) \neq 0\Big),$$

$$\mathrm{subsech}_{\kappa,\vartheta}(\varsigma, X) = \frac{1}{\mathrm{subcosh}_{\kappa,\vartheta}(\varsigma, X)}, \quad \Big(\mathrm{subcosh}_{\kappa,\vartheta}(\varsigma, X) \neq 0\Big),$$

where $\kappa, \vartheta, X \in \mathbb{C}$.

Theorem 8.12 ([4]) *Let $\kappa, \vartheta, X \in \mathbb{C}$. Then,*

$$\mho_{\kappa,\vartheta}(\varsigma, X) = \mathit{subcosh}_{\kappa,\vartheta}(\varsigma, X) + \mathit{subsinh}_{\kappa,\vartheta}(\varsigma, X).$$

In particular, we have the following [4]:

$$\mathrm{subsin}_{1,1}(\varsigma, X) = \sin(\varsigma X),$$

$$\mathrm{subcos}_{1,1}(\varsigma, X) = \cos(\varsigma X),$$

$$\mathrm{subtan}_{1,1}(\varsigma, X) = \tan(\varsigma X),$$

$$\mathrm{subcotan}_{1,1}(\varsigma, X) = \mathrm{cotan}(\varsigma X),$$

$$\mathrm{subcsc}_{1,1}(\varsigma, X) = \csc(\varsigma X),$$

$$\mathrm{subsec}_{1,1}(\varsigma, X) = \sec(\varsigma X),$$

$$\mathrm{subsinh}_{1,1}(\varsigma, X) = \sinh(\varsigma X),$$

$$\mathrm{subcosh}_{1,1}(\varsigma, X) = \cosh(\varsigma X),$$

$$\mathrm{subtanh}_{1,1}(\varsigma, X) = \tanh(\varsigma X),$$

$$\mathrm{subcotanh}_{1,1}(\varsigma, X) = \mathrm{cotanh}(\varsigma X),$$

$$\mathrm{subcsch}_{1,1}(\varsigma, X) = \mathrm{csch}(\varsigma X),$$

$$\mathrm{subsech}_{1,1}(\varsigma, X) = \mathrm{sech}(\varsigma X),$$

$$\mathrm{subsin}_1(\varsigma, X) = \sin(\varsigma X),$$

$$\mathrm{subcos}_1(\varsigma, X) = \cos(\varsigma X),$$

$$\mathrm{subtan}_1(\varsigma, X) = \tan(\varsigma X),$$

$$\mathrm{subcotan}_1(\varsigma, X) = \mathrm{cotan}(\varsigma X),$$

$$\mathrm{subcsc}_1(\varsigma, X) = \csc(\varsigma X),$$

$$\mathrm{subsec}_1(\varsigma, X) = \sec(\varsigma X),$$

$$\mathrm{subsinh}_1(\varsigma, X) = \sinh(\varsigma X),$$

$$\mathrm{subcosh}_1(\varsigma, X) = \cosh(\varsigma X),$$

$$\mathrm{subtanh}_1(\varsigma, X) = \tanh(\varsigma X),$$

$$\mathrm{subcotanh}_1(\varsigma, X) = \mathrm{cotanh}(\varsigma X),$$

$$\mathrm{subcsch}_1(\varsigma, X) = \mathrm{csch}(\varsigma X),$$

$$\mathrm{subsech}_1(\varsigma, X) = \mathrm{sech}(\varsigma X).$$

Also, we have that [4],

$$\mathrm{subsin}^2_\kappa(\varsigma, X) + \mathrm{subcos}^2_\kappa(\varsigma, X) = 1,$$

and

$$\mathrm{subcosh}^2_\kappa(\varsigma, X) - \mathrm{subsinh}^2_\kappa(\varsigma, X) = 1.$$

8.5 Fox Type Functions

Consider the infinite contour $\mathscr{L}$ which has one of the following forms:

- $\mathscr{L} = \mathscr{L}_{-\infty}$ is a left loop starting at $-\infty$ and ending at $-\infty$, enclosing all the poles of $\Gamma(Y)$.
- $\mathscr{L} = \mathscr{L}_{+\infty}$ is a left loop starting at $+\infty$ and ending at $+\infty$, enclosing all the poles of $\Gamma(d_j - Y)$, $j = 1, \ldots, s$, situated in a horizontal strip starting at the point $+\infty + i\mathcal{P}_1$ and terminating at the point $+\infty + i\mathcal{P}_2$ with $-\infty < \mathcal{P}_1 < \mathcal{P}_2 < +\infty$, and $d_j \in \mathbb{C}$.
- $\mathscr{L} = \mathscr{L}_{i\hbar\infty}$ is a contour starting at the point $\hbar - i\infty$ and terminating at the point $\hbar + i\infty$ where $\hbar \in \mathbb{R}$.

Fox's H function [15] is defined as

$$
{}_{s}^{v}\mathbb{H}_{r}^{w}\left[X \middle| \begin{matrix}(d_j, D_j)_{1,s} \\ (e_j, E_j)_{1,r}\end{matrix}\right] := \frac{1}{2\pi i}\int_{\mathscr{L}} \mathscr{O}(Y) X^Y dY, \tag{8.1}
$$

where $i^2 = -1$, $X \in \mathbb{C}\backslash\{0\}$, $X^Y = \exp(Y[\log|X| + i \ \arg(X)])$, $\log|X|$ denotes the natural logarithm of $|X|$, and $\arg(X)$ is not necessarily the principal value. For convenience,

$$
\mathscr{O}(Y) := \frac{\prod_{j=1}^{v} \Gamma(e_j - E_j Y) \prod_{j=1}^{w} \Gamma(1 - d_j + D_j Y)}{\prod_{j=v+1}^{r} \Gamma(1 - e_j + E_j Y) \prod_{j=w+1}^{s} \Gamma(d_j - D_j Y)},
$$

an empty product is interpreted as 1, and the integers v, w, s, r satisfy the inequalities $0 \le w \le s$ and $1 \le v \le r$. Assume the coefficients

$$
D_j > 0 \ (j = 1, \ldots, s) \quad \text{and} \quad E_j > 0 \ (j = 1, \ldots, r),
$$

and the complex parameters

$$
d_j \ (j = 1, \ldots, s) \quad \text{and} \quad e_j \ (j = 1, \ldots, r)
$$

are constrained such that no poles of integrand in (8.1) coincide, and $\mathscr{L}$ is a suitable contour of the Mellin-Barnes type (in the complex Y-plane) which separates the poles of one product from those of the other. Further, if we assume

$$
\ell := \sum_{j=1}^{w} D_j - \sum_{j=w+1}^{s} D_j + \sum_{j=1}^{v} E_j - \sum_{j=Q+1}^{r} E_j > 0,
$$

then the integral in (8.1) converges absolutely and defines the $\mathbb{H}$-function as an analytic in the sector:

$$
|\arg(X)| < \frac{1}{2}\ell\pi,
$$

the point $X = 0$ being excluded. Actually, the $\mathbb{H}$-function ia well-defined and is an analytic function of X also when either

$$\mathscr{E} := \sum_{j=1}^{s} D_j - \sum_{j=1}^{r} E_j < 0 \quad \text{and} \quad 0 < |X| < \infty,$$

or

$$\mathscr{E} = 0 \quad \text{and} \quad 0 < |X| < R := \prod_{j=1}^{s} D_j^{-D_j} \prod_{j=1}^{r} E_j^{E_j}.$$

The $\mathbb{H}$ function was introduced by Fox in 1961.

In particular, we have [15–17],

▷ The exponential function:

$${}_0\mathbb{H}_0[X] = \exp(X) = \sum_{k=0}^{\infty} \frac{X^k}{\Gamma(k)}, \quad X \in \mathbb{C}.$$

▷ The Mittag-Leffler function (generalized exponential function):

$${}_0\mathbb{H}_1[e_1; X] = \sum_{k=0}^{\infty} \frac{X^k}{\Gamma(1 + e_1 k)}, \quad e_1 \in \mathbb{C}, \ \Re(e_1) > 0, \ X \in \mathbb{C}.$$

▷ Hypergeometric function (the Gaussian Hypergeometric series):

$${}_2\mathbb{H}_1[d_1, d_2; e_1; X] = \sum_{w=0}^{\infty} \frac{(d_1)_w (d_2)_w}{(e_1)_w} \frac{X^w}{w!} = \frac{\Gamma(e_1)}{\Gamma(d_1)\Gamma(d_2)} \sum_{w=0}^{\infty} \frac{\Gamma(d_1 + w)\Gamma(d_2 + w)}{\Gamma(e_1 + w)} \frac{X^w}{w!},$$

where $d_1, d_2, e_1 \in \mathbb{C}$, $\Re(d_1), \Re(d_2), \Re(e_1) > 0$.

The Hypergeometric function can be represented in terms of the Mellin-Barnes integral of the form

$${}_2\mathbb{H}_1[d_1, d_2; e_1; X] = \frac{\Gamma(e_1)}{\Gamma(d_1)\Gamma(d_2)} \frac{1}{2\pi i} \int_{\mathscr{Z}} \frac{\Gamma(Y)\Gamma(d_1 - Y)\Gamma(d_2 - Y)}{\Gamma(e_1 - Y)} (-X)^{-Y} dY,$$

where $e_1 \neq -\wp$, $\wp \in \mathbb{N}_0 = \{0, 1, 2, 3, \ldots\}$.

▷ The Wright function (the Bessel-Maitland function):

$${}_1\mathbb{H}_1[d_1; e_1; X] = \sum_{n=0}^{\infty} \frac{\varsigma^n}{n!\Gamma(d_1 n + e_1)},$$

where $d_1 > -1$, $e_1, X \in \mathbb{C}$.

▷ Fox-Wright function (the generalized Wright function) [18]:

Consider positive vectors $\mathbf{D} = (D_1, \dots, D_s)$, $\mathbf{E} = (E_1, \dots, E_r)$, and complex vectors $\mathbf{d} = (d_1, \dots, d_s)$, and $\mathbf{E} = (e_1, \dots, e_r)$. The Fox-Wright function or the generalized Wright function is defined by the series

$$ {}_s\mathbb{H}_r\left[X\left|\begin{matrix}(d_1,D_1),\dots,(d_s,D_s)\\(e_1,E_1),\dots,(e_r,E_r)\end{matrix}\right.\right] = {}_s\mathbb{H}_r\left[X\left|\begin{matrix}(\mathbf{d},\mathbf{D})\\(\mathbf{e},\mathbf{E})\end{matrix}\right.\right] = \sum_{n=0}^{\infty}\frac{\Gamma(\mathbf{D}n+\mathbf{d})}{\Gamma(\mathbf{E}n+\mathbf{e})}\frac{X^n}{n!}, \tag{8.2}$$

where

$$\Gamma(\mathbf{D}n+\mathbf{d}) = \prod_{j=1}^{s}\Gamma(Dn+d), \tag{8.3}$$

and $\Gamma(\mathbf{E}n+\mathbf{e})$ is defined similarly.

The series (8.2) has a nonzero radius of convergence if

$$\mathscr{N} := \sum_{j=1}^{r}E_j - \sum_{j=1}^{s}D_j \geq -1. \tag{8.4}$$

If $\mathscr{N} > -1$ the series converges for all finite values of X to an entire function and if $\mathscr{N} = -1$, its radius of convergence equals

$$\mathscr{M} := \prod_{k=1}^{s}D_k^{-D_k}\prod_{j=1}^{r}E_j^{E_j}. \tag{8.5}$$

Convergence on the boundary $|X| = \mathscr{M}$ depends on the value of

$$\mathscr{W} := \sum_{j=1}^{r}e_j - \sum_{k=1}^{s}d_k + \frac{s-r-1}{2}. \tag{8.6}$$

Note that, (8.2) converges absolutely for $|X| = \mathscr{M}$ if $\Re(\mathscr{W}) > 0$.

The function ${}_s\mathbb{H}_r$ is an extension of the generalized hypergeometric function (which we will define it later). ${}_1\mathbb{H}_1$ and ${}_0\mathbb{H}_1$ are the Wright (the Bessel-Maitland) function and Mittag-Leffler function with $D_1 = d_1 = 1$, respectively.

▹ Meijer $\mathbb{G}$-function [19] :

The Meijer $\mathbb{G}$-function is a special case of the $\mathbb{H}$-function

$$
\begin{aligned}
&{}_{s}^{v}\mathbb{G}_{r}^{w}\left[X\left|\begin{matrix} d_1,\dots,d_s \\ e_1,\dots,e_r \end{matrix}\right.\right] \\
&\equiv {}_{s}^{v}\mathbb{H}_{r}^{w}\left[X\left|\begin{matrix} (d_1,1),\dots,(d_s,1) \\ (e_1,1),\dots,(e_r,1) \end{matrix}\right.\right] \\
&= \frac{1}{2\pi i}\int_{\mathscr{L}} \mathscr{O}'(Y) X^{-Y} dY, \qquad (8.7)
\end{aligned}
$$

where

$$
\mathscr{O}'(Y) := \frac{\prod_{j=1}^{v} \Gamma(e_j + Y) \prod_{i=1}^{w} \Gamma(1 - d_i - Y)}{\prod_{i=w+1}^{s} \Gamma(d_j + Y) \prod_{j=v+1}^{r} \Gamma(1 - d_j - Y)}. \qquad (8.8)
$$

Here, $X^{-Y} = \exp(-Y[\log|X| + i \ \arg(X)])$, $X \neq 0$ and $i^2 = -1$, and also $\log|X|$ represents the natural logarithm of $|X|$ and $\arg(X)$ is not necessarily the principle value.

Notice that, an empty product in (8.8), if it occurs, is taken to be one and the poles

$$
e_{j\wp} = -(e_j + \wp), \quad j = 1, \dots, v, \quad \wp \in \mathbb{N}_0, \qquad (8.9)
$$

of the gamma functions $\Gamma(e_j + Y)$ and the poles

$$
d_{i\wp'} = 1 - d_i + \wp', \quad j = 1, \dots, w, \quad \wp' \in \mathbb{N}_0, \qquad (8.10)
$$

of the gamma functions $\Gamma(1 - d_i - Y)$ do not coincide, that is

$$
e_j + \wp \neq d_i - \wp' - 1, \quad i = 1, \dots, w, \quad j = 1, \dots, v, \quad \wp, \wp' \in \mathbb{N}_0. \qquad (8.11)
$$

Note that contour $\mathscr{L}$ is one of the contours defined before which separates all poles $e_{j\wp}$ in (8.9) to the left and all poles $d_{i\wp}$ in (8.10) to the right of $\mathscr{L}$.

▷ $\mathbb{G}$-function (generalized Hypergeometric function) [19]:

The generalized hypergeometric function is defined by the following generalized hypergeometric series

$$
{}_{s}\mathbb{H}_{r}[d_1, \dots, d_s; e_1, \dots, e_s; X] = \sum_{k=0}^{\infty} \frac{\prod_{i=1}^{s} (d_i)_k}{\prod_{j=1}^{r} (e_j)_k} \frac{X^k}{k!}, \qquad (8.12)
$$

where $X \in \mathbb{C}$, $s, r \in \mathbb{N}_0$, and $d_i, e_j \in \mathbb{C}$, $i = 1, \dots, s$ and $j = 1, \dots, r$. For $z \in \mathbb{C}$, we have

$$
\begin{aligned}
&(z)_0 = 1, \quad z \neq 0, \\
&(z)_k = z(z+1)\dots(z+k-1), \quad k \in \mathbb{N}.
\end{aligned}
$$

If $d_j \neq -\wp$, $j = 1, \dots, r$ and $\wp \in \mathbb{N}_0$, then the generalized hypergeometric series (8.12) can be represented in terms of the Mellin-Barnes integral of the form

$$ {}_s\mathbb{H}_r[d_1,\ldots,d_s;e_1,\ldots,e_r;X] = \frac{\prod_{j=1}^{r}\Gamma(e_j)}{\prod_{i=1}^{s}\Gamma(d_i)}\frac{1}{2\pi i}\int_{\mathscr{L}}\frac{\Gamma(Y)\prod_{i=1}^{s}\Gamma(d_i-Y)}{\prod_{j=1}^{r}\Gamma(e_j-Y)}(-X)^{-Y}dY,\quad X\neq 0, $$

where $e_j \neq 0, -1, -2, \ldots$ $j = 1, \ldots r$, $e_j \neq 0, -1, -2, \ldots$ $j = 1, \ldots s$, and with the contour $\mathscr{L}$.

Such a formula leads to the representation of (8.12) via the Meijer $\mathbb{G}$-function as follows;

$$ {}_s\mathbb{H}_r[d_1,\ldots,d_s;e_1,\ldots,e_r;X] = \frac{\prod_{j=1}^{r}\Gamma(e_j)}{\prod_{i=1}^{s}\Gamma(d_i)}{}^{1}_{s}\mathbb{G}^{\mathsf{E}}_{r+1}\left[-X\middle|^{1-d_1,\ldots,1-d_s}_{0,1-e_1,\ldots,1-e_r}\right]. $$

References

1. Rabotnov YN (1948) The equilibrium of an elastic medium with after-effect. Prikl Mat Mekh (J Appl Math Mech) 12(1):53–62. (Reprinted in Fract Calc Appl Anal 17(3):684–696 (1948))
2. Yang XJ, Gao F, Ju Y (2020) General fractional derivatives with applications in viscoelasticity. Academic, New York
3. Mathai AM, Haubold HJ (2008) Special functions for applied scientists. Springer, New York
4. Yang X-J (2021) Theory and applications of special functions for scientists and engineers. Springer
5. Miller KS, Ross B (1993) An introduction to the fractional calculus and fractional differential equations. Wiley, New York
6. Hartley TT, Lorenzo CF (1998) A solution to the fundamental linear fractional order differential equation. NASA/TP-1998-208693. (Dec 1998)
7. Hartley TT, Lorenzo CF (2016) The fractional trigonometry: with applications to fractional differential equations and science. Wiley, New York
8. Hartley TT, Lorenzo CF (2008) Generalized functions for the fractional calculus. Crit Rev Biomed Eng 36(1):39–55
9. Kohlrausch R (1854) Theorie des elektrischen rckstandes in der leidener flasche. Ann Phys 167(2):179–214
10. Williams G, Watts DC (1970) Non-symmetrical dielectric relaxation behaviour arising from a simple empirical decay function. Trans Faraday Soc 66:80–85
11. Chung SH, Stevens JR (1991) Time-dependent correlation and the evaluation of the stretched exponential or Kohlrausch-Williams-Watts function. Am J Phys 59(11):1024–1030
12. Alvarez F, Alegra A, Colmenero J (1991) Relationship between the time-domain Kohlrausch-Williams-Watts and frequency-domain Havriliak-Negami relaxation functions. Phys Rev B 44(14):7306
13. Yoshioka S, Tajima S, Aso Y, Kojima S (2003) Inactivation and aggregation of β-galactosidase in lyophilized formulation described by Kohlrausch-Williams-Watts stretched exponential function. Pharm Res 20(10):1655–1660
14. Friedrich CHR (1991) Relaxation and retardation functions of the Maxwell model with fractional derivatives. Rheol Acta 30(2):151–158
15. Rezaei Aderyani S, Saadati R, Li C, Rassias TM, Park C (2023) Special functions and multi-stability of the Jensen type random operator equation in C^*-algebras via fixed point. J Inequal Appl 2023(1):1–24
16. Rezaei Aderyani S, Saadati R, Rassias TM, Park C (2022) Best approximation of $(\mathcal{G}_1, \mathcal{G}_2)$-random operator inequality in matrix Menger Banach algebras with application of stochastic Mittag-Leffler and $\mathbb{H}$-Fox control functions. J Inequal Appl

17. O'Regan D, Aderyani SR, Saadati R, Li C (2023) Stability results and parametric delayed Mittag-Leffler matrices in symmetric fuzzy-random spaces with application. Symmetry 15(10):1880
18. Karp DB, Prilepkina EG (2019) The Fox-Wright function near the singularity and branch cut.
19. Kilbas Anatoly A, Saxena Ram K, Saigo Megumi, Trujillo Juan J (2016) The generalized hypergeometric function as the Meijer G-function. Analysis 36(1):1–14

Chapter 9
Preliminaries

At first, we recall some definitions and results which will be used later on in the book. Then, we provide a new interpretation of the Ulam type stability with the application of classical, well–known special functions.

9.1 Weighted Spaces

In this part, we present several weighted spaces of continuous functions under different weighted norms [1].

Let $[\alpha, \beta]$ $(0 \leq \alpha < \beta < \infty)$ be an interval on $\mathbb{R}_+$, and $C[\alpha, \beta]$ be the space of continuous functions $h : [\alpha, \beta] \longrightarrow \mathbb{R}$ with the following norm

$$\|h\|_{C[\alpha,\beta]} = \max_{\alpha \leq \tau \leq \beta} |h(\tau)|.$$

The weighted space $C_{1-\mathcal{C};\xi}[\alpha, \beta]$ of continuous functions h on $(\alpha, \beta]$ is defined by

$$C_{1-\mathcal{C};\xi}[\alpha, \beta] = \left\{h : (\alpha, \beta] \longrightarrow \mathbb{R};\ (\xi(\tau) - \xi(\alpha))^{1-\mathcal{C}} h(\tau) \in C[\alpha, \beta]\right\}, \quad 0 \leq \mathcal{C} < 1,$$

with the norm below

$$\|h\|_{C_{1-\mathcal{C};\xi}[\alpha,\beta]} = \max_{\tau \in [\alpha,\beta]} \left|(\xi(\tau) - \xi(\alpha))^{1-\mathcal{C}} h(\tau)\right|.$$

The weighted space $C^n_{\mathcal{C};\xi}[\alpha, \beta]$ of continuous functions h on $(\alpha, \beta]$ is denoted by

$$C^n_{\mathcal{C};\xi}[\alpha, \beta] = \left\{h : (\alpha, \beta] \longrightarrow \mathbb{R};\ h(\tau) \in C^{n-1}[\alpha, \beta];\ h^{(n)}(\tau) \in C_{\mathcal{C};\xi}[\alpha, \beta]\right\}, \quad 0 \leq \mathcal{C} < 1,$$

S. Rezaei Aderyani et al., *Towards Ulam Type Multi Stability Analysis*, Studies in Systems, Decision and Control 523, https://doi.org/10.1007/978-3-031-55564-0_9

with the norm

$$\|h\|_{C^n_{C;\xi}[\alpha,\beta]} = \sum_{k=0}^{n-1} \|h^{(k)}\|_{C[\alpha,\beta]} + \|h^{(n)}\|_{C_{C;\xi}[\alpha,\beta]}.$$

In particular when $n = 0$, we have that $C^0_{C;\xi}[\alpha, \beta] = C_{C;\xi}[\alpha, \beta]$.

9.2 Fractional Calculus

Fractional calculus is the field of mathematical analysis which deals with the investigation and applications of integrals and derivatives of arbitrary order. It is an old topic since, starting from some speculations of G.W. Leibniz (1695–1697) and L. Euler (1730), it has been developed up to nowadays. In fact the idea of generalizing the notion of derivative to non integer order, in particular to the order 1/2, is contained in the correspondence of Leibniz with Bernoulli, L'Hôpital and Wallis. Euler took the first step by observing that the result of the evaluation of the derivative of the power function has a meaning for non–integer order due to his Gamma function.

There are different approaches to the fractional calculus available in the literature. Some of the most important ones are presented in the following parts.

9.2.1 *Hilfer's Fractional Derivatives and Related Theorems*

There is no unique answer to the question of how to define a derivative to a given fractional order. The most well–known and solidly established fractional derivative operators are those of Riemann–Liouville and Caputo [2], each of which has its own advantages and disadvantages compared with the other: for example, the Riemann–Liouville derivative is always continuous and analytic with respect to the order of differentiation, while the Caputo derivative gives rise to classical initial conditions rather than fractional ones. Hilfer invented an operator which is intermediate between Riemann–Liouville and Caputo, and which is now called the Hilfer fractional derivative.

In this part, we present the fractional derivative in the Hilfer sense and discuss its properties.

Definition 9.1 ([3]) Let (α, β) $(-\infty \leq \alpha < \beta \leq \infty)$ be an interval on $\mathbb{R}$. Consider the increasing monotone map $\xi(\mathfrak{B}_\circ) > 0$ on $(\alpha, \beta]$, which has a continuous derivative $\xi'(\mathfrak{B}_\circ)$ on (α, β). The fractional integrals of a function μ, with respect to ξ, on $[\alpha, \beta]$ are defined as

$$I_{\alpha^+}^{\mathcal{A};\xi}\mu(\mathfrak{B}_\circ) = \frac{1}{\Gamma(\mathcal{A})}\int_\alpha^{\mathfrak{B}_\circ} \xi'(\tau)(\xi(\mathfrak{B}_\circ) - \xi(\tau))^{\mathcal{A}-1}\mu(\tau)d\tau,$$

where $\mathcal{A} > 0$ and Γ is the Gamma function.

Definition 9.2 ([3]) Let $\mathcal{A} \in (n-1, n)$ with $n \in \mathbb{N}$, and let $\mu, \xi \in C^n[\alpha, \beta]$ be two mappings such that ξ is increasing and $\xi'(\mathfrak{B}_\circ) \neq 0$, for every $\mathfrak{B}_\circ \in [\alpha, \beta]$. Then, the ξ–Hilfer fractional derivative ${}^{\mathscr{H}}D_{0^+}^{\mathcal{A},\mathcal{B};\xi}(.)$ of a mapping μ of order $\mathcal{A}$ and type $0 \leq \mathcal{B} \leq 1$ is defined as

$${}^{\mathscr{H}}D_{\alpha^+}^{\mathcal{A},\mathcal{B};\xi}\mu(\mathfrak{B}_\circ) = I_{\alpha^+}^{\mathcal{B}(n-\mathcal{A});\xi}\left(\frac{1}{\xi'(\mathfrak{B}_\circ)}\frac{d}{d\mathfrak{B}_\circ}\right)^n I_{\alpha^+}^{(1-\mathcal{B})(n-\mathcal{A});\xi}\mu(\mathfrak{B}_\circ).$$

Theorem 9.1 ([4, 5]) *Let $\mu \in C^1[0, \beta]$, $\mathcal{A} \in (0, 1)$ and $\mathcal{B} \in [0, 1]$. Then, we have*

$${}^{\mathscr{H}}D_{0^+}^{\mathcal{A},\mathcal{B};\xi}I_{0^+}^{\mathcal{A};\xi}\mu(\mathfrak{B}_\circ) = \mu(\mathfrak{B}_\circ).$$

Theorem 9.2 ([4, 5]) *Let $\mu \in C^1[0, \beta]$, $\mathcal{A} \in (0, 1)$ and $\mathcal{B} \in [0, 1]$. Then, we have*

$$I_{0^+}^{\mathcal{A};\xi}{}^{\mathscr{H}}D_{0^+}^{\mathcal{A},\mathcal{B};\xi}\mu(\mathfrak{B}_\circ) = \mu(\mathfrak{B}_\circ) - \frac{(\xi(\mathfrak{B}_\circ) - \xi'(0))^{\mathcal{C}-1}}{\Gamma(\mathcal{C})}I_{0^+}^{(1-\mathcal{B})(1-\mathcal{A});\xi}\mu(0),$$

where $\mathcal{C} = \mathcal{A} + \mathcal{B}(1 - \mathcal{A})$.

Lemma 9.1 ([4, 5]) *Let $\hbar_1, \hbar_2 > 0$. If $\Psi(\tau) = (\xi(\tau) - \xi(\alpha))^{\hbar_2-1}$, then, we have*

$$I_{\alpha^+}^{\hbar_1;\xi}\Psi(\tau) = \frac{\Gamma(\hbar_2)}{\Gamma(\hbar_2 + \hbar_1)}(\xi(\tau) - \xi(\alpha))^{\hbar_1+\hbar_2-1}.$$

9.2.2 The Hadamard Fractional Derivatives

Apart from the most well–known Riemann-Liouville and Caputo fractional integral and derivative, there are other definitions of fractional integrals and derivatives, such as the Hadamard fractional integral and derivative. The fundamental distinction between the Hadamard integral and the Riemann-Liouville or Caputo integral is the type of kernel used; the Hadamard integral contains a logarithmic function, which was first developed by Hadamard in 1892 [6], whereas the Riemann-Liouville integral uses a power function.

Another distinction is that Hadamard fractional calculus is more suitable for describing phenomena unrelated to dilation on the semi-axis, while Riemann–Liouville fractional calculus is better appropriate to describe abnormal convection and diffusion phenomena. In igneous rocks, there is a creep phenomenon in the rheology and super slow kinetics. The Lomnitz logarithmic creep law describes it, and the Hadamard fractional calculus can more clearly illuminate its mathematical underpinnings [7, 8]. In addition, the Hadamard fractional calculus can also be used to describe a wide variety of material mechanics issues, including fracture analysis [9].

In this part, we introduce the definitions of the ϕ–Hadamard–type fractional integral and the ϕ–Hadamard–type fractional derivative of Caputo type.

Definition 9.3 ([10]) Let (α, β) be an interval on the real line $\mathbb{R}$, and $\phi(\varsigma)$ be a non-decreasing and positive monotone function on $(\alpha, \beta]$, having a continuous derivative $\phi'(\varsigma)$ (denote first derivative as $\frac{d}{d\varsigma}\phi(\varsigma)$ on (α, β)). The left–side ϕ–Hadamard–type fractional integral with order $\kappa > 0$ and parameter $\vartheta > 0$, of an integrable function $\eta(\varsigma)$, regarding $\phi(\varsigma)$, on $[\alpha, \beta]$ is defined as

$$ {}^{H}_{\alpha^+}I^{\kappa,\vartheta}_{\phi(\varsigma)}\eta(\varsigma) = \frac{1}{\Gamma(\kappa)}\int_{\alpha}^{\varsigma} \mathcal{Q}(\varsigma, \iota)\eta(\iota)d\iota, \tag{9.1} $$

where

$$ \mathcal{Q}(\varsigma, \iota) = \left(\frac{\phi(\iota)}{\phi(\varsigma)}\right)^{\vartheta}\left(\log\frac{\phi(\varsigma)}{\phi(\iota)}\right)^{\kappa-1}\frac{\phi'(\iota)}{\phi(\iota)}, $$

for every $\varsigma \in [\alpha, \beta]$ and Γ is the Gamma function.

Definition 9.4 ([10]) Let $n - 1 < \kappa < n$ with $n \in \mathbb{N}$. Let $\mathcal{J} = [\alpha, \beta]$ be an interval such that $-\infty \leq \alpha < \beta \leq \infty$, and let $\eta, \phi \in C^n[\alpha, \beta]$ be two functions such that ϕ is non–decreasing and $\phi'(\varsigma) \neq 0$, for every $\varsigma \in \mathcal{J}$. The ϕ–Hadamard–type fractional derivative of Caputo type ${}^{HC}_{\alpha^+}D^{\kappa,\vartheta}_{\phi(\varsigma)}\eta(\varsigma)$ of a function η of order $\kappa > 0$ and parameter $\vartheta > 0$ is defined as

$$ {}^{HC}_{\alpha^+}D^{\kappa,\vartheta}_{\phi(\varsigma)}\eta(\varsigma) = {}^{H}_{\alpha^+}I^{(n-\kappa),\vartheta}_{\phi(\varsigma)}\ {}^{H}_{\alpha^+}D^{n,\vartheta}_{\phi(\varsigma)}\eta(\varsigma), \qquad \varsigma \in [\alpha, \beta] \tag{9.2} $$

where

$$ {}^{H}_{\alpha^+}D^{n,\vartheta}_{\phi(\varsigma)}\eta(\varsigma) = \phi(\varsigma)^{-\vartheta}\left(\frac{\phi(\varsigma)}{\phi'(\varsigma)}.\frac{d}{d\varsigma}\right)^n\left[\phi(\varsigma)^{\vartheta}\eta(\varsigma)\right]. $$

9.3 Normed Algebras

Let X be a vector space on field $\mathbb{C}$. A normed space X in which, for all $x, y \in X$,

$$ xy \in X, $$

and

$$ \|xy\| \leq \|x\|\|y\|, $$

is called a complex normed algebra [11].

A complete normed algebra is called a Banach algebra. Moreover, if there exists a unit element e, such that

$$ex = xe = x,$$

for all $x \in X$, then, $\|e\| = 1$, and X is called a unital Banach algebra.

Definition 9.5 ([11]) Let X, Y be Banach algebras. A $\mathbb{C}$–linear mapping $H : X \to Y$, is called a homomorphism in Banach algebras, if H satisfies

$$H(xy) = H(x)H(y),$$

for all $x, y \in X$.

A $\mathbb{C}$–linear mapping $\delta : X \to X$, is called a derivation on X, if δ satisfies

$$\delta(x, y) = \delta(x)y + x\delta(y),$$

for all $x, y \in X$.

Definition 9.6 ([12]) Let X be a complex Banach algebra. A $\mathbb{C}$–linear mapping $\delta : X \to X$ is called an antiderivation, if it satisfies

$$\delta(x)\delta(y) = \delta(\delta(x)y) + \delta(x\delta(y)),$$

for every $x, y \in X$.

Example 9.1 Consider the two examples below:

(1) Let Q_m be the collection of all polynomials of degree m with complex coefficients and

$$Q = \{q_m \in Q_m | q(0) = 0, m \in \mathbb{N}\}.$$

Define $\delta : Q \to Q$ by

$$\delta(\sum_{k=1}^{n} b_k x^k) = \sum_{k=1}^{n} \frac{b_k}{k} x^k,$$

and $\delta(0) = 0$. Then, δ is an antiderivation.

(2) Consider the collection of all continuous functions on $\mathbb{R}$, represented by $C(\mathbb{R})$. Define $\delta : C(\mathbb{R}) \to C(\mathbb{R})$ by

$$\delta(g(x)) = \int_0^x g(t)dt,$$

for every $t \in \mathbb{R}$. Then, δ is an antiderivation.

Definition 9.7 ([13]) Let X and Y be two complex Banach algebras. We say a $\mathbb{C}$–trilinear mapping $H : X^3 \to Y$ is a permuting tri–homomorphism, if H satisfies the following conditions:

$$H(x_1x_2, x_3x_4, x_5x_6) = H(x_1, x_3, x_5)H(x_2, x_4, x_6),$$
$$H(y_{\sigma(1)}, y_{\sigma(2)}, y_{\sigma(3)}) = H(y_1, y_2, y_3),$$

for every $x_1, x_2, x_3, x_4, y_1, y_2, y_3 \in X$, and for every permutation $(\sigma(1), \sigma(2), \sigma(3))$ of $(1, 2, 3)$.

Definition 9.8 ([13]) Let X be a complex Banach algebra. We say a tri–additive mapping $\delta : X^3 \to X$ is a permuting tri–derivation on X, if δ satisfies the following conditions:

$$\delta(x_1, x_2, x_3, x_4) = \delta(x_1, x_3, x_4)x_2 + x_1\delta(x_2, x_3, x_4),$$
$$\delta(y_{\sigma(1)}, y_{\sigma(2)}, y_{\sigma(3)}) = \delta(y_1, y_2, y_3),$$

for every $x_1, x_2, x_3, x_4, y_1, y_2, y_3 \in X$, and for every permutation $(\sigma(1), \sigma(2), \sigma(3))$ of $(1, 2, 3)$.

9.4 C*–Algebras

Let U be a Banach algebra. Then, an involution on U is a mapping $u \to u^*$, from U into U which satisfies the following conditions:

(1) $u^{**} = u$, for all $u \in U$;
(2) $(\alpha u + \beta v)^* = \overline{\alpha}u^* + \overline{\beta}v^*$;
(3) $(uv)^* = v^*u^*$, for all $u, v \in U$.

If, in addition, $\|u^*u\| = \|u\|^2$, for all $u \in U$, then, U is a C*–algebra [11].

Definition 9.9 ([11]) Let U and V be C*–algebras. A $\mathbb{C}$–linear mapping $H : U \to V$ is called a homomorphism in C*–algebras if H satisfies

$$H(xy) = H(x)H(y),$$

and

$$H(x^*) = H(x)^*,$$

for all $x, y \in U$.

A $\mathbb{C}$–linear mapping $\delta : U \to U$, is called a derivation on U if δ satisfies

$$\delta(xy) = \delta(x)y + x\delta(y),$$

for all $x, y \in U$.

Suppose that $\mathcal{A}$ is a complex Banach $*$–algebra. Let $\mathbb{C}$–linear mapping $\delta : D(\delta) \to \mathcal{A}$ be a derivation on $\mathcal{A}$, where $D(\delta)$ is the domain of δ and $D(\delta)$ is dense in $\mathcal{A}$. If δ satisfies the additional condition

$$\delta(a^*) = \delta(a)^*,$$

for all $a \in \mathcal{A}$, then, δ is called a $*$–derivation on $\mathcal{A}$ [11].

It is well–known that, if $\mathcal{A}$ is a C*–algebra and $D(\delta)$ is $\mathcal{A}$, then, the derivation δ is bounded.

A C*–algebra $\mathcal{C}$ endowed with the Lie product

$$[x, y] := \frac{xy - xy}{2},$$

on $\mathcal{C}$ is called a Lie C*–algebra [11].

Definition 9.10 ([11]) Let A and B be Lie C*–algebras. A $\mathbb{C}$–linear mapping $H : A \to B$ is called a Lie C*–algebra homomorphism if

$$H([x, y]) = [H(x), H(y)],$$

for all $x, y \in A$.

A $\mathbb{C}$–linear mapping $\delta : A \to A$ is called a Lie derivation if

$$\delta([x, y]) = [\delta(x), y] + [x, \delta(y)],$$

for all $x, y \in A$.

9.5 C*–Ternary Algebras

Ternary algebraic structures appear more or less naturally in various domains of theoretical and mathematical physics, for example, the quark model inspired a particular brand of ternary algebraic system. One such attempt has been proposed by Nambu in 1973 and is now known under the name of "Nambu mechanics" [14, 15].

A C*–ternary algebra [11] is a complex Banach space A, equipped with a ternary product $(x, y, z) \to [x, y, z]$ of A^3 into A, which is $\mathbb{C}$–linear in the outer variables, conjugate $\mathbb{C}$–linear in the middle variable and associative in the sense that

$$[x, y, [z, w, v]] = [x, [w, z, y], v] = [[x, y, z], w, v],$$

and satisfies

$$\|[x, y, z]\| \leq \| \|x\| \cdot \|y\| \cdot \|z\|,$$

and

$$\|[x, x, x]\| \leq \| \|x\|^3.$$

If C*–ternary algebra $(A, [., ., .])$ has the identity, i.e., an element $e \in A$, such that

$$x = [x, e, e] = [e, e, x],$$

for all $x \in A$, then it is routine to verify that A, endowed with $x \circ y := [x, e, y]$ and $x^* := [e, x, e]$ is a unital C*–algebra. Conversely, if $(A, \circ)$ is a unital C*–algebra, then, $[x, y, z] := x \circ y^* \circ z$ makes A into a C*–ternary algebra.

Definition 9.11 ([11]) Let A and B be two C*–ternary algebras. A $\mathbb{C}$–linear mapping $H : A \to B$ is called a C*–ternary algebra homomorphism if,

$$H([x, y, z]) = [H(x), H(y), H(z)],$$

for all $x, y, z \in A$.

A $\mathbb{C}$–linear mapping $\delta : A \to A$ is called a C*–ternary derivation if,

$$\delta([x, y, z]) = [\delta(x), y, z] + [x, \delta(y), z] = [x, y, \delta(z)],$$

for all $x, y, z \in A$.

9.6 Classical Operators in Probability Measure Spaces

Consider Banach algebras $\mho_1$ and $\mho_2$. Let $(\pitchfork, \curlywedge, d)$ be a probability measure space with the Borel measurable spaces $(\mho_1, \mathfrak{B}_{\mho_1})$ and $(\mho_2, \mathfrak{B}_{\mho_2})$. An operator is a map $f : \pitchfork \times \mho_1 \to \mho_2$ such that $\{\leftthreetimes : f(\leftthreetimes, \mathsf{S}) \in \mathcal{C}\} \in \curlywedge$, for all S in $\mho_1$ and $\mathcal{C} \in \mathfrak{B}_{\mho_2}$ (see [16]).

The operator f is linear if,

$$f(\leftthreetimes, \alpha \mathsf{S}_1 + \beta \mathsf{S}_2) = \alpha f(\leftthreetimes, \mathsf{S}_1) + \beta f(\leftthreetimes, \mathsf{S}_2),$$

where $\mathsf{S}_1, \mathsf{S}_2 \in \mho_1$, $\alpha, \beta \in \mathbb{R}$ and $\leftthreetimes \in \pitchfork$.

The operator f is bounded, if we can find a $Q(\leftthreetimes) > 0$, such that

$$d(f(\leftthreetimes, \mathsf{S}_1), f(\leftthreetimes, \mathsf{S}_2)) \leq Q(\leftthreetimes) d(\mathsf{S}_1 - \mathsf{S}_2),$$

where $\mathsf{S}_1, \mathsf{S}_2 \in \mho_1$ and $\leftthreetimes \in \pitchfork$.

9.7 Generalized Metric Spaces and Fixed Point Theory

Speaking of the stability of a functional equation we follow a question raised in 1940 by Ulam, concerning approximate homomorphisms of groups [17]. The first partial answer (in the case of Cauchy's functional equation in Banach spaces) to Ulam's question was given by Hyers [18]. The method used by Hyers in [18] (quite often called the direct method) has been successfully applied for study of the Ulam type stability of large variety of equations, but unfortunately, as it was shown in [19], it does not work in numerous significant cases. Apart from it, there are also several other efficient approaches to the Ulam type stability, using different tools, for example, the method of invariant means [20], the method based on sandwich theorems [21], the method using the concept of shadowing [22], the Laplace transform method [23] and the Fourier transform method [24].

In this section, we discuss the fixed point method, which is the second most popular technique of proving the Ulam type stability. It was used for the first time by Baker [25] who applied a variant of Banach's fixed point theorem to obtain the Hyers–Ulam stability of a functional equation in a single variable. The Banach fixed point theorem [26] (also known as the Banach contraction principle) is an important tool in the theory of metric spaces because it guarantees the existence and uniqueness of fixed points of certain self mappings of metric spaces and provides a constructive method to find those fixed points. The theorem is named after Banach (1892–1945) and was first stated by him in 1922.

Theorem 9.3 (Banach [26]) *Let $(\mathcal{G}, d)$ be a complete metric space and $d : \mathcal{G} \to \mathcal{G}$ be a contraction, i.e., there exists $\mathcal{Z} \in [0, 1)$, such that*

$$d(\propto \Psi, \propto \Phi) \leq \mathcal{Z} d(\Psi, \Phi),$$

for all $\Psi, \Phi \in \mathcal{G}$. Then, there exists a unique $\Upsilon \in \mathcal{G}$, such that $\propto \Upsilon = \propto$.

Moreover, for all $\Phi \in \mathcal{G}$,

$$\lim_{m \to \infty} \propto^m \Phi = \Upsilon,$$

and, in fact, for all $\Phi \in \mathcal{G}$,

$$d(\Phi, \Upsilon) \leq \frac{1}{1 - \mathcal{Z}} d(\Phi, \propto \Phi).$$

We now propose an alternative fixed point theorem from the literature:

Theorem 9.4 ([27, 28]) *Let $(\mathcal{G}, d)$ be a complete metric space and $\propto: \mathcal{G} \to \mathcal{G}$ be strictly contractive, i.e.,*

$$d(\propto \Psi, \propto \Phi) \leq \mathcal{Z} d(\Psi, \Phi), \qquad \Psi, \Phi \in \mathcal{G}$$

for some Lipschitz constants $\mathcal{Z} < 1$. Then, we have that,

(1) *the mapping $\propto$ has a unique fixed point $\Psi^* = \propto \Psi^*$;*

(2) *the fixed point Ψ^* is globally attractive, i.e.,*

$$\lim_{m \to \infty} \propto^m \Psi = \Psi^*,$$

for any starting point $\Psi \in \mathcal{G}$;

(3) *the following three inequalities hold:*

$$d(\propto^m \Psi, \Psi^*) \leq \mathcal{Z}^m d(\Psi, \Psi^*),$$
$$d(\propto^m \Psi, \Psi^*) \leq \frac{1}{1-\mathcal{Z}} d(\propto^m \Psi, \propto^{m+1} \Psi),$$
$$d(\Psi, \Psi^*) \leq \frac{1}{1-\mathcal{Z}} d(\Psi, \propto \Psi),$$

for all nonnegative integers m and all $\Psi \in \mathcal{G}$.

Now, we present the vector valued generalized metric spaces and a generalized form of the alternative fixed point theorem.

Let $m \in \mathbb{N}$. Consider two vectors $\lambda = (\lambda_1, \ldots, \lambda_m)$, and $\gamma = (\gamma_1, \ldots, \gamma_m)$, with the partial order relation

$$\lambda \leq \gamma \Longleftrightarrow \lambda_i \leq \gamma_i, \quad i = 1, \cdots, m.$$

Also, $\lambda < \gamma$, denotes that $\lambda \leq \gamma$ and $\lambda \neq \gamma$; $\lambda \ll \gamma$ and $\lambda_i < \gamma_i$, for every $i = 1, \cdots, m$. Besides, we have that

$$\lambda \to 0 \Longleftrightarrow \lambda_i \to 0, \quad i = 1, \cdots, m.$$

Following Luxemburg [29], the concept of a vector valued generalized complete metric space may be introduced as in this quotation:

Definition 9.12 ([29, 30]) Suppose $\mathcal{G}$ is a nonempty set and $d : \mathcal{G}^2 \to [0, +\infty]^m$ (with $m \in \mathbb{N}$), is a given mapping. We say that d is a vector valued generalized complete meter on $\mathcal{G}$ if the following conditions satisfy:

(1) for every $(\Psi, \Phi) \in \mathcal{G} \times \mathcal{G}$, we have,

$$d(\Psi, \Phi) = \underbrace{(0, \cdots, 0)}_{m} \Longleftrightarrow \Psi = \Phi;$$

(2) for every $(\Psi, \Phi) \in \mathcal{G} \times \mathcal{G}$, we have the following symmetry,

$$d(\Phi, \Psi) = d(\Psi, \Phi) \Longleftrightarrow \Psi = \Phi;$$

(3) for every $\Psi, \Phi, \Upsilon \in \mathcal{G}$, the triangle inequality holds,

$$d(\Psi, \Upsilon) + d(\Upsilon, \Phi) \succeq d(\Phi, \Psi).$$

This concept differs from the usual concept of a complete metric space by the fact that not every two points in $\mathcal{G}$ have necessarily a finite distance. One might call such a space a vector valued generalized complete metric space.

Next, Diaz and Margolis [31] proved a theorem of the alternative for any contraction mapping $\propto$ on a generalized complete metric space $(\mathcal{G}, d)$. The conclusion of the theorem, speaking in general terms, asserts that: either all consecutive pairs of the sequence of successive approximations (starting from an element x_0 of $\mathcal{G}$) are infinitely far apart or the sequence of successive approximations, with initial element x_0 converges to a fixed point of $\propto$ (what particular fixed point depends, in general, on the initial element x_0).

Theorem 9.5 (Diaz–Margolis theorem [30, 31]) *Suppose* $d : \mathcal{G}^2 \to [0, +\infty]^m$, $m \in \mathbb{N}$, *and* $(\mathcal{G}, d)$ *is a vector valued generalized complete metric space. Suppose* $\propto: \mathcal{G} \to \mathcal{G}$ *is a strictly contractive mapping with Lipschitz constant* $\mathcal{Z} < 1$. *Then, for any given element* $\Psi \in \mathcal{G}$, *either*

$$d(\propto^n \Psi, \propto^{n+1} \Psi) = \underbrace{(+\infty, \cdots, +\infty)}_{m},$$

for any $n \in \mathbb{N} \cup \{0\}$ *or there exists an* $n_0 \in \mathbb{N}$ *such that*

(1) $d(\propto^n \Psi, \propto^{n+1} \Psi) \leq \underbrace{(+\infty, \cdots, +\infty)}_{m}, \quad \forall n \geq n_0;$

(2) *the fixed point* Φ^* *of* $\propto$ *is a convergent point of the sequence* $\{\propto^n \Psi\}$;

(3) Φ^* *is the unique fixed point of* $\propto$ *in the set* $\mathcal{Q} = \{\Phi \in \mathcal{G} \mid d(\propto^{n_0} \Psi, \Phi) \leq \underbrace{(+\infty, \cdots, +\infty)}_{m}\}$;

(4) $d(\Phi, \Phi^*) \leq \frac{1}{1-\mathcal{Z}} d(\Phi, \propto \Phi)$, *for all* $\Phi \in \mathcal{Q}$.

In the remainder of the book, we follow Radu's approach [28] and make use of a theorem of Diaz and Margolis.

9.8 Multi Stability Analysis

The main target of this section is to present a new concept of the Ulam type stability, i.e., multi–stability, through the classical, well–known special functions and to get the best approximation error estimates by different approaches of perturbation stability in various classical and fuzzy spaces. This stability allows us to obtain diverse approximations depending on the special functions that are initially chosen and to evaluate optimal stability and minimal error, which enable us to gain a unique optimum solution of both functional and fractional equations.

9.8.1 The Matrix–Valued Special Functions

Through the special functions given in the previous sections, we define the following diagonal matrix (a square matrix in which every element except the main diagonal elements is zero), to use in the next chapters of the book. This will help us to present a novel view on the Ulam type stability with application of special functions to provide a comprehensive discussion of optimizing the different types of the Ulam stabilities.

$$\mathfrak{W}(X) := \tag{9.3}$$

$$\mathrm{diag}\Big[\ {}_2\mathbb{F}_1(\mathsf{A},\mathsf{B};\mathsf{C};X),\ {}_2Supersin_1(\mathsf{A},\mathsf{B};\mathsf{C};X),\ {}_2Supercos_1(\mathsf{A},\mathsf{B};\mathsf{C};X),$$
$${}_2Supertan_1(\mathsf{A},\mathsf{B};\mathsf{C};X),\ {}_2Supercot_1(\mathsf{A},\mathsf{B};\mathsf{C};X),\ {}_2Supersec_1(\mathsf{A},\mathsf{B};\mathsf{C};X),$$
$${}_2Supercosec_1(\mathsf{A},\mathsf{B};\mathsf{C};X),\ {}_2Supersinh_1(\mathsf{A},\mathsf{B};\mathsf{C};X),\ {}_2Supercosh_1(\mathsf{A},\mathsf{B};\mathsf{C};X),$$
$${}_2Supertanh_1(\mathsf{A},\mathsf{B};\mathsf{C};X),\ {}_2Supercoth_1(\mathsf{A},\mathsf{B};\mathsf{C};X),\ {}_2Supersech_1(\mathsf{A},\mathsf{B};\mathsf{C};X),$$
$${}_2Supercosech_1(\mathsf{A},\mathsf{B};\mathsf{C};X),\ {}_p\mathbb{F}_q(\mathsf{A}_1,\cdots,\mathsf{A}_p;\mathsf{C}_1,\cdots,\mathsf{C}_p;X),$$
$${}_pSupersin_q(\mathsf{A}_1,\cdots,\mathsf{A}_p;\mathsf{C}_1,\cdots,\mathsf{C}_p;X),\ {}_pSupercos_q(\mathsf{A}_1,\cdots,\mathsf{A}_p;\mathsf{C}_1,\cdots,\mathsf{C}_p;X),$$

$${}_pSupertan_q(\mathsf{A}_1,\cdots,\mathsf{A}_p;\mathsf{C}_1,\cdots,\mathsf{C}_p;X),\ {}_pSupercot_q(\mathsf{A}_1,\cdots,\mathsf{A}_p;\mathsf{C}_1,\cdots,\mathsf{C}_p;X),$$
$${}_pSupersec_q(\mathsf{A}_1,\cdots,\mathsf{A}_p;\mathsf{C}_1,\cdots,\mathsf{C}_p;X),\ {}_pSupercosec_q(\mathsf{A}_1,\cdots,\mathsf{A}_p;\mathsf{C}_1,\cdots,\mathsf{C}_p;X),$$
$${}_pSupersinh_q(\mathsf{A}_1,\cdots,\mathsf{A}_p;\mathsf{C}_1,\cdots,\mathsf{C}_p;X),\ {}_pSupercosh_q(\mathsf{A}_1,\cdots,\mathsf{A}_p;\mathsf{C}_1,\cdots,\mathsf{C}_p;X),$$
$${}_pSupertanh_q(\mathsf{A}_1,\cdots,\mathsf{A}_p;\mathsf{C}_1,\cdots,\mathsf{C}_p;X),\ {}_pSupercoth_q(\mathsf{A}_1,\cdots,\mathsf{A}_p;\mathsf{C}_1,\cdots,\mathsf{C}_p;X),$$
$${}_pSupersech_q(\mathsf{A}_1,\cdots,\mathsf{A}_p;\mathsf{C}_1,\cdots,\mathsf{C}_p;X),\ {}_pSupercosech_q(\mathsf{A}_1,\cdots,\mathsf{A}_p;\mathsf{C}_1,\cdots,\mathsf{C}_p;X),$$
$${}_p\mathbb{W}_q(X),\ {}_p\sin_q(X),\ {}_p\cos_q(X),\ {}_p\tan_q(X),\ {}_p\cot_q(X),\ {}_p\sec_q(X),\ {}_p\mathrm{cosec}_q(X),$$

$${}_p\sinh_q(X),\ {}_p\cosh_q(X),\ {}_p\tanh_q(X),\ {}_p\coth_q(X),\ {}_p\mathrm{sech}_q(X),\ {}_p\mathrm{cosech}_q(X),$$
$${}_0\mathbb{W}_1(X),\ {}_0\sin_1(X),\ {}_0\cos_1(X),\ {}_0\tan_1(X),\ {}_0\cot_1(X),\ {}_0\sec_1(X),\ {}_0\mathrm{cosec}_1(X),$$
$${}_0\sinh_1(X),\ {}_0\cosh_1(X),\ {}_0\tanh_1(X),\ {}_0\coth_1(X),\ {}_0\mathrm{sech}_1(X),\ {}_0\mathrm{cosech}_1(X),$$
$${}_p\mathbb{B}_q(X),\ [{}_p\mathbb{W}_q]^t(X),\ [{}_p\sin_q]^t(X),\ [{}_p\cos_q]^t(X),\ [{}_p\sinh_q]^t(X),\ [{}_p\cosh_q]^t(X),\ \mathbb{E}_\kappa(X),$$
$$\mathbb{E}_\kappa(X^\kappa),\ \mathbb{E}_\kappa(X^\vartheta),\ \mathbb{E}_\kappa(X^{2\vartheta}),\ supersin_\kappa(X^\kappa),\ supercos_\kappa(X^\kappa),\ supertan_\kappa(X^\kappa),$$
$$supercotan_\kappa(X^\kappa),\ supercsc_\kappa(X^\kappa),\ supersec_\kappa(X^\kappa),\ supersinh_\kappa(X^\kappa),\ supercosh_\kappa(X^\kappa),$$

$$supertanh_\kappa(X^\kappa),\ supercoth_\kappa(X^\kappa),\ supercsch_\kappa(X^\kappa),\ supersech_\kappa(X^\kappa),$$
$$presin_\kappa(X),\ precos_\kappa(X),\ prtan_\kappa(X),\ precotan_\kappa(X),\ precsc_\kappa(X),\ presec_\kappa(X),$$
$$presin_\kappa(X^\vartheta),\ precos_\kappa(X^\vartheta),\ prtan_\kappa(X^\vartheta),\ precotan_\kappa(X^\vartheta),\ precsc_\kappa(X^\vartheta),\ presec_\kappa(X^\vartheta),$$
$$presinh_\kappa(X),\ precosh_\kappa(X),\ prtanh_\kappa(X),\ precotanh_\kappa(X),\ precsch_\kappa(X),\ presech_\kappa(X),$$
$$presinh_\kappa(X^\vartheta),\ precosh_\kappa(X^\vartheta),\ prtanh_\kappa(X^\vartheta),\ precotanh_\kappa(X^\vartheta),\ precsch_\kappa(X^\vartheta),$$
$$presech_\kappa(X^\vartheta),\ \mathbb{M}^\kappa_{m,r}(X),\ \{\mathbb{E}_\kappa\}^t(X),\ \{presin_\kappa\}^t(X),\ \{precos_\kappa\}^t(X),$$

$$\{presinh_{\kappa}\}^t(X), \{precosh_{\kappa}\}^t(X), \{presin_{\kappa}\}^t(X^{\vartheta}), \{precos_{\kappa}\}^t(X^{\vartheta}),$$
$$\{presinh_{\kappa}\}^t(X^{\vartheta}), \{precosh_{\kappa}\}^t(X^{\vartheta}), \{presin_{\kappa}\}^t(X^{\kappa}), \{precos_{\kappa}\}^t(X^{\kappa}),$$
$$\{presinh_{\kappa}\}^t(X^{\kappa}), \{precosh_{\kappa}\}^t(X^{\kappa}), \mathbb{E}_{\kappa,\vartheta}(X), X^{\vartheta-1}\mathbb{E}_{\kappa,\vartheta}(X^{\kappa}),$$
$$supersin_{\kappa,\vartheta}(X^{\kappa}), supercos_{\kappa,\vartheta}(X^{\kappa}), supertan_{\kappa,\vartheta}(X^{\kappa}), supercotan_{\kappa,\vartheta}(X^{\kappa}),$$
$$supercsc_{\kappa,\vartheta}(X^{\kappa}), supersec_{\kappa,\vartheta}(X^{\kappa}), supersinh_{\kappa,\vartheta}(X^{\kappa}), supercosh_{\kappa,\vartheta}(X^{\kappa}),$$
$$supertanh_{\kappa,\vartheta}(X^{\kappa}), supercotanh_{\kappa,\vartheta}(X^{\kappa}), supercsch_{\kappa,\vartheta}(X^{\kappa}), supersech_{\kappa,\vartheta}(X^{\kappa}),$$

$$presin_{\kappa,\vartheta}(X), precos_{\kappa,\vartheta}(X), pretan_{\kappa,\vartheta}(X), precotan_{\kappa,\vartheta}(X), precsc_{\kappa,\vartheta}(X),$$
$$presec_{\kappa,\vartheta}(X), presinh_{\kappa,\vartheta}(X), precosh_{\kappa,\vartheta}(X), pretanh_{\kappa,\vartheta}(X), precotanh_{\kappa,\vartheta}(X),$$
$$precsch_{\kappa,\vartheta}(X), presech_{\kappa,\vartheta}(X), presin_{\kappa,\vartheta}(X^{\gamma}), precos_{\kappa,\vartheta}(X^{\gamma}), pretan_{\kappa,\vartheta}(X^{\gamma}),$$
$$precotan_{\kappa,\vartheta}(X^{\gamma}), precsc_{\kappa,\vartheta}(X^{\gamma}), presec_{\kappa,\vartheta}(X^{\gamma}), presinh_{\kappa,\vartheta}(X^{\gamma}), precosh_{\kappa,\vartheta}(X^{\gamma}),$$
$$pretanh_{\kappa,\vartheta}(X^{\gamma}), precotanh_{\kappa,\vartheta}(X^{\gamma}), precsch_{\kappa,\vartheta}(X^{\gamma}), presech_{\kappa,\vartheta}(X^{\gamma}),$$

$$Tresin_{\kappa,\vartheta}(X), Trecos_{\kappa,\vartheta}(X), Tretan_{\kappa,\vartheta}(X), Trecotan_{\kappa,\vartheta}(X), Trecsc_{\kappa,\vartheta}(X),$$
$$Tresec_{\kappa,\vartheta}(X), Tresinh_{\kappa,\vartheta}(X), Trecosh_{\kappa,\vartheta}(X), Tretanh_{\kappa,\vartheta}(X), Trecotanh_{\kappa,\vartheta}(X),$$
$$Trecsch_{\kappa,\vartheta}(X), Tresech_{\kappa,\vartheta}(X), Gresin_{\kappa,\vartheta}(X^{\gamma}), Grecos_{\kappa,\vartheta}(X^{\gamma}), Gretan_{\kappa,\vartheta}(X^{\gamma}),$$
$$Grecotan_{\kappa,\vartheta}(X^{\gamma}), Grecsc_{\kappa,\vartheta}(X^{\gamma}), Gresec_{\kappa,\vartheta}(X^{\gamma}), Gresinh_{\kappa,\vartheta}(X^{\gamma}), Grecosh_{\kappa,\vartheta}(X^{\gamma}),$$
$$Gretanh_{\kappa,\vartheta}(X^{\gamma}), Grecotanh_{\kappa,\vartheta}(X^{\gamma}), Grecsch_{\kappa,\vartheta}(X^{\gamma}), Gresech_{\kappa,\vartheta}(X^{\gamma}),$$

$$supersin_{1,\vartheta}(X^{\kappa}), supercos_{1,\vartheta}(X^{\kappa}), supertan_{1,\vartheta}(X^{\kappa}), supercotan_{1,\vartheta}(X^{\kappa}),$$
$$supercsc_{1,\vartheta}(X^{\kappa}), supersec_{1,\vartheta}(X^{\kappa}), supersinh_{1,\vartheta}(X^{\kappa}), supercosh_{1,\vartheta}(X^{\kappa}),$$
$$supertanh_{1,\vartheta}(X^{\kappa}), supercotanh_{1,\vartheta}(X^{\kappa}), supercsch_{1,\vartheta}(X^{\kappa}), supersech_{1,\vartheta}(X^{\kappa}),$$
$$presin_{1,\vartheta}(X), precos_{1,\vartheta}(X), pretan_{1,\vartheta}(X), precotan_{1,\vartheta}(X),$$
$$precsc_{1,\vartheta}(X), presec_{1,\vartheta}(X), presinh_{1,\vartheta}(X), precosh_{1,\vartheta}(X),$$
$$pretanh_{1,\vartheta}(X), precotanh_{1,\vartheta}(X), precsch_{1,\vartheta}(X), presech_{1,\vartheta}(X),$$

$$precsc_{1,\vartheta}(X), presec_{1,\vartheta}(X), Gresin_{1,\vartheta}(X^{\gamma}), Grecos_{1,\vartheta}(X^{\gamma}),$$
$$Gretan_{1,\vartheta}(X^{\gamma}), Grecotan_{1,\vartheta}(X^{\gamma}), Grecsc_{1,\vartheta}(X^{\gamma}), Gresec_{1,\vartheta}(X^{\gamma}),$$
$$Gresinh_{1,\vartheta}(X^{\gamma}), Grecosh_{1,\vartheta}(X^{\gamma}), Gretanh_{1,\vartheta}(X^{\gamma}), Grecotanh_{1,\vartheta}(X^{\gamma}),$$
$$Grecsch_{1,\vartheta}(X^{\gamma}), Gresech_{1,\vartheta}(X^{\gamma}), \{\mathbb{E}_{\kappa,\vartheta}\}^t(X), \{\mathbb{E}_{\kappa,\vartheta}\}^t(X^{\kappa}),$$
$$\{\mathbb{E}_{\kappa,\vartheta}\}^t(X^{\gamma}), \{supersin_{\kappa,\vartheta}\}^t(X^{\kappa}), \{supercos_{\kappa,\vartheta}\}^t(X^{\kappa}), \{supersinh_{\kappa,\vartheta}\}^t(X^{\kappa}),$$
$$\{supercosh_{\kappa,\vartheta}\}^t(X^{\kappa}), \{presin_{\kappa,\vartheta}\}^t(X^{\gamma}), \{precos_{\kappa,\vartheta}\}^t(X^{\gamma}), \{presinh_{\kappa,\vartheta}\}^t(X^{\gamma}),$$

$$\{precosh_{\kappa,\vartheta}\}^t(X^\gamma), \{presin_{\kappa,\vartheta}\}^t(X), \{precos_{\kappa,\vartheta}\}^t(X), \{presinh_{\kappa,\vartheta}\}^t(X),$$
$$\{precosh_{\kappa,\vartheta}\}^t(X), \{Tresin_{\kappa,\vartheta}\}^t(X), \{Trecos_{\kappa,\vartheta}\}^t(X), \{Tresinh_{\kappa,\vartheta}\}^t(X),$$
$$\{Trecosh_{\kappa,\vartheta}\}^t(X), \{Gresin_{\kappa,\vartheta}\}^t(X^\gamma), \{Grecos_{\kappa,\vartheta}\}^t(X^\gamma), \{Gresinh_{\kappa,\vartheta}\}^t(X^\gamma),$$
$$\{Grecosh_{\kappa,\vartheta}\}^n(X^\gamma), \mathbb{E}^\gamma_{\kappa,\vartheta}(X), supersin^\gamma_{\kappa,\vartheta}(X^\kappa), supercos^\gamma_{\kappa,\vartheta}(X^\kappa),$$
$$supertan^\gamma_{\kappa,\vartheta}(X^\kappa), supercotan^\gamma_{\kappa,\vartheta}(X^\kappa), supercsc^\gamma_{\kappa,\vartheta}(X^\kappa), supersec^\gamma_{\kappa,\vartheta}(X^\kappa),$$
$$supersinh^\gamma_{\kappa,\vartheta}(X^\kappa), supercosh^\gamma_{\kappa,\vartheta}(X^\kappa), supertanh^\gamma_{\kappa,\vartheta}(X^\kappa), supercotanh^\gamma_{\kappa,\vartheta}(X^\kappa),$$

$$supercsch^\gamma_{\kappa,\vartheta}(X^\kappa), supersech^\gamma_{\kappa,\vartheta}(X^\kappa), presin^\gamma_{\kappa,\vartheta}(X), precos^\gamma_{\kappa,\vartheta}(X),$$
$$pretan^\gamma_{\kappa,\vartheta}(X), precotan^\gamma_{\kappa,\vartheta}(X), precsc^\gamma_{\kappa,\vartheta}(X), presec^\gamma_{\kappa,\vartheta}(X),$$
$$presinh^\gamma_{\kappa,\vartheta}(X), precosh^\gamma_{\kappa,\vartheta}(X), pretanh^\gamma_{\kappa,\vartheta}(X), precotanh^\gamma_{\kappa,\vartheta}(X),$$
$$precsch^\gamma_{\kappa,\vartheta}(X), presech^\gamma_{\kappa,\vartheta}(X), presin^\gamma_{\kappa,\vartheta}(X^\zeta), precos^\gamma_{\kappa,\vartheta}(X^\zeta),$$
$$pretan^\gamma_{\kappa,\vartheta}(X^\zeta), precotan^\gamma_{\kappa,\vartheta}(X^\zeta), precsc^\gamma_{\kappa,\vartheta}(X^\zeta), presec^\gamma_{\kappa,\vartheta}(X^\zeta),$$
$$presinh^\gamma_{\kappa,\vartheta}(X^\zeta), precosh^\gamma_{\kappa,\vartheta}(X^\zeta), pretanh^\gamma_{\kappa,\vartheta}(X^\zeta), precotanh^\gamma_{\kappa,\vartheta}(X^\zeta),$$

$$precsch^\gamma_{\kappa,\vartheta}(X^\zeta), presech^\gamma_{\kappa,\vartheta}(X^\zeta), Tresin^\gamma_{\kappa,\vartheta}(X), Trecos^\gamma_{\kappa,\vartheta}(X),$$
$$Tretan^\gamma_{\kappa,\vartheta}(X), Trecotan^\gamma_{\kappa,\vartheta}(X), Trecsc^\gamma_{\kappa,\vartheta}(X), Tresec^\gamma_{\kappa,\vartheta}(X),$$
$$Tresinh^\gamma_{\kappa,\vartheta}(X), Trecosh^\gamma_{\kappa,\vartheta}(X), Tretanh^\gamma_{\kappa,\vartheta}(X), Trecotanh^\gamma_{\kappa,\vartheta}(X),$$
$$Trecsch^\gamma_{\kappa,\vartheta}(X), Tresech^\gamma_{\kappa,\vartheta}(X), Gresin^\gamma_{\kappa,\vartheta}(X^\zeta), Grecos^\gamma_{\kappa,\vartheta}(X^\zeta),$$
$$Gretan^\gamma_{\kappa,\vartheta}(X^\zeta), Grecotan^\gamma_{\kappa,\vartheta}(X^\zeta), Grecsc^\gamma_{\kappa,\vartheta}(X^\zeta), Gresec^\gamma_{\kappa,\vartheta}(X^\zeta),$$
$$Gresinh^\gamma_{\kappa,\vartheta}(X^\zeta), Grecosh^\gamma_{\kappa,\vartheta}(X^\zeta), Gretanh^\gamma_{\kappa,\vartheta}(X^\zeta), Grecotanh^\gamma_{\kappa,\vartheta}(X^\zeta),$$

$$Grecsch^\gamma_{\kappa,\vartheta}(X^\zeta), Gresech^\gamma_{\kappa,\vartheta}(X^\zeta), \{\mathbb{E}^\gamma_{\kappa,\vartheta}\}^t(X), \{supersin^\gamma_{\kappa,\vartheta}\}^t(X^\kappa),$$
$$\{supercos^\gamma_{\kappa,\vartheta}\}^t(X^\kappa), \{supersinh^\gamma_{\kappa,\vartheta}\}^t(X^\kappa), \{supercosh^\gamma_{\kappa,\vartheta}\}^t(X^\kappa),$$
$$\{presin^\gamma_{\kappa,\vartheta}\}^t(X), \{precos^\gamma_{\kappa,\vartheta}\}^t(X), \{presinh^\gamma_{\kappa,\vartheta}\}^t(X), \{precosh^\gamma_{\kappa,\vartheta}\}^t(X),$$
$$\{presin^\gamma_{\kappa,\vartheta}\}^t(X^\zeta), \{precos^\gamma_{\kappa,\vartheta}\}^t(X^\zeta), \{presinh^\gamma_{\kappa,\vartheta}\}^t(X^\zeta), \{precosh^\gamma_{\kappa,\vartheta}\}^t(X^\zeta),$$
$$\{Tresin^\gamma_{\kappa,\vartheta}\}^t(X), \{Trecos^\gamma_{\kappa,\vartheta}\}^t(X), \{Tresinh^\gamma_{\kappa,\vartheta}\}^t(X), \{Trecosh^\gamma_{\kappa,\vartheta}\}^t(X),$$
$$\{Gresin^\gamma_{\kappa,\vartheta}\}^t(X^\zeta), \{Grecos^\gamma_{\kappa,\vartheta}\}^t(X^\zeta), \{Gresinh^\gamma_{\kappa,\vartheta}\}^t(X^\zeta), \{Grecosh^\gamma_{\kappa,\vartheta}\}^t(X^\zeta),$$

$$supersin^\gamma_{1,\vartheta}(X), supercos^\gamma_{1,\vartheta}(X), supersinh^\gamma_{1,\vartheta}(X), supercosh^\gamma_{1,\vartheta}(X),$$
$$presin^\gamma_{1,\vartheta}(X), precos^\gamma_{1,\vartheta}(X), presinh^\gamma_{1,\vartheta}(X), precosh^\gamma_{1,\vartheta}(X),$$
$$Gresin^\gamma_{1,\vartheta}(X^\zeta), Grecos^\gamma_{1,\vartheta}(X^\zeta), Gresinh^\gamma_{1,\vartheta}(X^\zeta), Grecosh^\gamma_{1,\vartheta}(X^\zeta),$$
$$supersin^\gamma_{\kappa,1}(X^\kappa), supercos^\gamma_{\kappa,1}(X^\kappa), supersinh^\gamma_{\kappa,1}(X^\kappa), supercosh^\gamma_{\kappa,1}(X^\kappa),$$
$$presin^\gamma_{\kappa,1}(X), precos^\gamma_{\kappa,1}(X), presinh^\gamma_{\kappa,1}(X), precosh^\gamma_{\kappa,1}(X),$$
$$\Psi_\psi(X), \Psi\sin_\psi(X), \Psi\cos_\psi(X), \Psi\tan_\psi(X), \Psi\mathrm{cotan}_\psi(X), \Psi\sec_\psi(X), \Psi\csc_\psi(X),$$

$\Psi_{\psi}(\varsigma, X)$, $\Psi\sin_{\psi}(\varsigma, X)$, $\Psi\cos_{\psi}(\varsigma, X)$, $\Psi\tan_{\psi}(\varsigma, X)$, $\Psi\mathrm{cotan}_{\psi}(\varsigma, X)$, $\Psi\sec_{\psi}(\varsigma, X)$, $\Psi\csc_{\psi}(\varsigma, X)$, $\Psi\sinh_{\psi}(X)$, $\Psi\cosh_{\psi}(X)$, $\Psi\tanh_{\psi}(X)$, $\Psi\mathrm{cotanh}_{\psi}(X)$, $\Psi\mathrm{sech}_{\psi}(X)$, $\Psi\mathrm{csch}_{\psi}(X)$, $\Psi\sinh_{\psi}(\varsigma, X)$, $\Psi\cosh_{\psi}(\varsigma, X)$, $\Psi\tanh_{\psi}(\varsigma, X)$, $\Psi\mathrm{cotanh}_{\psi}(\varsigma, X)$, $\Psi\mathrm{sech}_{\psi}(\varsigma, X)$, $\Phi^{\phi}_{\psi}(X)$, $\Phi^{\phi}_{\psi}(\varsigma, X)$, $\Phi\sin^{\phi}_{\psi}(X)$, $\Phi\cos^{\phi}_{\psi}(X)$, $\Phi\tan^{\phi}_{\psi}(X)$, $\Phi\mathrm{cotan}^{\phi}_{\psi}(X)$, $\Phi\csc^{\phi}_{\psi}(X)$, $\Phi\sec^{\phi}_{\psi}(X)$, $\Phi\sin^{\phi}_{\psi}(\varsigma, X)$, $\Phi\cos^{\phi}_{\psi}(\varsigma, X)$, $\Phi\tan^{\phi}_{\psi}(\varsigma, X)$, $\Phi\mathrm{cotan}^{\phi}_{\psi}(\varsigma, X)$, $\Phi\csc^{\phi}_{\psi}(\varsigma, X)$, $\Phi\sec^{\phi}_{\psi}(\varsigma, X)$, $\Phi\sinh^{\phi}_{\psi}(X)$, $\Phi\cosh^{\phi}_{\psi}(X)$, $\Phi\tanh^{\phi}_{\psi}(X)$, $\Phi\mathrm{cotanh}^{\phi}_{\psi}(X)$, $\Phi\mathrm{csch}^{\phi}_{\psi}(X)$, $\Phi\mathrm{sech}^{\phi}_{\psi}(X)$, $\Phi\sinh^{\phi}_{\psi}(\varsigma, X)$, $\Phi\cosh^{\phi}_{\psi}(\varsigma, X)$, $\Phi\tanh^{\phi}_{\psi}(\varsigma, X)$,

$\Phi\mathrm{cotanh}^{\phi}_{\psi}(\varsigma, X)$, $\Phi\mathrm{csch}^{\phi}_{\psi}(\varsigma, X)$, $\Phi\mathrm{sech}^{\phi}_{\psi}(\varsigma, X)$, $\Upsilon_{\kappa}(X)$, $\Upsilon_{\kappa}(\varsigma, X)$, $\Upsilon\sin_{\kappa}(X)$, $\Upsilon\cos_{\kappa}(X)$, $\Upsilon\tan_{\kappa}(X)$, $\Upsilon\mathrm{cotan}_{\kappa}(X)$, $\Upsilon\csc_{\kappa}(X)$, $\Upsilon\sec_{\kappa}(X)$, $\Upsilon\sin_{\kappa}(\varsigma, X)$, $\Upsilon\cos_{\kappa}(\varsigma, X)$, $\Upsilon\tan_{\kappa}(\varsigma, X)$, $\Upsilon\mathrm{cotan}_{\kappa}(X)$, $\Upsilon\csc_{\kappa}(\varsigma, X)$, $\Upsilon\sec_{\kappa}(\varsigma, X)$, $\Upsilon\sinh_{\kappa}(X)$, $\Upsilon\cosh_{\kappa}(X)$, $\Upsilon\tanh_{\kappa}(X)$, $\Upsilon\mathrm{cotanh}_{\kappa}(X)$, $\Upsilon\mathrm{csch}_{\kappa}(X)$, $\Upsilon\mathrm{sech}_{\kappa}(X)$, $\Upsilon\sinh_{\kappa}(\varsigma, X)$, $\Upsilon\cosh_{\kappa}(\varsigma, X)$, $\Upsilon\tanh_{\kappa}(\varsigma, X)$, $\Upsilon\mathrm{cotanh}_{\kappa}(X)$, $\Upsilon\mathrm{csch}_{\kappa}(\varsigma, X)$, $\Upsilon\mathrm{sech}_{\kappa}(\varsigma, X)$, $\Lambda_{\kappa}(X)$, $\Lambda_{\kappa}(\varsigma, X)$, $\Lambda\sin_{\kappa}(X)$, $\Lambda\cos_{\kappa}(X)$, $\Lambda\tan_{\kappa}(X)$, $\Lambda\mathrm{cotan}_{\kappa}(X)$, $\Lambda\csc_{\kappa}(X)$, $\Lambda\sec_{\kappa}(X)$,

$\Lambda\sin_{\kappa}(\varsigma, X)$, $\Lambda\cos_{\kappa}(\varsigma, X)$, $\Lambda\tan_{\kappa}(\varsigma, X)$, $\Lambda\mathrm{cotan}_{\kappa}(\varsigma, X)$, $\Lambda\csc_{\kappa}(\varsigma, X)$, $\Lambda\sec_{\kappa}(\varsigma, X)$, $\Lambda\sinh_{\kappa}(X)$, $\Lambda\cosh_{\kappa}(X)$, $\Lambda\tanh_{\kappa}(X)$, $\Lambda\mathrm{cotanh}_{\kappa}(X)$, $\Lambda\mathrm{csch}_{\kappa}(X)$, $\Lambda\mathrm{sech}_{\kappa}(X)$, $\Lambda\sinh_{\kappa}(\varsigma, X)$, $\Lambda\cosh_{\kappa}(\varsigma, X)$, $\Lambda\tanh_{\kappa}(\varsigma, X)$, $\Lambda\mathrm{cotanh}_{\kappa}(\varsigma, X)$, $\Lambda\mathrm{csch}_{\kappa}(\varsigma, X)$, $\Lambda\mathrm{sech}_{\kappa}(\varsigma, X)$, $\Theta_{\kappa}(X)$, $\Theta_{\kappa}(\varsigma, X)$, $\Theta\sin_{\kappa}(X)$, $\Theta\cos_{\kappa}(X)$, $\Theta\tan_{\kappa}(X)$, $\Theta\mathrm{cotan}_{\kappa}(X)$, $\Theta\csc_{\kappa}(X)$, $\Theta\sec_{\kappa}(X)$, $\Theta\sin_{\kappa}(\varsigma, X)$, $\Theta\cos_{\kappa}(\varsigma, X)$, $\Theta\tan_{\kappa}(\varsigma, X)$, $\Theta\mathrm{cotan}_{\kappa}(\varsigma, X)$, $\Theta\csc_{\kappa}(\varsigma, X)$, $\Theta\sec_{\kappa}(\varsigma, X)$, $\Theta\sinh_{\kappa}(X)$, $\Theta\cosh_{\kappa}(X)$, $\Theta\tanh_{\kappa}(X)$, $\Theta\mathrm{cotanh}_{\kappa}(X)$, $\Theta\mathrm{csch}_{\kappa}(X)$, $\Theta\mathrm{sech}_{\kappa}(X)$,

$\Theta\sinh_{\kappa}(\varsigma, X)$, $\Theta\cosh_{\kappa}(\varsigma, X)$, $\Theta\tanh_{\kappa}(\varsigma, X)$, $\Theta\mathrm{cotanh}_{\kappa}(\varsigma, X)$, $\Theta\mathrm{csch}_{\kappa}(\varsigma, X)$, $\Theta\mathrm{sech}_{\kappa}(\varsigma, X)$, $\Omega_{\kappa}(X)$, $\Omega_{\kappa}(\varsigma, X)$, $\Omega\sin_{\kappa}(X)$, $\Omega\cos_{\kappa}(X)$, $\Omega\tan_{\kappa}(X)$, $\Omega\mathrm{cotan}_{\kappa}(X)$, $\Omega\csc_{\kappa}(X)$, $\Omega\sec_{\kappa}(X)$, $\Omega\sin_{\kappa}(\varsigma, X)$, $\Omega\cos_{\kappa}(\varsigma, X)$, $\Omega\tan_{\kappa}(\varsigma, X)$, $\Omega\mathrm{cotan}_{\kappa}(\varsigma, X)$, $\Omega\csc_{\kappa}(\varsigma, X)$, $\Omega\sec_{\kappa}(\varsigma, X)$, $\Omega\sinh_{\kappa}(X)$, $\Omega\cosh_{\kappa}(X)$, $\Omega\tanh_{\kappa}(X)$, $\Omega\mathrm{cotanh}_{\kappa}(X)$, $\Omega\mathrm{csch}_{\kappa}(X)$, $\Omega\mathrm{sech}_{\kappa}(X)$, $\Omega\sinh_{\kappa}(\varsigma, X)$, $\Omega\cosh_{\kappa}(\varsigma, X)$, $\Omega\tanh_{\kappa}(\varsigma, X)$, $\Omega\mathrm{cotanh}_{\kappa}(\varsigma, X)$, $\Omega\mathrm{csch}_{\kappa}(\varsigma, X)$, $\Omega\mathrm{sech}_{\kappa}(\varsigma, X)$, $\Omega_{\kappa,\vartheta}(X)$, $\Omega_{\kappa,\vartheta}(\varsigma, X)$, $\Omega\sin_{\kappa,\vartheta}(X)$, $\Omega\cos_{\kappa,\vartheta}(X)$, $\Omega\tan_{\kappa,\vartheta}(X)$, $\Omega\mathrm{cotan}_{\kappa,\vartheta}(X)$, $\Omega\csc_{\kappa,\vartheta}(X)$, $\Omega\sec_{\kappa,\vartheta}(X)$, $\Omega\sin_{\kappa,\vartheta}(\varsigma, X)$, $\Omega\cos_{\kappa,\vartheta}(\varsigma, X)$, $\Omega\tan_{\kappa,\vartheta}(\varsigma, X)$, $\Omega\mathrm{cotan}_{\kappa,\vartheta}(\varsigma, X)$,

$\Omega\csc_{\kappa,\vartheta}(\varsigma, X)$, $\Omega\sec_{\kappa,\vartheta}(\varsigma, X)$, $\Omega\sinh_{\kappa,\vartheta}(X)$, $\Omega\cosh_{\kappa,\vartheta}(X)$, $\Omega\tanh_{\kappa,\vartheta}(X)$, $\Omega\mathrm{cotanh}_{\kappa,\vartheta}(X)$, $\Omega\mathrm{csch}_{\kappa,\vartheta}(X)$, $\Omega\mathrm{sech}_{\kappa,\vartheta}(X)$, $\Omega\sinh_{\kappa,\vartheta}(\varsigma, X)$, $\Omega\cosh_{\kappa,\vartheta}(\varsigma, X)$, $\Omega\tanh_{\kappa,\vartheta}(\varsigma, X)$, $\Omega\mathrm{cotanh}_{\kappa,\vartheta}(\varsigma, X)$, $\Omega\mathrm{csch}_{\kappa,\vartheta}(\varsigma, X)$, $\Omega\mathrm{sech}_{\kappa,\vartheta}(\varsigma, X)$, $\mho_{\kappa}(X)$, $\mho_{\kappa}(\varsigma, X)$, $\mathrm{subsin}_{\kappa}(X)$, $\mathrm{subcos}_{\kappa}(X)$, $\mathrm{subtan}_{\kappa}(X)$, $\mathrm{subcotan}_{\kappa}(X)$, $\mathrm{subcsc}_{\kappa}(X)$, $\mathrm{subsec}_{\kappa}(X)$, $\mathrm{subsin}_{\kappa}(\varsigma, X)$, $\mathrm{subcos}_{\kappa}(\varsigma, X)$, $\mathrm{subtan}_{\kappa}(\varsigma, X)$, $\mathrm{subcotan}_{\kappa}(\varsigma, X)$, $\mathrm{subcsc}_{\kappa}(\varsigma, X)$, $\mathrm{subsec}_{\kappa}(\varsigma, X)$, $\mathrm{subsinh}_{\kappa}(X)$, $\mathrm{subcosh}_{\kappa}(X)$, $\mathrm{subtanh}_{\kappa}(X)$, $\mathrm{subcotanh}_{\kappa}(X)$,

$\text{subcsch}_\kappa(X)$, $\text{subsech}_\kappa(X)$, $\text{subsinh}_\kappa(\varsigma, X)$, $\text{subcosh}_\kappa(\varsigma, X)$,
$\text{subtanh}_\kappa(\varsigma, X)$, $\text{subcotanh}_\kappa(\varsigma, X)$, $\text{subcsch}_\kappa(\varsigma, X)$, $\text{subcsch}_\kappa(\varsigma, X)$,
$\mho_{\kappa,\vartheta}(X)$, $\mho_{\kappa,\vartheta}(\varsigma, X)$, $\text{subsin}_{\kappa,\vartheta}(X)$, $\text{subcosin}_{\kappa,\vartheta}(X)$, $\text{subtan}_{\kappa,\vartheta}(X)$,
$\text{subcotan}_{\kappa,\vartheta}(X)$, $\text{subcsc}_{\kappa,\vartheta}(X)$, $\text{subsec}_{\kappa,\vartheta}(X)$, $\text{subsin}_{\kappa,\vartheta}(\varsigma, X)$, $\text{subcosin}_{\kappa,\vartheta}(\varsigma, X)$,
$\text{subtan}_{\kappa,\vartheta}(\varsigma, X)$, $\text{subcotan}_{\kappa,\vartheta}(\varsigma, X)$, $\text{subcsc}_{\kappa,\vartheta}(\varsigma, X)$, $\text{subsec}_{\kappa,\vartheta}(\varsigma, X)$,

$\text{subsinh}_{\kappa,\vartheta}(X)$, $\text{subcosinh}_{\kappa,\vartheta}(X)$, $\text{subtanh}_{\kappa,\vartheta}(X)$, $\text{subcotanh}_{\kappa,\vartheta}(X)$, $\text{subcsch}_{\kappa,\vartheta}(X)$,
$\text{subsech}_{\kappa,\vartheta}(X)$, $\text{subsinh}_{\kappa,\vartheta}(\varsigma, X)$, $\text{subcosinh}_{\kappa,\vartheta}(\varsigma, X)$, $\text{subtanh}_{\kappa,\vartheta}(\varsigma, X)$,
$\text{subcotanh}_{\kappa,\vartheta}(\varsigma, X)$, $\text{subcsch}_{\kappa,\vartheta}(\varsigma, X)$, $\text{subsech}_{\kappa,\vartheta}(\varsigma, X)$,

$$ {}^{v}_{s}\mathbb{H}^{w}_{r}\left[X\left|\begin{matrix}(d_j,D_j)_{1,s}\\(e_j,E_j)_{1,r}\end{matrix}\right.\right], {}^{v}_{s}\mathbb{H}^{w}_{r}\left[X\left|\begin{matrix}(d_1,1),\ldots,(d_s,1)\\(e_1,1),\ldots,(e_r,1)\end{matrix}\right.\right], {}_{s}\mathbb{H}_{r}\left[X\left|\begin{matrix}(d_1,D_1),\ldots,(d_s,D_s)\\(e_1,E_1),\ldots,(e_r,E_r)\end{matrix}\right.\right], $$

$$ {}_{s}\mathbb{H}_{r}[d_1,\ldots,d_s;e_1,\ldots,e_s;X]\ \Big], $$

where $X, \kappa, \vartheta, \varsigma, \gamma, \zeta, \phi, \psi, e_i, E_i, d_j, D_j, \mathsf{A}, \mathsf{C} \in \mathbb{C}$, $1 \le j \le s$, $1 \le i \le r$, $s, r, p, q \in \mathbb{N}_0$, $0 \le w \le s$, $1 \le v \le r$, and $t \in \mathbb{N}$.

9.8.2 The Concept of the Multi Stability

The concept of stability for a functional equation arises when we replace the functional equation by an inequality which acts as a perturbation of the initial equation. Over the last years, the stability results of functional equations have been strongly developed for getting approximate solution of the perturbed equation that is close to the exact solution. Very significant contributions about this topic were introduced by Ulam and Hyers in 1940–1941, and this type of stability called Hyers–Ulam stability [17, 18]. In 1978, the improvement of Hyers–Ulam stability provided by Rassias, led to the development of what is now known as Hyers–Ulam–Rassias stability [32]. In 1998, Alsina and Ger [33] established the Hyers–Ulam stability of differential equations by replacing functional equation. Thereafter, authors defined the types of Hyers–Ulam–Rassias–Mittag–Leffler stability of fractional differential equations to prove that every mapping from this type can be somehow approximated by an exact solution of the considered equation [34–36].

The recent growing interest in the Mittag–Leffler function and its various generalizations [37] is mainly due to their close relations to the Fractional Calculus (the theory of integrals and derivatives of non–integer order) and especially to fractional problems which come from applications. The special functions–along with the Mittag–Leffler function–including the functions of Wright type, the functions of hypergeometric type and others [38] which often appear in solutions of various types of equations with fractional operators, play a prominent role in the theory of the partial differential equations of the fractional order that are actively used nowadays for modeling of many phenomena [39–41].

As we all know, the main difficulty to find exact solution of such equation is very crucial, and the form of the exact solution (if it exists) is often so arduous that it is not appropriate for numerical calculation [42–44]. In view of this, it is imperative to discuss approximate solution and ask whether it lies near the exact solution. Mostly, we say that a fractional differential equation is stable in the Ulam sense if, for every solution of the fractional differential equation, there exists an approximate solution of the perturbed equation that is close to it.

To clarify the issue, let us give the definition of Ulam type stability of an operator equation [45]. Let $(\mathcal{X}, \| \,.\, \|)$ be a Banach space, $\mathcal{Y}$ be a nonempty subset of $\mathcal{X}$ and $T : \mathcal{Y} \longrightarrow \mathcal{X}$ be an operator. Consider the operator equation

$$T(x) = 0, \quad x \in \mathcal{Y}, \tag{9.4}$$

and the inequation

$$\|T(y)\| \leq \varepsilon, \quad \forall \varepsilon > 0, \; y \in \mathcal{Y}. \tag{9.5}$$

The Eq. (9.4) is called Hyers–Ulam stable if for each solution μ of the inequality (9.5), there exists a solution v of the operator Eq. (9.4) such that,

$$\|\mu - v\| \leq c\varepsilon,$$

where $c > 0$ is a constant depended on T.

Now, the Eq. (9.4) is called Hyers–Ulam–Rassias stable, if there exists a continuous function $\varphi : \mathbb{R} \longrightarrow \mathbb{R}$ such that for each $\varepsilon > 0$ and for each solution μ of the inequality

$$\|T(y)\| \leq \varepsilon\varphi(y), \quad \forall \varepsilon > 0, \; y \in \mathcal{Y}.$$

there exists a solution v of the operator Eq. (9.4) with a real number $c > 0$, such that,

$$\|\mu - v\| \leq c\varepsilon\varphi(y).$$

The Eq. (9.4) has Hyers–Ulam–Rassias–Mittag–Leffler stability or Hyers–Ulam–Rassias–Wright stability if the above statement is also true when we replace the control function φ by the Mittag–Leffler function or the Wright function, respectively.

In this section, we consider a diagonal matrix of special functions given in Sect. 9.8.1, as a controller to present a new concept of stability, namely, multi stability. This stability allows us to obtain different approximations depending on the different special functions that are initially chosen and to evaluate optimal stability and minimal error which enable us to get the best approximate solution.

Consider the diagonal matrix below

$$\mathrm{diag}[\imath_1, \cdots, \imath_n]_{n\times n} = \begin{bmatrix} \zeta_1 & 0 & \dots & 0 \\ 0 & \zeta_2 & \ddots & \vdots \\ \vdots & \ddots & \ddots & 0 \\ 0 & \dots & 0 & \zeta_n \end{bmatrix}_{n\times n},$$

with the following partial order relation

$$\imath := \mathrm{diag}[\imath_1, \cdots, \imath_n] \preceq \,\text{¸} := \mathrm{diag}[\text{¸}_1, \cdots, \text{¸}_n] \Leftrightarrow \imath_i \preceq \text{¸}_i,$$

for every $1 \leq i \leq n$ and $n \in \mathbb{N}$.

Besides, $\imath \prec$ ¸ denotes that $\imath \preceq$ ¸ and $\imath \neq$ ¸; $\imath \ll$ ¸ and $\zeta_i < \xi_i$, for every $1 \leq i \leq n$ and $n \in \mathbb{N}$.

Consider the matrix valued control function $\mathfrak{W}[X]$, as follows:

$$\mathfrak{W}[X] := \mathrm{diag}\left[\varphi_1^{\circledS}, \cdots, \varphi_n^{\circledS}(X)\right]_{n\times n},$$

where every $\underbrace{\varphi_i^{\circledS}}_{i=1,\cdots,n}$, with $n \in \mathbb{N}$, denotes a particular special function defined in Sect. 9.8.1 and the natural number n represents the numbers of special functions that we intend to consider.

Let a mapping Θ from vector space U to normed linear space V have Hyers–Ulam–Rassias stability. If we replace the control function of Hyers–Ulam–Rassias stability with $\mathfrak{W}[X]$, we say Θ has multi stability property.

Forthespecialcaseofmultiplestabilityi.e.Mittag–Leffler–Gauss-Hypergeometric–Bessel–Maitland–Fox stability, Mittag–Leffler–Supertrigonometric stability, Mittag–Leffler–Superhyperbolic stability and the others, we refer to [30, 46–48].

References

1. Boulfoul A, Tellab B, Abdellouahab N, Zennir K (2021) Existence and uniqueness results for initial value problem of nonlinear fractional integro-differential equation on an unbounded domain in a weighted Banach space. Math Methods Appl Sci 44(5):3509–3520
2. Alruwaily Y, Almaghamsi L, Karthikeyan K, El-hady ES (2023) Existence and uniqueness for a coupled system of fractional equations involving Riemann-Liouville and Caputo derivatives with coupled Riemann-Stieltjes integro-multipoint boundary conditions. AIMS Math 8:10067–10094
3. Dhayal R, Zhu Q (2023) Stability and controllability results of ψ-Hilfer fractional integro-differential systems under the influence of impulses. Chaos Solitons Fractals 168:113105
4. Kilbas AA, Srivastava HM, Trujillo JJ (2006) Theory and applications of fractional equations. Elsevier, Amsterdam
5. de Vanterler C, Sousa J, Capelas de Olivera E (2018) On the Ψ-Hilfer fractional derivative. Commun Nonlinear Sci Numer Simulat 60:72–91

6. Hadamard J (1892) Essai sur letude des fonctions donnees par leur developpment de Taylor. J Math Pures Appl 4(8):101–186
7. Lomnitz C (1956) Creep measurements in igneous rocks. J Geol 64:473–479
8. Mainardi F, Spada G (2012) On the viscoelastic characterization of the Jeffreys-Lomnitz law of creep. Rheol Acta 51:783–791
9. Ahmad B, Alsaedi A, Ntouyas S, Tariboon J (2017) Hadamard-type fractional differential equations, inclusions and inequalities. Springer International Publishing, pp 1–6
10. Ahmad B, Alsaedi A, Ntouyas SK, Tariboon J (2017) Hadamard-type fractional differential equations, inclusions and inequalities. Springer International Publishing, pp 3–11
11. Cho YJ, Park C, Rassias TM, Saadati R, Cho YJ, Park C, Rassias TM, Saadati R (2015) Stability of functional equations in Banach algebras. Springer International Publishing, pp 19–50
12. Miller JB (1966) Averaging and reynolds operators on Banach algebras I. Representation by derivations and antiderivations. J Math Anal Appl 14(3):527–548
13. Park C (2020) Permuting triderivations and permuting trihomomorphisms in complex Banach algebras. arXiv:2009.10346
14. Takhtajan L (1994) On foundation of the generalized Nambu mechanics. Commun Math Phys 160:295–315
15. Zettl H (1983) A characterization of ternary rings of operators. Adv Math 48:117–143
16. Saadati R, Vaezpour SM (2008) Linear operators in finite dimensional probabilistic normed spaces. J Math Anal Appl 346(2):446–450
17. Ulam SM (1960) A collection of mathematical problems. Interscience, New York
18. Hyers DH (1941) On the stability of the linear functional equation. Proc Natl Acad Sci USA 27:222–224
19. Kaiser Z, Pales Z (2005) An example of a stable functional equation when the Hyers method does not work. JIPAM. J Inequal Pure Appl Math 6(1, article 14):11
20. Szekélyhidi L (1982) Note on a stability theorem. Can Math Bull 25(4):500–501
21. Pales Z (1998) Generalized stability of the Cauchy functional equation. Aequationes Mathematicae 56(3):222–232
22. Tabor J, Tabor J (2007) General stability of functional equations of linear type. J Math Anal Appl 328(1):192–200
23. Razaei H, Jung S, Rassias TM (2013) Laplace transform and Hyers-Ulam stability of linear differential equation. J Math Anal Appl 403:244–251
24. Mohanapriya A, Park C, Ganesh A, Govindan V (2020) Mittag-Leffler-Hyers-Ulam stability of differential equation using Fourier transform. Adv Differ Equ 2020:1–16
25. Baker JA (1991) The stability of certain functional equations. Proc Am Math Soc 112(3):729–732
26. Banach S (1922) Sur les operations dans les ensembles abstraits et leur application aux euations integrales. Fundam Math 3:133–181
27. Căadariu L, Radu V (2003) Fixed points and the stability of Jensen's functional equation. J Inequal Pure Appl Math 4(1), Article 4
28. Radu V (2003) The fixed point alternative and the stability of functional equations. Fixed Point Theory 4:91–96
29. Luxemburg WAJ (1958) On the convergence of successive approximations in the theory of ordinary differential equations, II. Koninkl, Nederl. Akademie van Wetenschappen, Amsterdam, Proc Ser A (5)61. (Indag Math 20:540–546 (1958))
30. Rezaei Aderyani S, Saadati R, Li C, Rassias TM, Park C (2023) Special functions and multi-stability of the Jensen type random operator equation in C^*-algebras via fixed point. J Inequal Appl 2023(1):1–24
31. Diaz JB, Margolis B (1968) A fixed point theorem of the alternative, for contractions on a generalized complete metric space
32. Rassias ThM (1978) On the stability of the linear mapping in Banach spaces. Proc Am Math Soc 72:297–300
33. Alsina C, Ger R (1998) On some inequalities and stability results related to the exponential function. J Inequal Appl 2:373–380

34. Wu G, Baleanu D (2018) Stability analysis of impulsive fractional difference equations. Fract Calc Appl Anal 21:354–375
35. Wu G, Baleanu D, Huang L (2018) Novel Mittag-Leffler stability of linear fractional delay difference equations with impluse. Appl Math Lett 82:71–78
36. Wu G, Abdeljawad T, Liu J, Baleanu D, Wu K (2019) Mittag-Leffler stability analysis of fractional discrete-time neural networks via fixed point technique. Nonlinear Anal Model Control 24:919–936
37. Gorenflo R, Kilbas AA, Mainardi F, Rogosin SV (2020) Mittag-Leffler functions, related topics and applications. Springer, Berlin, p 540
38. Yang XJ (2021) Theory and applications of special functions for scientists and engineers. Springer, Singapore
39. Aderyani SR, Saadati R, O'Regan D, Alshammari FS (2022) Describing water wave propagation using the $\frac{G'}{G^2}$-expansion method. Mathematics 11(1):191
40. Aderyani SR, Saadati R, O'Regan D, Alshammari FS (2022) Existence, uniqueness and stability analysis with the multiple exp function method for NPDEs. Mathematics 10(21):4151
41. Aderyani SR, Saadati R, Vahidi J (2022) Multiple exp-function method to solve the nonlinear space-time fractional partial differential symmetric regularized long wave (SRLW) equation and the (1+ 1)-dimensional Benjami-Ono equation. Int J Modern Phys B 2350213
42. Aderyani SR, Saadati R, Vahidi J, Allahviranloo T (2022) The exact solutions of the conformable time-fractional modified nonlinear Schrödinger equation by the Trial equation method and modified Trial equation method. Adv Math Phys
43. Aderyani SR, Saadati R, Vahidi J, Mlaiki N, Abdeljawad T (2022) The exact solutions of conformable time-fractional modified nonlinear Schrödinger equation by Direct algebraic method and Sine-Gordon expansion method. AIMS Math 7(6):10807–10827
44. Aderyani SR, Saadati R, Vahidi J, Gómez-Aguilar JF (2022) The exact solutions of conformable time-fractional modified nonlinear Schrödinger equation by first integral method and functional variable method. Optic Quantum Electron 54(4):218
45. Addai E, Zhang L, Asamoah JK, Essel JF (2023) A fractional order age-specific smoke epidemic model. Appl Math Model 119:99–118
46. Aderyani SR, Saadati R, Abdeljawad T, Mlaiki N (2022) Multi-stability of non homogenous vector-valued fractional differential equations in matrix-valued Menger spaces. Alexandria Eng J 61(12):10913–10923
47. Aderyani SR, Saadati R (2023) Stability and controllability results by n-ary aggregation functions in matrix valued fuzzy n-normed spaces. Inf Sci 119265
48. Aderyani SR, Saadati R, O'Regan D, Alshammari FS (2023) Fuzzy approximate solutions of matrix-valued fractional differential equations by fuzzy control functions. Mathematics 11(6):1386

Chapter 10
Multi Stability of the Jensen Type Random Operator Equations in C^*–Algebras

In this chapter, we investigate the multi stability of homomorphisms and derivations in both C^*–algebras and Lie C^*–algebras for the following random operator equation [1] via the fixed point method,

$$\mu f\left(\leftthreetimes, \frac{x+y}{2}\right) + \mu f\left(\leftthreetimes, \frac{x-y}{2}\right) = f(\leftthreetimes, \mu x).$$

In particular for $\mu = 1$, the above equation turns out to be Jensen's random operator equation.

10.1 Multi Stability of Homomorphisms in C^*–Algebras

Throughout this section, let A be a C^*–algebra with norm $\|\cdot\|_A$ and that B be a C^*–algebra with norm $\|\cdot\|_B$.

For a given mapping $f : \pitchfork \times A \to B$, we define

$$D_\mu f(\leftthreetimes, x, y) := \mu f\left(\leftthreetimes, \frac{x+y}{2}\right) + \mu f\left(\leftthreetimes, \frac{x-y}{2}\right) - f(\leftthreetimes, \mu x),$$

for every $\mu \in \mathbb{T}^1 := \{\nu \in \mathbb{C} : |\nu| = 1\}$ and every $x, y \in A$ and $\leftthreetimes \in \pitchfork$.

Notice that a $\mathbb{C}$–linear operator $H : \pitchfork \times A \to B$ is a homomorphism [1] in C^*–algebras if H satisfies $H(\leftthreetimes, xy) = H(\leftthreetimes, x)H(\leftthreetimes, y)$ and $H(\leftthreetimes, x^*) = H(\leftthreetimes, x)^*$, for every $x, y \in A$ and $\leftthreetimes \in \pitchfork$.

We now investigate the multi stability of homomorphisms in C^*–algebras for the functional equation $D_\mu f(\leftthreetimes, x, y) = 0$.

Theorem 10.1 *Let $n \in \mathbb{N}$, and $f : \pitchfork \times A \to B$ be a mapping for which there exist functions* $\underbrace{\varphi_i}_{i=1,\ldots,n} : A^2 \to [0, \infty)$, *such that*

S. Rezaei Aderyani et al., *Towards Ulam Type Multi Stability Analysis*, Studies in Systems, Decision and Control 523, https://doi.org/10.1007/978-3-031-55564-0_10

$$\operatorname{diag}\left[\|D_\mu f(\lambda, x, y)\|_B, \ldots, \|D_\mu f(\lambda, x, y)\|_B\right]_{n\times n} \tag{10.1}$$
$$\preceq \operatorname{diag}\left[\varphi_1(x, y), \ldots, \varphi_n(x, y)\right]_{n\times n},$$

$$\operatorname{diag}\left[\|f(\lambda, xy) - f(\lambda, x)f(\lambda, y)\|_B, \ldots, \|f(\lambda, xy) - f(\lambda, x)f(\lambda, y)\|_B\right]_{n\times n} \tag{10.2}$$
$$\preceq \operatorname{diag}\left[\varphi_1(x, y), \ldots, \varphi_n(x, y)\right]_{n\times n},$$

and

$$\operatorname{diag}\left[\|f(\lambda, x^*) - f(\lambda, x)^*\|_B, \ldots, \|f(\lambda, x^*) - f(\lambda, x)^*\|_B\right]_{n\times n} \tag{10.3}$$
$$\preceq \operatorname{diag}\left[\varphi_1(x, x), \ldots, \varphi_n(x, x)\right]_{n\times n},$$

for every $\mu \in \mathbb{T}^1$ *and every* $x, y \in A$ *and* $\lambda \in \Omega$. *If there exist constants* $\underbrace{L_i}_{i=1,\ldots,n} < 1$, *such that* $\underbrace{\varphi_i}_{i=1,\ldots,n}(x, 0) \le 2\underbrace{L_i\varphi_i}_{i=1,\ldots,n}(\frac{x}{2}, 0)$, *for every* $x \in A$. *Then, there exists a unique* C^**–algebra homomorphism* $H : \Omega \times A \to B$, *such that*

$$\operatorname{diag}\left[\|f(\lambda, x) - H(\lambda, x)\|_B, \ldots, \|f(\lambda, x) - H(\lambda, x)\|_B\right]_{n\times n} \tag{10.4}$$
$$\preceq \operatorname{diag}\left[\frac{L_1}{1-L_1}\varphi_1(x, 0), \ldots, \frac{L_n}{1-L_n}\varphi_n(x, 0)\right]_{n\times n},$$

for every $x \in A$ *and* $\lambda \in \Omega$.

Proof Consider the set

$$X := \{g : \Omega \times A \to B\},$$

and define the generalized metric d on X, as follows:

$$d(g, h) := \inf\Bigg\{(C_1, \ldots, C_n) \in \mathbb{R}_+ : \operatorname{diag}\Big[\|g(\lambda, x) - h(\lambda, x)\|_B, \ldots,$$
$$\|g(\lambda, x) - h(\lambda, x)\|_B\Big]_{n\times n}$$
$$\preceq \left[C_1, \ldots, C_n\right]^T_{n\times 1} \operatorname{diag}\left[\varphi_1(x, 0), \ldots, \varphi_n(x, 0)\right]_{n\times n}, \quad \forall x \in A,\ \lambda \in \Omega\Bigg\}.$$

Then, we can easily prove that (X, d) is complete [1].

Consider the linear mapping $J : X \to X$, such that

$$Jg(\lambda, x) := \frac{1}{2} g(\lambda, 2x),$$

for every $x \in A$ and $\lambda \in \mathbb{T}$.

According to Theorem 3.1 of [2], we have that

$$d(Jg, Jh) \preceq (L_1, \ldots, L_n) d(g, h),$$

for every $g, h \in X$.

Setting $\mu = 1$ and $y = 0$ in (10.1), we have

$$\begin{aligned} &\operatorname{diag}\left[\|2f(\lambda, \frac{x}{2}) - f(\lambda, x)\|_B, \ldots, \|2f(\lambda, \frac{x}{2}) - f(\lambda, x)\|_B\right]_{n\times n} \quad (10.5)\\ &\preceq \operatorname{diag}\left[\varphi_1(x, 0), \ldots, \varphi_n(x, 0)\right]_{n\times n}, \end{aligned}$$

for every $x \in A$ and $\lambda \in \mathbb{T}$. So, we get

$$\begin{aligned} &\operatorname{diag}\left[\|f(\lambda, x) - \frac{1}{2} f(\lambda, 2x)\|_B, \ldots, \|f(\lambda, x) - \frac{1}{2} f(\lambda, 2x)\|_B\right]_{n\times n}\\ &\preceq \operatorname{diag}\left[\frac{1}{2}\varphi_1(2x, 0), \ldots, \frac{1}{2}\varphi_n(2x, 0)\right]_{n\times n}\\ &\preceq \operatorname{diag}\left[L_1\varphi_1(x, 0), \ldots, L_n\varphi_n(x, 0)\right]_{n\times n}, \end{aligned}$$

for every $x \in A$ and $\lambda \in \mathbb{T}$. Hence, $d(f, Jf) \preceq (L_1, \ldots, L_n)$.

According to Theorem 9.5, there exists a mapping $H : \mathbb{T} \times A \to B$, such that

(1) H is a fixed point of J, i.e.,

$$H(\lambda, 2x) = 2H(\lambda, x), \quad (10.6)$$

for every $x \in A$ and $\lambda \in \mathbb{T}$. The mapping H is a unique fixed point of J in the set

$$Y = \{g \in X : d(f, g) < \overbrace{(\infty, \ldots, \infty)}^{n}\}.$$

This concludes that H is a unique mapping satisfying the equality (10.6), such taht there exists $(C_1, \ldots, C_n) \in (0, \infty)^n$ satisfying

$$\text{diag}\left[\|H(\leftthreetimes, x) - f(\leftthreetimes, x)\|_B, \ldots, \|H(\leftthreetimes, x) - f(\leftthreetimes, x)\|_B\right]_{n\times n}$$
$$\preceq \text{diag}\left[C_1\varphi_1(x, 0), \ldots, C_n\varphi_n(x, 0)\right]_{n\times n}$$

for every $x \in A$ and $\leftthreetimes \in \pitchfork$.

(2) $d(J^k f, H) \to \overbrace{(0, \ldots, 0)}^{n}$ as $k \to \infty$. This concludes the following equality

$$\lim_{k\to\infty} \frac{f(\leftthreetimes, 2^k x)}{2^k} = H(\leftthreetimes, x), \tag{10.7}$$

for every $x \in A$ and $\leftthreetimes \in \pitchfork$.

(3) $d(f, H) \le (\frac{1}{1-L_1}, \ldots, \frac{1}{1-L_n})d(f, Jf)$, which implies the inequality below

$$d(f, H) \le \left(\frac{L_1}{1 - L_1}, \ldots, \frac{L_n}{1 - L_n}\right).$$

This claims that the inequality (10.4) holds.

According to the inequality (10.1) and the equality (10.7), we have

$$\text{diag}\left[\left\|H\left(\leftthreetimes, \frac{x+y}{2}\right) + H\left(\leftthreetimes, \frac{x-y}{2}\right) - H(\leftthreetimes, x)\right\|_B, \ldots, \left\|H\left(\leftthreetimes, \frac{x+y}{2}\right)\right.\right.$$
$$\left.\left. + H\left(\leftthreetimes, \frac{x-y}{2}\right) - H(\leftthreetimes, x)\right\|_B\right]_{n\times n}$$
$$= \text{diag}\left[\lim_{k\to\infty} \frac{1}{2^k}\|f(\leftthreetimes, 2^{k-1}(x+y)) + f(\leftthreetimes, 2^{k-1}(x-y)) - f(\leftthreetimes, 2^k x)\|_B, \ldots,\right.$$
$$\left.\lim_{k\to\infty} \frac{1}{2^k}\|f(\leftthreetimes, 2^{k-1}(x+y)) + f(\leftthreetimes, 2^{k-1}(x-y)) - f(\leftthreetimes, 2^k x)\|_B\right]_{n\times n}$$
$$\preceq \text{diag}\left[\lim_{k\to\infty} \frac{1}{2^k}\varphi_1(2^k x, 2^k y), \ldots, \lim_{k\to\infty} \frac{1}{2^k}\varphi_n(2^k x, 2^k y)\right]_{n\times n}$$
$$= \text{diag}\left[0, \ldots, 0\right]_{n\times n},$$

for every $x, y \in A$ and $\leftthreetimes \in \pitchfork$. So, we get

$$H(\leftthreetimes, \frac{x+y}{2}) + H(\leftthreetimes, \frac{x-y}{2}) = H(\leftthreetimes, x), \tag{10.8}$$

for every $x, y \in A$ and $\leftthreetimes \in \pitchfork$. Letting $z = \frac{x+y}{2}$ and $w = \frac{x-y}{2}$ in the equality (10.8), we have

$$H(\leftthreetimes, z) + H(\leftthreetimes, w) = H(\leftthreetimes, z + w),$$

for every $z, w \in A$ and $\leftthreetimes \in \pitchfork$. So the mapping $H : \pitchfork \times A \to B$ is Cauchy additive, i.e., $H(\leftthreetimes, z + w) = H(\leftthreetimes, z) + H(\leftthreetimes, w)$, for every $z, w \in A$ and $\leftthreetimes \in \pitchfork$.

Letting $y = x$ in the inequality (10.1), we have

$$\mu f(\leftthreetimes, x) = f(\leftthreetimes, \mu x),$$

for every $\mu \in \mathbb{T}^1$ and every $x \in A$ and $\leftthreetimes \in \pitchfork$. Similarly, we have

$$\mu H(\leftthreetimes, x) = H(\leftthreetimes, \mu x),$$

for every $\mu \in \mathbb{T}^1$ and every $x \in A$ and $\leftthreetimes \in \pitchfork$. Thus, one can prove that the mapping $H : \pitchfork \times A \to B$ is $\mathbb{C}$–linear.

According to the inequality (10.2), we have

$$\begin{aligned}
&\operatorname{diag}\Big[\|H(\leftthreetimes, xy) - H(\leftthreetimes, x)H(\leftthreetimes, y)\|_B, \ldots, \|H(\leftthreetimes, xy) - H(\leftthreetimes, x)H(\leftthreetimes, y)\|_B\Big]_{n\times n} \\
&= \operatorname{diag}\Big[\lim_{k\to\infty} \frac{1}{4^k}\|f(\leftthreetimes, 4^k xy) - f(\leftthreetimes, 2^k x) f(\leftthreetimes, 2^k y)\|_B, \ldots, \lim_{k\to\infty} \frac{1}{4^k}\|f(\leftthreetimes, 4^k xy) \\
&\quad - f(\leftthreetimes, 2^k x) f(\leftthreetimes, 2^k y)\|_B\Big]_{n\times n} \\
&\preceq \operatorname{diag}\Big[\lim_{k\to\infty} \frac{1}{4^k}\varphi_1(2^k x, 2^k y), \ldots, \lim_{k\to\infty} \frac{1}{4^k}\varphi_n(2^k x, 2^k y)\Big]_{n\times n} \\
&\preceq \operatorname{diag}\Big[\lim_{k\to\infty} \frac{1}{2^k}\varphi_1(2^k x, 2^k y), \ldots, \lim_{k\to\infty} \frac{1}{2^k}\varphi_n(2^k x, 2^k y)\Big]_{n\times n} \\
&= \operatorname{diag}\Big[0, \ldots, 0\Big]_{n\times n},
\end{aligned}$$

for every $x, y \in A$ and $\leftthreetimes \in \pitchfork$. So we get

$$H(\leftthreetimes, xy) = H(\leftthreetimes, x)H(\leftthreetimes, y),$$

for every $x, y \in A$ and $\leftthreetimes \in \pitchfork$.

According to the inequality (10.3), we have

$$\begin{aligned}
&\operatorname{diag}\Big[\|H(\leftthreetimes, x^*) - H(\leftthreetimes, x)^*\|_B, \ldots, \|H(\leftthreetimes, x^*) - H(\leftthreetimes, x)^*\|_B\Big]_{n\times n} \\
&= \operatorname{diag}\Big[\lim_{k\to\infty} \frac{1}{2^k}\|f(\leftthreetimes, 2^k x^*) - f(\leftthreetimes, 2^k x)^*\|_B, \ldots, \lim_{k\to\infty} \frac{1}{2^k}\|f(\leftthreetimes, 2^k x^*) \\
&\quad - f(\leftthreetimes, 2^k x)^*\|_B\Big]_{n\times n} \\
&\preceq \operatorname{diag}\Big[\lim_{k\to\infty} \frac{1}{2^k}\varphi_1(2^k x, 2^k x), \ldots, \lim_{k\to\infty} \frac{1}{2^k}\varphi_n(2^k x, 2^k x)\Big]_{n\times n}
\end{aligned}$$

$$= \operatorname{diag}\Big[0, \ldots, 0\Big]_{n\times n},$$

for every $x \in A$ and $\leftthreetimes \in \pitchfork$. So we get

$$H(\leftthreetimes, x^*) = H(\leftthreetimes, x)^*,$$

for every $x \in A$ and $\leftthreetimes \in \pitchfork$.

Thus, $H : \pitchfork \times A \to B$ is a C^*–algebra homomorphism satisfying the inequality (10.4), as desired.

Theorem 10.2 *Let $n \in \mathbb{N}$, and $f : \pitchfork \times A \to B$ be a mapping for which there exist functions $\underbrace{\varphi_i}_{i=1,\ldots,n} : A^2 \to [0, \infty)$ satisfying the inequalities* (10.1), (10.2) *and* (10.3). *Furthermore, if there exist constants $\underbrace{L_i}_{i=1,\ldots,n} < 1$ such that $\underbrace{\varphi_i}_{i=1,\ldots,n}(x, 0) \leq \frac{1}{2}\underbrace{L_i\varphi_i}_{i=1,\ldots,n}(2x, 0)$, for every $x \in A$. Then, there exists a unique C^*–algebra homomorphism $H : \pitchfork \times A \to B$, such that*

$$\operatorname{diag}\Big[\|f(\leftthreetimes, x) - H(\leftthreetimes, x)\|_B, \ldots, \|f(\leftthreetimes, x) - H(\leftthreetimes, x)\|_B\Big]_{n\times n} \tag{10.9}$$
$$\preceq \operatorname{diag}\left[\frac{L_1}{2-2L_1}\varphi_1(x, 0), \ldots, \frac{L_n}{2-2L_n}\varphi_n(x, 0)\right]_{n\times n},$$

for every $x \in A$ and $\leftthreetimes \in \pitchfork$.

Proof We consider the linear mapping $J : X \to X$, such that

$$Jg(\leftthreetimes, x) := 2g\left(\leftthreetimes, \frac{x}{2}\right),$$

for every $x \in A$ and $\leftthreetimes \in \pitchfork$.

It follows from the inequality (10.5) that

$$\left[\|f(\leftthreetimes, x) - 2f\left(\leftthreetimes, \frac{x}{2}\right)\|_B, \ldots, \|f(\leftthreetimes, x) - 2f\left(\leftthreetimes, \frac{x}{2}\right)\|_B\right]_{n\times n}$$
$$\preceq \left[\varphi_1(\frac{x}{2}, 0), \ldots, \varphi_n\left(\frac{x}{2}, 0\right)\right]_{n\times n}$$
$$\preceq \left[\frac{L_1}{2}\varphi_1(x, 0), \ldots, \frac{L_n}{2}\varphi_n(x, 0)\right]_{n\times n},$$

for every $x \in A$ and $\leftthreetimes \in \pitchfork$. Hence, $d(f, Jf) \leq (\frac{L_1}{2}, \ldots, \frac{L_n}{2})$.

According to Theorem 9.5, there exists a mapping $H : \pitchfork \times A \to B$, such that

(1) The map H is a fixed point of J, i.e.,

$$H(\lambda, 2x) = 2H(\lambda, x), \tag{10.10}$$

for every $x \in A$. Moreover, the mapping H is a unique fixed point of J in the set

$$Y = \{g \in X : d(f, g) < \underbrace{(\infty, \ldots, \infty)}_{n}\}.$$

This implies that H is a unique mapping satisfying the equality (10.10), such that there exists $(C_1, \ldots, C_n) \in (0, \infty)^n$ satisfying

$$\begin{aligned}&\operatorname{diag}\Big[\|H(\lambda, x) - f(\lambda, x)\|_B, \ldots, \|H(\lambda, x) - f(\lambda, x)\|_B\Big]_{n\times n}\\ &\preceq \operatorname{diag}\Big[C_1\varphi_1(x, 0), \ldots, C_n\varphi_n(x, 0)\Big]_{n\times n}\end{aligned}$$

for every $x \in A$ and $\lambda \in \pitchfork$.

(2) $d(J^k f, H) \to \underbrace{(0, \ldots, 0)}_{n}$ as $k \to \infty$. This deduces the equality below

$$\lim_{k\to\infty} 2^k f\left(\lambda, \frac{x}{2^k}\right) = H(\lambda, x),$$

for every $x \in A$ and $\lambda \in \pitchfork$.

(3) $d(f, H) \leq (\frac{1}{1-L_1}, \ldots, \frac{1}{1-L_n})d(f, Jf)$ claims the inequality

$$d(f, H) \leq \left(\frac{L_1}{2 - 2L_1}, \ldots, \frac{L_n}{2 - 2L_n}\right),$$

which infers that the inequality (10.9) holds.

The rest of the proof is similar to the proof of Theorem 10.1.

Theorem 10.3 *Let $n \in \mathbb{N}$, and $f : \pitchfork \times A \to B$ be an odd mapping for which there exist functions $\underbrace{\varphi_i}_{i=1,\ldots,n} : A^2 \to [0, \infty)$ satisfying the inequalities* (10.1), (10.2) *and* (10.3). *Moreover, if there exist constants $\underbrace{L_i}_{i=1,\ldots,n} < 1$, such that $\underbrace{\varphi_i}_{i=1,\ldots,n}(x, 3x) \leq 2\underbrace{L_i\varphi_i}_{i=1,\ldots,n}(\frac{x}{2}, \frac{3x}{2})$, for every $x \in A$. Then, there exists a unique C^*–algebra homomorphism $H : \pitchfork \times A \to B$, such that*

$$\operatorname{diag}\left[\|f(\leftthreetimes, x) - H(\leftthreetimes, x)\|_B, \ldots, \|f(\leftthreetimes, x) - H(\leftthreetimes, x)\|_B \right]_{n\times n} \tag{10.11}$$
$$\preceq \operatorname{diag}\left[\frac{1}{2-2L_1}\varphi_1(x, 3x), \ldots, \frac{1}{2-2L_n}\varphi_n(x, 3x) \right]_{n\times n},$$

for every $x \in A$ *and* $\leftthreetimes \in \pitchfork$.

Proof Consider the set below

$$X := \{g : \pitchfork \times A \to B\},$$

and introduce the generalized metric d on X, as follows:

$$d(g, h) := \inf\Bigg\{(C_1, \ldots, C_n) \in \mathbb{R}_+ : \operatorname{diag}[\|g(\leftthreetimes, x) - h(\leftthreetimes, x)\|_B, \ldots,$$
$$\|g(\leftthreetimes, x) - h(\leftthreetimes, x)\|_B]_{n\times n}$$
$$\preceq \left[C_1, \ldots, C_n \right]^T_{n\times 1} \operatorname{diag}\left[\varphi_1(x, 3x), \ldots, \varphi_n(x, 3x) \right]_{n\times n}, \quad \forall x \in A,\ \leftthreetimes \in \pitchfork \Bigg\}.$$

We can easily prove that the space (X, d) is complete [1].

Now, we consider the linear mapping $J : X \to X$, such that

$$Jg(\leftthreetimes, x) := \frac{1}{2} g(\leftthreetimes, 2x),$$

for every $x \in A$ and $\leftthreetimes \in \pitchfork$.

Based on Theorem 3.1 of [2], we get

$$d(Jg, Jh) \preceq (L_1, \ldots, L_n) d(g, h),$$

for every $g, h \in X$.

Letting $\mu = 1$ and relpacing y by $3x$ in the inequality (10.1), we have

$$\operatorname{diag}\left[\|f(\leftthreetimes, 2x) - 2f(\leftthreetimes, x)\|_B, \ldots, \|f(\leftthreetimes, 2x) - 2f(\leftthreetimes, x)\|_B \right]_{n\times n} \tag{10.12}$$
$$\preceq \operatorname{diag}\left[\varphi_1(x, 3x), \ldots, \varphi_n(x, 3x) \right]_{n\times n},$$

for every $x \in A$ and $\leftthreetimes \in \pitchfork$. So we get

$$\operatorname{diag}\left[\|f(\lambda, x) - \frac{1}{2}f(\lambda, 2x)\|_B, \ldots, \|f(\lambda, x) - \frac{1}{2}f(\lambda, 2x)\|_B\right]_{n\times n}$$
$$\preceq \operatorname{diag}\left[\frac{1}{2}\varphi_1(x, 3x), \ldots, \frac{1}{2}\varphi_n(x, 3x)\right]_{n\times n},$$

for every $x \in A$ and $\lambda \in \mathbb{T}$. Hence, $d(f, Jf) \leq (\frac{1}{2}, \ldots, \frac{1}{2})$.

According to Theorem 9.5, there exists a mapping $H : \mathbb{T} \times A \to B$, such that

(1) The map H is a fixed point of J, i.e.,

$$H(\lambda, 2x) = 2H(\lambda, x), \tag{10.13}$$

for every $x \in A$ and $\lambda \in \mathbb{T}$. In addition, the mapping H is a unique fixed point of J in the set below

$$Y = \{g \in X : d(f, g) < \underbrace{(\infty, \ldots, \infty)}_{n}\}.$$

This concludes that H is a unique mapping satisfying the equality (10.13), such that there exists $(C_1, \ldots, C_n) \in (0, \infty)^n$ satisfying

$$\operatorname{diag}\left[\|H(\lambda, x) - f(\lambda, x)\|_B, \ldots, \|H(\lambda, x) - f(\lambda, x)\|_B\right]_{n\times n}$$
$$\preceq \operatorname{diag}\left[C_1\varphi_1(x, 3x), \ldots, C_n\varphi_n(x, 3x)\right]_{n\times n},$$

for every $x \in A$ and $\lambda \in \mathbb{T}$.

(2) $d(J^n f, H) \to \underbrace{(0, \ldots, 0)}_{n}$ as $n \to \infty$. This infers the following equality

$$\lim_{n\to\infty} \frac{f(\lambda, 2^n x)}{2^n} = H(\lambda, x),$$

for every $x \in A$ and $\lambda \in \mathbb{T}$.

(3) $d(f, H) \leq (\frac{1}{1-L_1}, \ldots, \frac{1}{1-L_n})d(f, Jf)$ confirms the inequality

$$d(f, H) \leq \left(\frac{1}{2-2L_1}, \ldots, \frac{1}{2-2L_n}\right).$$

Therefore, the inequality (10.11) holds.

The rest follows immediately from the proof of Theorem 10.1.

Theorem 10.4 *Let $n \in \mathbb{N}$, and $f : \pitchfork \times A \to B$ be an odd mapping for which there exist functions $\underbrace{\varphi_i}_{i=1,\dots,n} : A^2 \to [0, \infty)$ satisfying the inequalities* (10.1), (10.2) *and* (10.3). *In addition, if there exist constants $\underbrace{L_i}_{i=1,\dots,n} < 1$, such that $\underbrace{\varphi_i}_{i=1,\dots,n}(x, 3x) \leq \frac{1}{2}\underbrace{L_i\varphi_i}_{i=1,\dots,n}(2x, 6x)$, for every $x \in A$. Then, there exists a unique C^*–algebra homomorphism $H : \pitchfork \times A \to B$, such that, we get*

$$\operatorname{diag}\left[\|f(\leftthreetimes, x) - H(\leftthreetimes, x)\|_B, \dots, \|f(\leftthreetimes, x) - H(\leftthreetimes, x)\|_B\right]_{n\times n} \tag{10.14}$$
$$\preceq \operatorname{diag}\left[\frac{L_1}{2-2L_1}\varphi_1(x, 3x), \dots, \frac{L_n}{2-2L_n}\varphi_n(x, 3x)\right]_{n\times n},$$

for every $x \in A$ and $\leftthreetimes \in \pitchfork$.

Proof We consider the linear mapping $J : X \to X$, such that

$$Jg(\leftthreetimes, x) := 2g\left(\leftthreetimes, \frac{x}{2}\right),$$

for every $x \in A$ and $\leftthreetimes \in \pitchfork$.

According to (10.12), we have

$$\begin{aligned}
&\operatorname{diag}\left[\|f(\leftthreetimes, x) - 2f\left(\leftthreetimes, \frac{x}{2}\right)\|_B, \dots, \|f(\leftthreetimes, x) - 2f\left(\leftthreetimes, \frac{x}{2}\right)\|_B\right]_{n\times n}\\
&\preceq \operatorname{diag}\left[\varphi_1\left(\frac{x}{2}, \frac{3x}{2}\right), \dots, \varphi_n\left(\frac{x}{2}, \frac{3x}{2}\right)\right]_{n\times n}\\
&\preceq \operatorname{diag}\left[\frac{L_1}{2}\varphi_1(x, 3x), \dots, \frac{L_n}{2}\varphi_n(x, 3x)\right]_{n\times n},
\end{aligned}$$

for every $x \in A$ and $\leftthreetimes \in \pitchfork$. Hence, $d(f, Jf) \leq (\frac{L_1}{2}, \dots, \frac{L_n}{2})$.

Using Theorem 9.5, there exists a mapping $H : \pitchfork \times A \to B$, such that

(1) The map H is a fixed point of J, i.e.,

$$H(\leftthreetimes, 2x) = 2H(\leftthreetimes, x), \tag{10.15}$$

for every $x \in A$ and $\leftthreetimes \in \pitchfork$. Further, the mapping H is a unique fixed point of J in the set below

$$Y = \{g \in X : d(f, g) < \underbrace{(\infty, \dots, \infty)}_{n}\}.$$

This indicates that H is a unique mapping satisfying the quality (10.15), such that there exists $(C_1, \dots, C_n) \in (0, \infty)^n$ satisfying the following inequality

$$\operatorname{diag}\Big[\|H(\leftthreetimes, x) - f(\leftthreetimes, x)\|_B, \dots, \|H(\leftthreetimes, x) - f(\leftthreetimes, x)\|_B\Big]_{n\times n} \preceq \operatorname{diag}\Big[C_1\varphi_1(x, 3x), \dots, C_n\varphi_n(x, 3x)\Big]_{n\times n}$$

for every $x \in A$ and $\leftthreetimes \in \pitchfork$.

(2) The condition $d(J^k f, H) \to \underbrace{(0, \dots, 0)}_{n}$, as $k \to \infty$, derives the equality below

$$\lim_{k\to\infty} 2^k f\left(\leftthreetimes, \frac{x}{2^k}\right) = H(\leftthreetimes, x),$$

for every $x \in A$ and $\leftthreetimes \in \pitchfork$.

(3) The inequality $d(f, H) \le (\frac{1}{1-L_1}, \dots, \frac{1}{1-L_n})d(f, Jf)$, implies the following inequality

$$d(f, H) \le \left(\frac{L_1}{2-2L_1}, \dots, \frac{L_n}{2-2L_n}\right),$$

which concludes that the inequality (10.14) holds.

The rest of the proof follows from the proof of Theorem 10.1.

10.2 Multi Stability of Derivations in C^*–Algebras

In this part, we present the multi stability of derivations on C^*–algebras for the functional equation $D_\mu f(\leftthreetimes, x, y) = 0$.

Throughout the section, we let A be a C^*–algebra with norm $\|\cdot\|_A$.

We recall that a $\mathbb{C}$–linear operator $\delta : \pitchfork \times A \to A$ is a derivation [1] on A, if δ satisfies $\delta(\leftthreetimes, xy) = \delta(\leftthreetimes, x)y + x\delta(\leftthreetimes, y)$, for every $x, y \in A$ and $\leftthreetimes \in \pitchfork$.

Theorem 10.5 *Let $n \in \mathbb{N}$, and $f : \pitchfork \times A \to A$ be a mapping for which there exist functions $\underbrace{\varphi_i}_{i=1,\dots,n} : A^2 \to [0, \infty)$, such that*

$$\operatorname{diag}\Big[\|D_\mu f(\leftthreetimes, x, y)\|_A, \dots, \|D_\mu f(\leftthreetimes, x, y)\|_A\Big]_{n\times n} \preceq \operatorname{diag}\Big[\varphi_1(x, y), \dots, \varphi_n(x, y)\Big], \tag{10.16}$$

$$\begin{aligned}
&\operatorname{diag}\Big[\|f(\leftthreetimes, xy) - f(\leftthreetimes, x)y - xf(\leftthreetimes, y)\|_A, \ldots, \|f(\leftthreetimes, xy) \\
&\quad - f(\leftthreetimes, x)y - xf(\leftthreetimes, y)\|_A\Big]_{n\times n} \\
&\preceq \operatorname{diag}\Big[\varphi_1(x, y), \ldots, \varphi_n(x, y)\Big]_{n\times n},
\end{aligned} \tag{10.17}$$

for every $\mu \in \mathbb{T}^1$ *and every* $x, y \in A$, $\leftthreetimes \in \pitchfork$. *Additionally, let there exist constants* $\underbrace{L_i}_{i=1,\ldots,n} < 1$, *such that* $\underbrace{\varphi_i}_{i=1,\ldots,n}(x, 0) \leq 2 \underbrace{L_i\varphi_i}_{i=1,\ldots,n}(\frac{x}{2}, 0)$, *for every* $x \in A$. *Then, there exists a unique derivation* $\delta : \pitchfork \times A \to A$ *satisfying*

$$\begin{aligned}
&\operatorname{diag}\Big[\|f(\leftthreetimes, x) - \delta(\leftthreetimes, x)\|_A, \ldots, \|f(\leftthreetimes, x) - \delta(\leftthreetimes, x)\|_A\Big]_{n\times n} \\
&\preceq \operatorname{diag}\Big[\frac{L_1}{1-L_1}\varphi_1(x, 0), \ldots, \frac{L_n}{1-L_n}\varphi_n(x, 0)\Big]_{n\times n},
\end{aligned} \tag{10.18}$$

for every $x \in A$ *and* $\leftthreetimes \in \pitchfork$.

Proof By the same reasoning as the proof of Theorem 10.1, there exists a unique evolutive $\mathbb{C}$–linear mapping $\delta : \pitchfork \times A \to A$ satisfying the inequality (10.18). The mapping $\delta : \pitchfork \times A \to A$ is defined by

$$\delta(\leftthreetimes, x) = \lim_{k\to\infty} \frac{f(\leftthreetimes, 2^k x)}{2^k},$$

for every $x \in A$ and $\leftthreetimes \in \pitchfork$.

Applying the inequality (10.17), we get

$$\begin{aligned}
&\operatorname{diag}\Big[\|\delta(\leftthreetimes, xy) - \delta(\leftthreetimes, x)y - x\delta(\leftthreetimes, y)\|_A, \ldots, \|\delta(\leftthreetimes, xy) \\
&\quad - \delta(\leftthreetimes, x)y - x\delta(\leftthreetimes, y)\|_A\Big]_{n\times n} \\
&= \operatorname{diag}\Big[\lim_{k\to\infty} \frac{1}{4^k}\|f(\leftthreetimes, 4^k xy) - f(\leftthreetimes, 2^k x)\cdot 2^k y - 2^k x f(\leftthreetimes, 2^k y)\|_A, \ldots, \\
&\qquad \lim_{k\to\infty} \frac{1}{4^k}\|f(\leftthreetimes, 4^k xy) - f(\leftthreetimes, 2^k x)\cdot 2^k y - 2^k x f(2^k y)\|_A\Big]_{n\times n} \\
&\preceq \operatorname{diag}\Big[\lim_{k\to\infty} \frac{1}{4^k}\varphi_1(2^k x, 2^k y), \ldots, \lim_{k\to\infty} \frac{1}{4^k}\varphi_n(2^k x, 2^k y)\Big]_{n\times n} \\
&\preceq \operatorname{diag}\Big[\lim_{k\to\infty} \frac{1}{2^k}\varphi_1(2^k x, 2^k y), \ldots, \lim_{k\to\infty} \frac{1}{2^k}\varphi_n(2^k x, 2^k y)\Big]_{n\times n}
\end{aligned}$$

$$= \operatorname{diag}\Big[0, \ldots, 0\Big]_{n\times n}$$

for every $x, y \in A$ and $\leftthreetimes \in \pitchfork$. So we get

$$\delta(\leftthreetimes, xy) = \delta(\leftthreetimes, x)y + x\delta(\leftthreetimes, y)$$

for every $x, y \in A$ and $\leftthreetimes \in \pitchfork$. Thus, $\delta : \pitchfork \times A \to A$ is a derivation satisfying the inequality (10.18).

Theorem 10.6 *Let $n \in \mathbb{N}$, and $f : \pitchfork \times A \to A$ be a mapping for which there exist functions $\underbrace{\varphi_i}_{i=1,\ldots,n} : A^2 \to [0, \infty)$ satisfying the inequalities* (10.16) *and* (10.17). *Also, let there exist constants $\underbrace{L_i}_{i=1,\ldots,n} < 1$, such that $\underbrace{\varphi_i}_{i=1,\ldots,n}(x, 0) \leq \frac{1}{2}\underbrace{L_i\varphi_i}_{i=1,\ldots,n}(2x, 0)$, for every $x \in A$. Then, there exists a unique derivation $\delta : \pitchfork \times A \to A$, such that*

$$\operatorname{diag}\Big[\|f(\leftthreetimes, x) - \delta(\leftthreetimes, x)\|_A, \ldots, \|f(\leftthreetimes, x) - \delta(\leftthreetimes, x)\|_A\Big]_{n\times n} \tag{10.19}$$
$$\preceq \operatorname{diag}\Big[\frac{L_1}{2 - 2L_1}\varphi_1(x, 0), \ldots, \frac{L_n}{2 - 2L_n}\varphi_n(x, 0)\Big]_{n\times n},$$

for every $x \in A$ and $\leftthreetimes \in \pitchfork$.

Proof The proof is similar to the proofs of Theorems 10.2 and 10.5.

Remark 10.1 For the inequalities controlled by the product of powers of norms, one can obtain the results which are similar to Theorems 10.3 and 10.4.

10.3 Multi Stability of Lie Homomorphisms in Lie C^*–Algebras

In this part, we prove the multi stability of Lie homomorphisms in Lie C^*–algebras for the functional equation $D_\mu f(x, y) = 0$.

A C^*–algebra $\mathcal{C}$, endowed with the Lie product $[x, y] := \frac{xy-yx}{2}$ on $\mathcal{C}$, is called a Lie C^*–algebra [1].

Definition 10.1 ([1]) Let A and B be two Lie C^*–algebras. A $\mathbb{C}$–linear operator $H : \pitchfork \times A \to B$ is called a Lie C^*–algebra homomorphism, if $H(\leftthreetimes, [x, y]) = [H(\leftthreetimes, x), H(\leftthreetimes, y)]$, for every $x, y \in A$.

Throughout this section, we let A be a Lie C^*–algebra with norm $\|\cdot\|_A$ and B be a Lie C^*–algebra with norm $\|\cdot\|_B$.

Theorem 10.7 *Let $n \in \mathbb{N}$, and $f : \pitchfork \times A \to B$ be a mapping for which there exist functions* $\underbrace{\varphi_i}_{i=1,\ldots,n} : A^2 \to [0, \infty)$, *satisfying* (10.1) *and*

$$\begin{aligned}
&\operatorname{diag}\Bigg[\|f(\leftthreetimes, [x, y]) - [f(\leftthreetimes, x), f(\leftthreetimes, y)]\|_B, \ldots, \|f(\leftthreetimes, [x, y]) \\
&\quad - [f(\leftthreetimes, x), f(\leftthreetimes, y)]\|_B\Bigg]_{n\times n} \\
&\preceq \operatorname{diag}\Bigg[\varphi_1(x, y), \ldots, \varphi_n(x, y)\Bigg]_{n\times n}
\end{aligned} \tag{10.20}$$

for every $x, y \in A$, and $\leftthreetimes \in \pitchfork$. Furthermore, if there exist constants $\underbrace{L_i}_{i=1,\ldots,n} < 1$, *such that* $\underbrace{\varphi_i}_{i=1,\ldots,n}(x, 0) \le 2 \underbrace{L_i\varphi_i}_{i=1,\ldots,n}(\frac{x}{2}, 0)$, *for every $x \in A$. Then, there exists a unique Lie C^*–algebra homomorphism $H : \pitchfork \times A \to B$ satisfying the inequality* (10.4).

Proof By the same arguments as the proof of Theorem 10.1, there exists a unique $\mathbb{C}$–linear mapping $\delta : \pitchfork \times A \to A$ satisfying the inequality (10.4). Consider the mapping $H : \pitchfork \times A \to B$ defined by

$$H(\leftthreetimes, x) = \lim_{k\to\infty} \frac{f(\leftthreetimes, 2^k x)}{2^k},$$

for every $x \in A$ and $\leftthreetimes \in \pitchfork$.

Utilizing the inequality (10.20), we get

$$\begin{aligned}
&\operatorname{diag}\Bigg[\|H(\leftthreetimes, [x, y]) - [H(\leftthreetimes, x), H(\leftthreetimes, y)]\|_B, \ldots, \|H(\leftthreetimes, [x, y]) \\
&\quad - [H(\leftthreetimes, x), H(\leftthreetimes, y)]\|_B\Bigg]_{n\times n} \\
&= \operatorname{diag}\Bigg[\lim_{k\to\infty} \frac{1}{4^k}\|f(\leftthreetimes, 4^k[x, y]) - [f(\leftthreetimes, 2^k x), f(\leftthreetimes, 2^k y)]\|_B, \ldots, \\
&\qquad \lim_{k\to\infty} \frac{1}{4^k}\|f(\leftthreetimes, 4^k[x, y]) - [f(\leftthreetimes, 2^k x), f(\leftthreetimes, 2^k y)]\|_B\Bigg]_{n\times n} \\
&\preceq \operatorname{diag}\Bigg[\lim_{k\to\infty} \frac{1}{4^k}\varphi_1(2^k x, 2^k y), \ldots, \lim_{k\to\infty} \frac{1}{4^k}\varphi_n(2^k x, 2^k y)\Bigg]_{n\times n} \\
&\preceq \operatorname{diag}\Bigg[\lim_{k\to\infty} \frac{1}{2^k}\varphi_1(2^k x, 2^k y), \ldots, \lim_{k\to\infty} \frac{1}{2^k}\varphi_n(2^k x, 2^k y)\Bigg]_{n\times n} \\
&= \operatorname{diag}\Bigg[0, \ldots, 0\Bigg]_{n\times n},
\end{aligned}$$

for every $x, y \in A$ and $\leftthreetimes \in \pitchfork$. Hence, we get

$$H(\leftthreetimes, [x, y]) = [H(\leftthreetimes, x), H(\leftthreetimes, y)],$$

for every $x, y \in A$ and $\leftthreetimes \in \pitchfork$.

In summary, $H : \pitchfork \times A \to B$ is a Lie C^*–algebra homomorphism satisfying the inequality (10.4), as desired.

Theorem 10.8 *Let $n \in \mathbb{N}$, and $f : \pitchfork \times A \to B$ be a mapping for which there exist functions $\underbrace{\varphi_i}_{i=1,\dots,n} : A^2 \to [0, \infty)$ satisfying the inequalities* (10.1) *and* (10.20). *Further, if there exist constants $\underbrace{L_i}_{i=1,\dots,n} < 1$, such that $\underbrace{\varphi_i}_{i=1,\dots,n}(x, 0) \leq \frac{1}{2}\underbrace{L_i\varphi_i}_{i=1,\dots,n}(2x, 0)$ for every $x \in A$. Then, there exists a unique Lie C^*–algebra homomorphism $H : \pitchfork \times A \to B$ satisfying the inequality* (10.9).

Proof The proof follows similarly from the proofs of Theorems 10.2 and 10.3.

Remark 10.2 For the inequalities controlled by the product of powers of norms, one can derive similar results to Theorems 10.3 and 10.4.

10.4 Multi Stability of Lie Derivations in Lie C^*–Algebras

In this section, we prove the multi stability of Lie derivations on Lie C^*–algebras for the functional equation $D_\mu f(\leftthreetimes, x, y) = 0$.

Definition 10.2 ([1]) Let A be a Lie C^*–algebra. A $\mathbb{C}$–linear operator $\delta : \pitchfork \times A \to A$ is called a Lie derivation if $\delta(\leftthreetimes, [x, y]) = [\delta(\leftthreetimes, x), y] + [x, \delta(\leftthreetimes, y)]$, for every $x, y \in A$ and $\leftthreetimes \in \pitchfork$.

Throughout this section, let A be a Lie C^*–algebra with norm $\|\cdot\|_A$.

Theorem 10.9 *Let $n \in \mathbb{N}$, and $f : \pitchfork \times A \to A$ be a mapping for which there exist functions $\underbrace{\varphi_i}_{i=1,\dots,n} : A^2 \to [0, \infty)$ satisfying the inequality* (10.16) *and*

$$\begin{aligned}
&\operatorname{diag}\Bigg[\|f(\leftthreetimes, [x, y]) - [f(\leftthreetimes, x), y] - [x, f(\leftthreetimes, y)]\|_A, \dots, \\
&\qquad \|f(\leftthreetimes, [x, y]) - [f(\leftthreetimes, x), y] - [x, f(\leftthreetimes, y)]\|_A\Bigg]_{n\times n} \\
&\preceq \operatorname{diag}\Bigg[\varphi_1(x, y), \dots, \varphi_n(x, y)\Bigg]_{n\times n}
\end{aligned} \tag{10.21}$$

for every $x, y \in A$, *and* $\curlywedge \in \pitchfork$. *Besides there exist constants* $\underbrace{L_i}_{i=1,\ldots,n} < 1$, *such that* $\underbrace{\varphi_i}_{i=1,\ldots,n}(x, 0) \leq 2 \underbrace{L_i\varphi_i}_{i=1,\ldots,n}(\frac{x}{2}, 0)$, *for every* $x \in A$. *Then, there exists a unique Lie derivation* $\delta : \pitchfork \times A \to A$ *satisfying the inequality* (10.18).

Proof By the same reasoning as the proof of Theorem 10.1, there exists a unique evolutive $\mathbb{C}$–linear mapping $\delta : \pitchfork \times A \to A$ satisfying the inequality (10.18). The mapping $\delta : \pitchfork \times A \to A$ is further defined by

$$\delta(\curlywedge, x) = \lim_{n\to\infty} \frac{f(\curlywedge, 2^n x)}{2^n},$$

for every $x \in A$ and $\curlywedge \in \pitchfork$.

According to the inequality (10.21), we get

$$\begin{aligned}
&\operatorname{diag}\Big[\|\delta(\curlywedge, [x, y]) - [\delta(\curlywedge, x), y] - [x, \delta(\curlywedge, y)]\|_A, \ldots, \|\delta(\curlywedge, [x, y]) \\
&\quad - [\delta(\curlywedge, x), y] - [x, \delta(\curlywedge, y)]\|_A\Big]_{n\times n} \\
&= \operatorname{diag}\Big[\lim_{k\to\infty}\frac{1}{4^k}\|f(\curlywedge, 4^k[x, y]) - [f(\curlywedge, 2^k x), 2^k y] - [2^k x, f(\curlywedge, 2^k y)]\|_A, \ldots, \\
&\quad \lim_{k\to\infty}\frac{1}{4^k}\|f(\curlywedge, 4^k[x, y]) - [f(\curlywedge, 2^k x), 2^k y] - [2^k x, f(\curlywedge, 2^k y)]\|_A\Big]_{n\times n} \\
&\leq \operatorname{diag}\Big[\lim_{k\to\infty}\frac{1}{4^k}\varphi(2^k x, 2^k y), \ldots, \lim_{k\to\infty}\frac{1}{4^k}\varphi(2^k x, 2^k y)\Big]_{n\times n} \\
&\leq \operatorname{diag}\Big[\lim_{k\to\infty}\frac{1}{2^k}\varphi(2^k x, 2^k y), \ldots, \lim_{k\to\infty}\frac{1}{2^k}\varphi(2^k x, 2^k y)\Big]_{n\times n} \\
&= \operatorname{diag}\Big[0, \ldots, 0\Big]_{n\times n}
\end{aligned}$$

for $x, y \in A$ and $\curlywedge \in \pitchfork$. So we get

$$\delta(\curlywedge, [x, y]) = [\delta(\curlywedge, x), y] + [x, \delta(\curlywedge, y)],$$

for every $x, y \in A$ and $\curlywedge \in \pitchfork$. Thus $\delta : \pitchfork \times A \to A$ is a derivation satisfying (10.18).

Theorem 10.10 *Let* $n \in \mathbb{N}$, *and* $f : \pitchfork \times A \to A$ *be a mapping for which there exist functions* $\underbrace{\varphi_i}_{i=1,\ldots,n} : A^2 \to [0, \infty)$ *satisfying the inequalities* (10.16) *and* (10.21). *In addition, if there exist constants* $\underbrace{L_i}_{i=1,\ldots,n} < 1$, *such that* $\underbrace{\varphi_i}_{i=1,\ldots,n}(x, 0) \leq \frac{1}{2}\underbrace{L_i\varphi_i}_{i=1,\ldots,n}(2x, 0)$, *for every* $x \in A$. *Then, there exists a unique Lie derivation* $\delta : \pitchfork \times A \to A$ *satisfying the inequality* (10.19).

Proof The proof is similar to the proofs of Theorems 10.2 and 10.3.

Remark 10.3 For the inequalities controlled by the product of powers of norms, one can obtain similar results to Theorems 10.3 and 10.4.

10.5 Application

Here, consider the diagonal matrix $\mathfrak{W}$ defined in Sect. 9.8.1.

Note 10.1 ([1]) For every $n \in \mathbb{N}$, and every $\theta > 0$, if we have,

$$\operatorname{diag}\Big[\varphi_1(x, y), \ldots, \varphi_n(x, y)\Big]_{n\times n} := \theta\mathfrak{W}(\|x\|_A^r + \|y\|_A^r),$$

then, the following are satisfied:

- for every $r < 1$, there exist constants $\underbrace{L_i}_{i=1,\ldots,n} < 1$, such that

$$\underbrace{\varphi_i}_{i=1,\ldots,n}(x, 0) \le 2\underbrace{L_i\varphi_i}_{i=1,\ldots,n}\left(\frac{x}{2}, 0\right),$$

- for every $r > 1$, there exist constants $\underbrace{L_i}_{i=1,\ldots,n} < 1$, such that

$$\underbrace{\varphi_i}_{i=1,\ldots,n}(x, 0) \le \frac{1}{2}\underbrace{L_i\varphi_i}_{i=1,\ldots,n}(2x, 0).$$

Corollary 10.1 *Let $n \in \mathbb{N}$, $r < 1$, $\theta > 0$ and $f : ⋔ \times A \to B$ be a mapping, such that*

$$\begin{aligned}&\operatorname{diag}\Big[\|D_\mu f(\curlywedge, x, y)\|_B, \ldots, \|D_\mu f(\curlywedge, x, y)\|_B\Big]_{n\times n}\\ &\preceq \theta\mathfrak{W}(\|x\|_A^r + \|y\|_A^r),\end{aligned} \tag{10.22}$$

$$\begin{aligned}&\operatorname{diag}\Big[\|f(\curlywedge, xy) - f(\curlywedge, x)f(\curlywedge, y)\|_B, \ldots, \|f(\curlywedge, xy) - f(\curlywedge, x)f(\curlywedge, y)\|_B\Big]_{n\times n}\\ &\preceq \theta\mathfrak{W}(\|x\|_A^r + \|y\|_A^r),\end{aligned} \tag{10.23}$$

$$\operatorname{diag}\left[\|f(\leftthreetimes, x^*) - f(\leftthreetimes, x)^*\|_B, \ldots, \|f(\leftthreetimes, x^*) - f(\leftthreetimes, x)^*\|_B\right]_{n\times n} \preceq \theta\mathfrak{W}(2\|x\|_A^r), \tag{10.24}$$

for every $\mu \in \mathbb{T}^1$ *and every* $x, y \in A$ *and* $\leftthreetimes \in \pitchfork$. *Then, there exists a unique* C^*–*algebra homomorphism* $H : \pitchfork \times A \to B$, *such that,*

$$\operatorname{diag}\left[\|f(\leftthreetimes, x) - H(\leftthreetimes, x)\|_B, \ldots, \|f(\leftthreetimes, x) - H(\leftthreetimes, x)\|_B\right]_{n\times n} \preceq \frac{2^r\theta}{2-2^r}\mathfrak{W}(\|x\|_A^r), \tag{10.25}$$

for every $x \in A$ *and* $\leftthreetimes \in \pitchfork$.

Proof The proof follows from Theorem 10.1 by taking

$$\operatorname{diag}\left[\varphi_1(x, y), \ldots, \varphi_n(x, y)\right]_{n\times n} := \theta\mathfrak{W}(\|x\|_A^r + \|y\|_A^r),$$

for every $x, y \in A$. Setting $\underbrace{L_i}_{i=1,\ldots,n} = 2^{r-1}$, we have the desired result.

Corollary 10.2 *Let* $n \in \mathbb{N}$, $r > 2$, $\theta > 0$ *and* $f : \pitchfork \times A \to B$ *be a mapping satisfying the inequalities* (10.22), (10.23) *and* (10.24). *Then, there exists a unique* C^*–*algebra homomorphism* $H : \pitchfork \times A \to B$, *such that,*

$$\operatorname{diag}\left[\|f(\leftthreetimes, x) - H(\leftthreetimes, x)\|_B, \ldots, \|f(\leftthreetimes, x) - H(\leftthreetimes, x)\|_B\right]_{n\times n} \preceq \frac{\theta}{2^r-2}\mathfrak{W}(\|x\|_A^r), \tag{10.26}$$

for every $x \in A$ *and* $\leftthreetimes \in \pitchfork$.

Proof The proof follows from Theorem 10.2 by taking

$$\operatorname{diag}\left[\varphi_1(x, y), \ldots, \varphi_{522}(x, y)\right]_{n\times n} := \theta\mathfrak{W}(\|x\|_A^r + \|y\|_A^r),$$

for every $x, y \in A$. Choosing $\underbrace{L_i}_{i=1,\ldots,n} = 2^{1-r}$, we get the desired result.

Corollary 10.3 *Let* $n \in \mathbb{N}$, $r < \frac{1}{2}$, $\theta > 0$ *and* $f : \pitchfork \times A \to B$ *be an odd mapping satisfying the inequalities below*

$$\operatorname{diag}\left[\|D_\mu f(\leftthreetimes, x, y)\|_B, \ldots, \|D_\mu f(\leftthreetimes, x, y)\|_B\right]_{n\times n} \tag{10.27}$$
$$\preceq \theta \cdot \mathfrak{W}(\|x\|_A^r \cdot \|y\|_A^r),$$

$$\operatorname{diag}\left[\|f(\leftthreetimes, xy) - f(\leftthreetimes, x) f(\leftthreetimes, y)\|_B, \ldots, \|f(\leftthreetimes, xy) - f(\leftthreetimes, x) f(\leftthreetimes, y)\|_B\right]_{n\times n} \tag{10.28}$$
$$\preceq \theta \cdot \mathfrak{W}(\|x\|_A^r \cdot \|y\|_A^r),$$

$$\operatorname{diag}\left[\|f(\leftthreetimes, x^*) - f(\leftthreetimes, x)^*\|_B, \ldots, \|f(\leftthreetimes, x^*) - f(\leftthreetimes, x)^*\|_B\right]_{n\times n} \tag{10.29}$$
$$\preceq \theta \cdot \mathfrak{W}(\|x\|_A^{2r}),$$

for every $\mu \in \mathbb{T}^1$ *and every* $x, y \in A$ *and* $\leftthreetimes \in ⫛$. *Then, there exists a unique* C^**–algebra homomorphism* $H : ⫛ \times A \to B$, *such that*

$$\operatorname{diag}\left[\|f(\leftthreetimes, x) - H(\leftthreetimes, x)\|_B, \ldots, \|f(\leftthreetimes, x) - H(\leftthreetimes, x)\|_B\right]_{n\times n} \tag{10.30}$$
$$\preceq \frac{3^r\theta}{2 - 2^{2r}} \mathfrak{W}(\|x\|_A^{2r}),$$

for every $x \in A$ *and* $\leftthreetimes \in ⫛$.

Proof The proof follows from Theorem 10.3 by taking

$$\operatorname{diag}\left[\varphi_1(x, y), \ldots, \varphi_n(x, y)\right]_{n\times n} := \theta \cdot \mathfrak{W}(\|x\|_A^r \cdot \|y\|_A^r),$$

for every $x, y \in A$. Picking $\underbrace{L_i}_{i=1,\ldots,n} = 2^{2r-1}$, we come to the desired result.

Corollary 10.4 *Let* $n \in \mathbb{N}$, $r > 1$, $\theta > 0$ *and* $f : ⫛ \times A \to B$ *be an odd mapping satisfying the inequalities* (10.27), (10.28), *and* (10.29). *Then, there exists a unique* C^**–algebra homomorphism* $H : ⫛ \times A \to B$, *such that*

$$\operatorname{diag}\left[\|f(\leftthreetimes, x) - H(\leftthreetimes, x)\|_B, \ldots, \|f(\leftthreetimes, x) - H(\leftthreetimes, x)\|_B\right]_{n\times n} \tag{10.31}$$
$$\preceq \frac{\theta}{2^{2r} - 2} \mathfrak{W}(\|x\|_A^{2r}),$$

for every $x \in A$ *and* $\leftthreetimes \in ⫛$.

Proof The proof can be derived from Theorem 10.4 by taking

$$\operatorname{diag}\Big[\varphi_1(x, y), \ldots, \varphi_n(x, y)\Big]_{n\times n} := \theta \cdot \mathfrak{W}(\|x\|_A^r \cdot \|y\|_A^r),$$

for every $x, y \in A$. Letting $\underbrace{L_i}_{i=1,\ldots,n} = 2^{1-2r}$, we get the desired result.

Corollary 10.5 *Let $n \in \mathbb{N}$, $r < 1$, $\theta > 0$ and $f : \pitchfork \times A \to A$ be a mapping, such that*

$$\begin{aligned}&\operatorname{diag}\Big[\|D_\mu f(\leftthreetimes, x, y)\|_A, \ldots, \|D_\mu f(\leftthreetimes, x, y)\|_A\Big]_{n\times n} \\ &\preceq \theta\mathfrak{W}(\|x\|_A^r + \|y\|_A^r),\end{aligned} \tag{10.32}$$

$$\begin{aligned}&\operatorname{diag}\Big[\|f(\leftthreetimes, xy) - f(\leftthreetimes, x)y - xf(\leftthreetimes, y)\|_A, \ldots, \|f(\leftthreetimes, xy) \\ &\qquad - f(\leftthreetimes, x)y - xf(\leftthreetimes, y)\|_A\Big]_{n\times n} \\ &\preceq \theta\mathfrak{W}(\|x\|_A^r + \|y\|_A^r),\end{aligned} \tag{10.33}$$

for every $\mu \in \mathbb{T}^1$ and every $x, y \in A$ and $\leftthreetimes \in \pitchfork$. Then, there exists a unique derivation $\delta : \pitchfork \times A \to A$, such that

$$\begin{aligned}&\operatorname{diag}\Big[\|f(\leftthreetimes, x) - \delta(\leftthreetimes, x)\|_A, \ldots, \|f(\leftthreetimes, x) - \delta(\leftthreetimes, x)\|_A\Big]_{n\times n} \\ &\preceq \frac{2^r\theta}{2-2^r}\mathfrak{W}(\|x\|_A^r),\end{aligned} \tag{10.34}$$

for every $x \in A$ and $\leftthreetimes \in \pitchfork$.

Proof The proof follows from Theorem 10.5 by taking

$$\operatorname{diag}\Big[\varphi_1(x, y), \ldots, \varphi_n(x, y)\Big]_{n\times n} := \theta\mathfrak{W}(\|x\|_A^r + \|y\|_A^r),$$

for every $x, y \in A$. Setting $\underbrace{L_i}_{i=1,\ldots,n} = 2^{r-1}$, we come to the conclusion.

Corollary 10.6 *Let $n \in \mathbb{N}$, $r > 2$, $\theta > 0$ and $f : \pitchfork \times A \to A$ be a mapping satisfying the inequalities* (10.32) *and* (10.33)*. Then, there exists a unique derivation $\delta : \pitchfork \times A \to A$, such that*

$$\operatorname{diag}\left[\|f(\leftthreetimes, x)-\delta(\leftthreetimes, x)\|_A, \ldots, \|f(\leftthreetimes, x)-\delta(\leftthreetimes, x)\|_A\right]_{n\times n} \tag{10.35}$$
$$\preceq \frac{\theta}{2^r-2}\mathfrak{W}(\|x\|_A^r),$$

for every $x \in A$ *and* $\leftthreetimes \in \pitchfork$.

Proof The proof is directly concludes from Theorem 10.6 by taking

$$\operatorname{diag}\left[\varphi_1(x, y), \ldots, \varphi_n(x, y)\right]_{n\times n} := \theta\mathfrak{W}(\|x\|_A^r + \|y\|_A^r)$$

for every $x, y \in A$. Choosing $\underbrace{L_i}_{i=1,\ldots,n} = 2^{1-r}$, we get the desired result.

Remark 10.4 For the inequalities controlled by the product of powers of norms, one can obtain similar results to Corollaries 10.3 and 10.4.

Corollary 10.7 *Let* $n \in \mathbb{N}$, $r < 1$, $\theta > 0$ *and* $f : \pitchfork \times A \to B$ *be a mapping satisfying the inequality* (10.22), *such that*

$$\operatorname{diag}\left[\|f(\leftthreetimes, [x, y]) - [f(\leftthreetimes, x), f(\leftthreetimes, y)]\|_B, \ldots, \|f(\leftthreetimes, [x, y]) - [f(\leftthreetimes, x), f(\leftthreetimes, y)]\|_B\right]_{n\times n} \tag{10.36}$$
$$\preceq \theta\mathfrak{W}(\|x\|_A^r + \|y\|_A^r),$$

for every $x, y \in A$ *and* $\leftthreetimes \in \pitchfork$. *Then, there exists a unique Lie* C^**–algebra homomorphism* $H : \pitchfork \times A \to B$ *satisfying the inequality* (10.25).

Proof The proof follows from Theorem 10.9 by taking

$$\operatorname{diag}\left[\varphi_1(x, y), \ldots, \varphi_n(x, y)\right]_{n\times n} := \theta\mathfrak{W}(\|x\|_A^r + \|y\|_A^r),$$

for every $x, y \in A$. Then, setting $\underbrace{L_i}_{i=1,\ldots,n} = 2^{r-1}$, we deduce the desired result.

Corollary 10.8 *Let* $n \in \mathbb{N}$, $r > 2$, $\theta > 0$ *and* $f : \pitchfork \times A \to B$ *be a mapping satisfying the inequalities* (10.22) *and* (10.36). *Then, there exists a unique Lie* C^**–algebra homomorphism* $H : \pitchfork \times A \to B$ *satisfying the inequality* (10.26).

Proof The proof follows from Theorem 10.10 by taking

$$\operatorname{diag}\left[\varphi_1(x, y), \ldots, \varphi_n(x, y)\right]_{n\times n} := \theta\mathfrak{W}(\|x\|_A^r + \|y\|_A^r),$$

for every $x, y \in A$. Letting $\underbrace{L_i}_{i=1,\ldots,n} = 2^{1-r}$, we get the desired result.

Corollary 10.9 *Let $n \in \mathbb{N}$, $r < 1$, $\theta > 0$ and $f : \pitchfork \times A \to A$ be a mapping satisfying the inequality* (10.32), *such taht*

$$\operatorname{diag}\left[\|f(\leftthreetimes, [x, y]) - [f(\leftthreetimes, x), y] - [x, f(\leftthreetimes, y)]\|_A, \ldots, \|f(\leftthreetimes, [x, y]) - [f(\leftthreetimes, x), y] - [x, f(\leftthreetimes, y)]\|_A \right]_{n\times n} \preceq \theta\mathfrak{W}(\|x\|_A^r + \|y\|_A^r), \tag{10.37}$$

for every $x, y \in A$ and $\leftthreetimes \in \pitchfork$. Then, there exists a unique Lie derivation $\delta : \pitchfork \times A \to A$ satisfying the inequality (10.34).

Proof The proof is directly from Theorem 10.9 by taking

$$\operatorname{diag}\left[\varphi_1(x, y), \ldots, \varphi_n(x, y)\right]_{n\times n} := \theta\mathfrak{W}(\|x\|_A^r + \|y\|_A^r),$$

for every $x, y \in A$. Then, setting $\underbrace{L_i}_{i=1,\ldots,n} = 2^{r-1}$, we get the desired result.

Corollary 10.10 *Let $n \in \mathbb{N}$, $r > 2$, $\theta > 0$ and $f : \pitchfork \times A \to A$ be a mapping satisfying the inequalities* (10.32) *and* (10.37). *Then, there exists a unique Lie derivation $\delta : \pitchfork \times A \to A$ satisfying the inequality* (10.35).

Proof It follows from Theorem 10.10 by taking

$$\operatorname{diag}\left[\varphi_1(x, y), \ldots, \varphi_n(x, y)\right]_{n\times n} := \theta\mathfrak{W}(\|x\|_A^r + \|y\|_A^r),$$

for every $x, y \in A$. Then choosing $\underbrace{L_i}_{i=1,\ldots,n} = 2^{1-r}$, we imply the desired result.

We refer the interested readers to [3–8].

References

1. Rezaei Aderyani S, Saadati R, Li C, Rassias TM, Park C (2023) Special functions and multi-stability of the Jensen type random operator equation in C^*-algebras via fixed point. J Inequal Appl 2023(1):1–24
2. Cadariu L, Radu V (2003) Fixed points and the stability of Jensen's functional equation. J Inequal Pure Appl Math 4(1), Art. ID 4 (2003)
3. Aderyani SR, Saadati R, O'Regan D, Alshammari FS (2022) Multi-super-stability of antiderivations in Banach algebras. AIMS Math 7(11):20143–20163

4. Aderyani SR, Saadati R, Feckan M (2021) The Cadariu-Radu method for existence, uniqueness and gauss hypergeometric stability of w-Hilfer fractional differential equations. Mathematics 9:1408
5. Aderyani SR, Saadati R, O'Regan D, Abdeljawad T (2022) UHML stability of a class of Δ-Hilfer FDEs via CRM. AIMS Math 7(4):5910–5919
6. Aderyani SR, Saadati R, O'Regan D (2022) The Cadariu-Radu method for existence, uniqueness and Gauss Hypergeometric stability of a class of Ξ-Hilfer fractional differential equations. Int J Nonlinear Sci Numer Simul 24(8)
7. Aderyani SR, Saadati R, O'Regan D, Alshammari FS (2023) Application of aggregated control functions for approximating $\mathscr{C}$-Hilfer fractional differential equations. AIMS Math 8(11):28010–28032
8. Aderyani SR, Saadati R, Rassias TM, Srivastava HM (2023) Existence, uniqueness and the multi-stability results for a w-Hilfer fractional differential equation. Axioms 12(7):681

Chapter 11
Super Multi Stability of Additive Functional Inequalities in Banach Algebras

In this chapter, we first study the multi stability results for the functional inequality below in complex Banach algebras by the fixed point method. Next, we present the concept of antiderivations in Banach algebras. Finally, through the fixed point theorem, we study super multi stability of antiderivations in Banach algebras, associated with the following inequality [1]:

$$\begin{aligned}
&\operatorname{diag}\Bigg[\Big\| f_1(\leftthreetimes, \varepsilon+\zeta+\varsigma) - f_1(\leftthreetimes, \varepsilon+\varsigma) - f_1(\leftthreetimes, \varepsilon+\zeta-\varsigma) \\
&\qquad + f_1(\leftthreetimes, \varepsilon-\varsigma))\Big\|, \ldots, \\
&\Big\| f_n(\leftthreetimes, \varepsilon+\zeta+\varsigma) - f_n(\leftthreetimes, \varepsilon+\varsigma) - f_n(\leftthreetimes, \varepsilon+\zeta-\varsigma) + f_n(\leftthreetimes, \varepsilon-\varsigma))\Big\|\Bigg]_{n\times n} \qquad (11.1)\\
&\preceq \operatorname{diag}\Bigg[\Big\| \theta_1\Big(f_1(\leftthreetimes, \varepsilon-\zeta+\varsigma) - f_1(\leftthreetimes, \varepsilon+\varsigma) - f_1(\leftthreetimes, \varepsilon-\zeta-\varsigma) \\
&\qquad + f_1(\leftthreetimes, \varepsilon-\varsigma)\Big)\Big\|, \ldots, \\
&\Big\| \theta_n\Big(f_n(\leftthreetimes, \varepsilon-\zeta+\varsigma) - f_n(\leftthreetimes, \varepsilon+\varsigma) - f_n(\leftthreetimes, \varepsilon-\zeta-\varsigma) + f_n(\leftthreetimes, \varepsilon-\varsigma)\Big)\Big\|\Bigg]_{n\times n},
\end{aligned}$$

for every $\varepsilon, \zeta, \varsigma \in \mho_1$, $\leftthreetimes \in \pitchfork$, and every $\theta_1, \ldots, \theta_n \in \mathbb{C}\backslash\{0\}$, with $|\theta_1|, \ldots, |\theta_n| < 1$.

Throughout this section, let $\mho_1$ be a complex Banach algebra and that $\theta_1, \ldots, \theta_n \in \mathbb{C}\backslash\{0\}$ with $|\theta_1|, \ldots, |\theta_n| < 1$. In addition, consider the control functions $\varphi_{j_i} : (\mho_1)^3 \to [0, \infty)$, for every $1 \leq i \leq n$, $1 \leq j_i \leq n$, and $n \in \mathbb{N}$.

Note 11.1 Let $\operatorname{diag}[\varphi_{j_1}, \ldots, \varphi_{j_n}]$ be a diagonal matrix of control functions φ_{j_i}, where $1 \leq i \leq n$, $1 \leq j_i \leq n$, and $n \in \mathbb{N}$. Then, $\varphi_{j_1}(\varphi_{j_n})$ represents an element at the 1th(nth) row and 1th(nth) column of the matrix and φ_{j_i} demonstrates the j_i^{th} given control function.

Note 11.2 The functional inequality (11.1) has the multiple stability results if, for every $n \in \mathbb{N}$, and every $j_i \in \mathbb{N}$, we get $f_1 = f_2 = \ldots = f_{n-1} = f_n$ and $\varphi_{j_1} \neq \varphi_{j_2} \neq \cdots \neq \varphi_{j_{n-1}} \neq \varphi_{j_n}$.

S. Rezaei Aderyani et al., *Towards Ulam Type Multi Stability Analysis*, Studies in Systems, Decision and Control 523, https://doi.org/10.1007/978-3-031-55564-0_11

11.1 Multi Stability of Additive Functional Inequalities in Complex Banach Algebras

In this section, we study the multi stability of the additive $(\theta_1, \ldots, \theta_n)$–functional inequality (11.1) in complex Banach algebras.

Lemma 11.1 *Let $n \in \mathbb{N}$, and $\underbrace{f_i}_{i=1,\ldots,n} : \pitchfork \times \mho_1 \to \mho_1$ be mappings satisfying $\underbrace{f_i}_{i=1,\ldots,n}(\leftthreetimes, 0) = 0$ and the functional inequality* (11.1), *for every $\varepsilon, \zeta, \varsigma \in \mho_1$, and $\leftthreetimes \in \pitchfork$. Then, the mappings $\underbrace{f_i}_{i=1,\ldots,n} : \pitchfork \times \mho_1 \to \mho_1$, are additive (the usual definition is at the end of the proof).*

Proof Let $\underbrace{f_i}_{i=1,\ldots,n} : \pitchfork \times \mho_1 \to \mho_1$ satisfy the functional inequality (11.1). Replacing ζ by $-\zeta$ in the functional inequality (11.1), we get

$$\begin{aligned}
&\operatorname{diag}\Bigg[\Big\| f_1(\leftthreetimes, \varepsilon - \zeta + \varsigma) - f_1(\leftthreetimes, \varepsilon + \varsigma) - f_1(\leftthreetimes, \varepsilon - \zeta - \varsigma) + f_1(\leftthreetimes, \varepsilon - \varsigma)\Big\|, \\
&\ldots, \Big\| f_n(\leftthreetimes, \varepsilon - \zeta + \varsigma) - f_n(\leftthreetimes, \varepsilon + \varsigma) - f_n(\leftthreetimes, \varepsilon - \zeta - \varsigma) + f_n(\leftthreetimes, \varepsilon - \varsigma)\Big\|\Bigg] \\
&\preceq \operatorname{diag}\Bigg[\Big\|\theta_1\Big(f_1(\leftthreetimes, \varepsilon + \zeta + \varsigma) - f_1(\leftthreetimes, \varepsilon + \varsigma) - f_1(\leftthreetimes, \varepsilon + \zeta - \varsigma) \\
&\qquad + f_1(\leftthreetimes, \varepsilon - \varsigma)\Big)\Big\|, \ldots, \\
&\Big\|\theta_n\Big(f_n(\leftthreetimes, \varepsilon + \zeta + \varsigma) - f_n(\leftthreetimes, \varepsilon + \varsigma) - f_n(\leftthreetimes, \varepsilon + \zeta - \varsigma) + f_n(\leftthreetimes, \varepsilon - \varsigma)\Big)\Big\|\Bigg]
\end{aligned} \tag{11.2}$$

for every $\varepsilon, \zeta, \varsigma \in \mho_1$, and $\leftthreetimes \in \pitchfork$. According to the inequalities (11.1) and (11.2), we have

$$\begin{aligned}
&\operatorname{diag}\Bigg[\| f_1(\leftthreetimes, \varepsilon + \zeta + \varsigma) - f_1(\leftthreetimes, \varepsilon + \varsigma) - f_1(\leftthreetimes, \varepsilon + \zeta - \varsigma) + f_1(\leftthreetimes, \varepsilon - \varsigma)\Big\|, \\
&\ldots, \Big\| f_n(\leftthreetimes, \varepsilon + \zeta + \varsigma) - f_n(\leftthreetimes, \varepsilon + \varsigma) - f_n(\leftthreetimes, \varepsilon + \zeta - \varsigma) + f_n(\leftthreetimes, \varepsilon - \varsigma)\Big\|\Bigg] \\
&\preceq \operatorname{diag}\Bigg[\Big\|\theta_1^2\Big(f_1(\leftthreetimes, \varepsilon + \zeta + \varsigma) - f_1(\leftthreetimes, \varepsilon + \varsigma) - f_1(\leftthreetimes, \varepsilon + \zeta - \varsigma) \\
&\qquad + f_1(\leftthreetimes, \varepsilon - \varsigma)\Big)\Big\|, \ldots, \\
&\Big\|\theta_n^2\Big(f_n(\leftthreetimes, \varepsilon + \zeta + \varsigma) - f_n(\leftthreetimes, \varepsilon + \varsigma) - f_n(\leftthreetimes, \varepsilon + \zeta - \varsigma) + f_n(\leftthreetimes, \varepsilon - \varsigma)\Big)\Big\|\Bigg]
\end{aligned}$$

and so

$$f_i(\curlywedge, \varepsilon + \zeta + \varsigma) - f_i(\curlywedge, \varepsilon + \varsigma) - f_i(\curlywedge, \varepsilon + \zeta - \varsigma) \\ + f_i(\curlywedge, \varepsilon - \varsigma) = 0, \quad i = 1, \ldots, n \tag{11.3}$$

for every $\varepsilon, \zeta, \varsigma \in \mho_1$, and $\curlywedge \in \pitchfork$, since $\underbrace{|\theta_i|}_{i=1,\ldots,n} < 1$.

Letting $\varsigma = \varepsilon$ in the equality (11.3),

$$f_i(\curlywedge, 2\varepsilon + \zeta) - f_i(\curlywedge, 2\varepsilon) - f_i(\curlywedge, \zeta) = 0, \quad i = 1, \ldots, n$$

for every $\varepsilon, \zeta \in \mho_1$, and $\curlywedge \in \pitchfork$. Thus, $\underbrace{f_i}_{i=1,\ldots,n}$ are additive.

Theorem 11.1 *Let* $n \in \mathbb{N}$, *and* $(\underbrace{\varphi_{j_1}}_{1\le j_1\le n}, \ldots, \underbrace{\varphi_{j_n}}_{1\le j_n\le n}) : (\mho_1 \times \mho_1 \times \mho_1)^n \to [0, \infty)^n$, *be functions, such that there exist* $(\mathcal{T}_1, \ldots, \mathcal{T}_n) < \underbrace{(1, \ldots, 1)}_{n}$ *with*

$$\begin{aligned} &\operatorname{diag}\left[\underbrace{\varphi_{j_1}}_{1\le j_1\le n}\left(\frac{\varepsilon}{2}, \frac{\zeta}{2}, \frac{\varsigma}{2}\right), \ldots, \underbrace{\varphi_{j_n}}_{1\le j_n\le n}\left(\frac{\varepsilon}{2}, \frac{\zeta}{2}, \frac{\varsigma}{2}\right)\right] \\ &\preceq \operatorname{diag}\left[\frac{\mathcal{T}_1}{2}\underbrace{\varphi_{j_1}}_{1\le j_1\le n}(\varepsilon, \zeta, \varsigma), \ldots, \frac{\mathcal{T}_n}{2}\underbrace{\varphi_{j_n}}_{1\le j_n\le n}(\varepsilon, \zeta, \varsigma)\right], \end{aligned} \tag{11.4}$$

for every $\varepsilon, \zeta, \varsigma \in \mho_1$. *Let* $\underbrace{f_i}_{1\le i\le n} : \pitchfork \times \mho_1 \to \mho_1$ *be mappings satisfying* $\underbrace{f_i}_{1\le i\le n}(\curlywedge, 0) =$ 0 *and*

$$\begin{aligned} &\operatorname{diag}\Bigg[\Big\| f_1(\curlywedge, \varepsilon + \zeta + \varsigma) - f_1(\curlywedge, \varepsilon + \varsigma) - f_1(\curlywedge, \varepsilon + \zeta - \varsigma) + f_1(\curlywedge, \varepsilon - \varsigma)\Big\| \\ &\qquad + |\theta_1| \underbrace{\varphi_{j_1}}_{1\le j_1\le n}(\varepsilon, -\zeta, \varsigma), \ldots, \\ &\quad \Big\| f_n(\curlywedge, \varepsilon + \zeta + \varsigma) - f_n(\curlywedge, \varepsilon + \varsigma) - f_n(\curlywedge, \varepsilon + \zeta - \varsigma) + f_n(\curlywedge, \varepsilon - \varsigma)\Big\| \\ &\qquad + |\theta_n| \underbrace{\varphi_{j_n}}_{1\le j_n\le n}(\varepsilon, -\zeta, \varsigma)\Bigg] \\ &\preceq \operatorname{diag}\Bigg[\Big\|\theta_1\Big(f_1(\curlywedge, \varepsilon - \zeta + \varsigma) - f_1(\curlywedge, \varepsilon + \varsigma) - f_1(\curlywedge, \varepsilon - \zeta - \varsigma) \\ &\qquad + f_1(\curlywedge, \varepsilon - \varsigma)\Big)\Big\| + \underbrace{\varphi_{j_1}}_{1\le j_1\le n}(\varepsilon, \zeta, \varsigma), \ldots, \end{aligned}$$

$$\left\|\theta_n\left(f_n(\curlywedge, \varepsilon-\zeta+\varsigma)-f_n(\curlywedge, \varepsilon+\varsigma)-f_n(\curlywedge, \varepsilon-\zeta-\varsigma)+f_n(\curlywedge, \varepsilon-\varsigma)\right)\right\|$$

$$+\underbrace{\varphi_{j_n}}_{1\leq j_n\leq n}(\varepsilon, \zeta, \varsigma)\Bigg], \tag{11.5}$$

for each $\varepsilon, \zeta, \varsigma \in \mho_1$ *and* $\curlywedge \in ⋔$. *Then, there exist unique additive mappings* $\underbrace{f_i'}_{1\leq i\leq n} : ⋔ \times\mho_1 \to \mho_1$, *such that*

$$\operatorname{diag}\left[\left\|f_1(\curlywedge, \varepsilon)-f_1'(\curlywedge, \varepsilon)\right\|, \ldots, \left\|f_n(\curlywedge, \varepsilon)-f_n'(\curlywedge, \varepsilon)\right\|\right]_{n\times n}$$

$$\preceq \operatorname{diag}\left[\frac{\mathcal{T}_1}{2(1-\mathcal{T}_1)}\underbrace{\varphi_{j_1}}_{1\leq j_1\leq n}\left(\frac{\varepsilon}{2}, \varepsilon, \frac{\varepsilon}{2}\right), \ldots, \frac{\mathcal{T}_n}{2(1-\mathcal{T}_n)}\underbrace{\varphi_{j_n}}_{1\leq j_n\leq n}\left(\frac{\varepsilon}{2}, \varepsilon, \frac{\varepsilon}{2}\right)\right]_{n\times n}, \tag{11.6}$$

for every $\varepsilon \in \mho_1$, *and* $\curlywedge \in ⋔$.

Proof Replacing ζ by $-\zeta$ in (11.5), we get

$$\operatorname{diag}\left[\left\|f_1(\curlywedge, \varepsilon-\zeta+\varsigma)-f_1(\curlywedge, \varepsilon+\varsigma)-f_1(\curlywedge, \varepsilon-\zeta-\varsigma)+f_1(\curlywedge, \varepsilon-\varsigma)\right\|\right.$$

$$+|\theta_1|\underbrace{\varphi_{j_1}}_{1\leq j_1\leq n}(\varepsilon, \zeta, \varsigma), \ldots,$$

$$\left\|f_n(\curlywedge, \varepsilon-\zeta+\varsigma)-f_n(\curlywedge, \varepsilon+\varsigma)-f_n(\curlywedge, \varepsilon-\zeta-\varsigma)+f_n(\curlywedge, \varepsilon-\varsigma)\right\|$$

$$\left.+|\theta_n|\underbrace{\varphi_{j_n}}_{1\leq j_n\leq n}(\varepsilon, \zeta, \varsigma)\right]$$

$$\preceq \operatorname{diag}\left[\left\|\theta_1(f_1(\curlywedge, \varepsilon+\zeta+\varsigma)-f_1(\curlywedge, \varepsilon+\varsigma)-f_1(\curlywedge, \varepsilon+\zeta-\varsigma)\right.\right.$$

$$\left.+f_1(\curlywedge, \varepsilon-\varsigma))\right\|+\underbrace{\varphi_{j_1}}_{1\leq j_1\leq n}(\varepsilon, -\zeta, \varsigma), \ldots,$$

$$\left\|\theta_n(f_n(\curlywedge, \varepsilon+\zeta+\varsigma)-f_n(\curlywedge, \varepsilon+\varsigma)-f_n(\curlywedge, \varepsilon+\zeta-\varsigma)+f_n(\curlywedge, \varepsilon-\varsigma))\right\|$$

$$\left.+\underbrace{\varphi_{j_n}}_{1\leq j_n\leq n}(\varepsilon, -\zeta, \varsigma)\right], \tag{11.7}$$

for every $\varepsilon, \zeta, \varsigma \in \mho_1$, and $\leftthreetimes \in \pitchfork$. According to the inequalities (11.5) and (11.7), we have

$$\begin{aligned}&\operatorname{diag}\Bigg[\Big\|f_1(\leftthreetimes, \varepsilon+\zeta+\varsigma) - f_1(\leftthreetimes, \varepsilon+\varsigma) - f_1(\leftthreetimes, \varepsilon+\zeta-\varsigma) + f_1(\leftthreetimes, \varepsilon-\varsigma)\Big\|, \ldots,\\ &\quad \Big\|f_n(\leftthreetimes, \varepsilon+\zeta+\varsigma) - f_n(\leftthreetimes, \varepsilon+\varsigma) - f_n(\leftthreetimes, \varepsilon+\zeta-\varsigma) + f_n(\leftthreetimes, \varepsilon-\varsigma)\Big\|\Bigg]_{n\times n}\\ &\preceq \operatorname{diag}\Bigg[\underbrace{\varphi_{j_1}}_{1\le j_1\le n}(\varepsilon,\zeta,\varsigma), \ldots, \underbrace{\varphi_{j_n}}_{1\le j_n\le n}(\varepsilon,\zeta,\varsigma)\Bigg]_{n\times n},\end{aligned} \tag{11.8}$$

for every $\varepsilon, \zeta, \varsigma \in \mho_1$, and $\leftthreetimes \in \pitchfork$.

Letting $\varepsilon = \varsigma = \frac{\sigma}{2}$ and $\zeta = \sigma$ in the inequality (11.8), we get

$$\begin{aligned}&\operatorname{diag}\Bigg[\Big\|f_1(\leftthreetimes, 2\sigma) - 2f_1(\leftthreetimes, \sigma)\Big\|, \ldots, \Big\|f_n(\leftthreetimes, 2\sigma) - 2f_n(\leftthreetimes, \sigma)\Big\|\Bigg]_{n\times n}\\ &\preceq \operatorname{diag}\Bigg[\underbrace{\varphi_{j_1}}_{1\le j_1\le n}\left(\frac{\sigma}{2},\sigma,\frac{\sigma}{2}\right), \ldots, \underbrace{\varphi_{j_n}}_{1\le j_n\le n}\left(\frac{\sigma}{2},\sigma,\frac{\sigma}{2}\right)\Bigg]_{n\times n},\end{aligned} \tag{11.9}$$

for every $\sigma \in \mho_1$, and $\leftthreetimes \in \pitchfork$.

Let $\hbar = (\hbar_1, \ldots, \hbar_n)$ and $\hbar' = (\hbar'_1, \ldots, \hbar'_n)$. Now, consider the set

$$\nabla := \{\hbar : (\pitchfork \times \mho_1)^n \to \mho_1^n : \hbar(\leftthreetimes, 0) = \overbrace{(0, \ldots, 0)}^{n}\},$$

and define the generalized metric d on ∇ by

$$\begin{aligned}d(\hbar, \hbar') = \inf\Bigg\{&(\mu_1, \ldots, \mu_n) \in \mathbb{R}^n_+ :\\ &\operatorname{diag}\Big[\|\hbar_1(\leftthreetimes, \varepsilon) - \hbar'_1(\leftthreetimes, \varepsilon)\|, \ldots \|\hbar_n(\leftthreetimes, \varepsilon) - \hbar'_n(\leftthreetimes, \varepsilon)\|\Big]\\ &\preceq \operatorname{diag}\Bigg[\mu_1 \underbrace{\varphi_{j_1}}_{1\le j_1\le n}\left(\frac{\varepsilon}{2},\varepsilon,\frac{\varepsilon}{2}\right), \ldots, \mu_n \underbrace{\varphi_{j_n}}_{1\le j_n\le n}\left(\frac{\varepsilon}{2},\varepsilon,\frac{\varepsilon}{2}\right)\Bigg], \forall \varepsilon \in \mho_1, \leftthreetimes \in \pitchfork\Bigg\},\end{aligned}$$

where $\inf \emptyset = \underbrace{(+\infty, \ldots, +\infty)}_{n}$.

We can easily prove that (∇, d) is complete [1].

Let $\mathscr{L} := (\mathscr{L}_1, \ldots, \mathscr{L}_n)$. Now, we consider the linear mapping $\mathscr{L} : \nabla \to \nabla$, such that

$$\mathscr{L}_i \hbar_i(\leftthreetimes, \varepsilon) := 2\hbar_i\left(\leftthreetimes, \frac{\varepsilon}{2}\right), \quad i = 1, \ldots, n,$$

for every $\varepsilon \in \mho_1$, and $\leftthreetimes \in \pitchfork$.

Let $\hbar, \hbar' \in \nabla$ be given, such that $d(\hbar, \hbar') = (\varepsilon_1, \ldots, \varepsilon_n)$. Then, we have

$$\begin{aligned}
&\mathrm{diag}\Big[\ \|\hbar_1(\curlywedge, \varepsilon) - \hbar'_1(\curlywedge, \varepsilon)\|, \ldots, \|\hbar_n(\curlywedge, \varepsilon) - \hbar'_n(\curlywedge, \varepsilon)\|\ \Big] \\
&\preceq \mathrm{diag}\Big[\varepsilon_1 \underbrace{\varphi_{j_1}}_{1\le j_1\le n}\left(\tfrac{\varepsilon}{2}, \varepsilon, \tfrac{\varepsilon}{2}\right), \ldots, \varepsilon_n \underbrace{\varphi_{j_n}}_{1\le j_n\le n}\left(\tfrac{\varepsilon}{2}, \varepsilon, \tfrac{\varepsilon}{2}\right)\Big],
\end{aligned}$$

for every $\varepsilon \in \mho_1$, and $\curlywedge \in \pitchfork$. Hence, we get

$$\begin{aligned}
&\mathrm{diag}\Big[\ \|\mathscr{L}_1\hbar_1(\curlywedge, \varepsilon) - \mathscr{L}_1\hbar'_1(\curlywedge, \varepsilon)\|, \ldots, \|\mathscr{L}_n\hbar_n(\curlywedge, \varepsilon) - \mathscr{L}_n\hbar'_n(\curlywedge, \varepsilon)\|\ \Big] \\
&= \mathrm{diag}\Big[\ \left\|2\hbar_1\left(\curlywedge, \frac{\varepsilon}{2}\right) - 2\hbar'_1\left(\curlywedge, \frac{\varepsilon}{2}\right)\right\|, \ldots, \left\|2\hbar_n\left(\curlywedge, \frac{\varepsilon}{2}\right) - 2\hbar'_n\left(\curlywedge, \frac{\varepsilon}{2}\right)\right\|\ \Big] \\
&\preceq \mathrm{diag}\Big[2\varepsilon_1 \underbrace{\varphi_{j_1}}_{1\le j_1\le n}\left(\frac{\varepsilon}{4}, \frac{\varepsilon}{2}, \frac{\varepsilon}{4}\right), \ldots, 2\varepsilon_n \underbrace{\varphi_{j_n}}_{1\le j_n\le n}\left(\frac{\varepsilon}{4}, \frac{\varepsilon}{2}, \frac{\varepsilon}{4}\right)\Big] \\
&\preceq \mathrm{diag}\Big[\mathcal{T}_1\varepsilon_1 \underbrace{\varphi_{j_1}}_{1\le j_1\le n}\left(\frac{\varepsilon}{2}, \varepsilon, \frac{\varepsilon}{2}\right), \ldots, \mathcal{T}_n\varepsilon_n \underbrace{\varphi_{j_n}}_{1\le j_n\le n}\left(\frac{\varepsilon}{2}, \varepsilon, \frac{\varepsilon}{2}\right)\Big],
\end{aligned}$$

for every $\varepsilon \in \mho_1$, and $\curlywedge \in \pitchfork$. Thus, $d(\hbar, \hbar') = \underbrace{(\varepsilon_1, \ldots, \varepsilon_n)}_{n}$, implies that

$$d(\mathscr{L}\hbar(\curlywedge, \varepsilon), \mathscr{L}\hbar'(\curlywedge, \varepsilon)) \preceq (\mathcal{T}_1\varepsilon_1, \ldots, \mathcal{T}_n\varepsilon_n).$$

Hence, we have

$$d(\mathscr{L}\hbar(\varepsilon), \mathscr{L}\hbar'(\varepsilon)) \preceq (\mathcal{T}_1, \ldots, \mathcal{T}_n)d(\hbar, \hbar'),$$

for every $\hbar, \hbar' \in \nabla$. According to the inequality (11.9), we get

$$\begin{aligned}
&\mathrm{diag}\Big[\ \left\|f_1(\curlywedge, \varepsilon) - 2f_1(\curlywedge, \frac{\varepsilon}{2})\right\|, \ldots, \left\|f_n(\curlywedge, \varepsilon) - 2f_n(\curlywedge, \frac{\varepsilon}{2})\right\|\ \Big]_{n\times n} \\
&\preceq \mathrm{diag}\Big[\ \underbrace{\varphi_{j_1}}_{1\le j_1\le n}\left(\frac{\varepsilon}{4}, \frac{\varepsilon}{2}, \frac{\varepsilon}{4}\right), \ldots, \underbrace{\varphi_{j_n}}_{1\le j_n\le n}\left(\frac{\varepsilon}{4}, \frac{\varepsilon}{2}, \frac{\varepsilon}{4}\right)\Big]_{n\times n} \\
&\preceq \mathrm{diag}\Big[\ \frac{\mathcal{T}_1}{2} \underbrace{\varphi_{j_1}}_{1\le j_1\le n}\left(\frac{\varepsilon}{2}, \varepsilon, \frac{\varepsilon}{2}\right), \ldots, \frac{\mathcal{T}_n}{2} \underbrace{\varphi_{j_n}}_{1\le j_n\le n}\left(\frac{\varepsilon}{2}, \varepsilon, \frac{\varepsilon}{2}\right)\Big]_{n\times n},
\end{aligned}$$

for every $\varepsilon \in \mho_1$ and $\curlywedge \in \pitchfork$. So $d(f, \mathscr{L}f) \preceq (\frac{\mathcal{T}_1}{2}, \ldots, \frac{\mathcal{T}_n}{2})$, for every $f := (f_1, \ldots, f_n)$.

According to Theorem 9.5, there exist mappings $\underbrace{f_i'}_{1\leq i\leq n} : \mho_1 \to \mho_1 (i = 1, \ldots, n)$ satisfying the following:

(1) The maps $\underbrace{f_i'}_{1\leq i\leq n}$ are fixed points of $\underbrace{\mathscr{L}_i}_{1\leq i\leq n}$, i.e.

$$\underbrace{f_i'}_{1\leq i\leq n}(\curlywedge, \varepsilon) = 2 \underbrace{f_i'}_{1\leq i\leq n}\left(\curlywedge, \frac{\varepsilon}{2}\right), \tag{11.10}$$

for every $\varepsilon \in \mho_1$, and $\curlywedge \in \pitchfork$. The mapping $f' := (f_1', \ldots, f_n')$ is a unique fixed point of $\mathscr{L} := (\mathscr{L}_1, \ldots, \mathscr{L}_n)$ in the set $\Bbbk = \{(\hbar_1, \cdots, \hbar_n) \in \nabla : d\Big((f_1, \cdots, f_n), (\hbar_1, \cdots, \hbar_n)\Big) < \underbrace{(\infty, \cdots, \infty)}_{n}\}$. This implies that f' is a unique mapping satisfying the equality (11.10), such that, there exist $\mu_1, \ldots, \mu_n \in (0, \infty)$ satisfying the inequality below

$$\begin{aligned}&\operatorname{diag}\Big[\|f_1(\curlywedge, \varepsilon) - f_1'(\curlywedge, \varepsilon)\|, \ldots, \|f_n(\curlywedge, \varepsilon) - f_n'(\curlywedge, \varepsilon)\|\Big]\\ &\preceq \operatorname{diag}\Big[\mu_1 \underbrace{\varphi_{j_1}}_{1\leq j_1\leq n}\left(\frac{\varepsilon}{2}, \varepsilon, \frac{\varepsilon}{2}\right), \ldots, \mu_n \underbrace{\varphi_{j_n}}_{1\leq j_n\leq n}\left(\frac{\varepsilon}{2}, \varepsilon, \frac{\varepsilon}{2}\right)\Big],\end{aligned}$$

for every $\varepsilon \in \mho_1$, and $\curlywedge \in \pitchfork$.

(2) Since $\lim_{m\to\infty} d\Big((\mathscr{L}_1^m f_1, \ldots, \mathscr{L}_n^m f_n), (f_1', \ldots, f_n')\Big) = \underbrace{(0, \ldots, 0)}_{n}$, therefore,

$$\lim_{m\to\infty} 2^m f_i\left(\curlywedge, \frac{\varepsilon}{2^m}\right) = f_i'(\curlywedge, \varepsilon), \ \ \forall i = 1, \ldots, n, \tag{11.11}$$

for every $\varepsilon \in \mho_1$, and $\curlywedge \in \pitchfork$.

(3) $d\Big((f_1, \ldots, f_n), (f_1', \ldots, f_n')\Big) \preceq \underbrace{(\frac{1}{1-\mathcal{T}_1}, \ldots, \frac{1}{1-\mathcal{T}_n})}_{n} d\Big((f_1, \ldots, f_n), (\mathscr{L}_1 f_1, \ldots, \mathscr{L}_n f_n)\Big)$, which implies that

$$\begin{aligned}&\operatorname{diag}\Big[\Big\|f_1(\curlywedge, \varepsilon) - f_1'(\curlywedge, \varepsilon)\Big\|, \ldots, \Big\|f_n(\curlywedge, \varepsilon) - f_n'(\curlywedge, \varepsilon)\Big\|\Big]_{n\times n}\\ &\preceq \operatorname{diag}\left[\frac{\mathcal{T}_1}{2(1-\mathcal{T}_1)} \underbrace{\varphi_{j_1}}_{1\leq j_1\leq n}\left(\frac{\varepsilon}{2}, \varepsilon, \frac{\varepsilon}{2}\right), \ldots, \frac{\mathcal{T}_n}{2(1-\mathcal{T}_n)} \underbrace{\varphi_{j_n}}_{1\leq j_n\leq n}\left(\frac{\varepsilon}{2}, \varepsilon, \frac{\varepsilon}{2}\right)\right]_{n\times n},\end{aligned}$$

for every $\varepsilon \in \mho_1$, and $\lambda \in \pitchfork$. According to the inequalities (11.4) and (11.5), we have

$$
\begin{aligned}
&\operatorname{diag}\Big[\Big\| f_1'(\lambda, \varepsilon+\zeta+\varsigma) - f_1'(\lambda, \varepsilon+\varsigma) - f_1'(\lambda, \varepsilon+\zeta-\varsigma) \\
&\quad + f_1'(\lambda, \varepsilon-\varsigma)\Big\|, \ldots, \Big\| f_n'(\lambda, \varepsilon+\zeta+\varsigma) - f_n'(\lambda, \varepsilon+\varsigma) \\
&\quad - f_n'(\lambda, \varepsilon+\zeta-\varsigma) + f_n'(\lambda, \varepsilon-\varsigma)\Big\|\Big] \\
&= \operatorname{diag}\Big[\lim_{m\to\infty} 2^m \Big\| f_1\left(\lambda, \frac{\varepsilon+\zeta+\varsigma}{2^m}\right) - f_1\left(\lambda, \frac{\varepsilon+\varsigma}{2^m}\right) \\
&\quad - f_1\left(\lambda, \frac{\varepsilon+\zeta-\varsigma}{2^m}\right) + f_1\left(\lambda, \frac{\varepsilon-\varsigma}{2^m}\right)\Big\| \quad, \ldots, \\
&\lim_{m\to\infty} 2^m \Big\| f_n\left(\lambda, \frac{\varepsilon+\zeta+\varsigma}{2^m}\right) - f_n\left(\lambda, \frac{\varepsilon+\varsigma}{2^m}\right) \\
&\quad - f_n\left(\lambda, \frac{\varepsilon+\zeta-\varsigma}{2^m}\right) + f_n\left(\lambda, \frac{\varepsilon-\varsigma}{2^m}\right)\Big\|\Big] \\
&\preceq \operatorname{diag}\Big[\lim_{m\to\infty} 2^m |\theta_1| \Big\| f_1\left(\lambda, \frac{\varepsilon-\zeta+\varsigma}{2^m}\right) - f_1\left(\lambda, \frac{\varepsilon+\varsigma}{2^m}\right) \\
&\quad - f_1\left(\lambda, \frac{\varepsilon-\zeta-\varsigma}{2^m}\right) + f_1\left(\lambda, \frac{\varepsilon-\varsigma}{2^m}\right)\Big\| \\
&\quad + \lim_{m\to\infty} 2^m \left(\underbrace{\varphi_{j_1}}_{1\le j_1\le n}\left(\frac{\varepsilon}{2^m}, \frac{\zeta}{2^m}, \frac{\varsigma}{2^m}\right) - \theta_1 \underbrace{\varphi_{j_1}}_{1\le j_1\le n}\left(\frac{\varepsilon}{2^m}, \frac{-\zeta}{2^m}, \frac{\varsigma}{2^m}\right)\right) \quad, \ldots, \\
&\lim_{m\to\infty} 2^m |\theta_n| \Big\| f_n\left(\lambda, \frac{\varepsilon-\zeta+\varsigma}{2^m}\right) - f_n\left(\lambda, \frac{\varepsilon+\varsigma}{2^m}\right) \\
&\quad - f_n\left(\lambda, \frac{\varepsilon-\zeta-\varsigma}{2^m}\right) + f_n\left(\lambda, \frac{\varepsilon-\varsigma}{2^m}\right)\Big\| \\
&\quad + \lim_{m\to\infty} 2^m \left(\underbrace{\varphi_{j_n}}_{1\le j_n\le n}\left(\frac{\varepsilon}{2^m}, \frac{\zeta}{2^m}, \frac{\varsigma}{2^m}\right) - \theta_n \underbrace{\varphi_{j_n}}_{1\le j_n\le n}\left(\frac{\varepsilon}{2^m}, \frac{-\zeta}{2^m}, \frac{\varsigma}{2^m}\right)\right)\Big] \\
&\preceq \operatorname{diag}\Big[\Big\| \theta_n (f_1'(\lambda, \varepsilon-\zeta+\varsigma) - f_1'(\lambda, \varepsilon+\varsigma) - f_1'(\lambda, \varepsilon-\zeta-\varsigma) \\
&\quad + f_1'(\lambda, \varepsilon-\varsigma))\Big\|, \ldots, \Big\| \theta_n (f_n'(\lambda, \varepsilon-\zeta+\varsigma) - f_n'(\lambda, \varepsilon+\varsigma) \\
&\quad - f_n'(\lambda, \varepsilon-\zeta-\varsigma) + f_n'(\lambda, \varepsilon-\varsigma))\Big\|\Big],
\end{aligned}
$$

for every $\varepsilon, \zeta, \varsigma \in \mho_1$, and $\leftthreetimes \in \pitchfork$. According to Lemma 11.1, the mappings $\underbrace{f_i'}_{i=1,\dots,n}$, are additive.

11.2 Super Multi Stability of Antiderivations in Banach Algebras

In this part, we first present the concept of antiderivations in Banach algebras and then investigate super multi stability of antiderivations associated to the functional inequality (11.1) in Banach algebras.

Definition 11.1 ([1]) Let $\mho_1$ be a complex Banach algebra. A $\mathbb{C}$-linear operator $\mathscr{G} : \pitchfork \times \mho_1 \to \mho_1$ is called an antiderivation if it satisfies

$$\mathscr{G}(\leftthreetimes, \varepsilon)\mathscr{G}(\leftthreetimes, \zeta) = \mathscr{G}(\leftthreetimes, \mathscr{G}(\leftthreetimes, \varepsilon)\zeta) + \mathscr{G}(\leftthreetimes, \varepsilon\mathscr{G}(\leftthreetimes, \zeta)),$$

for every $\varepsilon, \zeta \in \mho_1$ and $\leftthreetimes \in \pitchfork$.

Lemma 11.2 *[1, 2] Let $\mho_1$ be a complex Banach algebra and $f : \pitchfork \times \mho_1 \to \mho_1$ be an additive mapping, such that $f(\leftthreetimes, \mathfrak{J}\varepsilon) = \mathfrak{J} f(\leftthreetimes, \varepsilon)$, for every $\mathfrak{J} \in \mathbb{T}^1 := \{\eta \in \mathbb{C} : |\eta| = 1\}$, and every $\varepsilon \in \mho_1$. Then, f is $\mathbb{C}$–linear.*

Theorem 11.2 *Consider the control functions $\underbrace{\varphi_{j_1}}_{1\le j_1\le n}, \dots, \underbrace{\varphi_{j_n}}_{1\le j_n\le n} : \mho_1^3 \to [0, \infty)$, with $n \in \mathbb{N}$. Then, we have the following results:*

(i) If there exist $(\mathcal{T}_1, \dots, \mathcal{T}_n) < \underbrace{(1, \dots, 1)}_{n}$ satisfying the inequality below

$$\begin{aligned}
&\operatorname{diag}\left[\underbrace{\varphi_{j_1}}_{1\le j_1\le n}\left(\frac{\varepsilon}{\mathfrak{J}}, \frac{\zeta}{\mathfrak{J}}, \frac{\varsigma}{\mathfrak{J}}\right), \dots, \underbrace{\varphi_{j_n}}_{1\le j_n\le n}\left(\frac{\varepsilon}{\mathfrak{J}}, \frac{\zeta}{\mathfrak{J}}, \frac{\varsigma}{\mathfrak{J}}\right)\right] \\
&\preceq \operatorname{diag}\left[\frac{\mathcal{T}_1}{2}\underbrace{\varphi_{j_1}}_{1\le j_1\le n}(2\varepsilon, 2\zeta, 2\varsigma), \dots, \frac{\mathcal{T}_n}{2}\underbrace{\varphi_{j_n}}_{1\le j_n\le n}(2\varepsilon, 2\zeta, 2\varsigma)\right], \qquad (11.12)
\end{aligned}$$

and if $\underbrace{f_i}_{1\le i\le n} : \pitchfork \times \mho_1 \to \mho_1$, be mappings satisfying $\underbrace{f_i}_{1\le i\le n}(\leftthreetimes, 0) = 0$ and the following inequality,

$$\begin{aligned}
&\operatorname{diag}\Bigg[\Big\|\mathfrak{J} f_1(\leftthreetimes, \varepsilon + \zeta + \varsigma) - f_1(\leftthreetimes, \mathfrak{J}(\varepsilon + \varsigma)) - f_1(\leftthreetimes, \mathfrak{J}(\varepsilon + \zeta - \varsigma)) \\
&+ \mathfrak{J} f_1(\leftthreetimes, \varepsilon - \varsigma)\Big\| + |\theta_1|\underbrace{\varphi_{j_1}}_{1\le j_1\le n}(\varepsilon, -\zeta, \varsigma), \dots,
\end{aligned}$$

$$\Big\| \mathfrak{J} f_n(\leftthreetimes, \varepsilon+\zeta+\varsigma) - f_n(\leftthreetimes, \mathfrak{J}(\varepsilon+\varsigma)) - f_n(\leftthreetimes, \mathfrak{J}(\varepsilon+\zeta-\varsigma))$$

$$+ \mathfrak{J} f_n(\leftthreetimes, \varepsilon-\varsigma)\Big\| + |\theta_n| \underbrace{\varphi_{j_n}}_{1\le j_n \le n}(\varepsilon, -\zeta, \varsigma)\Big]$$

$$\preceq \operatorname{diag}\Big[\Big\|\theta_1(\mathfrak{J} f_1(\leftthreetimes, \varepsilon-\zeta+\varsigma) - f_1(\leftthreetimes, \mathfrak{J}(\varepsilon+\varsigma)) - f_1(\leftthreetimes, \mathfrak{J}(\varepsilon-\zeta-\varsigma))$$

$$+ \mathfrak{J} f_1(\leftthreetimes, \varepsilon-\varsigma))\Big\| + \underbrace{\varphi_{j_1}}_{1\le j_1 \le n}(\varepsilon, \zeta, \varsigma), \ldots,$$

$$\Big\|\theta_n(\mathfrak{J} f_n(\leftthreetimes, \varepsilon-\zeta+\varsigma) - f_n(\leftthreetimes, \mathfrak{J}(\varepsilon+\varsigma)) - f_n(\leftthreetimes, \mathfrak{J}(\varepsilon-\zeta-\varsigma))$$

$$+ \mathfrak{J} f_n(\leftthreetimes, \varepsilon-\varsigma))\Big\| + \underbrace{\varphi_{j_n}}_{1\le j_n \le n}(\varepsilon, \zeta, \varsigma)\Big], \tag{11.13}$$

for every $\mathfrak{J} \in \mathbb{T}^1$ *and every* $\varepsilon, \zeta, \varsigma \in \mho_1$, *and* $\leftthreetimes \in \pitchfork$. *Then, there exist unique* $\mathbb{C}$*-linear mappings* $\underbrace{\mathscr{G}_i}_{1\le i\le n} : \pitchfork \times \mho_1 \to \mho_1$, *such that*

$$\operatorname{diag}\Big[\Big\| f_1(\leftthreetimes, \varepsilon) - \mathscr{G}_1(\leftthreetimes, \varepsilon)\Big\|, \ldots, \Big\| f_n(\leftthreetimes, \varepsilon) - \mathscr{G}_n(\leftthreetimes, \varepsilon)\Big\|\Big]$$

$$\preceq \operatorname{diag}\Big[\frac{\mathcal{T}_1}{2(1-\mathcal{T}_1)} \underbrace{\varphi_{j_1}}_{1\le j_1 \le n}\Big(\frac{\varepsilon}{2}, \varepsilon, -\frac{\varepsilon}{2}\Big), \ldots, \frac{\mathcal{T}_n}{2(1-\mathcal{T}_n)} \underbrace{\varphi_{j_n}}_{1\le j_n \le n}\Big(\frac{\varepsilon}{2}, \varepsilon, -\frac{\varepsilon}{2}\Big)\Big], \tag{11.14}$$

for every $\varepsilon \in \mho_1$, *and* $\leftthreetimes \in \pitchfork$.

(ii) *In addition, if* $(\mathcal{T}_1, \ldots, \mathcal{T}_n) < \underbrace{(\frac{1}{2}, \ldots, \frac{1}{2})}_{n}$ *and* $\underbrace{f_i}_{1\le i\le n}$, *be continuous and satisfy* $\underbrace{f_i}_{1\le i\le n}(\leftthreetimes, 2\varepsilon) = 2 \underbrace{f_i}_{1\le i\le n}(\leftthreetimes, \varepsilon)$ *and the inequality below*

$$\operatorname{diag}\Big[\Big\| f_1(\leftthreetimes, \varepsilon) f_1(\leftthreetimes, \zeta) - f_1(\leftthreetimes, f_1(\leftthreetimes, \varepsilon)\zeta) - f_1(\leftthreetimes, \varepsilon f_1(\leftthreetimes, \zeta))\Big\|, \ldots,$$

$$\Big\| f_n(\leftthreetimes, \varepsilon) f_n(\leftthreetimes, \zeta) - f_n(\leftthreetimes, f_n(\leftthreetimes, \varepsilon)\zeta) - f_n(\leftthreetimes, \varepsilon f_n(\leftthreetimes, \zeta))\Big\|\Big]$$

$$\preceq \operatorname{diag}\Big[\underbrace{\varphi_{j_1}}_{1\le j_1 \le n}(\varepsilon, \zeta, \varepsilon), \ldots, \underbrace{\varphi_{j_n}}_{1\le j_n \le n}(\varepsilon, \zeta, \varepsilon)\Big], \tag{11.15}$$

for every $\varepsilon, \zeta \in \mho_1$ *and* $\leftthreetimes \in \pitchfork$. *Then,* $\underbrace{f_i}_{1\le i\le n} : \pitchfork \times \mho_1 \to \mho_1$ *are antiderivations.*

Proof By a similar method used in Theorem 11.1, the proof of (i) is straightforward. Now, we prove (ii).

(ii) Since $\underbrace{\mathscr{G}_i}_{1\le i\le n} = \underbrace{f_i}_{1\le i\le n}$, are continuous and $\mathbb{C}$–linear, we conclude from the inequalities (11.12) and (11.15) that

$$
\begin{aligned}
&\operatorname{diag}\Big[\|\mathscr{G}_1(\leftthreetimes,\varepsilon)\mathscr{G}_1(\leftthreetimes,\zeta)-\mathscr{G}_1(\leftthreetimes,\mathscr{G}_1(\leftthreetimes,\varepsilon)\zeta)-\mathscr{G}_1(\leftthreetimes,\varepsilon\mathscr{G}_1(\leftthreetimes,\zeta))\|\\
&\quad,\ldots,\|\mathscr{G}_n(\leftthreetimes,\varepsilon)\mathscr{G}_n(\leftthreetimes,\zeta)-\mathscr{G}_n(\leftthreetimes,\mathscr{G}_n(\leftthreetimes,\varepsilon)\zeta)-\mathscr{G}_n(\leftthreetimes,\varepsilon\mathscr{G}_n(\leftthreetimes,\zeta))\|\Big]\\
&=\operatorname{diag}\Big[\lim_{m\to\infty}4^m\Big\|\mathfrak{J}^m\Big(f_1\Big(\leftthreetimes,\frac{\varepsilon}{2^m\mathfrak{J}^m}\Big)f_1\Big(\leftthreetimes,\frac{\zeta}{2^m\mathfrak{J}^m}\Big)\\
&\quad-\mathscr{G}_1\Big(\leftthreetimes,f_1\Big(\leftthreetimes,\frac{\varepsilon}{2^m\mathfrak{J}^m}\Big)\frac{\zeta}{2^m\mathfrak{J}^m}\Big)-\mathscr{G}_1\Big(\leftthreetimes,\frac{\varepsilon}{2^m\mathfrak{J}^m}f_1\Big(\leftthreetimes,\frac{\zeta}{2^m\mathfrak{J}^m}\Big)\Big)\Big)\Big\|,\ldots,\\
&\lim_{m\to\infty}4^m\Big\|\mathfrak{J}^m\Big(f_n\Big(\leftthreetimes,\frac{\varepsilon}{2^m\mathfrak{J}^m}\Big)f_n\Big(\leftthreetimes,\frac{\zeta}{2^m\mathfrak{J}^m}\Big)-\mathscr{G}_n\Big(\leftthreetimes,f_n\Big(\leftthreetimes,\frac{\varepsilon}{2^m\mathfrak{J}^m}\Big)\frac{\zeta}{2^m\mathfrak{J}^m}\Big)\\
&\quad-\mathscr{G}_n\Big(\leftthreetimes,\frac{\varepsilon}{2^m\lambda^m}f_n\Big(\leftthreetimes,\frac{\zeta}{2^m\mathfrak{J}^m}\Big)\Big)\Big)\Big\|\Big]\\
&=\operatorname{diag}\Big[\lim_{m\to\infty}4^m\Big\|\mathfrak{J}^m\Big(f_1\Big(\leftthreetimes,\frac{\varepsilon}{2^m\mathfrak{J}^m}\Big)f_1\Big(\leftthreetimes,\frac{\zeta}{2^m\mathfrak{J}^m}\Big)\\
&\quad-f_1\Big(\leftthreetimes,f_1\Big(\leftthreetimes,\frac{\varepsilon}{2^m\mathfrak{J}^m}\Big)\frac{\zeta}{2^m\mathfrak{J}^m}\Big)-f_1\Big(\leftthreetimes,\frac{\varepsilon}{2^m\mathfrak{J}^m}f_1\Big(\leftthreetimes,\frac{\zeta}{2^m\mathfrak{J}^m}\Big)\Big)\Big)\Big\|,\ldots,\\
&\lim_{m\to\infty}4^m\Big\|\mathfrak{J}^m\Big(f_n\Big(\leftthreetimes,\frac{\varepsilon}{2^m\mathfrak{J}^m}\Big)f_n\Big(\leftthreetimes,\frac{\zeta}{2^m\mathfrak{J}^m}\Big)-f_n\Big(\leftthreetimes,f_1\Big(\leftthreetimes,\frac{\varepsilon}{2^m\mathfrak{J}^m}\Big)\frac{\zeta}{2^m\mathfrak{J}^m}\Big)\\
&\quad-f_n\Big(\leftthreetimes,\frac{\varepsilon}{2^m\mathfrak{J}^m}f_n\Big(\leftthreetimes,\frac{\zeta}{2^m\mathfrak{J}^m}\Big)\Big)\Big)\Big\|\Big]\\
&\preceq\operatorname{diag}\Big[\lim_{m\to\infty}2^{2m}\underbrace{\varphi_{j_1}}_{1\le j_1\le n}\Big(\frac{\varepsilon}{2^m\mathfrak{J}^m},\frac{\zeta}{2^m\mathfrak{J}^m},\frac{\varepsilon}{2^m\mathfrak{J}^m}\Big),\ldots,\\
&\lim_{m\to\infty}2^{2m}\underbrace{\varphi_{j_n}}_{1\le j_n\le n}\Big(\frac{\varepsilon}{2^m\mathfrak{J}^m},\frac{\zeta}{2^m\mathfrak{J}^m},\frac{\varepsilon}{2^m\mathfrak{J}^m}\Big)\Big]\\
&\preceq\operatorname{diag}\Big[\lim_{m\to\infty}(2\mathcal{T}_1)^m\underbrace{\varphi_{j_1}}_{1\le j_1\le n}(\varepsilon,\zeta,\varepsilon),\ldots,\lim_{m\to\infty}(2\mathcal{T}_n)^m\underbrace{\varphi_{j_n}}_{1\le j_n\le n}(\varepsilon,\zeta,\varepsilon)\Big],
\end{aligned}
$$

for every $\mathfrak{J} \in \mathbb{T}^1$ and every $\varepsilon, \zeta \in \mho_1$ and $\leftthreetimes \in \pitchfork$. Since $\underbrace{(2\mathcal{T}_1, \ldots, 2\mathcal{T}_n)}_{n} < \underbrace{(1, \ldots, 1)}_{n}$, the $\mathbb{C}$–linear mappings $\underbrace{\mathcal{G}_i}_{1 \le i \le n}$, are antiderivations. Thus, the mappings $\underbrace{f_i}_{1 \le i \le n} : \pitchfork \times \mho_1 \to \mho_1$ are antiderivations.

11.3 Super Multi Stability of Continuous Antiderivations in Banach Algebras

In this part, we investigate the super multi stability of continuous antiderivations in Banach algebras.

Theorem 11.3 *Consider* $\underbrace{\varphi_{j_1}}_{1 \le j_1 \le n}, \ldots, \underbrace{\varphi_{j_n}}_{1 \le j_n \le n} : \mho_1^3 \to [0, \infty)$, *with* $n \in \mathbb{N}$. *Then, we have the following results:*

(i) If there exist $(\mathcal{T}_1, \ldots, \mathcal{T}_n) \prec \overbrace{(1, \ldots, 1)}^{n}$ *satisfying the inequality below*

$$\operatorname{diag}\left[\underbrace{\varphi_{j_1}}_{1 \le j_1 \le n}\left(\frac{\varepsilon}{\mathfrak{J}}, \frac{\zeta}{\mathfrak{J}}, \frac{\varsigma}{\mathfrak{J}}\right), \ldots, \underbrace{\varphi_{j_n}}_{1 \le j_n \le n}\left(\frac{\varepsilon}{\mathfrak{J}}, \frac{\zeta}{\mathfrak{J}}, \frac{\varsigma}{\mathfrak{J}}\right)\right]$$
$$\preceq \operatorname{diag}\left[\frac{\mathcal{T}_1}{2}\underbrace{\varphi_{j_1}}_{1 \le j_1 \le n}(2\varepsilon, 2\zeta, 2\varsigma), \ldots, \frac{\mathcal{T}_n}{2}\underbrace{\varphi_{j_n}}_{1 \le j_n \le n}(2\varepsilon, 2\zeta, 2\varsigma)\right], \tag{11.16}$$

and if $\underbrace{f_i}_{1 \le i \le n} : \pitchfork \times \mho_1 \to \mho_1$, *be mappings satisfying* $\underbrace{f_i}_{1 \le i \le n}(\leftthreetimes, 0) = 0$ *and the following inequality*

$$\operatorname{diag}\Bigg[\Big\|\mathfrak{J} f_1(\leftthreetimes, \varepsilon + \zeta + \varsigma) - f_1(\leftthreetimes, \mathfrak{J}(\varepsilon + \varsigma)) - f_1(\leftthreetimes, \mathfrak{J}(\varepsilon + \zeta - \varsigma))$$
$$+ \mathfrak{J} f_1(\leftthreetimes, \varepsilon - \varsigma)\Big\| + |\theta_1| \underbrace{\varphi_{j_1}}_{1 \le j_1 \le n}(\varepsilon, -\zeta, \varsigma), \ldots,$$
$$\Big\|\mathfrak{J} f_n(\leftthreetimes, \varepsilon + \zeta + \varsigma) - f_n(\leftthreetimes, \mathfrak{J}(\varepsilon + \varsigma)) - f_n(\leftthreetimes, \mathfrak{J}(\varepsilon + \zeta - \varsigma))$$
$$+ \mathfrak{J} f_n(\leftthreetimes, \varepsilon - \varsigma)\Big\| + |\theta_n| \underbrace{\varphi_{j_n}}_{1 \le j_n \le n}(\varepsilon, -\zeta, \varsigma)\Bigg]$$
$$\preceq \operatorname{diag}\Bigg[\Big\|\theta_1(\mathfrak{J} f_1(\leftthreetimes, \varepsilon - \zeta + \varsigma) - f_1(\leftthreetimes, \mathfrak{J}(\varepsilon + \varsigma)) - f_1(\leftthreetimes, \mathfrak{J}(\varepsilon - \zeta - \varsigma))$$

$$+\mathfrak{J} f_1(\curlywedge, \varepsilon - \varsigma))\Big\| + \underbrace{\varphi_{j_1}}_{1\le j_1\le n}(\varepsilon, \zeta, \varsigma), \ldots,$$

$$\Big\|\theta_n(\mathfrak{J} f_n(\curlywedge, \varepsilon - \zeta + \varsigma) - f_n(\curlywedge, \mathfrak{J}(\varepsilon + \varsigma)) - f_n(\curlywedge, \mathfrak{J}(\varepsilon - \zeta - \varsigma))$$

$$+\mathfrak{J} f_n(\curlywedge, \varepsilon - \varsigma))\Big\| + \underbrace{\varphi_{j_n}}_{1\le j_n\le n}(\varepsilon, \zeta, \varsigma)\Big], \tag{11.17}$$

for every $\mathfrak{J} \in \mathbb{C} - \overline{\mathbb{T}}^1$ *and every* $\varepsilon, \zeta, \varsigma \in \mho_1$, *and* $\curlywedge \in \pitchfork$. *Then, there are unique* $\mathbb{C}$*–linear mappings* $\underbrace{\mathscr{G}_i}_{1\le i\le n} : \pitchfork \times \mho_1 \to \mho_1$, *such that*

$$\operatorname{diag}\Big[\ \|f_1(\curlywedge, \varepsilon) - \mathscr{G}_1(\curlywedge, \varepsilon)\|, \cdots, \|f_n(\curlywedge, \varepsilon) - \mathscr{G}_n(\curlywedge, \varepsilon)\|\Big]$$

$$\preceq \operatorname{diag}\Big[\ \frac{\mathcal{T}_1}{2(1-\mathcal{T}_1)}\underbrace{\varphi_{j_1}}_{1\le j_1\le n}\Big(\frac{\varepsilon}{2}, \varepsilon, -\frac{\varepsilon}{2}\Big), \cdots, \frac{\mathcal{T}_n}{2(1-\mathcal{T}_n)}\underbrace{\varphi_{j_n}}_{1\le j_n\le n}\Big(\frac{\varepsilon}{2}, \varepsilon, -\frac{\varepsilon}{2}\Big)\Big], \tag{11.18}$$

for every $\varepsilon \in \mho_1$, *and* $\curlywedge \in \pitchfork$.

(ii) *Furthermore, if* $(\mathcal{T}_1, \ldots, \mathcal{T}_n) \prec (\overbrace{\frac{1}{2}, \ldots, \frac{1}{2}}^{n})$, *and* $\underbrace{\varphi_{j_1}}_{1\le j_1\le n}, \ldots, \underbrace{\varphi_{j_n}}_{1\le j_n\le n}$ *be continuous functions and also* $\underbrace{f_i}_{1\le i\le n}(\curlywedge, 2\varepsilon) = 2\underbrace{f_i}_{1\le i\le n}(\curlywedge, \varepsilon)$ *and the inequality below*

$$\operatorname{diag}\Big[\|f_1(\curlywedge, \varepsilon) f_1(\curlywedge, \zeta) - f_1(\curlywedge, f_1(\curlywedge, \varepsilon)\zeta) - f_1(\curlywedge, \varepsilon f_1(\curlywedge, \zeta))\|, \ldots,$$

$$\|f_n(\curlywedge, \varepsilon) f_n(\curlywedge, \zeta) - f_n(\curlywedge, f_n(\curlywedge, \varepsilon)\zeta) - f_n(\curlywedge, \varepsilon f_n(\curlywedge, \zeta))\|\ \Big]$$

$$\preceq \operatorname{diag}\Big[\ \underbrace{\varphi_{j_1}}_{1\le j_1\le n}(\varepsilon, \zeta, \varepsilon), \ldots, \underbrace{\varphi_{j_n}}_{1\le j_n\le n}(\varepsilon, \zeta, \varepsilon)\Big],$$

for every ε *and* $\zeta \in \mho_1$, $\curlywedge \in \pitchfork$. *Then,* $\underbrace{f_i}_{1\le i\le n} : \pitchfork \times \mho_1 \to \mho_1$ *are continuous antiderivations.*

Proof Using the same reasoning as in the proof of Theorem 11.2, we obtain the desired result.

11.4 Application

Here, consider the control functions $\underbrace{\varphi_{j_i}^{Ⓢ}}_{\substack{1\leq j_i\leq n\\1\leq i\leq n}}$, defined in Sects. 9.8.1 and 9.8.2, with $n\in\mathbb{N}$. We now have the following results.

Corollary 11.1 *Let* $n\in\mathbb{N}$, *and* $\underbrace{f_i}_{1\leq i\leq n}:⫛\times\mho_1\to\mho_1$ *be mappings satisfying* $\underbrace{f_i}_{1\leq i\leq n}(\leftthreetimes,0)=0$ *and the inequality below*

$$
\begin{aligned}
&diag\Bigg[\bigg\|f_1(\leftthreetimes,\varepsilon+\zeta+\varsigma)-f_1(\leftthreetimes,\varepsilon+\varsigma)-f_1(\leftthreetimes,\varepsilon+\zeta-\varsigma)+f_1(\leftthreetimes,\varepsilon-\varsigma)\bigg\|\\
&\quad,\ldots,\bigg\|f_n(\leftthreetimes,\varepsilon+\zeta+\varsigma)-f_n(\leftthreetimes,\varepsilon+\varsigma)-f_n(\leftthreetimes,\varepsilon+\zeta-\varsigma)+f_n(\leftthreetimes,\varepsilon-\varsigma)\bigg\|\Bigg]\\
&\preceq diag\Bigg[\bigg\|\theta_1(f_1(\leftthreetimes,\varepsilon-\zeta+\varsigma)-f_1(\leftthreetimes,\varepsilon+\varsigma)-f_1(\leftthreetimes,\varepsilon-\zeta-\varsigma)\\
&\quad+f_1(\leftthreetimes,\varepsilon-\varsigma))\bigg\|+(1-|\theta_1|)\underbrace{\varphi_{j_1}^{Ⓢ}}_{1\leq j_1\leq n}(\|\varepsilon^2\|+\|\zeta^2\|+\|\varsigma^2\|),\ldots,\\
&\bigg\|\theta_n(f_n(\leftthreetimes,\varepsilon-\zeta+\varsigma)-f_n(\leftthreetimes,\varepsilon+\varsigma)-f_n(\leftthreetimes,\varepsilon-\zeta-\varsigma)+f_n(\leftthreetimes,\varepsilon-\varsigma))\bigg\|\\
&\quad+(1-|\theta_n|)\underbrace{\varphi_{j_n}^{Ⓢ}}_{1\leq j_n\leq n}(\|\varepsilon^2\|+\|\zeta^2\|+\|\varsigma^2\|)\Bigg],
\end{aligned}
$$

for every $\varepsilon,\zeta,\varsigma\in\mho_1$, *and* $\leftthreetimes\in⫛$. *Then, there are unique additive mappings* $\underbrace{f_i'}_{1\leq i\leq n}:⫛\times\mho_1\to\mho_1$, *such that*

$$
\begin{aligned}
&diag\Bigg[\|f_1(\leftthreetimes,\varepsilon)-f_1'(\leftthreetimes,\varepsilon)\|,\ldots,\|f_n(\leftthreetimes,\varepsilon)-f_n'(\leftthreetimes,\varepsilon)\|\Bigg]_{n\times n}\\
&\preceq diag\Bigg[\underbrace{\varphi_{j_1}^{Ⓢ}}_{1\leq j_1\leq n}(\|\varepsilon\|^2),\ldots,\underbrace{\varphi_{j_n}^{Ⓢ}}_{1\leq j_n\leq n}(\|\varepsilon\|^2)\Bigg],
\end{aligned}
$$

for every $\varepsilon\in\mho_1$ *and* $\leftthreetimes\in⫛$.

Proof The proof follows from Theorem 11.1 by letting

$$\mathrm{diag}\Big[\underbrace{\varphi_{j_1}}_{1\le j_1\le n}(\varepsilon,\zeta,\varsigma),\ldots,\underbrace{\varphi_{j_n}}_{1\le j_n\le n}(\varepsilon,\zeta,\varsigma)\Big]$$
$$:=\mathrm{diag}\Big[\underbrace{\varphi^{\circledS}_{j_1}}_{1\le j_1\le n}\Big(\|\varepsilon^2\|+\|\zeta^2\|+\|\varsigma^2\|\Big),\ldots,\underbrace{\varphi^{\circledS}_{j_n}}_{1\le j_n\le n}\Big(\|\varepsilon^2\|+\|\zeta^2\|+\|\varsigma^2\|\Big)\Big],$$

for every $\varepsilon,\zeta,\varsigma\in\mho_1$. Choosing $(\mathcal{T}_1,\ldots,\mathcal{T}_n)=\underbrace{(\frac{4}{7},\ldots,\frac{4}{7})}_{n}$, we obtain the desired result.

Corollary 11.2 *Let $n\in\mathbb{N}$, and $\underbrace{f_i}_{1\le i\le n}:\pitchfork\times\mho_1\to\mho_1$ be mappings satisfying $\underbrace{f_i}_{1\le i\le n}(\leftthreetimes,0)=0$ and the inequality below*

$$diag\Big[\Big\|f_1(\leftthreetimes,\varepsilon+\zeta+\varsigma)-f_1(\leftthreetimes,\varepsilon+\varsigma)-f_1(\leftthreetimes,\varepsilon+\zeta-\varsigma)+f_1(\leftthreetimes,\varepsilon-\varsigma)\Big\|$$
$$,\ldots,\Big\|f_n(\leftthreetimes,\varepsilon+\zeta+\varsigma)-f_n(\leftthreetimes,\varepsilon+\varsigma)-f_n(\leftthreetimes,\varepsilon+\zeta-\varsigma)+f_n(\leftthreetimes,\varepsilon-\varsigma)\Big\|\Big]$$
$$\preceq diag\Big[\Big\|\theta_1(f_1(\leftthreetimes,\varepsilon-\zeta+\varsigma)-f_1(\leftthreetimes,\varepsilon+\varsigma)-f_1(\leftthreetimes,\varepsilon-\zeta-\varsigma)$$
$$+f_1(\leftthreetimes,\varepsilon-\varsigma))\Big\|+(1-|\theta_1|)\underbrace{\varphi^{\circledS}_{j_1}}_{1\le j_1\le n}(\|\varepsilon\zeta\varsigma\|),\ldots,$$
$$\Big\|\theta_n(f_n(\leftthreetimes,\varepsilon-\zeta+\varsigma)-f_n(\leftthreetimes,\varepsilon+\varsigma)-f_n(\leftthreetimes,\varepsilon-\zeta-\varsigma)+f_n(\leftthreetimes,\varepsilon-\varsigma))\Big\|$$
$$+(1-|\theta_n|)\underbrace{\varphi^{\circledS}_{j_n}}_{1\le j_n\le n}(\|\varepsilon\zeta\varsigma\|)\Big]$$

for every $\varepsilon,\zeta,\varsigma\in\mho_1$ and $\leftthreetimes\in\pitchfork$. Then, there are unique additive mappings $\underbrace{f'_i}_{1\le i\le n}:\pitchfork\times\mho_1\to\mho_1$, such that

$$diag\Big[\|f_1(\leftthreetimes,\varepsilon)-f'_1(\leftthreetimes,\varepsilon)\|,\ldots,\|f_n(\leftthreetimes,\varepsilon)-f'_n(\leftthreetimes,\varepsilon)\|\Big]$$
$$\preceq diag\Big[\underbrace{\varphi^{\circledS}_{j_1}}_{1\le j_1\le n}(\|\varepsilon\|^3),\ldots,\underbrace{\varphi^{\circledS}_{j_n}}_{1\le j_n\le n}(\|\varepsilon\|^3)\Big],$$

for every $\varepsilon\in\mho_1$ and $\leftthreetimes\in\pitchfork$.

Proof The proof follows from Theorem 11.1 by letting

$$\begin{aligned}
&\operatorname{diag}\Big[\underbrace{\varphi_{j_1}}_{1\le j_1\le n}(\varepsilon,\zeta,\varsigma),\ldots,\underbrace{\varphi_{j_n}}_{1\le j_n\le n}(\varepsilon,\zeta,\varsigma)\Big]\\
&:=\operatorname{diag}\Big[\underbrace{\varphi^{\circledS}_{j_1}}_{1\le j_1\le n}(\|\varepsilon\zeta\varsigma\|),\ldots,\underbrace{\varphi^{\circledS}_{j_n}}_{1\le j_n\le n}(\|\varepsilon\zeta\varsigma\|)\Big],
\end{aligned}$$

for every $\varepsilon,\zeta,\varsigma\in\mho_1$ and $\leftthreetimes\in\pitchfork$. Choosing $(\mathcal{T}_1,\ldots,\mathcal{T}_n)=\underbrace{(\frac{8}{9},\ldots,\frac{8}{9})}_{n}$, we obtain the desired result.

Corollary 11.3 *Let $n\in\mathbb{N}$, and $\underbrace{f_i}_{1\le i\le n}:\pitchfork\times\mho_1\to\mho_1$, be odd mappings satisfying the inequality below*

$$\begin{aligned}
&diag\Big[\Big\|f_1(\leftthreetimes,\varepsilon+\zeta+\varsigma)-f_1(\leftthreetimes,\varepsilon+\varsigma)-f_1(\leftthreetimes,\varepsilon+\zeta-\varsigma)+f_1(\leftthreetimes,\varepsilon-\varsigma)\Big\|\\
&\quad,\ldots,\Big\|f_n(\leftthreetimes,\varepsilon+\zeta+\varsigma)-f_n(\leftthreetimes,\varepsilon+\varsigma)-f_n(\leftthreetimes,\varepsilon+\zeta-\varsigma)+f_n(\leftthreetimes,\varepsilon-\varsigma)\Big\|\Big]\\
&\preceq diag\Big[\Big\|\theta_1[f_1(\leftthreetimes,\varepsilon-\zeta+\varsigma)-f_1(\leftthreetimes,\varepsilon+\varsigma)-f_1(\leftthreetimes,\varepsilon-\zeta-\varsigma)\\
&\qquad+f_1(\leftthreetimes,\varepsilon-\varsigma)]\Big\|+(1-|\theta_1|)\underbrace{\varphi^{\circledS}_{j_1}}_{1\le j_1\le n}(\|\varepsilon\zeta\varsigma\|)\\
&\quad,\ldots,\Big\|\theta_n[f_n(\leftthreetimes,\varepsilon-\zeta+\varsigma)-f_n(\leftthreetimes,\varepsilon+\varsigma)-f_n(\leftthreetimes,\varepsilon-\zeta-\varsigma)\\
&\qquad+f_n(\leftthreetimes,\varepsilon-\varsigma)]\Big\|+(1-|\theta_n|)\underbrace{\varphi^{\circledS}_{j_n}}_{1\le j_n\le n}(\|\varepsilon\zeta\varsigma\|)\Big],
\end{aligned}\tag{11.19}$$

for every $\varepsilon,\zeta,\varsigma\in\mho_1$ and $\leftthreetimes\in\pitchfork$. Then, $\underbrace{f_i}_{1\le i\le n}$ are additive.

Proof Putting $\varepsilon=0$, in the inequality (11.19), we get

$$\begin{aligned}
&\operatorname{diag}\Big[\Big\|f_1(\leftthreetimes,\zeta+\varsigma)-f_1(\leftthreetimes,\varsigma)-f_1(\leftthreetimes,\zeta-\varsigma)+f_1(\leftthreetimes,-\varsigma)\Big\|,\ldots,\\
&\qquad\Big\|f_n(\leftthreetimes,\zeta+\varsigma)-f_n(\leftthreetimes,\varsigma)-f_n(\leftthreetimes,\zeta-\varsigma)+f_n(\leftthreetimes,-\varsigma)\Big\|\Big]\\
&\preceq\operatorname{diag}\Big[\Big\|\theta_1(f_1(\leftthreetimes,-\zeta+\varsigma)-f_1(\leftthreetimes,\varsigma)-f_1(\leftthreetimes,-\zeta-\varsigma)+f_1(\leftthreetimes,-\varsigma))\Big\|,\ldots,
\end{aligned}$$

$$\left\|\theta_n(f_n(\lambda, -\zeta + \varsigma) - f_n(\lambda, \varsigma) - f_n(\lambda, -\zeta - \varsigma) + f_n(\lambda, -\varsigma))\right\|\Big], \tag{11.20}$$

for every $\zeta, \varsigma \in \mho_1$ and $\lambda \in \pitchfork$. Replacing ζ by $-\zeta$ in the inequality (11.20), we have

$$\begin{aligned}
&\operatorname{diag}\Big[\left\| f_1(\lambda, -\zeta + \varsigma) - f_1(\lambda, \varsigma) - f_1(\lambda, -\zeta - \varsigma) + f_1(\lambda, -\varsigma)\right\|, \dots, \\
&\qquad \left\| f_n(\lambda, -\zeta + \varsigma) - f_n(\lambda, \varsigma) - f_n(\lambda, -\zeta - \varsigma) + f_n(\lambda, -\varsigma)\right\|\Big] \\
&\preceq \operatorname{diag}\Big[\left\|\theta_1[f_1(\lambda, \zeta + \varsigma) - f_1(\lambda, \varsigma) - f_1(\lambda, \zeta - \varsigma) + f_1(\lambda, -\varsigma)]\right\|, \dots, \\
&\qquad \left\|\theta_n[f_n(\lambda, \zeta + \varsigma) - f_n(\lambda, \varsigma) - f_n(\lambda, \zeta - \varsigma) + f_n(\lambda, -\varsigma)]\right\|\Big],
\end{aligned} \tag{11.21}$$

for every $\zeta, \varsigma \in \mho_1$ and $\lambda \in \pitchfork$. From the inequalities (11.20) and (11.21), it follows that

$$f_i(\lambda, \zeta + \varsigma) - f_i(\lambda, \varsigma) - f_i(\lambda, \zeta - \varsigma) + f_i(\lambda, -\varsigma) = 0, \quad i = 1, \dots, n,$$

for every $\zeta, \varsigma \in \mho_1$ and $\lambda \in \pitchfork$. Since $\underbrace{f_i}_{1 \le i \le n}$ are odd mappings, we get

$$f_i(\varsigma + \zeta) + f_i(\varsigma - \zeta) - 2f_i(\varsigma) = 0, \quad i = 1, \dots, n$$

for every $\zeta, \varsigma \in \mho_1$ and $\lambda \in \pitchfork$. Thus, the mappings $\underbrace{f_i}_{1 \le i \le n}$ are additive.

Corollary 11.4 *Let* $n \in \mathbb{N}$, *and* $\underbrace{f_i}_{1 \le i \le n} : \pitchfork \times \mho_1 \to \mho_1$, *be mappings satisfying* $\underbrace{f_i}_{1 \le i \le n}$ $(\lambda, 0) = 0$ *and the inequalities below*

$$\begin{aligned}
&diag\Big[\left\| f_1(\lambda, \varepsilon + \zeta + \varsigma) - f_1(\lambda, \varepsilon + \varsigma) - f_1(\lambda, \varepsilon + \zeta - \varsigma) + f_1(\lambda, \varepsilon - \varsigma)\right\| \\
&\quad , \dots, \left\| f_n(\lambda, \varepsilon + \zeta + \varsigma) - f_n(\lambda, \varepsilon + \varsigma) - f_n(\lambda, \varepsilon + \zeta - \varsigma) + f_n(\lambda, \varepsilon - \varsigma)\right\|\Big] \\
&\preceq diag\Big[\Big\|\theta_1(f_1(\lambda, \varepsilon - \zeta + \varsigma) - f_1(\lambda, \varepsilon + \varsigma) - f_1(\lambda, \varepsilon - \zeta - \varsigma) \\
&\quad + f_1(\lambda, \varepsilon - \varsigma))\Big\| + (1 - |\theta_1|) \underbrace{\varphi_{j_1}^{\circledS}}_{1 \le j_1 \le n} \left(\|\varepsilon^4 + \zeta^4 + \varsigma^4\|\right)
\end{aligned}$$

$$, \ldots, \left\| \theta_n(f_n(\curlywedge, \varepsilon - \zeta + \varsigma) - f_n(\curlywedge, \varepsilon + \varsigma) - f_n(\curlywedge, \varepsilon - \zeta - \varsigma) + f_n(\curlywedge, \varepsilon - \varsigma)) \right\| + (1 - |\theta_n|) \underbrace{\varphi_{j_n}^{\circledS}}_{1 \le j_n \le n} \left(\|\varepsilon^4 + \zeta^4 + \varsigma^4\| \right) \Big]$$

and

$$diag\Big[\left\| f_1(\curlywedge, \varepsilon) f_1(\curlywedge, \zeta) - f_1(\curlywedge, f_1(\curlywedge, \varepsilon)\zeta) - f_1(\curlywedge, \varepsilon f_1(\curlywedge, \zeta)) \right\|, \ldots, \left\| f_n(\curlywedge, \varepsilon) f_n(\curlywedge, \zeta) - f_n(\curlywedge, f_n(\curlywedge, \varepsilon)\zeta) - f_n(\curlywedge, \varepsilon f_n(\curlywedge, \zeta)) \right\| \Big]$$
$$\preceq diag\Big[\underbrace{\varphi_{j_1}^{\circledS}}_{1 \le j_1 \le n} \left(\|2\varepsilon^4 + \zeta^4\| \right), \ldots, \underbrace{\varphi_{j_n}^{\circledS}}_{1 \le j_n \le n} \left(\|2\varepsilon^4 + \zeta^4\| \right) \Big]_{n \times n},$$

for every $\varepsilon, \zeta, \varsigma \in \mho_1$ *and* $\curlywedge \in \pitchfork$. *If* $\underbrace{f_i}_{1 \le i \le n}(\curlywedge, 2\varepsilon) = 2 \underbrace{f_i}_{1 \le i \le n}(\curlywedge, \varepsilon)$, *for every* $\varepsilon, \zeta, \varsigma \in \mho_1$, *and* $\curlywedge \in \pitchfork$, *and* $\underbrace{f_i}_{1 \le i \le n}$, *be continuous, then, the mappings* $\underbrace{f_i}_{1 \le i \le n} : \pitchfork \times \mho_1 \to \mho_1$, *are antiderivations.*

Proof The proof follows from Theorem 11.2 by letting

$$\text{diag}\Big[\underbrace{\varphi_{j_1}}_{1 \le j_1 \le n}(\varepsilon, \zeta, \varsigma), \ldots, \underbrace{\varphi_{j_n}}_{1 \le j_n \le n}(\varepsilon, \zeta, \varsigma) \Big]$$
$$:= \text{diag}\Big[\underbrace{\varphi_{j_1}^{\circledS}}_{1 \le j_1 \le n} \left(\|\varepsilon^4 + \zeta^4 + \varsigma^4\| \right), \ldots, \underbrace{\varphi_{j_n}^{\circledS}}_{1 \le j_n \le n} \left(\|\varepsilon^4 + \zeta^4 + \varsigma^4\| \right) \Big]_{n \times n}$$

for every $\varepsilon, \zeta, \varsigma \in \mho_1$. Choosing $(\mathcal{T}_1, \ldots, \mathcal{T}_n) = (\overbrace{\frac{8}{17}, \ldots, \frac{8}{17}}^{n})$, we obtain the desired result.

We refer the interested readers to [3–8].

References

1. Aderyani SR, Saadati R, O'Regan D, Alshammari FS (2022) Multi-super-stability of antiderivations in Banach algebras. AIMS Math 7(11):20143–20163
2. Sayyari Y, Dehghanian M, Park C, Lee JR (2022) Stability of hyper homomorphisms and hyper derivations in complex Banach algebras. AIMS Math 7(6):10700–10710

3. Rezaei Aderyani S, Saadati R, Li C, Rassias TM, Park C (2023) Special functions and multi-stability of the Jensen type random operator equation in C^*-algebras via fixed point. J Inequal Appl 2023(1):1–24
4. Aderyani SR, Saadati R, Feckan M (2021) The Cadariu-Radu method for existence, uniqueness and gauss hypergeometric stability of w-Hilfer fractional differential equations. Mathematics 9:1408
5. Aderyani SR, Saadati R, O'Regan D, Abdeljawad T (2022) UHML stability of a class of Δ-Hilfer FDEs via CRM. AIMS Math 7(4):5910–5919
6. Aderyani SR, Saadati R, O'Regan D (2022) The Cadariu-Radu method for existence, uniqueness and Gauss Hypergeometric stability of a class of Ξ-Hilfer fractional differential equations. Int J Nonlinear Sci Numer Simul 24(8)
7. Aderyani SR, Saadati R, O'Regan D, Alshammari FS (2023) Application of aggregated control functions for approximating $\mathscr{C}$-Hilfer fractional differential equations. AIMS Math 8(11):28010–28032
8. Aderyani SR, Saadati R, Rassias TM, Srivastava HM (2023) Existence, uniqueness and the multi-stability results for a w-Hilfer fractional differential equation. Axioms 12(7):681

Chapter 12
Multi Stability of Tri–Additive Functional Inequalities in Unital C^*–Algebras

Considering the concept of multi stability, we stabilize the following tri–additive $(\boldsymbol{\Psi}_1, \dots, \boldsymbol{\Psi}_n)$–functional inequality [1] via the fixed point method,

$$
\begin{aligned}
&\operatorname{diag}\Big[\|\mathcal{L}_1(\lambda, \Psi_1 + \Psi_2, \Psi_3 - \Psi_4, \Psi_5 + \Psi_6) + \mathcal{L}_1(\lambda, \Psi_1 - \Psi_2, \Psi_3 + \Psi_4, \Psi_5 - \Psi_6) \\
&\qquad -2\mathcal{L}_1(\lambda, \Psi_1, \Psi_3, \Psi_5) + 2\mathcal{L}_1(\lambda, \Psi_1, \Psi_4, \Psi_6) \\
&\qquad\qquad -2\mathcal{L}_1(\lambda, \Psi_2, \Psi_3, \Psi_6) + 2\mathcal{L}_1(\lambda, \Psi_2, \Psi_4, \Psi_5)\| \\
&\quad \dots, \|\mathcal{L}_n(\lambda, \Psi_1 + \Psi_2, \Psi_3 - \Psi_4, \Psi_5 + \Psi_6) + \mathcal{L}_n(\lambda, \Psi_1 - \Psi_2, \Psi_3 + \Psi_4, \Psi_5 - \Psi_6) \\
&\qquad -2\mathcal{L}_n(\lambda, \Psi_1, \Psi_3, \Psi_5) + 2\mathcal{L}_n(\lambda, \Psi_1, \Psi_4, \Psi_6) \\
&\qquad\qquad -2\mathcal{L}_n(\lambda, \Psi_2, \Psi_3, \Psi_6) + 2\mathcal{L}_n(\lambda, \Psi_2, \Psi_4, \Psi_5)\| \Big] \\
&\preceq \operatorname{diag}\Big[\Big\| \boldsymbol{\Psi}_1 \Big[2\mathcal{L}_1\Big(\lambda, \frac{\Psi_1 + \Psi_2}{2}, \Psi_3 - \Psi_4, \Psi_5 + \Psi_6\Big) \\
&\qquad +2\mathcal{L}_1\Big(\lambda, \frac{\Psi_1 - \Psi_2}{2}, \Psi_3 + \Psi_4, \Psi_5 - \Psi_6\Big) \\
&\qquad -2\mathcal{L}_1(\lambda, \Psi_1, \Psi_3, \Psi_5) + 2\mathcal{L}_1(\lambda, \Psi_1, \Psi_4, \Psi_6) \\
&\qquad\qquad -2\mathcal{L}_1(\lambda, \Psi_2, \Psi_3, \Psi_6) + 2\mathcal{L}_1(\lambda, \Psi_2, \Psi_4, \Psi_5)]\|, \\
&\quad \dots, \Big\| \boldsymbol{\Psi}_n \Big[2\mathcal{L}_n\Big(\lambda, \frac{\Psi_1 + \Psi_2}{2}, \Psi_3 - \Psi_4, \Psi_5 + \Psi_6\Big) \\
&\qquad +2\mathcal{L}_n\Big(\lambda, \frac{\Psi_1 - \Psi_2}{2}, \Psi_3 + \Psi_4, \Psi_5 - \Psi_6\Big) \\
&\qquad -2\mathcal{L}_n(\lambda, \Psi_1, \Psi_3, \Psi_5) + 2\mathcal{L}_n(\lambda, \Psi_1, \Psi_4, \Psi_6) \\
&\qquad\qquad -2\mathcal{L}_n(\lambda, \Psi_2, \Psi_3, \Psi_6) + 2\mathcal{L}_n(\lambda, \Psi_2, \Psi_4, \Psi_5)]\| \Big],
\end{aligned}
\tag{12.1}
$$

where $n \in \mathbb{N}$, $\underbrace{\boldsymbol{\Psi}_i}_{i=1,\dots,n}$ $(\neq 0) \in \mathbb{C}$ are fixed numbers with $\underbrace{|\boldsymbol{\Psi}_i|}_{i=1,\dots,n} < 1$.

In addition, we get estimations for permuting tri–homomorphisms and permuting tri–derivations associated with the above inequality in the unital C^*–algebras and

S. Rezaei Aderyani et al., *Towards Ulam Type Multi Stability Analysis*, Studies in Systems, Decision and Control 523, https://doi.org/10.1007/978-3-031-55564-0_12

Banach algebras. As an application, we apply several aggregation maps on special functions to present an application of the multi stability.

In this chapter, let $\mathcal{G}$ and $\mathcal{Q}$ be complex normed spaces and complex Banach spaces, respectively. Also, let $\mathfrak{M}$ and $\mathfrak{N}$ be two complex Banach algebras.

Now, consider the definitions of permuting tri–homomorphisms and permuting tri–derivations, as follows:

Definition 12.1 ([1, 2]) Let $\mathfrak{M}$ and $\mathfrak{N}$ be two complex Banach algebras. We say a $\mathbb{C}$–trilinear operator $\mathcal{E} : \text{ሐ} \times \mathfrak{M}^3 \to \mathfrak{N}$ is a permuting tri–homomorphism, if $\mathcal{E}$ satisfies the following conditions:

$$\mathcal{E}(\leftthreetimes, \Psi_1\Psi_2, \Psi_3\Psi_4, \Psi_5\Psi_6) = \mathcal{E}(\leftthreetimes, \Psi_1, \Psi_3, \Psi_5)\mathcal{E}(\leftthreetimes, \Psi_2, \Psi_4, \Psi_6),$$
$$\mathcal{E}(\leftthreetimes, \Phi_{\delta(1)}, \Phi_{\delta(2)}, \Phi_{\delta(3)}) = \mathcal{E}(\leftthreetimes, \Phi_1, \Phi_2, \Phi_3),$$

for every $\Psi_1, \Psi_2, \Psi_3, \Psi_4, \Phi_1, \Phi_2, \Phi_3 \in \mathfrak{M}$, $\leftthreetimes \in \text{ሐ}$, and for every permutation $(\delta(1), \delta(2), \delta(3))$ of $(1, 2, 3)$.

Definition 12.2 ([1, 2]) Let $\mathfrak{M}$ be a complex Banach algebra. We say a tri–additive operator $\mathfrak{Z} : \text{ሐ} \times \mathfrak{M}^3 \to \mathfrak{M}$ is a permuting tri–derivation on $\mathfrak{M}$, if $\mathfrak{Z}$ satisfies the following conditions:

$$\mathfrak{Z}(\leftthreetimes, \Psi_1\Psi_2, \Psi_3, \Psi_4) = \mathfrak{Z}(\leftthreetimes, \Psi_1, \Psi_3, \Psi_4)\Psi_2 + \Psi_1\mathfrak{Z}(\leftthreetimes, \Psi_2, \Psi_3, \Psi_4),$$
$$\mathfrak{Z}(\leftthreetimes, \Phi_{\delta(1)}, \Phi_{\delta(2)}, \Phi_{\delta(3)}) = \mathfrak{Z}(\leftthreetimes, \Phi_1, \Phi_2, \Phi_3),$$

for every $\Psi_1, \Psi_2, \Psi_3, \Psi_4, \Phi_1, \Phi_2, \Phi_3 \in \mathfrak{M}$, $\leftthreetimes \in \text{ሐ}$, and for every permutation $(\delta(1), \delta(2), \delta(3))$ of $(1, 2, 3)$.

12.1 Multi Stability of Tri–Additive Functional Inequalities in Complex Banach Spaces

In this part, we investigate the multi stability results for the tri–additive $(\boldsymbol{\Psi}_1, \ldots, \boldsymbol{\Psi}_n)$–functional inequality (12.1) in complex Banach spaces.

Lemma 12.1 ([1, 3]) *Let $n \in \mathbb{N}$, and for every $\Psi_1, \Psi_2, \Psi_3, \Psi_4, \Psi_5, \Psi_6 \in \mathcal{G}$, and $\leftthreetimes \in \text{ሐ}$, the mappings $\underbrace{\mathcal{L}_i}_{i=1,\ldots,n} : \text{ሐ} \times \mathcal{G}^3 \to \mathcal{Q}$ satisfy $\underbrace{\mathcal{L}_i}_{i=1,\ldots,n} (\leftthreetimes, 0, \Psi_3, \Psi_5) = \underbrace{\mathcal{L}_i}_{i=1,\ldots,n} (\leftthreetimes, \Psi_1, 0, \Psi_5) = \underbrace{\mathcal{L}_i}_{i=1,\ldots,n} (\leftthreetimes, \Psi_1, \Psi_3, 0) = 0$ and the inequality (12.1). Then, $\underbrace{\mathcal{L}_i}_{i=1,\ldots,n} : \text{ሐ} \times \mathcal{G}^3 \to \mathcal{Q}$ are tri–additive.*

Theorem 12.1 *Let $n \in \mathbb{N}$, and $\underbrace{\varphi_i}_{i=1,\ldots,n} : \mathcal{G}^6 \to \mathbb{R}$ be functions, such that there exist $0 < \underbrace{\mathfrak{P}_i}_{i=1,\ldots,n} < 0.25$ with*

$$\underbrace{\varphi_i}_{i=1,\dots,n}\left(\frac{\Psi_1}{2},\frac{\Psi_2}{2},\frac{\Psi_3}{2},\frac{\Psi_4}{2},\frac{\Psi_5}{2},\frac{\Psi_6}{2}\right) < \frac{\overbrace{\mathfrak{P}_i}^{i=1,\dots,n}}{2}\underbrace{\varphi_i}_{i=1,\dots,n}(\Psi_1,\Psi_2,\Psi_3,\Psi_4,\Psi_5,\Psi_6) \quad (12.2)$$

and let $\underbrace{\mathcal{L}_i}_{i=1,\dots,n} :⋔ \times \mathcal{G}^3 \to \mathcal{Q}$ *be mappings satisfying* $\underbrace{\mathcal{L}_i}_{i=1,\dots,n}(\curlywedge,\Psi_1,0,\Psi_5) = \underbrace{\mathcal{L}_i}_{i=1,\dots,n}(\curlywedge,0,\Psi_3,\Psi_5) = \underbrace{\mathcal{L}_i}_{i=1,\dots,n}(\curlywedge,\Psi_1,\Psi_3,0) = 0$ *and the inequality below*

$$\begin{aligned}
diag\Big[&\|\mathcal{L}_1(\curlywedge,\Psi_1+\Psi_2,\Psi_3-\Psi_4,\Psi_5+\Psi_6)+\mathcal{L}_1(\curlywedge,\Psi_1-\Psi_2,\Psi_3+\Psi_4,\Psi_5-\Psi_6)\\
&-2\mathcal{L}_1(\curlywedge,\Psi_1,\Psi_3,\Psi_5)+2\mathcal{L}_1(\curlywedge,\Psi_1,\Psi_4,\Psi_6)\\
&-2\mathcal{L}_1(\curlywedge,\Psi_2,\Psi_3,\Psi_6)+2\mathcal{L}_1(\curlywedge,\Psi_2,\Psi_4,\Psi_5)\|,\\
&\dots,\|\mathcal{L}_n(\curlywedge,\Psi_1+\Psi_2,\Psi_3-\Psi_4,\Psi_5+\Psi_6)+\mathcal{L}_n(\curlywedge,\Psi_1-\Psi_2,\Psi_3+\Psi_4,\Psi_5-\Psi_6)\\
&-2\mathcal{L}_n(\curlywedge,\Psi_1,\Psi_3,\Psi_5)+2\mathcal{L}_n(\curlywedge,\Psi_1,\Psi_4,\Psi_6)\\
&-2\mathcal{L}_n(\curlywedge,\Psi_2,\Psi_3,\Psi_6)+2\mathcal{L}_n(\curlywedge,\Psi_2,\Psi_4,\Psi_5)\|\Big]\\
\preceq diag\Big[&\Big\|\boldsymbol{\Psi}_1\Big[2\mathcal{L}_1\Big(\curlywedge,\frac{\Psi_1+\Psi_2}{2},\Psi_3-\Psi_4,\Psi_5+\Psi_6\Big)\\
&+2\mathcal{L}_1\Big(\curlywedge,\frac{\Psi_1-\Psi_2}{2},\Psi_3+\Psi_4,\Psi_5-\Psi_6\Big)\\
&-2\mathcal{L}_1(\curlywedge,\Psi_1,\Psi_3,\Psi_5)+2\mathcal{L}_1(\curlywedge,\Psi_1,\Psi_4,\Psi_6)\\
&-2\mathcal{L}_1(\curlywedge,\Psi_2,\Psi_3,\Psi_6)+2\mathcal{L}_1(\curlywedge,\Psi_2,\Psi_4,\Psi_5)]\|\\
&+\varphi_1(\Psi_1,\Psi_2,\Psi_3,\Psi_4,\Psi_5,\Psi_6)\\
&,\dots,\Big\|\boldsymbol{\Psi}_n\Big[2\mathcal{L}_n\Big(\curlywedge,\frac{\Psi_1+\Psi_2}{2},\Psi_3-\Psi_4,\Psi_5+\Psi_6\Big)\\
&+2\mathcal{L}_n\Big(\curlywedge,\frac{\Psi_1-\Psi_2}{2},\Psi_3+\Psi_4,\Psi_5-\Psi_6\Big)\\
&-2\mathcal{L}_n(\curlywedge,\Psi_1,\Psi_3,\Psi_5)+2\mathcal{L}_n(\curlywedge,\Psi_1,\Psi_4,\Psi_6)\\
&-2\mathcal{L}_n(\curlywedge,\Psi_2,\Psi_3,\Psi_6)+2\mathcal{L}_n(\curlywedge,\Psi_2,\Psi_4,\Psi_5)]\|\\
&+\varphi_n(\Psi_1,\Psi_2,\Psi_3,\Psi_4,\Psi_5,\Psi_6)\Big],
\end{aligned} \quad (12.3)$$

for eavery $\Psi_1,\Psi_2,\Psi_3,\Psi_4,\Psi_5,\Psi_6 \in \mathcal{G}$, *and* $\curlywedge \in ⋔$. *Then, there exist unique tri–additive mappings* $\underbrace{\mathcal{L}'_i}_{i=1,\dots,n} :⋔ \times \mathcal{G}^3 \to \mathcal{Q}$ *satisfying the inequality below*

$$\begin{aligned}
diag\Big[&\|\mathcal{L}_1(\curlywedge,\Psi_1,\Psi_3,\Psi_5)-\mathcal{L}'_1(\curlywedge,\Psi_1,\Psi_3,\Psi_5)\|,\dots,\\
&\|\mathcal{L}_n(\curlywedge,\Psi_1,\Psi_3,\Psi_5)-\mathcal{L}'_n(\curlywedge,\Psi_1,\Psi_3,\Psi_5)\|\Big]\\
\preceq diag\Big[&\frac{1}{1-4\mathfrak{P}_1}\Big[\varphi_1\left(\Psi_1,0,\frac{\Psi_3}{2},\frac{\Psi_3}{2},\Psi_5,0\right)
\end{aligned} \quad (12.4)$$

$$+2\varphi_1\left(\frac{\Psi_1}{2},\frac{\Psi_1}{2},\frac{\Psi_3}{2},0,\Psi_5,0\right)+4\varphi_1\left(\frac{\Psi_1}{2},0,\frac{\Psi_3}{2},0,\frac{\Psi_5}{2},\frac{\Psi_5}{2}\right)\Bigg]$$

$$\ldots,\frac{1}{1-4\mathfrak{P}_n}\Bigg[\varphi_n\left(\Psi_1,0,\frac{\Psi_3}{2},\frac{\Psi_3}{2},\Psi_5,0\right)$$
$$+2\varphi_n\left(\frac{\Psi_1}{2},\frac{\Psi_1}{2},\frac{\Psi_3}{2},0,\Psi_5,0\right)+4\varphi_n\left(\frac{\Psi_1}{2},0,\frac{\Psi_3}{2},0,\frac{\Psi_5}{2},\frac{\Psi_5}{2}\right)\Bigg]\Bigg].$$

for every $\Psi_1,\Psi_3,\Psi_5\in\mathcal{G}$ *and* $\leftthreetimes\in\pitchfork$.

Proof Letting $\Psi_4=\Psi_6=0$ and $\Psi_2=\Psi_1$ in (12.3), we get

$$\operatorname{diag}\Bigg[\|\mathcal{L}_1(\leftthreetimes,2\Psi_1,\Psi_3,\Psi_5)-2\mathcal{L}_1(\leftthreetimes,\Psi_1,\Psi_3,\Psi_5)\|,\ldots,$$
$$\|\mathcal{L}_n(\leftthreetimes,2\Psi_1,\Psi_3,\Psi_5)-2\mathcal{L}_n(\leftthreetimes,\Psi_1,\Psi_3,\Psi_5)\|\Bigg]$$
$$\preceq\operatorname{diag}\Bigg[\varphi_1(\Psi_1,\Psi_1,\Psi_3,0,\Psi_5,0),\ldots,\varphi_n(\Psi_1,\Psi_1,\Psi_3,0,\Psi_5,0)\Bigg].$$

So we have that

$$\operatorname{diag}\Bigg[\|\mathcal{L}_1(\leftthreetimes,2\Psi_1,\Psi_3,2\Psi_5)-2\mathcal{L}_1(\leftthreetimes,\Psi_1,\Psi_3,2\Psi_5)\|,\ldots,$$
$$\|\mathcal{L}_n(\leftthreetimes,2\Psi_1,\Psi_3,2\Psi_5)-2\mathcal{L}_n(\leftthreetimes,\Psi_1,\Psi_3,2\Psi_5)\|\Bigg]$$
$$\preceq\operatorname{diag}\Bigg[\varphi_1(\Psi_1,\Psi_1,\Psi_3,0,2\Psi_5,0),\ldots,\varphi_n(\Psi_1,\Psi_1,\Psi_3,0,2\Psi_5,0)\Bigg],$$

for every $\Psi_1,\Psi_3,\Psi_5\in\mathcal{G}$ and $\leftthreetimes\in\pitchfork$.

Choosing $\Psi_2=\Psi_4=0$ and $\Psi_6=\Psi_5$ in the inequality (12.3), we have

$$\operatorname{diag}\Bigg[\|\mathcal{L}_1(\leftthreetimes,\Psi_1,\Psi_3,2\Psi_5)-2\mathcal{L}_1(\leftthreetimes,\Psi_1,\Psi_3,\Psi_5)\|,\ldots,$$
$$\|\mathcal{L}_n(\leftthreetimes,\Psi_1,\Psi_3,2\Psi_5)-2\mathcal{L}_n(\leftthreetimes,\Psi_1,\Psi_3,\Psi_5)\|\Bigg]$$
$$\preceq\operatorname{diag}\Bigg[\varphi_1(\Psi_1,0,\Psi_3,0,\Psi_5,\Psi_5),\ldots,\varphi_n(\Psi_1,0,\Psi_3,0,\Psi_5,\Psi_5)\Bigg],$$

which implies that

$$\operatorname{diag}\Bigg[\|\mathcal{L}_1(\leftthreetimes,2\Psi_1,\Psi_3,2\Psi_5)-4\mathcal{L}_1(\leftthreetimes,\Psi_1,\Psi_3,\Psi_5)\|,\ldots,\tag{12.5}$$
$$\|\mathcal{L}_n(\leftthreetimes,2\Psi_1,\Psi_3,2\Psi_5)-4\mathcal{L}_n(\leftthreetimes,\Psi_1,\Psi_3,\Psi_5)\|\Bigg]$$

$$\preceq \operatorname{diag}\Big[\varphi_1(\Psi_1, \Psi_1, \Psi_3, 0, 2\Psi_5, 0) + 2\varphi_1(\Psi_1, 0, \Psi_3, 0, \Psi_5, \Psi_5), \ldots, \varphi_n(\Psi_1, \Psi_1, \Psi_3, 0, 2\Psi_5, 0) + 2\varphi_n(\Psi_1, 0, \Psi_3, 0, \Psi_5, \Psi_5)\Big],$$

for every $\Psi_1, \Psi_3, \Psi_5 \in \mathcal{G}$ and $\curlywedge \in \pitchfork$.

Setting $\Psi_2 = \Psi_6 = 0$ and $\Psi_4 = \Psi_3$ in the inequality (12.3), we get

$$\operatorname{diag}\Big[\|\mathcal{L}_1(\curlywedge, \Psi_1, 2\Psi_3, \Psi_5) - 2\mathcal{L}_1(\curlywedge, \Psi_1, \Psi_3, \Psi_5)\|, \ldots, \|\mathcal{L}_n(\curlywedge, \Psi_1, 2\Psi_3, \Psi_5) - 2\mathcal{L}_n(\curlywedge, \Psi_1, \Psi_3, \Psi_5)\|\Big]$$
$$\preceq \operatorname{diag}\Big[\varphi_1(\Psi_1, 0, \Psi_3, \Psi_3, \Psi_5, 0), \ldots, \varphi_n(\Psi_1, 0, \Psi_3, \Psi_3, \Psi_5, 0)\Big].$$

Hence, we get

$$\operatorname{diag}\Big[\|\mathcal{L}_1(\curlywedge, 2\Psi_1, 2\Psi_3, 2\Psi_5) - 2\mathcal{L}_1(\curlywedge, 2\Psi_1, \Psi_3, 2\Psi_5)\|, \ldots, \|\mathcal{L}_n(\curlywedge, 2\Psi_1, 2\Psi_3, 2\Psi_5) - 2\mathcal{L}_n(\curlywedge, 2\Psi_1, \Psi_3, 2\Psi_5)\|\Big]$$
$$\preceq \operatorname{diag}\Big[\varphi_1(2\Psi_1, 0, \Psi_3, \Psi_3, 2\Psi_5, 0), \ldots, \varphi_n(2\Psi_1, 0, \Psi_3, \Psi_3, 2\Psi_5, 0)\Big], \tag{12.6}$$

for every $\Psi_1, \Psi_3, \Psi_5 \in \mathcal{G}$ and $\curlywedge \in \pitchfork$.

According to the inequalities (12.5) and (12.6), we have

$$\operatorname{diag}\Big[\|\mathcal{L}_1(\curlywedge, 2\Psi_1, 2\Psi_3, 2\Psi_5) - 8\mathcal{L}_1(\curlywedge, \Psi_1, \Psi_3, \Psi_5)\|, \ldots, \|\mathcal{L}_n(\curlywedge, 2\Psi_1, 2\Psi_3, 2\Psi_5) - 8\mathcal{L}_n(\curlywedge, \Psi_1, \Psi_3, \Psi_5)\|\Big] \tag{12.7}$$
$$\preceq \operatorname{diag}\Big[\varphi_1(2\Psi_1, 0, \Psi_3, \Psi_3, 2\Psi_5, 0) + 2\varphi_1(\Psi_1, \Psi_1, \Psi_3, 0, 2\Psi_5, 0) + 4\varphi_1(\Psi_1, 0, \Psi_3, 0, \Psi_5, \Psi_5), \ldots, \varphi_n(2\Psi_1, 0, \Psi_3, \Psi_3, 2\Psi_5, 0) + 2\varphi_n(\Psi_1, \Psi_1, \Psi_3, 0, 2\Psi_5, 0) + 4\varphi_n(\Psi_1, 0, \Psi_3, 0, \Psi_5, \Psi_5)\Big],$$

and so

$$\operatorname{diag}\Big[\Big\|\mathcal{L}_1(\curlywedge, \Psi_1, \Psi_3, \Psi_5) - 8\mathcal{L}_1\Big(\curlywedge, \frac{\Psi_1}{2}, \frac{\Psi_3}{2}, \frac{\Psi_5}{2}\Big)\Big\|, \ldots, \Big\|\mathcal{L}_n(\curlywedge, \Psi_1, \Psi_3, \Psi_5) - 8\mathcal{L}_n\Big(\curlywedge, \frac{\Psi_1}{2}, \frac{\Psi_3}{2}, \frac{\Psi_5}{2}\Big)\Big\|\Big] \tag{12.8}$$
$$\preceq \operatorname{diag}\Big[\varphi_1\Big(\Psi_1, 0, \frac{\Psi_3}{2}, \frac{\Psi_3}{2}, \Psi_5, 0\Big) + 2\varphi_1\Big(\frac{\Psi_1}{2}, \frac{\Psi_1}{2}, \frac{\Psi_3}{2}, 0, \Psi_5, 0\Big)$$

$$+4\varphi_1\left(\frac{\Psi_1}{2},0,\frac{\Psi_3}{2},0,\frac{\Psi_5}{2},\frac{\Psi_5}{2}\right),\ldots,$$
$$\varphi_n\left(\Psi_1,0,\frac{\Psi_3}{2},\frac{\Psi_3}{2},\Psi_5,0\right)+2\varphi_n\left(\frac{\Psi_1}{2},\frac{\Psi_1}{2},\frac{\Psi_3}{2},0,\Psi_5,0\right)$$
$$+4\varphi_n\left(\frac{\Psi_1}{2},0,\frac{\Psi_3}{2},0,\frac{\Psi_5}{2},\frac{\Psi_5}{2}\right)\Bigg],$$

for every $\Psi_1, \Psi_3, \Psi_5 \in \mathcal{G}$ and $\curlywedge \in ⫛$.

Consider the set

$$\Xi := \Bigg\{ \underbrace{\iota_i}_{i=1,\ldots,n} : ⫛ \times \mathcal{G}^3 \to \mathcal{Q} :$$
$$\underbrace{\iota_i}_{i=1,\ldots,n}(\curlywedge,\Psi_1,0,\Psi_5) = \underbrace{\iota_i}_{i=1,\ldots,n}(\curlywedge,0,\Psi_3,\Psi_5) = \underbrace{\iota_i}_{i=1,\ldots,n}(\curlywedge,\Psi_1,\Psi_3,0) = 0\Bigg\},$$

and introduce the vector valued generalized metric $\lambda : \Xi^n \times \Xi^n \longrightarrow [0,+\infty]^{3n}$, as follows:

$$\lambda\Big((\iota_1,\ldots,\iota_n),(\varsigma_1,\ldots,\varsigma_n)\Big) \tag{12.9}$$
$$=\Big(\lambda_1\Big((\iota_1,\ldots,\iota_n),(\varsigma_1,\ldots,\varsigma_n)\Big),\lambda_2\Big((\iota_1,\ldots,\iota_n),(\varsigma_1,\ldots,\varsigma_n)\Big),\lambda_3\Big((\iota_1,\ldots,\iota_n),(\varsigma_1,\ldots,\varsigma_n)\Big)\Big)$$
$$=\inf\Bigg\{(\mu_1,\mu_2,\mu_3)^n \in \mathbb{R}_+^{3n} : \mathrm{diag}\Big[\|\iota_1(\curlywedge,\Psi_1,\Psi_3,\Psi_5)-\varsigma_1(\curlywedge,\Psi_1,\Psi_3,\Psi_5)\|,\ldots,$$
$$\|\iota_n(\curlywedge,\Psi_1,\Psi_3,\Psi_5)-\varsigma_n(\curlywedge,\Psi_1,\Psi_3,\Psi_5)\|\Big]$$
$$\preceq \mathrm{diag}\Big[\mu_1\varphi_1\left(\Psi_1,0,\frac{\Psi_3}{2},\frac{\Psi_3}{2},\Psi_5,0\right)+\mu_2\varphi_1\left(\frac{\Psi_1}{2},\frac{\Psi_1}{2},\frac{\Psi_3}{2},0,\Psi_5,0\right)$$
$$+\mu_3\varphi_1\left(\frac{\Psi_1}{2},0,\frac{\Psi_3}{2},0,\frac{\Psi_5}{2},\frac{\Psi_5}{2}\right),\ldots,$$
$$\mu_1\varphi_n\left(\Psi_1,0,\frac{\Psi_3}{2},\frac{\Psi_3}{2},\Psi_5,0\right)+\mu_2\varphi_n\left(\frac{\Psi_1}{2},\frac{\Psi_1}{2},\frac{\Psi_3}{2},0,\Psi_5,0\right)$$
$$+\mu_3\varphi_n\left(\frac{\Psi_1}{2},0,\frac{\Psi_3}{2},0,\frac{\Psi_5}{2},\frac{\Psi_5}{2}\right)\Big],\ \forall\Psi_1,\Psi_3,\Psi_5\in\mathcal{G}\Bigg\},$$

where we denote $\inf \boldsymbol{\emptyset} = (\inf\emptyset, \inf\emptyset, \inf\emptyset) = (+\infty,+\infty,+\infty)$.

We prove that (Ξ^n, λ) is a generalized metric space. For the metric part, we prove the triangle inequality below:

$$\lambda\Big((\iota_1,\ldots,\iota_n),(\varsigma_1,\ldots,\varsigma_n)\Big)\le\lambda\Big((\iota_1,\ldots,\iota_n),(\upsilon_1,\ldots,\upsilon_n)\Big)+\lambda\Big((\upsilon_1,\ldots,\upsilon_n),(\varsigma_1,\ldots,\varsigma_n)\Big),$$

for every $(\iota_1,\ldots,\iota_n),(\varsigma_1,\ldots,\varsigma_n),(\upsilon_1,\ldots,\upsilon_n)\in\Xi^n$. Now, we have

$$
\begin{aligned}
&\lambda\Big((\iota_1,\ldots,\iota_n),(\varsigma_1,\ldots,\varsigma_n)\Big)\\
&=\Big(\lambda_1((\iota_1,\ldots,\iota_n),(\varsigma_1,\ldots,\varsigma_n)),\lambda_2((\iota_1,\ldots,\iota_n),(\varsigma_1,\ldots,\varsigma_n)),\lambda_3((\iota_1,\ldots,\iota_n),(\varsigma_1,\ldots,\varsigma_n))\Big)\\
&=\inf\Bigg\{(\mu_1,\mu_2,\mu_3)^n\in\mathbb{R}_+^{3n}:\operatorname{diag}\Big[\|\iota_1(\curlywedge,\Psi_1,\Psi_3,\Psi_5)-\varsigma_1(\curlywedge,\Psi_1,\Psi_3,\Psi_5)\|,\\
&\qquad\ldots\|\iota_n(\curlywedge,\Psi_1,\Psi_3,\Psi_5)-\varsigma_n(\curlywedge,\Psi_1,\Psi_3,\Psi_5)\|\Big]\\
&\qquad\preceq\operatorname{diag}\Big[\mu_1\varphi_1\left(\Psi_1,0,\frac{\Psi_3}{2},\frac{\Psi_3}{2},\Psi_5,0\right)+\mu_2\varphi_1\left(\frac{\Psi_1}{2},\frac{\Psi_1}{2},\frac{\Psi_3}{2},0,\Psi_5,0\right)\\
&\qquad+\mu_3\varphi_1\left(\frac{\Psi_1}{2},0,\frac{\Psi_3}{2},0,\frac{\Psi_5}{2},\frac{\Psi_5}{2}\right),\\
&\qquad\ldots,\mu_1\varphi_n\left(\Psi_1,0,\frac{\Psi_3}{2},\frac{\Psi_3}{2},\Psi_5,0\right)+\mu_2\varphi_n\left(\frac{\Psi_1}{2},\frac{\Psi_1}{2},\frac{\Psi_3}{2},0,\Psi_5,0\right)\\
&\qquad+\mu_3\varphi_n\left(\frac{\Psi_1}{2},0,\frac{\Psi_3}{2},0,\frac{\Psi_5}{2},\frac{\Psi_5}{2}\right)\Big],\quad\forall\Psi_1,\Psi_3,\Psi_5\in\mathcal{G}\Bigg\}
\end{aligned}
$$

$$
\begin{aligned}
&\le\inf\Bigg\{(\mu_1,\mu_2,\mu_3)^n\in\mathbb{R}_+^{3n}:\operatorname{diag}\Big[\|\iota_1(\curlywedge,\Psi_1,\Psi_3,\Psi_5)-\upsilon_1(\curlywedge,\Psi_1,\Psi_3,\Psi_5)\|\\
&\qquad+\|\upsilon_1(\curlywedge,\Psi_1,\Psi_3,\Psi_5)-\varsigma_1(\curlywedge,\Psi_1,\Psi_3,\Psi_5)\|,\ldots,\\
&\|\iota_n(\curlywedge,\Psi_1,\Psi_3,\Psi_5)-\upsilon_n(\curlywedge,\Psi_1,\Psi_3,\Psi_5)\|+\|\upsilon_n(\curlywedge,\Psi_1,\Psi_3,\Psi_5)-\varsigma_n(\curlywedge,\Psi_1,\Psi_3,\Psi_5)\|\Big]\\
&\qquad\preceq\operatorname{diag}\Big[\mu_1\varphi_1\left(\Psi_1,0,\frac{\Psi_3}{2},\frac{\Psi_3}{2},\Psi_5,0\right)+\mu_2\varphi_1\left(\frac{\Psi_1}{2},\frac{\Psi_1}{2},\frac{\Psi_3}{2},0,\Psi_5,0\right)\\
&\qquad+\mu_3\varphi_1\left(\frac{\Psi_1}{2},0,\frac{\Psi_3}{2},0,\frac{\Psi_5}{2},\frac{\Psi_5}{2}\right),\\
&\qquad\ldots,\mu_1\varphi_n\left(\Psi_1,0,\frac{\Psi_3}{2},\frac{\Psi_3}{2},\Psi_5,0\right)+\mu_2\varphi_n\left(\frac{\Psi_1}{2},\frac{\Psi_1}{2},\frac{\Psi_3}{2},0,\Psi_5,0\right)\\
&\qquad+\mu_3\varphi_n\left(\frac{\Psi_1}{2},0,\frac{\Psi_3}{2},0,\frac{\Psi_5}{2},\frac{\Psi_5}{2}\right)\Big],\quad\forall\Psi_1,\Psi_3,\Psi_5\in\mathcal{G}\Bigg\}
\end{aligned}
$$

$$
\begin{aligned}
&\le\inf\Bigg\{(\mu_1,\mu_2,\mu_3)^n\in\mathbb{R}_+^{3n}:\|\iota_1(\curlywedge,\Psi_1,\Psi_3,\Psi_5)-\upsilon_1(\curlywedge,\Psi_1,\Psi_3,\Psi_5)\|,\\
&\qquad\ldots,\|\iota_n(\curlywedge,\Psi_1,\Psi_3,\Psi_5)-\upsilon_n(\curlywedge,\Psi_1,\Psi_3,\Psi_5)\|\Big]\\
&\qquad\preceq\operatorname{diag}\Big[\mu_1\varphi_1\left(\Psi_1,0,\frac{\Psi_3}{2},\frac{\Psi_3}{2},\Psi_5,0\right)+\mu_2\varphi_1\left(\frac{\Psi_1}{2},\frac{\Psi_1}{2},\frac{\Psi_3}{2},0,\Psi_5,0\right)\\
&\qquad+\mu_3\varphi_1\left(\frac{\Psi_1}{2}0,\frac{\Psi_3}{2},0,\frac{\Psi_5}{2},\frac{\Psi_5}{2}\right),\\
&\qquad,\ldots,\mu_1\varphi_n\left(\Psi_1,0,\frac{\Psi_3}{2},\frac{\Psi_3}{2},\Psi_5,0\right)+\mu_2\varphi_n\left(\frac{\Psi_1}{2},\frac{\Psi_1}{2},\frac{\Psi_3}{2},0,\Psi_5,0\right)\\
&\qquad+\mu_3\varphi_n\left(\frac{\Psi_1}{2}0,\frac{\Psi_3}{2},0,\frac{\Psi_5}{2},\frac{\Psi_5}{2}\right)\Big],\quad\forall\Psi_1,\Psi_3,\Psi_5\in\mathcal{G}\Bigg\}
\end{aligned}
$$

$$+\inf\left\{(\mu_1,\mu_2,\mu_3)^n\in\mathbb{R}_+^{3n}:\|\upsilon_1(\curlywedge,\Psi_1,\Psi_3,\Psi_5)-\varsigma_1(\curlywedge,\Psi_1,\Psi_3,\Psi_5)\|,\right.$$
$$\left.\ldots,\|\upsilon_n(\curlywedge,\Psi_1,\Psi_3,\Psi_5)-\varsigma_n(\curlywedge,\Psi_1,\Psi_3,\Psi_5)\|\right]$$
$$\preceq \operatorname{diag}\left[\mu_1\varphi_1\left(\Psi_1,0,\frac{\Psi_3}{2},\frac{\Psi_3}{2},\Psi_5,0\right)+\mu_2\varphi_1\left(\frac{\Psi_1}{2},\frac{\Psi_1}{2},\frac{\Psi_3}{2},0,\Psi_5,0\right)\right.$$
$$+\mu_3\varphi_1\left(\frac{\Psi_1}{2},0,\frac{\Psi_3}{2},0,\frac{\Psi_5}{2},\frac{\Psi_5}{2}\right),$$
$$,\ldots\mu_1\varphi_n\left(\Psi_1,0,\frac{\Psi_3}{2},\frac{\Psi_3}{2},\Psi_5,0\right)+\mu_2\varphi_n\left(\frac{\Psi_1}{2},\frac{\Psi_1}{2},\frac{\Psi_3}{2},0,\Psi_5,0\right)$$
$$\left.+\mu_3\varphi_n\left(\frac{\Psi_1}{2},0,\frac{\Psi_3}{2},0,\frac{\Psi_5}{2},\frac{\Psi_5}{2}\right)\right],\ \ \forall\Psi_1,\Psi_3,\Psi_5\in\mathcal{G}\Big\}$$
$$=\lambda((\iota_1,\ldots,\iota_n),(\upsilon_1,\ldots,\upsilon_n))+\lambda((\upsilon_1,\ldots,\upsilon_n),(\varsigma_1,\ldots,\varsigma_n)). \tag{12.10}$$

Next, we deduce that (Ξ^n,λ) is complete. Let $(\underbrace{\omega_i}_{i=1,\ldots,n})_\iota$ be a Cauchy sequence in (Ξ,λ). Then, for every $\epsilon_1,\epsilon_2,\epsilon_3>0$, there exist an $\aleph_{\epsilon_1,\epsilon_2,\epsilon_3}\in\mathbb{N}$, such that

$$\lambda\Big(((\omega_1)_\jmath,\ldots,(\omega_n)_\jmath),((\omega_1)_\iota,\ldots,(\omega_n)_\iota)\Big)\preceq(\epsilon_1,\epsilon_2,\epsilon_3)^n,$$

for every $\jmath,\iota\geq\aleph_{\epsilon_1,\epsilon_2,\epsilon_3}$. From the relation (12.9), we get

$$\operatorname{diag}\Big[\|(\omega_1)_\jmath(\curlywedge,\Psi_1,\Psi_3,\Psi_5)-(\omega_1)_\iota(\curlywedge,\Psi_1,\Psi_3,\Psi_5)\|, \tag{12.11}$$
$$\ldots,\|(\omega_n)_\jmath(\curlywedge,\Psi_1,\Psi_3,\Psi_5)-(\omega_n)_\iota(\curlywedge,\Psi_1,\Psi_3,\Psi_5)\|\Big]$$
$$\preceq\operatorname{diag}\left[\epsilon_1\varphi_1\left(\Psi_1,0,\frac{\Psi_3}{2},\frac{\Psi_3}{2},\Psi_5,0\right)+\epsilon_2\varphi_1\left(\frac{\Psi_1}{2},\frac{\Psi_1}{2},\frac{\Psi_3}{2},0,\Psi_5,0\right)\right.$$
$$+\epsilon_3\varphi_1\left(\frac{\Psi_1}{2},0,\frac{\Psi_3}{2},0,\frac{\Psi_5}{2},\frac{\Psi_5}{2}\right),\ldots,$$
$$\epsilon_1\varphi_n\left(\Psi_1,0,\frac{\Psi_3}{2},\frac{\Psi_3}{2},\Psi_5,0\right)+\epsilon_2\varphi_n\left(\frac{\Psi_1}{2},\frac{\Psi_1}{2},\frac{\Psi_3}{2},0,\Psi_5,0\right)$$
$$\left.+\epsilon_n\varphi_n\left(\frac{\Psi_1}{2},0,\frac{\Psi_3}{2},0,\frac{\Psi_5}{2},\frac{\Psi_5}{2}\right)\right],$$

for every $\Psi_1,\Psi_3,\Psi_5\in\mathcal{G}$ and $\curlywedge\in\pitchfork$. If Ψ_1,Ψ_3,Ψ_5 be fixed, then, the inequality (12.11) implies that $(\underbrace{\omega_i}_{i=1,\ldots,n})_\iota(\curlywedge,\Psi_1,\Psi_3,\Psi_5)$ are Cauchy sequences in $\mathcal{Q}$. Since $\mathcal{Q}$ is complete, $(\underbrace{\omega_i}_{i=1,\ldots,n})_\iota(\curlywedge,\Psi_1,\Psi_3,\Psi_5)$ are convergent, for every $\Psi_1,\Psi_3,\Psi_5\in\mathcal{G}$ and $\curlywedge\in\pitchfork$. Thus, we can define a mapping $\underbrace{\omega_i}_{i=1,\ldots,n}$ by

$$\underbrace{\omega_i}_{i=1,\ldots,n}(\curlywedge, \Psi_1, \Psi_3, \Psi_5) := \lim_{\iota\to\infty}(\underbrace{\omega_i}_{i=1,\ldots,n})_\iota(\curlywedge, \Psi_1, \Psi_3, \Psi_5), \tag{12.12}$$

for every $\Psi_1, \Psi_3, \Psi_5 \in \mathcal{G}$ and $\curlywedge \in \pitchfork$.

It is easy to check $\underbrace{\omega_i}_{i=1,\ldots,n} \in \Xi$. Indeed, if we let $\iota \to \infty$, then it follows from the inequality (12.11) that

$$\text{diag}\Big[\|\omega_1(\curlywedge, \Psi_1, \Psi_3, \Psi_5) - (\omega_1)_\iota(\curlywedge, \Psi_1, \Psi_3, \Psi_5)\|, \ldots, \tag{12.13}$$
$$\|\omega_n(\curlywedge, \Psi_1, \Psi_3, \Psi_5) - (\omega_n)_\iota(\curlywedge, \Psi_1, \Psi_3, \Psi_5)\|\Big]$$
$$\preceq \text{diag}\Big[\epsilon_1\varphi_1\left(\Psi_1, 0, \frac{\Psi_3}{2}, \frac{\Psi_3}{2}, \Psi_5, 0\right) + \epsilon_2\varphi_1\left(\frac{\Psi_1}{2}, \frac{\Psi_1}{2}, \frac{\Psi_3}{2}, 0, \Psi_5, 0\right)$$
$$+\epsilon_3\varphi_1\left(\frac{\Psi_1}{2}, 0, \frac{\Psi_3}{2}, 0, \frac{\Psi_5}{2}, \frac{\Psi_5}{2}\right),$$
$$\ldots, \epsilon_1\varphi_n\left(\Psi_1, 0, \frac{\Psi_3}{2}, \frac{\Psi_3}{2}, \Psi_5, 0\right) + \epsilon_2\varphi_n\left(\frac{\Psi_1}{2}, \frac{\Psi_1}{2}, \frac{\Psi_3}{2}, 0, \Psi_5, 0\right)$$
$$+\epsilon_3\varphi_n\left(\frac{\Psi_1}{2}, 0, \frac{\Psi_3}{2}, 0, \frac{\Psi_5}{2}, \frac{\Psi_5}{2}\right)\Big], \tag{12.14}$$

for every $\Psi_1, \Psi_3, \Psi_5 \in \mathcal{G}$.

Considering the relation (12.9), we get

$$\lambda\Big((\omega_1, \ldots, \omega_n), ((\omega_1)_\iota, \ldots, (\omega_n)_\iota)\Big) \preceq (\epsilon_1, \epsilon_2, \epsilon_3)^n.$$

This implies that the Cauchy sequences $(\underbrace{\omega_i}_{i=1,\ldots,n})_\iota$ are convergent to $\underbrace{\omega_i}_{i=1,\ldots,n}$ in (Ξ, λ). Thus, (Ξ^n, λ) is complete.

We now define the linear mapping $\underbrace{\Gamma_i}_{i=1,\ldots,n} : \Xi \to \Xi$, as follows:

$$\underbrace{\Gamma_i}_{i=1,\ldots,n}(\underbrace{\iota_i}_{i=1,\ldots,n}(\curlywedge, \Psi_1, \Psi_3, \Psi_5)) := 8\underbrace{\iota_i}_{i=1,\ldots,n}\left(\curlywedge, \frac{\Psi_1}{2}, \frac{\Psi_3}{2}, \frac{\Psi_5}{2}\right),$$

for every $\Psi_1, \Psi_3, \Psi_5 \in \mathcal{G}$ and $\curlywedge \in \pitchfork$.

Let $\underbrace{\iota_i}_{i=1,\ldots,n}, \underbrace{\varsigma_i}_{i=1,\ldots,n} \in \Xi$ be given, such that $\lambda\Big((\iota_1, \ldots, \iota_n), (\varsigma_1, \ldots, \varsigma_n)\Big) = (\varepsilon_1, \varepsilon_2, \varepsilon_3)^n$. Then, we have

$$\begin{aligned}
&\operatorname{diag}\Big[\|\iota_1(\leftthreetimes,\Psi_1,\Psi_3,\Psi_5)-\varsigma_1(\leftthreetimes,\Psi_1,\Psi_3,\Psi_5)\|,\\
&\qquad\ldots,\|\iota_n(\leftthreetimes,\Psi_1,\Psi_3,\Psi_5)-\varsigma_n(\leftthreetimes,\Psi_1,\Psi_3,\Psi_5)\|\Big]\\
&\preceq \operatorname{diag}\Big[\epsilon_1\varphi_1\left(\Psi_1,0,\frac{\Psi_3}{2},\frac{\Psi_3}{2},\Psi_5,0\right)+\epsilon_2\varphi_1\left(\frac{\Psi_1}{2},\frac{\Psi_1}{2},\frac{\Psi_3}{2},0,\Psi_5,0\right)\\
&\qquad+\epsilon_3\varphi_1\left(\frac{\Psi_1}{2},0,\frac{\Psi_3}{2},0,\frac{\Psi_5}{2},\frac{\Psi_5}{2}\right),\\
&\qquad,\ldots\epsilon_1\varphi_n\left(\Psi_1,0,\frac{\Psi_3}{2},\frac{\Psi_3}{2},\Psi_5,0\right)+\epsilon_2\varphi_n\left(\frac{\Psi_1}{2},\frac{\Psi_1}{2},\frac{\Psi_3}{2},0,\Psi_5,0\right)\\
&\qquad+\epsilon_3\varphi_n\left(\frac{\Psi_1}{2},0,\frac{\Psi_3}{2},0,\frac{\Psi_5}{2},\frac{\Psi_5}{2}\right)\Big],
\end{aligned}$$

for every $\Psi_1,\Psi_3,\Psi_5\in\mathcal{G}$ and $\leftthreetimes\in\pitchfork$. Thus, we get

$$\begin{aligned}
&\operatorname{diag}\Big[\|\Gamma_1(\iota_1(\leftthreetimes,\Psi_1,\Psi_3,\Psi_5))-\Gamma_1(\varsigma_1(\leftthreetimes,\Psi_1,\Psi_3,\Psi_5))\|,\\
&\qquad\ldots,\|\Gamma_n(\iota_n(\leftthreetimes,\Psi_1,\Psi_3,\Psi_5))-\Gamma_n(\varsigma_n(\leftthreetimes,\Psi_1,\Psi_3,\Psi_5))\|\Big]\\
&=\operatorname{diag}\Big[\left\|8\iota_1\left(\leftthreetimes,\frac{\Psi_1}{2},\frac{\Psi_3}{2},\frac{\Psi_5}{2}\right)-8\varsigma_1\left(\leftthreetimes,\frac{\Psi_1}{2},\frac{\Psi_3}{2},\frac{\Psi_5}{2}\right)\right\|,\\
&\qquad\ldots,\left\|8\iota_n\left(\leftthreetimes,\frac{\Psi_1}{2},\frac{\Psi_3}{2},\frac{\Psi_5}{2}\right)-8\varsigma_n\left(\leftthreetimes,\frac{\Psi_1}{2},\frac{\Psi_3}{2},\frac{\Psi_5}{2}\right)\right\|\Big]\\
&\preceq\operatorname{diag}\Big[8\varepsilon_1\varphi_1\left(\frac{\Psi_1}{2},0,\frac{\Psi_3}{4},\frac{\Psi_3}{4},\frac{\Psi_5}{2},0\right)+8\varepsilon_2\varphi_1\left(\frac{\Psi_1}{4},\frac{\Psi_1}{4},\frac{\Psi_3}{4},0,\frac{\Psi_5}{2},0\right)\\
&\qquad+8\varepsilon_3\varphi_1\left(\frac{\Psi_1}{4},0,\frac{\Psi_3}{4},0,\frac{\Psi_5}{4},\frac{\Psi_5}{4}\right),\\
&\qquad\ldots,8\varepsilon_1\varphi_n\left(\frac{\Psi_1}{2},0,\frac{\Psi_3}{4},\frac{\Psi_3}{4},\frac{\Psi_5}{2},0\right)+8\varepsilon_2\varphi_n\left(\frac{\Psi_1}{4},\frac{\Psi_1}{4},\frac{\Psi_3}{4},0,\frac{\Psi_5}{2},0\right)\\
&\qquad+8\varepsilon_3\varphi_n\left(\frac{\Psi_1}{4},0,\frac{\Psi_3}{4},0,\frac{\Psi_5}{4},\frac{\Psi_5}{4}\right)\Big]\\
&\preceq\operatorname{diag}\Big[4\varepsilon_1\mathfrak{P}_1\varphi_1\left(\Psi_1,0,\frac{\Psi_3}{2},\frac{\Psi_3}{2},\Psi_5,0\right)+4\varepsilon_2\mathfrak{P}_1\varphi_1\left(\frac{\Psi_1}{2},\frac{\Psi_1}{2},\frac{\Psi_3}{2},0,\Psi_5,0\right)\\
&\qquad+4\varepsilon_3\mathfrak{P}_1\varphi_1\left(\frac{\Psi_1}{2},0,\frac{\Psi_3}{2},0,\frac{\Psi_5}{2},\frac{\Psi_5}{2}\right),\\
&\qquad\ldots,4\varepsilon_1\mathfrak{P}_n\varphi_n\left(\Psi_1,0,\frac{\Psi_3}{2},\frac{\Psi_3}{2},\Psi_5,0\right)+4\varepsilon_2\mathfrak{P}_n\varphi_n\left(\frac{\Psi_1}{2},\frac{\Psi_1}{2},\frac{\Psi_3}{2},0,\Psi_5,0\right)\\
&\qquad+4\varepsilon_3\mathfrak{P}_n\varphi_n\left(\frac{\Psi_1}{2},0,\frac{\Psi_3}{2},0,\frac{\Psi_5}{2},\frac{\Psi_5}{2}\right)\Big],
\end{aligned}\tag{12.15}$$

for every $\Psi_1,\Psi_3,\Psi_5\in\mathcal{G},\leftthreetimes\in\pitchfork$ and $\underbrace{(\mathfrak{P}_1,\ldots,\mathfrak{P}_n)}_{n}>\underbrace{(0,\ldots,0)}_{n}$. Thus, we have that

$$\lambda\Big((\Gamma_1(\iota_1),\ldots,\Gamma_n(\iota_n)),(\Gamma_1(\varsigma_1),\ldots,\Gamma_n(\varsigma_n))\Big)\preceq 4\underbrace{\Big(\mathfrak{P}_1(\varepsilon_1,\varepsilon_2,\varepsilon_3),\ldots,\mathfrak{P}_n(\varepsilon_1,\varepsilon_2,\varepsilon_3)\Big)}_{n},$$

for every $\underbrace{\iota_i}_{i=1,\ldots,n}, \underbrace{\varsigma_i}_{i=1,\ldots,n} \in \Xi$. This means that

$$\lambda\Big((\Gamma_1(\iota_1),\ldots,\Gamma_n(\iota_n)),(\Gamma_1(\varsigma_1),\ldots,\Gamma_n(\varsigma_n))\Big)\preceq 4(\mathfrak{P}_1,\ldots,\mathfrak{P}_n)\lambda\Big((\iota_1,\ldots,\iota_n),(\varsigma_1,\ldots,\varsigma_n)\Big),$$

for every $\underbrace{\iota_i}_{i=1,\ldots,n}, \underbrace{\varsigma_i}_{i=1,\ldots,n} \in \Xi$.

It follows from the inequality (12.8) that

$$\begin{aligned}
&\operatorname{diag}\Bigg[\left\|\mathcal{L}_1(\leftthreetimes,\Psi_1,\Psi_3,\Psi_5)-8\mathcal{L}_1\left(\leftthreetimes,\frac{\Psi_1}{2},\frac{\Psi_3}{2},\frac{\Psi_5}{2}\right)\right\|, \qquad (12.16)\\
&\qquad\ldots,\left\|\mathcal{L}_n(\leftthreetimes,\Psi_1,\Psi_3,\Psi_5)-8\mathcal{L}_n\left(\leftthreetimes,\frac{\Psi_1}{2},\frac{\Psi_3}{2},\frac{\Psi_5}{2}\right)\right\|\Bigg]\\
&\preceq \operatorname{dig}\Bigg[\varphi_1\left(\Psi_1,0,\frac{\Psi_3}{2},\frac{\Psi_3}{2},\Psi_5,0\right)+2\varphi_1\left(\frac{\Psi_1}{2},\frac{\Psi_1}{2},\frac{\Psi_3}{2},0,\Psi_5,0\right)\\
&\qquad+4\varphi_1\left(\frac{\Psi_1}{2},0,\frac{\Psi_3}{2},0,\frac{\Psi_5}{2},\frac{\Psi_5}{2}\right),\\
&\qquad\ldots,\varphi_n\left(\Psi_1,0,\frac{\Psi_3}{2},\frac{\Psi_3}{2},\Psi_5,0\right)+2\varphi_n\left(\frac{\Psi_1}{2},\frac{\Psi_1}{2},\frac{\Psi_3}{2},0,\Psi_5,0\right)\\
&\qquad+4\varphi_n\left(\frac{\Psi_1}{2},0,\frac{\Psi_3}{2},0,\frac{\Psi_5}{2},\frac{\Psi_5}{2}\right)\Bigg],
\end{aligned}$$

for every $\Psi_1,\Psi_3,\Psi_5\in\mathcal{G}$, $\leftthreetimes\in\pitchfork$, and $\underbrace{\mathcal{L}_i}_{i=1,\ldots,n}\in\Xi$. Thus, we get

$$\lambda\Big((\mathcal{L}_1,\ldots,\mathcal{L}_n),(\Gamma_1\mathcal{L}_1,\ldots,\Gamma_n\mathcal{L}_n)\Big)\preceq\underbrace{\Big((1,2,4),\ldots,(1,2,4)\Big)}_{n}.$$

Based on Theorem 9.5, we can find mappings $\underbrace{\mathcal{L}'_i}_{i=1,\ldots,n}:\pitchfork\times\mathcal{G}^3\to\mathcal{Q}$ satisfying the following:

(1) The maps $\underbrace{\mathcal{L}'_i}_{i=1,\ldots,n}$ are the fixed points of $\underbrace{\Gamma_i}_{i=1,\ldots,n}$, i.e.,

$$\underbrace{\Gamma_i}_{i=1,\ldots,n}\left(\underbrace{\mathcal{L}'_i}_{i=1,\ldots,n}(\leftthreetimes,\Psi_1,\Psi_3,\Psi_5)\right) \qquad (12.17)$$

$$= \underbrace{\mathcal{L}_i'}_{i=1,\ldots,n}(\curlywedge, \Psi_1, \Psi_3, \Psi_5)$$
$$:= 8 \underbrace{\mathcal{L}_i}_{i=1,\ldots,n}\left(\curlywedge, \frac{\Psi_1}{2}, \frac{\Psi_3}{2}, \frac{\Psi_5}{2}\right),$$

for every $\Psi_1, \Psi_3, \Psi_5 \in \mathcal{G}$ and $\curlywedge \in ⫛$. The mappings $\underbrace{\mathcal{L}_i'}_{i=1,\ldots,n}$ are unique fixed points of $\underbrace{\Gamma_i}_{i=1,\ldots,n}$ in the set below

$$\left\{ \underbrace{\mathcal{L}_i}_{i=1,\ldots,n} \in \Xi : \lambda\Big((\mathcal{L}_1, \ldots, \mathcal{L}_n), (\mathcal{L}_1', \ldots, \mathcal{L}_n')\Big) < (+\infty, +\infty, +\infty)^n \right\}.$$

This concludes that $\underbrace{\mathcal{L}_i'}_{i=1,\ldots,n}$ are unique mappings satisfying the relation (12.17) and also there exists $(\mu_1, \mu_2, \mu_3) \in (0, +\infty)^3$, such that

$$\begin{aligned}
&\operatorname{diag}\Big[\|\mathcal{L}_1(\curlywedge, \Psi_1, \Psi_3, \Psi_5) - \mathcal{L}_1'(\curlywedge, \Psi_1, \Psi_3, \Psi_5)\|, \\
&\qquad \ldots, \|\mathcal{L}_n(\curlywedge, \Psi_1, \Psi_3, \Psi_5) - \mathcal{L}_n'(\curlywedge, \Psi_1, \Psi_3, \Psi_5)\|\Big], \\
&\preceq \operatorname{diag}\Big[\mu_1\varphi_1\left(\Psi_1, 0, \frac{\Psi_3}{2}, \frac{\Psi_3}{2}, \Psi_5, 0\right) + \mu_2\varphi_1\left(\frac{\Psi_1}{2}, \frac{\Psi_1}{2}, \frac{\Psi_3}{2}, 0, \Psi_5, 0\right) \\
&\qquad + \mu_3\varphi_1\left(\frac{\Psi_1}{2}, 0, \frac{\Psi_3}{2}, 0, \frac{\Psi_5}{2}, \frac{\Psi_5}{2}\right), \\
&\qquad \ldots, \mu_1\varphi_n\left(\Psi_1, 0, \frac{\Psi_3}{2}, \frac{\Psi_3}{2}, \Psi_5, 0\right) + \mu_2\varphi_n\left(\frac{\Psi_1}{2}, \frac{\Psi_1}{2}, \frac{\Psi_3}{2}, 0, \Psi_5, 0\right) \\
&\qquad + \mu_3\varphi_n\left(\frac{\Psi_1}{2}, 0, \frac{\Psi_3}{2}, 0, \frac{\Psi_5}{2}, \frac{\Psi_5}{2}\right)\Big],
\end{aligned}$$

for every $\Psi_1, \Psi_3, \Psi_5 \in \mathcal{G}$ and $\curlywedge \in ⫛$.

(2) $\lambda\Big((\Gamma_1^k\mathcal{L}_1, \ldots, \Gamma_n^k\mathcal{L}_n), (\mathcal{L}_1', \ldots, \mathcal{L}_n')\Big) \to (0, 0, 0)^n$ as $k \to \infty$. This concludes the following equality

$$\underbrace{\mathcal{L}_i'}_{i=1,\ldots,n}(\curlywedge, \Psi_1, \Psi_3, \Psi_5) = \lim_{k\to\infty} 8^k \underbrace{\mathcal{L}_i}_{i=1,\ldots,n}\left(\curlywedge, \frac{\Psi_1}{2^k}, \frac{\Psi_3}{2^k}, \frac{\Psi_5}{2^k}\right),$$

for every $\Psi_1, \Psi_3, \Psi_5 \in \mathcal{G}$ and $\curlywedge \in ⫛$;

(3) $\lambda((\mathcal{L}_1,\ldots,\mathcal{L}_n),(\mathcal{L}'_1,\ldots,\mathcal{L}'_n)) \preceq (\frac{1}{1-4\mathfrak{P}_1},\ldots,\frac{1}{1-4\mathfrak{P}_n})\lambda((\mathcal{L}_1,\ldots,\mathcal{L}_n),(\Gamma_1 \mathcal{L}_1,\ldots,\Gamma_n\mathcal{L}_n))$. This implies that

$$\begin{aligned}
&\operatorname{diag}\Big[\|\mathcal{L}_1(\leftthreetimes,\Psi_1,\Psi_3,\Psi_5)-\mathcal{L}'_1(\leftthreetimes,\Psi_1,\Psi_3,\Psi_5)\|,\\
&\qquad \ldots,\|\mathcal{L}_n(\leftthreetimes,\Psi_1,\Psi_3,\Psi_5)-\mathcal{L}'_n(\leftthreetimes,\Psi_1,\Psi_3,\Psi_5)\|\Big]\\
&\preceq \operatorname{diag}\Bigg[\frac{1}{1-4\mathfrak{P}_1}\Bigg[\varphi_1\left(\Psi_1,0,\frac{\Psi_3}{2},\frac{\Psi_3}{2},\Psi_5,0\right)+2\varphi_1\left(\frac{\Psi_1}{2},\frac{\Psi_1}{2},\frac{\Psi_3}{2},0,\Psi_5,0\right)\\
&\qquad +4\varphi_1\left(\frac{\Psi_1}{2},0,\frac{\Psi_3}{2},0,\frac{\Psi_5}{2},\frac{\Psi_5}{2}\right)\Bigg],\\
&\qquad \ldots,\frac{1}{1-4\mathfrak{P}_n}\Bigg[\varphi_n\left(\Psi_1,0,\frac{\Psi_3}{2},\frac{\Psi_3}{2},\Psi_5,0\right)+2\varphi_n\left(\frac{\Psi_1}{2},\frac{\Psi_1}{2},\frac{\Psi_3}{2},0,\Psi_5,0\right)\\
&\qquad +4\varphi_n\left(\frac{\Psi_1}{2},0,\frac{\Psi_3}{2},0,\frac{\Psi_5}{2},\frac{\Psi_5}{2}\right)\Bigg]\Bigg],
\end{aligned}$$

for every $\Psi_1,\Psi_3,\Psi_5 \in \mathcal{G}$ and $\leftthreetimes \in \pitchfork$.

Now, let $\underbrace{\mathcal{E}_i}_{i=1,\ldots,n} : \pitchfork \times \mathcal{G}^3 \to \mathcal{Q}$ be other additive mappings satisfying the inequality (12.4). Then, we have

$$\begin{aligned}
&\operatorname{diag}\Big[\|\mathcal{E}_1(\leftthreetimes,\Psi_1,\Psi_3,\Psi_5)-\mathcal{L}'_1(\leftthreetimes,\Psi_1,\Psi_3,\Psi_5)\|,\\
&\qquad \ldots,\|\mathcal{E}_n(\leftthreetimes,\Psi_1,\Psi_3,\Psi_5)-\mathcal{L}'_n(\leftthreetimes,\Psi_1,\Psi_3,\Psi_5)\|\Big]\\
&=\operatorname{diag}\Bigg[\left\|8^q\mathcal{E}_1\left(\leftthreetimes,\frac{\Psi_1}{2^q},\frac{\Psi_3}{2^q},\frac{\Psi_5}{2^q}\right)-8^q\mathcal{L}'_1\left(\leftthreetimes,\frac{\Psi_1}{2^q},\frac{\Psi_3}{2^q},\frac{\Psi_5}{2^q}\right)\right\|,\ldots,\\
&\qquad \left\|8^q\mathcal{E}_n\left(\leftthreetimes,\frac{\Psi_1}{2^q},\frac{\Psi_3}{2^q},\frac{\Psi_5}{2^q}\right)-8^q\mathcal{L}'_n\left(\leftthreetimes,\frac{\Psi_1}{2^q},\frac{\Psi_3}{2^q},\frac{\Psi_5}{2^q}\right)\right\|\Bigg]\\
&\preceq \operatorname{diag}\Bigg[\left\|8^q\mathcal{E}_1\left(\leftthreetimes,\frac{\Psi_1}{2^q},\frac{\Psi_3}{2^q},\frac{\Psi_5}{2^q}\right)-8^q\mathcal{L}_1\left(\leftthreetimes,\frac{\Psi_1}{2^q},\frac{\Psi_3}{2^q},\frac{\Psi_5}{2^q}\right)\right\|\\
&\qquad +\left\|8^q\mathcal{L}'_1\left(\leftthreetimes,\frac{\Psi_1}{2^q},\frac{\Psi_3}{2^q},\frac{\Psi_5}{2^q}\right)-8^q\mathcal{L}_1\left(\leftthreetimes,\frac{\Psi_1}{2^q},\frac{\Psi_3}{2^q},\frac{\Psi_5}{2^q}\right)\right\|,\ldots,\\
&\qquad \left\|8^q\mathcal{E}_n\left(\leftthreetimes,\frac{\Psi_1}{2^q},\frac{\Psi_3}{2^q},\frac{\Psi_5}{2^q}\right)-8^q\mathcal{L}_n\left(\leftthreetimes,\frac{\Psi_1}{2^q},\frac{\Psi_3}{2^q},\frac{\Psi_5}{2^q}\right)\right\|\\
&\qquad +\left\|8^q\mathcal{L}'_n\left(\leftthreetimes,\frac{\Psi_1}{2^q},\frac{\Psi_3}{2^q},\frac{\Psi_5}{2^q}\right)-8^q\mathcal{L}_n\left(\leftthreetimes,\frac{\Psi_1}{2^q},\frac{\Psi_3}{2^q},\frac{\Psi_5}{2^q}\right)\right\|\Bigg]\\
&\preceq \operatorname{diag}\Bigg[\frac{1}{1-4\mathfrak{P}_1}8^q\Bigg[2\varphi_1\left(\frac{\Psi_1}{2^q},0,\frac{\Psi_3}{2^{q+1}},\frac{\Psi_3}{2^{q+1}},\frac{\Psi_5}{2^q},0\right)+4\varphi_1\left(\frac{\Psi_1}{2^{q+1}},\frac{\Psi_1}{2^{q+1}},\frac{\Psi_3}{2^{q+1}},0,\frac{\Psi_5}{2^q},0\right)\\
&\qquad +8\varphi_1\left(\frac{\Psi_1}{2^{q+1}},0,\frac{\Psi_3}{2^{q+1}},0,\frac{\Psi_5}{2^{q+1}},\frac{\Psi_5}{2^{q+1}}\right)\Bigg],\ldots,\\
&\frac{1}{1-4\mathfrak{P}_n}8^q\Bigg[2\varphi_n\left(\frac{\Psi_1}{2^q},0,\frac{\Psi_3}{2^{q+1}},\frac{\Psi_3}{2^{q+1}},\frac{\Psi_5}{2^q},0\right)+4\varphi_n\left(\frac{\Psi_1}{2^{q+1}},\frac{\Psi_1}{2^{q+1}},\frac{\Psi_3}{2^{q+1}},0,\frac{\Psi_5}{2^q},0\right)\\
&\qquad +8\varphi_n\left(\frac{\Psi_1}{2^{q+1}},0,\frac{\Psi_3}{2^{q+1}},0,\frac{\Psi_5}{2^{q+1}},\frac{\Psi_5}{2^{q+1}}\right)\Bigg]\Bigg],
\end{aligned}$$

which tends to zero as $q \to \infty$, for every $\Psi_1, \Psi_3, \Psi_5 \in \mathcal{G}$ and $\leftthreetimes \in \pitchfork$. So we can conclude that $\underbrace{\mathcal{E}_i}_{i=1,\dots,n}(\leftthreetimes, \Psi_1, \Psi_3, \Psi_5) = \underbrace{\mathcal{L}'_i}_{i=1,\dots,n}(\leftthreetimes, \Psi_1, \Psi_3, \Psi_5)$, for every $\Psi_1, \Psi_3, \Psi_5 \in \mathcal{G}$ and $\leftthreetimes \in \pitchfork$. This proves the uniqueness of $\underbrace{\mathcal{L}'_i}_{i=1,\dots,n}$.

Further, it follows from the inequality (12.3) that

$$
\begin{aligned}
\operatorname{diag}\Big[& \|\mathcal{L}'_1(\leftthreetimes, \Psi_1 + \Psi_2, \Psi_3 - \Psi_4, \Psi_5 + \Psi_6) + \mathcal{L}'_1(\leftthreetimes, \Psi_1 - \Psi_2, \Psi_3 + \Psi_4, \Psi_5 - \Psi_6) \\
& -2\mathcal{L}'_1(\leftthreetimes, \Psi_1, \Psi_3, \Psi_5) + 2\mathcal{L}'_1(\leftthreetimes, \Psi_1, \Psi_4, \Psi_6) \\
& -2\mathcal{L}'_1(\leftthreetimes, \Psi_2, \Psi_3, \Psi_6) + 2\mathcal{L}'_1(\leftthreetimes, \Psi_2, \Psi_4, \Psi_5)\|, \\
& \dots, \|\mathcal{L}'_n(\leftthreetimes, \Psi_1 + \Psi_2, \Psi_3 - \Psi_4, \Psi_5 + \Psi_6) + \mathcal{L}'_n(\leftthreetimes, \Psi_1 - \Psi_2, \Psi_3 + \Psi_4, \Psi_5 - \Psi_6) \\
& -2\mathcal{L}'_n(\leftthreetimes, \Psi_1, \Psi_3, \Psi_5) + 2\mathcal{L}'_n(\leftthreetimes, \Psi_1, \Psi_4, \Psi_6) \\
& -2\mathcal{L}'_n(\leftthreetimes, \Psi_2, \Psi_3, \Psi_6) + 2\mathcal{L}'_n(\leftthreetimes, \Psi_2, \Psi_4, \Psi_5)\| \Big]
\end{aligned}
$$

$$
\begin{aligned}
= \operatorname{diag}\Big[& \lim_{k\to\infty} \Big\| 8^k \Big(\mathcal{L}_1\Big(\leftthreetimes, \frac{\Psi_1 + \Psi_2}{2^k}, \frac{\Psi_3 - \Psi_4}{2^k}, \frac{\Psi_5 + \Psi_6}{2^k}\Big) \\
& + \mathcal{L}_1\Big(\leftthreetimes, \frac{\Psi_1 - \Psi_2}{2^k}, \frac{\Psi_3 + \Psi_4}{2^k}, \frac{\Psi_5 - \Psi_6}{2^k}\Big) \\
& - 2\mathcal{L}_1\Big(\leftthreetimes, \frac{\Psi_1}{2^k}, \frac{\Psi_3}{2^k}, \frac{\Psi_5}{2^k}\Big) + 2\mathcal{L}_1\Big(\leftthreetimes, \frac{\Psi_1}{2^k}, \frac{\Psi_4}{2^k}, \frac{\Psi_6}{2^k}\Big) \\
& - 2\mathcal{L}_1\Big(\leftthreetimes, \frac{\Psi_2}{2^k}, \frac{\Psi_3}{2^k}, \frac{\Psi_6}{2^k}\Big) + 2\mathcal{L}_1\Big(\leftthreetimes, \frac{\Psi_2}{2^k}, \frac{\Psi_4}{2^k}, \frac{\Psi_5}{2^k}\Big)\Big)\Big\|, \\
& \dots, \lim_{k\to\infty} \Big\| 8^k \Big(\mathcal{L}_n\Big(\leftthreetimes, \frac{\Psi_1 + \Psi_2}{2^k}, \frac{\Psi_3 - \Psi_4}{2^k}, \frac{\Psi_5 + \Psi_6}{2^k}\Big) \\
& + \mathcal{L}_n\Big(\leftthreetimes, \frac{\Psi_1 - \Psi_2}{2^k}, \frac{\Psi_3 + \Psi_4}{2^k}, \frac{\Psi_5 - \Psi_6}{2^k}\Big) \\
& - 2\mathcal{L}_n\Big(\leftthreetimes, \frac{\Psi_1}{2^k}, \frac{\Psi_3}{2^k}, \frac{\Psi_5}{2^k}\Big) + 2\mathcal{L}_n\Big(\leftthreetimes, \frac{\Psi_1}{2^k}, \frac{\Psi_4}{2^k}, \frac{\Psi_6}{2^k}\Big) \\
& - 2\mathcal{L}_n\Big(\leftthreetimes, \frac{\Psi_2}{2^k}, \frac{\Psi_3}{2^k}, \frac{\Psi_6}{2^k}\Big) + 2\mathcal{L}_n\Big(\leftthreetimes, \frac{\Psi_2}{2^k}, \frac{\Psi_4}{2^k}, \frac{\Psi_5}{2^k}\Big)\Big)\Big\| \Big],
\end{aligned}
$$

$$
\begin{aligned}
\preceq \operatorname{diag}\Big[& \lim_{k\to\infty} \Big\| 8^k \boldsymbol{\Psi}_1 \Big(2\mathcal{L}_1\Big(\leftthreetimes, \frac{\Psi_1 + \Psi_2}{2^{k+1}}, \frac{\Psi_3 - \Psi_4}{2^k}, \frac{\Psi_5 + \Psi_6}{2^k}\Big) \\
& + 2\mathcal{L}_1\Big(\leftthreetimes, \frac{\Psi_1 - \Psi_2}{2^{k+1}}, \frac{\Psi_3 + \Psi_4}{2^k}, \frac{\Psi_5 - \Psi_6}{2^k}\Big) \\
& - 2\mathcal{L}_1\Big(\leftthreetimes, \frac{\Psi_1}{2^k}, \frac{\Psi_3}{2^k}, \frac{\Psi_5}{2^k}\Big) + 2\mathcal{L}_1\Big(\leftthreetimes, \frac{\Psi_1}{2^k}, \frac{\Psi_4}{2^k}, \frac{\Psi_6}{2^k}\Big) \\
& - 2\mathcal{L}_1\Big(\leftthreetimes, \frac{\Psi_2}{2^k}, \frac{\Psi_3}{2^k}, \frac{\Psi_6}{2^k}\Big) + 2\mathcal{L}_1\Big(\leftthreetimes, \frac{\Psi_2}{2^k}, \frac{\Psi_4}{2^k}, \frac{\Psi_5}{2^k}\Big)\Big)\Big\| \\
& + \lim_{k\to\infty} 8^k \varphi_1\Big(\frac{\Psi_1}{2^k}, \frac{\Psi_2}{2^k}, \frac{\Psi_3}{2^k}, \frac{\Psi_4}{2^k}, \frac{\Psi_5}{2^k}, \frac{\Psi_6}{2^k}\Big), \dots,
\end{aligned}
$$

$$\lim_{k\to\infty}\Big\|8^k\boldsymbol{\Psi}_n\Big(2\mathcal{L}_n\Big(\leftthreetimes,\frac{\Psi_1+\Psi_2}{2^{k+1}},\frac{\Psi_3-\Psi_4}{2^k},\frac{\Psi_5+\Psi_6}{2^k}\Big)$$
$$+2\mathcal{L}_n\Big(\leftthreetimes,\frac{\Psi_1-\Psi_2}{2^{k+1}},\frac{\Psi_3+\Psi_4}{2^k},\frac{\Psi_5-\Psi_6}{2^k}\Big)$$
$$-2\mathcal{L}_n\Big(\leftthreetimes,\frac{\Psi_1}{2^k},\frac{\Psi_3}{2^k},\frac{\Psi_5}{2^k}\Big)+2\mathcal{L}_n\Big(\leftthreetimes,\frac{\Psi_1}{2^k},\frac{\Psi_4}{2^k},\frac{\Psi_6}{2^k}\Big)$$
$$-2\mathcal{L}_n\Big(\leftthreetimes,\frac{\Psi_2}{2^k},\frac{\Psi_3}{2^k},\frac{\Psi_6}{2^k}\Big)+2\mathcal{L}_n\Big(\leftthreetimes,\frac{\Psi_2}{2^k},\frac{\Psi_4}{2^k},\frac{\Psi_5}{2^k}\Big)\Big)\Big\|$$
$$+\lim_{k\to\infty}8^k\varphi_n\Big(\frac{\Psi_1}{2^k},\frac{\Psi_2}{2^k},\frac{\Psi_3}{2^k},\frac{\Psi_4}{2^k},\frac{\Psi_5}{2^k},\frac{\Psi_6}{2^k}\Big)\Big]$$
$$\preceq \operatorname{diag}\Big[\Big\|\boldsymbol{\Psi}_1\Big(2\mathcal{L}'_1\Big(\leftthreetimes,\frac{\Psi_1+\Psi_2}{2},\Psi_3-\Psi_4,\Psi_5+\Psi_6\Big)$$
$$+2\mathcal{L}'_1\Big(\leftthreetimes,\frac{\Psi_1-\Psi_2}{2},\Psi_3+\Psi_4,\Psi_5-\Psi_6\Big)$$
$$-2\mathcal{L}'_1(\leftthreetimes,\Psi_1,\Psi_3,\Psi_5)+2\mathcal{L}'_1(\leftthreetimes,\Psi_1,\Psi_4,\Psi_6)$$
$$-2\mathcal{L}'_1(\leftthreetimes,\Psi_2,\Psi_3,\Psi_6)+2\mathcal{L}'_1(\leftthreetimes,\Psi_2,\Psi_4,\Psi_5))\|,\ldots,$$
$$\Big\|\boldsymbol{\Psi}_n\Big(2\mathcal{L}'_n\Big(\leftthreetimes,\frac{\Psi_1+\Psi_2}{2},\Psi_3-\Psi_4,\Psi_5+\Psi_6\Big)$$
$$+2\mathcal{L}'_n\Big(\leftthreetimes,\frac{\Psi_1-\Psi_2}{2},\Psi_3+\Psi_4,\Psi_5-\Psi_6\Big)$$
$$-2\mathcal{L}'_n(\leftthreetimes,\Psi_1,\Psi_3,\Psi_5)+2\mathcal{L}'_n(\leftthreetimes,\Psi_1,\Psi_4,\Psi_6)$$
$$-2\mathcal{L}'_n(\leftthreetimes,\Psi_2,\Psi_3,\Psi_6)+2\mathcal{L}'_n(\leftthreetimes,\Psi_2,\Psi_4,\Psi_5))\|\Big],$$

for every $\Psi_1,\Psi_2,\Psi_3,\Psi_4,\Psi_5,\Psi_6\in\mathcal{G}$ and $\leftthreetimes\in\pitchfork$. Thus, we get

$$\operatorname{diag}\Big[\|\mathcal{L}'_1(\leftthreetimes,\Psi_1+\Psi_2,\Psi_3-\Psi_4,\Psi_5+\Psi_6)$$
$$+\mathcal{L}'_1(\leftthreetimes,\Psi_1-\Psi_2,\Psi_3+\Psi_4,\Psi_5-\Psi_6)$$
$$-2\mathcal{L}'_1(\leftthreetimes,\Psi_1,\Psi_3,\Psi_5)+2\mathcal{L}'_1(\leftthreetimes,\Psi_1,\Psi_4,\Psi_6)$$
$$-2\mathcal{L}'_1(\leftthreetimes,\Psi_2,\Psi_3,\Psi_6)+2\mathcal{L}'_1(\leftthreetimes,\Psi_2,\Psi_4,\Psi_5)\|,\ldots,$$
$$\|\mathcal{L}'_n(\leftthreetimes,\Psi_1+\Psi_2,\Psi_3-\Psi_4,\Psi_5+\Psi_6)$$
$$+\mathcal{L}'_n(\leftthreetimes,\Psi_1-\Psi_2,\Psi_3+\Psi_4,\Psi_5-\Psi_6)$$
$$-2\mathcal{L}'_n(\leftthreetimes,\Psi_1,\Psi_3,\Psi_5)+2\mathcal{L}'_n(\leftthreetimes,\Psi_1,\Psi_4,\Psi_6)$$
$$-2\mathcal{L}'_n(\leftthreetimes,\Psi_2,\Psi_3,\Psi_6)+2\mathcal{L}'_n(\leftthreetimes,\Psi_2,\Psi_4,\Psi_5)\|,\Big]$$
$$\preceq \operatorname{diag}\Big[\Big\|\boldsymbol{\Psi}_1\Big(2\mathcal{L}'_1\Big(\leftthreetimes,\frac{\Psi_1+\Psi_2}{2},\Psi_3-\Psi_4,\Psi_5+\Psi_6\Big)$$
$$+2\mathcal{L}'_1\Big(\leftthreetimes,\frac{\Psi_1-\Psi_2}{2},\Psi_3+\Psi_4,\Psi_5-\Psi_6\Big)$$
$$-2\mathcal{L}'_1(\leftthreetimes,\Psi_1,\Psi_3,\Psi_5)+2\mathcal{L}'_1(\leftthreetimes,\Psi_1,\Psi_4,\Psi_6)$$
$$-2\mathcal{L}'_1(\leftthreetimes,\Psi_2,\Psi_3,\Psi_6)+2\mathcal{L}'_1(\leftthreetimes,\Psi_2,\Psi_4,\Psi_5))\|,\ldots,$$

$$\Bigg\| \boldsymbol{\Psi}_n \Bigg(2\mathcal{L}'_n\Big(\leftthreetimes, \frac{\Psi_1+\Psi_2}{2}, \Psi_3-\Psi_4, \Psi_5+\Psi_6\Big) + 2\mathcal{L}'_n\Big(\leftthreetimes, \frac{\Psi_1-\Psi_2}{2}, \Psi_3+\Psi_4, \Psi_5-\Psi_6\Big) - 2\mathcal{L}'_n(\leftthreetimes, \Psi_1, \Psi_3, \Psi_5) + 2\mathcal{L}'_n(\leftthreetimes, \Psi_1, \Psi_4, \Psi_6) - 2\mathcal{L}'_n(\leftthreetimes, \Psi_2, \Psi_3, \Psi_6) + 2\mathcal{L}'_n(\leftthreetimes, \Psi_2, \Psi_4, \Psi_5)\Bigg)\Bigg\| \Bigg],$$

for every $\Psi_1, \Psi_2, \Psi_3, \Psi_4, \Psi_5, \Psi_6 \in \mathcal{G}$ and $\leftthreetimes \in ⫛$. Based on Lemma 12.1, we conclude that the mappings $\underbrace{\mathcal{L}'_i}_{i=1,\dots,n} : ⫛ \times \mathcal{G}^3 \to \mathcal{Q}$ are tri–additive.

Theorem 12.2 *Let $n \in \mathbb{N}$, and $\underbrace{\varphi_i}_{i=1,\dots,n} : \mathcal{G}^6 \to [0, \infty)$ be functions such that*

$$\underbrace{\varphi_i}_{i=1,\dots,n} (2\Psi_1, 2\Psi_2, 2\Psi_3, 2\Psi_4, 2\Psi_5, 2\Psi_6) < 2 \underbrace{\mathfrak{P}_i \varphi_i}_{i=1,\dots,n} (\Psi_1, \Psi_2, \Psi_3, \Psi_4, \Psi_5, \Psi_6),$$

for every $\Psi_1, \Psi_2, \Psi_3, \Psi_4, \Psi_5, \Psi_6 \in \mathcal{G}$, $\leftthreetimes \in ⫛$, and every $0 < \underbrace{\mathfrak{P}_i}_{i=1,\dots,n} < 4$. Let $\underbrace{\mathcal{L}_i}_{i=1,\dots,n} : ⫛ \times \mathcal{G}^3 \to \mathcal{Q}$ be mappings satisfying the inequality (12.3) and $\underbrace{\mathcal{L}_i}_{i=1,\dots,n} (\leftthreetimes, \Psi_1, 0, \Psi_5) = \underbrace{\mathcal{L}_i}_{i=1,\dots,n} (\leftthreetimes, 0, \Psi_3, \Psi_5) = \underbrace{\mathcal{L}_i}_{i=1,\dots,n} (\leftthreetimes, \Psi_1, \Psi_3, 0) = 0$, for every $\Psi_1, \Psi_2, \Psi_3, \Psi_4, \Psi_5, \Psi_6 \in \mathcal{G}$ and $\leftthreetimes \in ⫛$. Then, there exist unique tri–additive mappings $\underbrace{\mathcal{L}'_i}_{i=1,\dots,n} : ⫛ \times \mathcal{G}^3 \to \mathcal{Q}$ satisfying the inequality below

$$diag\Big[\|\mathcal{L}_1(\leftthreetimes, \Psi_1, \Psi_3, \Psi_5) - \mathcal{L}'_1(\leftthreetimes, \Psi_1, \Psi_3, \Psi_5)\|, \dots, \|\mathcal{L}_n(\leftthreetimes, \Psi_1, \Psi_3, \Psi_5) - \mathcal{L}'_n(\leftthreetimes, \Psi_1, \Psi_3, \Psi_5)\| \Big]$$

$$\preceq diag\Bigg[\frac{4}{4-\mathfrak{P}_1}\Big(\frac{1}{8}\varphi_1 (2\Psi_1, 0, \Psi_3, \Psi_3, 2\Psi_5, 0) + \frac{1}{4}\varphi_1 (\Psi_1, \Psi_1, \Psi_3, 0, 2\Psi_5, 0) + \frac{1}{2}\varphi_1 (\Psi_1, 0, \Psi_3, 0, \Psi_5, \Psi_5)\Big), \dots, \frac{4}{4-\mathfrak{P}_n}\Big(\frac{1}{8}\varphi_n (2\Psi_1, 0, \Psi_3, \Psi_3, 2\Psi_5, 0) + \frac{1}{4}\varphi_n (\Psi_1, \Psi_1, \Psi_3, 0, 2\Psi_5, 0) + \frac{1}{2}\varphi_n (\Psi_1, 0, \Psi_3, 0, \Psi_5, \Psi_5)\Big) \Bigg],$$

for every $\Psi_1, \Psi_3, \Psi_5 \in \mathcal{G}$ and $\leftthreetimes \in ⫛$.

Proof According to the inequality (12.7), we have that

$$\begin{aligned}&\operatorname{diag}\Bigg[\left\|\mathcal{L}_1(\curlywedge, \Psi_1, \Psi_3, \Psi_5) - \frac{1}{8}\mathcal{L}_1(\curlywedge, 2\Psi_1, 2\Psi_3, 2\Psi_5)\right\|, \ldots, \\ &\qquad\qquad \left\|\mathcal{L}_n(\curlywedge, \Psi_1, \Psi_3, \Psi_5) - \frac{1}{8}\mathcal{L}_n(\curlywedge, 2\Psi_1, 2\Psi_3, 2\Psi_5)\right\|\Bigg] \\ &\preceq \operatorname{diag}\Bigg[\frac{1}{8}(\varphi_1(2\Psi_1, 0, \Psi_3, \Psi_3, 2\Psi_5, 0) + 2\varphi_1(\Psi_1, \Psi_1, \Psi_3, 0, 2\Psi_5, 0) \\ &\qquad +4\varphi_1(\Psi_1, 0, \Psi_3, 0, \Psi_5, \Psi_5)), \ldots, \\ &\qquad \frac{1}{8}(\varphi_n(2\Psi_1, 0, \Psi_3, \Psi_3, 2\Psi_5, 0) + 2\varphi_n(\Psi_1, \Psi_1, \Psi_3, 0, 2\Psi_5, 0) \\ &\qquad +4\varphi_n(\Psi_1, 0, \Psi_3, 0, \Psi_5, \Psi_5))\Bigg],\end{aligned}$$

for every $\Psi_1, \Psi_3, \Psi_5 \in \mathcal{G}$ and $\curlywedge \in ⫛$.

We complete the proof by following a similar method in Theorem 12.1.

12.2 Multi Stability of Permuting Tri–Derivations in Complex Banach Algebras

In this part, we study the multi stability of permuting tri–derivations related to the tri–additive $(\boldsymbol{\Psi}_1, \ldots, \boldsymbol{\Psi}_n)$–functional inequality (12.1), in complex Banach algebras.

Lemma 12.2 *Let* $n \in \mathbb{N}$, *and* $\underbrace{\mathcal{L}_i}_{i=1,\ldots,n} : ⫛ \times \mathcal{G}^3 \to \mathcal{Q}$ *be tri–additive mappings satisfying*

$$\underbrace{\mathcal{L}_i}_{i=1,\ldots,n}(\curlywedge, \Lambda_1\Psi_1, \Lambda_2\Psi_3, \Lambda_3\Psi_5) = \Lambda_1\Lambda_2\Lambda_3 \underbrace{\mathcal{L}_i}_{i=1,\ldots,n}(\curlywedge, \Psi_1, \Psi_3, \Psi_5),$$

for every $\Psi_1, \Psi_3, \Psi_5 \in \Psi_5$, $\curlywedge \in ⫛$, *and* $\Lambda_1, \Lambda_2, \Lambda_3 \in \Delta^1 := \{\varpi \in \mathbb{C} \ : \ |\varpi| = 1\}$. *Then,* $\underbrace{\mathcal{L}_i}_{i=1,\ldots,n}$ *are* $\mathbb{C}$*–trilinear.*

Proof The proof follows from a similar method to the proof of Lemma 2.1 in [4].

Theorem 12.3 *Let* $n \in \mathbb{N}$, *and* $\underbrace{\varphi_i}_{i=1,\ldots,n} : \mathfrak{M}^6 \to \mathbb{R}$ *be functions such that there exist* $0 < \underbrace{\mathfrak{P}_i}_{i=1,\ldots,n} < 0.25$ *with*

$$\underbrace{\varphi_i}_{i=1,\ldots,n}\left(\frac{\Psi_1}{2},\frac{\Psi_2}{2},\frac{\Psi_3}{2},\frac{\Psi_4}{2},\frac{\Psi_5}{2},\frac{\Psi_6}{2}\right) < \frac{\overbrace{\mathfrak{P}_i}^{i=1,\ldots,n}}{2}\underbrace{\varphi_i}_{i=1,\ldots,n}(\Psi_1,\Psi_2,\Psi_3,\Psi_4,\Psi_5,\Psi_6), \tag{12.18}$$

and let $\underbrace{\mathcal{L}_i}_{i=1,\ldots,n} : ⋔ \times \mathfrak{M}^3 \to \mathfrak{M}$ *be mappings satisfying* $\underbrace{\mathcal{L}_i}_{i=1,\ldots,n}(\curlywedge, \Psi_1, 0, \Psi_5) = \underbrace{\mathcal{L}_i}_{i=1,\ldots,n}(\curlywedge, 0, \Psi_3, \Psi_5) = \underbrace{\mathcal{L}_i}_{i=1,\ldots,n}(\curlywedge, \Psi_1, \Psi_3, 0) = 0$ *and the inequality below*

$$\begin{aligned}
diag\Big[&\|\mathcal{L}_1(\curlywedge, \Lambda_1(\Psi_1+\Psi_2), \Lambda_2(\Psi_3-\Psi_4), \Lambda_3(\Psi_5+\Psi_6)) \\
&+\mathcal{L}_1(\curlywedge, \Lambda_1(\Psi_1-\Psi_2), \Lambda_2(\Psi_3+\Psi_4), \Lambda_3(\Psi_5-\Psi_6)) \\
&-\Lambda_1\Lambda_2\Lambda_3(2\mathcal{L}_1(\curlywedge, \Psi_1, \Psi_3, \Psi_5) - 2\mathcal{L}_1(\curlywedge, \Psi_1, \Psi_4, \Psi_6) \\
&+2\mathcal{L}_1(\curlywedge, \Psi_2, \Psi_3, \Psi_6) - 2\mathcal{L}_1(\curlywedge, \Psi_2, \Psi_4, \Psi_5))\|, \ldots, \\
&\|\mathcal{L}_n(\curlywedge, \Lambda_1(\Psi_1+\Psi_2), \Lambda_2(\Psi_3-\Psi_4), \Lambda_3(\Psi_5+\Psi_6)) \\
&+\mathcal{L}_n(\curlywedge, \Lambda_1(\Psi_1-\Psi_2), \Lambda_2(\Psi_3+\Psi_4), \Lambda_3(\Psi_5-\Psi_6)) \\
&-\Lambda_1\Lambda_2\Lambda_3(2\mathcal{L}_n(\curlywedge, \Psi_1, \Psi_3, \Psi_5) - 2\mathcal{L}_n(\curlywedge, \Psi_1, \Psi_4, \Psi_6) \\
&+2\mathcal{L}_n(\curlywedge, \Psi_2, \Psi_3, \Psi_6) - 2\mathcal{L}_n(\curlywedge, \Psi_2, \Psi_4, \Psi_5))\|\Big] \\
\preceq diag\Big[&\Big\|\boldsymbol{\Psi}_1\Big(2\mathcal{L}_1\Big(\curlywedge, \frac{\Psi_1+\Psi_2}{2}, \Psi_3-\Psi_4, \Psi_5+\Psi_6\Big) \\
&+2\mathcal{L}_1\Big(\curlywedge, \frac{\Psi_1-\Psi_2}{2}, \Psi_3+\Psi_4, \Psi_5-\Psi_6\Big) \\
&-2\mathcal{L}_1(\curlywedge, \Psi_1, \Psi_3, \Psi_5) + 2\mathcal{L}_1(\curlywedge, \Psi_1, \Psi_4, b) \\
&-2\mathcal{L}_1(\curlywedge, \Psi_2, \Psi_3, \Psi_6) + 2\mathcal{L}_1(\curlywedge, \Psi_2, \Psi_4, \Psi_5))\| \\
&+\varphi_1(\Psi_1, \Psi_2, \Psi_3, \Psi_4, \Psi_5, \Psi_6), \ldots, \\
&\Big\|\boldsymbol{\Psi}_n\Big(2\mathcal{L}_n\Big(\curlywedge, \frac{\Psi_1+\Psi_2}{2}, \Psi_3-\Psi_4, \Psi_5+\Psi_6\Big) \\
&+2\mathcal{L}_n\Big(\curlywedge, \frac{\Psi_1-\Psi_2}{2}, \Psi_3+\Psi_4, \Psi_5-\Psi_6\Big) \\
&-2\mathcal{L}_n(\curlywedge, \Psi_1, \Psi_3, \Psi_5) + 2\mathcal{L}_n(\curlywedge, \Psi_1, \Psi_4, b) \\
&-2\mathcal{L}_n(\curlywedge, \Psi_2, \Psi_3, \Psi_6) + 2\mathcal{L}_n(\curlywedge, \Psi_2, \Psi_4, \Psi_5))\| \\
&+\varphi_n(\Psi_1, \Psi_2, \Psi_3, \Psi_4, \Psi_5, \Psi_6)\Big],
\end{aligned} \tag{12.19}$$

for every $\Lambda_1, \Lambda_2, \Lambda_3 \in \Delta^1$ *and every* $\Psi_1, \Psi_2, \Psi_3, \Psi_4, \Psi_5, \Psi_6 \in \mathfrak{M}$ *and* $\curlywedge \in ⋔$. *Then, there exist unique* $\mathbb{C}$*–trilinear mappings* $\underbrace{\mathfrak{Z}_i}_{i=1,\ldots,n} : ⋔ \times \mathfrak{M}^3 \to \mathfrak{M}$ *satisfying the inequality below*

$$diag\Big[\ \|\mathcal{L}_1(\leftthreetimes,\Psi_1,\Psi_3,\Psi_5)-\mathfrak{Z}_1(\leftthreetimes,\Psi_1,\Psi_3,\Psi_5)\|,\ldots,\|\mathcal{L}_n(\leftthreetimes,\Psi_1,\Psi_3,\Psi_5)-\mathfrak{Z}_n(\leftthreetimes,\Psi_1,\Psi_3,\Psi_5)\|\ \Big]$$
$$\preceq diag\Big[\frac{1}{1-4\mathfrak{P}_1}\Big[\varphi_1\left(\Psi_1,0,\frac{\Psi_3}{2},\frac{\Psi_3}{2},\Psi_5,0\right)+2\varphi_1\left(\frac{\Psi_1}{2},\frac{\Psi_1}{2},\frac{\Psi_3}{2},0,\Psi_5,0\right)$$
$$+4\varphi_1\left(\frac{\Psi_1}{2},0,\frac{\Psi_3}{2},0,\frac{\Psi_5}{2},\frac{\Psi_5}{2}\right)\Big],\ldots,$$
$$\frac{1}{1-4\mathfrak{P}_n}\Big[\varphi_n\left(\Psi_1,0,\frac{\Psi_3}{2},\frac{\Psi_3}{2},\Psi_5,0\right)+2\varphi_n\left(\frac{\Psi_1}{2},\frac{\Psi_1}{2},\frac{\Psi_3}{2},0,\Psi_5,0\right)$$
$$+4\varphi_n\left(\frac{\Psi_1}{2},0,\frac{\Psi_3}{2},0,\frac{\Psi_5}{2},\frac{\Psi_5}{2}\right)\Big]\Big], \tag{12.20}$$

for every $\Psi_1,\Psi_3,\Psi_5\in\mathfrak{M}$, and $\leftthreetimes\in\pitchfork$.

In addition, if the mappings $\underbrace{\mathcal{L}_i}_{i=1,\ldots,n}:\pitchfork\times\mathfrak{M}^3\to\mathfrak{M}$ satisfy $\underbrace{\mathcal{L}_i}_{i=1,\ldots,n}(\leftthreetimes,\Psi_1,\Psi_3,\Psi_5)=2\underbrace{\mathcal{L}_i}_{i=1,\ldots,n}(\leftthreetimes,\Psi_1,\Psi_3,\Psi_5)$ and the following inequalities

$$diag\Big[\|\mathcal{L}_1(\leftthreetimes,\Psi_1\Psi_2,\Psi_3,\Psi_5)-\mathcal{L}_1(\leftthreetimes,\Psi_1,\Psi_3,\Psi_5)\Psi_2-\Psi_1\mathcal{L}_1(\leftthreetimes,\Psi_2,\Psi_3,\Psi_5)\|, \tag{12.21}$$
$$\ldots,\|\mathcal{L}_n(\leftthreetimes,\Psi_1\Psi_2,\Psi_3,\Psi_5)-\mathcal{L}_n(\leftthreetimes,\Psi_1,\Psi_3,\Psi_5)\Psi_2-\Psi_1\mathcal{L}_n(\leftthreetimes,\Psi_2,\Psi_3,\Psi_5)\|\ \Big]$$
$$\preceq diag\Big[\varphi_1(\Psi_1,\Psi_2,\Psi_3,\Psi_3,\Psi_5,\Psi_5),\ldots,\varphi_n(\Psi_1,\Psi_2,\Psi_3,\Psi_3,\Psi_5,\Psi_5)\Big],$$

and

$$diag\Big[\ \|\mathcal{L}_1(\leftthreetimes,\Phi_{\delta(1)},\Phi_{\delta(2)},\Phi_{\delta(3)})-\mathcal{L}_1(\leftthreetimes,\Phi_1,\Phi_2,\Phi_3)\| \tag{12.22}$$
$$\|\mathcal{L}_n(\leftthreetimes,\Phi_{\delta(1)},\Phi_{\delta(2)},\Phi_{\delta(3)})-\mathcal{L}_n(\leftthreetimes,\Phi_1,\Phi_2,\Phi_3)\|\ \Big]$$
$$\preceq diag\Big[\varphi_1(\Phi_1,\Phi_1,\Phi_2,\Phi_2,\Phi_3,\Phi_3),\ldots,\varphi_n(\Phi_1,\Phi_1,\Phi_2,\Phi_2,\Phi_3,\Phi_3)\ \Big],$$

for every $\Psi_1,\Psi_2,\Psi_3,\Phi_1,\Phi_2,\Phi_3,\Psi_5\in\mathfrak{M}$ and $\leftthreetimes\in\pitchfork$, and for every permutation $(\delta(1),\delta(2),\delta(3))$ of $(1,2,3)$, Then, the $\mathbb{C}$–trilinear mappings $\underbrace{\mathfrak{Z}_i}_{i=1,\ldots,n}:\pitchfork\times\mathfrak{M}^3\to\mathfrak{M}$ are permuting tri–derivations.

Proof Set $\Lambda_1=\Lambda_2=\Lambda_3=1$ in inequality (12.19). Then, the result follows directly from Theorem 12.1 and [5, Theorem 3.3].

Theorem 12.4 *Let $n\in\mathbb{N}$, and $\underbrace{\varphi_i}_{i=1,\ldots,n}:\mathfrak{M}^6\to[0,\infty)$ be functions such that there exist $0<\underbrace{\mathfrak{P}_i}_{i=1,\ldots,n}<4$ with*

$$\underbrace{\varphi_i}_{i=1,\ldots,n}(2\Psi_1,2\Psi_2,2\Psi_3,2\Psi_4,2\Psi_5,2\Psi_6)<2\underbrace{\mathfrak{P}_i\varphi_i}_{i=1,\ldots,n}(\Psi_1,\Psi_2,\Psi_3,\Psi_4,\Psi_5,\Psi_6), \tag{12.23}$$

for every $\Psi_1, \Psi_2, \Psi_3, \Psi_4, \Psi_5, \Psi_6 \in \mathfrak{M}$ *and* $\curlywedge \in \pitchfork$. *Let* $\underbrace{\mathcal{L}_i}_{i=1,\ldots,n} : \pitchfork \times \mathfrak{M}^3 \to \mathfrak{M}$ *be mappings satisfying the inequality (12.19) and* $\underbrace{\mathcal{L}_i}_{i=1,\ldots,n} (\curlywedge, \Psi_1, 0, \Psi_5) = \underbrace{\mathcal{L}_i}_{i=1,\ldots,n} (\curlywedge, 0, \Psi_3, \Psi_5) = \underbrace{\mathcal{L}_i}_{i=1,\ldots,n} (\curlywedge, \Psi_1, \Psi_3, 0) = 0$, *for every* $\Psi_1, \Psi_3, \Psi_5 \in \mathfrak{M}$ *and* $\curlywedge \in \pitchfork$. *Then, we can obtain unique* $\mathbb{C}$*–trilinear mappings* $\underbrace{\mathfrak{Z}_i}_{i=1,\ldots,n} : \pitchfork \times \mathfrak{M}^3 \to \mathfrak{M}$ *satisfying the inequality below*

$$
\begin{aligned}
&diag\Big[\, \|\mathcal{L}_1(\curlywedge, \Psi_1, \Psi_3, \Psi_5) - \mathfrak{Z}_1(\curlywedge, \Psi_1, \Psi_3, \Psi_5)\|, \ldots, \|\mathcal{L}_n(\curlywedge, \Psi_1, \Psi_3, \Psi_5) - \mathfrak{Z}_n(\curlywedge, \Psi_1, \Psi_3, \Psi_5)\| \Big] \\
&\preceq diag\Bigg[\frac{4}{4-\mathfrak{P}_1}\bigg[\frac{1}{8}\varphi_1\,(2\Psi_1, 0, \Psi_3, \Psi_3, 2\Psi_5, 0) + \frac{1}{4}\varphi_1\,(\Psi_1, \Psi_1, \Psi_3, 0, 2\Psi_5, 0) \\
&\qquad\qquad + \frac{1}{2}\varphi_1\,(\Psi_1, 0, \Psi_3, 0, \Psi_5, \Psi_5)\bigg], \ldots, \\
&\qquad \frac{4}{4-\mathfrak{P}_n}\bigg[\frac{1}{8}\varphi_n\,(2\Psi_1, 0, \Psi_3, \Psi_3, 2\Psi_5, 0) + \frac{1}{4}\varphi_n\,(\Psi_1, \Psi_1, \Psi_3, 0, 2\Psi_5, 0) \\
&\qquad\qquad + \frac{1}{2}\varphi_n\,(\Psi_1, 0, \Psi_3, 0, \Psi_5, \Psi_5)\bigg]\Bigg], \qquad (12.24)
\end{aligned}
$$

for every $\Psi_1, \Psi_3, \Psi_5 \in \mathfrak{M}$ *and* $\curlywedge \in \pitchfork$.

Also, if the mappings $\underbrace{\mathcal{L}_i}_{i=1,\ldots,n} : \pitchfork \times \mathfrak{M}^3 \to \mathfrak{M}$ *satisfy the inequalities (12.21) and (12.22) and* $\underbrace{\mathcal{L}_i}_{i=1,\ldots,n} (\curlywedge, 2\Psi_1, \Psi_3, \Psi_5) = 2 \underbrace{\mathcal{L}_i}_{i=1,\ldots,n} (\curlywedge, \Psi_1, \Psi_3, \Psi_5)$, *for every* $\Psi_1, \Psi_3, \Psi_5 \in \mathfrak{M}$ *and* $\curlywedge \in \pitchfork$, *then, the* $\mathbb{C}$*–trilinear mappings* $\underbrace{\mathfrak{Z}_i}_{i=1,\ldots,n} : \pitchfork \times \mathfrak{M}^3 \to \mathfrak{M}$ *are permuting tri-derivations.*

Proof The proof follows from a similar method to the proof of Theorem 12.3.

12.3 Multi Stability of Permuting Tri–Derivations in Unital C^*–Algebras

In this part, we study the multi stability of permuting tri–derivations related to the tri–additive $(\boldsymbol{\Psi}_1, \ldots, \boldsymbol{\Psi}_n)$–functional inequality (12.1), in unital C^*–algebras.

Here, let $\mathfrak{M}$ and $U(\mathfrak{M})$ be a unital C^*–algebra with unit e and a unitary group, respectively. We now have the following theorem:

Theorem 12.5 *Let $n \in \mathbb{N}$, and $\underbrace{\varphi_i}_{i=1,\ldots,n} : \mathfrak{M}^6 \to \mathbb{R}$ be functions satisfying the inequality* (12.18) *and $\underbrace{\mathcal{L}_i}_{i=1,\ldots,n} : \pitchfork \times \mathfrak{M}^3 \to \mathfrak{M}$ be mappings satisfying the inequality* (12.19) *and $\underbrace{\mathcal{L}_i}_{i=1,\ldots,n}(\leftthreetimes, \Psi_1, 0, \Psi_5) = \underbrace{\mathcal{L}_i}_{i=1,\ldots,n}(\leftthreetimes, 0, \Psi_3, \Psi_5) = \underbrace{\mathcal{L}_i}_{i=1,\ldots,n}(\leftthreetimes, \Psi_1, \Psi_3, 0) = 0$, for every $\Psi_1, \Psi_3, \Psi_5 \in \mathfrak{M}$ and $\leftthreetimes \in \pitchfork$. Then, there exist unique $\mathbb{C}$–trilinear mappings $\underbrace{\mathfrak{Z}_i}_{i=1,\ldots,n} : \pitchfork \times \mathfrak{M}^3 \to \mathfrak{M}$ satisfying the inequality* (12.20).

In addition, if the mappings $\underbrace{\mathcal{L}_i}_{i=1,\ldots,n} : \pitchfork \times \mathfrak{M}^3 \to \mathfrak{M}$ satisfy the inequality (12.22), *and $\underbrace{\mathcal{L}_i}_{i=1,\ldots,n}(\leftthreetimes, 2\beta, \Psi_3, \Psi_5) = 2\underbrace{\mathcal{L}_i}_{i=1,\ldots,n}(\leftthreetimes, \beta, \Psi_3, \Psi_5)$ and*

$$
\begin{aligned}
& diag\Big[\|\mathcal{L}_1(\leftthreetimes, \beta\Psi_2, \Psi_3, \Psi_5) - \mathcal{L}_1(\leftthreetimes, \beta, \Psi_3, \Psi_5)\Psi_2 - \beta\mathcal{L}_1(\leftthreetimes, \Psi_2, \Psi_3, \Psi_5)\|, \ldots, \\
& \qquad \|\mathcal{L}_n(\leftthreetimes, \beta\Psi_2, \Psi_3, \Psi_5) - \mathcal{L}_n(\leftthreetimes, \beta, \Psi_3, \Psi_5)\Psi_2 - \beta\mathcal{L}_n(\leftthreetimes, \Psi_2, \Psi_3, \Psi_5)\|\Big] \\
& \preceq diag\Big[\varphi_1(\beta, \Psi_2, \Psi_3, \Psi_3, \Psi_5, \Psi_5), \ldots, \varphi_n(\beta, \Psi_2, \Psi_3, \Psi_3, \Psi_5, \Psi_5)\Big],
\end{aligned}
\tag{12.25}
$$

for every $\Psi_2, \Psi_3, \Psi_5 \in \mathfrak{M}$ and $\leftthreetimes \in \pitchfork$, and every $\beta \in U(\mathfrak{M})$, then, the $\mathbb{C}$–trilinear mappings $\underbrace{\mathfrak{Z}_i}_{i=1,\ldots,n} : \pitchfork \times \mathfrak{M}^3 \to \mathfrak{M}$ are permuting tri-derivations.

Proof It follows directly from Theorem 12.3 and [5, Theorem 3.7].

Remark 12.1 By a similar method in the proof of Theorem 12.5, we can conclude that if the inequality (12.25) in Theorem 12.5 is replaced by the inequality below

$$
\begin{aligned}
& \mathrm{diag}\Big[\|\mathcal{L}_1(\leftthreetimes, \beta\alpha, \alpha_1, \alpha_2) - \mathcal{L}_1(\leftthreetimes, \beta, \alpha_1, \alpha_2)\alpha - \beta\mathcal{L}_1(\leftthreetimes, \alpha, \alpha_1, \alpha_2)\|, \ldots, \\
& \qquad \|\mathcal{L}_n(\leftthreetimes, \beta\alpha, \alpha_1, \alpha_2) - \mathcal{L}_n(\leftthreetimes, \beta, \alpha_1, \alpha_2)\alpha - \beta\mathcal{L}_n(\leftthreetimes, \alpha, \alpha_1, \alpha_2)\|\Big] \\
& \preceq \mathrm{diag}\Big[\varphi_1(\beta, \alpha, \alpha_1, \alpha_1, \alpha_2, \alpha_2), \ldots, \varphi_n(\beta, \alpha, \alpha_1, \alpha_1, \alpha_2, \alpha_2)\Big],
\end{aligned}
$$

for every $\beta, \alpha, \alpha_1, \alpha_2 \in U(\mathfrak{M})$, then, the $\mathbb{C}$–trilinear mappings $\underbrace{\mathfrak{Z}_i}_{i=1,\ldots,n} : \pitchfork \times \mathfrak{M}^3 \to \mathfrak{M}$ are permuting tri-derivations.

Theorem 12.6 *Let* $n \in \mathbb{N}$, *and* $\underbrace{\varphi_i}_{i=1,\ldots,n} : \mathfrak{M}^6 \to \mathbb{R}$ *be functions satisfying the inequality* (12.23) *and* $\underbrace{\mathcal{L}_i}_{i=1,\ldots,n} : \pitchfork \times \mathfrak{M}^3 \to \mathfrak{M}$ *be mappings satisfying the inequality* (12.19) *and* $\underbrace{\mathcal{L}_i}_{i=1,\ldots,n}(\curlywedge, \Psi_1, 0, \Psi_5) = \underbrace{\mathcal{L}_i}_{i=1,\ldots,n}(\curlywedge, 0, \Psi_3, \Psi_5) = \underbrace{\mathcal{L}_i}_{i=1,\ldots,n}(\curlywedge, \Psi_1, \Psi_3, 0) = 0$, *for every* $\Psi_1, \Psi_3, \Psi_5 \in \mathfrak{M}$ *and* $\curlywedge \in \pitchfork$. *Then, we can find unique* $\mathbb{C}$*–trilinear mappings* $\underbrace{\mathfrak{Z}_i}_{i=1,\ldots,n} : \pitchfork \times \mathfrak{M}^3 \to \mathfrak{M}$ *satisfying the inequality* (12.24).

In addition, if the mappings $\underbrace{\mathcal{L}_i}_{i=1,\ldots,n} : \pitchfork \times \mathfrak{M}^3 \to \mathfrak{M}$ *satisfy the inequalities* (12.22) *and* (12.25) *and* $\underbrace{\mathcal{L}_i}_{i=1,\ldots,n}(\curlywedge, 2\beta, \Psi_3, \Psi_5) = 2\underbrace{\mathcal{L}_i}_{i=1,\ldots,n}(\curlywedge, \beta, \Psi_3, \Psi_5)$, *for every* $\Psi_3, \Psi_5 \in \mathfrak{M}$, $\curlywedge \in \pitchfork$, *and every* $\beta \in U(\mathfrak{M})$, *then, the* $\mathbb{C}$*–trilinear mappings* $\underbrace{\mathfrak{Z}_i}_{i=1,\ldots,n} : \pitchfork \times \mathfrak{M}^3 \to \mathfrak{M}$ *are permuting tri-derivations.*

Proof By a similar method to the proof of Theorem 12.5, we can get the result.

12.4 Multi Stability of Permuting Tri–Homomorphisms in Complex Banach Algebras

In this section, we study permuting tri–homomorphisms dependending on the tri–additive $(\boldsymbol{\Psi}_1, \ldots, \boldsymbol{\Psi}_n)$–functional inequality (12.1) in complex Banach algebras.

Theorem 12.7 *Let* $n \in \mathbb{N}$, *and* $\underbrace{\varphi_i}_{i=1,\ldots,n} : \mathfrak{M}^6 \to \mathbb{R}$ *be functions such that there exist* $0 < \underbrace{\mathfrak{P}_i}_{i=1,\ldots,n} < 0.25$ *with*

$$\underbrace{\varphi_i}_{i=1,\ldots,n}\left(\frac{\Psi_1}{2}, \frac{\Psi_2}{2}, \frac{\Psi_3}{2}, \frac{\Psi_4}{2}, \frac{\Psi_5}{2}, \frac{\Psi_6}{2}\right) < \frac{\overbrace{\mathfrak{P}_i}^{i=1,\ldots,n}}{2}\underbrace{\varphi_i}_{i=1,\ldots,n}(\Psi_1, \Psi_2, \Psi_3, \Psi_4, \Psi_5, \Psi_6), \quad (12.26)$$

and let $\underbrace{\mathcal{L}_i}_{i=1,\ldots,n} : \pitchfork \times \mathfrak{M}^3 \to \mathfrak{N}$ *be mappings satisfying the inequality* (12.19) *and* $\underbrace{\mathcal{L}_i}_{i=1,\ldots,n}(\curlywedge, \Psi_1, 0, \Psi_5) = \underbrace{\mathcal{L}_i}_{i=1,\ldots,n}(\curlywedge, 0, \Psi_3, \Psi_5) = \underbrace{\mathcal{L}_i}_{i=1,\ldots,n}(\curlywedge, \Psi_1, \Psi_3, 0) = 0$, *for every* $\Psi_1, \Psi_3, \Psi_4 \in \mathfrak{M}$ *and* $\curlywedge \in \pitchfork$. *Then, there exist unique* $\mathbb{C}$*–trilinear mappings* $\underbrace{\mathcal{E}_i}_{i=1,\ldots,n} : \pitchfork \times \mathfrak{M}^3 \to \mathfrak{N}$ *satisfying the inequality below*

$$diag\Bigg[\|\mathcal{L}_1(\curlywedge, \Psi_1, \Psi_3, \Psi_5) - \mathcal{E}_1(\curlywedge, \Psi_1, \Psi_3, \Psi_5)\|, \ldots, \quad (12.27)$$

$$\|\mathcal{L}_n(\curlywedge, \Psi_1, \Psi_3, \Psi_5) - \mathcal{E}_n(\curlywedge, \Psi_1, \Psi_3, \Psi_5)\| \Bigg]$$

$$\preceq diag\Bigg[\frac{1}{1-4\mathfrak{P}_1}\Bigg[\varphi_1\left(\Psi_1, 0, \frac{\Psi_3}{2}, \frac{\Psi_3}{2}, \Psi_5, 0\right) + 2\varphi_1\left(\frac{\Psi_1}{2}, \frac{\Psi_1}{2}, \frac{\Psi_3}{2}, 0, \Psi_5, 0\right)$$

$$+4\varphi_1\left(\frac{\Psi_1}{2}, 0, \frac{\Psi_3}{2}, 0, \frac{\Psi_5}{2}, \frac{\Psi_5}{2}\right)\Bigg], \ldots,$$

$$\frac{1}{1-4\mathfrak{P}_n}\Bigg[\varphi_n\left(\Psi_1, 0, \frac{\Psi_3}{2}, \frac{\Psi_3}{2}, \Psi_5, 0\right) + 2\varphi_n\left(\frac{\Psi_1}{2}, \frac{\Psi_1}{2}, \frac{\Psi_3}{2}, 0, \Psi_5, 0\right)$$

$$+4\varphi_n\left(\frac{\Psi_1}{2}, 0, \frac{\Psi_3}{2}, 0, \frac{\Psi_5}{2}, \frac{\Psi_5}{2}\right)\Bigg]\Bigg],$$

for every $\Psi_1, \Psi_3, \Psi_5 \in \mathfrak{M}$ *and* $\curlywedge \in ⫛$.

Moreover, if the mappings $\underbrace{\mathcal{L}_i}_{i=1,\ldots,n} : ⫛ \times \mathfrak{M}^3 \to \mathfrak{N}$ *satisfy the inequality (12.22) and the inequality below*

$$diag\Bigg[\|\mathcal{L}_1(\curlywedge, \Psi_1\Psi_2, \Psi_3\Psi_4, \Psi_5\Psi_6) - \mathcal{L}_1(\curlywedge, \Psi_1, \Psi_3, \Psi_5)\mathcal{L}_1(\curlywedge, \Psi_2, \Psi_4, \Psi_6)\| \quad (12.28)$$

$$\|\mathcal{L}_n(\curlywedge, \Psi_1\Psi_2, \Psi_3\Psi_4, \Psi_5\Psi_6) - \mathcal{L}_n(\curlywedge, \Psi_1, \Psi_3, \Psi_5)\mathcal{L}_n(\curlywedge, \Psi_2, \Psi_4, \Psi_6)\| \Bigg]$$

$$\preceq diag\Bigg[\varphi_1(\Psi_1, \Psi_2, \Psi_3, \Psi_3, \Psi_5, \Psi_5), \ldots, \varphi_n(\Psi_1, \Psi_2, \Psi_3, \Psi_3, \Psi_5, \Psi_5) \Bigg],$$

for every $\Psi_1, \Psi_2, \Psi_3, \Psi_4, \Psi_5, \Psi_6 \in \mathfrak{M}$ *and* $\curlywedge \in ⫛$*, then, the* $\mathbb{C}$*–trilinear mappings* $\underbrace{\mathcal{E}_i}_{i=1,\ldots,n} : ⫛ \times \mathfrak{M}^3 \to \mathfrak{N}$ *are permuting tri–homomorphisms.*

Proof It follows directly from Theorem 12.3 and [5, Theorem 4.1].

Theorem 12.8 *Let* $n \in \mathbb{N}$*, and* $\underbrace{\varphi_i}_{i=1,\ldots,n} : \mathfrak{M}^6 \to [0, \infty)$ *be functions satisfying the inequality* (12.23)*, for* $0 < \underbrace{\mathfrak{P}_i}_{i=1,\ldots,n} < 4$*, and let* $\underbrace{\mathcal{L}_i}_{i=1,\ldots,n} : ⫛ \times \mathfrak{M}^3 \to \mathfrak{N}$ *be mappings satisfying the inequality* (12.19) *and* $\underbrace{\mathcal{L}_i}_{i=1,\ldots,n}(\curlywedge, \Psi_1, 0, \Psi_5) = \underbrace{\mathcal{L}_i}_{i=1,\ldots,n}(\curlywedge, 0, \Psi_3, \Psi_5) = \underbrace{\mathcal{L}_i}_{i=1,\ldots,n}(\curlywedge, \Psi_1, \Psi_3, 0) = 0$*, for every* $\Psi_1, \Psi_3, \Psi_4 \in \mathfrak{M}$ *and* $\curlywedge \in ⫛$*. Then, there exist unique* $\mathbb{C}$*–trilinear mappings* $\underbrace{\mathcal{E}_i}_{i=1,\ldots,n} : ⫛ \times \mathfrak{M}^3 \to \mathfrak{N}$ *satisfying the inequality below*

$$diag\Bigg[\|\mathcal{L}_1(\curlywedge, \Psi_1, \Psi_3, \Psi_5) - \mathcal{E}_1(\curlywedge, \Psi_1, \Psi_3, \Psi_5)\|, \ldots, \quad (12.29)$$

$$\|\mathcal{L}_n(\leftthreetimes, \Psi_1, \Psi_3, \Psi_5) - \mathcal{E}_n(\leftthreetimes, \Psi_1, \Psi_3, \Psi_5)\| \Bigg]$$

$$\preceq diag\Bigg[\frac{4}{4-\mathfrak{P}_1}\Bigg[\frac{1}{8}\varphi_1\left(2\Psi_1, 0, \Psi_3, \Psi_3, 2\Psi_5, 0\right) + \frac{1}{4}\varphi_1\left(\Psi_1, \Psi_1, \Psi_3, 0, 2\Psi_5, 0\right)$$
$$+\frac{1}{2}\varphi_1\left(\Psi_1, 0, \Psi_3, 0, \Psi_5, \Psi_5\right)\Bigg], \ldots,$$
$$\frac{4}{4-\mathfrak{P}_n}\Bigg[\frac{1}{8}\varphi_n\left(2\Psi_1, 0, \Psi_3, \Psi_3, 2\Psi_5, 0\right) + \frac{1}{4}\varphi_n\left(\Psi_1, \Psi_1, \Psi_3, 0, 2\Psi_5, 0\right)$$
$$+\frac{1}{2}\varphi_n\left(\Psi_1, 0, \Psi_3, 0, \Psi_5, \Psi_5\right)\Bigg]\Bigg],$$

for every $\Psi_1, \Psi_3, \Psi_5 \in \mathfrak{M}$ *and* $\leftthreetimes \in \pitchfork$.

In addition, if the mappings $\underbrace{\mathcal{L}_i}_{i=1,\ldots,n} : \pitchfork \times \mathfrak{M}^3 \to \mathfrak{N}$ *satisfy the inequalities* (12.22) *and* (12.28)*, then, the* $\mathbb{C}$*–trilinear mappings* $\underbrace{\mathcal{E}_i}_{i=1,\ldots,n} : \pitchfork \times \mathfrak{M}^3 \to \mathfrak{N}$ *are permuting tri-homomorphisms.*

Proof By a similar method to the proof of Theorem 12.7, we can obtain the result.

12.5 Multi Stability of Permuting Tri–Homomorphisms in Unital C^*–Algebras

In this section, we study permuting tri–homomorphisms dependending on the tri–additive $(\boldsymbol{\Psi}_1, \ldots, \boldsymbol{\Psi}_n)$–functional inequality (12.1) in unital C^*–algebras.

Let $\mathfrak{M}$ and $U(\mathfrak{M})$ be a unital C^*–algebra with unit e and a unitary group, respectively. Now, we come to the following theorems:

Theorem 12.9 *Let* $\underbrace{\varphi_i}_{i=1,\ldots,n} : \pitchfork \times \mathfrak{M}^6 \to \mathbb{R}$ *be functions satisfying the inequality* (12.26) *and* $\underbrace{\mathcal{L}_i}_{i=1,\ldots,n} : \mathfrak{M}^3 \to \mathfrak{N}$ *be mappings satisfying the inequality* (12.19) *and* $\underbrace{\mathcal{L}_i}_{i=1,\ldots,n}(\leftthreetimes, \Psi_1, 0, \Psi_5) = \underbrace{\mathcal{L}_i}_{i=1,\ldots,n}(\leftthreetimes, 0, \Psi_3, \Psi_5) = \underbrace{\mathcal{L}_i}_{i=1,\ldots,n}(\leftthreetimes, \Psi_1, \Psi_3, 0) = 0$*, for every* $\Psi_1, \Psi_3, \Psi_5 \in \mathfrak{M}$ *and* $\leftthreetimes \in \pitchfork$*. Then, there exist unique* $\mathbb{C}$*–trilinear mappings* $\underbrace{\mathcal{E}_i}_{i=1,\ldots,n} : \pitchfork \times \mathfrak{M}^3 \to \mathfrak{N}$ *satisfying the inequality* (12.27).

Also, if the mappings $\underbrace{\mathcal{L}_i}_{i=1,\ldots,n} : \pitchfork \times \mathfrak{M}^3 \to \mathfrak{N}$ *satisfy the inequality* (12.22) *and the inequality below*

$$diag\Big[\|\mathcal{L}_1(\curlywedge, \beta_1\alpha_1, \beta_2\alpha_2, \beta_3\alpha_3) - \mathcal{L}_1(\curlywedge, \beta_1, \beta_2, \beta_3)\mathcal{L}_1(\curlywedge, \alpha_1, \alpha_2, \alpha_3)\|, \ldots,$$
$$\|\mathcal{L}_n(\curlywedge, \beta_1\alpha_1, \beta_2\alpha_2, \beta_3\alpha_3) - \mathcal{L}_n(\curlywedge, \beta_1, \beta_2, \beta_3)\mathcal{L}_n(\curlywedge, \alpha_1, \alpha_2, \alpha_3)\| \Big]$$
$$\preceq diag\Big[\varphi_1(\beta_1, \alpha_1, \beta_2, \alpha_2, \beta_3, \alpha_3), \ldots, \varphi_n(\beta_1, \alpha_1, \beta_2, \alpha_2, \beta_3, \alpha_3) \Big], \tag{12.30}$$

for every $\beta_j, \alpha_j \in U(\mathfrak{M})$, $j = 1, 2, 3$, *then, the* $\mathbb{C}$*–trilinear mappings* $\underbrace{\mathcal{E}_i}_{i=1,\ldots,n} : \pitchfork \times \mathfrak{M}^3 \to \mathfrak{N}$ *are permuting tri-homomorphisms.*

Proof Using Theorem 12.3 and [5, Theorem 4.5], we prove the theorem.

Theorem 12.10 *Let* $n \in \mathbb{N}$, *and* $\underbrace{\varphi_i}_{i=1,\ldots,n} : \mathfrak{M}^6 \to \mathbb{R}$ *be functions satisfying the inequality* (12.23) *and* $\underbrace{\mathcal{L}_i}_{i=1,\ldots,n} : \pitchfork \times \mathfrak{M}^3 \to \mathfrak{N}$ *be mappings satisfying the inequality* (12.19) *and* $\underbrace{\mathcal{L}_i}_{i=1,\ldots,n}(\curlywedge, \Psi_1, 0, \Psi_5) = \underbrace{\mathcal{L}_i}_{i=1,\ldots,n}(\curlywedge, 0, \Psi_3, \Psi_5) = \underbrace{\mathcal{L}_i}_{i=1,\ldots,n}(\curlywedge, \Psi_1, \Psi_3, 0) = 0$, *for every* $\Psi_1, \Psi_3, \Psi_5 \in \mathfrak{M}$ *and* $\curlywedge \in \pitchfork$. *Then, there exist unique* $\mathbb{C}$*–trilinear mappings* $\underbrace{\mathcal{E}_i}_{i=1,\ldots,n} : \pitchfork \times \mathfrak{M}^3 \to \mathfrak{N}$ *satisfying the inequality* (12.29).

Besides, if the mappings $\underbrace{\mathcal{L}_i}_{i=1,\ldots,n} : \pitchfork \times \mathfrak{M}^3 \to \mathfrak{N}$ *satisfy the inequalities* (12.22) *and* (12.30), *then, the* $\mathbb{C}$*–trilinear mappings* $\underbrace{\mathcal{E}_i}_{i=1,\ldots,n} : \pitchfork \times \mathfrak{M}^3 \to \mathfrak{N}$ *are permuting tri–homomorphisms.*

Proof Using a similar method to the proof of Theorem 12.9, we can get the result.

12.6 Application

In this section, firstly we introduce the concept of aggregation maps. Next, we apply a small list of well–known aggregation maps on special functions to present an application of the concept of the multi stability.

12.6.1 An Introduction on Aggregation Maps

The central concept we are investigating in this section is that of aggregation, which refers to the process of combining and merging several (most often numerical) values into a single one. Perhaps the oldest example in this respect is the notion of arithmetic mean or average, which has been used during all the history of physics and all experimental sciences. Any function like the arithmetic mean computing a single output value from an (arbitrarily long) vector of input values is called an aggregation function.

Aggregation functions play an important role in many of the technological tasks scientists are faced with nowadays. They are specifically important in many problems related to the fusion of information. More generally, aggregation functions are widely used in pure mathematics (e.g., functional equations, theory of means and averages, measure and integration theory), applied mathematics (e.g., probability, statistics, decision mathematics), computer and engineering sciences (e.g., artificial intelligence, operations research, information theory, engineering design, pattern recognition and image analysis, data fusion, automated reasoning), economics and finance (e.g., game theory, voting theory, decision making), social sciences (e.g., representational measurement, mathematical psychology) as well as many other applied fields of physics and natural sciences. Thus, a main characteristic of the aggregation functions is that they are used in a large number of areas and disciplines.

The essence of aggregation is that the output value computed by the aggregation function should represent or synthesize "in some sense" all individual inputs, where quotes are put to emphasize the fact that the precise meaning of this expression is highly dependent on the context. In any case, defining or choosing the right class of aggregation functions for a specific problem is a difficult task, considering the huge variety of potential aggregation functions. In this respect, one could ask the following question: Consider a set of n values, lying in some real interval $[a, b]$ to be aggregated. Is any function from $[a, b]^n$ to $\mathbb{R}$ a candidate to be an aggregation function? Obviously not. On the other hand, it is not that easy to define the minimal set of properties a function should fulfill to be an aggregation function. A first natural requirement comes from the fact that the output should be a synthetic value. Then, if inputs are supposed to lie in the interval $[a, b]$, the output should also lie in this interval. Moreover, if all input values are identical to the lower bound a, then the output should also be a, and similarly for the case of the upper bound b. This defines a boundary condition. A second natural requirement is nondecreasing monotonicity. It means that if some of the input values increase, the representative output value should reflect this increase, or at worst, stay constant. These two requirements are commonly accepted in the field, and we adopt them as the basic definition of an aggregation function.

Thus defined, the class of aggregation functions is huge, making the problem of choosing the right function (or family) for a given application a difficult one. Besides this practical consideration, the study of the main classes of aggregation functions, their properties and their relationships, is so complex and rich that it becomes a mathematical topic of its own. The content of this section will offer to the reader a solid mathematical analysis of aggregation functions, able to answer to both mathematical and practical concerns. This section establishes the basic material to start with, essentially a precise definition of an aggregation function, on the basis of the above discussion, the first elementary and well–known examples of aggregation functions, and the list of all the conventional notation used in the section.

In the present section, the symbol |, when used in an interval notation, denotes either [or], indifferently. For instance, $|a, b|$ denotes any real interval with a and b as endpoints.

We assume throughout that the variables of any aggregation function have a common domain $\mathbb{I}$ which is a nonempty real interval, bounded or not. In some exceptional cases, $\mathbb{I}$ will represent a nonempty interval of the extended real number system $\overline{\mathbb{R}} := [-\infty, +\infty]$. This latter case will always be mentioned explicitly and then we will often observe the convention that $0 \times \infty = 0$ and $+\infty + (-\infty) = -\infty$. Also, the notation $\operatorname{int}(\mathbb{I})$ denotes the interior of $\mathbb{I}$, that is, the corresponding open interval, and $\overline{\mathbb{I}}$, denotes its (set) closure.

Let n be any nonzero natural integer and set $[n] := \{1, \ldots, n\}$. We will often use bold symbols to denote n-tuples. For instance $\operatorname{diag}[x_1, \ldots, x_n]_{n\times n}$ will often be written $\mathbf{X}$.

Let us now introduce the concept of aggregation function in a formal way. We shall make a distinction between aggregation functions having a definite number of arguments and extended aggregation functions defined for any number of arguments. Let us first examine the simple case $\mathbb{I} = [0, 1]$, which is very often considered in the literature. An aggregation function in $\operatorname{diag}[[0, 1], \ldots, [0, 1]]_{n\times n}$ is merely a function $\mathsf{A}^{(n)} : \operatorname{diag}[[0, 1], \ldots, [0, 1]]_{n\times n} \longrightarrow [0, 1]$ that is nondecreasing in each variable and that fulfills the boundary conditions

$$\mathsf{A}^{(n)}\Big(\operatorname{diag}[0, \ldots, 0]_{n\times n}\Big) = 0, \quad \text{and} \quad \mathsf{A}^{(n)}\Big(\operatorname{diag}[1, \ldots, 1]_{n\times n}\Big) = 1.$$

In the general case, we have the next definition:

Definition 12.3 ([6]) An aggregation function in $\operatorname{diag}[\mathbb{I}, \ldots, \mathbb{I}]_{n\times n}$ is a function $\mathsf{A}^{(n)} : \operatorname{diag}[\mathbb{I}, \ldots, \mathbb{I}]_{n\times n} \longrightarrow \mathbb{I}$ that

(i) is nondecreasing (in each variable),
(ii) fulfills the boundary conditions

$$\inf_{\mathbf{X}\in\mathbb{I}^n} \mathsf{A}^{(n)}(\mathbf{X}) = \inf \mathbb{I}, \quad \text{and} \quad \sup_{\mathbf{X}\in\mathbb{I}^n} \mathsf{A}^{(n)}(\mathbf{X}) = \sup \mathbb{I}. \tag{12.31}$$

The integer n represents the arity of the aggregation function, that is, the number of its variables. When no confusion can arise, the aggregation functions will simply be written A instead of $\mathsf{A}^{(n)}$. To give a first instance, the arithmetic mean, defined by

$$\mathsf{AM}^{(n)}(\mathbf{X}) := \frac{1}{n}\sum_{i=1}^{n} x_i, \tag{12.32}$$

is clearly an aggregation function in any domain $\operatorname{diag}[\mathbb{I}, \ldots, \mathbb{I}]_{n\times n}$.

A specific case is the aggregation of a singleton, i.e., the unary function $\mathsf{A}^{(1)} : \mathbb{I} \longrightarrow \mathbb{I}$. As the aggregation of a singleton is not an actual aggregation, the following convention is often considered:

$$\mathsf{A}^{(1)}(x) = x, \tag{12.33}$$

for all $x \in \mathbb{I}$. Unless otherwise specified, we will adopt this convention throughout. For the sake of clarity, we will often use the letter F(or $\mathsf{G}, \mathsf{H}, \ldots$) to designate an arbitrary function of several variables and the letter A for an aggregation function. The range of any function F will be denoted $\operatorname{ran}(\mathsf{F})$.

Remark 12.2 (i) It is easy to see that any function $\mathsf{F} : \operatorname{diag}[\mathbb{I}, \ldots, \mathbb{I}]_{n\times n} \longrightarrow \mathbb{R}$ fulfilling conditions (i) and (ii) of Definition 12.3 is such that $\operatorname{ran}(\mathsf{F}) \subseteq \overline{\mathbb{I}}$. However, by definition, any aggregation function A in $\operatorname{diag}[\mathbb{I}, \ldots, \mathbb{I}]_{n\times n}$ fulfills $\operatorname{ran}(\mathsf{A}) \subseteq \mathbb{I}$, which is a much more natural property.

(ii) Note that any n-tuple $\mathbf{X} = \operatorname{diag}[x_1, \ldots, x_n]_{n\times n} \in \operatorname{diag}[\mathbb{I}, \ldots, \mathbb{I}]_{n\times n}$, whose coordinates are to be aggregated by an aggregation function A, can itself be considered as a function from $[n]$ to $\mathbb{I}$. In this case the aggregation function A is simply regarded as a functional from $\operatorname{diag}[\mathbb{I}, \ldots, \mathbb{I}]_{[n]\times[n]}$ to $\mathbb{I}$. Instances of such functionals are given by integrals constructed from capacities (see Chap. 5 of [6]).

Definition 12.4 ([6]) The diagonal section of any function $\mathsf{F} : \operatorname{diag}[\mathbb{I}, \ldots, \mathbb{I}]_{n\times n} \longrightarrow \overline{\mathbb{R}}$ is the unary function $\delta_{\mathsf{F}} : \mathbb{I} \longrightarrow \overline{\mathbb{R}}$ defined as $\delta_{\mathsf{F}}(x) = \mathsf{F}\Big(\operatorname{diag}[x, \ldots, x]_{n\times n}\Big)$, for all $x \in \mathbb{I}$.

The boundary conditions given in Definition 12.3 show that, generally, an aggregation function cannot be defined without specifying its domain. The following proposition shows that, for a nondecreasing function $\mathsf{F} : \operatorname{diag}[\mathbb{I}, \ldots, \mathbb{I}]_{n\times n} \longrightarrow \overline{\mathbb{R}}$ to be an aggregation function in $\operatorname{diag}[\mathbb{I}, \ldots, \mathbb{I}]_{n\times n}$, the condition $\operatorname{ran}(\mathsf{F}) \subseteq \mathbb{I}$ and the boundary conditions (12.31) are required only for the diagonal section δ_{F}.

Proposition 12.1 ([6]) *A nondecreasing function* $\mathsf{F} : \operatorname{diag}[\mathbb{I}, \ldots, \mathbb{I}]_{n\times n} \longrightarrow \overline{\mathbb{R}}$ *is an aggregation function in* $\operatorname{diag}[\mathbb{I}, \ldots, \mathbb{I}]_{n\times n}$ *if and only if* $\operatorname{ran}(\delta_{\mathsf{F}}) \subseteq \mathbb{I}$ *holds with*

$$\inf_{x\in\mathbb{I}} \delta_{\mathsf{F}(x)} = \inf \mathbb{I}, \quad \textit{and} \quad \sup_{x\in\mathbb{I}} \delta_{\mathsf{F}(x)} = \sup \mathbb{I}.$$

The following result provides an easy way to determine each subdomain $\operatorname{diag}[\mathbb{J}, \ldots, \mathbb{J}]_{n\times n}$ of $\operatorname{diag}[\mathbb{I}, \ldots, \mathbb{I}]_{n\times n}$, with $\mathbb{J} \subseteq \mathbb{I} \subseteq \mathbb{R}$, in which a given nondecreasing function $\mathsf{F} : \operatorname{diag}[\mathbb{I}, \ldots, \mathbb{I}]_{n\times n} \to \overline{\mathbb{R}}$ is an aggregation function.

Proposition 12.2 ([6]) *Let* $\mathsf{F} : \operatorname{diag}[\mathbb{I}, \ldots, \mathbb{I}]_{n\times n} \longrightarrow \overline{\mathbb{R}}$ *be a nondecreasing function and let* $a := \inf \mathbb{I}$ *and* $b := \sup \mathbb{I}$. *Then,*

$$\inf_{x\in\mathbb{I}} \delta_{\mathsf{F}}(x) = a \Longleftrightarrow \begin{cases} \delta_{\mathsf{F}}(a) = a, & \textit{if } a \in \mathbb{I}, \\ \lim_{x\to a^+} \delta\mathsf{F}(x) = a, & \textit{if } a \notin \mathbb{I}, \end{cases}$$

and

$$\inf_{x\in\mathbb{I}} \delta_{\mathsf{F}}(x) = b \Longleftrightarrow \begin{cases} \delta_{\mathsf{F}}(b) = b, & \textit{if } b \in \mathbb{I}, \\ \lim_{x\to b^-} \delta\mathsf{F}(x) = a, & \textit{if } b \notin \mathbb{I}, \end{cases}$$

Example 12.1 Consider the product function

$$\Pi(\mathbf{X}) := \prod_{i=1}^{n} x_i. \tag{12.34}$$

Let us restrict this function to $\mathrm{diag}[\mathbb{I}, \ldots, \mathbb{I}]_{n\times n} = \mathrm{diag}[[0, +\infty], \ldots, [0, \infty]]_{n\times n}$ in which it is continuous and nondecreasing. To ensure the existence of the product, we adopt the convention that $0 \times \infty = 0$. If $n > 1$, the solutions of the equation $\delta_\Pi(x) = x$ in $[0, \infty]$ are 0, 1, and ∞. Therefore, according to Propositions 12.1 and 12.2, the product function, when restricted to $\mathrm{diag}[[0, \infty], \ldots, [0, \infty]]_{n\times n}$ is an aggregation function in the subdomains $\mathrm{diag}[|0, 1|, \ldots |0, 1|]_{n\times n}$, $\mathrm{diag}[|1, \infty|, \ldots, |1, \infty|]_{n\times n}$, $\mathrm{diag}[|0, \infty|, \ldots, |0, \infty|]_{n\times n}$, and only in these.

Example 12.2 Consider the function

$$\mathsf{F}(\mathbf{X}) := \frac{(\sum_i x_i)^2 - \sum_i x_i^2}{\sum_i x_i}.$$

Let us restrict this function to $\mathrm{diag}[\mathbb{I}, \ldots \mathbb{I}]_{n\times n} = \mathrm{diag}[(0, \infty), \ldots, (0, \infty)]_{n\times n}$ in which it is continuous and nondecreasing. If $n > 2$, the equation $\delta_{\mathsf{F}}(x) = x$ has no solution in $(0, \infty)$. Hence, the restriction of F to $\mathrm{diag}[(0, \infty), \ldots, (0, \infty)]_{n\times n}$ is an aggregation function but not in a proper subdomain. For $n = 2$ any subdomain can be considered (and then F is the harmonic mean).

We have the immediate corollary:

Corollary 12.1 ([6]) *Let* $\mathsf{F} : \mathit{diag}[\mathbb{I}, \ldots \mathbb{I}]_{n\times n} \longrightarrow \overline{\mathbb{R}}$ *be nondecreasing. If* F *is idempotent, i.e.,* $\delta_{\mathsf{F}(x)} = x$*, for all* $x \in \mathbb{I}$*, then it is an aggregation function in* $\mathit{diag}[\mathbb{I}, \ldots \mathbb{I}]_{n\times n}$ *and even in any subinterval* $\mathit{diag}[\mathbb{J}, \ldots \mathbb{J}]_{n\times n}$ *of* $\mathit{diag}[\mathbb{I}, \ldots \mathbb{I}]_{n\times n}$*, with* $\mathbb{J} \subseteq \mathbb{I}$.

We now introduce the concept of extended aggregation function. Denote by $\mathbb{N}$ the set of strictly positive integers.

Definition 12.5 ([6]) An extended function in $\bigcup_{n\in\mathbb{N}} \mathrm{diag}[\mathbb{I}, \ldots \mathbb{I}]_{n\times n}$ is a mapping

$$\mathsf{F} : \bigcup_{n\in\mathbb{N}} \mathrm{diag}[\mathbb{I}, \ldots \mathbb{I}]_{n\times n} \longrightarrow \overline{\mathbb{R}}.$$

An extended aggregation function in $\bigcup_{n\in\mathbb{N}} \mathrm{diag}[\mathbb{I}, \ldots \mathbb{I}]_{n\times n}$ is an extended function A in $\bigcup_{n\in\mathbb{N}} \mathrm{diag}[\mathbb{I}, \ldots \mathbb{I}]_{n\times n}$ whose restriction $\mathsf{A}^{(n)} := \mathsf{A}|_{\mathrm{diag}[\mathbb{I},\ldots\mathbb{I}]_{n\times n}}$ to $\mathrm{diag}[\mathbb{I}, \ldots \mathbb{I}]_{n\times n}$ is an aggregation function in $\mathrm{diag}[\mathbb{I}, \ldots \mathbb{I}]_{n\times n}$ for any $n \in \mathbb{N}$.

Since an extended function is defined for any number of arguments, it can also be regarded as a sequence $(\mathsf{F}^{(n)})_{n\in\mathbb{N}}$, whose nth element is an n-ary function $\mathsf{F}^{(n)}$: $\text{diag}[\mathbb{I},\ldots\mathbb{I}]_{n\times n} \longrightarrow \overline{\mathbb{R}}$ When no confusion can arise, an extended function will simply be called a function.

For example, the arithmetic mean as an extended aggregation function is the sequence $(\mathsf{A}^{(n)})_{n\in\mathbb{N}}$, where $\mathsf{A}^{(n)}$ is defined by (12.32) for all $n \in \mathbb{N}$.

Remark 12.3 In an arbitrary extended function, for different n and m the functions $\mathsf{F}^{(n)}$ and $\mathsf{F}^{(m)}$ need not be related to each other. For instance, $\mathsf{F}^{(n)}$ could be the arithmetic mean while $\mathsf{F}^{(m)}$ could be the product. However, as we can see in Chapter 2 in [6], they can be related in a relevant way by means of certain grouping properties, such as associativity or decomposability.

For illustration and subsequent use we now give a small list of well–known aggregation functions, as follows:

- The arithmetic mean function $\mathsf{AM} : \text{diag}[\mathbb{I},\ldots,\mathbb{I}]_{n\times n} \longrightarrow \mathbb{I}$ and the geometric mean function $\mathsf{GM} : \text{diag}[\mathbb{I},\ldots,\mathbb{I}]_{n\times n} \longrightarrow \mathbb{I}$ are respectively defined by

$$\mathsf{AM}(\mathbf{X}) := \frac{1}{n}\sum_{i=1}^{n} x_i, \tag{12.35}$$

$$\mathsf{GM}(\mathbf{X}) := (\prod_{i=1}^{n} x_i)^{\frac{1}{n}}. \tag{12.36}$$

  Note that, when $n > 1$, the geometric mean is not an aggregation function in every domain.We must consider a domain $\text{diag}[\mathbb{I},\ldots,\mathbb{I}]_{n\times n}$ such that $\mathbb{I} \subseteq [0,\infty]$.
- For any $k \in [n]$, the projection function $\mathsf{P}_k : \text{diag}[\mathbb{I},\ldots,\mathbb{I}]_{n\times n} \longrightarrow \mathbb{I}$ and the order statistic function $\mathsf{OS}_k : \text{diag}[\mathbb{I},\ldots,\mathbb{I}]_{n\times n} \longrightarrow \mathbb{I}$ associated with the kth argument, are respectively defined by

$$\mathsf{P}_k(\mathbf{X}) := x_k, \tag{12.37}$$
$$\mathsf{OS}_k(\mathbf{X}) := (x)_k, \tag{12.38}$$

  where $(x)_k$ is the kth lowest coordinate of x, that is,

$$x_{(1)} \leqslant \cdots \leqslant x_{(k)} \leqslant \cdots x_{(n)}.$$

  The projections onto the first and the last coordinates are defined as

$$\mathsf{P}_F(\mathbf{X}) := \mathsf{P}_1(\mathbf{X}) = x_1, \tag{12.39}$$
$$\mathsf{P}_L(\mathbf{X}) := \mathsf{P}_n(\mathbf{X}) = x_n. \tag{12.40}$$

Similarly, the extreme order statistics x_1 and x_n are respectively the minimum and maximum functions

$$\mathsf{Min}(\mathbf{X}) := \mathsf{OS}_1(\mathbf{X}) = \min\{x_1, \ldots, x_n\}, \tag{12.41}$$
$$\mathsf{Max}(\mathbf{X}) := \mathsf{OS}_n(\mathbf{X}) = \max\{x_1, \ldots, x_n\}, \tag{12.42}$$

which will sometimes be written by means of the lattice operations $\vee$ and $\wedge$, respectively, that is,

$$\mathsf{Min}(\mathbf{X}) = \bigwedge_{i=1}^{n} x_i, \quad \text{and} \quad \mathsf{Max}(\mathbf{X}) = \bigvee_{i=1}^{n} x_i.$$

Note that OS_k can be written in terms of only minima and maxima as follows

$$\mathsf{OS}_k(\mathbf{X}) = \bigwedge_{\substack{K\subseteq[n]\\|K|=k}} \bigvee_{i\in K} x_i = \bigvee_{\substack{K\subseteq[n]\\|K|=n-k+1}} \bigwedge_{i\in K} x_i.$$

Also, the median of an odd number of values $\mathrm{diag}[x_1, \ldots, x_{2k-1}]_{(2k-1)\times(2k-1)}$ is simply defined by

$$\mathsf{Med}\Big(\mathrm{diag}[x_1, \ldots, x_{2k-1}]_{(2k-1)\times(2k-1)}\Big) = x_{(k)},$$

which can be rewritten as

$$\mathsf{Med}\Big(\mathrm{diag}[x_1, \ldots, x_{2k-1}]_{(2k-1)\times(2k-1)}\Big) = \bigwedge_{\substack{K\subseteq[2k-1]\\|K|=k}} \bigvee_{i\in K} x_i = \bigvee_{\substack{K\subseteq[2k-1]\\|K|=k}} \bigwedge_{i\in K} x_i.$$

For instance, we have

$$\begin{aligned}\mathsf{Med}(\mathrm{diag}[x_1, x_2, x_3]_{3\times3}) &= (x_1 \wedge x_2) \vee (x_1 \wedge x_3) \vee (x_2 \wedge x_3)\\ &= (x_1 \vee x_2) \wedge (x_1 \vee x_3) \wedge (x_2 \vee x_3).\end{aligned}$$

For an even number of values $\mathrm{diag}[x_1, \ldots, x_{2k}]$, the median is defined by

$$\mathsf{Med}\Big(\mathrm{diag}[x_1, \ldots, x_{2k}]_{2k\times2k}\Big) := \mathsf{AM}\Big(\mathrm{diag}[x_{(k)}, x_{(k+1)}]_{2\times2}\Big) = \frac{x_{(k)} + x_{(k+1)}}{2}.$$

For any $\alpha \in \mathbb{I}$, we also define the α-median, $\mathsf{Med}_\alpha : \mathrm{diag}[\mathbb{I}, \ldots, \mathbb{I}]_{n\times n} \longrightarrow \mathbb{I}$, by

$$\mathsf{Med}_\alpha(\mathbf{X}) = \mathsf{Med}\Big(\mathrm{diag}[x_1, \ldots, x_n, \underbrace{\alpha, \ldots, \alpha}_{n-1}]_{(2n-1)\times(2n-1)}\Big)$$

$$= \mathsf{Med}(\mathsf{Min}(\mathsf{X}), \alpha, \mathsf{Max}(\mathsf{X})). \quad (12.43)$$

- For any nonempty subset $K \subseteq [n]$, the partial minimum $\mathsf{Min}_k : \text{diag}[\mathbb{I}, \ldots, \mathbb{I}]_{n\times n} \longrightarrow \mathbb{I}$ and the partial maximum $\mathsf{Max}_k : \text{diag}[\mathbb{I}, \ldots, \mathbb{I}]_{n\times n} \longrightarrow \mathbb{I}$, associated with K, are respectively defined by

$$\mathsf{Min}_k(\mathbf{X}) := \bigwedge_{i\in K} x_i, \quad (12.44)$$

$$\mathsf{Max}_k(\mathbf{X}) := \bigvee_{i\in K} x_i. \quad (12.45)$$

- For any weight vector $\mathbf{W} = \text{diag}[w_1, \ldots, w_n]_{n\times n} \in \text{diag}[[0, 1], \ldots, [0, 1]]_{n\times n}$ such that $\sum_{i=1}^{n} w_i = 1$, the weighted arithmetic mean function $\mathsf{WAM}_\mathbf{w} : \text{diag}[\mathbb{I}, \ldots, \mathbb{I}]_{n\times n} \longrightarrow \mathbb{I}$ and the ordered weighted averaging function $\mathsf{OWA}_\mathbf{w} : \text{diag}[\mathbb{I}, \ldots, \mathbb{I}]_{n\times n} \longrightarrow \mathbb{I}$, associated with $\mathbf{W}$, are respectively defined by

$$\mathsf{WAM}_\mathbf{w}(\mathbf{X}) := \sum_{i=1}^{n} w_i x_i, \quad (12.46)$$

$$\mathsf{OWA}_\mathbf{w}(\mathbf{X}) := \sum_{i=1}^{n} w_i x_{(i)}. \quad (12.47)$$

- The sum $\sum : \text{diag}[\overline{\mathbb{R}}, \ldots, \overline{\mathbb{R}}]_{n\times n} \longrightarrow \mathbb{R}$ and product $\Pi : \text{diag}[\overline{\mathbb{R}}, \ldots, \overline{\mathbb{R}}]_{n\times n} \longrightarrow \mathbb{R}$ functions are respectively defined by

$$\sum(\mathbf{X}) := \sum_{i=1}^{n} x_i, \quad (12.48)$$

$$\Pi(\mathbf{X}) := \prod_{i=1}^{n} x_i. \quad (12.49)$$

  Note that, when $n > 1$, these latter two functions are not aggregation functions in every domain. For the sum, we need to consider the domains $\text{diag}[|-\infty, \infty|, \ldots, |-\infty, \infty|]_{n\times n}$, $\text{diag}[|0, \infty|, \ldots, |0, \infty|]_{n\times n}$, $\text{diag}[|-\infty, 0|, \ldots, |-\infty, 0|]_{n\times n}$. For the product, when restricted to $\text{diag}[[0, \infty], \ldots, [0, \infty]]_{n\times n}$ we have to consider the domains $\text{diag}[|0, 1|, \ldots, |0, 1|]_{n\times n}$, $\text{diag}[|1, \infty|, \ldots, |1, \infty|]_{n\times n}$, $\text{diag}[|0, \infty|, \ldots, |0, \infty|]_{n\times n}$.
- Assume $\mathbb{I} = [a, b]$ is a closed interval. The smallest and the greatest aggregation functions in $\text{diag}[[a, b], \ldots, [a, b]]_{n\times n}$ are respectively defined by

$$\mathsf{A}_\perp(\mathbf{x}) = \begin{cases} b \text{ if } x_i = b, & \text{for all } i \in [n], \\ a & otherwise. \end{cases}$$

$$\mathsf{A}_{\top}(\mathbf{x}) = \begin{cases} a \text{ if } x_i = a, & \text{for all } i \in [n], \\ b & otherwise. \end{cases}$$

By definition, we have $\mathsf{A}_{\perp} \leq \mathsf{A} \leq \mathsf{A}_{\top}$ for any aggregation function $\mathsf{A} : \text{diag}[[a, b], \ldots, [a, b]]_{n\times n} \to [a, b]$.

Clearly, the constant function $K_c : \text{diag}[\mathbb{I}, \ldots, \mathbb{I}]_{n\times n} \to \mathbb{I}$, given by

$$K_c(\mathbf{x}) := c,$$

where $c \in \mathbb{I}$ is a fixed constant, is a trivial example of a function that is not an aggregation function (unless $\mathbb{I} = \{c\}$).

12.6.2 Multi Stability with Application of Aggregate Special Functions

In this part, using the idea of aggregation maps and special functions, we present an application of the multi stability.

Consider the following aggregation maps defined in Sect. 12.6.1, as follows:

$$\begin{aligned}&\mathsf{AM}(\mathbf{X}), \mathsf{GM}(\mathbf{X}), \mathsf{P}_k(\mathbf{X}), \mathsf{OS}_k(\mathbf{X}), \mathsf{P}_F(\mathbf{X}), \mathsf{P}_L(\mathbf{X}), \mathsf{Min}(\mathbf{X}), \mathsf{Max}(\mathbf{X}), \mathsf{Med}(\mathbf{X}), \mathsf{Med}_\alpha(\mathbf{X}),\\ &\mathsf{Min}_k(\mathbf{X}), \mathsf{Max}_k(\mathbf{X}), \mathsf{WAM}_{\mathbf{W}}(\mathbf{X}), \mathsf{OWA}_{\mathbf{W}}(\mathbf{X}), \sum(\mathbf{X}), \prod(\mathbf{X}), \mathsf{A}_{\perp}(\mathbf{X}), \mathsf{A}_{\top}(\mathbf{X}),\end{aligned}$$

where k, α and $\mathbf{W}$ are defined in the previous part.

Let $\underbrace{\mathsf{AG}_i}_{\substack{1\leq i\leq n\\ n\in\mathbb{N}}}(\mathbf{X})$, denotes the ith given aggregation maps, (here, $n = 18$). In addition, let $\mathfrak{W}(X) = \text{diag}[\varphi_1^{\circledS}(X), \ldots, \varphi_m^{\circledS}(X)]$, with $m \in \mathbb{N}$, be the matrix valued controller of given special functions defined in Sects. 9.8.1 and 9.8.2.

We now have the following results of the multi stability with the application of aggregation maps and special functions.

Note 12.1 Here, we have the multi stability results, when $\mathcal{L}_1 = \cdots = \mathcal{L}_{n\in\mathbb{N}}$.

Corollary 12.2 *Let* $\underbrace{\Theta_i}_{i=1,\ldots,n} \geq 0$ *and* $\underbrace{r_i}_{i=1,\ldots,n} > 3$ *be in* $\mathbb{R}$ *and* $\underbrace{\mathcal{L}_i}_{i=1,\ldots,n} : \pitchfork \times \mathcal{G}^3 \to \mathcal{Q}$ *be mappings satisfying* $\underbrace{\mathcal{L}_i}_{i=1,\ldots,n}(\curlywedge, \Psi_1, 0, \Psi_5) = \underbrace{\mathcal{L}_i}_{i=1,\ldots,n}(\curlywedge, 0, \Psi_3, \Psi_5) = \underbrace{\mathcal{L}_i}_{i=1,\ldots,n}(\curlywedge, \Psi_1, \Psi_3, 0) = 0$ *and*

$$\begin{aligned} diag\Big[&\|\mathcal{L}_1(\curlywedge, \Psi_1 + \Psi_2, \Psi_3 - \Psi_4, \Psi_5 + \Psi_6) + \mathcal{L}_1(\curlywedge, \Psi_1 - \Psi_2, \Psi_3 + \Psi_4, \Psi_5 - \Psi_6) \\ &\quad -2\mathcal{L}_1(\curlywedge, \Psi_1, \Psi_3, \Psi_5) + 2\mathcal{L}_1(\curlywedge, \Psi_1, \Psi_4, \Psi_6) \\ &\qquad -2\mathcal{L}_1(\curlywedge, \Psi_2, \Psi_3, \Psi_6) + 2\mathcal{L}_1(\curlywedge, \Psi_2, \Psi_4, \Psi_5)\|, \ldots, \end{aligned} \tag{12.50}$$

$$\|\mathcal{L}_n(\leftthreetimes, \Psi_1+\Psi_2, \Psi_3-\Psi_4, \Psi_5+\Psi_6)+\mathcal{L}_n(\leftthreetimes, \Psi_1-\Psi_2, \Psi_3+\Psi_4, \Psi_5-\Psi_6)$$
$$-2\mathcal{L}_n(\leftthreetimes, \Psi_1, \Psi_3, \Psi_5)+2\mathcal{L}_n(\leftthreetimes, \Psi_1, \Psi_4, \Psi_6)$$
$$-2\mathcal{L}_n(\leftthreetimes, \Psi_2, \Psi_3, \Psi_6)+2\mathcal{L}_n(\leftthreetimes, \Psi_2, \Psi_4, \Psi_5)\|\Big]$$
$$\preceq diag\Bigg[\Bigg\|\boldsymbol{\Psi}_1\left(2\mathcal{L}_1\left(\leftthreetimes, \frac{\Psi_1+\Psi_2}{2}, \Psi_3-\Psi_4, \Psi_5+\Psi_6\right)\right.$$
$$+2\mathcal{L}_1\left(\leftthreetimes, \frac{\Psi_1-\Psi_2}{2}, \Psi_3+\Psi_4, \Psi_5-\Psi_6\right)$$
$$-2\mathcal{L}_1(\leftthreetimes, \Psi_1, \Psi_3, \Psi_5)+2\mathcal{L}_1(\leftthreetimes, \Psi_1, \Psi_4, \Psi_6)$$
$$\left.-2\mathcal{L}_1(\leftthreetimes, \Psi_2, \Psi_3, \Psi_6)+2\mathcal{L}_1(\leftthreetimes, \Psi_2, \Psi_4, \Psi_5)\right)\|$$
$$+\Theta_1 \mathrm{AG}_1\left(\mathfrak{W}\left(\|\Psi_1\|^{r_1}+\|\Psi_2\|^{r_1}+\|\Psi_3\|^{r_1}+\|\Psi_4\|^{r_1}+\|\Psi_5\|^{r_1}+\|\Psi_6\|^{r_1}\right)\right),$$
$$\ldots, \Bigg\|\boldsymbol{\Psi}_n\left(2\mathcal{L}_n\left(\leftthreetimes, \frac{\Psi_1+\Psi_2}{2}, \Psi_3-\Psi_4, \Psi_5+\Psi_6\right)\right.$$
$$+2\mathcal{L}_n\left(\leftthreetimes, \frac{\Psi_1-\Psi_2}{2}, \Psi_3+\Psi_4, \Psi_5-\Psi_6\right)$$
$$-2\mathcal{L}_n(\leftthreetimes, \Psi_1, \Psi_3, \Psi_5)+2\mathcal{L}_n(\leftthreetimes, \Psi_1, \Psi_4, \Psi_6)$$
$$\left.-2\mathcal{L}_n(\leftthreetimes, \Psi_2, \Psi_3, \Psi_6)+2\mathcal{L}_n(\leftthreetimes, \Psi_2, \Psi_4, \Psi_5)\right)\|$$
$$+\Theta_n \mathrm{AG}_n\left(\mathfrak{W}\left(\|\Psi_1\|^{r_n}+\|\Psi_2\|^{r_n}+\|\Psi_3\|^{r_n}+\|\Psi_4\|^{r_n}+\|\Psi_5\|^{r_n}+\|\Psi_6\|^{r_n}\right)\right)\Bigg],$$

for every $\Psi_1, \Psi_2, \Psi_3, \Psi_4, \Psi_5, \Psi_6 \in \mathcal{G}$ *and* $\leftthreetimes \in \pitchfork$. *Then, there exist unique tri–additive mappings* $\underbrace{\mathcal{L}_i'}_{i=1,\ldots,n} : \pitchfork \times \mathcal{G}^3 \to \mathcal{Q}$ *satisfying the inequality below*

$$diag\Big[\|\mathcal{L}_1(\leftthreetimes, \Psi_1, \Psi_3, \Psi_5)-\mathcal{L}_1'(\leftthreetimes, \Psi_1, \Psi_3, \Psi_5)\|, \ldots, \tag{12.51}$$
$$\|\mathcal{L}_n(\leftthreetimes, \Psi_1, \Psi_3, \Psi_5)-\mathcal{L}_n'(\leftthreetimes, \Psi_1, \Psi_3, \Psi_5)\|\Big]$$
$$\preceq diag\left[\frac{\Theta_1}{2^{r_1}-2^3}\mathrm{AG}_1\left(\mathfrak{W}\left((2^{r_1}+8)\|\Psi_1\|^{r_1}+8\|\Psi_3\|^{r_1}+(2+2^{r_1+1}+8)\|\Psi_5\|^{r_1}\right)\right),\right.$$
$$\left.\ldots, \frac{\Theta_n}{2^{r_n}-2^3}\mathrm{AG}_n\left(\mathfrak{W}\left((2^{r_n}+8)\|\Psi_1\|^{r_n}+8\|\Psi_3\|^{r_n}+(2+2^{r_n+1}+8)\|\Psi_5\|^{r_n}\right)\right)\right],$$

for every $\Psi_1, \Psi_3, \Psi_5 \in \mathcal{G}$ *and* $\leftthreetimes \in \pitchfork$.

Proof We define the control function $\underbrace{\varphi_i}_{i=1,\ldots,n}$ by the coefficient

$$\varphi_i(\Psi_1, \Psi_2, \Psi_3, \Psi_4, \Psi_5, \Psi_6) \tag{12.52}$$
$$=\Theta_i \mathrm{AG}_i\left(\mathfrak{W}\left(\|\Psi_1\|^{r_i}+\|\Psi_2\|^{r_i}+\|\Psi_3\|^{r_i}+\|\Psi_4\|^{r_i}+\|\Psi_5\|^{r_i}+\|\Psi_6\|^{r_i}\right)\right), \quad i=1,\ldots,n,$$

for every Ψ_1, ..., $\Psi_6 \in \mathcal{G}$ and $\leftthreetimes \in \pitchfork$. Letting $\underbrace{\mathfrak{P}_i}_{i=1,\ldots,n} = 2^{1-\overbrace{r_i}^{i=1,\ldots,n}}$, the result follows immediately from Theorem 12.1.

for all $\Psi_1, \Psi_3, \Psi_5 \in \mathcal{G}, \leftthreetimes \in \pitchfork$.
for all $\Psi_1, \Psi_3, \Psi_5 \in \mathcal{G}, \leftthreetimes \in \pitchfork$.

Corollary 12.3 *Let* $\underbrace{\Theta_i}_{i=1,\dots,n} \geq 0$ *and* $\underbrace{r_i}_{i=1,\dots,n} < 3$ *be in* $\mathbb{R}$, *and* $\underbrace{\mathcal{L}_i}_{i=1,\dots,n} : \pitchfork \times \mathcal{G}^3 \to \mathcal{Q}$ *be mappings satisfying* $\underbrace{\mathcal{L}_i}_{i=1,\dots,n}(\leftthreetimes, \Psi_1, 0, \Psi_5) = \underbrace{\mathcal{L}_i}_{i=1,\dots,n}(\leftthreetimes, 0, \Psi_3, \Psi_5) = \underbrace{\mathcal{L}_i}_{i=1,\dots,n}(\leftthreetimes, \Psi_1, \Psi_3, 0) = 0$ *and the inequality* (12.50). *Then, we can find unique tri–additive mappings* $\underbrace{\mathcal{L}'_i}_{i=1,\dots,n} : \pitchfork \times \mathcal{G}^3 \to \mathcal{Q}$ *satisfying the inequality below*

$$
\begin{aligned}
& diag\Big[\|\mathcal{L}_1(\leftthreetimes, \Psi_1, \Psi_3, \Psi_5) - \mathcal{L}'_1(\leftthreetimes, \Psi_1, \Psi_3, \Psi_5)\|, \dots, \\
& \qquad \|\mathcal{L}_n(\leftthreetimes, \Psi_1, \Psi_3, \Psi_5) - \mathcal{L}'_n(\leftthreetimes, \Psi_1, \Psi_3, \Psi_5)\|\Big] \\
& \preceq diag\Big[\frac{\Theta_1}{8 - 2^{r_1}} \mathrm{AG}_1\Big(\mathfrak{W}\Big((2^{r_1} + 8)\|\Psi_1\|^{r_1} + 8\|\Psi_3\|^{r_1} + (2^{r_1} + 2^{r_1+1} + 8)\|\Psi_5\|^{r_1}\Big)\Big), \\
& \qquad \dots, \frac{\Theta_n}{8 - 2^{r_n}} \mathrm{AG}_n\Big(\mathfrak{W}\Big((2^{r_n} + 8)\|\Psi_1\|^{r_n} + 8\|\Psi_3\|^{r_n} + (2^{r_n} + 2^{r_n+1} + 8)\|\Psi_5\|^{r_n}\Big)\Big)\Big],
\end{aligned}
$$

for every $\Psi_1, \Psi_3, \Psi_5 \in \mathcal{G}$ *and* $\leftthreetimes \in \pitchfork$.

Proof Put $\underbrace{\mathfrak{P}_i}_{i=1,\dots,n} = 2^{\overbrace{r_i}^{i=1,\dots,n}} - 1$. It follows immediately from the relation (12.52) and Theorem 12.2.

Corollary 12.4 *Let* $\underbrace{\Theta_i}_{i=1,\dots,n} \geq 0$ *and* $\underbrace{r_i}_{i=1,\dots,n} > 4$ *be in* $\mathbb{R}$, *and* $\underbrace{\mathcal{L}_i}_{i=1,\dots,n} : \pitchfork \times \mathfrak{M}^3 \to \mathfrak{M}$ *be mappings satisfying* $\underbrace{\mathcal{L}_i}_{i=1,\dots,n}(\leftthreetimes, \Psi_1, 0, \Psi_5) = \underbrace{\mathcal{L}_i}_{i=1,\dots,n}(\leftthreetimes, 0, \Psi_3, \Psi_5) = \underbrace{\mathcal{L}_i}_{i=1,\dots,n}(\leftthreetimes, \Psi_1, \Psi_3, 0) = 0$ *and the following inequality*

$$
\begin{aligned}
& diag\Big[\|\mathcal{L}_1(\leftthreetimes, \Lambda_1(\Psi_1 + \Psi_2), \Lambda_2(\Psi_3 - \Psi_4), \Lambda_3(\Psi_5 + \Psi_6)) \\
& \qquad + \mathcal{L}_1(\leftthreetimes, \Lambda_1(\Psi_1 - \Psi_2), \Lambda_2(\Psi_3 + \Psi_4), \Lambda_3(\Psi_5 - \Psi_6)) \\
& \qquad - \Lambda_1\Lambda_2\Lambda_3(2\mathcal{L}_1(\leftthreetimes, \Psi_1, \Psi_3, \Psi_5) - 2\mathcal{L}_1(\leftthreetimes, \Psi_1, \Psi_4, \Psi_6) \\
& \qquad + 2\mathcal{L}_1(\leftthreetimes, \Psi_2, \Psi_3, \Psi_6) - 2\mathcal{L}_1(\leftthreetimes, \Psi_2, \Psi_4, \Psi_5))\|, \dots, \\
& \qquad \|\mathcal{L}_n(\leftthreetimes, \Lambda_1(\Psi_1 + \Psi_2), \Lambda_2(\Psi_3 - \Psi_4), \Lambda_3(\Psi_5 + \Psi_6)) \\
& \qquad + \mathcal{L}_n(\leftthreetimes, \Lambda_1(\Psi_1 - \Psi_2), \Lambda_2(\Psi_3 + \Psi_4), \Lambda_3(\Psi_5 - \Psi_6)) \\
& \qquad - \Lambda_1\Lambda_2\Lambda_3(2\mathcal{L}_n(\leftthreetimes, \Psi_1, \Psi_3, \Psi_5) - 2\mathcal{L}_n(\leftthreetimes, \Psi_1, \Psi_4, \Psi_6) \\
& \qquad + 2\mathcal{L}_n(\leftthreetimes, \Psi_2, \Psi_3, \Psi_6) - 2\mathcal{L}_n(\leftthreetimes, \Psi_2, \Psi_4, \Psi_5))\|\Big] \\
& \preceq diag\Big[\Big\|\boldsymbol{\Psi}_1\Big(2\mathcal{L}_1\Big(\leftthreetimes, \frac{\Psi_1 + \Psi_2}{2}, \Psi_3 - \Psi_4, \Psi_5 + \Psi_6\Big)
\end{aligned}
\tag{12.53}
$$

$$
\begin{aligned}
&+2\mathcal{L}_1\left(\leftthreetimes, \frac{\Psi_1-\Psi_2}{2}, \Psi_3+\Psi_4, \Psi_5-\Psi_6\right) \\
&\quad -2\mathcal{L}_1(\leftthreetimes, \Psi_1, \Psi_3, \Psi_5)+2\mathcal{L}_1(\leftthreetimes, \Psi_1, \Psi_4, \Psi_6) \\
&\qquad -2\mathcal{L}_1(\leftthreetimes, \Psi_2, \Psi_3, \Psi_6)+2\mathcal{L}_1(\leftthreetimes, \Psi_2, \Psi_4, \Psi_5))\| \\
&+\Theta_1 \mathrm{AG}_1\left(\mathfrak{W}\left(\|\Psi_1\|^{r_1}+\|\Psi_2\|^{r_1}+\|\Psi_3\|^{r_1}+\|\Psi_4\|^{r_1}+\|\Psi_5\|^{r_1}+\|\Psi_6\|^{r_1}\right)\right), \\
&\ldots, \left\|\Psi_n\left(2\mathcal{L}_n\left(\leftthreetimes, \frac{\Psi_1+\Psi_2}{2}, \Psi_3-\Psi_4, \Psi_5+\Psi_6\right)\right.\right. \\
&+2\mathcal{L}_n\left(\leftthreetimes, \frac{\Psi_1-\Psi_2}{2}, \Psi_3+\Psi_4, \Psi_5-\Psi_6\right) \\
&\quad -2\mathcal{L}_n(\leftthreetimes, \Psi_1, \Psi_3, \Psi_5)+2\mathcal{L}_n(\leftthreetimes, \Psi_1, \Psi_4, \Psi_6) \\
&\qquad -2\mathcal{L}_n(\leftthreetimes, \Psi_2, \Psi_3, \Psi_6)+2\mathcal{L}_n(\leftthreetimes, \Psi_2, \Psi_4, \Psi_5))\| \\
&+\Theta_n \mathrm{AG}_n\left(\mathfrak{W}\left(\|\Psi_1\|^{r_n}+\|\Psi_2\|^{r_n}+\|\Psi_3\|^{r_n}+\|\Psi_4\|^{r_n}+\|\Psi_5\|^{r_n}+\|\Psi_6\|^{r_n}\right)\right)\Big],
\end{aligned}
$$

for every $\Lambda_1, \Lambda_2, \Lambda_3 \in \Delta^1$ *and every* $\Psi_1, \Psi_2, \Psi_3, \Psi_4, \Psi_5, \Psi_6 \in \mathfrak{M}$ *and* $\leftthreetimes \in \pitchfork$. *Then, there exist unique* $\mathbb{C}$*–trilinear mappings* $\underbrace{\mathfrak{Z}_i}_{i=1,\ldots,n} : \pitchfork \times \mathfrak{M}^3 \to \mathfrak{M}$ *satisfying the inequality below*

$$
\begin{aligned}
&diag\Big[\ \|\mathcal{L}_1(\leftthreetimes, \Psi_1, \Psi_3, \Psi_5)-\mathfrak{Z}_1(\leftthreetimes, \Psi_1, \Psi_3, \Psi_5)\|, \ldots, \\
&\qquad \|\mathcal{L}_n(\leftthreetimes, \Psi_1, \Psi_3, \Psi_5)-\mathfrak{Z}_n(\leftthreetimes, \Psi_1, \Psi_3, \Psi_5)\|\ \Big] \\
&\preceq diag\Big[\ \frac{\Theta_1}{2^{r_1}-2^3} \mathrm{AG}_1\left(\mathfrak{W}\left((2^{r_1}+8)\|\Psi_1\|^{r_1}+8\|\Psi_3\|^{r_1}+(2^{r_1}+2^{r_1+1}+8)\|\Psi_5\|^{r_1}\right)\right), \\
&\ldots, \frac{\Theta_n}{2^{r_n}-2^3} \mathrm{AG}_n\left(\mathfrak{W}\left((2^{r_n}+8)\|\Psi_1\|^{r_n}+8\|\Psi_3\|^{r_n}+(2^{r_n}+2^{r_n+1}+8)\|\Psi_5\|^{r_n}\right)\right)\Big],
\end{aligned}
\tag{12.54}
$$

for every $\Psi_1, \Psi_3, \Psi_5 \in \mathfrak{M}$ *and* $\leftthreetimes \in \pitchfork$.

Also, if the mappings $\underbrace{\mathcal{L}_i}_{i=1,\ldots,n} : \pitchfork \times \mathfrak{M}^3 \to \mathfrak{M}$ *satisfy* $\underbrace{\mathcal{L}_i}_{i=1,\ldots,n}(\leftthreetimes, 2\Psi_1, \Psi_3, \Psi_5) = 2\underbrace{\mathcal{L}_i}_{i=1,\ldots,n}(\leftthreetimes, \Psi_1, \Psi_3, \Psi_5)$ *and the inequalities below*

$$
\begin{aligned}
&diag\Big[\ \|\mathcal{L}_1(\leftthreetimes, \Psi_1\Psi_2, \Psi_3, \Psi_5)-\mathcal{L}_1(\leftthreetimes, \Psi_1, \Psi_3, \Psi_5)\Psi_2-\Psi_1\mathcal{L}_1(\leftthreetimes, \Psi_2, \Psi_3, \Psi_5)\|, \\
&\ldots, \|\mathcal{L}_n(\leftthreetimes, \Psi_1\Psi_2, \Psi_3, \Psi_5)-\mathcal{L}_n(\leftthreetimes, \Psi_1, \Psi_3, \Psi_5)\Psi_2-\Psi_1\mathcal{L}_n(\leftthreetimes, \Psi_2, \Psi_3, \Psi_5)\|\ \Big] \\
&\preceq diag\Big[\ \Theta_1 \mathrm{AG}_1\left(\mathfrak{W}\left(\|\Psi_1\|^{r_1}+\|\Psi_2\|^{r_1}+\|\Psi_3\|^{r_1}+\|\Psi_4\|^{r_1}+\|\Psi_5\|^{r_1}+\|\Psi_6\|^{r_1}\right)\right), \\
&\ldots, \Theta_n \mathrm{AG}_n\left(\mathfrak{W}\left(\|\Psi_1\|^{r_n}+\|\Psi_2\|^{r_n}+\|\Psi_3\|^{r_n}+\|\Psi_4\|^{r_n}+\|\Psi_5\|^{r_n}+\|\Psi_6\|^{r_n}\right)\right)\Big],
\end{aligned}
\tag{12.55}
$$

and

$$diag\Big[\|\mathcal{L}_1(\curlywedge, \Phi_{\delta(1)}, \Phi_{\delta(2)}, \Phi_{\delta(3)}) - \mathcal{L}_1(\curlywedge, \Phi_1, \Phi_2, \Phi_3)\| \tag{12.56}$$

$$\|\mathcal{L}_n(\curlywedge, \Phi_{\delta(1)}, \Phi_{\delta(2)}, \Phi_{\delta(3)}) - \mathcal{L}_n(\curlywedge, \Phi_1, \Phi_2, \Phi_3)\| \Big]$$

$$\preceq diag\Big[\Theta_1 \mathrm{AG}_1\Big(\mathfrak{W}\Big(\|\Psi_1\|^{r_1} + \|\Psi_2\|^{r_1} + \|\Psi_3\|^{r_1} + \|\Psi_4\|^{r_1} + \|\Psi_5\|^{r_1} + \|\Psi_6\|^{r_1}\Big)\Big),$$

$$\ldots, \Theta_n \mathrm{AG}_n\Big(\mathfrak{W}\Big(\|\Psi_1\|^{r_n} + \|\Psi_2\|^{r_n} + \|\Psi_3\|^{r_n} + \|\Psi_4\|^{r_n} + \|\Psi_5\|^{r_n} + \|\Psi_6\|^{r_n}\Big)\Big)\Big],$$

for every $\Psi_1, \Psi_2, \Psi_3, \Phi_1, \Phi_2, \Phi_3, \Psi_5 \in \mathfrak{M}$ *and* $\curlywedge \in \pitchfork$, *and for every permutation* $(\delta(1), \delta(2), \delta(3))$ *of* $(1, 2, 3)$, *then, the* $\mathbb{C}$*–trilinear mappings* $\underbrace{\mathfrak{Z}_i}_{i=1,\ldots,n} : \pitchfork \times \mathfrak{M}^3 \to \mathfrak{M}$ *are permuting tri–derivations.*

Proof Put $\underbrace{\mathfrak{P}_i}_{i=1,\ldots,n} = 2^{1-\overbrace{r_i}^{i=1,\ldots,n}}$. It is the coefficient that defines the control function $\underbrace{\varphi_i}_{i=1,\ldots,n}$ by (12.52). Then, we apply Theorem 12.3 to complete the proof.

Corollary 12.5 *Let* $\underbrace{\Theta_i}_{i=1,\ldots,n} \geq 0$ *and* $\underbrace{r_i}_{i=1,\ldots,n} < 3$ *be in* $\mathbb{R}$, *and* $\underbrace{\mathcal{L}_i}_{i=1,\ldots,n} : \pitchfork \times \mathfrak{M}^3 \to \mathfrak{M}$ *be mappings satisfying the inequality* (12.53) *and* $\underbrace{\mathcal{L}_i}_{i=1,\ldots,n}(\curlywedge, \Psi_1, 0, \Psi_5) = \underbrace{\mathcal{L}_i}_{i=1,\ldots,n}(\curlywedge, 0, \Psi_3, \Psi_5) = \underbrace{\mathcal{L}_i}_{i=1,\ldots,n}(\curlywedge, \Psi_1, \Psi_3, 0) = 0$, *for every* $\Psi_1, \Psi_3, \Psi_5 \in \mathfrak{M}$ *and* $\curlywedge \in \pitchfork$. *Then, we can find unique* $\mathbb{C}$*–trilinear mappings* $\underbrace{\mathfrak{Z}_i}_{i=1,\ldots,n} : \pitchfork \times \mathfrak{M}^3 \to \mathfrak{M}$ *satisfying the inequality below*

$$diag\Big[\|\mathcal{L}_1(\curlywedge, \Psi_1, \Psi_3, \Psi_5) - \mathfrak{Z}_1(\curlywedge, \Psi_1, \Psi_3, \Psi_5)\|, \ldots, \tag{12.57}$$

$$\|\mathcal{L}_n(\curlywedge, \Psi_1, \Psi_3, \Psi_5) - \mathfrak{Z}_n(\curlywedge, \Psi_1, \Psi_3, \Psi_5)\| \Big]$$

$$\preceq diag\Big[\frac{\Theta_1}{8 - 2^{r_1}} \mathrm{AG}_1\Big(\mathfrak{W}\Big((2^{r_1} + 8)\|\Psi_1\|^{r_1} + 8\|\Psi_3\|^{r_1} + (2 + 2^{r_1+1} + 8)\|\Psi_5\|^{r_1}\Big)\Big),$$

$$\ldots, \frac{\Theta_n}{8 - 2^{r_n}} \mathrm{AG}_n\Big(\mathfrak{W}\Big((2^{r_n} + 8)\|\Psi_1\|^{r_n} + 8\|\Psi_3\|^{r_n} + (2 + 2^{r_n+1} + 8)\|\Psi_5\|^{r_n}\Big)\Big)\Big],$$

for every $\Psi_1, \Psi_3, \Psi_5 \in \mathfrak{M}$ *and* $\curlywedge \in \pitchfork$.

Also, if the mappings $\underbrace{\mathcal{L}_i}_{i=1,\ldots,n} : \pitchfork \times \mathfrak{M}^3 \to \mathfrak{M}$ *satisfy the inequalities* (12.55) *and* (12.56) *and* $\underbrace{\mathcal{L}_i}_{i=1,\ldots,n}(\curlywedge, 2\Psi_1, \Psi_3, \Psi_5) = 2\underbrace{\mathcal{L}_i}_{i=1,\ldots,n}(\curlywedge, \Psi_1, \Psi_3, \Psi_5)$, *for every* $\Psi_1, \Psi_3,$

$\Psi_5 \in \mathfrak{M}$ *and* $\leftthreetimes \in ⫛$, *then, the* $\mathbb{C}$*–trilinear mapping* $\underbrace{\mathfrak{Z}_i}_{i=1,\dots,n} : ⫛ \times \mathfrak{M}^3 \to \mathfrak{M}$ *are permuting tri–derivations.*

Proof Put $\underbrace{\mathfrak{P}_i}_{i=1,\dots,n} = 2^{\overbrace{r_i}^{i=1,\dots,n} - 1}$. It is the coefficient that defines the control function $\underbrace{\varphi_i}_{i=1,\dots,n}$ by (12.52). Then we apply Theorem 12.4 to complete the proof.

Corollary 12.6 *Let* $\underbrace{\Theta_i}_{i=1,\dots,n} \geq 0$ *and* $\underbrace{r_i}_{i=1,\dots,n} > 3$ *be in* $\mathbb{R}$, *and* $\underbrace{\mathcal{L}_i}_{i=1,\dots,n} : ⫛ \times \mathfrak{M}^3 \to \mathfrak{M}$ *be mappings satisfying the inequality* (12.53) *and* $\underbrace{\mathcal{L}_i}_{i=1,\dots,n} (\leftthreetimes, \Psi_1, 0, a) = \underbrace{\mathcal{L}_i}_{i=1,\dots,n} (\leftthreetimes, 0, \Psi_3, \Psi_5) = \underbrace{\mathcal{L}_i}_{i=1,\dots,n} (\leftthreetimes, \Psi_1, \Psi_3, 0) = 0$, *for every* $\Psi_1, \Psi_3, \Psi_5 \in \mathfrak{M}$ *and* $\leftthreetimes \in ⫛$. *Then, there exist unique* $\mathbb{C}$*–trilinear mappings* $\underbrace{\mathfrak{Z}_i}_{i=1,\dots,n} : ⫛ \times \mathfrak{M}^3 \to \mathfrak{M}$ *satisfying the inequality* (12.54).

Furthermore, if the mappings $\underbrace{\mathcal{L}_i}_{i=1,\dots,n} : ⫛ \times \mathfrak{M}^3 \to \mathfrak{M}$ *satisfy the inequality* (12.56) *and* $\underbrace{\mathcal{L}_i}_{i=1,\dots,n} (\leftthreetimes, 2\beta, \Psi_3, \Psi_5) = 2 \underbrace{\mathcal{L}_i}_{i=1,\dots,n} (\leftthreetimes, \beta, \Psi_3, \Psi_5)$ *and the inequality below*

$$
\begin{aligned}
& diag\Big[\|\mathcal{L}_1(\leftthreetimes, \beta\Psi_2, \Psi_3, \Psi_5) - \mathcal{L}_1(\leftthreetimes, \beta, \Psi_3, \Psi_5)\Psi_2 - \beta\mathcal{L}_1(\leftthreetimes, \Psi_2, \Psi_3, \Psi_5)\|, \dots, \\
& \qquad \|\mathcal{L}_n(\leftthreetimes, \beta\Psi_2, \Psi_3, \Psi_5) - \mathcal{L}_n(\leftthreetimes, \beta, \Psi_3, \Psi_5)\Psi_2 - \beta\mathcal{L}_n(\leftthreetimes, \Psi_2, \Psi_3, \Psi_5)\| \Big] \\
& \preceq diag\Big[\Theta_1 \mathrm{AG}_1\Big(\mathfrak{W}\Big(1 + \|\Psi_2\|^{r_1} + 2\|\Psi_3\|^{r_1} + 2\|\Psi_5\|^{r_1}\Big)\Big), \dots, \\
& \qquad \Theta_n \mathrm{AG}_n\Big(\mathfrak{W}\Big(1 + \|\Psi_2\|^{r_n} + 2\|\Psi_3\|^{r_n} + 2\|\Psi_5\|^{r_n}\Big)\Big) \Big],
\end{aligned} \tag{12.58}
$$

for every $\beta \in U(\mathfrak{M})$ *and every* $\Psi_2, \Psi_3, \Psi_5 \in \mathfrak{M}$ *and* $\leftthreetimes \in ⫛$, *then, the* $\mathbb{C}$*–trilinear mappings* $\underbrace{\mathfrak{Z}_i}_{i=1,\dots,n} : ⫛ \times \mathfrak{M}^3 \to \mathfrak{M}$ *are permuting tri–derivations.*

Proof It is the coefficient that defines the control function $\underbrace{\varphi_i}_{i=1,\dots,n}$ by the relation (12.52). Then, we complete the proof using Theorem 12.5.

Corollary 12.7 *Let* $\underbrace{\Theta_i}_{i=1,\dots,n} \geq 0$ *and* $\underbrace{r_i}_{i=1,\dots,n} < 3$ *be in* $\mathbb{R}$, *and* $\underbrace{\mathcal{L}_i}_{i=1,\dots,n} : ⫛ \times \mathfrak{M}^3 \to \mathfrak{M}$ *be mappings satisfying the inequality* (12.53) *and* $\underbrace{\mathcal{L}_i}_{i=1,\dots,n} (\leftthreetimes, \Psi_1, 0, \Psi_5) = \underbrace{\mathcal{L}_i}_{i=1,\dots,n}$

$(\curlywedge, 0, \Psi_3, \Psi_5) = \underbrace{\mathcal{L}_i}_{i=1,\dots,n} (\curlywedge, \Psi_1, \Psi_3, 0) = 0$, *for every* $\Psi_1, \Psi_3, \Psi_5 \in \mathfrak{M}$ *and* $\curlywedge \in \pitchfork$.
Then, there exist unique $\mathbb{C}$*–trilinear mappings* $\underbrace{\mathfrak{Z}_i}_{i=1,\dots,n} : \pitchfork \times \mathfrak{M}^3 \to \mathfrak{M}$ *satisfying the inequality* (12.57).

Also, if the mappings $\underbrace{\mathcal{L}_i}_{i=1,\dots,n} : \mathfrak{M}^3 \to \mathfrak{M}$ *satisfy the inequalities* (12.56), *and* (12.58) *and* $\underbrace{\mathcal{L}_i}_{i=1,\dots,n} (\curlywedge, 2\beta, \Psi_3, \Psi_5) = 2 \underbrace{\mathcal{L}_i}_{i=1,\dots,n} (\curlywedge, \beta, \Psi_3, \Psi_5)$, *for every* $\beta \in U(\mathfrak{M})$ *and every* $\Psi_3, \Psi_5 \in \mathfrak{M}$ *and* $\curlywedge \in \pitchfork$, *then, the* $\mathbb{C}$*–trilinear mappings* $\underbrace{\mathfrak{Z}_i}_{i=1,\dots,n} : \pitchfork \times \mathfrak{M}^3 \to \mathfrak{M}$ *are permuting tri–derivations.*

Proof It is the coefficient that defines the control function $\underbrace{\varphi_i}_{i=1,\dots,n}$ by the relation (12.52). Then, we complete the proof by Theorem 12.6.

Corollary 12.8 *Let* $\underbrace{\Theta_i}_{i=1,\dots,n} \geq 0$ *and* $\underbrace{r_i}_{i=1,\dots,n} > 3$ *be in* $\mathbb{R}$, *and* $\mathcal{L} : \mathfrak{M}^3 \to \mathfrak{N}$ *be mappings satisfying the inequality* (12.53) *and* $\underbrace{\mathcal{L}_i}_{i=1,\dots,n} (\curlywedge, \Psi_1, 0, \Psi_5) = \underbrace{\mathcal{L}_i}_{i=1,\dots,n} (\curlywedge, 0, \Psi_3, \Psi_5) = \underbrace{\mathcal{L}_i}_{i=1,\dots,n} (\curlywedge, \Psi_1, \Psi_3, 0) = 0$, *for every* $\Psi_1, \Psi_2, \Psi_3, \Psi_4, \Psi_5, \Psi_6 \in \mathfrak{M}$ *and* $\curlywedge \in \pitchfork$.
Then, there exist unique $\mathbb{C}$*–trilinear mappings* $\underbrace{\mathcal{E}_i}_{i=1,\dots,n} : \pitchfork \times \mathfrak{M}^3 \to \mathfrak{N}$ *satisfying the inequality below*

$$
\begin{aligned}
&diag\Big[\|\mathcal{L}_1(\curlywedge, \Psi_1, \Psi_3, \Psi_5) - \mathcal{E}_1(\curlywedge, \Psi_1, \Psi_3, \Psi_4)\|, \dots, \\
&\qquad \|\mathcal{L}_n(\curlywedge, \Psi_1, \Psi_3, \Psi_5) - \mathcal{E}_n(\curlywedge, \Psi_1, \Psi_3, \Psi_4)\|\Big], \\
&\preceq diag\Big[\frac{\Theta_1}{2^{r_1} - 2^3}\mathrm{AG}_1\Big(\mathfrak{W}\Big((2^{r_1} + 8)\|\Psi_1\|^{r_1} + 8\|\Psi_3\|^{r_1} + (2^{r_1} + 2^{r_1+1} + 8)\|\Psi_5\|^{r_1}\Big)\Big), \\
&\qquad \dots, \frac{\Theta_n}{2^{r_n} - 2^3}\mathrm{AG}_n\Big(\mathfrak{W}\Big((2^{r_n} + 8)\|\Psi_1\|^{r_n} + 8\|\Psi_3\|^{r_n} + (2^{r_n} + 2^{r_n+1} + 8)\|\Psi_5\|^{r_n}\Big)\Big)\Big],
\end{aligned}
\tag{12.59}
$$

for every $\Psi_1, \Psi_3, \Psi_5 \in \mathfrak{M}$ *and* $\curlywedge \in \pitchfork$.

Besides, if the mappings $\underbrace{\mathcal{L}_i}_{i=1,\dots,n} : \pitchfork \times \mathfrak{M}^3 \to \mathfrak{N}$ *satisfy the inequality* (12.56) *and the inequality below*

$$
\begin{aligned}
&diag\Big[\|\mathcal{L}_1(\curlywedge, \Psi_1\Psi_2, \Psi_3\Psi_4, \Psi_5\Psi_6) - \mathcal{L}_1(\curlywedge, \Psi_1, \Psi_3, \Psi_5)\mathcal{L}_1(\curlywedge, \Psi_2, \Psi_4, \Psi_6)\|, \\
&\qquad \dots, \|\mathcal{L}_n(\curlywedge, \Psi_1\Psi_2, \Psi_3\Psi_4, \Psi_5\Psi_6) - \mathcal{L}_n(\curlywedge, \Psi_1, \Psi_3, \Psi_5)\mathcal{L}_n(\curlywedge, \Psi_2, \Psi_4, \Psi_6)\|\Big] \\
&\preceq diag\Big[\Theta_1\mathrm{AG}_1\Big(\mathfrak{W}\Big(\|\Psi_1\|^{r_1} + \|\Psi_2\|^{r_1} + \|\Psi_3\|^{r_1} + \|\Psi_4\|^{r_1} + \|\Psi_5\|^{r_1} + \|\Psi_6\|^{r_1}\Big),
\end{aligned}
\tag{12.60}
$$

$$\ldots, \Theta_n \mathrm{AG}_n\left(\mathfrak{W}\left(\|\Psi_1\|^{r_n}+\|\Psi_2\|^{r_n}+\|\Psi_3\|^{r_n}+\|\Psi_4\|^{r_n}+\|\Psi_5\|^{r_n}+\|\Psi_6\|^{r_n}\right)\right],$$

for every $\Psi_1, \Psi_2, \Psi_3, \Psi_4, \Psi_5, \Psi_6 \in \mathfrak{M}$ *and* $\leftthreetimes \in ⫙$, *then, the* $\mathbb{C}$*–trilinear mappings* $\underbrace{\mathcal{E}_i}_{i=1,\ldots,n} : ⫙ \times \mathfrak{M}^3 \to \mathfrak{N}$ *are permuting tri–homomorphisms.*

Proof It is the coefficient that defines the control function $\underbrace{\varphi_i}_{i=1,\ldots,n}$ by the relation (12.52). Then, we complete the proof by Theorem 12.7.

for all $\Psi_1, \Psi_3, \Psi_5 \in \mathfrak{M}$, $\leftthreetimes \in ⫙$.

Corollary 12.9 *Let* $\underbrace{\Theta_i}_{i=1,\ldots,n} \geq 0$ *and* $\underbrace{r_i}_{i=1,\ldots,n} < 3$ *be in* $\mathbb{R}$, *and* $\underbrace{\mathcal{L}_i}_{i=1,\ldots,n} : ⫙ \times \mathfrak{M}^3 \to \mathfrak{N}$ *be mappings satisfying the inequality* (12.53) *and* $\underbrace{\mathcal{L}_i}_{i=1,\ldots,n}(\leftthreetimes, \Psi_1, 0, \Psi_5) = \underbrace{\mathcal{L}_i}_{i=1,\ldots,n}(\leftthreetimes, 0, \Psi_3, \Psi_5) = \underbrace{\mathcal{L}_i}_{i=1,\ldots,n}(\leftthreetimes, \Psi_1, \Psi_3, 0) = 0$, *for every* $\Psi_1, \Psi_2, \Psi_3, \Psi_4, \Psi_5, \Psi_6 \in \mathfrak{M}$ *and* $\leftthreetimes \in ⫙$. *Then, there exist unique* $\mathbb{C}$*–trilinear mappings* $\underbrace{\mathcal{E}_i}_{i=1,\ldots,n} : ⫙ \times \mathfrak{M}^3 \to \mathfrak{N}$ *satisfying the inequality below*

$$\begin{aligned} diag\Big[& \|\mathcal{L}_1(\leftthreetimes, \Psi_1, \Psi_3, \Psi_4) - \mathcal{E}_1(\leftthreetimes, \Psi_1, \Psi_3, \Psi_4)\|, \\ & \ldots, \|\mathcal{L}_n(\leftthreetimes, \Psi_1, \Psi_3, \Psi_4) - \mathcal{E}_n(\leftthreetimes, \Psi_1, \Psi_3, \Psi_4)\|\Big], \\ \preceq diag\Big[& \frac{\Theta_1}{8-2^{r_1}} \mathrm{AG}_1\left(\mathfrak{W}\left((2^{r_1}+8)\|\Psi_1\|^{r_1} + 8\|\Psi_3\|^{r_1} + (2^{r_1}+2^{r_1+1}+8)\|\Psi_5\|^{r_1}\right)\right), \\ & \ldots, \frac{\Theta_n}{8-2^{r_n}} \mathrm{AG}_n\left(\mathfrak{W}\left((2^{r_n}+8)\|\Psi_1\|^{r_n} + 8\|\Psi_3\|^{r_n} + (2^{r_n}+2^{r_n+1}+8)\|\Psi_5\|^{r_n}\right)\right)\Big], \end{aligned} \tag{12.61}$$

for every $\Psi_1, \Psi_3, \Psi_5 \in \mathfrak{M}$ *and* $\leftthreetimes \in ⫙$.

Also, if the mappings $\underbrace{\mathcal{L}_i}_{i=1,\ldots,n} : ⫙ \times \mathfrak{M}^3 \to \mathfrak{N}$ *satisfy the inequalities* (12.56) *and* (12.60), *then, the* $\mathbb{C}$*–trilinear mappings* $\underbrace{\mathcal{E}_i}_{i=1,\ldots,n} : ⫙ \times \mathfrak{M}^3 \to \mathfrak{N}$ *are permuting tri–homomorphisms.*

Proof It is the coefficient that defines the control function $\underbrace{\varphi_i}_{i=1,\ldots,n}$ by the relation (12.52). Then, we complete the proof by Theorem 12.8.

Corollary 12.10 *Let* $\underbrace{\Theta_i}_{i=1,\dots,n} \geq 0$ *and* $\underbrace{r_i}_{i=1,\dots,n} > 6$ *be in* $\mathbb{R}$, *and* $\underbrace{\mathcal{L}_i}_{i=1,\dots,n} : ⋔ \times \mathfrak{M}^3 \to \mathfrak{N}$ *be mappings satisfying the inequality* (12.53) *and* $\underbrace{\mathcal{L}_i}_{i=1,\dots,n} (\curlywedge, \Psi_1, 0, \Psi_5) = \underbrace{\mathcal{L}_i}_{i=1,\dots,n} (\curlywedge, 0, \Psi_3, \Psi_5) = \underbrace{\mathcal{L}_i}_{i=1,\dots,n} (\curlywedge, \Psi_1, \Psi_3, 0) = 0$, *for every* $\Psi_1, \Psi_2, \Psi_3, \Psi_4, \Psi_5, \Psi_6 \in \mathfrak{M}$ *and* $\curlywedge \in ⋔$. *Then, there exist unique* $\mathbb{C}$*–trilinear mappings* $\underbrace{\mathcal{E}_i}_{i=1,\dots,n} : ⋔ \times \mathfrak{M}^3 \to \mathfrak{N}$ *satisfying the inequality* (12.59).

Also, if the mappings $\underbrace{\mathcal{L}_i}_{i=1,\dots,n} : ⋔ \times \mathfrak{M}^3 \to \mathfrak{N}$ *satisfy the inequality* (12.56) *and the inequality below*

$$diag\Big[\|\mathcal{L}_1(\curlywedge, \beta_1\alpha_1, \beta_2\alpha_2, \beta_3\alpha_3) - \mathcal{L}_1(\curlywedge, \beta_1, \beta_2, \beta_3)\mathcal{L}_1(\curlywedge, \alpha_1, \alpha_2, \alpha_3)\|, \dots, \tag{12.62}$$
$$\|\mathcal{L}_n(\curlywedge, \beta_1\alpha_1, \beta_2\alpha_2, \beta_3\alpha_3) - \mathcal{L}_n(\curlywedge, \beta_1, \beta_2, \beta_3)\mathcal{L}_n(\curlywedge, \alpha_1, \alpha_2, \alpha_3)\|\Big]$$
$$\preceq diag\Big[6\Theta_1 \mathrm{AG}_1\Big(\mathfrak{W}\Big(\|\Psi_1\|^{r_1} + \|\Psi_2\|^{r_1} + \|\Psi_3\|^{r_1} + \|\Psi_4\|^{r_1} + \|\Psi_5\|^{r_1} + \|\Psi_6\|^{r_1}\Big)\Big), \dots,$$
$$6\Theta_n \mathrm{AG}_n\Big(\mathfrak{W}\Big(\|\Psi_1\|^{r_n} + \|\Psi_2\|^{r_n} + \|\Psi_3\|^{r_n} + \|\Psi_4\|^{r_n} + \|\Psi_5\|^{r_n} + \|\Psi_6\|^{r_n}\Big)\Big)\Big],$$

for every $\beta_j, \alpha_j \in U(\mathfrak{M})$, $j = 1, 2, 3$, *then, the* $\mathbb{C}$*–trilinear mappings* $\underbrace{\mathcal{E}_i}_{i=1,\dots,n} : ⋔ \times \mathfrak{M}^3 \to \mathfrak{N}$ *are permuting tri–homomorphisms.*

Proof It is the coefficient that defines the control function $\underbrace{\varphi_i}_{i=1,\dots,n}$ by the relation (12.52). Then, we complete the proof by Theorem 12.9.

Corollary 12.11 *Let* $\underbrace{\Theta_i}_{i=1,\dots,n} \geq 0$ *and* $\underbrace{r_i}_{i=1,\dots,n} < 3$ *be in* $\mathbb{R}$, *and* $\underbrace{\mathcal{L}_i}_{i=1,\dots,n} : ⋔ \times \mathfrak{M}^3 \to \mathfrak{N}$ *be mappings satisfying the inequality* (12.53) *and* $\underbrace{\mathcal{L}_i}_{i=1,\dots,n} (\curlywedge, \Psi_1, 0, \Psi_5) = \underbrace{\mathcal{L}_i}_{i=1,\dots,n} (\curlywedge, 0, \Psi_3, \Psi_5) = \underbrace{\mathcal{L}_i}_{i=1,\dots,n} (\curlywedge, \Psi_1, \Psi_3, 0) = 0$, *for every* $\Psi_1, \Psi_2, \Psi_3, \Psi_4, \Psi_5, \Psi_6 \in \mathfrak{M}$ *and* $\curlywedge \in ⋔$. *Then, we can find unique* $\mathbb{C}$*–trilinear mappings* $\underbrace{\mathcal{E}_i}_{i=1,\dots,n} : ⋔ \times \mathfrak{M}^3 \to \mathfrak{N}$ *satisfying the inequality* (12.61).

Also, if the mappings $\underbrace{\mathcal{L}_i}_{i=1,\dots,n} : ⋔ \times \mathfrak{M}^3 \to \mathfrak{N}$ *satisfy the inequalities* (12.56) *and* (12.62), *then, the* $\mathbb{C}$*–trilinear mappings* $\underbrace{\mathcal{E}_i}_{i=1,\dots,n} : ⋔ \times \mathfrak{M}^3 \to \mathfrak{N}$ *are permuting tri–homomorphisms.*

Proof It is the coefficient that defines the control function $\underbrace{\varphi_i}_{i=1,\ldots,n}$ by the relation (12.52). Then, we complete the proof by Theorem 12.10.

We refer the interested readers to [7–11].

References

1. Aderyani SR, Ahadi A, Saadati R, Srivastava HM (2024) Aggregate special functions to approximate permuting tri-homomorphisms and permuting tri-derivations associated with a tri-additive ψ-functional inequality in Banach algebras. Acta Math Sci Ser B (Engl. Ed.) 44(1):311–338
2. Öztürk MA (1999) Permuting tri-derivations in prime and semi-prime rings. East Asian Math J 15:177–190
3. Park C (2020) Permuting triderivations and permuting trihomomorphisms in complex Banach algebras. arXiv:2009.10346
4. Bae J, Park W (2010) Approximate bi-homomorphisms and bi-derivations in C^* -ternary algebras. Bull Korean Math Soc 47:195–209
5. Park C, Jin Y, Shin D, Zhang X, Govindan V (2020) Permuting triderivations and permuting tri-homomorphisms in Banach algebras. Rocky Mountain J Math 50(5):1793–1806. MR4170688
6. Grabisch M, Marichal JL, Mesiar R, Pap E (2009) Aggregation functions, vol 127. Cambridge University Press
7. Aderyani SR, Saadati R, O'Regan D, Alshammari FS (2023) Application of aggregated control functions for approximating $\mathscr{C}$-Hilfer fractional differential equations. AIMS Math 8(11):28010–28032
8. ORegan D, Aderyani SR, Saadati R, Li C (2023) Stability results and parametric delayed Mittag-Leffler matrices in symmetric fuzzy-random spaces with application. Symmetry 15(10):1880
9. Aderyani SR, Saadati R, ORegan D, Li C (2023) On a new approach for stability and controllability analysis of functional equations. Mathematics 11(16):3458
10. Aderyani SR, Saadati R, Rassias TM, Srivastava HM (2023) Existence, uniqueness and the multi-stability results for a W-Hilfer fractional differential equation. Axioms 12(7):681
11. Aderyani SR, Saadati R (2023) Stability and controllability results by n-ary aggregation functions in matrix valued fuzzy n-normed spaces. Inf Sci 119265

Chapter 13
Multi Stability of Fractional–Order Systems I

Applying the fixed point theory, we study the existence, uniqueness and the multi stability of the following ξ–Hilfer fractional differential equations defined on compact domains [1, 2]. Next, we generalize the main results for unbounded domains. To illustrate the main result for a fractional system, we present an example.

Now, consider the fractional–order systems I, as follows:

$$\begin{cases} {}^{\mathscr{H}}D_{0^+}^{\mathcal{A},\mathcal{B};\xi} g(\tau) = F(\tau, g(\tau), g(\mu(\tau))), & \tau \in (0, p], \\ I_{0^+}^{1-\mathcal{C};\xi} g(0^+) = \mathcal{E}_\circ, & \mathcal{E}_\circ \in \mathbb{R}, \end{cases} \tag{13.1}$$

where ${}^{\mathscr{H}}D_{0^+}^{\mathcal{A},\mathcal{B};\xi}(.)$ is the ξ–Hilfer fractional derivative of order $0 < \mathcal{A} \leq 1$ and type $0 \leq \mathcal{B} \leq 1$, $I_{0^+}^{1-\mathcal{C};\xi}(.)$ is the Riemann–Liouville fractional integral of order $1 - \mathcal{C}$, $\mathcal{C} = \mathcal{A} + \mathcal{B}(1 - \mathcal{A})$ with respect to the mapping ξ, and $F : (0, p] \times \mathbb{R}^2 \longrightarrow \mathbb{R}$ is a given mapping.

13.1 Multi Stability of Hilfer Fractional Systems I on Compact Domains

In this part, we prove the existence, uniqueness and the multi stability results of the ξ–Hilfer fractional systems (13.1) defined on compact domains by the fixed point theorem.

For $F \in C((0, p] \times \mathbb{R}^2, \mathbb{R})$ and $\underbrace{\theta_i}_{i=1,\ldots,n} > 0$, with $n \in \mathbb{N}$, consider the following fractional equations

$${}^{\mathscr{H}}D_{0^+}^{\mathcal{A},\mathcal{B};\xi} g(\tau) = F(\tau, g(\tau), g(\mu(\tau))), \qquad \tau \in (0, p], \tag{13.2}$$

$$I_{0^+}^{1-\mathcal{C};\xi} g(0^+) = \mathcal{E}_\circ, \qquad \mathcal{E}_\circ \in \mathbb{R}, \tag{13.3}$$

S. Rezaei Aderyani et al., *Towards Ulam Type Multi Stability Analysis*, Studies in Systems, Decision and Control 523, https://doi.org/10.1007/978-3-031-55564-0_13

and consider the inequality below

$$\operatorname{diag}\Bigg[\Big|{}^{\mathscr{H}}D_{0^+}^{\mathcal{A},\mathcal{B};\xi} f(\tau) - F(\tau, f(\tau), f(\mu(\tau)))\Big|, \ldots, \Big|{}^{\mathscr{H}}D_{0^+}^{\mathcal{A},\mathcal{B};\xi} f(\tau) - F(\tau, f(\tau), f(\mu(\tau)))\Big|\Bigg]_{n\times n} \tag{13.4}$$
$$\preceq \operatorname{diag}\Bigg[\theta_1\, \varphi_1^{\circledS}\Big(((\xi(\tau)-\xi(0))^{\mathcal{A}}\Big), \ldots, \theta_n \varphi_n^{\circledS}\Big(((\xi(\tau)-\xi(0))^{\mathcal{A}}\Big)\Bigg]_{n\times n}, \quad \tau \in (0, p],$$

where $\underbrace{\varphi_i^{\circledS}}_{i=1,\ldots,n} \in C((0, p], \mathbb{R}_+)$ are special control functions defined in Sects. 9.8.1 and 9.8.2.

Definition 13.1 Equations (13.2) and (13.3) have the multi stability with respect to

$$\operatorname{diag}[\varphi_1^{\circledS}((\xi(\tau)-\xi(0))^{\mathcal{A}}), \ldots, \varphi_n^{\circledS}((\xi(\tau)-\xi(0))^{\mathcal{A}})],$$

if there exist $\underbrace{c_i}_{i=1,\ldots,n} > 0$, such that, for every $\underbrace{\theta_i}_{i=1,\ldots,n} > 0$ and every solution $f \in C_{1-\mathcal{C};\xi}(0, p]$ to the inequality (13.4) and $I_{0^+}^{1-\mathcal{C};\xi} f(0^+) = \mathcal{E}_\circ$, there exists a solution $g \in C_{1-\mathcal{C};\xi}(0, p]$ to the Eqs. (13.2) and (13.3) with

$$\operatorname{diag}\Big[|f(\tau)-g(\tau)|, \ldots, |f(\tau)-g(\tau)|\Big]_{n\times n}$$
$$\preceq \operatorname{diag}\Big[c_1\,\theta_1\,\varphi_1^{\circledS}((\xi(\tau)-\xi(0))^{\mathcal{A}}), \ldots, c_n\,\theta_n\,\varphi_n^{\circledS}((\xi(\tau)-\xi(0))^{\mathcal{A}})\Big]_{n\times n},$$

where $\tau \in (0, p]$.

Lemma 13.1 *Consider the continuous mapping $F : (0, p] \times \mathbb{R} \times \mathbb{R} \longrightarrow \mathbb{R}$. Then, the fractional equations* (13.2) *and* (13.3) *are equivalent to*

$$g(\tau) = \frac{(\xi(\tau)-\xi(0))^{\mathcal{C}-1}}{\Gamma(\mathcal{C})}\mathcal{E}_\circ + \frac{1}{\Gamma(\mathcal{A})}\int_0^\tau \xi'(\tau')(\xi(\tau)-\xi(\tau'))^{\mathcal{A}-1}$$
$$F(\tau', g(\tau'), g(\mu(\tau')))d\tau', \tag{13.5}$$

for every $\tau \in (0, p]$.

Proof To see this note applying the fractional integral operator $I_{0^+}^{\mathcal{A};\xi}(.)$ on both sides of the fractional quation (13.1) and using Theorem 9.2, we get the integral equation (13.5).

On the other hand, if g satisfies the integral equation (13.5), then g satisfies the fractional equation (13.1). However, applying the fractional derivative ${}^{\mathscr{H}}D_{0^+}^{\mathcal{A},\mathcal{B};\xi}(.)$ on both sides of the integral equation (13.5), we have

$$
\begin{aligned}
{}^{\mathscr{H}}D_{0^+}^{\mathcal{A},\mathcal{B};\xi} f(\tau) = {}^{\mathscr{H}}D_{0^+}^{\mathcal{A},\mathcal{B};\xi}\left[\frac{(\xi(\tau)-\xi(0))^{\mathcal{C}-1}}{\Gamma(\mathcal{C})}\mathcal{E}_\circ\right] \\
+ {}^{\mathscr{H}}D_{0^+}^{\mathcal{A},\mathcal{B};\xi} I_{0^+}^{\mathcal{A};\xi} F(\tau', f(\tau'), f(\mu(\tau'))).
\end{aligned}
$$

Making use of Theorem 9.1 and

$$
{}^{\mathscr{H}}D_{0^+}^{\mathcal{A},\mathcal{B};\xi}(\xi(\tau)-\xi(0))^{\mathcal{C}-1} = 0, \quad 0 < \mathcal{C} < 1,
$$

we conclude that $g(\tau)$ satisfies the initial value problem equation (13.1) if and only if $g(\tau)$ satisfies the integral equation (13.5).

Remark 13.1 ([1, 2]) Let $f \in C_{1-\mathcal{C};\xi}(0, p]$ be a solution of inequality (13.4) and $I_{0^+}^{1-\mathcal{C};\xi} f(0^+) = \mathcal{E}_\circ$. Then, f is a solution of the following integral inequality

$$
\begin{aligned}
&\operatorname{diag}\Bigg[\Bigg|f(\tau) - \frac{(\xi(\tau)-\xi(0))^{\mathcal{C}-1}}{\Gamma(\mathcal{C})}\mathcal{E}_\circ \\
&\qquad - \frac{1}{\Gamma(\mathcal{A})}\int_0^\tau \xi'(s)(\xi(\tau)-\xi(s))^{\mathcal{A}-1} F(s, f(s), f(\mu(s)))ds\Bigg|, \\
&\quad \ldots, \Bigg|f(\tau) - \frac{(\xi(\tau)-\xi(0))^{\mathcal{C}-1}}{\Gamma(\mathcal{C})}\mathcal{E}_\circ \\
&\qquad - \frac{1}{\Gamma(\mathcal{A})}\int_0^\tau \xi'(s)(\xi(\tau)-\xi(s))^{\mathcal{A}-1} F(s, f(s), f(\mu(s)))ds\Bigg|\Bigg]_{n\times n} \\
&\preceq \operatorname{diag}\left[\theta_1\, \varphi_1^{\circledS}((\xi(\tau)-\xi(0))^{\mathcal{A}}), \ldots, \theta_n\, \varphi_n^{\circledS}((\xi(\tau)-\xi(0))^{\mathcal{A}})\right]_{n\times n},
\end{aligned}
$$

where $\underbrace{\theta_i}_{i=1,\ldots,n} > 0$.

We only prove the following special cases:

$$
\begin{aligned}
&\Bigg|f(\tau) - \frac{(\xi(\tau)-\xi(0))^{\mathcal{C}-1}}{\Gamma(\mathcal{C})}\mathcal{E}_\circ - \frac{1}{\Gamma(\mathcal{A})}\int_0^\tau \xi'(s)(\xi(\tau)-\xi(s))^{\mathcal{A}-1} \\
&\quad F(s, f(s), f(\mu(s)))ds\Bigg| \\
&\le \frac{\theta}{\Gamma(\mathcal{A})}\int_0^\tau \xi'(s)(\xi(\tau)-\xi(s))^{\mathcal{A}-1} {}_2\mathbb{F}_1(\mathsf{A}, \mathsf{B}, \mathsf{C}; (\xi(s)-\xi(0))^{\mathcal{A}})ds \\
&\le \frac{\theta}{\Gamma(\mathcal{A})}\int_0^\tau \xi'(s)(\xi(\tau)-\xi(s))^{\mathcal{A}-1} \frac{\Gamma(\mathsf{C})}{\Gamma(\mathsf{A})\Gamma(\mathsf{B})}\sum_{k=0}^{\infty}\frac{\Gamma(\mathsf{A}+k)\Gamma(\mathsf{B}+k)}{\Gamma(\mathsf{C}+k)}
\end{aligned}
$$

$$\frac{(\xi(s)-\xi(0))^{k\mathcal{A}}}{k!}ds$$

$$= \frac{\theta}{\Gamma(\mathcal{A})}\frac{\Gamma(\mathsf{C})}{\Gamma(\mathsf{A})\Gamma(\mathsf{B})}\sum_{k=0}^{\infty}\frac{\Gamma(\mathsf{A}+k)\Gamma(\mathsf{B}+k)}{\Gamma(\mathsf{C}+k)}\frac{1}{k!}\int_0^{\tau}(\xi(\tau)-\xi(s))^{\mathcal{A}-1}$$

$$(\xi(s)-\xi(0))^{k\mathcal{A}}d\xi(s)$$

$$= \frac{\theta}{\Gamma(\mathcal{A})}\frac{\Gamma(\mathsf{C})}{\Gamma(\mathsf{A})\Gamma(\mathsf{B})}\sum_{k=0}^{\infty}\frac{\Gamma(\mathsf{A}+k)\Gamma(\mathsf{B}+k)}{\Gamma(\mathsf{C}+k)}\frac{1}{k!}\int_0^{\xi(\tau)-\xi(0)}(\xi(\tau)-\xi(0)-u)^{\mathcal{A}-1}$$

$$(u)^{k\mathcal{A}}du(u=\xi(s)-\xi(0))$$

$$\leq \frac{\theta}{\Gamma(\mathcal{A})}\frac{\Gamma(\mathsf{C})}{\Gamma(\mathsf{A})\Gamma(\mathsf{B})}\sum_{k=0}^{\infty}\frac{\Gamma(\mathsf{A}+k)\Gamma(\mathsf{B}+k)}{\Gamma(\mathsf{C}+k)}\frac{1}{k!}(\xi(\tau)-\xi(0))^{\mathcal{A}-1}$$

$$\int_0^{\xi(\tau)-\xi(0)}(1-\frac{u}{\xi(\tau)-\xi(0)})^{\mathcal{A}-1}u^{k\mathcal{A}}du$$

$$= \frac{\theta}{\Gamma(\mathcal{A})}\frac{\Gamma(\mathsf{C})}{\Gamma(\mathsf{A})\Gamma(\mathsf{B})}\sum_{k=0}^{\infty}\frac{\Gamma(\mathsf{A}+k)\Gamma(\mathsf{B}+k)}{\Gamma(\mathsf{C}+k)}\frac{1}{k!}(\xi(\tau)-\xi(0))^{(k+1)\mathcal{A}}$$

$$\int_0^1(1-v)^{\mathcal{A}-1}v^{k\mathcal{A}}dv\left(v=\frac{u}{\xi(\tau)-\xi(0)}\right)$$

$$= \frac{\theta}{\Gamma(\mathcal{A})}\frac{\Gamma(\mathsf{C})}{\Gamma(\mathsf{A})\Gamma(\mathsf{B})}\sum_{k=0}^{\infty}\frac{\Gamma(\mathsf{A}+k)\Gamma(\mathsf{B}+k)}{\Gamma(\mathsf{C}+k)}\frac{1}{k!}(\xi(\tau)-\xi(0))^{(k+1)\mathcal{A}}$$

$$\frac{\Gamma(k\mathcal{A}+1)\Gamma(\mathcal{A})}{\Gamma((k+1)\mathcal{A}+1)}$$

$$\leq \theta\frac{\Gamma(\mathsf{C})}{\Gamma(\mathsf{A})\Gamma(\mathsf{B})}\sum_{k=0}^{\infty}\frac{\Gamma(\mathsf{A}+k)\Gamma(\mathsf{B}+k)}{\Gamma(\mathsf{C}+k)}\frac{(\xi(\tau)-\xi(0))^{k\mathcal{A}}}{k!}$$

$$= \theta\ {}_2\mathbb{F}_1(\mathsf{A},\mathsf{B},\mathsf{C};(\xi(\tau)-\xi(0))^{\mathcal{A}}).$$

and

$$\left|f(\tau)-\frac{(\xi(\tau)-\xi(0))^{\mathcal{C}-1}}{\Gamma(\mathcal{C})}\mathcal{E}_{\circ}-\frac{1}{\Gamma(\mathcal{A})}\int_0^{\tau}\xi'(s)(\xi(\tau)-\xi(s))^{\mathcal{A}-1}\right.$$

$$\left.F(s,f(s),f(\mu(s)))ds\right|$$

$$\leq \frac{\theta}{\Gamma(\mathcal{A})}\int_0^{\tau}\xi'(s)(\xi(\tau)-\xi(s))^{\mathcal{A}-1}\mathbb{E}_{\mathcal{A}}\left((\xi(s)-\xi(0))^{\mathcal{A}}\right)ds$$

$$\leq \frac{\theta}{\Gamma(\mathcal{A})}\int_0^{\tau} \xi'(s)(\xi(\tau)-\xi(s))^{\mathcal{A}-1}\sum_{k=0}^{\infty}\frac{(\xi(s)-\xi(0))^{k\mathcal{A}}}{\Gamma(k\mathcal{A}+1)}ds$$

$$= \frac{\theta}{\Gamma(\mathcal{A})}\sum_{k=0}^{\infty}\frac{1}{\Gamma(k\mathcal{A}+1)}\int_0^{\tau}(\xi(\zeta)-\xi(s))^{\mathcal{A}-1}(\xi(s)-\xi(0))^{k\mathcal{A}}d\xi(s)$$

$$= \frac{\theta}{\Gamma(\mathcal{A})}\sum_{k=0}^{\infty}\frac{1}{\Gamma(k\mathcal{A}+1)}\int_0^{\xi(\tau)-\xi(0)}(\xi(\tau)-\xi(0)-u)^{\mathcal{A}-1}(u)^{k\mathcal{A}}du$$

$(u=\xi(s)-\xi(0))$

$$\leq \frac{\theta}{\Gamma(\mathcal{A})}\sum_{k=0}^{\infty}\frac{1}{\Gamma(k\mathcal{A}+1)}(\xi(\tau)-\xi(0))^{\mathcal{A}-1}\int_0^{\xi(\tau)-\xi(0)}(1-\frac{u}{\xi(\tau)-\xi(0)})^{\mathcal{A}-1}u^{k\mathcal{A}}du$$

$$= \frac{\theta}{\Gamma(\mathcal{A})}\sum_{k=0}^{\infty}\frac{1}{\Gamma(k\mathcal{A}+1)}(\xi(\tau)-\xi(0))^{(k+1)\mathcal{A}}\int_0^{1}(1-v)^{\mathcal{A}-1}v^{k\mathcal{A}}dv$$

$$\left(v=\frac{u}{\xi(\tau)-\xi(0)}\right)$$

$$= \frac{\theta}{\Gamma(\mathcal{A})}\sum_{k=0}^{\infty}\frac{1}{\Gamma(k\mathcal{A}+1)}(\xi(\tau)-\xi(0))^{(k+1)\mathcal{A}}\frac{\Gamma(k\mathcal{A}+1)\Gamma(\mathcal{A})}{\Gamma((k+1)\mathcal{A}+1)}$$

$$\leq \theta\sum_{n=0}^{\infty}\frac{(\xi(\tau)-\xi(0))^{n\mathcal{A}}}{\Gamma(n\mathcal{A}+1)}$$

$$= \theta\mathbb{E}_{\mathcal{A}}\Big((\xi(\tau)-\xi(0))^{\mathcal{A}}\Big).$$

where ${}_2\mathbb{F}_1(\mathsf{A},\mathsf{B},\mathsf{C};.)$ and $\mathbb{E}_{\mathcal{A}}(.)$ denote the hypergeometric function and the one–parameter mittag–Leffler function, with $\theta>0$, $\mathsf{A},\mathsf{B},\mathsf{C}\in\mathbf{C}$, $\Re(\mathsf{A}),\Re(\mathsf{B}),\Re(\mathsf{C})>0$, and $\mathcal{A}\in(0,1]$.

Making use of the Diaz–Margolis theorem (i.e., Theorem 9.5), we obtain the multi stability results for the fractional system (13.1).

Let us suppose the following axioms are satisfied:

($\mathfrak{J}_1$) $F\in C((0,p]\times\mathbb{R}^2,\mathbb{R})$, $\mu\in C([0,p],[0,p])$, $\mu(\tau)\leq\tau,\tau\geq 0$.

($\mathfrak{J}_2$) There exists $0<\Theta<\frac{1}{2}$ such that

$$\Big|F(\tau,S_1,S_2)-F(\tau,R_1,R_2)\Big|\leq\Theta\sum_{j=1}^{2}|S_j-R_j|,$$

$$\text{for every}\quad \tau\in(0,p],\, S_j,R_j\in\mathbb{R},\, j=1,2.$$

Theorem 13.1 *Let the assumptions* ($\mathfrak{J}_1$) *and* ($\mathfrak{J}_2$) *be satisfied. If* f *in* $C_{1-\mathcal{C};\xi}(0,p]$ *satisfies the inequality* (13.4) *and* $I_{0^+}^{1-\mathcal{C};\xi}f(0^+)=\mathcal{E}_\circ$*, thus, we can find a unique* *function g satisfying the fractional equations* (13.2) *and* (13.3), *such that*

$$
\begin{aligned}
& diag\Big[|g(\tau)-f(\tau)|,\dots,|g(\tau)-f(\tau)|\Big]_{n\times n} \\
& \preceq diag\Big[\frac{\theta_1}{1-2\Theta}\varphi_1^{\circledS}((\xi(\tau)-\xi(0))^{\mathcal{A}}),\dots,\frac{\theta_n}{1-2\Theta}\varphi_n^{\circledS}((\xi(\tau)-\xi(0))^{\mathcal{A}})\Big]_{n\times n},
\end{aligned}
\tag{13.6}
$$

for every $\tau \in (0, p]$ *and* $n \in \mathbb{N}$.

Proof Set $\Xi = C_{1-\mathcal{C};\xi}(0, p]$ and define a mapping $\delta : \Xi \times \Xi \to [0, \infty]^n$ by

$$
\begin{aligned}
\delta(\rho,\varrho) = \inf\Bigg\{ & (\mathtt{C}_1,\dots,\mathtt{C}_n) \geq \underbrace{(0,\dots,0)}_{n} : \mathrm{diag}\Big[|\rho(\tau)-\varrho(\tau)| \\
& ,\dots,|\rho(\tau)-\varrho(\tau)|\Big]_{n\times n} \\
& \preceq \mathrm{diag}\Big[\mathtt{C}_1\theta_1\varphi_1^{\circledS}((\xi(\tau)-\xi(0))^{\mathcal{A}}),\dots,\mathtt{C}_n\theta_n\varphi_n^{\circledS}((\xi(\tau)-\xi(0))^{\mathcal{A}})\Big]_{n\times n}\Bigg\},
\end{aligned}
\tag{13.7}
$$

for every $\tau \in (0, p]$ and $\underbrace{\theta_i}_{1\leq i\leq n} > 0$.

We prove that (Ξ, δ) is a complete $[0, \infty]^n$–valued metric space. For the metric part, it is enough to assume that $\delta(\rho, \varrho) > \delta(\rho, \nu) + \delta(\nu, \varrho)$, for some ρ, ϱ and $\nu \in \Xi$. Thus, we can find $\tau_\circ \in (0, p]$ with

$$
\begin{aligned}
& \mathrm{diag}\Big[|\rho(\tau_\circ)-\varrho(\tau_\circ)|,\dots,|\rho(\tau_\circ)-\varrho(\tau_\circ)|\Big]_{n\times n} \\
& \succ (\delta(\rho,\nu)+\delta(\nu,\varrho))\mathrm{diag}\Big[\theta_1\varphi_1^{\circledS}((\xi(\tau_\circ)-\xi(0))^{\mathcal{A}}) \\
& ,\dots,\theta_n\varphi_n^{\circledS}((\xi(\tau_\circ)-\xi(0))^{\mathcal{A}})\Big]_{n\times n}.
\end{aligned}
$$

Thus, from the definition of δ, we get

$$
\begin{aligned}
& \mathrm{diag}\Big[|\rho(\tau_\circ)-\varrho(\tau_\circ)|,\dots,|\rho(\tau_\circ)-\varrho(\tau_\circ)|\Big]_{n\times n} \\
& \succ \mathrm{diag}\Big[|\rho(\tau_\circ)-\nu(\tau_\circ)|+|(\nu(\tau_\circ)-\varrho(\tau_\circ)| \\
& ,\dots,|\rho(\tau_\circ)-\nu(\tau_\circ)|+|(\nu(\tau_\circ)-\varrho(\tau_\circ)|\Big]_{n\times n},
\end{aligned}
$$

that contradicts. Next, we prove that the metric space (Ξ, δ) is complete. Let ω_k be a Cauchy sequence in (Ξ, δ). Thus, for every $\underbrace{\epsilon_i}_{i=1,\dots,n} > 0$ we can find $\underbrace{\aleph_{\epsilon_i}}_{i=1,\dots,n} \in \mathbb{N}$ such that $\delta(\omega_m, \omega_k) \preceq (\epsilon_1, \dots, \epsilon_n)$, for every $m, k \geq \underbrace{\aleph_{\epsilon_i}}_{i=1,\dots,n}$. In view of the relation (13.7), we have that

$$\begin{aligned} &\operatorname{diag}\Big[|\omega_m(\tau) - \omega_k(\tau)|, \dots, |\omega_m(\tau) - \omega_k(\tau)|\Big]_{n\times n} \\ &\preceq \operatorname{diag}\Big[\epsilon_1\theta_1\varphi_1^{Ⓢ}((\xi(\tau) - \xi(0))^{\mathcal{A}}), \dots, \epsilon_n\theta_n\varphi_n^{Ⓢ}((\xi(\tau) - \xi(0))^{\mathcal{A}})\Big]_{n\times n}, \end{aligned} \tag{13.8}$$

for every $\tau \in (0, p]$. Let $\tau \in (0, p]$ be arbitrary and fixed, from the inequality (13.8), we can conclude in the reals that the sequence $\{\omega_k(\tau)\}$ is Cauchy and convergence. Thus, we can define a mapping ω by

$$\omega(\tau) := \lim_{k\to\infty} \omega_k(\tau), \quad (\tau \in (0, p]).$$

Considering the inequality (13.8) and letting $m \to \infty$, we get

$$\begin{aligned} &\operatorname{diag}\Big[|\omega(\tau) - \omega_k(\tau)|, \dots, |\omega(\tau) - \omega_k(\tau)|\Big]_{n\times n} \\ &\preceq \operatorname{diag}\Big[\epsilon_1\theta_1\varphi_1^{Ⓢ}((\xi(\tau) - \xi(0))^{\mathcal{A}}), \dots, \epsilon_n\theta_n\varphi_n^{Ⓢ}((\xi(\tau) - \xi(0))^{\mathcal{A}})\Big]_{n\times n}. \end{aligned}$$

Making use of the relation (13.7), we have

$$\delta(\omega, \omega_k) \preceq (\epsilon_1, \dots, \epsilon_n).$$

This implies that the Cauchy sequence $\{\omega_k\}$ converges to ω in (Ξ, δ). Thus, the metric space (Ξ, δ) is complete.

Based on Lemma 13.1, we have that the fractional system (13.1) is equivalent to the the integral equation (13.5). Now, consider the self–mapping $\propto: \Xi \longrightarrow \Xi$ such that for $\rho \in \Xi$,

$$\begin{aligned} \propto(\rho(\tau)) = \frac{(\xi(\tau) - \xi(0))^{\mathcal{C}-1}}{\Gamma(\mathcal{C})}\mathcal{E}_\circ + \frac{1}{\Gamma(\mathcal{A})}\int_0^\tau \xi'(\tau')(\xi(\tau) - \xi(\tau'))^{\mathcal{A}-1} \\ F(\tau', \rho(\tau'), \rho(\mu(\tau')))d\tau', \end{aligned} \tag{13.9}$$

for every $\tau \in (0, p]$.

For $\rho \in \Xi$, we have

$$
\begin{aligned}
&|\propto(\rho(\tau)) - \propto(\rho(\tau_0))| \\
&= \Bigg| \frac{(\xi(\tau)-\xi(0))^{\mathcal{C}-1}}{\Gamma(\mathcal{C})}\mathcal{E}_\circ + \frac{1}{\Gamma(\mathcal{A})}\int_0^{\tau} \xi'(\tau')(\xi(\tau)-\xi(\tau'))^{\mathcal{A}-1} \\
&\quad F(\tau', \rho(\tau'), \rho(\mu(\tau')))d\tau' \\
&\quad - \frac{(\xi(\tau_0)-\xi(0))^{\mathcal{C}-1}}{\Gamma(\mathcal{C})}\mathcal{E}_\circ - \frac{1}{\Gamma(\mathcal{A})}\int_0^{\tau_0} \xi'(\tau')(\xi(\tau_0)-\xi(\tau'))^{\mathcal{A}-1} \\
&\quad F(\tau', \rho(\tau'), \rho(\mu(\tau')))d\tau' \Bigg| \\
&\quad \longrightarrow 0
\end{aligned}
$$

as $\tau \longrightarrow \tau_0$, which implies the the continuity property of $\propto \rho$ and so $\propto \rho \in \Xi$.

Next, we show that the self-mapping $\propto$ is contractive on Ξ. Consider $\propto$: $\Xi \longrightarrow \Xi$ defined in the relation (13.9). Let $\rho, \varrho \in C_{1-\mathcal{C};\xi}(0, p]$ and $\delta(\rho(\tau), \varrho(\tau)) \preceq (\Bbbk_1, \ldots, \Bbbk_n)$, with $\underbrace{\Bbbk_i}_{i=1,\ldots,n} \in [0, \infty]$. Then, for every $\tau \in (0, p]$, we get

$$
\begin{aligned}
&\operatorname{diag}\Big[|\rho(\tau)-\varrho(\tau)|, \ldots, |\rho(\tau)-\varrho(\tau)|\Big]_{n\times n} \\
&\preceq \operatorname{diag}\Big[\Bbbk_1\theta_1\, \varphi_1^{\circledS}((\xi(\tau)-\xi(0))^{\mathcal{A}}), \ldots, \Bbbk_n\theta_n\, \varphi_n^{\circledS}((\xi(\tau)-\xi(0))^{\mathcal{A}})\Big].
\end{aligned}
$$

Using Remark 13.1, for all $\tau \in (0, p]$, we have

$$
\begin{aligned}
&\operatorname{diag}\Big[\Big|\propto(\rho(\tau)) - \propto(\varrho(\tau))\Big|, \ldots, \Big|\propto(\rho(\tau)) - \propto(\varrho(\tau))\Big|\Big]_{n\times n} \\
&\preceq \operatorname{diag}\Big[\frac{1}{\Gamma(\mathcal{A})}\int_0^{\tau} \xi'(\tau')(\xi(\tau)-\xi(\tau'))^{\mathcal{A}-1}\Big|F(\tau', \rho(\tau'), \rho(\mu(\tau'))) \\
&\quad - F(\tau', \varrho(\tau'), \varrho(\mu(\tau')))\Big|d\tau', \ldots, \\
&\quad \frac{1}{\Gamma(\mathcal{A})}\int_0^{\tau} \xi'(\tau')(\xi(\tau)-\xi(\tau'))^{\mathcal{A}-1}\Big|F(\tau', \rho(\tau'), \rho(\mu(\tau'))) \\
&\quad - F(\tau', \varrho(\tau'), \varrho(\mu(\tau')))\Big|d\tau'\Big]_{n\times n}
\end{aligned}
$$

$$\preceq \operatorname{diag}\Bigg[\frac{\Theta}{\Gamma(\mathcal{A})}\int_0^{\tau}\xi'(\tau')(\xi(\tau)-\xi(\tau'))^{\mathcal{A}-1}\Big[|\rho(\tau')-\varrho(\tau')|$$

$$+|\rho(\mu(\tau'))-\varrho(\mu(\tau'))|\Big]d\tau',\ldots,$$

$$\frac{\Theta}{\Gamma(\mathcal{A})}\int_0^{\tau}\xi'(\tau')(\xi(\tau)-\xi(\tau'))^{\mathcal{A}-1}\Big[|\rho(\tau')-\varrho(\tau')|+|\rho(\mu(\tau'))$$

$$-\varrho(\mu(\tau'))|\Big]d\tau'\Bigg]_{n\times n}$$

$$\preceq \operatorname{diag}\Bigg[\frac{2\Theta}{\Gamma(\mathcal{A})}\int_0^{\tau}\xi'(\tau')(\xi(\tau)-\xi(\tau'))^{\mathcal{A}-1}\varphi_1^{\circledS}((\xi(\tau')-\xi(0))^{\mathcal{A}})d\tau',\ldots,$$

$$\frac{2\Theta}{\Gamma(\mathcal{A})}\int_0^{\tau}\xi'(\tau')(\xi(\tau)-\xi(\tau'))^{\mathcal{A}-1}\varphi_n^{\circledS}((\xi(\tau')-\xi(0))^{\mathcal{A}})d\tau'\Bigg]_{n\times n}$$

$$\preceq \operatorname{diag}\Big[2\Theta\mathbb{k}_1\theta_1\varphi_1^{\circledS}((\xi(\tau)-\xi(0))^{\mathcal{A}}),\ldots,2\Theta\mathbb{k}_n\theta_n\varphi_n^{\circledS}((\xi(\tau)-\xi(0))^{\mathcal{A}})\Big]_{n\times n}.$$

Then, we have

$$\delta(\propto\rho,\propto\varrho)\preceq\underbrace{(2\Theta,\ldots,2\Theta)}_{n}\ \delta(\rho,\varrho).$$

Since $0<\Theta<\frac{1}{2}$, we can conclude the contractively property of $\propto$.

Let $f\in\Xi$. Since $\propto f\in\Xi$, we have that

$$\operatorname{diag}\Big[\Big|\propto(f(\tau))-f(\tau)\Big|,\ldots,\Big|\propto(f(\tau))-f(\tau)\Big|\Big]_{n\times n}$$

$$\preceq \operatorname{diag}\Bigg[\Bigg|f(\tau)-\frac{(\xi(\tau)-\xi(0))^{\mathcal{C}-1}}{\Gamma(\mathcal{C})}\mathcal{E}_\circ-\frac{1}{\Gamma(\mathcal{A})}\int_0^{\tau}\xi'(\tau')(\xi(\tau)-\xi(0))^{\mathcal{A}-1}$$

$$F(\tau',f(\tau'),f(\mu(\tau')))d\tau'\Bigg|,$$

$$\ldots,\Bigg|f(\tau)-\frac{(\xi(\tau)-\xi(0))^{\mathcal{C}-1}}{\Gamma(\mathcal{C})}\mathcal{E}_\circ-\frac{1}{\Gamma(\mathcal{A})}\int_0^{\tau}\xi'(\tau')(\xi(\tau)-\xi(0))^{\mathcal{A}-1}$$

$$F(\tau',f(\tau'),f(\mu(\tau')))d\tau'\Bigg|\Bigg]_{n\times n}$$

$$\preceq \operatorname{diag}\Big[\theta_1\,\varphi_1^{\circledS}((\xi(\tau)-\xi(0))^{\mathcal{A}}),\ldots,\theta_n\,\varphi_n^{\circledS}((\xi(\tau)-\xi(0))^{\mathcal{A}})\Big],$$

for every $\tau \in (0, p]$, which implies that

$$\delta(\propto f, f) \leq \underbrace{(1, \ldots, 1)}_{n}. \tag{13.10}$$

and hence for every $k \in \mathbb{N}$, we have $\delta(\propto^k f, \propto^{k+1} f) \leq \underbrace{(+\infty, \ldots, +\infty)}_{n}$. Now, we can apply the second part of Theorem 9.5 and so we can find a unique map $g \in \{\tilde{\sigma} \in \Xi : \delta(\propto f, \tilde{\sigma}) < \underbrace{(+\infty, \ldots, +\infty)}_{n}\}$ such that $\propto g = g$ and so

$$g(\tau) = \frac{(\xi(\tau) - \xi(0))^{\mathcal{C}-1}}{\Gamma(\mathcal{C})}\mathcal{E}_\circ + \frac{1}{\Gamma(\mathcal{A})}\int_0^\tau \xi'(\tau')(\xi(\tau) - \xi(\tau'))^{\mathcal{A}-1}$$
$$F(\tau', g(\tau'), g(\mu(\tau')))d\tau',$$

for every $\tau \in (0, p]$, where $I_{0+}^{1-\mathcal{C};\xi} g(0^+) = \mathcal{E}_\circ \in \mathbb{R}$.

In view of Theorem 9.5 and the inequality (13.10), we have

$$\delta(g, f) \leq \underbrace{\left(\frac{1}{1-2\Theta}, \ldots, \frac{1}{1-2\Theta}\right)}_{n} \delta(\propto f, f)$$
$$\leq \underbrace{\left(\frac{1}{1-2\Theta}, \ldots, \frac{1}{1-2\Theta}\right)}_{n},$$

which implies the inequality (13.6).

13.2 Multi Stability of Hilfer Fractional Systems I on Unbounded Domains

In Theorem 13.1, we proved the multi stability of the ξ–Hilfer fractional differential system (13.1) on bounded domains. Now, we extend our result to unbounded domains. Here, let $\mathsf{S} = C_{1-\mathcal{C};\xi}(\mathbb{R})$.

Theorem 13.2 *Let the assumptions* ($\mathfrak{J}_1$) *and* ($\mathfrak{J}_2$) *be hold. If* f *in* S *satisfies the inequality* (13.4) *and* $I_{0+}^{1-\mathcal{C};\xi} f(0^+) = \mathcal{E}_\circ$, *thus, we can find a unique function* g *satisfying in the fractional equations* (13.2)–(13.3) *such that satisfies the inequality* (13.6) *for every* $\tau \in \mathbb{R}$.

Proof We are going to show the result for $\mathbb{R}$. By the same way, we can prove the theorem for $(-\infty, 0]$ or $[0, +\infty)$.

For every $k \in \mathbb{N}$, Let $\mathsf{P}_k = [\mathfrak{P} - k, \mathfrak{P} + k]$. As stated by Theorem 13.1, we can find a unique mapping $g_k \in C_{1-\mathcal{C};\xi}(\mathsf{P}_k, \mathbb{R})$, such that

$$ {}^{\mathscr{H}}D_{0^+}^{\mathcal{A},\mathcal{B};\xi} g_k(\tau) = F(\tau, g_k(\tau), g_k(\mu(\tau))), \tag{13.11}$$

$$ I_{0^+}^{1-\mathcal{C};\xi} g_k(0^+) = \mathcal{E}_\circ, \qquad \mathcal{E}_\circ \in \mathbb{R}, \tag{13.12}$$

and

$$\begin{aligned} &\mathrm{diag}\Big[|g_k(\tau) - f(\tau)|, \ldots, |g_k(\tau) - f(\tau)| \Big]_{n\times n} \\ &\preceq \mathrm{diag}\Big[\frac{\theta_1}{1-2\Theta} \varphi_1^{\circledS}((\xi(\tau) - \xi(0))^{\mathcal{A}}), \ldots, \frac{\theta_n}{1-2\Theta} \varphi_n^{\circledS}((\xi(\tau) - \xi(0))^{\mathcal{A}}) \Big], \end{aligned} \tag{13.13}$$

for every $\tau \in \mathsf{P}_k$. The uniqueness of g_k implies that if $\tau \in \mathsf{P}_k$, then,

$$ g_k(\tau) = g_{k+1}(\tau) = g_{k+2}(\tau) = \cdots . \tag{13.14}$$

For every $\tau \in \mathbb{R}$, we define $k(\tau) \in \mathbb{N}$ as follows:

$$ k(\tau) = \min\{k \in \mathbb{N} \mid \tau \in \mathsf{P}_k\}. $$

In addition, we consider a mapping g given by

$$ g(\tau) = g_{k(\tau)}(\tau), \quad \tau \in \mathbb{R}, $$

and we claim that $g \in \mathsf{S}$. For a real number $\tau_1 \in \mathbb{R}$, we consider $k_1 = k(\tau_1) \in \mathbb{N}$. Thus, $\tau_1 \in (\mathsf{P}_{k_1+1})^\circ$ (the interior) and we can find an $\epsilon > 0$ such that $g(\tau) = g_{k+1}(\tau)$ for all τ with $\tau_1 - \epsilon < \tau < \tau_1 + \epsilon$.

Now, we want to show that g satisfies the relations (13.2), (13.3) and (13.6), for every $\tau \in \mathbb{R}$. For every $\tau \in \mathbb{R}$, consider $k(\tau) \in \mathbb{N}$. Thus, $\tau \in \mathsf{P}_{k(\tau)}$ and it concludes from the fractional equations (13.11) and (13.12) that

$$\begin{aligned} g(\tau) &= g_{k(\tau)}(\tau) \\ &= \frac{(\xi(\tau) - \xi(0))^{\mathcal{C}-1}}{\Gamma(\mathcal{C})} \mathcal{E}_\circ \\ &\quad + \frac{1}{\Gamma(\mathcal{A})} \int_0^\tau \xi'(\tau')(\xi(\tau) - \xi(\tau'))^{\mathcal{A}-1} F(\tau', g_{k(\tau')}(\tau'), g_{k(\tau')}(\mu(\tau')))d\tau', \\ &= \frac{(\xi(\tau) - \xi(0))^{\mathcal{C}-1}}{\Gamma(\mathcal{C})} \mathcal{E}_\circ \end{aligned}$$

$$+\frac{1}{\Gamma(\mathcal{A})}\int_0^\tau \xi'(\tau')(\xi(\tau)-\xi(\tau'))^{\mathcal{A}-1}F(\tau',g(\tau'),g(\mu(\tau')))d\tau',$$

where the above equality holds true because $k(\tau') \leq k(\tau)$, for every $\tau' \in \mathsf{P}_{k(\tau)}$ and it concludes from the relation (13.14) that

$$g(\tau') = g_{k(\tau')}(\tau') = g_{k(\tau)}(\tau').$$

Since $g(\tau) = g_{k(\tau)}(\tau)$ and $\tau \in \mathsf{P}_{k(\tau)}$, for every $\tau \in \mathbb{R}$, the inequality (13.13) implies that

$$\begin{aligned}
&\operatorname{diag}\Big[\,|g(\tau)-f(\tau)|,\ldots,|g(\tau)-f(\tau)|\,\Big]_{n\times n}\\
&=\operatorname{diag}\Big[\,|g_{k(\tau)}(\tau)-f(\tau)|,\ldots,|g_{k(\tau)}(\tau)-f(\tau)|\,\Big]_{n\times n}\\
&\preceq \operatorname{diag}\Big[\frac{\theta_1}{1-2\Theta}\varphi_1^{\circledS}((\xi(\tau)-\xi(0))^{\mathcal{A}}),\ldots,\frac{\theta_n}{1-2\Theta}\varphi_n^{\circledS}((\xi(\tau)-\xi(0))^{\mathcal{A}})\Big].
\end{aligned}$$

In the end, we claim that g is unique. Let $g' \in \mathsf{S}$ be another function, which satisfies the relations (13.2), (13.3) and (13.6), for every $\tau \in \mathbb{R}$. Since $g|_{\mathsf{P}_{k(\tau)}} (= g_{k(\tau)})$ and $g'|_{\mathsf{P}_{k(\tau)}}$ both satisfy (13.2), (13.3) and (13.6), for every $\tau \in \mathsf{P}_{k(\tau)}$, the uniqueness of $g_{k(\tau)} = g|_{\mathsf{P}_{k(\tau)}}$ implies that

$$g(\tau) = g|_{\mathsf{P}_{k(\tau)}}(\tau) = g'|_{\mathsf{P}_{k(\tau)}}(\tau) = g'(\tau),$$

as required.

13.3 Application

In this part, as an application, we apply our main result to prove an example.

Example 13.1 Consider the following fractional system

$$\begin{cases}
{}^{\mathscr{H}}D_{0^+}^{\frac{1}{4},\frac{3}{4};e^{\tau}} g(\tau) = \frac{1}{4}\sin\left(g\left(\frac{3}{4}\tau\right)\right) + \frac{1}{4}\operatorname{arccot}(g(\tau)), & \tau \in (0,1],\\
I_{0^+}^{\frac{3}{16};e^{\varsigma}} g(0^+) = g_0,
\end{cases} \tag{13.15}$$

and the inequality below

$$
\begin{aligned}
&\operatorname{diag}\Bigg[\left|{}^{\mathscr{H}}D_{0^+}^{\frac{1}{4},\frac{3}{4};e^{\tau}}g(\tau)-F\left(\tau,g(\tau),g\left(\frac{3}{4}\tau\right)\right)\right|,\ldots,\\
&\qquad\qquad\left|{}^{\mathscr{H}}D_{0^+}^{\frac{1}{4},\frac{3}{4};e^{\tau}}g(\tau)-F\left(\tau,g(\tau),g\left(\frac{3}{4}\tau\right)\right)\right|\Bigg]_{n\times n}\\
&\preceq\operatorname{diag}\left[\theta_1\,\varphi_1^{\circledS}((e^{\tau}-1)^{\frac{1}{4}}),\ldots,\theta_n\,\varphi_n^{\circledS}((e^{\tau}-1)^{\frac{1}{4}})\right].
\end{aligned}
$$

Notice that in fractional equations (13.2) and (13.3), we let $\mathcal{A}=\frac{1}{4}$, $\mathcal{B}=\frac{3}{4}$, then $\mathcal{C}=\mathcal{A}+\mathcal{B}(1-\mathcal{A})=\frac{13}{16}$, and $p=1$, $\xi(.)=e^{\cdot}$, $\mu(.)=\frac{3}{4}(.)$, $F(.,g(.),g(\mu(.)))=\frac{1}{4}\sin(g(\frac{3}{4}(.)))+\frac{1}{4}\text{arccot}(g(.))$, and $\Theta=\frac{1}{4}$.

We want to show that $F(.,g(.),g(\mu(.)))$ satisfies in assumption $(\mathfrak{J}_2)$:

$$
\begin{aligned}
&\left|\frac{1}{4}\sin(g_1)+\frac{1}{4}\text{arccot}(g_1)-\frac{1}{4}\sin(g_2)-\frac{1}{4}\text{arccot}(g_2)\right|\\
&\leq\left|\frac{1}{4}\sin(g_1)-\frac{1}{4}\sin(g_2)\right|+\left|\frac{1}{4}\text{arccot}(g_1)-\frac{1}{4}\text{arccot}(g_2)\right|\\
&\leq\left|\frac{1}{4}(g_1-g_2)\right|+\left|\frac{1}{4}(\text{arccot}(\mathfrak{C}))'(g_1-g_2)\right|\\
&\leq\left|\frac{1}{4}(g_1-g_2)\right|+\left|\frac{1}{4}\frac{-1}{1+\mathfrak{C}^2}(g_1-g_2)\right|\\
&\leq\left|\frac{1}{4}(g_1-g_2)\right|+\left|\frac{1}{4}(g_1-g_2)\right|\\
&\leq\frac{1}{2}\left|g_1-g_2\right|,
\end{aligned}
$$

where $g_1\leq\mathfrak{C}\leq g_2$. thus, the function F is a contraction.

Now, Theorem 13.1 implies that problem (13.15) has a unique solution and also is multi stable with

$$
\begin{aligned}
&\operatorname{diag}\Big[|g(\tau)-f(\tau)|,\ldots,|g(\tau)-f(\tau)|\Big]_{n\times n}\\
&\preceq\operatorname{diag}\left[2\theta_1\varphi_1^{\circledS}((e^{\tau}-1)^{\frac{1}{4}}),\ldots,2\theta_n\varphi_n^{\circledS}((e^{\tau}-1)^{\frac{1}{4}})\right],\qquad\tau\in(0,1].
\end{aligned}
$$

We refer the interested readers to [3–6].

References

1. Aderyani SR, Saadati R, Feckan M (2021) The Cadariu-Radu method for existence, uniqueness and gauss hypergeometric stability of W-Hilfer fractional differential equations. Mathematics 9:1408
2. Aderyani SR, Saadati R, O'Regan D, Abdeljawad T (2022) UHML stability of a class of Δ-Hilfer FDEs via CRM. AIMS Math 7(4):5910–5919
3. Aderyani SR, Saadati R, O'Regan D, Alshammari FS (2022) Multi-super-stability of antiderivations in Banach algebras. AIMS Math 7(11):20143–20163
4. Aderyani SR, Saadati R, O'Regan D, Alshammari FS (2023) Application of aggregated control functions for approximating $\mathscr{C}$-Hilfer fractional differential equations. AIMS Math 8(11):28010–28032
5. Aderyani SR, Saadati R, Rassias TM, Srivastava HM (2023) Existence, Uniqueness and the Multi-Stability Results for a w-Hilfer Fractional Differential Equation. Axioms 12(7):681
6. Rezaei Aderyani S, Saadati R, Li C, Rassias TM, Park C (2023) Special functions and multi-stability of the Jensen type random operator equation in C^*-algebras via fixed point. J Inequal Appl 2023(1):1–24

Chapter 14
Multi Stability of Fractional–Order System II

Applying the fixed point theory, we study the existence, uniqueness and the multi stability of the following ξ–Hilfer fractional differential equations defined on compact domains. Next, we generalize the main results for unbounded domains. An example is given to illustrate the main result for a fractional system.

Now, consider the fractional–order systems II, as follows [1]:

$$\begin{cases} {}^{\mathscr{H}}D_{\lambda^+}^{\mathcal{A},\mathcal{B}:\xi}\Phi(\zeta) = F(\zeta, \Phi(\zeta), {}^{\mathscr{H}}D_{\lambda^+}^{\mathcal{A},\mathcal{B}:\xi}\Phi(\zeta)), \\ I_{\lambda^+}^{1-\mathcal{C}:\xi}\Phi(\lambda^+) = J_\lambda, \end{cases} \tag{14.1}$$

where ${}^{\mathscr{H}}D_{\lambda^+}^{\mathcal{A},\mathcal{B}:\xi}(.)$ is the ξ–Hilfer fractional derivative of order $0 < \mathcal{A} \leq 1$ and type $0 \leq \mathcal{B} \leq 1$, $I_{\lambda^+}^{1-\mathcal{C}:\xi}(.)$ is the Riemann–Liouville fractional integral of order $1 - \mathcal{C}$, $\mathcal{C} = \mathcal{A} + \mathcal{B}(1 - \mathcal{A})$, with respect to the function ξ, and $F : \Delta \times \mathbb{R}^2 \longrightarrow \mathbb{R}$ is a given function. Here, assume $\zeta \in \Delta = [\lambda, D]$ with $D > \lambda$ and $J_\lambda \in \mathbb{R}$.

14.1 Multi Stability of Hilfer Fractional Systems II on Compact Domains

In this part, we prove the existence, uniqueness and the multi stability results of the ξ–Hilfer fractional system (14.1) defined on compact domains by the fixed point theorem.

For $F : \Delta \times \mathbb{R}^2 \to \mathbb{R}$ and $\theta > 0$, consider the following fractional equations

$${}^{\mathscr{H}}D_{\lambda^+}^{\mathcal{A},\mathcal{B}:\xi}\Phi(\zeta) = F(\zeta, \Phi(\zeta), {}^{\mathscr{H}}D_{\lambda^+}^{\mathcal{A},\mathcal{B}:\xi}\Phi(\zeta)), \tag{14.2}$$

$$I_{\lambda^+}^{\mathcal{C}:\xi}\Phi(\lambda^+) = J_\lambda, \qquad J_\lambda \in \mathbb{R}, \tag{14.3}$$

and the inequality below

S. Rezaei Aderyani et al., *Towards Ulam Type Multi Stability Analysis*, Studies in Systems, Decision and Control 523, https://doi.org/10.1007/978-3-031-55564-0_14

$$\begin{aligned}
&\operatorname{diag}\Bigg[\left|{}^{\mathscr{H}}D_{\lambda^+}^{\mathcal{A},\mathcal{B};\xi}\Psi(\zeta) - F(\zeta,\Psi(\zeta),{}^{\mathscr{H}}D_{\lambda^+}^{\mathcal{A},\mathcal{B};\xi}\Psi(\zeta))\right|,\cdots, \\
&\qquad\left|{}^{\mathscr{H}}D_{\lambda^+}^{\mathcal{A},\mathcal{B};\xi}\Psi(\zeta) - F(\zeta,\Psi(\zeta),{}^{\mathscr{H}}D_{\lambda^+}^{\mathcal{A},\mathcal{B};\xi}\Psi(\zeta))\right|\Bigg]_{n\times n} \\
&\preceq \operatorname{diag}\Bigg[\theta_1\varphi_1^{\circledS}\Big((\xi(\zeta)-\xi(0))^{\mathcal{A}}\Big),\cdots,\theta_n\varphi_n^{\circledS}\Big((\xi(\zeta)-\xi(0))^{\mathcal{A}}\Big)\Bigg],
\end{aligned} \tag{14.4}$$

where $\zeta \in \Delta, n \in \mathbb{N}$ and $\underbrace{\varphi_i^{\circledS}}_{i=1,\cdots,n} \in C(\Delta,\mathbb{R}_+)$ are special control functions defined in Sects. 9.8.1 and 9.8.2.

We now define the concept of the multi stability for fractional system (14.1).

Definition 14.1 Equations (14.2)–(14.3) have the multi stability with respect to

$$\operatorname{diag}\Bigg[\varphi_1^{\circledS}\Big((\xi(\zeta)-\xi(0))^{\mathcal{A}}\Big),\cdots,\varphi_n^{\circledS}\Big((\xi(\zeta)-\xi(0))^{\mathcal{A}}\Big)\Bigg],$$

if there exist $\underbrace{c_i}_{i=1,\cdots,n} > 0$, such that for every $\underbrace{\theta_i}_{i=1,\cdots,n} > 0$ and every solution $\Psi \in C_{1-\mathcal{C};\xi}(\Delta,\mathbb{R})$ to the inequality (14.4) and $I_{\lambda^+}^{1-\mathcal{C};\xi}\Psi(\lambda^+) = J_\lambda$, there exists a solution $\Phi \in C_{1-\mathcal{C};\xi}(\Delta,\mathbb{R})$ to fractional equations (14.2)–(14.3) with

$$\begin{aligned}
&\operatorname{diag}\Big[|\Psi(\zeta)-\Phi(\zeta)|,\cdots,|\Psi(\zeta)-\Phi(\zeta)|\Big]_{n\times n} \\
&\preceq \operatorname{diag}\Bigg[c_1\theta_1\varphi_1^{\circledS}\Big((\xi(\zeta)-\xi(0))^{\mathcal{A}}\Big),\cdots,c_n\theta_n\varphi_n^{\circledS}\Big((\xi(\zeta)-\xi(0))^{\mathcal{A}}\Big)\Bigg],
\end{aligned}$$

for every $\zeta \in \Delta$.

Lemma 14.1 *Consider a continuous function $F : \Delta \times \mathbb{R}^2 \to \mathbb{R}$. Then, the Eq.* (14.1) *is equivalent to*

$$\Phi(\zeta) = \frac{(\xi(\zeta)-\xi(\lambda))^{\mathcal{C}-1}}{\Gamma(\mathcal{C})}J_\lambda + I_{\lambda^+}^{\mathcal{A};\xi}G(\zeta), \tag{14.5}$$

where $G \in C(\Delta,\mathbb{R})$ satisfies the functional equation below

$$G_\Phi(\zeta) = F\Big(\zeta,\Phi(\zeta),G_\Phi(\zeta)\Big).$$

Proof To see this note applying $I_{\lambda^+}^{\mathcal{A};\xi}(.)$ on both sides of the fractional equation (14.1) and using Theorem 9.2, we have

$$\Phi(\zeta) - \frac{(\xi(\zeta)-\xi(\lambda))^{\mathcal{C}-1}}{\Gamma(\mathcal{C})}I_{\lambda^+}^{(1-\mathcal{A})(1-\mathcal{B});\xi}J_\lambda = I_{\lambda^+}^{\mathcal{A};\xi}G(\zeta).$$

Thus, we get

$$\Phi(\zeta) = \frac{(\xi(\zeta) - \xi(\lambda))^{\mathcal{C}-1}}{\Gamma(\mathcal{C})} J_\lambda + I_{\lambda^+}^{\mathcal{A}:\xi} G(\zeta). \tag{14.6}$$

On the other hand, if Φ satisfies the equality (14.6), then, it satisfies the fractional equation (14.1). To see this, apply ${}^{\mathscr{H}}D_{\lambda^+}^{\mathcal{A},\mathcal{B}:\xi}(.)$ on both sides of Eq. (14.6). Now, according to Theorem 9.1, we get

$${}^{\mathscr{H}}D_{\lambda^+}^{\mathcal{A},\mathcal{B}:\xi}\Phi(\zeta) = {}^{\mathscr{H}}D_{\lambda^+}^{\mathcal{A},\mathcal{B}:\xi}\frac{(\xi(\zeta) - \xi(\lambda))^{\mathcal{C}-1}}{\Gamma(\mathcal{C})} J_\lambda + {}^{\mathscr{H}}D_{\lambda^+}^{\mathcal{A},\mathcal{B}:\xi} I_{\lambda^+}^{\mathcal{A}:\xi} G(\zeta) = G(\zeta),$$

where for $\mathcal{C} \in (0, 1)$, we used ${}^{\mathscr{H}}D_{\lambda^+}^{\mathcal{A},\mathcal{B}:\xi}(\xi(\zeta) - \xi(\lambda))^{\mathcal{C}-1} = 0$. We conclude that $\Phi(\zeta)$ satisfies the fractional equation (14.1) if and only if $\Phi(\zeta)$ satisfies (14.5).

Remark 14.1 ([1]) Let $\Psi \in C_{1-\mathcal{C}:\xi}(\Delta, \mathbb{R})$ be a solution of the inequality (14.4) and $I_{\lambda^+}^{1-\mathcal{C}:\xi}\Psi(\lambda^+) = J_\lambda$. Then, Ψ is a solution of the following integral inequality

$$\begin{aligned}
&\operatorname{diag}\Bigg[\left|\Psi(\zeta) - \frac{(\xi(\zeta) - \xi(0))^{\mathcal{C}-1}}{\Gamma(\mathcal{C})} J_\lambda - \frac{1}{\Gamma(\mathcal{A})}\int_0^\zeta \xi'(\mathcal{A})(\xi(\zeta) - \xi(s))^{\mathcal{A}-1} G_\Psi(s)ds\right|, \cdots,\\
&\qquad \left|\Psi(\zeta) - \frac{(\xi(\zeta) - \xi(0))^{\mathcal{C}-1}}{\Gamma(\mathcal{C})} J_\lambda - \frac{1}{\Gamma(\mathcal{A})}\int_0^\zeta \xi'(\mathcal{A})(\xi(\zeta) - \xi(s))^{\mathcal{A}-1} G_\Psi(s)ds\right|\Bigg]_{n\times n}\\
&\preceq \operatorname{diag}\left[\theta_1\varphi_1^{\circledS}\Big((\xi(\zeta) - \xi(0))^{\mathcal{A}}\Big), \cdots, \theta_n\varphi_n^{\circledS}\Big((\xi(\zeta) - \xi(0))^{\mathcal{A}}\Big)\right],
\end{aligned}$$

where $G_\Psi \in C(\Delta, \mathbb{R})$ satisfies the functional equation below

$$G_\Psi(\zeta) = F\Big(\zeta, \Psi(\zeta), G_\Psi(\zeta)\Big). \tag{14.7}$$

Also, you can see Remmark 13.1.

Let us suppose the following axioms are satisfied:

$(\mathcal{F}_1)$ $F : \Delta \times \mathbb{R}^2 \to \mathbb{R}$ is continuous.

$(\mathcal{F}_2)$ There exists $0 < \Theta_1, \Theta_2$, with $0 < \dfrac{\Theta_1}{(1 - \Theta_2)} < 1$ such that

$$|F(\zeta, \kappa, \upsilon) - F(\zeta, \overline{\kappa}, \overline{\upsilon})| \leq \Theta_1|\kappa - \overline{\kappa}| + \Theta_2|\upsilon - \overline{\upsilon}| \quad \text{for every} \quad \kappa, \upsilon, \overline{\kappa}, \overline{\upsilon} \in \mathbb{R} \text{ and } \zeta \in \Delta.$$

Theorem 14.1 *Let the assumptions* $(\mathcal{F}_1)$ *and* $(\mathcal{F}_2)$ *be satisfied. If* $\Psi \in C_{1-\mathcal{C}:\xi}(\Delta, \mathbb{R})$ *satisfies the inequality* (14.4) *and* $I_{\lambda^+}^{1-\mathcal{C}:\xi}\Psi(\lambda^+) = J_\lambda$*, then, there exists a unique function* Φ *satisfying the fractional equations* (14.2)–(14.3) *such that*

$$daig\left[|\Phi(\zeta) - \Psi(\zeta)|, \cdots, |\Phi(\zeta) - \Psi(\zeta)| \right]_{n\times n} \tag{14.8}$$
$$\preceq daig\left[\frac{\theta_1}{1 - \frac{\Theta_1}{(1-\Theta_2)}} \varphi_1^{\circledS}\left((\xi(\zeta) - \xi(0))^{\mathcal{A}} \right), \cdots, \frac{\theta_n}{1 - \frac{\Theta_1}{(1-\Theta_2)}} \varphi_n^{\circledS}\left((\xi(\zeta) - \xi(0))^{\mathcal{A}} \right) \right],$$

for every $\zeta \in \Delta$.

Proof Set $\mathcal{H} = C_{1-C;\xi}(\Delta, \mathbb{R})$ and define a mapping $\mathcal{P} : \mathcal{H} \times \mathcal{H} \to [0, \infty]^n$ with $n \in \mathbb{N}$ by

$$\mathcal{P}(\rho, \varrho) = \inf \Big\{ (\mathtt{C}_1, \cdots, \mathtt{C}_n) \geq 0 : \mathrm{diag}\left[|\rho(\zeta) - \varrho(\zeta)|, \cdots, |\rho(\zeta) - \varrho(\zeta)| \right]_{n\times n} \tag{14.9}$$
$$\preceq \mathrm{diag}\left[\mathtt{C}_1\theta_1\varphi_1^{\circledS}\left((\xi(\zeta) - \xi(0))^{\mathcal{A}} \right), \cdots, \mathtt{C}_n\theta_n\varphi_n^{\circledS}\left((\xi(\zeta) - \xi(0))^{\mathcal{A}} \right) \right] \Big\}.$$

We show that $(\mathcal{H}, \mathcal{P})$ is a complete generalized metric space. For the metric part it is enough to assume that $\mathcal{P}(\rho, \varrho) > \mathcal{P}(\rho, \nu) + \mathcal{P}(\nu, \varrho)$, for some ρ, ϱ and $\nu \in \mathcal{H}$. Then, there exists $\zeta_\circ \in \Delta$ with

$$\mathrm{diag}\left[|\rho(\zeta_\circ) - \varrho(\zeta_\circ)|, \cdots, |\rho(\zeta_\circ) - \varrho(\zeta_\circ)| \right]_{n\times n}$$
$$\succ (\mathcal{P}(\rho, \nu) + \mathcal{P}(\nu, \varrho))\mathrm{diag}\left[\theta_1\varphi_1^{\circledS}\left((\xi(\zeta_\circ) - \xi(0))^{\mathcal{A}} \right), \cdots, \theta_n\varphi_n^{\circledS}\left((\xi(\zeta_\circ) - \xi(0))^{\mathcal{A}} \right) \right].$$

Hence, from the definition of $\mathcal{P}$, we get

$$|\rho(\zeta_\circ) - \varrho(\zeta_\circ)|$$
$$> |\rho(\zeta_\circ) - \nu(\zeta_\circ)| + |(\nu(\zeta_\circ) - \varrho(\zeta_\circ)|,$$

that contradicts. Next, we prove that $(\mathcal{H}, \mathcal{P})$ is complete. Assume ω_k is a Cauchy sequence in $(\mathcal{H}, \mathcal{P})$. Thus, for every $\underbrace{\epsilon_i}_{i=1,\cdots,n} > 0$, there exists an $\underbrace{\aleph_{\epsilon_i}}_{i=1,\cdots,n} \in \mathbb{N}$ such that $\mathcal{P}(\omega_m, \omega_k) \leq (\epsilon_1, \cdots, \epsilon_n)$, for every $m, k \geq \underbrace{\aleph_{\epsilon_i}}_{i=1,\cdots,n}$. Making use of the relation (14.9), we have

$$\mathrm{diag}\left[|\omega_m(\zeta) - \omega_k(\zeta)|, \cdots, |\omega_m(\zeta) - \omega_k(\zeta)| \right]_{n\times n}$$
$$\preceq \mathrm{diag}\left[\epsilon_1\theta_1\varphi_1^{\circledS}\left((\xi(\zeta) - \xi(0))^{\mathcal{A}} \right), \cdots, \epsilon_n\theta_n\varphi_n^{\circledS}\left((\xi(\zeta) - \xi(0))^{\mathcal{A}} \right) \right], \tag{14.10}$$

for every $\zeta \in \Delta$. If ζ is fixed, the inequality (14.10) implies that $\{\omega_k(\zeta)\}$ is a Cauchy sequence in $\mathbb{R}$. Since $\mathbb{R}$ is complete, $\{\omega_k(\zeta)\}$ converges for every $\zeta \in \Delta$. Thus, we can define a function ω by

$$\omega(\zeta) := \lim_{k\to\infty} \omega_k(\zeta), \quad (\zeta \in \Delta). \tag{14.11}$$

It is easy to check $\omega \in \mathcal{H}$. If we let $m \to \infty$, it follows from the inequality (14.10) that

$$\begin{aligned} &\operatorname{diag}\left[|\omega(\zeta)-\omega_k(\zeta)|,\cdots,|\omega(\zeta)-\omega_k(\zeta)|\right]_{n\times n} \\ &\preceq \operatorname{diag}\left[\epsilon_1\theta_1\varphi_1^{\circledS}\Big((\xi(\zeta)-\xi(0))^{\mathcal{A}}\Big),\cdots,\epsilon_n\theta_n\varphi_n^{\circledS}\Big((\xi(\zeta)-\xi(0))^{\mathcal{A}}\Big)\right]. \end{aligned} \tag{14.12}$$

Considering the relation (14.9), we get

$$\mathcal{P}(\omega,\omega_k) \preceq (\epsilon_1,\cdots,\epsilon_n).$$

This implies that the Cauchy sequence $\{\omega_k\}$ is convergent to ω in $(\mathcal{H},\mathcal{P})$. Therefore, $(\mathcal{H},\mathcal{P})$ is complete.

In view of Lemma 14.1, we have that the fractional equations (14.2)–(14.3) are equivalent to the Eq. (14.5). Now, consider $\propto: \mathcal{H} \longrightarrow \mathcal{H}$ such that for $\rho \in \mathcal{H}$,

$$\begin{aligned} &\propto(\rho(\zeta)) \\ &= \frac{(\xi(\zeta)-\xi(0))^{\mathcal{C}-1}}{\Gamma(\mathcal{C})}J_\lambda + \frac{1}{\Gamma(\mathcal{A})}\int_0^\zeta \xi'(\mathcal{A})(\xi(\zeta)-\xi(\backepsilon))^{\mathcal{A}-1}G_\rho(\backepsilon)d\backepsilon, \end{aligned} \tag{14.13}$$

in which $G_\rho \in C(\Delta,\mathbb{R})$ satisfies the functional equation:

$$G_\rho(\zeta) = F\Big(\zeta,\rho(\zeta),G_\rho(\zeta)\Big).$$

For $\rho \in \mathcal{H}$, we have

$$\begin{aligned} &|\propto(\rho(\zeta)) - \propto(\rho(\zeta_0))| \\ &= \Bigg|\frac{(\xi(\zeta)-\xi(0))^{\mathcal{C}-1}}{\Gamma(\mathcal{C})}J_\lambda + \frac{1}{\Gamma(\mathcal{A})}\int_0^\zeta \xi'(\mathcal{A})(\xi(\zeta)-\xi(\backepsilon))^{\mathcal{A}-1}G_\rho(\backepsilon)d\backepsilon \\ &\qquad -\frac{(\xi(\zeta_0)-\xi(0))^{\mathcal{C}-1}}{\Gamma(\mathcal{C})}J_\lambda + \frac{1}{\Gamma(\mathcal{A})}\int_0^\zeta \xi'(\mathcal{A})(\xi(\zeta_0)-\xi(\backepsilon))^{\mathcal{A}-1}G_\rho(\backepsilon)d\backepsilon\Bigg| \\ &\longrightarrow 0, \quad \text{as } \zeta \longrightarrow \zeta_0, \end{aligned}$$

so $\propto: \mathcal{H} \longrightarrow \mathcal{H}$ is continuous.

Next, we show that the self–mapping $\propto$ is contractive on $\mathcal{H}$. Consider $\propto: \mathcal{H} \longrightarrow \mathcal{H}$ defined in (14.13). Let $\rho, \varrho \in C_{1-\mathcal{C};\xi}(\Delta,\mathbb{R})$ and $\mathcal{P}(\rho(\zeta),\varrho(\zeta)) \preceq (\Bbbk_1,\cdots,\Bbbk_n)$, with $\underbrace{\Bbbk_i}_{1\le i\le n} \in [0,\infty]$. Then for every $\zeta \in \Delta$, we get

$$\operatorname{diag}\Big[|\rho(\zeta)-\varrho(\zeta)|,\cdots,|\rho(\zeta)-\varrho(\zeta)|\Big]_{n\times n}$$
$$\preceq \operatorname{diag}\Big[\Bbbk_1\theta_1\varphi_1^{\circledS}\Big((\xi(\zeta)-\xi(0))^{\mathcal{A}}\Big),\cdots,\Bbbk_n\theta_n\varphi_n^{\circledS}\Big((\xi(\zeta)-\xi(0))^{\mathcal{A}}\Big)\Big].$$

For all $\zeta\in\Delta$, we have

$$\Big|\propto(\rho(\zeta))-\propto(\varrho(\zeta))\Big| \tag{14.14}$$
$$\leq\frac{1}{\Gamma(\mathcal{A})}\int_0^{\zeta}\xi'(\mathcal{A})(\xi(\zeta)-\xi(\ni))^{\mathcal{A}-1}|G_\rho(\ni)-G_\varrho(\ni)|d\ni,$$

where

$$G_\rho(\zeta)=F(\zeta,\rho(\zeta),G_\rho(\zeta)),$$

and

$$G_\varrho(\zeta)=F(\zeta,\varrho(\zeta),G_\varrho(\zeta)).$$

Using hypothesis $(\mathcal{F}_2)$ we have, for any $\zeta\in\Delta$,

$$|G_\rho(\zeta)-G_\varrho(\zeta)|\leq\Theta_1|\rho(\zeta)-\varrho(\zeta)|+\Theta_2|G_\rho(\zeta)-G_\varrho(\zeta)|,$$

which can be written as

$$|G_\rho(\zeta)-G_\varrho(\zeta)|\leq\frac{\Theta_1}{1-\Theta_2}|\rho(\zeta)-\varrho(\zeta)|. \tag{14.15}$$

Now using Remark 14.1, and the inequalities (14.14) and (14.15), we have

$$\operatorname{diag}\Big[\Big|\propto(\rho(\zeta))-\propto(\varrho(\zeta))\Big|,\cdots,\Big|\propto(\rho(\zeta))-\propto(\varrho(\zeta))\Big|\Big]_{n\times n}$$
$$\preceq\operatorname{diag}\Big[\frac{\Theta_1}{(1-\Theta_2)\Gamma(\mathcal{A})}\int_0^{\zeta}\xi'(\mathcal{A})(\xi(\zeta)-\xi(\ni))^{\mathcal{A}-1}|\rho(\ni)-\varrho(\ni)|d\ni,\cdots,$$
$$\frac{\Theta_1}{(1-\Theta_2)\Gamma(\mathcal{A})}\int_0^{\zeta}\xi'(\mathcal{A})(\xi(\zeta)-\xi(\ni))^{\mathcal{A}-1}|\rho(\ni)-\varrho(\ni)|d\ni\Big]_{n\times n}$$
$$\preceq\operatorname{diag}\Big[\frac{\Bbbk_1\theta_1\Theta_1}{(1-\Theta_2)\Gamma(\mathcal{A})}\int_0^{\zeta}\xi'(\mathcal{A})(\xi(\zeta)-\xi(\ni))^{\mathcal{A}-1}\varphi_1^{\circledS}\Big((\xi(\ni)-\xi(0))^{\mathcal{A}}\Big)d\ni,\cdots,$$
$$\frac{\Bbbk_n\theta_n\Theta_n}{(1-\Theta_2)\Gamma(\mathcal{A})}\int_0^{\zeta}\xi'(\mathcal{A})(\xi(\zeta)-\xi(\ni))^{\mathcal{A}-1}\varphi_n^{\circledS}\Big((\xi(\ni)-\xi(0))^{\mathcal{A}}\Big)d\ni\Big]$$
$$\preceq\operatorname{diag}\Big[\frac{\Bbbk_1\theta_1\Theta_1}{(1-\Theta_2)}\varphi_1^{\circledS}\Big((\xi(\zeta)-\xi(0))^{\mathcal{A}}\Big),\cdots,\frac{\Bbbk_n\theta_n\Theta_1}{(1-\Theta_2)}\varphi_n^{\circledS}\Big((\xi(\zeta)-\xi(0))^{\mathcal{A}}\Big)\Big].$$

Then, we have

$$\mathcal{P}(\propto(\rho), \propto(\varrho)) \leq \underbrace{(\frac{\Theta_1}{(1-\Theta_2)}, \cdots, \frac{\Theta_1}{(1-\Theta_2)})}_{n} \ \mathcal{P}(\rho, \varrho). \tag{14.16}$$

Now, since $0 < \dfrac{\Theta_1}{(1-\Theta_2)} < 1$, we can conclude the contractively property of $\propto$.

Let $\Psi \in \mathcal{H}$. Since $\propto(\Psi) \in \mathcal{H}$, we have that

$$\begin{aligned}
&\operatorname{diag}\Big[\Big|\propto(\Psi(\zeta)) - \Psi(\zeta)\Big|, \cdots, \Big|\propto(\Psi(\zeta)) - \Psi(\zeta)\Big|\Big]_{n\times n} \\
&\preceq \operatorname{diag}\Big[\Big|\Psi(\zeta) - \frac{(\xi(\zeta)-\xi(0))^{\mathcal{C}-1}}{\Gamma(\mathcal{C})}J_\lambda - \frac{1}{\Gamma(\mathcal{A})}\int_0^\zeta \xi'(\mathcal{A})(\xi(\zeta)-\xi(\ni))^{\mathcal{A}-1}G_\Psi(\ni)d\ni\Big|, \cdots, \\
&\qquad \Big|\Psi(\zeta) - \frac{(\xi(\zeta)-\xi(0))^{\mathcal{C}-1}}{\Gamma(\mathcal{C})}J_\lambda - \frac{1}{\Gamma(\mathcal{A})}\int_0^\zeta \xi'(\mathcal{A})(\xi(\zeta)-\xi(\ni))^{\mathcal{A}-1}G_\Psi(\ni)d\ni\Big|\Big]_{n\times n} \\
&\preceq \operatorname{diag}\Big[\theta_1\varphi_1^{\circledS}\Big((\xi(\zeta)-\xi(0))^{\mathcal{A}}\Big), \cdots, \theta_n\varphi_n^{\circledS}\Big((\xi(\zeta)-\xi(0))^{\mathcal{A}}\Big)\Big],
\end{aligned}$$

for $\zeta \in \Delta$, which implies that

$$\mathcal{P}(\propto(\Psi), \Psi) \leq \underbrace{(1, \cdots, 1)}_{n}. \tag{14.17}$$

Hence, for every $k \in \mathbb{N}$ we have $\mathcal{P}(\propto^k(\Psi), \propto^{k+1}(\Psi)) < \underbrace{(+\infty, \cdots, +\infty)}_{n}$.

Now, we can apply the second part of Theorem 9.5 and so we can find a unique map $\Phi \in \{\tilde{\sigma} \in \mathcal{H} : \mathcal{P}(\propto \Psi, \tilde{\sigma}) < \underbrace{(+\infty, \cdots, +\infty)}_{n}\}$ such that $\propto \Phi = \Phi$ and so

$$\Phi(\zeta) = \frac{(\xi(\zeta)-\xi(0))^{\mathcal{C}-1}}{\Gamma(\mathcal{C})}J_\lambda + \frac{1}{\Gamma(\mathcal{A})}\int_0^\zeta \xi'(\ni)(\xi(\zeta)-\xi(\ni))^{\mathcal{A}-1}G_\Phi(\ni)d\ni, \tag{14.18}$$

for every $\zeta \in \Delta$, in which $G_\Phi \in C(\Delta, \mathbb{R})$ satisfies the functional equation (14.1) and $I_{\lambda^+}^{1-\mathcal{C};\xi}\Phi(\lambda^+) = J_\lambda \in \mathbb{R}$.

Using Theorem 9.5 and the inequality (14.17), we have

$$\begin{aligned}
\mathcal{P}(\Phi, \Psi) &\leq \underbrace{\left(\frac{1}{1-\frac{\Theta_1}{(1-\Theta_2)}}, \cdots, \frac{1}{1-\frac{\Theta_1}{(1-\Theta_2)}}\right)}_{n}\mathcal{P}(\propto(\Psi), \Psi) \\
&\leq \underbrace{\left(\frac{1}{1-\frac{\Theta_1}{(1-\Theta_2)}}, \cdots, \left(\frac{1}{1-\frac{\Theta_1}{(1-\Theta_2)}}\right)\right.}_{n},
\end{aligned}$$

which implies (14.8).

14.2 Multi Stability of Hilfer Fractional Systems II on Unbounded Domains

We prove the next theorem for $\mathbb{R}$. Similar reasoning establishes the result for $(-\infty, 0]$ and $[0, +\infty)$. Let $\mathsf{S} = C_{1-\mathcal{C};\xi}(\mathbb{R})$.

Theorem 14.2 *Let the assumptions* $(\mathcal{F}_1)$ *and* $(\mathcal{F}_2)$ *be satisfied. If* Ψ *in* S *satisfies the inequality* (14.4) *and* $I_{\lambda^+}^{1-\mathcal{C};\xi}\Psi(\lambda^+) = J_\lambda$*, then, there exists a unique function* Φ *satisfying the fractional equations* (14.2)–(14.3) *such that the inequality* (14.8) *is satisfied for every* $\zeta \in \mathbb{R}$.

Proof For every $k \in \mathbb{N}$, we consider $\mathsf{P}_k = [\mathfrak{I} - k, \mathfrak{I} + k]$. From Theorem 14.1, there is a unique function $\Phi_k \in C_{1-\mathcal{C};\xi}(\mathsf{P}_k)$, such that

$$^{\mathscr{H}}D_{\lambda^+}^{\mathcal{A},\mathcal{B};\xi}\Phi_k(\zeta) = \mathcal{C}(\zeta, \Phi_k(\zeta), \Phi_k(\eta(\zeta))), \qquad \zeta \in \mathsf{P}_k, \tag{14.19}$$

$$I_{\lambda^+}^{1-\mathcal{C};\xi}\Phi_k(\lambda^+) = J_\lambda, \qquad J_\lambda \in \mathbb{R}, \tag{14.20}$$

and

$$\begin{aligned}&\operatorname{diag}\Big[\,|\Phi_k(\zeta) - \Psi(\zeta)|, \cdots, |\Phi_k(\zeta) - \Psi(\zeta)|\,\Big]_{n\times n}\\ &\preceq \operatorname{diag}\left[\frac{\theta_1}{1-\dfrac{\Theta_1}{(1-\Theta_2)}}\varphi_1^{\circledS}\Big((\xi(\zeta)-\xi(0))^{\mathcal{A}}\Big), \cdots, \frac{\theta_n}{1-\dfrac{\Theta_1}{(1-\Theta_2)}}\varphi_n^{\circledS}\Big((\xi(\zeta)-\xi(0))^{\mathcal{A}}\Big)\right],\end{aligned} \tag{14.21}$$

for every $\zeta \in \mathsf{P}_k$ and $n \in \mathbb{N}$, The uniqueness of Φ_k implies that if $\zeta \in \mathsf{P}_k$, then,

$$\Phi_k(\zeta) = \Phi_{k+1}(\zeta) = \Phi_{k+2}(\zeta) = \cdots. \tag{14.22}$$

Define $k(\zeta) \in \mathbb{N}$ as

$$k(\zeta) = \min\{k \in \mathbb{N} \mid \zeta \in \mathsf{P}_k\}.$$

Moreover, define a function Φ, by

$$\Phi(\zeta) = \Phi_{k(\zeta)}(\zeta), \quad \zeta \in \mathbb{R},$$

and we claim that $\Phi \in \mathsf{S}$. For a real number ζ_1, we consider the integer $k_1 = k(\zeta_1)$. Thus, ζ_1 belongs to the interior of P_{k_1+1} and there exists an $\epsilon > 0$ that $\Phi(\zeta) = \Phi_{k_1+1}(\zeta)$, for every ζ with $\zeta_1 - \epsilon < \zeta < \zeta_1 + \epsilon$. Next, we want to prove that Φ satisfies the fractional equations (14.2) and (14.3) and the inequality (14.8), for every $\zeta \in \mathbb{R}$. For every $\zeta \in \mathbb{R}$, consider the integer $k(\zeta)$. Thus $\zeta \in \mathsf{P}_{k(\zeta)}$ and we conclude from the Eqs. (14.19) and (14.20) that

$$\begin{aligned}\Phi(\zeta) &= \Phi_{k(\zeta)}(\zeta) \\ &= \frac{(\xi(\zeta)-\xi(0))^{\mathcal{C}-1}}{\Gamma(\mathcal{C})}J_\lambda + \frac{1}{\Gamma(\mathcal{A})}\int_0^\zeta \xi'(\mathcal{A})(\xi(\zeta)-\xi(\ni))^{\mathcal{A}-1}G_{\Phi_{n(\zeta)}}(\ni)d\ni \\ &= \frac{(\xi(\zeta)-\xi(0))^{\mathcal{C}-1}}{\Gamma(\mathcal{C})}J_\lambda + \frac{1}{\Gamma(\mathcal{A})}\int_0^\zeta \xi'(\mathcal{A})(\xi(\zeta)-\xi(\ni))^{\mathcal{A}-1}G_{\Phi}(\ni)d\ni,\end{aligned} \tag{14.23}$$

where

$$G_\Phi(\zeta) = F(\zeta, \Phi(\zeta), G_\Phi(\zeta))$$

and

$$G_{\Phi_{k(\zeta)}}(\zeta) = F(\zeta, \Phi_{k(\zeta)}(\zeta), G_{\Phi_{k(\zeta)}}(\zeta))$$

are in $C(\Delta, \mathbb{R})$. The above equality is true because $k(\ni) \le k(\zeta)$, for every $\ni \in \mathsf{P}_{k(\zeta)}$ and we conclude from the relation (14.22) that

$$\Phi(\ni) = \Phi_{k(\ni)}(\ni) = \Phi_{k(\zeta)}(\ni). \tag{14.24}$$

Since $\Phi(\zeta) = \Phi_{k(\zeta)}(\zeta)$ and $\zeta \in \mathsf{P}_{k(\zeta)}$, for every $\zeta \in \mathbb{R}$, the inequality (14.21) implies that

$$\begin{aligned}&\operatorname{diag}\Big[|\Phi(\zeta)-\Psi(\zeta)|, \cdots, |\Phi(\zeta)-\Psi(\zeta)|\Big]_{n\times n} \\ &= \operatorname{diag}\Big[|\Phi_{k(\zeta)}(\zeta)-\Psi(\zeta)|, \cdots, |\Phi_{k(\zeta)}(\zeta)-\Psi(\zeta)|\Big]_{n\times n} \\ &\preceq \operatorname{diag}\left[\frac{\theta_1}{1-\frac{\Theta_1}{(1-\Theta_2)}}\varphi_1^{\circledS}\Big((\xi(\zeta)-\xi(0))^{\mathcal{A}}\Big), \cdots, \frac{\theta_n}{1-\frac{\Theta_1}{(1-\Theta_2)}}\varphi_n^{\circledS}\Big((\xi(\zeta)-\xi(0))^{\mathcal{A}}\Big)\right].\end{aligned} \tag{14.25}$$

Finally, we claim that Φ is unique. Let $\Phi' \in \mathsf{S}$ be another function, which satisfies the Eqs. (14.2), (14.3) and the inequality (14.8), for every $\zeta \in \mathbb{R}$. Since $\Phi|_{\mathsf{P}_{k(\zeta)}} (= \Phi_{k(\zeta)})$ and $\Phi'|_{\mathsf{P}_{k(\zeta)}}$ both satisfy (14.2), (14.3) and (14.8), for every $\zeta \in \mathsf{P}_{k(\zeta)}$, the uniqueness of $\Phi_{k(\zeta)} = \Phi|_{\mathsf{P}_{k(\zeta)}}$ implies that

$$\Phi(\zeta) = \Phi|_{\mathsf{P}_{k(\zeta)}}(\zeta) = \Phi'|_{\mathsf{P}_{k(\zeta)}}(\zeta) = \Phi'(\zeta),$$

as required.

14.3 Application

Example 14.1 Consider the fractional system (14.1), for $\xi(\zeta) = \zeta^2$, $\lambda = 0$, $D = 1$, $\mathcal{A} = \mathcal{B} = \frac{1}{2}$, $\Phi : [0, 1] \to \mathbb{R}$ and the continuous function $F : [0, 1] \times \mathbb{R}^2 \to \mathbb{R}$

given by

$$F(\zeta, u, w) = \frac{1}{5}{}_2\mathbb{F}_1(\mathsf{A}, \mathsf{B}, \mathsf{B}; \zeta)u + \frac{1}{8}w,$$

and we have

$$\begin{cases} {}^{\mathscr{H}}D_{0^+}^{\frac{1}{2},\frac{1}{2};\zeta^2}\Phi(\zeta) = \frac{1}{5}{}_2\mathbb{F}_1\left(\mathsf{A}, \mathsf{B}, \mathsf{B}; \zeta\right)\Phi(\zeta) + \frac{1}{8}{}^{\mathscr{H}}D_{0^+}^{\frac{1}{2},\frac{1}{2};\zeta^2}\Phi(\zeta), \\ I_{0^+}^{\frac{1}{4};\zeta^2}\Phi(0^+) = J_0 \in \mathbb{R}, \end{cases} \tag{14.26}$$

where $\mathsf{A} \in \mathbb{R}^+$ and $\mathsf{B} \in \mathbb{R}$, and ${}_2\mathbb{F}_1(\mathsf{A}, \mathsf{B}, \mathsf{B}; .)$ is the Hypergeometric function defined in Chap. 2.

For every $\kappa, \upsilon, \overline{\kappa}, \overline{\upsilon} \in \mathbb{R}$ and $\zeta \in [0, 1]$, we get

$$\begin{aligned} &|F(\zeta, \kappa, \upsilon) - F(\zeta, \overline{\kappa}, \overline{\upsilon})| \\ =&\left|\frac{1}{5}{}_2\mathbb{F}_1\left(\mathsf{A}, \mathsf{B}, \mathsf{B}; \zeta\right)\kappa + \frac{1}{8}{}^{\mathscr{H}}D_{0^+}^{\frac{1}{2},\frac{1}{2};\zeta^2}\upsilon - \frac{1}{5}{}_2\mathbb{F}_1\left(\mathsf{A}, \mathsf{B}, \mathsf{B}; \zeta\right)\overline{\kappa} - \frac{1}{8}{}^{\mathscr{H}}D_{0^+}^{\frac{1}{2},\frac{1}{2};\zeta^2}\overline{\upsilon}\right| \\ \leq&\frac{1}{5}{}_2\mathbb{F}_1\left(\mathsf{A}, \mathsf{B}, \mathsf{B}; \zeta\right)|\kappa - \overline{\kappa}| + \frac{1}{8}|\upsilon - \overline{\upsilon}|. \end{aligned}$$

Thus, the hypothesis $(\mathcal{F}_2)$ is satisfied with $\Theta_1 = \frac{1}{5}{}_2\mathbb{F}_1\left(\mathsf{A}, \mathsf{B}, \mathsf{B}; \zeta\right)$ and $\Theta_2 = \frac{1}{8}$. On the other hand, the condition

$$\frac{8}{35}{}_2\mathbb{F}_1\left(\mathsf{A}, \mathsf{B}, \mathsf{B}; \zeta\right) < 1,$$

is satisfied. Now, Theorem 14.1 implies that problem (14.26) has a unique solution and also is the multi stable with

$$\begin{aligned} &\operatorname{diag}\Big[|\Phi(\zeta) - \Psi(\zeta)|, \cdots, |\Phi(\zeta) - \Psi(\zeta)|\Big]_{n\times n} \\ &\preceq \operatorname{diag}\left[\frac{\theta_1}{1 - \frac{8}{35}{}_2\mathbb{F}_1\left(\mathsf{A}, \mathsf{B}, \mathsf{B}; \zeta\right)}\varphi_1^{\circledS}(\zeta), \cdots, \frac{\theta_n}{1 - \frac{8}{35}{}_2\mathbb{F}_1\left(\mathsf{A}, \mathsf{B}, \mathsf{B}; \zeta\right)}\varphi_n^{\circledS}(\zeta)\right]_{n\times n}, \end{aligned}$$

where $\zeta \in [0, 1]$.

We refer the interested readers to [2–7].

References

1. Aderyani SR, Saadati R, ORegan D (2022) The Cadariu-Radu method for existence, uniqueness and Gauss Hypergeometric stability of a class of Ξ-Hilfer fractional differential equations. Int J Nonlinear Sci Numer Simul
2. Aderyani SR, Saadati R, O'Regan D, Alshammari FS (2022) Multi-super-stability of antiderivations in Banach algebras. AIMS Math 7(11):20143–20163
3. Aderyani SR, Saadati R, ORegan D, Alshammari FS (2023) Application of aggregated control functions for approximating $\mathscr{C}$-Hilfer fractional differential equations. AIMS Math 8(11):28010–28032
4. Aderyani SR, Saadati R, Rassias TM, Srivastava HM (2023) Existence, uniqueness and the multi-stability results for a w-Hilfer fractional differential equation. Axioms 12(7):681
5. Rezaei Aderyani S, Saadati R, Li C, Rassias TM, Park C (2023) Special functions and multi-stability of the Jensen type random operator equation in C^*-algebras via fixed point. J Inequalities Appl 2023(1):1–24
6. Aderyani SR, Saadati R, Feckan M (2021) The Cadariu-Radu method for existence. Uniqueness and Gauss hypergeometric stability of W-Hilfer fractional differential equations. Mathematics 9:1408
7. Aderyani SR, Saadati R, ORegan D, Abdeljawad T (2022) UHML stability of a class of Δ-Hilfer FDEs via CRM. AIMS Math 7(4):5910–5919

Chapter 15
Fuzzy Topological Spaces

Fuzzy mathematics as fuzzy sets and fuzzy logics is a new branch of mathematics which has been found multiple uses across numerous fields such as control systems, engineering, image processing, power engineering, industrial automation, robotics, consumer electronics, and optimization. Fuzzy concept is considered to be an important aspect in the field of mathematics that interprets the uncertainties, and deals with the unreliable information and vagueness of data. Fuzzy set theory is a particularly useful tool to describe situations in which the data are imprecise or vague. The main goal of this chapter is to introduce basic and standard properties often required for fuzzy spaces.

15.1 Generalized Triangular Norms

The purpose of this section is to supply the basic definitions of triangular norms that are required for a proper reading of the rest of the volume.

The history of triangular norms started with the paper "Statistical metrics" by Karl Menger [1]. The main idea of Menger was to construct metric spaces where probability distributions rather than numbers are used in order to describe the distance between two elements of the space in question. Triangular norms (t–norms for short) naturally came into the picture in the course of the generalization of the classical triangle inequality to this more general setting. Consequently, the first field where t–norms played a major role was the theory of probabilistic metric spaces (as statistical metric spaces were called after 1964). Berthold Schweizer and Abe Sklar in [2–4] provided the axioms of t–norms, as they are used today, and a redefinition of statistical metric spaces given in [5] led to a rapid development of the field.

Definition 15.1 ([6]) A triangular norm (shortly, t–norm) is a binary operation on the unit interval [0, 1], i.e., a function $\bigotimes_{TN} : [0, 1] \times [0, 1] \to [0, 1]$, such that, the following four axioms are satisfied:

S. Rezaei Aderyani et al., *Towards Ulam Type Multi Stability Analysis*, Studies in Systems, Decision and Control 523, https://doi.org/10.1007/978-3-031-55564-0_15

(T1) $\mathcal{F} \bigotimes_{\text{TN}} \mathcal{H} = \mathcal{H} \bigotimes_{\text{TN}} \mathcal{F}$, for every $\mathcal{F}, \mathcal{H} \in [0, 1]$, (commutativity);

(T2) $\mathcal{F} \bigotimes_{\text{TN}} (\mathcal{P} \bigotimes_{\text{TN}} \mathcal{H}) = (\mathcal{F} \bigotimes_{\text{TN}} \mathcal{P}) \bigotimes_{\text{TN}} \mathcal{H}$, for every $\mathcal{F}, \mathcal{H}, \mathcal{P} \in [0, 1]$ (associativity);

(T3) $\mathcal{F} \bigotimes_{\text{TN}} 1 = \mathcal{F}$, for every $\mathcal{F} \in [0, 1]$, (boundary condition);

(T4) $\mathcal{F} \bigotimes_{\text{TN}} \mathcal{P} \leq \mathcal{F}' \bigotimes_{\text{TN}} \mathcal{P}'$, whenever $\mathcal{F} \leq \mathcal{F}'$ and $\mathcal{P} \leq \mathcal{P}'$, for every $\mathcal{F}, \mathcal{F}', \mathcal{P}, \mathcal{P}' \in [0, 1]$, (monotonicity).

The commutativity of (T1), the boundary condition (T3), and the monotonicity (T4) imply that, for every t–norm $\bigotimes_{\text{TN}}$ and $\mathcal{F} \in [0, 1]$, the following boundary conditions are also satisfied:

$$\mathcal{F} \bigotimes_{\text{TN}} 1 = 1 \bigotimes_{\text{TN}} \mathcal{F} = \mathcal{F},$$

$$\mathcal{F} \bigotimes_{\text{TN}} 0 = 0 \bigotimes_{\text{TN}} \mathcal{F} = 0,$$

and so all the t–norms coincide with the boundary of the unit square $[0, 1]^2$.

The monotonicity of a t–norm $\bigotimes_{\text{TN}}$ in the second component (T4) is, together with the commutativity (T1), equivalent to the (joint) monotonicity in both components, that is,

$$\mathcal{F} \bigotimes_{\text{TN}} \mathcal{P} \leq \mathcal{F}' \bigotimes_{\text{TN}} \mathcal{P}', \tag{15.1}$$

whenever $\mathcal{F} \leq \mathcal{F}'$ and $\mathcal{P} \leq \mathcal{P}'$.

Basic examples are the Lukasiewicz t–norm $\bigotimes_{\text{TN}}^{L}$:

$$\mathcal{F} \bigotimes_{\text{TN}}^{L} \mathcal{H} := \max\{\mathcal{F} + \mathcal{H} - 1, 0\},$$

for every $\mathcal{F}, \mathcal{H} \in [0, 1]$, and the t–norms $\bigotimes_{\text{TN}}^{P}$, $\bigotimes_{\text{TN}}^{M}$ and $\bigotimes_{\text{TN}}^{D}$ are defined as follows:

$$\mathcal{F} \bigotimes_{\text{TN}}^{P} \mathcal{H} := \mathcal{F}\mathcal{H} \text{ (: the product t–norm)},$$

$$\mathcal{F} \bigotimes_{\text{TN}}^{M} \mathcal{H} := \wedge\{\mathcal{F}, \mathcal{H}\} \text{ (: the minimum t–norm)},$$

and

$$\mathcal{F} \bigotimes_{\text{TN}}^{D} \mathcal{H} := \begin{cases} \min\{\mathcal{F}, \mathcal{H}\}, & \text{if } \max\{\mathcal{F}, \mathcal{H}\} = 1, \\ 0, & \text{otherwise}, \end{cases} \text{ (: the drastic t–norm)},$$

for every $\mathcal{F}, \mathcal{H} \in [0, 1]$.

If, for every two t–norms $\bigotimes^{1}_{\mathrm{TN}}$ and $\bigotimes^{2}_{\mathrm{TN}}$, the inequality $\mathcal{F} \bigotimes^{1}_{\mathrm{TN}} \mathcal{H} \leq \mathcal{F} \bigotimes^{2}_{\mathrm{TN}} \mathcal{H}$ holds for every $\mathcal{F}, \mathcal{H} \in [0, 1]$, then, we say that $\bigotimes^{1}_{\mathrm{TN}}$ is weaker than $\bigotimes^{2}_{\mathrm{TN}}$ or, equivalently, $\bigotimes^{2}_{\mathrm{TN}}$ is stronger than $\bigotimes^{1}_{\mathrm{TN}}$.

From (15.1), it follows that, for every $\mathcal{F}, \mathcal{H} \in [0, 1]$,

$$\mathcal{F} \bigotimes_{\mathrm{TN}} \mathcal{H} \leq \mathcal{F} \bigotimes_{\mathrm{TN}} 1 = \mathcal{F}, \quad \text{and}$$
$$\mathcal{F} \bigotimes_{\mathrm{TN}} \mathcal{H} \leq 1 \bigotimes_{\mathrm{TN}} \mathcal{H} = \mathcal{H}.$$

Since $\mathcal{F} \bigotimes_{\mathrm{TN}} \mathcal{H} \geq 0 = \mathcal{F} \bigotimes^{D}_{\mathrm{TN}} \mathcal{H}$, for every $\mathcal{F}, \mathcal{H} \in [0, 1]$, holds trivially, for any t–norm $\bigotimes_{\mathrm{TN}}$, we have

$$\bigotimes^{D}_{\mathrm{TN}} \leq \bigotimes_{\mathrm{TN}} \leq \bigotimes^{M}_{\mathrm{TN}},$$

that is, $\bigotimes^{D}_{\mathrm{TN}}$ is weaker and $\bigotimes^{M}_{\mathrm{TN}}$ is stronger than any other t–norms. Also, since $\bigotimes^{L}_{\mathrm{TN}} \leq \bigotimes^{P}_{\mathrm{TN}}$, we obtain the following ordering for four basic t–norms:

$$\bigotimes^{D}_{\mathrm{TN}} \leq \bigotimes^{L}_{\mathrm{TN}} \leq \bigotimes^{P}_{\mathrm{TN}} \leq \bigotimes^{M}_{\mathrm{TN}}.$$

Proposition 15.1 ([6]) *We have the following results:*

- *The minimum $\bigotimes^{M}_{\mathrm{TN}}$ is the only t–norm satisfying $\mathcal{F} \bigotimes_{\mathrm{TN}} \mathcal{F} = \mathcal{F}$, for every $\mathcal{F} \in [0, 1]$.*
- *The weakest t–norm $\bigotimes^{D}_{\mathrm{TN}}$ is the only t–norm satisfying $\mathcal{F} \bigotimes_{\mathrm{TN}} \mathcal{F} = 0$, for every $\mathcal{F} \in [0, 1]$.*

Proposition 15.2 ([6]) *A t–norm $\bigotimes_{\mathrm{TN}}$ is continuous if and only if it is continuous in its first component, i.e., for all $\mathcal{H} \in [0, 1]$, if the one–place function*

$$\bigotimes_{\mathrm{TN}}(., \mathcal{H}) : [0, 1] \longrightarrow [0, 1], \quad \mathcal{F} \mapsto \mathcal{F} \bigotimes_{\mathrm{TN}} \mathcal{H},$$

is continuous.

For example, the minimum $\bigotimes^{M}_{\mathrm{TN}}$ and Lukasiewicz t–norm $\bigotimes^{L}_{\mathrm{TN}}$ are continuous, but the t–norm $\bigotimes^{\Delta}_{\mathrm{TN}}$ defined by

$$\mathcal{F} \bigotimes^{\Delta}_{\mathrm{TN}} \mathcal{H} := \begin{cases} \frac{\mathcal{F}\mathcal{H}}{2} & \text{if } \max\{\mathcal{F}, \mathcal{H}\} < 1, \\ \mathcal{F}\mathcal{H}, & \text{otherwise.} \end{cases}$$

for every $\mathcal{F}, \mathcal{H} \in [0, 1]$, is not continuous.

Definition 15.2 ([6]) We have the following definitions:

- A t–norm $\bigotimes_{\mathrm{TN}}$ is said to be strictly monotone if

$$\mathcal{F} \bigotimes_{\mathrm{TN}} \mathcal{H} < \mathcal{F} \bigotimes_{\mathrm{TN}} \mathcal{P},$$

  whenever $\mathcal{F} \in [0, 1]$ and $\mathcal{H} < \mathcal{P}$.
- A t–norm $\bigotimes_{\mathrm{TN}}$ is said to be strict if it is continuous and strictly monotone.

For example, the t–norm $\bigotimes_{\mathrm{TN}}^{\Delta}$ is strictly monotone, but the minimum $\bigotimes_{\mathrm{TN}}^{M}$ and Lukasiewicz t–norm $\bigotimes_{\mathrm{TN}}^{L}$ are not strictly monotone.

Proposition 15.3 ([6]) *A t–norm* $\bigotimes_{\mathrm{TN}}$ *is strictly monotone if and only if*

$$\mathcal{F} \bigotimes_{\mathrm{TN}} \mathcal{H} = \mathcal{F} \bigotimes_{\mathrm{TN}} \mathcal{P},\ \mathcal{F} > 0 \Longrightarrow \mathcal{H} = \mathcal{P}.$$

Now, we extend the concept of triangular norms mentioned above on diagonal matrices.

Let $n \in \mathbb{N}$, and consider the square matrix diag$\mathcal{M}_n$ on closed interval $[0, 1]$ defined by

$$\text{diag}\mathcal{M}_n([0, 1]) = \left\{ \begin{bmatrix} \mathscr{H}_1 & & \\ & \ddots & \\ & & \mathscr{H}_n \end{bmatrix}_{n\times n} = \text{diag}[\mathscr{H}_1, \ldots, \mathscr{H}_n],\ \mathscr{H}_1, ..., \mathscr{H}_n \in [0, 1] \right\},$$

where diag$\mathcal{M}_n([0, 1])$ is equipped with the partial order relation:

$$\begin{aligned} &\mathscr{H} := \text{diag}[\mathscr{H}_1, \ldots, \mathscr{H}_n], \mathscr{G} := \text{diag}[\mathscr{G}_1, \ldots, \mathscr{G}_n] \in \text{diag}\mathcal{M}_n([0, 1]), \\ &\mathscr{H} \preceq \mathscr{G} \Longleftrightarrow \mathscr{H}_i \leq \mathscr{G}_i, \quad \text{for every } i \in \mathbb{N}. \end{aligned}$$

Also, $\mathscr{H} \prec \mathscr{G}$ denotes that $\mathscr{H} \preceq \mathscr{G}$ and $\mathscr{H} \neq \mathscr{G}$; $\mathscr{H} \ll \mathscr{G}$ and $\mathscr{H}_{\mathrm{i}} < \mathscr{G}_{\mathrm{i}}$, for any $i \in \mathbb{N}$.

We will often use the bold symbols **1** and **0** to denote the following an identify matrix (square unit matrix) and a square zero matrix given by

$$\mathbf{1} := \begin{bmatrix} 1 & & \\ & \ddots & \\ & & 1 \end{bmatrix}_{n\times n} = \text{diag}[1, \ldots, 1],$$

and

$$\mathbf{0} := \begin{bmatrix} 0 & & \\ & \ddots & \\ & & 0 \end{bmatrix}_{n\times n} = \text{diag}[0, \ldots, 0].$$

We now present generalized triangular norm $\bigotimes_{\text{GTN}}$ on diag$\mathcal{M}_n([0, 1])$, as follows:

Definition 15.3 ([7, 8]) A generalized triangular norm (in short, generalized t–norm) on diag$\mathcal{M}_n([0, 1])$ is an operation $\bigotimes_{\text{GTN}}$: diag$\mathcal{M}_n([0, 1]) \times$ diag$\mathcal{M}_n([0, 1]) \to$ diag$\mathcal{M}_n([0, 1])$ satisfying the following conditions:

(1) $(\forall \mathscr{H} \in \text{diag}\mathcal{M}_n([0, 1]))(\mathscr{H} \bigotimes_{\text{GTN}} \mathbf{1}) = \mathscr{H})$ (boundary condition);

(2) $(\forall(\mathscr{H}, \mathscr{G}) \in (\text{diag}\mathcal{M}_n([0, 1]))^2)(\mathscr{H} \bigotimes_{\text{GTN}} \mathscr{G} = \mathscr{G} \bigotimes_{\text{GTN}} \mathscr{H})$ (commutativity);

(3) $(\forall(\mathscr{H}, \mathscr{G}, \mathscr{K}) \in (\text{diag}\mathcal{M}_n([0, 1])^3)(\mathscr{H} \bigotimes_{\text{GTN}}(\mathscr{G} \bigotimes_{\text{GTN}} \mathscr{K}) = (\mathscr{H} \bigotimes_{\text{GTN}} \mathscr{G}) \bigotimes_{\text{GTN}} \mathscr{K})$ (associativity);

(4) $(\forall(\mathscr{H}, \mathscr{H}', \mathscr{G}, \mathscr{G}') \in (\text{diag}\mathcal{M}_n([0, 1]^4)(\mathscr{H} \preceq \mathscr{H}' \text{ and } \mathscr{G} \preceq \mathscr{G}' \Longrightarrow \mathscr{H} \bigotimes_{\text{GTN}} \mathscr{G} \preceq \mathscr{H}' \bigotimes_{\text{GTN}} \mathscr{G}'$ (monotonicity).

Note 15.1 ([7, 8]) For every $\mathscr{H}, \mathscr{G} \in \text{diag}\mathcal{M}_n([0, 1])$ and every sequences $\{\mathscr{H}_k\}$ and $\{\mathscr{G}_k\}$ converging to $\mathscr{H}$ and $\mathscr{G}$, let we have

$$\lim_k(\mathscr{H}_k \bigotimes_{\text{GTN}} \mathscr{G}_k) = \mathscr{H} \bigotimes_{\text{GTN}} \mathscr{G},$$

then, the generalized triangular norm $\bigotimes_{\text{GTN}}$ on diag$\mathcal{M}_n([0, 1])$ is a continuous generalized triangular norm.

Here, we generalize triangular norms $\bigotimes^P_{\text{TN}}$, $\bigotimes^M_{\text{TN}}$ and $\bigotimes^L_{\text{TN}}$ on diag$\mathcal{M}_n([0, 1])$, as follows:

(1) Define $\bigotimes^P_{\text{GTN}}$: diag$\mathcal{M}_n([0, 1]) \times$ diag$\mathcal{M}_n([0, 1]) \to$ diag$\mathcal{M}_n([0, 1])$, such that,

$$\begin{aligned}\mathscr{H} \bigotimes^P_{\text{GTN}} \mathscr{G} &= \text{diag}[\mathscr{H}_1, \ldots, \mathscr{H}_n] \bigotimes^P_{\text{GTN}} \text{diag}[\mathscr{G}_1, \ldots, \mathscr{G}_n] \\ &= \text{diag}[\mathscr{H}_1.\mathscr{G}_1, \ldots, \mathscr{H}_n.\mathscr{G}_n],\end{aligned}$$

then, $\bigotimes^P_{\text{GTN}}$ is a product continuous generalized triangular norm.

(2) Define $\bigotimes^M_{\text{GTN}}$: diag$\mathcal{M}_n([0, 1]) \times$ diag$\mathcal{M}_n([0, 1]) \to$ diag$\mathcal{M}_n([0, 1])$, such that,

$$\begin{aligned}\mathscr{H} \bigotimes^M_{\text{GTN}} \mathscr{G} &= \text{diag}[\mathscr{H}_1, \ldots, \mathscr{H}_n] \bigotimes^M_{\text{GTN}} \text{diag}[\mathscr{G}_1, \ldots, \mathscr{G}_n] \\ &= \text{diag}[\min\{\mathscr{H}_1, \mathscr{G}_1\}, \cdots, \min\{\mathscr{H}_n, \mathscr{G}_n\}],\end{aligned}$$

then, $\bigotimes_{\text{GTN}}^{M}$ is a minimum continuous generalized triangular norm.

(3) Define $\bigotimes_{\text{GTN}}^{L} : \text{diag}\mathcal{M}_n([0,1]) \times \text{diag}\mathcal{M}_n([0,1]) \to \text{diag}\mathcal{M}_n([0,1])$, such that,

$$\begin{aligned}\mathscr{H} \overset{L}{\underset{\text{GTN}}{\bigotimes}} \mathscr{G} &= \text{diag}[\mathscr{H}_1, \cdots, \mathscr{H}_n] \overset{L}{\underset{\text{GTN}}{\bigotimes}} \text{diag}[\mathscr{G}_1, \cdots, \mathscr{G}_n] \\ &= \text{diag}[\max\{\mathscr{H}_1 + \mathscr{G}_1 - 1, 0\}, \cdots, \max\{\mathscr{H}_n + \mathscr{G}_n - 1, 0\}],\end{aligned}$$

then, $\bigotimes_{\text{GTN}}^{L}$ is a Lukasiewicz continuous generalized triangular norm.

Now, we present some numerical examples of product generalized t–norm, minimum generalized t–norm and Lukasiewicz generalized t–norm and compare the results.

$$\text{diag}\left[\frac{1}{2}, \frac{1}{5}, 1\right] \overset{M}{\underset{\text{GTN}}{\bigotimes}} \text{diag}\left[\frac{3}{10}, \frac{7}{10}, 0\right] = \begin{bmatrix} \frac{1}{2} & & \\ & \frac{1}{5} & \\ & & 1 \end{bmatrix} \overset{M}{\underset{\text{GTN}}{\bigotimes}} \begin{bmatrix} \frac{3}{10} & & \\ & \frac{7}{10} & \\ & & 0 \end{bmatrix} = \begin{bmatrix} \frac{3}{10} & & \\ & \frac{1}{5} & \\ & & 0 \end{bmatrix},$$

$$\text{diag}\left[\frac{1}{2}, \frac{1}{5}, 1\right] \overset{P}{\underset{\text{GTN}}{\bigotimes}} \text{diag}\left[\frac{3}{10}, \frac{7}{10}, 0\right] = \begin{bmatrix} \frac{1}{2} & & \\ & \frac{1}{5} & \\ & & 1 \end{bmatrix} \overset{P}{\underset{\text{GTN}}{\bigotimes}} \begin{bmatrix} \frac{3}{10} & & \\ & \frac{7}{10} & \\ & & 0 \end{bmatrix} = \begin{bmatrix} \frac{3}{20} & & \\ & \frac{7}{50} & \\ & & 0 \end{bmatrix},$$

and

$$\text{diag}\left[\frac{1}{2}, \frac{1}{5}, 1\right] \overset{L}{\underset{\text{GTN}}{\bigotimes}} \text{diag}\left[\frac{3}{10}, \frac{7}{10}, 0\right] = \begin{bmatrix} \frac{1}{2} & & \\ & \frac{1}{5} & \\ & & 1 \end{bmatrix} \overset{L}{\underset{\text{GTN}}{\bigotimes}} \begin{bmatrix} \frac{3}{10} & & \\ & \frac{7}{10} & \\ & & 0 \end{bmatrix} = \begin{bmatrix} 0 & & \\ & 0 & \\ & & 0 \end{bmatrix}.$$

Also, since

$$\text{diag}\left[\frac{3}{10}, \frac{1}{5}, 0\right] \succeq \text{diag}\left[\frac{3}{20}, \frac{7}{50}, 0\right] \succeq \text{diag}\,[0, 0, 0]\,,$$

we have that the minimum generalized t–norm $\bigotimes_{\text{GTN}}^{M}$ is stronger than any other generalized t–norms, as given below:

$$\text{diag}\left[\frac{1}{2}, \frac{1}{5}, 1\right] \overset{M}{\underset{\text{GTN}}{\bigotimes}} \text{diag}\left[\frac{3}{10}, \frac{7}{10}, 0\right]$$

$$\succeq \operatorname{diag}\left[\frac{1}{2}, \frac{1}{5}, 1\right] \underset{\text{GTN}}{\overset{P}{\bigotimes}} \operatorname{diag}\left[\frac{3}{10}, \frac{7}{10}, 0\right]$$

$$\succeq \operatorname{diag}\left[\frac{1}{2}, \frac{1}{5}, 1\right] \underset{\text{GTN}}{\overset{L}{\bigotimes}} \operatorname{diag}\left[\frac{3}{10}, \frac{7}{10}, 0\right].$$

15.2 Matrix Valued Fuzzy Normed Spaces

In this section, we recall some definitions and results as fuzzy sets, fuzzy normed spaces which will be used later on in the book.

The classical set theory is built on the fundamental concept of "set" of which an individual is either a member or not a member. A sharp, crisp, and unambiguous distinction exists between a member and a nonmember for any well–defined "set" of entities in this theory, and there is a very precise and clear boundary to indicate if an entity belongs to the set. In other words, when one asks the question "Is this entity a member of that set?" The answer is either "yes" or "no." This is true for both the deterministic and the stochastic cases. In probability and statistics, one may ask a question like "What is the probability of this entity being a member of that set?" In this case, although an answer could be like "The probability for this entity to be a member of that set is 90%," the final outcome (i.e., conclusion) is still either "it is" or "it is not" a member of the set. The chance for one to make a correct prediction as "it is a member of the set" is 90%, which does not mean that it has 90% membership in the set and in the meantime it possesses 10% non–membership. Namely, in the classical set theory, it is not allowed that an element is in a set and not in the set at the same time. Thus, many real–world application problems cannot be described and handled by the classical set theory, including all those involving elements with only partial membership of a set. On the contrary, fuzzy set theory accepts partial memberships, and, therefore, in a sense generalizes the classical set theory to some extent. Fuzzy set theory was proposed by Zadeh in 1965 as an extension of the classical notion of a set [9]. With the proposed methodology, Zadeh introduced a mathematic method with which decision–making using fuzzy descriptions of some information becomes possible [10].

In mathematics, fuzzy sets or uncertain sets are sets whose elements have degrees of membership. In classical set theory, the membership of elements in a set is assessed in binary terms (true and false) according to a bivalent condition–an element either belongs or does not belong to the set. By contrast, fuzzy set theory permits the gradual assessment of the membership of elements in a set with the aid of a membership function valued in the real unit interval [0,1]. Fuzzy sets generalize classical bivalent sets or crisp sets, since the indicator functions or characteristic functions of classical sets that map elements of the subsets of X to one, and all other elements to zero, are special cases of the membership functions of fuzzy sets, if the latter only takes values 0 or 1.

In the sequel, we shall adopt usual terminology, notation, and conventions of the theory of fuzzy normed spaces.

Consider the vector space $\mathcal{W}$. Let Λ denotes the space of all fuzzy sets, that is, the space of all mappings $\mathcal{N} : \mathcal{W} \times (0, +\infty) \to [0, 1]$ satisfying the following axioms:

- $\mathcal{N}(\mathcal{V}, .) : (0, +\infty) \to [0, 1]$ is continuous, for every $\mathcal{V} \in \mathcal{W}$;
- $\mathcal{N}(\mathcal{V}, .)$ is non–decreasing, for every $\mathcal{V} \in \mathcal{W}$;
- $\lim_{\mathcal{T}\to+\infty} \mathcal{N}(\mathcal{V}, \mathcal{T}) = 1$, and $\lim_{\mathcal{T}\to 0} \mathcal{N}(\mathcal{V}, \mathcal{T}) = 0$, for every $\mathcal{V} \in \mathcal{W}$ and $\mathcal{T} > 0$.

Definition 15.4 ([7, 8]) Consider the continuous t–norm $\bigotimes_{\mathrm{TN}}$, the vector space $\mathcal{W}$ and the fuzzy set $\mathcal{N} : \mathcal{W} \times (0, +\infty) \to [0, 1]$. The triple $(\mathcal{W}, \mathcal{N}, \bigotimes_{\mathrm{TN}})$ is called a fuzzy normed space if,

(C-1) $\mathcal{N}(\mathcal{V}, \mathcal{T}) > 0$, for every $\mathcal{T} \in (0, +\infty)$ and $\mathcal{V} \in \mathcal{W}$;
(C-2) $\mathcal{N}(\mathcal{V}, \mathcal{T}) = 1$, for every $\mathcal{T} \in (0, +\infty)$, if and only if $\mathcal{V} = 0$;
(C-3) $\mathcal{N}(\hbar\mathcal{V}, \mathcal{T}) = \mathcal{N}\left(\mathcal{V}, \frac{\mathcal{T}}{|\hbar|}\right)$, for every $\mathcal{T} \in (0, +\infty)$, $\mathcal{V} \in \mathcal{W}$ and $\hbar \in \mathbb{C}$ with $\hbar \neq 0$;
(C-4) $\mathcal{N}(\mathcal{V} + \mathcal{V}', \mathcal{T} + \mathcal{T}') \geq \mathcal{N}(\mathcal{V}, \mathcal{T}) \bigotimes_{\mathrm{TN}} \mathcal{N}(\mathcal{V}', \mathcal{T}')$, for every $\mathcal{V}, \mathcal{V}' \in \mathcal{W}$ and $\mathcal{T}, \mathcal{T}' \in (0, +\infty)$.

The concept of fuzzy metric spaces was initially introduced by Kaleva and Seikkla in 1984 [11], who proved a fixed point theorem for such spaces. In 1992, Felbin [12] introduced the concept of fuzzy norm and showed that every finite dimensional normed linear space has a completion. In 1994, Cheng and Mordenson [13] presented another idea of fuzzy norm on a linear space and in this situation the corresponding fuzzy metric is of Kramosil and Michalek type. In 2002, Xiao and Zhu [14] modified the definition of fuzzy norm and studied the topological properties of fuzzy normed linear spaces. In 2003, Bag and Samanta [15] introduced a new concept of fuzzy norm and they studied the properties of finite dimensional fuzzy normed linear space. Other approaches for fuzzy normed spaces can be found in [16–19].

We now present some examples of fuzzy normed spaces by means of special functions given in Chap. 1.

Example 15.1 Consider the one–parameter Mittag–Leffler function, as follows:

$$\mathbb{E}_{\kappa}\left(-\frac{\|\mathcal{V}\|}{\mathcal{T}}\right) = \sum_{k=0}^{\infty} \frac{\left(-\frac{\|\mathcal{V}\|}{\mathcal{T}}\right)^k}{\Gamma(1+\kappa k)}, \quad \|\mathcal{V}\| < \frac{1}{n}, \ n \in \mathbb{N},$$

where $\kappa \in [0, 1]$, $\mathcal{V} \in \mathcal{W}$ and $\mathcal{T} \in (0, +\infty)$.

Here, we want to show in the following four steps that $\left(\mathcal{W}, \mathbb{E}_{\kappa}\left(-\frac{\|\mathcal{V}\|}{\mathcal{T}}\right), \bigotimes_{\mathrm{TN}}^{M}\right)$ is a fuzzy normed space.

(1) If $\kappa \in [0, 1]$, then $\mathbb{E}_\kappa(0) = 1$ and $\lim_{\mathcal{V}\to-\infty} \mathbb{E}_\kappa(\mathcal{V}) = 0$, therefore we can conclude that $\mathbb{E}_\kappa$ is an increasing function for all $\kappa \in [0, 1]$, and also we have that $\mathbb{E}_\kappa \in [0, 1]$.

(2) It is straight forward to show $\mathbb{E}_\kappa\left(-\frac{\|\mathcal{V}\|}{\mathcal{T}}\right) = 1$, for every $\mathcal{T} \in (0, +\infty)$, if and only if $\mathcal{V} = 0$.

(3) For any $\mathcal{V} \in \mathcal{W}$, $\mathcal{T} \in (0, +\infty)$ and $0 \neq \hbar \in \mathbb{C}$, we have

$$\begin{aligned}\mathbb{E}_\kappa\left(-\frac{\|\hbar\mathcal{V}\|}{\mathcal{T}}\right) &= \sum_{k=0}^{\infty} \frac{\left(-\frac{\|\hbar\mathcal{V}\|}{\mathcal{T}}\right)^k}{\Gamma(1+\kappa k)} \\ &= \sum_{k=0}^{\infty} \frac{\left(-\frac{\|\mathcal{V}\|}{\frac{\mathcal{T}}{|\hbar|}}\right)^k}{\Gamma(1+\kappa k)} \\ &= \mathbb{E}_\kappa\left(-\frac{\|\mathcal{V}\|}{\frac{\mathcal{T}}{|\hbar|}}\right).\end{aligned}$$

(4) Let $\mathbb{E}_\kappa\left(-\frac{\|\mathcal{V}\|}{\mathcal{T}}\right) \leq \mathbb{E}_\kappa\left(-\frac{\|\mathcal{V}'\|}{\mathcal{T}'}\right)$. Then, we have $\frac{\|\mathcal{V}'\|}{\mathcal{T}'} \leq \frac{\|\mathcal{V}\|}{\mathcal{T}}$, for any $\mathcal{V}, \mathcal{V}' \in \mathcal{W}$ and $\mathcal{T}, \mathcal{T}' \in (0, +\infty)$. Now, if $\mathcal{V} = \mathcal{V}'$, we have $\mathcal{T} \leq \mathcal{T}'$. Thus, otherwise, we have that

$$\begin{aligned}\frac{\|\mathcal{V}\|}{\mathcal{T}} + \frac{\|\mathcal{V}\|}{\mathcal{T}} &\geq \frac{\|\mathcal{V}\|}{\mathcal{T}} + \frac{\|\mathcal{V}'\|}{\mathcal{T}'} \\ &\geq 2\frac{\|\mathcal{V}\|}{\mathcal{T}+\mathcal{T}'} + 2\frac{\|\mathcal{V}'\|}{\mathcal{T}+\mathcal{T}'} \\ &\geq 2\frac{\|\mathcal{V}+\mathcal{V}'\|}{\mathcal{T}+\mathcal{T}'},\end{aligned}$$

then, $\frac{\|\mathcal{V}\|}{\mathcal{T}} \geq \frac{\|\mathcal{V}+\mathcal{V}'\|}{\mathcal{T}+\mathcal{T}'}$. But $-\frac{\|\mathcal{V}\|}{\mathcal{T}} \leq -\frac{\|\mathcal{V}+\mathcal{V}'\|}{\mathcal{T}+\mathcal{T}'}$, and also

$$\sum_{k=0}^{\infty} \frac{\left(-\frac{\|\mathcal{V}\|}{\mathcal{T}}\right)^k}{\Gamma(1+\kappa k)} \leq \sum_{k=0}^{\infty} \frac{\left(-\frac{\|\mathcal{V}+\mathcal{V}'\|}{\mathcal{T}+\mathcal{T}'}\right)^k}{\Gamma(1+\kappa k)}, \tag{15.2}$$

which implies that

$$\mathbb{E}_\kappa\left(-\frac{\|\mathcal{V}\|}{\mathcal{T}}\right) \leq \mathbb{E}_\kappa\left(-\frac{\|\mathcal{V}+\mathcal{V}'\|}{\mathcal{T}+\mathcal{T}'}\right).$$

Hence we have,

$$\mathbb{E}_{\kappa}\left(-\frac{\|\mathcal{V}+\mathcal{V}'\|}{\mathcal{T}+\mathcal{T}'}\right) \geq \min\left\{\mathbb{E}_{\kappa}\left(-\frac{\|\mathcal{V}\|}{\mathcal{T}}\right), \mathbb{E}_{\kappa}\left(-\frac{\|\mathcal{V}'\|}{\mathcal{T}'}\right)\right\},$$

for any $\mathcal{V}, \mathcal{V}' \in \mathcal{W}$ and $\mathcal{T}, \mathcal{T}' \in (0, +\infty)$. Thus,

$$\mathcal{N}(\mathcal{V}, \mathcal{T}) = \mathbb{E}_{\kappa}\left(-\frac{\|\mathcal{V}\|}{\mathcal{T}}\right),$$

defines a fuzzy norm and $(\mathcal{W}, \mathcal{N}, \bigotimes_{\mathrm{TN}}^{M})$ is a fuzzy normed space, for any $\mathcal{V} \in \mathcal{W}$, $\mathcal{T} \in (0, +\infty)$ and $\kappa \in [0, 1]$; here $(\mathcal{W}, \|.\|)$ is a normed linear space.

Example 15.2 Consider the following increasing Hypergeometric function given by

$${}_2\mathbb{F}_1\left(\mathsf{A}, \mathsf{B}; \mathsf{C}; -\frac{\|\mathcal{V}\|}{\mathcal{T}}\right) = \sum_{k=0}^{\infty} \frac{(\mathsf{A})_k(\mathsf{B})_k}{(\mathsf{C})_k} \frac{\left(-\frac{\|\mathcal{V}\|}{\mathcal{T}}\right)^k}{k!}, \quad \|\mathcal{V}\| < \frac{1}{n}, \quad n \in \mathbb{N}$$

where $\mathsf{A}, \mathsf{B}, \mathsf{C} \in [0, +\infty]$, $\mathcal{V} \in \mathcal{W}$, and $\mathcal{T} \in (0, +\infty)$.

Here, we want to show in the following steps that $\left(\mathcal{W}, {}_2\mathbb{F}_1\left(\mathsf{A}, \mathsf{B}; \mathsf{C}; -\frac{\|\mathcal{V}\|}{\mathcal{T}}\right), \bigotimes_{\mathrm{TN}}^{M}\right)$ is a fuzzy normed space.

(1) It is easy to show that ${}_2\mathbb{F}_1\left(\mathsf{A}, \mathsf{B}; \mathsf{C}; -\frac{\|\mathcal{V}\|}{\mathcal{T}}\right) = 1$, for every $\mathcal{T} \in (0, +\infty)$, if and only if $\mathcal{V} = 0$.

(2) For any $\mathcal{V} \in \mathcal{W}$, $\mathcal{T} \in (0, +\infty)$, and $0 \neq \hbar \in \mathbb{C}$, we have that

$$\begin{aligned}
{}_2\mathbb{F}_1\left(\mathsf{A}, \mathsf{B}; \mathsf{C}; -\frac{\|\hbar\mathcal{V}\|}{\mathcal{T}}\right) &= \sum_{k=0}^{\infty} \frac{(\mathsf{A})_k(\mathsf{B})_k}{(\mathsf{C})_k} \frac{-\frac{\|\hbar\mathcal{V}\|^k}{\mathcal{T}}}{k!} \\
&= \sum_{k=0}^{\infty} \frac{(\mathsf{A})_k(\mathsf{B})_k}{(\mathsf{C})_k} \frac{-\frac{\|\mathcal{V}\|^k}{\frac{\mathcal{T}}{|\hbar|}}}{k!} \\
&= {}_2\mathbb{F}_1\left(\mathsf{A}, \mathsf{B}; \mathsf{C}; -\frac{\|\mathcal{V}\|}{\frac{\mathcal{T}}{|\hbar|}}\right).
\end{aligned}$$

(3) Let ${}_2\mathbb{F}_1\left(\mathsf{A}, \mathsf{B}; \mathsf{C}; -\frac{\|\mathcal{V}\|}{\mathcal{T}}\right) \leq {}_2\mathbb{F}_1\left(\mathsf{A}, \mathsf{B}; \mathsf{C}; -\frac{\|\mathcal{V}'\|}{\mathcal{T}'}\right)$. Then, we have $\frac{\|\mathcal{V}'\|}{\mathcal{T}'} \leq \frac{\|\mathcal{V}\|}{\mathcal{T}}$, for any $\mathcal{V}, \mathcal{V}' \in \mathcal{W}$ and $\mathcal{T}, \mathcal{T}' \in (0, +\infty)$. Now, if $\mathcal{V} = \mathcal{V}'$, we have $\mathcal{T} \leq \mathcal{T}'$. Thus, otherwise, we have

$$\frac{\|\mathcal{V}\|}{\mathcal{T}}+\frac{\|\mathcal{V}\|}{\mathcal{T}}\geq\frac{\|\mathcal{V}\|}{\mathcal{T}}+\frac{\|\mathcal{V}'\|}{\mathcal{T}'}$$
$$\geq 2\frac{\|\mathcal{V}\|}{\mathcal{T}+\mathcal{T}'}+2\frac{\|\mathcal{V}'\|}{\mathcal{T}+\mathcal{T}'}$$
$$\geq 2\frac{\|\mathcal{V}+\mathcal{V}'\|}{\mathcal{T}+\mathcal{T}'},$$

then, $\frac{\|\mathcal{V}\|}{\mathcal{T}}\geq\frac{\|\mathcal{V}+\mathcal{V}'\|}{\mathcal{T}+\mathcal{T}'}$. But $-\frac{\|\mathcal{V}\|}{\mathcal{T}}\leq-\frac{\|\mathcal{V}+\mathcal{V}'\|}{\mathcal{T}+\mathcal{T}'}$, and also

$$\sum_{k=0}^{\infty}\frac{(\mathsf{A})_k(\mathsf{B})_k}{(\mathsf{C})_k}\frac{\left(-\frac{\|\mathcal{V}\|}{\mathcal{T}}\right)^k}{k!}\leq\sum_{k=0}^{\infty}\frac{(\mathsf{A})_k(\mathsf{B})_k}{(\mathsf{C})_k}\frac{\left(-\frac{\|\mathcal{V}+\mathcal{V}'\|}{\mathcal{T}+\mathcal{T}'}\right)^k}{k!}, \tag{15.3}$$

which implies that

$$_2\mathbb{F}_1\left(\mathsf{A},\mathsf{B};\mathsf{C};-\frac{\|\mathcal{V}\|}{\mathcal{T}}\right)\leq{}_2\mathbb{F}_1\left(\mathsf{A},\mathsf{B};\mathsf{C};-\frac{\|\mathcal{V}+\mathcal{V}'\|}{\mathcal{T}+\mathcal{T}'}\right).$$

Hence we have,

$$_2\mathbb{F}_1\left(\mathsf{A},\mathsf{B};\mathsf{C};-\frac{\|\mathcal{V}+\mathcal{V}'\|}{\mathcal{T}+\mathcal{T}'}\right)\geq\min\left\{{}_2\mathbb{F}_1\left(\mathsf{A},\mathsf{B};\mathsf{C};-\frac{\|\mathcal{V}\|}{\mathcal{T}}\right),{}_2\mathbb{F}_1\left(\mathsf{A},\mathsf{B};\mathsf{C};-\frac{\|\mathcal{V}'\|}{\mathcal{T}'}\right)\right\},$$

for any $\mathcal{V},\mathcal{V}'\in\mathcal{W}$ and $\mathcal{T},\mathcal{T}'\in(0,+\infty)$. Thus,

$$_2\mathbb{F}_1\left(\mathsf{A},\mathsf{B};\mathsf{C};-\frac{\|\mathcal{V}\|}{\mathcal{T}}\right),$$

denotes a fuzzy norm and $(\mathcal{W},{}_2\mathbb{F}_1\left(\mathsf{A},\mathsf{B};\mathsf{C};-\frac{\|\mathcal{V}\|}{\mathcal{T}}\right),\bigotimes_{\mathrm{TN}}^{M})$ is a fuzzy normed space, for all $\mathcal{V}\in\mathcal{W}$, $\mathcal{T}\in(0,+\infty)$ and $\mathsf{A},\mathsf{B},\mathsf{C}\in[0,+\infty]$; here $(\mathcal{W},\|.\|)$ is a normed linear space.

Example 15.3 Consider the following increasing Wright function

$$\mathbb{K}\left(\vartheta,b,-\frac{\|\mathcal{V}\|}{\mathcal{T}}\right)=\sum_{k=0}^{\infty}\frac{\left(-\frac{\|\mathcal{V}\|}{\mathcal{T}}\right)^k}{k!\Gamma(\vartheta k+b)},\ \|\mathcal{V}\|<\frac{1}{n},\ n\in\mathbb{N}$$

where $\vartheta,b\in[0,+\infty]$, $\mathcal{V}\in\mathcal{W}$ and $\mathcal{T}\in(0,+\infty)$.

Here, we want to show in the following steps that $\left(\mathcal{W},\mathbb{K}\left(\vartheta,b,-\frac{\|\mathcal{V}\|}{\mathcal{T}}\right),\bigotimes_{\mathrm{TN}}^{M}\right)$ is a fuzzy normed space.

(1) It is easy to show that $\mathbb{K}\left(\vartheta, b, -\frac{\|\mathcal{V}\|}{\mathcal{T}}\right) = 1$, for every $\mathcal{T} \in (0, +\infty)$, if and only if $\mathcal{V} = 0$.

(2) For any $\mathcal{V} \in \mathcal{W}$, $\mathcal{T} \in (0, +\infty)$, and $0 \neq \hbar \in \mathbb{C}$, we have that

$$\begin{aligned}\mathbb{K}\left(\vartheta, b, -\frac{\|\hbar\mathcal{V}\|}{\mathcal{T}}\right) &= \sum_{k=0}^{\infty} \frac{-\frac{\|\hbar\mathcal{V}\|^k}{\mathcal{T}}}{k!\Gamma(\vartheta k + b)} \\ &= \sum_{k=0}^{\infty} \frac{-\frac{\|\mathcal{V}\|^k}{\frac{\mathcal{T}}{|\hbar|}}}{k!\Gamma(\vartheta k + b)} \\ &= \mathbb{K}\left(\vartheta, b, -\frac{\|\mathcal{V}\|}{\frac{\mathcal{T}}{|\hbar|}}\right).\end{aligned}$$

(3) Let $\mathbb{K}\left(\vartheta, b, -\frac{\|\mathcal{V}\|}{\mathcal{T}}\right) \leq \mathbb{K}\left(\vartheta, b, -\frac{\|\mathcal{V}'\|}{\mathcal{T}'}\right)$. Then, we have $\frac{\|\mathcal{V}'\|}{\mathcal{T}'} \leq \frac{\|\mathcal{V}\|}{\mathcal{T}}$, for any $\mathcal{V}, \mathcal{V}' \in \mathcal{W}$ and $\mathcal{T}, \mathcal{T}' \in (0, +\infty)$. Now, if $\mathcal{V} = \mathcal{V}'$, we have $\mathcal{T} \leq \mathcal{T}'$. Thus, otherwise, we have that

$$\begin{aligned}\frac{\|\mathcal{V}\|}{\mathcal{T}} + \frac{\|\mathcal{V}\|}{\mathcal{T}} &\geq \frac{\|\mathcal{V}\|}{\mathcal{T}} + \frac{\|\mathcal{V}'\|}{\mathcal{T}'} \\ &\geq 2\frac{\|\mathcal{V}\|}{\mathcal{T} + \mathcal{T}'} + 2\frac{\|\mathcal{V}'\|}{\mathcal{T} + \mathcal{T}'} \\ &\geq 2\frac{\|\mathcal{V} + \mathcal{V}'\|}{\mathcal{T} + \mathcal{T}'},\end{aligned}$$

then, $\frac{\|\mathcal{V}\|}{\mathcal{T}} \geq \frac{\|\mathcal{V} + \mathcal{V}'\|}{\mathcal{T} + \mathcal{T}'}$. But $-\frac{\|\mathcal{V}\|}{\mathcal{T}} \leq -\frac{\|\mathcal{V} + \mathcal{V}'\|}{\mathcal{T} + \mathcal{T}'}$, and also

$$\sum_{k=0}^{\infty} \frac{\left(-\frac{\|\mathcal{V}\|}{\mathcal{T}}\right)^k}{k!\Gamma(\vartheta k + b)} \leq \sum_{k=0}^{\infty} \frac{\left(-\frac{\|\mathcal{V} + \mathcal{V}'\|}{\mathcal{T} + \mathcal{T}'}\right)^k}{k!\Gamma(\vartheta k + b)}, \tag{15.4}$$

which implies that

$$\mathbb{K}\left(\vartheta, b, -\frac{\|\mathcal{V}\|}{\mathcal{T}}\right) \leq \mathbb{K}\left(\vartheta, b, -\frac{\|\mathcal{V} + \mathcal{V}'\|}{\mathcal{T} + \mathcal{T}'}\right).$$

Hence we have that

$$\mathbb{K}\left(\vartheta, b, -\frac{\|\mathcal{V}+\mathcal{V}'\|}{\mathcal{T}+\mathcal{T}'}\right) \geq \min\left\{\mathbb{K}\left(\vartheta, b, -\frac{\|\mathcal{V}\|}{\mathcal{T}}\right), \mathbb{K}\left(\vartheta, b, -\frac{\|\mathcal{V}'\|}{\mathcal{T}'}\right)\right\},$$

for any $\mathcal{V}, \mathcal{V}' \in \mathcal{W}$ and $\mathcal{T}, \mathcal{T}' \in (0, +\infty)$. Thus,

$$\mathbb{K}\left(\vartheta, b, -\frac{\|\mathcal{V}\|}{\mathcal{T}}\right),$$

denotes a fuzzy norm and $\left(\mathcal{W}, \mathbb{K}\left(\vartheta, b, -\frac{\|\mathcal{V}\|}{\mathcal{T}}\right), \bigotimes_{\mathrm{TN}}^{M}\right)$ is a fuzzy normed space, for all $\mathcal{V} \in \mathcal{W}, \mathcal{T} \in (0, +\infty)$ and $b, \vartheta \in [0, +\infty]$; here $(\mathcal{W}, \|.\|)$ is a normed linear space.

Let us generalize the concept of fuzzy normed spaces on diagonal matrices $\mathrm{diag}\mathcal{M}_{n\in\mathbb{N}}([0, 1])$ defined in Section 3.1.1.

Consider the vector space $\mathcal{W}$. Let Λ^* denotes the space of all matrix valued fuzzy sets, that is, the space of all mappings $\mathcal{N} : \mathcal{W} \times (0, +\infty) \longrightarrow \mathrm{diag}\mathcal{M}_n([0, 1])$ satisfying the following axioms:

- $\mathcal{N}(\mathcal{V}, .) : (0, +\infty) \to [0, 1]$ is continuous, for every $\mathcal{V} \in \mathcal{W}$;
- $\mathcal{N}(\mathcal{V}, .)$ is non–decreasing, for every $\mathcal{V} \in \mathcal{W}$;
- $\lim_{\mathcal{T}\to+\infty} (\mathcal{V}, \mathcal{T}) = \mathbf{1}$, and $\lim_{\mathcal{T}\to 0} \mathcal{N}(\mathcal{V}, \mathcal{T}) = \mathbf{0}$, for every $\mathcal{V} \in \mathcal{W}$ and $\mathcal{T} > 0$.

In Λ^*, we define "$\preceq$" as follows:

$$\mathcal{N} \preceq \mathcal{N}' \iff \mathcal{N}(\mathcal{V}, \mathcal{T}) \preceq \mathcal{N}'(\mathcal{V}, \mathcal{T}'),$$

for every $\mathcal{T}, \mathcal{T}' > 0$ and every $\mathcal{V} \in \mathcal{W}$.

Definition 15.5 ([8, 20]) Consider the continuous generalized triangular norm $\bigotimes_{\mathrm{GTN}}$, a vector space $\mathcal{W}$ and matrix valued fuzzy set $\mathcal{N} : \mathcal{W} \times (0, +\infty) \longrightarrow \mathrm{diag}\mathcal{M}_n([0, 1])$. In this case,we define a matrix valued fuzzy normed space (shortly, MVFN space) $(\mathcal{W}, \mathcal{N}, \bigotimes_{\mathrm{GTN}})$ as,

(D-1) $\mathcal{N}(\mathcal{V}, \mathcal{T}) = \mathbf{1}$, for every $\mathcal{T} > 0$ if and only if $\mathcal{V} = 0$;

(D-2) $\mathcal{N}(\hbar\mathcal{V}, \mathcal{T}) = \mathcal{N}(\mathcal{V}, \frac{\mathcal{T}}{|\hbar|})$, for every $\mathcal{V} \in \mathcal{W}, \mathcal{T} > 0$, and $\hbar \in \mathbb{C}$ with $\hbar \neq 0$;

(D-3) $\mathcal{N}(\mathcal{V} + \mathcal{V}', \mathcal{T} + \mathcal{T}') \succeq \mathcal{N}(\mathcal{V}, \mathcal{T}) \bigotimes_{\mathrm{GTN}} \mathcal{N}(\mathcal{V}', \mathcal{T}')$, for every $\mathcal{V}, \mathcal{V}' \in \mathcal{W}$ and $\mathcal{T}, \mathcal{T}' \geq 0$;

(D-4) $\lim_{\mathcal{T}\to+\infty} \mathcal{N}(\mathcal{V}, \mathcal{T}) = \mathbf{1}$, for every $\mathcal{V} \in \mathcal{W}$.

Note 15.2 A complete MVFN space is called a matrix valued fuzzy Banach space (or MVFB space).

Example 15.4 ([20]) The matrix valued fuzzy set $\mathcal{N}$ given by

$$\mathcal{N}(\mathcal{V}, \mathcal{T}) = \mathrm{diag}\left[\exp(-\frac{\|\mathcal{V}\|}{\mathcal{T}}), \frac{\mathcal{T}}{\mathcal{T} + \|\mathcal{V}\|}\right], \quad \mathcal{V} \in \mathcal{W}, \ \mathcal{T} > 0,$$

is a matrix valued fuzzy norm, and the triple $(\mathcal{W}, \mathcal{N}, \bigotimes_{\mathrm{GTN}}^{M})$ is a matrix valued fuzzy normed space; here, $(\mathcal{W}, \|\cdot\|)$ denotes a linear normed space.

Note 15.3 Throughout the Sections 3.2–3.7, we let $\bigotimes_{\mathrm{TN}} := \bigotimes_{\mathrm{TN}}^{M}$ and $\bigotimes_{\mathrm{GTN}} := \bigotimes_{\mathrm{GTN}}^{M}$.

15.3 Matrix Valued Fuzzy Normed Algebras

Fuzzy normed algebra was first introduced by Sadeqi and Amiripour [21] in 2007. From the notion of fuzzy normed space to the concept of fuzzy normed algebra it is one step that we present in the following definition.

Definition 15.6 ([21, 22]) A matrix valued fuzzy normed algebra (in short, MVFN–algebra) $(\mathcal{W}, \mathcal{N}, \bigotimes_{\mathrm{GTN}}, \bigoplus_{\mathrm{GTN}})$ is a matrix valued fuzzy normed space $(\mathcal{W}, \mathcal{N}, \bigotimes_{\mathrm{GTN}})$ with algebraic structure such that

(D-5) $\mathcal{N}(\mathcal{V}\mathcal{V}', \mathcal{T}\mathcal{T}') \succeq \mathcal{N}(\mathcal{V}, \mathcal{T}) \bigoplus_{\mathrm{GTN}} \mathcal{N}(\mathcal{V}', \mathcal{T}')$, for every $\mathcal{V}, \mathcal{V}' \in \mathcal{W}$ and every $\mathcal{T}', \mathcal{T} > 0$. In which $\bigoplus_{\mathrm{GTN}}$ is a continuous generalized triangular norm.

Note 15.4 A complete matrix valued fuzzy normed algebra is called a matrix valued fuzzy Banach algebra (or MVFB–algebra).

Example 15.5 Every normed algebra $(\mathcal{W}, \|\cdot\|)$ defines a matrix valued fuzzy normed algebra $(\mathcal{W}, \mathcal{N}, \bigotimes_{\mathrm{GTN}}^{M}, \bigotimes_{\mathrm{GTN}}^{P})$, where

$$\mathcal{N}(\mathcal{V}, \vec{\mathcal{T}}) = \operatorname{diag}\left[\frac{\mathcal{T}_1}{\mathcal{T}_1 + \|\mathcal{V}\|}, \frac{\mathcal{T}_2}{\mathcal{T}_2 + \|\mathcal{V}\|}\right],$$

for every $\vec{\mathcal{T}} > \vec{0}$, if and only if

$$\|\mathcal{V}\mathcal{V}'\| \le \|\mathcal{V}\|\|\mathcal{V}'\| + \mathcal{T}_i'\|\mathcal{V}'\| + \mathcal{T}_i\|\mathcal{V}\|,$$

for every $\mathcal{V}, \mathcal{V}' \in (\mathcal{W}, \|\cdot\|)$, $\mathcal{T}_i, \mathcal{T}_i' > 0$ and $i = 1, 2$.

This space is called the matrix valued induced fuzzy normed algebra.

Note that the last example is held with $(\mathcal{W}, \mathcal{N}, \bigotimes_{\mathrm{GTN}}^{P}, \bigotimes_{\mathrm{GTN}}^{P})$.

Note 15.5 Throughout the Sects. 3.2–3.7, we let $\bigotimes_{\mathrm{TN}} := \bigotimes_{\mathrm{TN}}^{M}$, $\bigoplus_{\mathrm{TN}} := \bigoplus_{\mathrm{TN}}^{M}$, $\bigotimes_{\mathrm{GTN}} := \bigotimes_{\mathrm{GTN}}^{M}$, and $\bigoplus_{\mathrm{GTN}} := \bigoplus_{\mathrm{GTN}}^{M}$.

15.4 Matrix Valued Fuzzy C–⋄–Algebras

Consider the MVFB–algebra $(\mathcal{W}, \mathcal{N}, \bigotimes_{\mathrm{GTN}}, \bigoplus_{\mathrm{GTN}})$. An involution on $\mathcal{W}$ is a mapping $\mathcal{V} \to \mathcal{V}^{\diamond}$ from $\mathcal{W}$ into $\mathcal{W}$ satisfying the following axioms:

(1) $\mathcal{V}^{\diamond\diamond} = \mathcal{V}$, for every $\mathcal{V} \in \mathcal{W}$;
(2) $(\alpha\mathcal{V} + \beta\mathcal{V})^{\diamond} = \bar{\alpha}\mathcal{V}^{\diamond} + \bar{\beta}\mathcal{V}^{\diamond}$, for every $\mathcal{V} \in \mathcal{W}$;
(3) $(\mathcal{V}\mathcal{V}')^{\diamond} = \mathcal{V}'^{\diamond}\mathcal{V}^{\diamond}$, for every $\mathcal{V}, \mathcal{V}' \in \mathcal{W}$.

Then, $\mathcal{W}$ is called a matrix valued fuzzy Banach–⋄–algebra (in short, MVFB–⋄–algebra) [23, 24].

Moreover, if $\mathcal{N}(\mathcal{V}^{\diamond}\mathcal{V}, \mathcal{T}) = \mathcal{N}(\mathcal{V}^2, \mathcal{T})$, for every $\mathcal{V} \in \mathcal{W}$ and $\mathcal{T} > 0$, then $\mathcal{W}$ is called a matrix valued fuzzy C–⋄–algebra (in short, MVFC–⋄–algebra) [23, 24].

Besides, if there exists a unit element e, such that

$$e\mathcal{V} = \mathcal{V}e = \mathcal{V},$$

for all $\mathcal{V} \in \mathcal{W}$, then, $(\mathcal{W}, \mathcal{N}, \bigotimes_{\text{GTN}}, \bigoplus_{\text{GTN}})$ is called a unital MVFC–⋄–algebra with unit e and the unitary group $U(\mathcal{W}) = \{t \in \mathcal{W} : t^{\diamond}t = tt^{\diamond} = e\}$.

Definition 15.7 ([23]) Consider the MVFC–⋄–algebra $(\mathcal{W}, \mathcal{N}, \bigotimes_{\text{GTN}}, \bigoplus_{\text{GTN}})$. A $\mathbb{C}$–linear fuzzy mapping $\delta : \mathcal{W} \to \mathcal{W}$ is a derivation if,

$$\delta(\mathcal{V}_1\mathcal{V}_2) = \delta(\mathcal{V}_1)\mathcal{V}_2 + \mathcal{V}_1\delta(\mathcal{V}_2),$$

for every $\mathcal{V}_1, \mathcal{V}_2 \in \mathcal{W}$.

A $\mathbb{C}$–linear fuzzy mapping $H : \mathcal{W} \to \mathcal{W}$ is a homomorphism if,

$$H(\mathcal{V}_1\mathcal{V}_2) = H(\mathcal{V}_1)H(\mathcal{V}_2),$$

and

$$H(\mathcal{V}_1^{\diamond}) = H(\mathcal{V}_1)^{\diamond},$$

for every $\mathcal{V}_1, \mathcal{V}_2 \in \mathcal{W}$.

Definition 15.8 ([24]) Let $(\mathcal{W}, \mathcal{N}, \bigotimes_{\text{GTN}}, \bigoplus_{\text{GTN}})$ be an MVFC–⋄–algebra. We say a tri–additive mapping $\delta : \mathcal{W}^3 \to \mathcal{W}$ is a fuzzy permuting tri–derivation on $\mathcal{W}$ if δ satisfies the following conditions:

$$\delta(\mathcal{V}_1\mathcal{V}_2, \mathcal{V}_3, \mathcal{V}_4) = \delta(\mathcal{V}_1, \mathcal{V}_3, \mathcal{V}_4)\mathcal{V}_2 + \mathcal{V}_1\delta(\mathcal{V}_2, \mathcal{V}_3, \mathcal{V}_4),$$
$$\delta(\mathcal{V}'_{\sigma(1)}, \mathcal{V}'_{\sigma(2)}, \mathfrak{X}'_{\sigma(3)}) = \delta(\mathcal{V}'_1, \mathcal{V}'_2, \mathcal{V}'_3)$$

for all $\mathcal{V}_1, \mathcal{V}_2, \mathcal{V}_3, \mathcal{V}_4, \mathcal{V}'_1, \mathcal{V}'_2, \mathcal{V}'_3 \in \mathcal{W}$ and for all permutation $(\sigma(1), \sigma(2), \sigma(3))$ of $(1, 2, 3)$.

Definition 15.9 ([24]) Let $(\mathcal{W}_1, \mathcal{N}, \bigotimes_{\text{GTN}}, \bigoplus_{\text{GTN}})$ and $(\mathcal{W}_2, \mathcal{N}, \bigotimes_{\text{GTN}}, \bigoplus_{\text{GTN}})$ be two MVFC–⋄–algebras. We say a $\mathbb{C}$–trilinear mapping $H : \mathcal{W}_1^3 \to \mathcal{W}_2$ is a *permuting tri-homomorphism* if H satisfies the following conditions:

$$H(\mathcal{V}_1\mathcal{V}_2, \mathcal{V}_3\mathcal{V}_4, \mathcal{V}_5\mathcal{V}_6) = H(\mathcal{V}_1, \mathcal{V}_3, \mathcal{V}_5)H(\mathcal{V}_2, \mathcal{V}_4, \mathcal{V}_6),$$
$$H(\mathcal{V}'_{\sigma(1)}, \mathcal{V}'_{\sigma(2)}, \mathcal{V}'_{\sigma(3)}) = H(\mathcal{V}'_1, \mathcal{V}'_2, \mathcal{V}'_3)$$

for all $\mathcal{V}_1, \mathcal{V}_2, \mathcal{V}_3, \mathcal{V}_4, \mathcal{V}_1', \mathcal{V}_2', \mathcal{V}_3' \in \mathcal{W}_1$ and for all permutation $(\sigma(1), \sigma(2), \sigma(3))$ of $(1, 2, 3)$.

15.5 Matrix Valued Fuzzy C–⋄–Ternary Algebras

Consider the MVFB–algebra $(\mathcal{W}, \mathscr{N}, \bigotimes_{\mathrm{GTN}}, \bigoplus_{\mathrm{GTN}})$. A matrix valued fuzzy C–⋄–ternary algebra (in short, MVFC–⋄–ternary algebra) [23] is an MVFC–⋄–algebra $\mathcal{W}$, equipped with a ternary product $(\mathcal{V}, \mathcal{F}, \mathcal{H}) \mapsto [\mathcal{V}, \mathcal{F}, \mathcal{H}]$ of $\mathcal{W}^3$ into $\mathcal{W}$, which is $\mathbb{C}$–linear in the outer variables, conjugate $\mathbb{C}$–linear in the middle variable, and associative in the sense that

$$[\mathcal{V}, \mathcal{F}, [\mathcal{H}, \mathcal{G}, \mathcal{K}]] = [\mathcal{V}, [\mathcal{G}, \mathcal{H}, \mathcal{F}], \mathcal{K}] = [[\mathcal{V}, \mathcal{F}, \mathcal{H}], \mathcal{G}, \mathcal{K}],$$

and satisfies

$$\|[\mathcal{V}, \mathcal{F}, \mathcal{H}]\| \leq \|\mathcal{V}\|.\|\mathcal{F}\|.\|\mathcal{H}\|,$$

and

$$\|[\mathcal{V}, \mathcal{V}, \mathcal{V}]\| = \|\mathcal{V}\|^3,$$

for every $\mathcal{V}, \mathcal{G}, \mathcal{H}, \mathcal{F}, \mathcal{K} \in \mathcal{W}$, in which $(\mathcal{W}, \|.\|)$ is a normed algebra.

Definition 15.10 ([23]) Let $(\mathcal{W}, \mathscr{N}, \bigotimes_{\mathrm{GTN}}, \bigoplus_{\mathrm{GTN}})$ be an MVFC–⋄–ternary algebra. A $\mathbb{C}$–linear fuzzy mapping $\delta : \mathcal{W} \to \mathcal{W}$ is a FC–⋄–ternary derivation if

$$\delta([\mathcal{V}, \mathcal{F}, \mathcal{H}]) = [\delta(\mathcal{V}), \mathcal{F}, \mathcal{H}] + [\mathcal{V}, \delta(\mathcal{F}), \mathcal{H}] + [\mathcal{V}, \mathcal{F}, \delta(\mathcal{H})],$$

for every $\mathcal{V}, \mathcal{F}, \mathcal{H} \in \mathcal{W}$.

A $\mathbb{C}$–linear fuzzy mapping $H : \mathcal{W} \to \mathcal{W}$ is a FC–⋄–ternary homomorphism if

$$H([\mathcal{V}, \mathcal{F}, \mathcal{H}]) = [H(\mathcal{V}), H(\mathcal{F}), H(\mathcal{H})],$$

for every $\mathcal{V}, \mathcal{F}, \mathcal{H} \in \mathcal{W}$.

15.6 Fuzzy Operators in Probability Measure Spaces

Consider the MVFB–algebras $(\mho_1, \mathscr{N}, \bigotimes_{\mathrm{GTN}}, \bigoplus_{\mathrm{GTN}})$ and $(\mho_2, \mathscr{N}, \bigotimes_{\mathrm{GTN}}, \bigoplus_{\mathrm{GTN}})$. Let $(\pitchfork, \curlywedge, \mathscr{N})$ be a probability measure space with the Borel measurable spaces $(\mho_1, \mathfrak{B}_{\mho_1})$ and $(\mho_2, \mathfrak{B}_{\mho_2})$. A map $f : \pitchfork \times \mho_1 \to \mho_2$ is a fuzzy operator if $\{\leftthreetimes : f(\leftthreetimes, \mathsf{S}) \in \mathcal{C}\} \in \curlywedge$, for every S in $\mho_1$ and $\mathcal{C} \in \mathfrak{B}_{\mho_2}$.

The fuzzy operator f is linear if,

$$f(\lambda, \alpha S_1 + \beta S_2) = \alpha f(\lambda, S_1) + \beta f(\lambda, S_2),$$

where $S_1, S_2 \in \mho_1, \alpha, \beta \in \mathbb{R}$ and $\lambda \in \pitchfork$.

The fuzzy operator f is bounded if we can find a $Q(\lambda) > 0$, such that

$$\mathcal{N}\Big(f(\lambda, S_1) - f(\lambda, S_2), Q(\lambda)\mathcal{T}\Big) \geq \mathcal{N}(S_1 - S_2, \mathcal{T}),$$

where$S_1, S_2 \in \mho_1, \lambda \in \pitchfork$ and $\mathcal{T} > 0$.

References

1. Menger K, Menger K (2003) Statistical metrics. Sel Math 2:433–435
2. Schweizer B, Sklar A (1958) Espaces métriques aléatories. Comptes Rendus Hebdomadaires Des Seances De L Academie Des Sciences 247(23):2092–2094
3. Schweizer B, Sklar A (1960) Statistical metric spaces. Pac J Math 10(1):313–334
4. Schweizer B, Sklar A, Thorp E (1960) The metrization of statistical metric spaces
5. Sherstnev AN (1962) Random normed spaces. questions of completeness. Uchenye Zapiski Kazanskogo Universiteta. Seriya Fiziko-Matematicheskie Nauki 122(4):3–20
6. Saadati R (2016) Random operator theory. Academic Press
7. Aderyani SR, Saadati R, Abdeljawad T, Mlaiki N (2022) Multi-stability of non homogenous vector-valued fractional differential equations in matrix-valued Menger spaces. Alex Eng J 61(12):10913–10923
8. Aderyani SR, Saadati R (2023) Stability and controllability results by n-ary aggregation functions in matrix valued fuzzy n-normed spaces. Inf Sci 119265
9. Zadeh L (1965) Fuzzy sets. Inf Control 8:338–353
10. Bellman RE, Zadeh LA (1970) Decision-making in a fuzzy environment. Manag Sci 17(4): B-141
11. Kaleva O, Seikkala S (1984) On fuzzy metric spaces. Fuzzy Sets Syst 12:215–229
12. Felbin C (1992) Finite dimensional fuzzy normed linear space. Fuzzy Sets Syst 48:239–248
13. Cheng SC, Mordeson JN (1994) Fuzzy linear operator and fuzzy normed linear spaces. Bull Calcutta Math Soc 86:429–436
14. Xiao J, Zhu X (2002) On linearly topological structure and property of fuzzy normed linear space. Fuzzy Sets Syst 125:153–161
15. Bag T, Samanta SK (2003) Finite dimensional fuzzy normed linear spaces. J Fuzzy Math 11:687–705
16. Saadati R, Vaezpour SM (2005) Some results on fuzzy Banach spaces. J Appl Math Comput 17:475–484
17. Bag T, Samanta SK (2008) A comparative study of fuzzy norms on a linear space. Fuzzy Sets Syst 159:670–684
18. Alegre C, Romaguera S (2010) Characterizations of fuzzy metrizable topological vector spaces and their asymmetric generalization in terms of fuzzy (quasi-)norms. Fuzzy Sets Syst 161:2182–2192
19. Golet I (2010) On generalized fuzzy normed spaces and coincidence point theorems. Fuzzy Sets Syst 161:1138–1144
20. Aderyani SR, Saadati R, Allahviranloo T, Abbasbandy S, Catak M (2023) Fuzzy approximation of a fractional Lorenz system and a fractional financial crisis. Iran J Fuzzy Syst 20(7):27–36
21. Sadeqi I, Amiripour A (2007, August). Fuzzy Banach algebra. In: First joint congress on fuzzy and intelligent systems, vol 1. Ferdorwsi University of Mashhad, Iran

22. Cho YJ, Rassias TM, Saadati R (2018) Operator theory and fixed points in fuzzy normed algebras and applications. In: Fuzzy Operator theory in mathematical analysis, pp 339–346
23. Aderyani SR, Saadati R (2022). Approximation of derivation–homomorphism fuzzy functional inequalities in matrix valued FC-$\diamond$-algebras. In: 2022 9th Iranian Joint Congress on Fuzzy and Intelligent Systems (CFIS), pp 1–6. IEEE. (Mar 2022)
24. Aderyani SR, Saadati R, Mesiar R (2022) Estimation of permuting tri-homomorphisms and permuting tri-derivations associated with the tri-additive Υ-random operator inequality in matrix MB-algebra. Int J Gen Syst 51(6):547–569

Chapter 16
Multi Stability of Pseudo Stochastic Tri–Additive Fuzzy Operator Inequalities in MVFC–⋄–Algebras

In this chapter, we study the multi stability of the following pseudo stochastic tri–additive Υ–fuzzy operator inequality [1] in MVFB–algebras by the fixed point method.

$$\begin{aligned}
&\mathscr{N}\Big(\mathcal{Q}(\lambda, \mathfrak{X}_1+\mathfrak{X}_2, \mathfrak{X}_3-\mathfrak{X}_4, \mathfrak{X}_5-\mathfrak{X}_6)+\mathcal{Q}(\lambda, \mathfrak{X}_1-\mathfrak{X}_2, \mathfrak{X}_3+\mathfrak{X}_4, \mathfrak{X}_5-\mathfrak{X}_6) \\
&\quad -2\mathcal{Q}(\lambda, \mathfrak{X}_1, \mathfrak{X}_3, \mathfrak{X}_5)+2\mathcal{Q}(\lambda, \mathfrak{X}_1, \mathfrak{X}_4, \mathfrak{X}_6)-2\mathcal{Q}(\lambda, \mathfrak{X}_2, \mathfrak{X}_3, \mathfrak{X}_6)+2\mathcal{Q}(\lambda, \mathfrak{X}_2, \mathfrak{X}_4, \mathfrak{X}_5), \mathcal{T}\Big) \\
&\geq \mathscr{N}\Big(\Upsilon[2\mathcal{Q}(\lambda, \frac{\mathfrak{X}_1+\mathfrak{X}_2}{2}, \mathfrak{X}_3-\mathfrak{X}_4, \mathfrak{X}_5+\mathfrak{X}_6)+2\mathcal{Q}(\lambda, \frac{\mathfrak{X}_1-\mathfrak{X}_2}{2}, \mathfrak{X}_3+\mathfrak{X}_4, \mathfrak{X}_5-\mathfrak{X}_6) \\
&\quad -2\mathcal{Q}(\lambda, \mathfrak{X}_1, \mathfrak{X}_3, \mathfrak{X}_5)+2\mathcal{Q}(\lambda, \mathfrak{X}_1, \mathfrak{X}_4, \mathfrak{X}_6)-2\mathcal{Q}(\lambda, \mathfrak{X}_2, \mathfrak{X}_3, \mathfrak{X}_6)+2\mathcal{Q}(\lambda, \mathfrak{X}_2, \mathfrak{X}_4, \mathfrak{X}_5)], \mathcal{T}\Big),
\end{aligned} \tag{16.1}$$

where $0 \neq \Upsilon \in \mathbb{C}$ is fixed with $|\Upsilon| < 1$, $\lambda \in ⋔$ and $\mathcal{T} > 0$. In addition, we study the multi stability of permuting tri–homomorphisms and permuting tri–derivations in MVFB–algebras and unital MVFC–⋄–algebras associated with the tri–additive Υ–fuzzy operator inequality (16.1).

Definition 16.1 ([1]) Let $(\mathfrak{M}, \mathscr{N}, \bigotimes_{\text{GTN}}, \bigoplus_{\text{GTN}})$ be an MVFB–algebra. We say a tri–additive operator $\mathfrak{Z} : ⋔ \times \mathfrak{M}^3 \to \mathfrak{M}$ is a *permuting tri-derivation* on $\mathfrak{M}$ if $\mathfrak{Z}$ satisfies the following conditions:

$$\mathfrak{Z}(\lambda, \mathfrak{X}_1\mathfrak{X}_2, \mathfrak{X}_3, \mathfrak{X}_4) = \mathfrak{Z}(\lambda, \mathfrak{X}_1, \mathfrak{X}_3, \mathfrak{X}_4)\mathfrak{X}_2 + \mathfrak{X}_1\mathfrak{Z}(\lambda, \mathfrak{X}_2, \mathfrak{X}_3, \mathfrak{X}_4),$$
$$\mathfrak{Z}(\lambda, \mathfrak{X}'_{\delta(1)}, \mathfrak{X}'_{\delta(2)}, \mathfrak{X}'_{\delta(3)}) = \mathfrak{Z}(\lambda, \mathfrak{X}'_1, \mathfrak{X}'_2, \mathfrak{X}'_3)$$

for all $\mathfrak{X}_1, \mathfrak{X}_2, \mathfrak{X}_3, \mathfrak{X}_4, \mathfrak{X}'_1, \mathfrak{X}'_2, \mathfrak{X}'_3 \in \mathfrak{M}$, $\lambda \in ⋔$ and for all permutation $(\delta(1), \delta(2), \delta(3))$ of $(1, 2, 3)$.

Definition 16.2 ([1]) Let $(\mathfrak{M}, \mathscr{N}, \bigotimes_{\text{GTN}}, \bigoplus_{\text{GTN}})$ and $(\mathfrak{N}, \mathscr{N}, \bigotimes_{\text{GTN}}, \bigoplus_{\text{GTN}})$ be two MVFB–algebra. We say a $\mathbb{C}$–trilinear operator $\mathcal{E} : ⋔ \times \mathfrak{M}^3 \to \mathfrak{N}$ is a permuting tri-homomorphism if $\mathcal{E}$ satisfies the following conditions:

S. Rezaei Aderyani et al., *Towards Ulam Type Multi Stability Analysis*, Studies in Systems, Decision and Control 523, https://doi.org/10.1007/978-3-031-55564-0_16

$$\mathcal{E}(\curlywedge, \mathfrak{X}_1\mathfrak{X}_2, \mathfrak{X}_3\mathfrak{X}_4, \mathfrak{X}_5\mathfrak{X}_6) = \mathcal{E}(\curlywedge, \mathfrak{X}_1, \mathfrak{X}_3, \mathfrak{X}_5)\mathcal{E}(\curlywedge, \mathfrak{X}_2, \mathfrak{X}_4, \mathfrak{X}_6),$$
$$\mathcal{E}(\curlywedge, \mathfrak{X}'_{\delta(1)}, \mathfrak{X}'_{\delta(2)}, \mathfrak{X}'_{\delta(3)}) = \mathcal{E}(\curlywedge, \mathfrak{X}'_1, \mathfrak{X}'_2, \mathfrak{X}'_3)$$

for all $\mathfrak{X}_1, \mathfrak{X}_2, \mathfrak{X}_3, \mathfrak{X}_4, \mathfrak{X}'_1, \mathfrak{X}'_2, \mathfrak{X}'_3 \in \mathfrak{M}$, $\curlywedge \in \pitchfork$ and for all permutation $(\delta(1), \delta(2), \delta(3))$ of $(1, 2, 3)$.

16.1 Multi Stability of Tri–Additive Functional Inequalities in MVFB–Algebras

In this part, we study the multi stability of the tri–additive Υ–functional inequality (16.1) in MVFB–algebras by an alternative fixed point theorem.

Throughout this section, we consider the MVFB–algebras $(\mho_1, \mathcal{N}, \bigotimes_{\mathrm{GTN}}, \bigoplus_{\mathrm{GTN}})$, and $(\mho_2, \mathcal{N}, \bigotimes_{\mathrm{GTN}}, \bigoplus_{\mathrm{GTN}})$.

Lemma 16.1 *Let $\mathcal{Q} : \pitchfork \times \mho_1^3 \to \mho_2$ be a fuzzy operator satisfying the functional inequality (16.1), for every $\mathfrak{X}_i \in \mho_1, i = 1, \dots, 6$ and $\curlywedge \in \pitchfork$. Let $\mathcal{Q}(\curlywedge, 0, \mathfrak{X}_3, \mathfrak{X}_5) = \mathcal{Q}(\curlywedge, \mathfrak{X}_1, 0, \mathfrak{X}_5) = \mathcal{Q}(\curlywedge, \mathfrak{X}_1, \mathfrak{X}_3, 0) = \mathcal{Q}(\curlywedge, \mathfrak{X}_1, 0, 0) = \mathcal{Q}(\curlywedge, \mathfrak{X}_2, 0, \mathfrak{X}_5) = \mathcal{Q}(\curlywedge, \mathfrak{X}_2, \mathfrak{X}_3, 0) = 0$, for every $\mathfrak{X}_i \in \mho_1, i = 1, 2, 3, 5$ and $\curlywedge \in \pitchfork$. Then, fuzzy operator $\mathcal{Q} : \pitchfork \times \mho_1^3 \to \mho_2$ is tri–additive.*

Proof Putting $\mathfrak{X}_1 = \mathfrak{X}_2$ and $\mathfrak{X}_4 = \mathfrak{X}_6 = 0$ in functional inequality (16.1), we have that

$$\mathcal{Q}(\curlywedge, 2\mathfrak{X}_1, \mathfrak{X}_3, \mathfrak{X}_5) = 2\mathcal{Q}(\curlywedge, \mathfrak{X}_1, \mathfrak{X}_3, \mathfrak{X}_5),$$

for every $\mathfrak{X}_1, \mathfrak{X}_3, \mathfrak{X}_5 \in \mho_1$ and $\curlywedge \in \pitchfork$.

Thus, we get

$$\begin{aligned} &\mathcal{N}\Big(\mathcal{Q}(\curlywedge, \mathfrak{X}_1 + \mathfrak{X}_2, \mathfrak{X}_3 - \mathfrak{X}_4, \mathfrak{X}_5 - \mathfrak{X}_6) + \mathcal{Q}(\curlywedge, \mathfrak{X}_1 - \mathfrak{X}_2, \mathfrak{X}_3 + \mathfrak{X}_4, \mathfrak{X}_5 - \mathfrak{X}_6) \\ &\quad -2\mathcal{Q}(\curlywedge, \mathfrak{X}_1, \mathfrak{X}_3, \mathfrak{X}_5) + 2\mathcal{Q}(\curlywedge, \mathfrak{X}_1, \mathfrak{X}_4, \mathfrak{X}_6) - 2\mathcal{Q}(\curlywedge, \mathfrak{X}_2, \mathfrak{X}_3, \mathfrak{X}_6) + 2\mathcal{Q}(\curlywedge, \mathfrak{X}_2, \mathfrak{X}_4, \mathfrak{X}_5), \mathcal{T}\Big) \\ &\geq \mathcal{N}\Big(\Upsilon[\mathcal{Q}(\curlywedge, \mathfrak{X}_1 + \mathfrak{X}_2, \mathfrak{X}_3 - \mathfrak{X}_4, \mathfrak{X}_5 + \mathfrak{X}_6) + \mathcal{Q}(\curlywedge, \mathfrak{X}_1 - \mathfrak{X}_2, \mathfrak{X}_3 + \mathfrak{X}_4, \mathfrak{X}_5 - \mathfrak{X}_6) \\ &\quad -2\mathcal{Q}(\curlywedge, \mathfrak{X}_1, \mathfrak{X}_3, \mathfrak{X}_5) + 2\mathcal{Q}(\curlywedge, \mathfrak{X}_1, \mathfrak{X}_4, \mathfrak{X}_6) - 2\mathcal{Q}(\curlywedge, 3, \mathfrak{X}_2, \mathfrak{X}_3, \mathfrak{X}_6) + 2\mathcal{Q}(\curlywedge, \mathfrak{X}_2, \mathfrak{X}_4, \mathfrak{X}_5)], \mathcal{T}\Big), \end{aligned} \tag{16.2}$$

and so

$$\begin{aligned} &\mathcal{Q}(\curlywedge, \mathfrak{X}_1 + \mathfrak{X}_2, \mathfrak{X}_3 - \mathfrak{X}_4, \mathfrak{X}_5 - \mathfrak{X}_6) + \mathcal{Q}(\curlywedge, \mathfrak{X}_1 - \mathfrak{X}_2, \mathfrak{X}_3 + \mathfrak{X}_4, \mathfrak{X}_5 - \mathfrak{X}_6) \\ &-2\mathcal{Q}(\curlywedge, \mathfrak{X}_1, \mathfrak{X}_3, \mathfrak{X}_5) + 2\mathcal{Q}(\curlywedge, \mathfrak{X}_1, \mathfrak{X}_4, \mathfrak{X}_6) - 2\mathcal{Q}(\curlywedge, \mathfrak{X}_2, \mathfrak{X}_3, \mathfrak{X}_6) + 2\mathcal{Q}(\curlywedge, \mathfrak{X}_2, \mathfrak{X}_4, \mathfrak{X}_5) = 0 \end{aligned} \tag{16.3}$$

for every $\mathfrak{X}_i \in \mho_1, i = 1, \dots, 6$, $\curlywedge \in \pitchfork$ and $\mathcal{T} > 0$. Now, putting $\mathfrak{X}_4 = \mathfrak{X}_6 = 0$ in (16.3), we have

$$
\begin{aligned}
&\mathcal{Q}(\leftthreetimes, \mathfrak{X}_1 + \mathfrak{X}_2, \mathfrak{X}_3, \mathfrak{X}_5) + \mathcal{Q}(\leftthreetimes, \mathfrak{X}_1 - \mathfrak{X}_2, \mathfrak{X}_3, \mathfrak{X}_5) \\
&= 2\mathcal{Q}(\leftthreetimes, \mathfrak{X}_1, \mathfrak{X}_3, \mathfrak{X}_5),
\end{aligned} \tag{16.4}
$$

and so

$$
\begin{aligned}
&\mathcal{Q}(\leftthreetimes, \mathfrak{X}, \mathfrak{X}_3, \mathfrak{X}_5) + \mathcal{Q}(\leftthreetimes, \mathfrak{X}', \mathfrak{X}_3, \mathfrak{X}_5) \\
&= 2\mathcal{Q}(\leftthreetimes, \frac{\mathfrak{X} + \mathfrak{X}'}{2}, \mathfrak{X}_3, \mathfrak{X}_5) = \mathcal{Q}(\leftthreetimes, \mathfrak{X} + \mathfrak{X}', \mathfrak{X}_3, \mathfrak{X}_5),
\end{aligned} \tag{16.5}
$$

for every $\mathfrak{X} = \mathfrak{X}_1 + \mathfrak{X}_2, \mathfrak{X}' = \mathfrak{X}_1 - \mathfrak{X}_2, \mathfrak{X}_3, \mathfrak{X}_4, \mathfrak{X}_5, \mathfrak{X}_6 \in \mho_1$ and $\leftthreetimes \in \pitchfork$. Since $|\Upsilon| < 1$, the fuzzy operator $\mathcal{Q} : \pitchfork \times \mho_1^3 \to \mho_2$ is additive in the second variable. Similarly, one can show that $\mathcal{Q} : \pitchfork \times \mho_1^3 \to \mho_2$ is additive both in the second and in the third variable. Thus, $\mathcal{Q} : \pitchfork \times \mho_1^3 \to \mho_2$ is tri–additive.

Theorem 16.1 *Let $n \in \mathbb{N}$, and $(\mho_1, \mathcal{N}, \bigotimes^M_{\mathrm{GTN}}, \bigotimes^M_{\mathrm{GTN}})$ be an MVFB–algebra, and $\underbrace{\varphi_i}_{i=1,\ldots,n} : \mho_1^6 \times (0, +\infty) \to [0, 1]$ be fuzzy sets such that there exist $\underbrace{\mathfrak{P}_i}_{i=1,\ldots,n} < 1$ with*

$$
\begin{aligned}
&\underbrace{\varphi_i}_{i=1,\ldots,n}\left(\frac{\mathfrak{X}_1}{2}, \frac{\mathfrak{X}_2}{2}, \frac{\mathfrak{X}_3}{2}, \frac{\mathfrak{X}_4}{2}, \frac{\mathfrak{X}_5}{2}, \frac{\mathfrak{X}_6}{2}, \mathcal{T}\right) \\
&\geq \underbrace{\varphi_i}_{i=1,\ldots,n}\left(\mathfrak{X}_1, \mathfrak{X}_2, \mathfrak{X}_3, \mathfrak{X}_4, \mathfrak{X}_5, \mathfrak{X}_6, \frac{2}{\underbrace{\mathfrak{P}_i}_{i=1,\ldots,n}}\mathcal{T}\right)
\end{aligned} \tag{16.6}
$$

and

$$
\lim_{j \to \infty} \underbrace{\varphi_i}_{i=1,\ldots,n}\left(\frac{\mathfrak{X}_1}{2^j}, \frac{\mathfrak{X}_2}{2^j}, \frac{\mathfrak{X}_3}{2^j}, \frac{\mathfrak{X}_4}{2^j}, \frac{\mathfrak{X}_5}{2^j}, \frac{\mathfrak{X}_6}{2^j}, \frac{\mathcal{T}}{2^j}\right) = 1, \tag{16.7}
$$

for every $\mathfrak{X}_i \in \mho_1, i = 1, \ldots, 6$ and $\mathcal{T} > 0$. Suppose the fuzzy operator $\mathcal{Q} : \pitchfork \times \mho_1^3 \to \mho_2$ satisfies $\mathcal{Q}(\leftthreetimes, 0, \mathfrak{X}_3, \mathfrak{X}_5) = \mathcal{Q}(\leftthreetimes, \mathfrak{X}_1, 0, \mathfrak{X}_5) = \mathcal{Q}(\leftthreetimes, \mathfrak{X}_1, \mathfrak{X}_3, 0) = \mathcal{Q}(\leftthreetimes, \mathfrak{X}_1, 0, 0) = \mathcal{Q}(\leftthreetimes, \mathfrak{X}_2, 0, \mathfrak{X}_5) = \mathcal{Q}(\leftthreetimes, \mathfrak{X}_2, \mathfrak{X}_3, 0) = 0$ and

$$
\begin{aligned}
&\mathcal{N}\Big(\mathcal{Q}(\leftthreetimes, \mathfrak{X}_1 + \mathfrak{X}_2, \mathfrak{X}_3 - \mathfrak{X}_4, \mathfrak{X}_5 - \mathfrak{X}_6) + \mathcal{Q}(\leftthreetimes, \mathfrak{X}_1 - \mathfrak{X}_2, \mathfrak{X}_3 + \mathfrak{X}_4, \mathfrak{X}_5 - \mathfrak{X}_6) \\
&\qquad -2\mathcal{Q}(\leftthreetimes, \mathfrak{X}_1, \mathfrak{X}_3, \mathfrak{X}_5) + 2\mathcal{Q}(\leftthreetimes, \mathfrak{X}_1, \mathfrak{X}_4, \mathfrak{X}_6) - 2\mathcal{Q}(\leftthreetimes, \mathfrak{X}_2, \mathfrak{X}_3, \mathfrak{X}_6) + 2\mathcal{Q}(\leftthreetimes, \mathfrak{X}_2, \mathfrak{X}_4, \mathfrak{X}_5), \mathcal{T}\Big) \\
&\geq \mathcal{N}\Big(\mathcal{G}|2\mathcal{Q}(\leftthreetimes, \frac{\mathfrak{X}_1 + \mathfrak{X}_2}{2}, \mathfrak{X}_3 - \mathfrak{X}_4, \mathfrak{X}_5 + \mathfrak{X}_6) + 2\mathcal{Q}(\leftthreetimes, \frac{\mathfrak{X}_1 - \mathfrak{X}_2}{2}, \mathfrak{X}_3 + \mathfrak{X}_4, \mathfrak{X}_5 - \mathfrak{X}_6) \\
&\qquad -2\mathcal{Q}(\leftthreetimes, \mathfrak{X}_1, \mathfrak{X}_3, \mathfrak{X}_5) + 2\mathcal{Q}(\leftthreetimes, \mathfrak{X}_1, \mathfrak{X}_4, \mathfrak{X}_6) - 2\mathcal{Q}(\leftthreetimes, \mathfrak{X}_2, \mathfrak{X}_3, \mathfrak{X}_6) + 2\mathcal{Q}(\leftthreetimes, \mathfrak{X}_2, \mathfrak{X}_4, \mathfrak{X}_5)|, \mathcal{T}\Big) \\
&\bigotimes_{\mathrm{GTN}} diag\Big[\varphi_1\Big(\mathfrak{X}_1, \mathfrak{X}_2, \mathfrak{X}_3, \mathfrak{X}_4, \mathfrak{X}_5, \mathfrak{X}_6, \mathcal{T}\Big), \ldots, \varphi_n\Big(\mathfrak{X}_1, \mathfrak{X}_2, \mathfrak{X}_3, \mathfrak{X}_4, \mathfrak{X}_5, \mathfrak{X}_6, \mathcal{T}\Big)\Big],
\end{aligned} \tag{16.8}
$$

for every $\mathfrak{X}_i \in \mho_1, i = 1, \ldots, 6$, $\leftthreetimes \in \pitchfork$ and $\mathcal{T} > 0$. Then, there exists a unique tri–additive mapping $\mathcal{Q}' : \pitchfork \times \mho_1^3 \to \mho_2$ satisfying

$$\mathcal{N}\left(\mathcal{Q}(\curlywedge, \mathfrak{X}_1, \mathfrak{X}_3, \mathfrak{X}_5) - \mathcal{Q}'(\curlywedge, \mathfrak{X}_1, \mathfrak{X}_3, \mathfrak{X}_5), \mathcal{T}\right) \tag{16.9}$$
$$\succeq diag\left[\varphi_1\left(\mathfrak{X}_1, 0, \frac{\mathfrak{X}_3}{2}, \frac{\mathfrak{X}_3}{2}, \mathfrak{X}_5, 0, (1-4\mathfrak{P}_1)\mathcal{T}\right) \bigotimes_{\text{TN}} \varphi_1\left(\frac{\mathfrak{X}_1}{2}, \frac{\mathfrak{X}_1}{2}, \frac{\mathfrak{X}_3}{2}, 0, \mathfrak{X}_5, 0, \frac{(1-4\mathfrak{P}_1)\mathcal{T}}{2}\right) \bigotimes_{\text{TN}} \varphi_1\left(\frac{\mathfrak{X}_1}{2}, 0, \frac{\mathfrak{X}_3}{2}, 0, \frac{\mathfrak{X}_5}{2}, \frac{\mathfrak{X}_5}{2}, \frac{(1-4\mathfrak{P}_1)\mathcal{T}}{4}\right), \ldots, \varphi_n\left(\mathfrak{X}_1, 0, \frac{\mathfrak{X}_3}{2}, \frac{\mathfrak{X}_3}{2}, \mathfrak{X}_5, 0, (1-4\mathfrak{P}_n)\mathcal{T}\right) \bigotimes_{\text{TN}} \varphi_n\left(\frac{\mathfrak{X}_1}{2}, \frac{\mathfrak{X}_1}{2}, \frac{\mathfrak{X}_3}{2}, 0, \mathfrak{X}_5, 0, \frac{(1-4\mathfrak{P}_n)\mathcal{T}}{2}\right) \bigotimes_{\text{TN}} \varphi_n\left(\frac{\mathfrak{X}_1}{2}, 0, \frac{\mathfrak{X}_3}{2}, 0, \frac{\mathfrak{X}_5}{2}, \frac{\mathfrak{X}_5}{2}, \frac{(1-4\mathfrak{P}_n)\mathcal{T}}{4}\right)\right],$$

for every $\mathfrak{X}_1, \mathfrak{X}_3, \mathfrak{X}_5 \in \mho_1$ *and* $\curlywedge \in \pitchfork$.

Proof Letting $\mathfrak{X}_4 = \mathfrak{X}_6 = 0$ and $\mathfrak{X}_2 = \mathfrak{X}_1$ in the inequality (16.8), we get

$$\mathcal{N}\left(\mathcal{Q}(\curlywedge, 2\mathfrak{X}_1, \mathfrak{X}_3, \mathfrak{X}_5) - 2\mathcal{Q}(\curlywedge, \mathfrak{X}_1, \mathfrak{X}_3, \mathfrak{X}_5), \mathcal{T}\right) \succeq \text{diag}\left[\varphi_1\left(\mathfrak{X}_1, \mathfrak{X}_1, \mathfrak{X}_3, 0, \mathfrak{X}_5, 0, \mathcal{T}\right), \ldots, \varphi_n\left(\mathfrak{X}_1, \mathfrak{X}_1, \mathfrak{X}_3, 0, \mathfrak{X}_5, 0, \mathcal{T}\right)\right],$$

and so

$$\mathcal{N}\left(\mathcal{Q}(\curlywedge, 2\mathfrak{X}_1, \mathfrak{X}_3, 2\mathfrak{X}_5) - 2\mathcal{Q}(\curlywedge, \mathfrak{X}_1, \mathfrak{X}_3, 2\mathfrak{X}_5), \mathcal{T}\right) \succeq \text{diag}\left[\varphi_1\left(\mathfrak{X}_1, \mathfrak{X}_1, \mathfrak{X}_3, 0, 2\mathfrak{X}_5, 0, \mathcal{T}\right), \ldots, \varphi_n\left(\mathfrak{X}_1, \mathfrak{X}_1, \mathfrak{X}_3, 0, 2\mathfrak{X}_5, 0, \mathcal{T}\right)\right],$$

for every $\mathfrak{X}_1, \mathfrak{X}_3, \mathfrak{X}_5 \in \mho_1$, $\curlywedge \in \pitchfork$ and $\mathcal{T} > 0$.

Letting $\mathfrak{X}_2 = \mathfrak{X}_4 = 0$ and $\mathfrak{X}_6 = \mathfrak{X}_5$ in the inequality (16.8), we get

$$\mathcal{N}\left(\mathcal{Q}(\curlywedge, \mathfrak{X}_1, \mathfrak{X}_3, 2\mathfrak{X}_5) - 2\mathcal{Q}(\curlywedge, \mathfrak{X}_1, \mathfrak{X}_3, \mathfrak{X}_5), \mathcal{T}\right) \succeq \text{diag}\left[\varphi_1\left(\mathfrak{X}_1, 0, \mathfrak{X}_3, 0, \mathfrak{X}_5, \mathfrak{X}_5, \mathcal{T}\right), \ldots, \varphi_n\left(\mathfrak{X}_1, 0, \mathfrak{X}_3, 0, \mathfrak{X}_5, \mathfrak{X}_5, \mathcal{T}\right)\right],$$

and so

$$\mathcal{N}\left(\mathcal{Q}(\leftthreetimes, 2\mathfrak{X}_1, \mathfrak{X}_3, 2\mathfrak{X}_5) - 4\mathcal{Q}(\leftthreetimes, \mathfrak{X}_1, \mathfrak{X}_3, \mathfrak{X}_5), \mathcal{T}\right) \tag{16.10}$$
$$\succeq \operatorname{diag}\left[\varphi_1\left(\mathfrak{X}_1, \mathfrak{X}_1, \mathfrak{X}_3, 0, 2\mathfrak{X}_5, 0, \mathcal{T}\right) \bigotimes_{\mathrm{TN}} \varphi_1\left(\mathfrak{X}_1, 0, \mathfrak{X}_3, 0, \mathfrak{X}_5, \mathfrak{X}_5, \frac{\mathcal{T}}{2}\right), \ldots, \right.$$
$$\left. \varphi_n\left(\mathfrak{X}_1, \mathfrak{X}_1, \mathfrak{X}_3, 0, 2\mathfrak{X}_5, 0, \mathcal{T}\right) \bigotimes_{\mathrm{TN}} \varphi_n\left(\mathfrak{X}_1, 0, \mathfrak{X}_3, 0, \mathfrak{X}_5, \mathfrak{X}_5, \frac{\mathcal{T}}{2}\right)\right],$$

for every $\mathfrak{X}_1, \mathfrak{X}_3, \mathfrak{X}_5 \in \mho_1$, $\leftthreetimes \in ⋔$ and $\mathcal{T} > 0$.

Letting $\mathfrak{X}_2 = \mathfrak{X}_6 = 0$ and $\mathfrak{X}_4 = \mathfrak{X}_3$ in the inequality (16.8), we get

$$\mathcal{N}\left(\mathcal{Q}(\leftthreetimes, \mathfrak{X}_1, 2\mathfrak{X}_3, \mathfrak{X}_5) - 2\mathcal{Q}(\leftthreetimes, \mathfrak{X}_1, \mathfrak{X}_3, \mathfrak{X}_5), \mathcal{T}\right)$$
$$\succeq \operatorname{diag}\left[\varphi_1\left(\mathfrak{X}_1, 0, \mathfrak{X}_3, \mathfrak{X}_3, \mathfrak{X}_5, 0, \mathcal{T}\right), \ldots, \varphi_n\left(\mathfrak{X}_1, 0, \mathfrak{X}_3, \mathfrak{X}_3, \mathfrak{X}_5, 0, \mathcal{T}\right)\right],$$

and so

$$\mathcal{N}\left(\mathcal{Q}(\leftthreetimes, 2\mathfrak{X}_1, 2\mathfrak{X}_3, 2\mathfrak{X}_5) - 2\mathcal{Q}(\leftthreetimes, 2\mathfrak{X}_1, \mathfrak{X}_3, 2\mathfrak{X}_5), \mathcal{T}\right) \tag{16.11}$$
$$\succeq \operatorname{diag}\left[\varphi_1\left(2\mathfrak{X}_1, 0, \mathfrak{X}_3, \mathfrak{X}_3, 2\mathfrak{X}_5, 0, \mathcal{T}\right), \ldots, \varphi_n\left(2\mathfrak{X}_1, 0, \mathfrak{X}_3, \mathfrak{X}_3, 2\mathfrak{X}_5, 0, \mathcal{T}\right)\right],$$

for every $\mathfrak{X}_1, \mathfrak{X}_3, \mathfrak{X}_5 \in \mho_1$, $\leftthreetimes \in ⋔$ and $\mathcal{T} > 0$.

Making use of the inequalities (16.10) and (16.11), we have that

$$\mathcal{N}\left(\mathcal{Q}(\leftthreetimes, 2\mathfrak{X}_1, 2\mathfrak{X}_3, 2\mathfrak{X}_5) - 8\mathcal{Q}(\leftthreetimes, \mathfrak{X}_1, \mathfrak{X}_3, \mathfrak{X}_5), \mathcal{T}\right) \tag{16.12}$$
$$\succeq \operatorname{diag}\left[\varphi_1\left(2\mathfrak{X}_1, 0, \Psi_3, \Psi_3, 2\Psi_5, 0, \mathcal{T}\right)\right.$$
$$\bigotimes_{\mathrm{TN}} \varphi_1\left(\mathfrak{X}_1, \mathfrak{X}_1, \mathfrak{X}_3, 0, 2\mathfrak{X}_5, 0, \frac{\mathcal{T}}{2}\right)$$
$$\bigotimes_{\mathrm{TN}} \varphi_1\left(\mathfrak{X}_1, 0, \mathfrak{X}_3, 0, \mathfrak{X}_5, \mathfrak{X}_5, \frac{\mathcal{T}}{4}\right), \ldots,$$
$$\varphi_n\left(2\mathfrak{X}_1, 0, \Psi_3, \Psi_3, 2\Psi_5, 0, \mathcal{T}\right)$$
$$\bigotimes_{\mathrm{TN}} \varphi_n\left(\mathfrak{X}_1, \mathfrak{X}_1, \mathfrak{X}_3, 0, 2\mathfrak{X}_5, 0, \frac{\mathcal{T}}{2}\right)$$
$$\left.\bigotimes_{\mathrm{TN}} \varphi_n\left(\mathfrak{X}_1, 0, \mathfrak{X}_3, 0, \mathfrak{X}_5, \mathfrak{X}_5, \frac{\mathcal{T}}{4}\right)\right],$$

and so

$$\mathcal{N}\left(\mathcal{Q}(\lambda, \mathfrak{X}_1, \mathfrak{X}_3, \mathfrak{X}_5) - 8\mathcal{Q}\left(\lambda, \frac{\mathfrak{X}_1}{2}, \frac{\mathfrak{X}_3}{2}, \frac{\mathfrak{X}_5}{2}\right), \mathcal{T}\right) \tag{16.13}$$
$$\succeq \operatorname{diag}\Bigg[\varphi_1\left(\mathfrak{X}_1, 0, \frac{\mathfrak{X}_3}{2}, \frac{\mathfrak{X}_3}{2}, \mathfrak{X}_5, 0, \mathcal{T}\right) \bigotimes_{\mathrm{TN}} \varphi_1\left(\frac{\mathfrak{X}_1}{2}, \frac{\mathfrak{X}_1}{2}, \frac{\mathfrak{X}_3}{2}, 0, \mathfrak{X}_5, 0, \frac{\mathcal{T}}{2}\right) \bigotimes_{\mathrm{TN}} \varphi_1\left(\frac{\mathfrak{X}_1}{2}, 0, \frac{\mathfrak{X}_3}{2}, 0, \frac{\mathfrak{X}_5}{2}, \frac{\mathfrak{X}_5}{2}, \frac{\mathcal{T}}{4}\right), \ldots,$$
$$\varphi_n\left(\mathfrak{X}_1, 0, \frac{\mathfrak{X}_3}{2}, \frac{\mathfrak{X}_3}{2}, \mathfrak{X}_5, 0, \mathcal{T}\right) \bigotimes_{\mathrm{TN}} \varphi_n\left(\frac{\mathfrak{X}_1}{2}, \frac{\mathfrak{X}_1}{2}, \frac{\mathfrak{X}_3}{2}, 0, \mathfrak{X}_5, 0, \frac{\mathcal{T}}{2}\right) \bigotimes_{\mathrm{TN}} \varphi_n\left(\frac{\mathfrak{X}_1}{2}, 0, \frac{\mathfrak{X}_3}{2}, 0, \frac{\mathfrak{X}_5}{2}, \frac{\mathfrak{X}_5}{2}, \frac{\mathcal{T}}{4}\right)\Bigg],$$

for every $\mathfrak{X}_1, \mathfrak{X}_3, \mathfrak{X}_5 \in \mho_1$, $\lambda \in \pitchfork$ and $\mathcal{T} > 0$.

Consider the set below

$$\Xi := \left\{\xi : \pitchfork \times \mho_1^3 \to \mho_2,\ \ \xi(\lambda, \mathfrak{X}_1, 0, \mathfrak{X}_5) = \xi(\lambda, 0, \mathfrak{X}_3, \mathfrak{X}_5) = \xi(\lambda, \mathfrak{X}_1, \mathfrak{X}_3, 0) = 0\right\},$$

and introduce the vector valued generalized metric $\lambda : \Xi \times \Xi \longrightarrow [0, +\infty]^{3n}$, as follows:

$$\lambda(\xi, \zeta) = \left(\lambda_1(\xi, \zeta), \lambda_2(\xi, \zeta), \lambda_3(\xi, \zeta)\right) \tag{16.14}$$
$$= \inf\Bigg\{(\mu_1, \mu_2, \mu_3)^n \in \mathbb{R}_+^{3n} : \mathcal{N}\left(\xi(\lambda, \mathfrak{X}_1, \mathfrak{X}_3, \mathfrak{X}_5) - \zeta(\lambda, \mathfrak{X}_1, \mathfrak{X}_3, \mathfrak{X}_5), \mathcal{T}\right)$$
$$\preceq \varphi_1\left(\mathfrak{X}_1, 0, \frac{\mathfrak{X}_3}{2}, \frac{\mathfrak{X}_3}{2}, \mathfrak{X}_5, 0, \frac{\mathcal{T}}{\mu_1}\right) \bigotimes_{\mathrm{TN}} \varphi_1\left(\frac{\mathfrak{X}_1}{2}, \frac{\mathfrak{X}_1}{2}, \frac{\mathfrak{X}_3}{2}, 0, \mathfrak{X}_5, 0, \frac{\mathcal{T}}{\mu_2}\right) \bigotimes_{\mathrm{TN}} \varphi_1\left(\frac{\mathfrak{X}_1}{2}, 0, \frac{\mathfrak{X}_3}{2}, 0, \frac{\mathfrak{X}_5}{2}, \frac{\mathfrak{X}_5}{2}, \frac{\mathcal{T}}{\mu_3}\right), \ldots,$$
$$\varphi_n\left(\mathfrak{X}_1, 0, \frac{\mathfrak{X}_3}{2}, \frac{\mathfrak{X}_3}{2}, \mathfrak{X}_5, 0, \frac{\mathcal{T}}{\mu_1}\right) \bigotimes_{\mathrm{TN}} \varphi_n\left(\frac{\mathfrak{X}_1}{2}, \frac{\mathfrak{X}_1}{2}, \frac{\mathfrak{X}_3}{2}, 0, \mathfrak{X}_5, 0, \frac{\mathcal{T}}{\mu_2}\right) \bigotimes_{\mathrm{TN}} \varphi_n\left(\frac{\mathfrak{X}_1}{2}, 0, \frac{\mathfrak{X}_3}{2}, 0, \frac{\mathfrak{X}_5}{2}, \frac{\mathfrak{X}_5}{2}, \frac{\mathcal{T}}{\mu_3}\right),$$
$$\mathfrak{X}_1, \mathfrak{X}_3, \mathfrak{X}_5 \in \mho_1, \lambda \in \pitchfork, \text{ and } \mathcal{T} > 0\Bigg\},$$

where, as usual, $\inf \boldsymbol{\emptyset} = (\inf \emptyset, \inf \emptyset, \inf \emptyset) = (+\infty, +\infty, +\infty)$. We now prove that $(\varXi, \lambda)$ is a complete generalized metric space. For the metric part, we show that $\lambda(\xi, \zeta) \leq \lambda(\xi, \varsigma) + \lambda(\varsigma, \zeta)$.

$$\begin{aligned}
\lambda(\xi, \zeta) &= \Big(\lambda_1(\xi, \zeta), \lambda_2(\xi, \zeta), \lambda_3(\xi, \zeta)\Big) \\
&= \inf \Bigg\{ (\mu_1, \mu_2, \mu_3)^n \in \mathbb{R}_+^{3n} : \mathscr{N}\Big(\xi(\curlywedge, \mathfrak{X}_1, \mathfrak{X}_3, \mathfrak{X}_5) - \zeta(\curlywedge, \mathfrak{X}_1, \mathfrak{X}_3, \mathfrak{X}_5), \mathcal{T}\Big) \\
&\qquad \succeq \operatorname{diag}\Bigg[\varphi_1\Big(\mathfrak{X}_1, 0, \frac{\mathfrak{X}_3}{2}, \frac{\mathfrak{X}_3}{2}, \mathfrak{X}_5, 0, \frac{\mathcal{T}}{\mu_1}\Big) \\
&\qquad\qquad \bigotimes_{\mathrm{TN}} \varphi_1\Big(\frac{\mathfrak{X}_1}{2}, \frac{\mathfrak{X}_1}{2}, \frac{\mathfrak{X}_3}{2}, 0, \mathfrak{X}_5, 0, \frac{\mathcal{T}}{\mu_2}\Big) \\
&\qquad\qquad \bigotimes_{\mathrm{TN}} \varphi_1\Big(\frac{\mathfrak{X}_1}{2}, 0, \frac{\mathfrak{X}_3}{2}, 0, \frac{\mathfrak{X}_5}{2}, \frac{\mathfrak{X}_5}{2}, \frac{\mathcal{T}}{\mu_3}\Big), \ldots, \\
&\qquad\quad \varphi_n\Big(\mathfrak{X}_1, 0, \frac{\mathfrak{X}_3}{2}, \frac{\mathfrak{X}_3}{2}, \mathfrak{X}_5, 0, \frac{\mathcal{T}}{\mu_1}\Big) \\
&\qquad\qquad \bigotimes_{\mathrm{TN}} \varphi_n\Big(\frac{\mathfrak{X}_1}{2}, \frac{\mathfrak{X}_1}{2}, \frac{\mathfrak{X}_3}{2}, 0, \mathfrak{X}_5, 0, \frac{\mathcal{T}}{\mu_2}\Big) \\
&\qquad\qquad \bigotimes_{\mathrm{TN}} \varphi_n\Big(\frac{\mathfrak{X}_1}{2}, 0, \frac{\mathfrak{X}_3}{2}, 0, \frac{\mathfrak{X}_5}{2}, \frac{\mathfrak{X}_5}{2}, \frac{\mathcal{T}}{\mu_3}\Big)\Bigg], \quad \forall \mathfrak{X}_1, \mathfrak{X}_3, \mathfrak{X}_5 \in \mho_1, \curlywedge \in \pitchfork \Bigg\}
\end{aligned}$$

$$\begin{aligned}
&\leq \inf \Bigg\{ (\mu_1, \mu_2, \mu_3)^n \in \mathbb{R}_+^{3n} : \mathscr{N}\Big(\xi(\curlywedge, \mathfrak{X}_1, \mathfrak{X}_3, \mathfrak{X}_5) - \varsigma(\curlywedge, \mathfrak{X}_1, \mathfrak{X}_3, \mathfrak{X}_5), \mathcal{T}\Big) \\
&\qquad \bigotimes_{\mathrm{TN}} \mathscr{N}\Big(\varsigma(\curlywedge, \mathfrak{X}_1, \mathfrak{X}_3, \mathfrak{X}_5) - \zeta(\curlywedge, \mathfrak{X}_1, \mathfrak{X}_3, \mathfrak{X}_5), \mathcal{T}\Big) \\
&\qquad \succeq \operatorname{diag}\Bigg[\varphi_1\Big(\Psi_1, 0, \frac{\mathfrak{X}_3}{2}, \frac{\mathfrak{X}_3}{2}, \mathfrak{X}_5, 0, \frac{\mathcal{T}}{\mu_1}\Big) \\
&\qquad\qquad \bigotimes_{\mathrm{TN}} \varphi_1\Big(\frac{\mathfrak{X}_1}{2}, \frac{\mathfrak{X}_1}{2}, \frac{\mathfrak{X}_3}{2}, 0, \mathfrak{X}_5, 0, \frac{\mathcal{T}}{\mu_2}\Big) \\
&\qquad\qquad \bigotimes_{\mathrm{TN}} \varphi_1\Big(\frac{\Psi_1}{2}, 0, \frac{\Psi_3}{2}, 0, \frac{\mathfrak{X}_5}{2}, \frac{\mathfrak{X}_5}{2}, \frac{\mathcal{T}}{\mu_3}\Big), \ldots, \\
&\qquad\quad \varphi_n\Big(\Psi_1, 0, \frac{\mathfrak{X}_3}{2}, \frac{\mathfrak{X}_3}{2}, \mathfrak{X}_5, 0, \frac{\mathcal{T}}{\mu_1}\Big) \\
&\qquad\qquad \bigotimes_{\mathrm{TN}} \varphi_n\Big(\frac{\mathfrak{X}_1}{2}, \frac{\mathfrak{X}_1}{2}, \frac{\mathfrak{X}_3}{2}, 0, \mathfrak{X}_5, 0, \frac{\mathcal{T}}{\mu_2}\Big) \\
&\qquad\qquad \bigotimes_{\mathrm{TN}} \varphi_n\Big(\frac{\Psi_1}{2}, 0, \frac{\Psi_3}{2}, 0, \frac{\mathfrak{X}_5}{2}, \frac{\mathfrak{X}_5}{2}, \frac{\mathcal{T}}{\mu_3}\Big)\Bigg], \quad \forall \mathfrak{X}_1, \mathfrak{X}_3, \mathfrak{X}_5 \in \mho_1, \curlywedge \in \pitchfork \Bigg\}
\end{aligned}$$

$$\begin{aligned}
&\leq \inf \Bigg\{ (\mu_1, \mu_2, \mu_3)^n \in \mathbb{R}_+^{3n} : \mathscr{N}\Big(\xi(\curlywedge, \mathfrak{X}_1, \mathfrak{X}_3, \mathfrak{X}_5) - \varsigma(\curlywedge, \mathfrak{X}_1, \mathfrak{X}_3, \mathfrak{X}_5), \mathcal{T}\Big) \\
&\qquad \succeq \operatorname{diag}\Bigg[\varphi_1\Big(\mathfrak{X}_1, 0, \frac{\mathfrak{X}_3}{2}, \frac{\mathfrak{X}_3}{2}, \mathfrak{X}_5, 0, \frac{\mathcal{T}}{\mu_1}\Big) \\
&\qquad\qquad \bigotimes_{\mathrm{TN}} \varphi_1\Big(\frac{\mathfrak{X}_1}{2}, \frac{\mathfrak{X}_1}{2}, \frac{\mathfrak{X}_3}{2}, 0, \mathfrak{X}_5, 0, \frac{\mathcal{T}}{\mu_2}\Big) \\
&\qquad\qquad \bigotimes_{\mathrm{TN}} \varphi_1\Big(\frac{\mathfrak{X}_1}{2}, 0, \frac{\mathfrak{X}_3}{2}, 0, \frac{\mathfrak{X}_5}{2}, \frac{\mathfrak{X}_5}{2}, \frac{\mathcal{T}}{\mu_3}\Big), \ldots, \\
&\qquad\quad \varphi_n\Big(\mathfrak{X}_1, 0, \frac{\mathfrak{X}_3}{2}, \frac{\mathfrak{X}_3}{2}, \mathfrak{X}_5, 0, \frac{\mathcal{T}}{\mu_1}\Big) \\
&\qquad\qquad \bigotimes_{\mathrm{TN}} \varphi_n\Big(\frac{\mathfrak{X}_1}{2}, \frac{\mathfrak{X}_1}{2}, \frac{\mathfrak{X}_3}{2}, 0, \mathfrak{X}_5, 0, \frac{\mathcal{T}}{\mu_2}\Big)
\end{aligned}$$

$$\bigotimes_{\mathrm{TN}} \varphi_n\left(\frac{\mathfrak{X}_1}{2}, 0, \frac{\mathfrak{X}_3}{2}, 0, \frac{\mathfrak{X}_5}{2}, \frac{\mathfrak{X}_5}{2}, \frac{\mathcal{T}}{\mu_3}\right), \ \forall \mathfrak{X}_1, \mathfrak{X}_3, \mathfrak{X}_5 \in \mho_1, \leftthreetimes \in \pitchfork \Bigg\},$$

$$\begin{aligned}
&+\inf\Bigg\{(\mu_1, \mu_2, \mu_3)^n \in \mathbb{R}_+^{3n} : \mathcal{N}\left(\varsigma(\leftthreetimes, \mathfrak{X}_1, \mathfrak{X}_3, \mathfrak{X}_5) - \zeta(\leftthreetimes, \mathfrak{X}_1, \mathfrak{X}_3, \mathfrak{X}_5), \mathcal{T}\right)\\
&\qquad \succeq \operatorname{diag}\Bigg[\varphi_1\left(\Psi_1, 0, \frac{\mathfrak{X}_3}{2}, \frac{\mathfrak{X}_3}{2}, \mathfrak{X}_5, 0, \frac{\mathcal{T}}{\mu_1}\right)\\
&\qquad\qquad \bigotimes_{\mathrm{TN}} \varphi_1\left(\frac{\mathfrak{X}_1}{2}, \frac{\mathfrak{X}_1}{2}, \frac{\mathfrak{X}_3}{2}, 0, \mathfrak{X}_5, 0, \frac{\mathcal{T}}{\mu_2}\right)\\
&\qquad\qquad \bigotimes_{\mathrm{TN}} \varphi_1\left(\frac{\mathfrak{X}_1}{2}, 0, \frac{\mathfrak{X}_3}{2}, 0, \frac{\mathfrak{X}_5}{2}, \frac{\mathfrak{X}_5}{2}, \frac{\mathcal{T}}{\mu_3}\right), \ldots,\\
&\qquad\qquad \varphi_n\left(\Psi_1, 0, \frac{\mathfrak{X}_3}{2}, \frac{\mathfrak{X}_3}{2}, \mathfrak{X}_5, 0, \frac{\mathcal{T}}{\mu_1}\right)\\
&\qquad\qquad \bigotimes_{\mathrm{TN}} \varphi_n\left(\frac{\mathfrak{X}_1}{2}, \frac{\mathfrak{X}_1}{2}, \frac{\mathfrak{X}_3}{2}, 0, \mathfrak{X}_5, 0, \frac{\mathcal{T}}{\mu_2}\right)\\
&\qquad\qquad \bigotimes_{\mathrm{TN}} \varphi_n\left(\frac{\mathfrak{X}_1}{2}, 0, \frac{\mathfrak{X}_3}{2}, 0, \frac{\mathfrak{X}_5}{2}, \frac{\mathfrak{X}_5}{2}, \frac{\mathcal{T}}{\mu_3}\right), \ \forall \mathfrak{X}_1, \mathfrak{X}_3, \mathfrak{X}_5 \in \mho_1, \leftthreetimes \in \pitchfork \Bigg\},\\
&= \lambda(\xi, \varsigma) + \lambda(\varsigma, \zeta).
\end{aligned} \tag{16.15}$$

Next, we prove that (Ξ, λ) is complete. Let ω_w be a Cauchy sequence in (Ξ, λ). Thus, for every $\epsilon_1, \epsilon_2, \epsilon_3 > 0$, there exists an $\aleph_{\epsilon_1,\epsilon_2,\epsilon_3} \in \mathbb{N}$ such that $\lambda(\omega_m, \omega_w) \leq (\epsilon_1, \epsilon_2, \epsilon_3)^n$, for every $m, w \geq \aleph_{\epsilon_1,\epsilon_2,\epsilon_3}$. According to (16.14), we have

$$\begin{aligned}
&\mathcal{N}\left(\omega_m(\leftthreetimes, \mathfrak{X}_1, \mathfrak{X}_3, \mathfrak{X}_5) - \omega_w(\leftthreetimes, \mathfrak{X}_1, \mathfrak{X}_3, \mathfrak{X}_5), \mathcal{T}\right)\\
&\succeq \operatorname{diag}\Bigg[\varphi_1\left(\mathfrak{X}_1, 0, \frac{\mathfrak{X}_3}{2}, \frac{\mathfrak{X}_3}{2}, \mathfrak{X}_5, 0, \frac{\mathcal{T}}{\epsilon_1}\right)\\
&\qquad \bigotimes_{\mathrm{TN}} \varphi_1\left(\frac{\mathfrak{X}_1}{2}, \frac{\mathfrak{X}_1}{2}, \frac{\mathfrak{X}_3}{2}, 0, \mathfrak{X}_5, 0, \frac{\mathcal{T}}{\epsilon_2}\right)\\
&\qquad\qquad \bigotimes_{\mathrm{TN}} \varphi_1\left(\frac{\mathfrak{X}_1}{2}, 0, \frac{\mathfrak{X}_3}{2}, 0, \frac{\mathfrak{X}_5}{2}, \frac{\mathfrak{X}_5}{2}, \frac{\mathcal{T}}{\epsilon_3}\right), \ldots,\\
&\qquad \varphi_n\left(\mathfrak{X}_1, 0, \frac{\mathfrak{X}_3}{2}, \frac{\mathfrak{X}_3}{2}, \mathfrak{X}_5, 0, \frac{\mathcal{T}}{\epsilon_1}\right)\\
&\qquad \bigotimes_{\mathrm{TN}} \varphi_n\left(\frac{\mathfrak{X}_1}{2}, \frac{\mathfrak{X}_1}{2}, \frac{\mathfrak{X}_3}{2}, 0, \mathfrak{X}_5, 0, \frac{\mathcal{T}}{\epsilon_2}\right)\\
&\qquad\qquad \bigotimes_{\mathrm{TN}} \varphi_n\left(\frac{\mathfrak{X}_1}{2}, 0, \frac{\mathfrak{X}_3}{2}, 0, \frac{\mathfrak{X}_5}{2}, \frac{\mathfrak{X}_5}{2}, \frac{\mathcal{T}}{\epsilon_3}\right)\Bigg],
\end{aligned} \tag{16.16}$$

for every $\mathfrak{X}_1, \mathfrak{X}_3, \mathfrak{X}_5 \in \mho_1$, $\leftthreetimes \in \pitchfork$ and $\mathcal{T} > 0$. If $\mathfrak{X}_1, \mathfrak{X}_3, \mathfrak{X}_5$ be fixed, the inequality (16.16) implies that the sequence $\{\omega_w(\leftthreetimes, \mathfrak{X}_1, \mathfrak{X}_3, \mathfrak{X}_5)\}$ is a Cauchy sequence in $\mho_1$. Since $\mho_1$ is complete, the sequence $\{\omega_w(\leftthreetimes, \mathfrak{X}_1, \mathfrak{X}_3, \mathfrak{X}_5)\}$ converges for every

$\mathfrak{X}_1, \mathfrak{X}_3, \mathfrak{X}_5 \in \mho_1$, and $\curlywedge \in \pitchfork$. Thus, we can define a function ω by

$$\omega(\curlywedge, \mathfrak{X}_1, \mathfrak{X}_2, \mathfrak{X}_3) := \lim_{w \to \infty} \omega_w(\curlywedge, \mathfrak{X}_1, \mathfrak{X}_3, \mathfrak{X}_5), \tag{16.17}$$

for every $\mathfrak{X}_1, \mathfrak{X}_3, \mathfrak{X}_5 \in \mho_1$ and $\curlywedge \in \pitchfork$.

It is easy to check $\omega \in \Xi$. If we let $m \to \infty$ it follows from the inequality (16.16) that

$$\begin{aligned} &\mathcal{N}\left(\omega(\curlywedge, \mathfrak{X}_1, \mathfrak{X}_3, \mathfrak{X}_5) - \omega_w(\curlywedge, \mathfrak{X}_1, \mathfrak{X}_3, \mathfrak{X}_5), \mathcal{T}\right) \\ &\succeq \operatorname{diag}\Bigg[\varphi_1\left(\mathfrak{X}_1, 0, \frac{\mathfrak{X}_3}{2}, \frac{\mathfrak{X}_3}{2}, \mathfrak{X}_5, 0, \frac{\mathcal{T}}{\epsilon_1}\right) \\ &\qquad \bigotimes_{\mathrm{TN}} \varphi_1\left(\frac{\mathfrak{X}_1}{2}, \frac{\mathfrak{X}_1}{2}, \frac{\mathfrak{X}_3}{2}, 0, \mathfrak{X}_5, 0, \frac{\mathcal{T}}{\epsilon_2}\right) \\ &\qquad\qquad \bigotimes_{\mathrm{TN}} \varphi_1\left(\frac{\mathfrak{X}_1}{2}, 0, \frac{\mathfrak{X}_3}{2}, 0, \frac{\mathfrak{X}_5}{2}, \frac{\mathfrak{X}_5}{2}, \frac{\mathcal{T}}{\epsilon_3}\right), \ldots, \\ &\qquad \varphi_n\left(\mathfrak{X}_1, 0, \frac{\mathfrak{X}_3}{2}, \frac{\mathfrak{X}_3}{2}, \mathfrak{X}_5, 0, \frac{\mathcal{T}}{\epsilon_1}\right) \\ &\qquad \bigotimes_{\mathrm{TN}} \varphi_n\left(\frac{\mathfrak{X}_1}{2}, \frac{\mathfrak{X}_1}{2}, \frac{\mathfrak{X}_3}{2}, 0, \mathfrak{X}_5, 0, \frac{\mathcal{T}}{\epsilon_2}\right) \\ &\qquad\qquad \bigotimes_{\mathrm{TN}} \varphi_n\left(\frac{\mathfrak{X}_1}{2}, 0, \frac{\mathfrak{X}_3}{2}, 0, \frac{\mathfrak{X}_5}{2}, \frac{\mathfrak{X}_5}{2}, \frac{\mathcal{T}}{\epsilon_3}\right)\Bigg], \end{aligned} \tag{16.18}$$

for every $\mathfrak{X}_1, \mathfrak{X}_3, \mathfrak{X}_5 \in \mho_1$ and $\curlywedge \in \pitchfork$.

Considering (16.14), we get

$$\lambda(\omega, \omega_w) \preceq (\epsilon_1, \epsilon_2, \epsilon_3)^n.$$

This implies that the Cauchy sequence $\{\omega_w\}$ is convergent to ω in (Ξ, λ). Therefore, (Ξ, λ) is complete.

We define the linear mapping $\Gamma : \Xi \to \Xi$, as follows:

$$\Gamma(\xi(\curlywedge, \mathfrak{X}_1, \mathfrak{X}_3, \mathfrak{X}_5)) := 8\xi\left(\curlywedge, \frac{\mathfrak{X}_1}{2}, \frac{\mathfrak{X}_3}{2}, \frac{\mathfrak{X}_5}{2}\right)$$

for every $\mathfrak{X}_1, \mathfrak{X}_3, \mathfrak{X}_5 \in \mho_1$ and $\curlywedge \in \pitchfork$.

Let $\xi, \zeta \in \Xi$ be given such that $\lambda(\xi, \zeta) = (\varepsilon_1, \varepsilon_2, \varepsilon_3)^n$. Then, we get

$$\begin{aligned} &\mathcal{N}\left(\xi(\curlywedge, \mathfrak{X}_1, \mathfrak{X}_3, \mathfrak{X}_5) - \zeta(\curlywedge, \mathfrak{X}_1, \mathfrak{X}_3, \mathfrak{X}_5), \mathcal{T}\right) \\ &\succeq \operatorname{diag}\Bigg[\varphi_1\left(\mathfrak{X}_1, 0, \frac{\mathfrak{X}_3}{2}, \frac{\mathfrak{X}_3}{2}, \mathfrak{X}_5, 0, \frac{\mathcal{T}}{\epsilon_1}\right) \end{aligned}$$

$$\bigotimes_{\text{TN}} \varphi_1\left(\frac{\mathfrak{X}_1}{2}, \frac{\mathfrak{X}_1}{2}, \frac{\mathfrak{X}_3}{2}, 0, \mathfrak{X}_5, 0, \frac{\mathcal{T}}{\epsilon_2}\right)$$
$$\bigotimes_{\text{TN}} \varphi_1\left(\frac{\mathfrak{X}_1}{2}, 0, \frac{\mathfrak{X}_3}{2}, 0, \frac{\mathfrak{X}_5}{2}, \frac{\mathfrak{X}_5}{2}, \frac{\mathcal{T}}{\epsilon_3}\right), \ldots,$$
$$\varphi_n\left(\mathfrak{X}_1, 0, \frac{\mathfrak{X}_3}{2}, \frac{\mathfrak{X}_3}{2}, \mathfrak{X}_5, 0, \frac{\mathcal{T}}{\epsilon_1}\right)$$
$$\bigotimes_{\text{TN}} \varphi_n\left(\frac{\mathfrak{X}_1}{2}, \frac{\mathfrak{X}_1}{2}, \frac{\mathfrak{X}_3}{2}, 0, \mathfrak{X}_5, 0, \frac{\mathcal{T}}{\epsilon_2}\right)$$
$$\bigotimes_{\text{TN}} \varphi_n\left(\frac{\mathfrak{X}_1}{2}, 0, \frac{\mathfrak{X}_3}{2}, 0, \frac{\mathfrak{X}_5}{2}, \frac{\mathfrak{X}_5}{2}, \frac{\mathcal{T}}{\epsilon_3}\right)\Bigg],$$

for every $\mathfrak{X}_1, \mathfrak{X}_3, \mathfrak{X}_5 \in \mho_1$, $\curlywedge \in \pitchfork$ and $\mathcal{T} > 0$.

Thus, we have that

$$\begin{aligned}
&\mathcal{N}\left(\Gamma(\xi(\curlywedge, \mathfrak{X}_1, \mathfrak{X}_3, \mathfrak{X}_5)) - \Gamma(\zeta(\curlywedge, \mathfrak{X}_1, \mathfrak{X}_3, \mathfrak{X}_5)), \mathcal{T}\right) \qquad (16.19)\\
= &\mathcal{N}\left(8\xi(\curlywedge, \frac{\mathfrak{X}_1}{2}, \frac{\mathfrak{X}_3}{2}, \frac{\mathfrak{X}_5}{2}) - 8\zeta(\curlywedge, \frac{\mathfrak{X}_1}{2}, \frac{\mathfrak{X}_3}{2}, \frac{\mathfrak{X}_5}{2}), \mathcal{T}\right)\\
\succeq &\operatorname{diag}\Bigg[\varphi_1\left(\frac{\mathfrak{X}_1}{2}, 0, \frac{\mathfrak{X}_3}{4}, \frac{\mathfrak{X}_3}{4}, \frac{\mathfrak{X}_5}{2}, 0, \frac{1}{8\varepsilon_1}\mathcal{T}\right)\\
&\bigotimes_{\text{TN}} \varphi_1\left(\frac{\mathfrak{X}_1}{4}, \frac{\mathfrak{X}_1}{4}, \frac{\mathfrak{X}_3}{4}, 0, \frac{\mathfrak{X}_5}{2}, 0, \frac{1}{8\varepsilon_2}\mathcal{T}\right)\\
&\bigotimes_{\text{TN}} \varphi_1\left(\frac{\mathfrak{X}_1}{4}, 0, \frac{\mathfrak{X}_3}{4}, 0, \frac{\mathfrak{X}_5}{4}, \frac{\mathfrak{X}_5}{4}, \frac{1}{8\varepsilon_3}\mathcal{T}\right), \ldots,\\
&\varphi_n\left(\frac{\mathfrak{X}_1}{2}, 0, \frac{\mathfrak{X}_3}{4}, \frac{\mathfrak{X}_3}{4}, \frac{\mathfrak{X}_5}{2}, 0, \frac{1}{8\varepsilon_1}\mathcal{T}\right)\\
&\bigotimes_{\text{TN}} \varphi_n\left(\frac{\mathfrak{X}_1}{4}, \frac{\mathfrak{X}_1}{4}, \frac{\mathfrak{X}_3}{4}, 0, \frac{\mathfrak{X}_5}{2}, 0, \frac{1}{8\varepsilon_2}\mathcal{T}\right)\\
&\bigotimes_{\text{TN}} \varphi_n\left(\frac{\mathfrak{X}_1}{4}, 0, \frac{\mathfrak{X}_3}{4}, 0, \frac{\mathfrak{X}_5}{4}, \frac{\mathfrak{X}_5}{4}, \frac{1}{8\varepsilon_3}\mathcal{T}\right)\Bigg]\\
\succeq &\operatorname{diag}\Bigg[\varphi_1\left(\mathfrak{X}_1, 0, \frac{\mathfrak{X}_3}{2}, \frac{\mathfrak{X}_3}{2}, \mathfrak{X}_5, 0, \frac{1}{4\mathfrak{P}_1\varepsilon_1}\mathcal{T}\right)\\
&\bigotimes_{\text{TN}} \varphi_1\left(\frac{\mathfrak{X}_1}{2}, \frac{\mathfrak{X}_1}{2}, \frac{\mathfrak{X}_3}{2}, 0, \mathfrak{X}_5, 0, \frac{1}{4\mathfrak{P}_1\varepsilon_2}\mathcal{T}\right)\\
&\bigotimes_{\text{TN}} \varphi_1\left(\frac{\mathfrak{X}_1}{2}, 0, \frac{\mathfrak{X}_3}{2}, 0, \frac{\mathfrak{X}_5}{2}, \frac{\mathfrak{X}_5}{2}, \frac{1}{4\mathfrak{P}_1\varepsilon_3}\mathcal{T}\right), \ldots,
\end{aligned}$$

$$\varphi_n\left(\mathfrak{X}_1, 0, \frac{\mathfrak{X}_3}{2}, \frac{\mathfrak{X}_3}{2}, \mathfrak{X}_5, 0, \frac{1}{4\mathfrak{P}_n\varepsilon_1}\mathcal{T}\right)$$
$$\bigotimes_{\mathrm{TN}} \varphi_n\left(\frac{\mathfrak{X}_1}{2}, \frac{\mathfrak{X}_1}{2}, \frac{\mathfrak{X}_3}{2}, 0, \mathfrak{X}_5, 0, \frac{1}{4\mathfrak{P}_n\varepsilon_2}\mathcal{T}\right)$$
$$\bigotimes_{\mathrm{TN}} \varphi_n\left(\frac{\mathfrak{X}_1}{2}, 0, \frac{\mathfrak{X}_3}{2}, 0, \frac{\mathfrak{X}_5}{2}, \frac{\mathfrak{X}_5}{2}, \frac{1}{4\mathfrak{P}_n\varepsilon_3}\mathcal{T}\right)\Bigg],$$

for every $\mathfrak{X}_1, \mathfrak{X}_3, \mathfrak{X}_5 \in \mho_1$, $\leftthreetimes \in \pitchfork$, and $\mathcal{T} > 0$. Then, we get

$$\lambda(\Gamma\xi, \Gamma\zeta) \leq \Big(4\mathfrak{P}_1(\varepsilon_1, \varepsilon_2, \varepsilon_3), \ldots, 4\mathfrak{P}_n(\varepsilon_1, \varepsilon_2, \varepsilon_3)\Big).$$

This means that

$$\lambda(\Gamma\xi, \Gamma\zeta) \leq (4\mathfrak{P}_1, \ldots, 4\mathfrak{P}_n)\lambda(\xi, \zeta)$$

for every $\xi, \zeta \in \Xi$.

It follows from (16.13) that

$$\mathcal{N}\left(\mathcal{Q}(\leftthreetimes, \mathfrak{X}_1, \mathfrak{X}_3, \mathfrak{X}_5) - 8\mathcal{Q}\left(\leftthreetimes, \frac{\mathfrak{X}_1}{2}, \frac{\mathfrak{X}_3}{2}, \frac{\mathfrak{X}_5}{2}\right), \mathcal{T}\right) \tag{16.20}$$
$$\succeq \operatorname{diag}\Bigg[\varphi_1\left(\mathfrak{X}_1, 0, \frac{\mathfrak{X}_3}{2}, \frac{\mathfrak{X}_3}{2}, \mathfrak{X}_5, 0, \mathcal{T}\right)$$
$$\bigotimes_{\mathrm{TN}} \varphi_1\left(\frac{\mathfrak{X}_1}{2}, \frac{\mathfrak{X}_1}{2}, \frac{\Psi_3}{2}, 0, \mathfrak{X}_5, 0, 2\mathcal{T}\right)$$
$$\bigotimes_{\mathrm{TN}} \varphi_1\left(\frac{\mathfrak{X}_1}{2}, 0, \frac{\mathfrak{X}_3}{2}, 0, \frac{\mathfrak{X}_5}{2}, \frac{\mathfrak{X}_5}{2}, 4\mathcal{T}\right), \ldots,$$
$$\varphi_n\left(\mathfrak{X}_1, 0, \frac{\mathfrak{X}_3}{2}, \frac{\mathfrak{X}_3}{2}, \mathfrak{X}_5, 0, \mathcal{T}\right)$$
$$\bigotimes_{\mathrm{TN}} \varphi_n\left(\frac{\mathfrak{X}_1}{2}, \frac{\mathfrak{X}_1}{2}, \frac{\Psi_3}{2}, 0, \mathfrak{X}_5, 0, 2\mathcal{T}\right)$$
$$\bigotimes_{\mathrm{TN}} \varphi_n\left(\frac{\mathfrak{X}_1}{2}, 0, \frac{\mathfrak{X}_3}{2}, 0, \frac{\mathfrak{X}_5}{2}, \frac{\mathfrak{X}_5}{2}, 4\mathcal{T}\right)\Bigg],$$

for every $\mathfrak{X}_1, \mathfrak{X}_3, \mathfrak{X}_5 \in \mho_1$, $\leftthreetimes \in \pitchfork$, $\mathcal{T} > 0$ and $\mathcal{Q} \in \Xi$. So $\lambda(\mathcal{Q}, \Gamma\mathcal{Q}) \leq (1, 2, 4)^n$.

In view of Theorem 2.20, we can find a mapping $\mathcal{Q}' : \pitchfork \times \mho_1^3 \to \mho_2$ such that,
(1) The map $\mathcal{Q}'$ is a fixed point of Γ, i.e.,

$$\Gamma(\mathcal{Q}'(\leftthreetimes, \mathfrak{X}_1, \mathfrak{X}_3, \mathfrak{X}_5)) = \mathcal{Q}'(\leftthreetimes, \mathfrak{X}_1, \mathfrak{X}_3, \mathfrak{X}_5) := 8\mathcal{Q}\left(\leftthreetimes, \frac{\mathfrak{X}_1}{2}, \frac{\mathfrak{X}_3}{2}, \frac{\mathfrak{X}_5}{2}\right) \tag{16.21}$$

for every $\mathfrak{X}_1, \mathfrak{X}_3, \mathfrak{X}_5 \in \mho_1$ and $\leftthreetimes \in \pitchfork$. The mapping $\mathcal{Q}'$ is a unique fixed point of Γ. We conclude that $\mathcal{Q}'$ is a unique mapping satisfying (16.21) such that there exists $(\mu_1, \mu_2, \mu_3) \in (0, \infty)^3$ satisfying

$$\begin{aligned}
&\mathcal{N}\Big(\mathcal{Q}(\leftthreetimes, \mathfrak{X}_1, \mathfrak{X}_3, \mathfrak{X}_5) - \mathcal{Q}'(\leftthreetimes, \mathfrak{X}_1, \mathfrak{X}_3, \mathfrak{X}_5), \mathcal{T}\Big)\\
&\succeq \operatorname{diag}\Big[\varphi_1\Big(\mathfrak{X}_1, 0, \frac{\mathfrak{X}_3}{2}, \frac{\mathfrak{X}_3}{2}, \mathfrak{X}_5, 0, \frac{\mathcal{T}}{\mu_1}\Big)\\
&\qquad \bigotimes_{\mathrm{TN}} \varphi_1\Big(\frac{\mathfrak{X}_1}{2}, \frac{\mathfrak{X}_1}{2}, \frac{\mathfrak{X}_3}{2}, 0, \mathfrak{X}_5, 0, \frac{\mathcal{T}}{\mu_2}\Big)\\
&\qquad\qquad \bigotimes_{\mathrm{TN}} \varphi_1\Big(\frac{\mathfrak{X}_1}{2}, 0, \frac{\mathfrak{X}_3}{2}, 0, \frac{\mathfrak{X}_5}{2}, \frac{\mathfrak{X}_5}{2}, \frac{\mathcal{T}}{\mu_3}\Big), \ldots,\\
&\quad \varphi_n\Big(\mathfrak{X}_1, 0, \frac{\mathfrak{X}_3}{2}, \frac{\mathfrak{X}_3}{2}, \mathfrak{X}_5, 0, \frac{\mathcal{T}}{\mu_1}\Big)\\
&\qquad \bigotimes_{\mathrm{TN}} \varphi_n\Big(\frac{\mathfrak{X}_1}{2}, \frac{\mathfrak{X}_1}{2}, \frac{\mathfrak{X}_3}{2}, 0, \mathfrak{X}_5, 0, \frac{\mathcal{T}}{\mu_2}\Big)\\
&\qquad\qquad \bigotimes_{\mathrm{TN}} \varphi_n\Big(\frac{\mathfrak{X}_1}{2}, 0, \frac{\mathfrak{X}_3}{2}, 0, \frac{\mathfrak{X}_5}{2}, \frac{\mathfrak{X}_5}{2}, \frac{\mathcal{T}}{\mu_3}\Big)\Big],
\end{aligned}$$

for every $\mathfrak{X}_1, \mathfrak{X}_3, \mathfrak{X}_5 \in \mho_1$, $\leftthreetimes \in \pitchfork$ and $\mathcal{T} > 0$.
(2) $\lambda(\Gamma^k \mathcal{L}, \mathcal{L}') \to (0, 0, 0)^n$ as $k \to \infty$. This implies the following equality

$$\mathcal{Q}'(\leftthreetimes, \mathfrak{X}_1, \mathfrak{X}_3, \mathfrak{X}_5) = \lim_{k\to\infty} 8^k \mathcal{Q}\Big(\leftthreetimes, \frac{\mathfrak{X}_1}{2^k}, \frac{\mathfrak{X}_3}{2^k}, \frac{\mathfrak{X}_5}{2^k}\Big)$$

for every $\mathfrak{X}_1, \mathfrak{X}_3, \mathfrak{X}_5 \in \mho_1$ and $\leftthreetimes \in \pitchfork$.
(3) $\lambda(\mathcal{Q}, \mathcal{Q}') \preceq (\frac{1}{1-4\mathfrak{P}_1}, \ldots, \frac{1}{1-4\mathfrak{P}_n})\lambda(\mathcal{Q}, \Gamma\mathcal{Q})$, which implies

$$\begin{aligned}
&\mathcal{N}\Big(\mathcal{Q}(\leftthreetimes, \mathfrak{X}_1, \mathfrak{X}_3, \mathfrak{X}_5) - \mathcal{Q}'(\leftthreetimes, \mathfrak{X}_1, \mathfrak{X}_3, \mathfrak{X}_5), \mathcal{T}\Big)\\
&\succeq \operatorname{diag}\Big[\varphi_1\Big(\mathfrak{X}_1, 0, \frac{\mathfrak{X}_3}{2}, \frac{\mathfrak{X}_3}{2}, \mathfrak{X}_5, 0, (1 - 4\mathfrak{P}_1)\mathcal{T}\Big)\\
&\qquad \bigotimes_{\mathrm{TN}} \varphi_1\Big(\frac{\mathfrak{X}_1}{2}, \frac{\mathfrak{X}_1}{2}, \frac{\mathfrak{X}_3}{2}, 0, \mathfrak{X}_5, 0, \frac{1 - 4\mathfrak{P}_1}{2}\mathcal{T}\Big)\\
&\qquad\qquad \bigotimes_{\mathrm{TN}} \varphi_1\Big(\frac{\mathfrak{X}_1}{2}, 0, \frac{\mathfrak{X}_3}{2}, 0, \frac{\mathfrak{X}_5}{2}, \frac{\mathfrak{X}_5}{2}, \frac{1 - 4\mathfrak{P}_1}{4}\mathcal{T}\Big), \ldots,\\
&\quad \varphi_n\Big(\mathfrak{X}_1, 0, \frac{\mathfrak{X}_3}{2}, \frac{\mathfrak{X}_3}{2}, \mathfrak{X}_5, 0, (1 - 4\mathfrak{P}_n)\mathcal{T}\Big)
\end{aligned}$$

$$\bigotimes_{\mathrm{TN}} \varphi_n\left(\frac{\mathfrak{X}_1}{2}, \frac{\mathfrak{X}_1}{2}, \frac{\mathfrak{X}_3}{2}, 0, \mathfrak{X}_5, 0, \frac{1-4\mathfrak{P}_n}{2}\mathcal{T}\right)$$
$$\bigotimes_{\mathrm{TN}} \varphi_n\left(\frac{\mathfrak{X}_1}{2}, 0, \frac{\mathfrak{X}_3}{2}, 0, \frac{\mathfrak{X}_5}{2}, \frac{\mathfrak{X}_5}{2}, \frac{1-4\mathfrak{P}_n}{4}\mathcal{T}\right)\Bigg],$$

for every $\mathfrak{X}_1, \mathfrak{X}_3, \mathfrak{X}_5 \in \mho_1$, $\curlywedge \in ⫛$ and $\mathcal{T} > 0$.

Now, let $\mathcal{Q}'' : ⫛ \times \mho_1^3 \to \mho_2$ be another additive mapping satisfying (16.9). Then, we have that

$$\begin{aligned}
&\mathcal{N}\Big(\mathcal{Q}''(\curlywedge, \mathfrak{X}_1, \mathfrak{X}_3, \mathfrak{X}_5) - \mathcal{Q}'(\curlywedge, \mathfrak{X}_1, \mathfrak{X}_3, \mathfrak{X}_5, \mathcal{T}\Big) \\
&= \mathcal{N}\left(8^q \mathcal{Q}''\left(\curlywedge, \frac{\mathfrak{X}_1}{2^q}, \frac{\mathfrak{X}_3}{2^q}, \frac{\mathfrak{X}_5}{2^q}\right) - 8^q \mathcal{Q}'\left(\curlywedge, \frac{\mathfrak{X}_1}{2^q}, \frac{\mathfrak{X}_3}{2^q}, \frac{\mathfrak{X}_5}{2^q}\right), \mathcal{T}\right) \\
&\succeq \mathcal{N}\left(8^q \mathcal{Q}''\left(\curlywedge, \frac{\mathfrak{X}_1}{2^q}, \frac{\mathfrak{X}_3}{2^q}, \frac{\mathfrak{X}_5}{2^q}\right) - 8^q \mathcal{Q}\left(\curlywedge, \frac{\mathfrak{X}_1}{2^q}, \frac{\mathfrak{X}_3}{2^q}, \frac{\mathfrak{X}_5}{2^q}\right), \mathcal{T}\right) \\
&\qquad \bigotimes_{\mathrm{GTN}} \mathcal{N}\left(8^q \mathcal{Q}'\left(\curlywedge, \frac{\mathfrak{X}_1}{2^q}, \frac{\mathfrak{X}_3}{2^q}, \frac{\mathfrak{X}_5}{2^q}\right) - 8^q \mathcal{Q}\left(\curlywedge, \frac{\mathfrak{X}_1}{2^q}, \frac{\mathfrak{X}_3}{2^q}, \frac{\mathfrak{X}_5}{2^q}\right), \mathcal{T}\right) \\
&\succeq \mathrm{diag}\Bigg[\varphi_1\left(\frac{\mathfrak{X}_1}{2^q}, 0, \frac{\mathfrak{X}_3}{2^{q+1}}, \frac{\mathfrak{X}_3}{2^{q+1}}, \frac{\mathfrak{X}_5}{2^q}, 0, \frac{1-4\mathfrak{P}_1}{2.8^q}\mathcal{T}\right) \\
&\qquad \bigotimes_{\mathrm{TN}} \varphi_1\left(\frac{\mathfrak{X}_1}{2^{q+1}}, \frac{\mathfrak{X}_1}{2^{q+1}}, \frac{\mathfrak{X}_3}{2^{q+1}}, 0, \frac{\mathfrak{X}_5}{2^q}, 0, \frac{1-4\mathfrak{P}_1}{4.8^q}\mathcal{T}\right) \\
&\qquad\qquad \bigotimes_{\mathrm{TN}} \varphi_1\left(\frac{\mathfrak{X}_1}{2^{q+1}}, 0, \frac{\mathfrak{X}_3}{2^{q+1}}, 0, \frac{\mathfrak{X}_5}{2^{q+1}}, \frac{\mathfrak{X}_5}{2^{q+1}}, \frac{1-4\mathfrak{P}_1}{8.8^q}\mathcal{T}\right), \ldots, \\
&\quad \varphi_n\left(\frac{\mathfrak{X}_1}{2^q}, 0, \frac{\mathfrak{X}_3}{2^{q+1}}, \frac{\mathfrak{X}_3}{2^{q+1}}, \frac{\mathfrak{X}_5}{2^q}, 0, \frac{1-4\mathfrak{P}_n}{2.8^q}\mathcal{T}\right) \\
&\qquad \bigotimes_{\mathrm{TN}} \varphi_n\left(\frac{\mathfrak{X}_1}{2^{q+1}}, \frac{\mathfrak{X}_1}{2^{q+1}}, \frac{\mathfrak{X}_3}{2^{q+1}}, 0, \frac{\mathfrak{X}_5}{2^q}, 0, \frac{1-4\mathfrak{P}_n}{4.8^q}\mathcal{T}\right) \\
&\qquad\qquad \bigotimes_{\mathrm{TN}} \varphi_n\left(\frac{\mathfrak{X}_1}{2^{q+1}}, 0, \frac{\mathfrak{X}_3}{2^{q+1}}, 0, \frac{\mathfrak{X}_5}{2^{q+1}}, \frac{\mathfrak{X}_5}{2^{q+1}}, \frac{1-4\mathfrak{P}_n}{8.8^q}\mathcal{T}\right)\Bigg],
\end{aligned}$$

which tends to **1** as $q \to \infty$, for every $\mathfrak{X}_1, \mathfrak{X}_3, \mathfrak{X}_5 \in \mho_1$, $\curlywedge \in ⫛$ and $\mathcal{T} > 0$. So we can conclude that $\mathcal{Q}''(\curlywedge, \mathfrak{X}_1, \mathfrak{X}_3, \mathfrak{X}_5) = \mathcal{Q}'(\curlywedge, \mathfrak{X}_1, \mathfrak{X}_3, \mathfrak{X}_5)$, for every $\mathfrak{X}_1, \mathfrak{X}_3, \mathfrak{X}_5 \in \mho_1$ and $\curlywedge \in ⫛$. This proves the uniqueness of $\mathcal{Q}'$.

It follows from (16.8) that

$$\begin{aligned}
&\mathcal{N}\Big(\mathcal{Q}'(\curlywedge, \mathfrak{X}_1 + \mathfrak{X}_2, \mathfrak{X}_3 - \mathfrak{X}_4, \mathfrak{X}_5 + \mathfrak{X}_6) + \mathcal{Q}'(\curlywedge, \mathfrak{X}_1 - \mathfrak{X}_2, \mathfrak{X}_3 + \mathfrak{X}_4, \mathfrak{X}_5 - \mathfrak{X}_6), \\
&\quad - 2\mathcal{Q}'(\curlywedge, \psi_1, \psi_3, \psi_5) + 2\mathcal{Q}'(\curlywedge, \mathfrak{X}_1, \mathfrak{X}_4, \mathfrak{X}_6) - 2\mathcal{Q}'(\curlywedge, \mathfrak{X}_2, \mathfrak{X}_3, \mathfrak{X}_6) + 2\mathcal{Q}'(\curlywedge, \mathfrak{X}_2, \mathfrak{X}_4, \mathfrak{X}_5), \mathcal{T}\Big) \\
&= \lim_{q\to\infty} \mathcal{N}\Big(8^q|\mathcal{Q}(\curlywedge, \frac{\mathfrak{X}_1 + \mathfrak{X}_2}{2^q}, \frac{\mathfrak{X}_3 - \mathfrak{X}_4}{2^q}, \frac{\mathfrak{X}_5 + \mathfrak{X}_6}{2^q}) + \mathcal{Q}(\curlywedge, \frac{\mathfrak{X}_1 - \mathfrak{X}_2}{2^q}, \frac{\mathfrak{X}_3 + \mathfrak{X}_4}{2^q}, \frac{\mathfrak{X}_5 - \mathfrak{X}_6}{2^q}) \\
&\qquad - 2\mathcal{Q}(\curlywedge, \frac{\mathfrak{X}_1}{2^q}, \frac{\mathfrak{X}_3}{2^q}, \frac{\mathfrak{X}_5}{2^q}) + 2\mathcal{Q}(\curlywedge, \frac{\mathfrak{X}_1}{2^q}, \frac{\mathfrak{X}_4}{2^q}, \frac{\mathfrak{X}_6}{2^q})
\end{aligned}$$

$$-2\mathcal{Q}(\curlywedge,\frac{\mathfrak{X}_2}{2^q},\frac{\mathfrak{X}_3}{2^q},\frac{\mathfrak{X}_6}{2^q})+2\mathcal{Q}(\curlywedge,\frac{\mathfrak{X}_2}{2^q},\frac{\mathfrak{X}_4}{2^q},\frac{\mathfrak{X}_5}{2^q})],\mathcal{T}\Big)$$

$$\succeq \lim_{q\to\infty}\mathcal{N}\Big(8^q\Upsilon[2\mathcal{Q}(\curlywedge,\frac{\mathfrak{X}_1+\mathfrak{X}_2}{2^{q+1}},\frac{\mathfrak{X}_3-\mathfrak{X}_4}{2^q},\frac{\mathfrak{X}_5+\mathfrak{X}_6}{2^q})+2\mathcal{Q}(\curlywedge,\frac{\mathfrak{X}_1-\mathfrak{X}_2}{2^{q+1}},\frac{\mathfrak{X}_3+\mathfrak{X}_4}{2^q},\frac{\mathfrak{X}_5-\mathfrak{X}_6}{2^q}\Big)$$

$$-2\mathcal{Q}\Big(\curlywedge,\frac{\mathfrak{X}_1}{2^q},\frac{\mathfrak{X}_3}{2^q},\frac{\mathfrak{X}_5}{2^q})+2\mathcal{Q}(\curlywedge,\frac{\mathfrak{X}_1}{2^q},\frac{\mathfrak{X}_4}{2^q},\frac{\mathfrak{X}_6}{2^q}\Big)$$

$$-2\mathcal{Q}(\curlywedge,\frac{\mathfrak{X}_2}{2^q},\frac{\mathfrak{X}_3}{2^q},\frac{\mathfrak{X}_6}{2^q})+2\mathcal{Q}(\curlywedge,\frac{\mathfrak{X}_2}{2^q},\frac{\mathfrak{X}_4}{2^q},\frac{\mathfrak{X}_5}{2^q})],\mathcal{T}\Big)$$

$$\bigotimes_{\mathrm{GTN}}\mathrm{diag}\Big[\lim_{q\to\infty}\varphi_1\Big(\frac{\mathfrak{X}_1}{2^q},\frac{\mathfrak{X}_2}{2^q},\frac{\mathfrak{X}_3}{2^q},\frac{\mathfrak{X}_4}{2^q},\frac{\mathfrak{X}_5}{2^q},\frac{\mathfrak{X}_6}{2^q},\frac{\mathcal{T}}{8^q}\Big),\ldots,$$

$$\lim_{q\to\infty}\varphi_n\Big(\frac{\mathfrak{X}_1}{2^q},\frac{\mathfrak{X}_2}{2^q},\frac{\mathfrak{X}_3}{2^q},\frac{\mathfrak{X}_4}{2^q},\frac{\mathfrak{X}_5}{2^q},\frac{\mathfrak{X}_6}{2^q},\frac{\mathcal{T}}{8^q}\Big)\Big]$$

$$\succeq \mathcal{N}\Big(\Upsilon[2\mathcal{Q}'(\curlywedge,\frac{\mathfrak{X}_1+\mathfrak{X}_2}{2},\mathfrak{X}_3-\mathfrak{X}_4,\mathfrak{X}_5+\mathfrak{X}_6)+2\mathcal{Q}'(\curlywedge,\frac{\mathfrak{X}_1-\mathfrak{X}_2}{2},\mathfrak{X}_3+\mathfrak{X}_4,\mathfrak{X}_5-\mathfrak{X}_6)$$

$$-2\mathcal{Q}'(\curlywedge,\mathfrak{X}_1,\mathfrak{X}_3,\mathfrak{X}_5)+2\mathcal{Q}'(\curlywedge,\mathfrak{X}_1,\mathfrak{X}_4,\mathfrak{X}_6)$$

$$-2\mathcal{Q}'(\curlywedge,\mathfrak{X}_2,\mathfrak{X}_3,\mathfrak{X}_6)+2\mathcal{Q}'(\curlywedge,\mathfrak{X}_2,\mathfrak{X}_4,\mathfrak{X}_5)],\mathcal{T}\Big),$$

for every $\mathfrak{X}_1,\mathfrak{X}_2,\mathfrak{X}_3,\mathfrak{X}_4,\mathfrak{X}_5,\mathfrak{X}_6\in\mho_1$, $\curlywedge\in\pitchfork$ and $\mathcal{T}>0$. So, we get

$$\mathcal{N}\Big(\mathcal{Q}'(\curlywedge,\mathfrak{X}_1+\mathfrak{X}_2,\mathfrak{X}_3-\mathfrak{X}_4,\mathfrak{X}_5+\mathfrak{X}_6)+\mathcal{Q}'(\curlywedge,\mathfrak{X}_1-\mathfrak{X}_2,\mathfrak{X}_3+\mathfrak{X}_4,\mathfrak{X}_5-\mathfrak{X}_6)$$

$$-2\mathcal{Q}'(\curlywedge,\mathfrak{X}_1,\mathfrak{X}_3,\mathfrak{X}_5)+2\mathcal{Q}'(\curlywedge,\mathfrak{X}_1,\mathfrak{X}_4,\mathfrak{X}_6)$$

$$-2\mathcal{Q}'(\curlywedge,\mathfrak{X}_2,\mathfrak{X}_3,\mathfrak{X}_6)+2\mathcal{Q}'(\curlywedge,\mathfrak{X}_2,\mathfrak{X}_4,\mathfrak{X}_5),\mathcal{T}\Big)$$

$$\succeq \mathcal{N}\Big(\Upsilon[2\mathcal{Q}'(\curlywedge,\frac{\mathfrak{X}_1+\mathfrak{X}_2}{2},\mathfrak{X}_3-\mathfrak{X}_4,\mathfrak{X}_5+\mathfrak{X}_6)+2\mathcal{Q}'(\curlywedge,\frac{\mathfrak{X}_1-\mathfrak{X}_2}{2},\mathfrak{X}_3+\mathfrak{X}_4,\mathfrak{X}_5-\mathfrak{X}_6)$$

$$-2\mathcal{Q}'(\curlywedge,\mathfrak{X}_1,\mathfrak{X}_3,\mathfrak{X}_5)+2\mathcal{Q}'(\curlywedge,\mathfrak{X}_1,\mathfrak{X}_4,\mathfrak{X}_6)$$

$$-2\mathcal{Q}'(\curlywedge,\mathfrak{X}_2,\mathfrak{X}_3,\mathfrak{X}_6)+2\mathcal{Q}'(\curlywedge,\mathfrak{X}_2,\mathfrak{X}_4,\mathfrak{X}_5)],\mathcal{T}\Big)$$

for every $\mathfrak{X}_1,\mathfrak{X}_2,\mathfrak{X}_3,\mathfrak{X}_4,\mathfrak{X}_5,\mathfrak{X}_6\in\mho_1$, $\curlywedge\in\pitchfork$ and $\mathcal{T}>0$. In view of Lemma 16.1, we conclude that the mapping $\mathcal{Q}':\pitchfork\times\mho_1^3\to\mho_2$ is tri–additive.

Theorem 16.2 *Let* $n\in\mathbb{N}$, *and* $\underbrace{\varphi_i}_{i=1,\ldots,n}:\mho_1^6\times(0,+\infty)\to[0,1]$ *be functions such that,*

$$\underbrace{\varphi_i}_{i=1,\ldots,n}\Big(2\mathfrak{X}_1,2\mathfrak{X}_2,2\mathfrak{X}_3,2\mathfrak{X}_4,2\mathfrak{X}_5,2\mathfrak{X}_6,\mathcal{T}\Big) \tag{16.22}$$

$$\geq \underbrace{\varphi_i}_{i=1,\ldots,n}\Big(\mathfrak{X}_1,\mathfrak{X}_2,\mathfrak{X}_3,\mathfrak{X}_4,\mathfrak{X}_5,\mathfrak{X}_6,\frac{\mathcal{T}}{2\underbrace{\mathfrak{P}_i}_{i=1,\ldots,n}}\Big),$$

for every $\mathfrak{X}_1, \mathfrak{X}_2, \mathfrak{X}_3, \mathfrak{X}_4, \mathfrak{X}_5, \mathfrak{X}_6 \in \mho_1, \leftthreetimes \in \pitchfork$, $\mathcal{T} > 0$ *and* $0 < \underbrace{\mathfrak{P}_i}_{i=1,\dots,n} < 1$.

Let $\mathcal{Q} : \pitchfork \times \mho_1^3 \to \mho_2$ *be a mapping satisfying* (16.8) *and* $\mathcal{Q}(\leftthreetimes, \mathfrak{X}_1, 0, \mathfrak{X}_5) = \mathcal{Q}(\leftthreetimes, 0, \mathfrak{X}_3, \mathfrak{X}_5)$
$= \mathcal{Q}(\leftthreetimes, \mathfrak{X}_1, \mathfrak{X}_3, 0) = 0$, *for every* $\mathfrak{X}_1, \mathfrak{X}_2, \mathfrak{X}_3, \mathfrak{X}_4, \mathfrak{X}_5, \mathfrak{X}_6 \in \mho_1$ *and* $\leftthreetimes \in \pitchfork$. *Then, we can find a unique tri–additive mapping* $\mathcal{Q}' : \pitchfork \times \mho_1^3 \to \mho_2$ *satisfying*

$$
\begin{aligned}
&\mathcal{N}\Big(\mathcal{Q}(\leftthreetimes, \mathfrak{X}_1, \mathfrak{X}_3, \mathfrak{X}_5) - \mathcal{Q}'(\leftthreetimes, \mathfrak{X}_1, \mathfrak{X}_3, \mathfrak{X}_5), \mathcal{T}\Big)\\
&\succeq diag\Big[\varphi_1\Big(\mathfrak{X}_1, 0, \mathfrak{X}_3, \mathfrak{X}_3, 2\mathfrak{X}_5, 0, (8 - 2\mathfrak{P}_1)\mathcal{T}\Big)\\
&\qquad \bigotimes_{\mathrm{TN}} \varphi_1\Big(\mathfrak{X}_1, \mathfrak{X}_1, \mathfrak{X}_3, 0, 2\mathfrak{X}_5, 0, (4 - \mathfrak{P}_1)\mathcal{T}\Big)\\
&\qquad\qquad \bigotimes_{\mathrm{TN}} \varphi_1\Big(\mathfrak{X}_1, 0, \mathfrak{X}_3, 0, \mathfrak{X}_5, \mathfrak{X}_5, \frac{4 - \mathfrak{P}_1}{2}\mathcal{T}\Big), \dots,\\
&\quad \varphi_n\Big(\mathfrak{X}_1, 0, \mathfrak{X}_3, \mathfrak{X}_3, 2\mathfrak{X}_5, 0, (8 - 2\mathfrak{P}_n)\mathcal{T}\Big)\\
&\qquad \bigotimes_{\mathrm{TN}} \varphi_n\Big(\mathfrak{X}_1, \mathfrak{X}_1, \mathfrak{X}_3, 0, 2\mathfrak{X}_5, 0, (4 - \mathfrak{P}_n)\mathcal{T}\Big)\\
&\qquad\qquad \bigotimes_{\mathrm{TN}} \varphi_n\Big(\mathfrak{X}_1, 0, \mathfrak{X}_3, 0, \mathfrak{X}_5, \mathfrak{X}_5, \frac{4 - \mathfrak{P}_n}{2}\mathcal{T}\Big)\Big],
\end{aligned}
$$

for every $\mathfrak{X}_1, \mathfrak{X}_3, \mathfrak{X}_5 \in \mho_1, \leftthreetimes \in \pitchfork$.

Proof Based on (16.12), we have that

$$
\begin{aligned}
&\mathcal{N}\Big(\mathcal{Q}(\leftthreetimes, \mathfrak{X}_1, \mathfrak{X}_3, \mathfrak{X}_5) - \frac{1}{8}\mathcal{Q}(\leftthreetimes, 2\mathfrak{X}_1, 2\mathfrak{X}_3, 2\mathfrak{X}_5), \mathcal{T}\Big)\\
&\succeq \mathrm{diag}\Big[\varphi_1\Big(2\mathfrak{X}_1, 0, \mathfrak{X}_3, \mathfrak{X}_3, 2\mathfrak{X}_5, 0, 8\mathcal{T}\Big)\\
&\qquad \bigotimes_{\mathrm{TN}} \varphi_1\Big(\mathfrak{X}_1, \mathfrak{X}_1, \mathfrak{X}_3, 0, 2\mathfrak{X}_5, 0, 4\mathcal{T}\Big)\\
&\qquad\qquad \bigotimes_{\mathrm{TN}} \varphi_1\Big(\mathfrak{X}_1, 0, \mathfrak{X}_3, 0, \mathfrak{X}_5, \mathfrak{X}_5, 2\mathcal{T}\Big), \dots,\\
&\quad \varphi_n\Big(2\mathfrak{X}_1, 0, \mathfrak{X}_3, \mathfrak{X}_3, 2\mathfrak{X}_5, 0, 8\mathcal{T}\Big)\\
&\qquad \bigotimes_{\mathrm{TN}} \varphi_n\Big(\mathfrak{X}_1, \mathfrak{X}_1, \mathfrak{X}_3, 0, 2\mathfrak{X}_5, 0, 4\mathcal{T}\Big)\\
&\qquad\qquad \bigotimes_{\mathrm{TN}} \varphi_n\Big(\mathfrak{X}_1, 0, \mathfrak{X}_3, 0, \mathfrak{X}_5, \mathfrak{X}_5, 2\mathcal{T}\Big)\Big],
\end{aligned}
$$

for every $\mathfrak{X}_1, \mathfrak{X}_3, \mathfrak{X}_5 \in \mho_1, \leftthreetimes \in \pitchfork$ and $\mathcal{T} > 0$.

By using a similar method in the proof of Theorem 16.1, the proof will be completed.

16.2 Multi Stability of Permuting Tri–Derivations in Complex MVFB–Algebras

In this section, we study the multi stability of permuting tri–derivations in complex MVFB–algebras associated with the tri–additive Υ–functional inequality (16.1).

Throughout this section, we consider MVFN–space $(\mho_1, \mathscr{N}, \bigotimes_{\mathrm{GTN}})$, MVFB–space $(\mho_2, \mathscr{N}, \bigotimes_{\mathrm{GTN}})$ and MVFB–algebras $(\mathfrak{M}, \mathscr{N}, \bigotimes_{\mathrm{GTN}}, \bigoplus_{\mathrm{GTN}})$.

Lemma 16.2 *Let $\mathcal{Q} : ⫛ \times \mho_1^3 \to \mho_2$ be a tri–additive mapping satisfying*

$$\mathcal{Q}(\curlywedge, \Lambda_1\mathfrak{X}_1, \Lambda_2\mathfrak{X}_3, \Lambda_3\mathfrak{X}_5) = \Lambda_1\Lambda_2\Lambda_3\mathcal{Q}(\curlywedge, \mathfrak{X}_1, \mathfrak{X}_3, \mathfrak{X}_5),$$

for every $\mathfrak{X}_1, \mathfrak{X}_3, \mathfrak{X}_5 \in \mho_1, \curlywedge \in ⫛$ and $\Lambda_1, \Lambda_2, \Lambda_3 \in \Delta^1 := \{\varpi \in \mathbb{C} \ : \ |\varpi| = 1\}$. Then, $\mathcal{Q}$ is $\mathbb{C}$–trilinear.

Proof It follows from a similar method in the proof of Lemma 2.1 in [2] and [1].

Theorem 16.3 *Let $n \in \mathbb{N}$, and $\underbrace{\varphi_i}_{i=1,\dots,n} : \mathfrak{M}^6 \times (0, +\infty) \to [0, 1]$ be fuzzy sets such that there exist $0 < \underbrace{\mathfrak{P}_i}_{i=1,\dots,n} < 1$ with*

$$\underbrace{\varphi_i}_{i=1,\dots,n}\left(\frac{\mathfrak{X}_1}{2}, \frac{\mathfrak{X}_2}{2}, \frac{\mathfrak{X}_3}{2}, \frac{\mathfrak{X}_4}{2}, \frac{\mathfrak{X}_5}{2}, \frac{\mathfrak{X}_6}{2}, \mathcal{T}\right) \geq \underbrace{\varphi_i}_{i=1,\dots,n}\left(\mathfrak{X}_1, \mathfrak{X}_2, \mathfrak{X}_3, \mathfrak{X}_4, \mathfrak{X}_5, \mathfrak{X}_6, \frac{2}{\mathfrak{P}}\mathcal{T}\right), \tag{16.23}$$

and let $\mathcal{Q} : ⫛ \times \mathfrak{M}^3 \to \mathfrak{M}$ be a mapping satisfying $\mathcal{Q}(\curlywedge, \mathfrak{X}_1, 0, \mathfrak{X}_5) = \mathcal{Q}(\curlywedge, 0, \mathfrak{X}_3, \mathfrak{X}_5) = \mathcal{Q}(\curlywedge, \mathfrak{X}_1, \mathfrak{X}_3, 0) = 0$ and

$$\begin{aligned}
&\mathscr{N}\Big(\mathcal{Q}(\curlywedge, \Lambda_1(\mathfrak{X}_1 + \mathfrak{X}_2), \Lambda_2(\mathfrak{X}_3 - \mathfrak{X}_4), \Lambda_3(\mathfrak{X}_5 + \mathfrak{X}_6)) \\
&\quad + \mathcal{Q}(\curlywedge, \Lambda_1(\mathfrak{X}_1 - \mathfrak{X}_2), \Lambda_2(\mathfrak{X}_3 + \mathfrak{X}_4), \Lambda_3(\mathfrak{X}_5 - \mathfrak{X}_6)) \\
&\quad - \Lambda_1\Lambda_2\Lambda_3(2\mathcal{Q}(\curlywedge, \mathfrak{X}_1, \mathfrak{X}_3, \mathfrak{X}_5) - 2\mathcal{Q}(\curlywedge, \mathfrak{X}_1, \mathfrak{X}_4, \mathfrak{X}_6) \\
&\quad + 2\mathcal{Q}(\curlywedge, \mathfrak{X}_2, \mathfrak{X}_3, \mathfrak{X}_6) - 2\mathcal{Q}(\curlywedge, \mathfrak{X}_2, \mathfrak{X}_4, \mathfrak{X}_5)), \mathcal{T}\Big) \\
&\geq \mathscr{N}\Big(\Upsilon[2\mathcal{Q}(\curlywedge, \frac{\mathfrak{X}_1 + \mathfrak{X}_2}{2}, \mathfrak{X}_3 - \mathfrak{X}_4, \mathfrak{X}_5 + \mathfrak{X}_6) + 2\mathcal{Q}(\curlywedge, \frac{\mathfrak{X}_1 - \mathfrak{X}_2}{2}, \mathfrak{X}_3 + \mathfrak{X}_4, \mathfrak{X}_5 - \mathfrak{X}_6) \\
&\quad - 2\mathcal{Q}(\curlywedge, \mathfrak{X}_1, \mathfrak{X}_3, \mathfrak{X}_5) + 2\mathcal{Q}(\curlywedge, \mathfrak{X}_1, \mathfrak{X}_4, \mathfrak{X}_6) \\
&\quad - 2\mathcal{Q}(\curlywedge, \mathfrak{X}_2, \mathfrak{X}_3, \mathfrak{X}_6) + 2\mathcal{Q}(\curlywedge, \mathfrak{X}_2, \mathfrak{X}_4, \mathfrak{X}_5))], \mathcal{T}\Big) \\
&\quad \bigotimes_{\mathrm{GTN}} diag\Big[\varphi_1\Big(\mathfrak{X}_1, \mathfrak{X}_2, \mathfrak{X}_3, \mathfrak{X}_4, \mathfrak{X}_5, \mathfrak{X}_6, \mathcal{T}\Big), \dots, \varphi_n\Big(\mathfrak{X}_1, \mathfrak{X}_2, \mathfrak{X}_3, \mathfrak{X}_4, \mathfrak{X}_5, \mathfrak{X}_6, \mathcal{T}\Big)\Big],
\end{aligned} \tag{16.24}$$

for every $\Lambda_1, \Lambda_2, \Lambda_3 \in \Delta^1$, $\mathfrak{X}_1, \mathfrak{X}_2, \mathfrak{X}_3, \mathfrak{X}_4, \mathfrak{X}_5, \mathfrak{X}_6 \in \mathfrak{M}$, $\leftthreetimes \in \pitchfork$ *and* $\mathcal{T} > 0$. *Then, we can find a unique* $\mathbb{C}$*–trilinear mapping* $\mathfrak{Z} : \pitchfork \times \mathfrak{M}^3 \to \mathfrak{M}$ *satisfying*

$$\begin{aligned}
&\mathcal{N}\Big(\mathcal{Q}(\leftthreetimes, \mathfrak{X}_1, \mathfrak{X}_3, \mathfrak{X}_5) - \mathfrak{Z}(\leftthreetimes, \mathfrak{X}_1, \mathfrak{X}_3, \mathfrak{X}_5), \mathcal{T}\Big) \qquad (16.25)\\
&\succeq diag\Bigg[\varphi_1\Big(\mathfrak{X}_1, 0, \frac{\mathfrak{X}_3}{2}, \frac{\mathfrak{X}_3}{2}, \mathfrak{X}_5, 0, (1-4\mathfrak{P}_1)\mathcal{T}\Big)\\
&\qquad \bigotimes_{\mathrm{TN}} \varphi_1\Big(\frac{\mathfrak{X}_1}{2}, \frac{\mathfrak{X}_1}{2}, \frac{\mathfrak{X}_3}{2}, 0, \mathfrak{X}_5, 0, \frac{(1-4\mathfrak{P}_1)}{2}\mathcal{T}\Big)\\
&\qquad\qquad \bigotimes_{\mathrm{TN}} \varphi_1\Big(\frac{\mathfrak{X}_1}{2}, 0, \frac{\mathfrak{X}_3}{2}, 0, \frac{\mathfrak{X}_5}{2}, \frac{\mathfrak{X}_5}{2}, \frac{(1-4\mathfrak{P}_1)}{4}\mathcal{T}\Big), \ldots,\\
&\qquad \varphi_n\Big(\mathfrak{X}_1, 0, \frac{\mathfrak{X}_3}{2}, \frac{\mathfrak{X}_3}{2}, \mathfrak{X}_5, 0, (1-4\mathfrak{P}_n)\mathcal{T}\Big)\\
&\qquad \bigotimes_{\mathrm{TN}} \varphi_n\Big(\frac{\mathfrak{X}_1}{2}, \frac{\mathfrak{X}_1}{2}, \frac{\mathfrak{X}_3}{2}, 0, \mathfrak{X}_5, 0, \frac{(1-4\mathfrak{P}_n)}{2}\mathcal{T}\Big)\\
&\qquad\qquad \bigotimes_{\mathrm{TN}} \varphi_n\Big(\frac{\mathfrak{X}_1}{2}, 0, \frac{\mathfrak{X}_3}{2}, 0, \frac{\mathfrak{X}_5}{2}, \frac{\mathfrak{X}_5}{2}, \frac{(1-4\mathfrak{P}_n)}{4}\mathcal{T}\Big)\Bigg],
\end{aligned}$$

for every $\mathfrak{X}_1, \mathfrak{X}_3, \mathfrak{X}_5 \in \mathfrak{M}$, $\leftthreetimes \in \pitchfork$ *and* $\mathcal{T} > 0$, *where* $\underbrace{\varphi_i}_{1 \le i \le n}$ *are given in Theorem 16.1.*

Also, if the mapping $\mathcal{Q} : \pitchfork \times \mathfrak{M}^3 \to \mathfrak{M}$ *satisfies* $\mathcal{Q}(\leftthreetimes, \mathfrak{X}_1, \mathfrak{X}_3, \mathfrak{X}_5) = 2\mathcal{Q}(\leftthreetimes, \mathfrak{X}_1, \mathfrak{X}_3, \mathfrak{X}_5)$ *and*

$$\begin{aligned}
&\mathcal{N}\Big(\mathcal{Q}(\leftthreetimes, \mathfrak{X}_1\mathfrak{X}_2, \mathfrak{X}_3, \mathfrak{X}_5) - \mathcal{Q}(\leftthreetimes, \mathfrak{X}_1, \mathfrak{X}_3, \mathfrak{X}_5)\mathfrak{X}_2 - \mathfrak{X}_1\mathcal{Q}(\leftthreetimes, \mathfrak{X}_2, \mathfrak{X}_3, \mathfrak{X}_5), \mathcal{T}\Big) \qquad (16.26)\\
&\succeq diag\Big[\varphi_1\Big(\mathfrak{X}_1, \mathfrak{X}_2, \mathfrak{X}_3, \mathfrak{X}_3, \mathfrak{X}_5, \mathfrak{X}_5, \mathcal{T}\Big), \ldots, \varphi_n\Big(\mathfrak{X}_1, \mathfrak{X}_2, \mathfrak{X}_3, \mathfrak{X}_3, \mathfrak{X}_5, \mathfrak{X}_5, \mathcal{T}\Big)\Big],
\end{aligned}$$

and

$$\begin{aligned}
&\mathcal{N}\Big(\mathcal{Q}(\leftthreetimes, \mathfrak{X}'_{\delta(1)}, \mathfrak{X}'_{\delta(2)}, \mathfrak{X}'_{\delta(3)}) - \mathcal{Q}(\leftthreetimes, \mathfrak{X}'_1, \mathfrak{X}'_2, \mathfrak{X}'_3), \mathcal{T}\Big) \qquad (16.27)\\
&\succeq \varphi\Big(\mathfrak{X}'_1, \mathfrak{X}'_1, \mathfrak{X}'_2, \mathfrak{X}'_2, \mathfrak{X}'_3, \mathfrak{X}'_3, \mathcal{T}\Big)
\end{aligned}$$

for every $\mathfrak{X}_1, \mathfrak{X}_2, \mathfrak{X}_3, \mathfrak{X}'_1, \mathfrak{X}'_2, \mathfrak{X}'_3, \mathfrak{X}_5 \in \mathfrak{M}$, $\leftthreetimes \in \pitchfork$, $\mathcal{T} > 0$, *and for every permutation* $(\delta(1), \delta(2), \delta(3))$ *of* $(1, 2, 3)$, *then, the* $\mathbb{C}$*–trilinear mapping* $\mathfrak{Z} : \pitchfork \times \mathfrak{M}^3 \to \mathfrak{M}$ *is a permuting tri–derivation.*

Proof Let $\Lambda_1 = \Lambda_2 = \Lambda_3 = 1$ in (16.24). Theorem 16.1 and [3, Theorem 3.3] prove the theorem.

Theorem 16.4 *Let $n \in \mathbb{N}$, and* $\underbrace{\varphi_i}_{i=1,\ldots,n} : \mathfrak{M}^6 \times (0, +\infty) \to [0, 1]$ *be fuzzy sets such that there exist* $0 < \underbrace{\mathfrak{P}_i}_{i=1,\ldots,n} < 1$ *with*

$$\underbrace{\varphi_i}_{i=1,\ldots,n}\left(2\mathfrak{X}_1, 2\mathfrak{X}_2, 2\mathfrak{X}_3, 2\mathfrak{X}_4, 2\mathfrak{X}_5, 2\mathfrak{X}_6, \mathcal{T}\right) > \underbrace{\varphi_i}_{i=1,\ldots,n}\left(\mathfrak{X}_1, \mathfrak{X}_2, \mathfrak{X}_3, \mathfrak{X}_4, \mathfrak{X}_5, \mathfrak{X}_6, \frac{\mathcal{T}}{2\underbrace{\mathfrak{P}_i}_{i=1,\ldots,n}}\right), \tag{16.28}$$

for every $\mathfrak{X}_1, \mathfrak{X}_2, \mathfrak{X}_3, \mathfrak{X}_4, \mathfrak{X}_5, \mathfrak{X}_6 \in \mathfrak{M}$ and $\mathcal{T} > 0$. Let $\mathcal{Q} : \pitchfork \times \mathfrak{M}^3 \to \mathfrak{M}$ be a mapping satisfying (16.24) *and $\mathcal{Q}(\leftthreetimes, \mathfrak{X}_1, 0, \mathfrak{X}_5) = \mathcal{Q}(\leftthreetimes, 0, \mathfrak{X}_3, \mathfrak{X}_5) = \mathcal{Q}(\leftthreetimes, \mathfrak{X}_1, \mathfrak{X}_3, 0) = 0$, for every $\mathfrak{X}_1, \mathfrak{X}_3, \mathfrak{X}_5 \in \mathfrak{M}$ and $\leftthreetimes \in \pitchfork$. Then, we can find a unique $\mathbb{C}$–trilinear mapping $\mathfrak{Z} : \pitchfork \times \mathfrak{M}^3 \to \mathfrak{M}$ satisfying*

$$\begin{aligned}
&\mathcal{N}\left(\mathcal{Q}(\leftthreetimes, \mathfrak{X}_1, \mathfrak{X}_3, \mathfrak{X}_5) - \mathfrak{Z}(\leftthreetimes, \mathfrak{X}_1, \mathfrak{X}_3, \mathfrak{X}_5), \mathcal{T}\right) \\
&\succeq diag\Bigg[\varphi_1\left(2\mathfrak{X}_1, 0, \mathfrak{X}_3, \mathfrak{X}_3, 2\mathfrak{X}_5, 0, 2(4 - \mathfrak{P}_1)\mathcal{T}\right) \\
&\qquad \bigotimes \varphi_1\left(\mathfrak{X}_1, \mathfrak{X}_1, \mathfrak{X}_3, 0, 2\mathfrak{X}_5, 0, (4 - \mathfrak{P}_1)\mathcal{T}\right) \\
&\qquad\qquad \bigotimes \varphi_1\left(\mathfrak{X}_1, 0, \mathfrak{X}_3, 0, \mathfrak{X}_5, \mathfrak{X}_5, \frac{(4 - \mathfrak{P}_1)}{2}\mathcal{T}\right), \ldots, \\
&\quad \varphi_n\left(2\mathfrak{X}_1, 0, \mathfrak{X}_3, \mathfrak{X}_3, 2\mathfrak{X}_5, 0, 2(4 - \mathfrak{P}_n)\mathcal{T}\right) \\
&\qquad \bigotimes \varphi_n\left(\mathfrak{X}_1, \mathfrak{X}_1, \mathfrak{X}_3, 0, 2\mathfrak{X}_5, 0, (4 - \mathfrak{P}_n)\mathcal{T}\right) \\
&\qquad\qquad \bigotimes \varphi_n\left(\mathfrak{X}_1, 0, \mathfrak{X}_3, 0, \mathfrak{X}_5, \mathfrak{X}_5, \frac{(4 - \mathfrak{P}_n)}{2}\mathcal{T}\right)\Bigg],
\end{aligned} \tag{16.29}$$

for every $\mathfrak{X}_1, \mathfrak{X}_3, \mathfrak{X}_5 \in \mathfrak{M}$ and $\leftthreetimes \in \pitchfork$.

Also, if the mapping $\mathcal{Q} : \pitchfork \times \mathfrak{M}^3 \to \mathfrak{M}$, satisfies (16.26), (16.27) *and $\mathcal{Q}(\leftthreetimes, 2\mathfrak{X}_1, \mathfrak{X}_3, \mathfrak{X}_5) = 2\mathcal{Q}(\leftthreetimes, \mathfrak{X}_1, \mathfrak{X}_3, \mathfrak{X}_5)$, for every $\mathfrak{X}_1, \mathfrak{X}_3, \mathfrak{X}_5 \in \mathfrak{M}$, and $\leftthreetimes \in \pitchfork$, then, the $\mathbb{C}$–trilinear mapping $\mathfrak{Z} : \pitchfork \times \mathfrak{M}^3 \to \mathfrak{M}$ is a permuting tri–derivation.*

Proof It follows from a similar method in the proof of Theorem 16.3.

16.3 Multi Stability of Permuting Tri–Derivations in Unital MVFC–⋄–Algebras

In this part, we study the multi stability of permuting tri–derivations in unital MVFC–⋄–algebras associated with the tri–additive Υ–functional inequality (16.1).

From now on, let $(\mathfrak{M}, \mathscr{N}, \bigotimes_{\text{GTN}}, \bigoplus_{\text{GTN}})$ and $U(\mathfrak{M}) := \{t \in \mathfrak{M} : t^{\diamond}t = tt^{\diamond} = e\}$ be a unital MVFC–⋄–algebra with unit e and unitary group, respectively.

Theorem 16.5 *Let* $n \in \mathbb{N}$, *and* $\underbrace{\varphi_i}_{i=1,\ldots,n} : \mathfrak{M}^6 \times (0, +\infty) \to [0, 1]$ *be fuzzy sets satisfying* (16.23) *and* $\mathcal{Q} : ⫛ \times \mathfrak{M}^3 \to \mathfrak{M}$ *be a mapping satisfying* (16.24) *and* $\mathcal{Q}(\leftthreetimes, \mathfrak{X}_1, 0, \mathfrak{X}_5) = \mathcal{Q}(\leftthreetimes, 0, \mathfrak{X}_3, \mathfrak{X}_5) = \mathcal{Q}(\leftthreetimes, \mathfrak{X}_1, \mathfrak{X}_3, 0) = 0$, *for every* $\mathfrak{X}_1, \mathfrak{X}_3, \mathfrak{X}_5 \in \mathfrak{M}$ *and* $\leftthreetimes \in ⫛$. *Then, we can find a unique* $\mathbb{C}$*–trilinear mapping* $\mathfrak{Z} : ⫛ \times \mathfrak{M}^3 \to \mathfrak{M}$ *satisfying* (16.25).

Also, if the mapping $\mathcal{Q} : ⫛ \times \mathfrak{M}^3 \to \mathfrak{M}$ *satisfies* (16.27), *and* $\mathcal{Q}(\leftthreetimes, 2\beta, \mathfrak{X}_3, \mathfrak{X}_5) = 2\mathcal{Q}(\leftthreetimes, \beta, \mathfrak{X}_3, \mathfrak{X}_5)$ *and*

$$\mathscr{N}\Big(\mathcal{Q}(\leftthreetimes, \beta\mathfrak{X}_2, \mathfrak{X}_3, \mathfrak{X}_5) - \mathcal{Q}(\leftthreetimes, \beta, \mathfrak{X}_3, \mathfrak{X}_5)\mathfrak{X}_2 - \beta\mathcal{Q}(\leftthreetimes, \mathfrak{X}_2, \mathfrak{X}_3, \mathfrak{X}_5), \mathcal{T}\Big) \tag{16.30}$$
$$\succeq diag\Big[\varphi_1\Big(\beta, \mathfrak{X}_2, \mathfrak{X}_3, \mathfrak{X}_3, \mathfrak{X}_5, \mathfrak{X}_5\Big), \ldots, \varphi_n\Big(\beta, \mathfrak{X}_2, \mathfrak{X}_3, \mathfrak{X}_3, \mathfrak{X}_5, \mathfrak{X}_5\Big)\Big],$$

for every $\mathfrak{X}_2, \mathfrak{X}_3, \mathfrak{X}_5 \in \mathfrak{M}$, $\leftthreetimes \in ⫛$, $\mathcal{T} > 0$ *and every* $\beta \in U(\mathfrak{M})$, *then, the* $\mathbb{C}$*–trilinear mapping* $\mathfrak{Z} : ⫛ \times \mathfrak{M}^3 \to \mathfrak{M}$ *is a permuting tri–derivation.*

Proof Theorem 16.3 and [3, Theorem 3.7] prove the theorem.

Remark 16.1 By a similar method in the proof of Theorem 16.5, we can conclude that if (16.30) in Theorem 16.5 is replaced by

$$\mathscr{N}\Big(\mathcal{Q}(\leftthreetimes, \beta\alpha, \alpha_1, \alpha_2) - \mathcal{Q}(\leftthreetimes, \beta, \alpha_1, \alpha_2)\alpha - \beta\mathcal{Q}(\leftthreetimes, \alpha, \alpha_1, \alpha_2), \mathcal{T}\Big)$$
$$\succeq \text{diag}\Big[\varphi_1\Big(\beta, \alpha, \alpha_1, \alpha_1, \alpha_2, \alpha_2, \mathcal{T}\Big), \ldots, \varphi_n\Big(\beta, \alpha, \alpha_1, \alpha_1, \alpha_2, \alpha_2, \mathcal{T}\Big)\Big],$$

for every $\beta, \alpha, \alpha_1, \alpha_2 \in U(\mathfrak{M})$, then, the $\mathbb{C}$–trilinear mapping $\mathfrak{Z} : ⫛ \times \mathfrak{M}^3 \to \mathfrak{M}$ is a permuting tri–derivation.

Theorem 16.6 *Let* $n \in \mathbb{N}$, *and* $\overbrace{\varphi_i}^{i=1,\ldots,n} : \mathfrak{M}^6 \times (0, +\infty) \to [0, 1]$ *be functions satisfying* (16.28) *and* $\mathcal{Q} : ⫛ \times \mathfrak{M}^3 \to \mathfrak{M}$ *be a mapping satisfying* (16.24) *and* $\mathcal{Q}(\leftthreetimes, \mathfrak{X}_1, 0, \mathfrak{X}_5) = \mathcal{Q}(\leftthreetimes, 0, \mathfrak{X}_3, \mathfrak{X}_5) = \mathcal{Q}(\mathfrak{X}_1, \mathfrak{X}_3, 0) = 0$, *for every* $\mathfrak{X}_1, \mathfrak{X}_3, \mathfrak{X}_5 \in \mathfrak{M}$, *and* $\leftthreetimes \in ⫛$. *Then, we can find a unique* $\mathbb{C}$*–trilinear mapping* $\mathfrak{Z} : ⫛ \times \mathfrak{M}^3 \to \mathfrak{M}$ *satisfying* (16.29).

Also, if the mapping $\mathcal{Q} : ⋔ \times \mathfrak{M}^3 \to \mathfrak{M}$ *satisfies* (16.27), (16.30) *and* $\mathcal{Q}(\curlywedge, 2\beta, \mathfrak{X}_3, \mathfrak{X}_5) = 2\mathcal{Q}(\curlywedge, \beta, \mathfrak{X}_3, \mathfrak{X}_5)$*, for every* $\mathfrak{X}_3, \mathfrak{X}_5 \in \mathfrak{M}, \curlywedge \in ⋔$ *and every* $\beta \in U(\mathfrak{M})$*, then, the* $\mathbb{C}$*–trilinear mapping* $\mathfrak{Z} : ⋔ \times \mathfrak{M}^3 \to \mathfrak{M}$ *is a permuting tri–derivation.*

Proof Using a similar method in the proof of Theorem 16.5, it follows.

16.4 Multi Stability of Permuting Tri–Homomorphisms in Complex MVFB–Algebras

In this section, we study the multi stability of permuting tri–homomorphisms in complex MVFB–algebras associated with the tri–additive Υ–functional inequality (16.1).

Throughout this section, we consider MVFB–algebras $(\mathfrak{M}, \mathscr{N}, \bigotimes_{\mathrm{GTN}}, \bigoplus_{\mathrm{GTN}})$ and $(\mathfrak{N}, \mathscr{N}, \bigotimes_{\mathrm{GTN}}, \bigoplus_{\mathrm{GTN}})$.

Theorem 16.7 *Let* $n \in \mathbb{N}$*, and* $\underbrace{\varphi_i}_{i=1,\dots,n} : \mathfrak{M}^6 \times (0, +\infty) \to [0, 1]$ *be fuzzy sets such that there exist* $0 < \underbrace{\mathfrak{P}_i}_{i=1,\dots,n} < 1$ *with*

$$\underbrace{\varphi_i}_{i=1,\dots,n}\left(\frac{\mathfrak{X}_1}{2}, \frac{\mathfrak{X}_2}{2}, \frac{\mathfrak{X}_3}{2}, \frac{\mathfrak{X}_4}{2}, \frac{\mathfrak{X}_5}{2}, \frac{\mathfrak{X}_6}{2}, \mathcal{T}\right) > \underbrace{\varphi_i}_{i=1,\dots,n}\left(\mathfrak{X}_1, \mathfrak{X}_2, \mathfrak{X}_3, \mathfrak{X}_4, \mathfrak{X}_5, \Psi_6, \frac{2}{\underbrace{\mathfrak{P}_i}_{i=1,\dots,n}}\mathcal{T}\right), \tag{16.31}$$

and let $\mathcal{Q} : ⋔ \times \mathfrak{M}^3 \to \mathfrak{N}$ *be a mapping satisfying* (16.24) *and* $\mathcal{Q}(\curlywedge, \mathfrak{X}_1, 0, \mathfrak{X}_5) = \mathcal{Q}(\curlywedge, 0, \mathfrak{X}_3, \mathfrak{X}_5) = \mathcal{Q}(\curlywedge, \mathfrak{X}_1, \mathfrak{X}_3, 0) = 0$*, for every* $\mathfrak{X}_1, \mathfrak{X}_3, \mathfrak{X}_4 \in \mathfrak{M}$*, and* $\curlywedge \in ⋔$*. Then, we can find a unique* $\mathbb{C}$*–trilinear mapping* $\mathcal{E} : ⋔ \times \mathfrak{M}^3 \to \mathfrak{N}$ *satisfying*

$$\begin{aligned}
&\mathscr{N}\Big(\mathcal{Q}(\curlywedge, \mathfrak{X}_1, \mathfrak{X}_3, \mathfrak{X}_5) - \mathcal{E}(\curlywedge, \mathfrak{X}_1, \mathfrak{X}_3, \mathfrak{X}_5), \mathcal{T}\Big) \\
&\succeq diag\Bigg[\varphi_1\left(\mathfrak{X}_1, 0, \frac{\mathfrak{X}_3}{2}, \frac{\mathfrak{X}_3}{2}, \mathfrak{X}_5, 0, (1 - 4\mathfrak{P}_1)\mathcal{T}\right) \\
&\qquad \bigotimes_{\mathrm{TN}} \varphi_1\left(\frac{\mathfrak{X}_1}{2}, \frac{\mathfrak{X}_1}{2}, \frac{\mathfrak{X}_3}{2}, 0, \mathfrak{X}_5, 0, \frac{(1 - 4\mathfrak{P}_1)}{2}\mathcal{T}\right) \\
&\qquad\qquad \bigotimes_{\mathrm{TN}} \varphi_1\left(\frac{\mathfrak{X}_1}{2}, 0, \frac{\mathfrak{X}_3}{2}, 0, \frac{\Psi_5}{2}, \frac{\mathfrak{X}_5}{2}, \frac{(1 - 4\mathfrak{P}_1)}{4}\mathcal{T}\right), \dots, \\
&\quad \varphi_n\left(\mathfrak{X}_1, 0, \frac{\mathfrak{X}_3}{2}, \frac{\mathfrak{X}_3}{2}, \mathfrak{X}_5, 0, (1 - 4\mathfrak{P}_n)\mathcal{T}\right)
\end{aligned} \tag{16.32}$$

$$\bigotimes_{\mathrm{TN}} \varphi_n\left(\frac{\mathfrak{X}_1}{2}, \frac{\mathfrak{X}_1}{2}, \frac{\mathfrak{X}_3}{2}, 0, \mathfrak{X}_5, 0, \frac{(1-4\mathfrak{P}_n)}{2}\mathcal{T}\right)$$
$$\bigotimes_{\mathrm{TN}} \varphi_n\left(\frac{\mathfrak{X}_1}{2}, 0, \frac{\mathfrak{X}_3}{2}, 0, \frac{\Psi_5}{2}, \frac{\mathfrak{X}_5}{2}, \frac{(1-4\mathfrak{P}_n)}{4}\mathcal{T}\right)\Bigg],$$

for every $\mathfrak{X}_1, \mathfrak{X}_3, \mathfrak{X}_5 \in \mathfrak{M}, \leftthreetimes \in ⫛$ *and* $\mathcal{T} > 0$, *where* $\underbrace{\varphi_i}_{i=1,\ldots,n}$ *are given in Theorem 16.1.*

Also, if the mapping $\mathcal{Q} : ⫛ \times \mathfrak{M}^3 \to \mathfrak{N}$ *satisfies* (16.27) *and*

$$\mathcal{N}\Big(\mathcal{Q}(\leftthreetimes, \mathfrak{X}_1\mathfrak{X}_2, \mathfrak{X}_3\mathfrak{X}4, \mathfrak{X}_5\mathfrak{X}_6) - \mathcal{Q}(\leftthreetimes, \mathfrak{X}_1, \mathfrak{X}_3, \mathfrak{X}_5)\mathcal{Q}(\leftthreetimes, \mathfrak{X}_2, \mathfrak{X}_4, \mathfrak{X}_6), \mathcal{T}\Big) \tag{16.33}$$
$$\succeq diag\Big[\varphi_1\Big(\mathfrak{X}_1, \mathfrak{X}_2, \mathfrak{X}_3, \mathfrak{X}_3, \mathfrak{X}_5, \mathfrak{X}_5, \mathcal{T}\Big), \ldots, \varphi_n\Big(\mathfrak{X}_1, \mathfrak{X}_2, \mathfrak{X}_3, \mathfrak{X}_3, \mathfrak{X}_5, \mathfrak{X}_5, \mathcal{T}\Big)\Big],$$

for every $\mathfrak{X}_1, \mathfrak{X}_2, \mathfrak{X}_3, \mathfrak{X}_4, \mathfrak{X}_5, \mathfrak{X}_6 \in \mathfrak{M}, \leftthreetimes \in ⫛$ *and* $\mathcal{T} > 0$, *then, the* $\mathbb{C}$*–trilinear mapping* $\mathcal{E} : ⫛ \times \mathfrak{M}^3 \to \mathfrak{N}$ *is a permuting tri–homomorphism.*

Proof Theorem 16.3 and [3, Theorem 4.1] prove the theorem.

Theorem 16.8 *Let* $n \in \mathbb{N}$, *and* $\overbrace{\varphi_i}^{i=1,\ldots,n} : \mathfrak{M}^6 \times (0, +\infty) \to [0, 1]$ *be functions satisfying* (16.28) *and* $\mathcal{Q} : ⫛ \times \mathfrak{M}^3 \to \mathfrak{N}$ *be a mapping satisfying* (16.24) *and* $\mathcal{Q}(\leftthreetimes, \mathfrak{X}_1, 0, \mathfrak{X}_5) = \mathcal{Q}(\leftthreetimes, 0, \mathfrak{X}_3, \mathfrak{X}_5) = \mathcal{Q}(\leftthreetimes, \mathfrak{X}_1, \mathfrak{X}_3, 0) = 0$, *for every* $\mathfrak{X}_1, \mathfrak{X}_3, \mathfrak{X}_4 \in \mathfrak{M}$, *and* $\leftthreetimes \in ⫛$. *Then, we can find a unique* $\mathbb{C}$*–trilinear mapping* $\mathcal{E} : ⫛ \times \mathfrak{M}^3 \to \mathfrak{N}$ *satisfying*

$$\mathcal{N}\Big(\mathcal{Q}(\leftthreetimes, \mathfrak{X}_1, \mathfrak{X}_3, \mathfrak{X}_5) - \mathcal{E}(\leftthreetimes, \mathfrak{X}_1, \mathfrak{X}_3, \mathfrak{X}_5), \mathcal{T}\Big) \tag{16.34}$$
$$\succeq diag\Big[\varphi_1\Big(2\mathfrak{X}_1, 0, \mathfrak{X}_3, \mathfrak{X}_3, 2\mathfrak{X}_5, 0, 2(4-\mathfrak{P}_1)\mathcal{T}\Big)$$
$$\bigotimes_{\mathrm{TN}} \varphi_1\Big(\mathfrak{X}_1, \mathfrak{X}_1, \mathfrak{X}_3, 0, 2\mathfrak{X}_5, 0, (4-\mathfrak{P}_1)\mathcal{T}\Big)$$
$$\bigotimes_{\mathrm{TN}} \varphi_1\Big(\mathfrak{X}_1, 0, \mathfrak{X}_3, 0, \mathfrak{X}_5, \mathfrak{X}_5, \frac{(4-\mathfrak{P}_1)}{2}\mathcal{T}\Big), \ldots,$$
$$\varphi_n\Big(2\mathfrak{X}_1, 0, \mathfrak{X}_3, \mathfrak{X}_3, 2\mathfrak{X}_5, 0, 2(4-\mathfrak{P}_n)\mathcal{T}\Big)$$
$$\bigotimes_{\mathrm{TN}} \varphi_n\Big(\mathfrak{X}_1, \mathfrak{X}_1, \mathfrak{X}_3, 0, 2\mathfrak{X}_5, 0, (4-\mathfrak{P}_n)\mathcal{T}\Big)$$
$$\bigotimes_{\mathrm{TN}} \varphi_n\Big(\mathfrak{X}_1, 0, \mathfrak{X}_3, 0, \mathfrak{X}_5, \mathfrak{X}_5, \frac{(4-\mathfrak{P}_n)}{2}\mathcal{T}\Big)\Big], \tag{16.35}$$

for every $\mathfrak{X}_1, \mathfrak{X}_3, \mathfrak{X}_5 \in \mathfrak{M}, \leftthreetimes \in ⫛$ *and* $\mathcal{T} > 0$.

Also, if the mapping $\mathcal{Q} : ⋔ \times \mathfrak{M}^3 \to \mathfrak{N}$ *satisfies* (16.27) *and* (16.33), *then, the* $\mathbb{C}$*–trilinear mapping* $\mathcal{E} : ⋔ \times \mathfrak{M}^3 \to \mathfrak{N}$ *is a permuting tri–homomorphism.*

Proof It follows from Theorem 16.7.

16.5 Multi Stability of Permuting Tri–Homomorphisms in Unital MVFC–⋄–Algebras

In this part, we study the multi stability of permuting tri–homomorphisms in unital MVFC–⋄–algebras associated with the tri-additive Υ–functional inequality (16.1).

From now on, assume that $(\mathfrak{M}, \mathcal{N}, \bigotimes_{\text{GTN}}, \bigoplus_{\text{GTN}})$ is a unital MVFC–⋄–algebra with unit e and a unitary group $U(\mathfrak{M}) := \{t \in \mathfrak{M} : t^\diamond t = tt^\diamond = e\}$, and that $(\mathfrak{N}, \mathcal{N}, \bigotimes_{\text{GTN}}, \bigoplus_{\text{GTN}})$ is a unital MVFC–⋄–algebra.

Theorem 16.9 *Let* $n \in \mathbb{N}$, *and* $\overbrace{\varphi_i}^{i=1,\ldots,n} : \mathfrak{M}^6 \times (0, +\infty) \to [0, 1]$ *be functions satisfying* (16.31) *and* $\mathcal{Q} : ⋔ \times \mathfrak{M}^3 \to \mathfrak{N}$ *be a mapping satisfying* (16.24) *and* $\mathcal{Q}(\curlywedge, \mathfrak{X}_1, 0, \mathfrak{X}_5) = \mathcal{Q}(\curlywedge, 0, \mathfrak{X}_3, \mathfrak{X}_5) = \mathcal{Q}(\curlywedge, \mathfrak{X}_1, \mathfrak{X}_3, 0) = 0$, *for every* $\mathfrak{X}_1, \mathfrak{X}_3, \mathfrak{X}_5 \in \mathfrak{M}$ *and* $\curlywedge \in ⋔$. *Then, we can find a unique* $\mathbb{C}$*–trilinear mapping* $\mathcal{E} : ⋔ \times \mathfrak{M}^3 \to \mathfrak{N}$ *satisfying* (16.32).

Also, if the mapping $\mathcal{Q} : ⋔ \times \mathfrak{M}^3 \to \mathfrak{N}$ *satisfies* (16.27) *and*

$$\mathcal{N}\Big(\mathcal{Q}(\curlywedge, \beta_1\alpha_1, \beta_2\alpha_2, \beta_3\alpha_3) - \mathcal{Q}(\curlywedge, \beta_1, \beta_2, \beta_3)\mathcal{Q}(\curlywedge, \alpha_1, \alpha_2, \alpha_3), \mathcal{T}\Big) \quad (16.36)$$
$$\succeq diag\Big[\varphi_1\Big(\beta_1, \alpha_1, \beta_2, \alpha_2, \beta_3, \alpha_3, \mathcal{T}\Big), \ldots, \varphi_n\Big(\beta_1, \alpha_1, \beta_2, \alpha_2, \beta_3, \alpha_3, \mathcal{T}\Big)\Big],$$

for every $\beta_i, \alpha_i \in U(\mathfrak{M}), i = 1, 2, 3$, *and* $\mathcal{T} > 0$, *then, the* $\mathbb{C}$*–trilinear mapping* $\mathcal{E} : ⋔ \times \mathfrak{M}^3 \to \mathfrak{N}$ *is a permuting tri–homomorphism.*

Proof Theorem 16.3 and [3, Theorem 4.5] prove the theorem.

Theorem 16.10 *Let* $n \in \mathbb{N}$, *and* $\overbrace{\varphi_i}^{i=1,\ldots,n} : \mathfrak{M}^6 \times (0, +\infty) \to [0, 1]$ *be functions satisfying* (16.28) *and* $\mathcal{Q} : ⋔ \times \mathfrak{M}^3 \to \mathfrak{N}$ *be a mapping satisfying* (16.24) *and* $\mathcal{Q}(\curlywedge, \mathfrak{X}_1, 0, \mathfrak{X}_5) = \mathcal{Q}(\curlywedge, 0, \mathfrak{X}_3, \mathfrak{X}_5) = \mathcal{Q}(\curlywedge, \mathfrak{X}_1, \mathfrak{X}_3, 0) = 0$, *for every* $\mathfrak{X}_1, \mathfrak{X}_3, \mathfrak{X}_5 \in \mathfrak{M}$ *and* $\curlywedge \in ⋔$. *Then, we can find a unique* $\mathbb{C}$*–trilinear mapping* $\mathcal{E} : ⋔ \times \mathfrak{M}^3 \to \mathfrak{N}$ *satisfying* (16.34).

Also, if the mapping $\mathcal{Q} : ⋔ \times \mathfrak{M}^3 \to \mathfrak{N}$ *satisfies* (16.27) *and* (16.36), *then, the* $\mathbb{C}$*–trilinear mapping* $\mathcal{E} : ⋔ \times \mathfrak{M}^3 \to \mathfrak{N}$ *is a permuting tri–homomorphism.*

Proof It follows from Theorem 16.9.

16.6 Application

In this section, consider the control functions $\underbrace{\varphi_i^{\circledS}}_{1\le i\le n}$ defined in Sects. 9.8.1 and 9.8.2, for every $n \in \mathbb{N}$.

Corollary 16.1 *Consider the MVFB–algebras* $(\mho_1, \mathcal{N}, \bigotimes_{\mathrm{GTN}}, \bigoplus_{\mathrm{GTN}})$, *and* $(\mho_2, \mathcal{N}, \bigotimes_{\mathrm{GTN}}, \bigoplus_{\mathrm{GTN}})$. *Let* $n \in \mathbb{N}$, *and* $\underbrace{r_i}_{i=1,\dots,n} > 3$ *and* $\underbrace{\Omega_i}_{i=1,\dots,n} > 0$ *be in* $\mathbb{R}$ *and* $\mathcal{Q} : \pitchfork \times \mho_1^3 \to \mho_2$ *be a mapping satisfying* $\mathcal{Q}(\leftthreetimes, \mathfrak{X}_1, 0, \mathfrak{X}_5) = \mathcal{Q}(\leftthreetimes, 0, \mathfrak{X}_3, \mathfrak{X}_5) = \mathcal{Q}(\leftthreetimes, \mathfrak{X}_1, \mathfrak{X}_3, 0) = 0$ *and*

$$\begin{aligned}
&\mathcal{N}\Big(\mathcal{Q}(\leftthreetimes, \mathfrak{X}_1+\mathfrak{X}_2, \mathfrak{X}_3-\mathfrak{X}_4, \mathfrak{X}_5+\mathfrak{X}_6)+\mathcal{Q}(\leftthreetimes, \mathfrak{X}_1-\mathfrak{X}_2, \mathfrak{X}_3+\mathfrak{X}_4, \mathfrak{X}_5-\mathfrak{X}_6)\\
&\qquad -2\mathcal{Q}(\leftthreetimes, \mathfrak{X}_1, \mathfrak{X}_3, \mathfrak{X}_5)+2\mathcal{Q}(\leftthreetimes, \mathfrak{X}_1, \mathfrak{X}_4, \mathfrak{X}_6)\\
&\qquad\qquad -2\mathcal{Q}(\leftthreetimes, \mathfrak{X}_2, \mathfrak{X}_3, \mathfrak{X}_6)+2\mathcal{Q}(\leftthreetimes, \mathfrak{X}_2, \mathfrak{X}_4, \mathfrak{X}_5), \mathcal{T}\Big)\\
&\succeq \mathcal{N}\Big(\Upsilon[2\mathcal{Q}(\leftthreetimes, \frac{\mathfrak{X}_1+\mathfrak{X}_2}{2}, \mathfrak{X}_3-\mathfrak{X}_4, \mathfrak{X}_5+\mathfrak{X}_6)+2\mathcal{Q}(\leftthreetimes, \frac{\mathfrak{X}_1-\mathfrak{X}_2}{2}, \mathfrak{X}_3+\mathfrak{X}_4, \mathfrak{X}_5-\mathfrak{X}_6)\\
&\qquad -2\mathcal{Q}(\leftthreetimes, \mathfrak{X}_1, \mathfrak{X}_3, \mathfrak{X}_5)+2\mathcal{Q}(\leftthreetimes, \mathfrak{X}_1, \mathfrak{X}_4, \mathfrak{X}_6)\\
&\qquad\qquad -2\mathcal{Q}(\leftthreetimes, \mathfrak{X}_2, \mathfrak{X}_3, \mathfrak{X}_6)+2\mathcal{Q}(\leftthreetimes, \mathfrak{X}_2, \mathfrak{X}_4, \mathfrak{X}_5)], \mathcal{T}\Big)\\
&\bigotimes_{\mathrm{GTN}} diag\Big[\varphi_1^{\circledS}\Big(\frac{-\Omega_1[\|\mathfrak{X}_1\|^{r_1}+\|\mathfrak{X}_2\|^{r_1}+\|\mathfrak{X}_3\|^{r_1}+\|\mathfrak{X}_4\|^{r_1}+\|\mathfrak{X}_5\|^{r_1}+\|\mathfrak{X}_6\|^{r_1}]}{\mathcal{T}}\Big), \dots,\\
&\qquad \varphi_n^{\circledS}\Big(\frac{-\Omega_n[\|\mathfrak{X}_1\|^{r_n}+\|\mathfrak{X}_2\|^{r_n}+\|\mathfrak{X}_3\|^{r_n}+\|\mathfrak{X}_4\|^{r_n}+\|\mathfrak{X}_5\|^{r_n}+\|\mathfrak{X}_6\|^{r_n}]}{\mathcal{T}}\Big)\Big].
\end{aligned}\tag{16.37}$$

for every $\mathfrak{X}_1, \mathfrak{X}_2, \mathfrak{X}_3, \mathfrak{X}_4, \mathfrak{X}_5, \mathfrak{X}_6 \in \mho_1$, $\leftthreetimes \in \pitchfork$ *and* $\mathcal{T} > 0$. *Then, we can find a unique tri–additive mapping* $\mathcal{Q}' : \pitchfork \times \mho_1^3 \to \mho_2$ *satisfying*

$$\begin{aligned}
&\mathcal{N}\Big(\mathcal{Q}(\leftthreetimes, \mathfrak{X}_1, \mathfrak{X}_3, \mathfrak{X}_5)-\mathcal{Q}'(\leftthreetimes, \mathfrak{X}_1, \mathfrak{X}_3, \mathfrak{X}_5), \mathcal{T}\Big)\\
&\succeq diag[\sigma_1, \dots, \sigma_n],
\end{aligned}\tag{16.38}$$

in which

$$\begin{aligned}
&diag[\sigma_1, \dots, \sigma_n]\\
&:= diag\Big[\varphi_1^{\circledS}\Big(\frac{-\Omega_1[\|\mathfrak{X}_1\|^{r_1}+2\frac{\|\mathfrak{X}_3\|^{r_1}}{2^{r_1}}+\|\mathfrak{X}_5\|^{r_1}]}{(1-4\mathfrak{P}_1)\mathcal{T}}\Big)\\
&\qquad \bigotimes_{\mathrm{TN}} \varphi_1^{\circledS}\Big(\frac{-\Omega_1[\frac{2\|\mathfrak{X}_1\|^{r_1}}{2^{r_1}}+2\frac{\|\mathfrak{X}_3\|^{r_1}}{2^{r_1}}+\|\mathfrak{X}_5\|^{r_1}]}{\frac{(1-4\mathfrak{P}_1)}{2}\mathcal{T}}\Big)
\end{aligned}\tag{16.39}$$

$$\bigotimes_{\mathrm{TN}} \varphi_1^{\circledS}\left(\frac{-\Omega_1[\frac{\|\mathfrak{X}_1\|^{r_1}}{2^{r_1}}+\frac{\|\mathfrak{X}_3\|^{r_1}}{2^{r_1}}+\frac{\|\mathfrak{X}_5\|^{r_1}}{2^{r_1}}]}{\frac{(1-4\mathfrak{P}_1)}{4}\mathcal{T}}\right),\ldots,$$

$$\varphi_n^{\circledS}\left(\frac{-\Omega_n[\|\mathfrak{X}_1\|^{r_n}+2\frac{\|\mathfrak{X}_3\|^{r_n}}{2^{r_n}}+\|\mathfrak{X}_5\|^{r_n}]}{(1-4\mathfrak{P}_n)\mathcal{T}}\right)$$

$$\bigotimes_{\mathrm{TN}} \varphi_n^{\circledS}\left(\frac{-\Omega_n[\frac{2\|\mathfrak{X}_1\|^{r_n}}{2^{r_n}}+2\frac{\|\mathfrak{X}_3\|^{r_n}}{2^{r_n}}+\|\mathfrak{X}_5\|^{r_n}]}{\frac{(1-4\mathfrak{P}_n)}{2}\mathcal{T}}\right)$$

$$\bigotimes_{\mathrm{TN}} \varphi_n^{\circledS}\left(\frac{-\Omega_n[\frac{\|\mathfrak{X}_1\|^{r_n}}{2^{r_n}}+\frac{\|\mathfrak{X}_3\|^{r_n}}{2^{r_n}}+\frac{\|\mathfrak{X}_5\|^{r_n}}{2^{r_n}}]}{\frac{(1-4\mathfrak{P}_n)}{4}\mathcal{T}}\right)\Bigg],$$

for every $\underbrace{\mathfrak{P}_i}_{i=1,\ldots,n} = 2^{\overbrace{r_i}^{i=1,\ldots,n}-1}$, $\mathfrak{X}_1, \mathfrak{X}_3, \mathfrak{X}_5 \in \mho_1$, $\lambda \in \pitchfork$ *and* $\mathcal{T} > 0$.

Proof We define the control function $\underbrace{\varphi_i}_{1\le i\le n}$ by the coefficient

$$\varphi_i(\mathfrak{X}_1,\mathfrak{X}_2,\mathfrak{X}_3,\mathfrak{X}_4,\mathfrak{X}_5,\mathfrak{X}_6) := \varphi_i^{\circledS}\left(\frac{-\Omega_i[\|\mathfrak{X}_1\|^{r_i}+\|\mathfrak{X}_2\|^{r_n}+\|\mathfrak{X}_3\|^{r_n}+\|\mathfrak{X}_4\|^{r_i}+\|\mathfrak{X}_5\|^{r_i}+\|\mathfrak{X}_6\|^{r_i}]}{\mathcal{T}}\right),\quad i=1,\ldots,n, \tag{16.40}$$

for every $\mathfrak{X}_1, \mathfrak{X}_2, \mathfrak{X}_3, \mathfrak{X}_4, \mathfrak{X}_5, \mathfrak{X}_6 \in \mho_1$, $\lambda \in \pitchfork$ and $\mathcal{T} > 0$. Letting $\underbrace{\mathfrak{P}_i}_{i=1,\ldots,n} = 2^{\overbrace{r_i}^{i=1,\ldots,n}-1}$,

the result follows immediately from Theorem 16.1.

Corollary 16.2 *Consider the MVFB–algebras* $(\mho_1, \mathcal{N}, \bigotimes_{\mathrm{GTN}}, \bigoplus_{\mathrm{GTN}})$, *and* $(\mho_2, \mathcal{N}, \bigotimes_{\mathrm{GTN}}, \bigoplus_{\mathrm{GTN}})$. *Let* $n \in \mathbb{N}$, *and* $\underbrace{r_i}_{i=1,\ldots,n} < 3$ *and* $\underbrace{\Omega_i}_{i=1,\ldots,n} > 0$ *be in* $\mathbb{R}$, *and* $\mathcal{Q} : \pitchfork \times \mho_1^3 \to \mho_2$ *be a mapping satisfying* $\mathcal{Q}(\lambda, \mathfrak{X}_1, 0, \mathfrak{X}_5) = \mathcal{Q}(\lambda, 0, \mathfrak{X}_3, \mathfrak{X}_5) = \mathcal{Q}(\lambda, \mathfrak{X}_1, \mathfrak{X}_3, 0) = 0$ *and the inequality* (16.37). *Then, we can find a unique tri–additive mapping* $\mathcal{Q}' : \pitchfork \times \mho_1^3 \to \mho_2$ *satisfying*

$$\mathcal{N}\Big(\mathcal{Q}(\lambda, \mathfrak{X}_1, \mathfrak{X}_3, \mathfrak{X}_5) - \mathcal{Q}'(\lambda, \mathfrak{X}_1, \mathfrak{X}_3, \mathfrak{X}_5), \mathcal{T}\Big) \succeq diag[\sigma', \ldots, \sigma'] \tag{16.41}$$

in which

$$
\begin{aligned}
& diag[\sigma', \ldots, \sigma'] \\
& := diag\Bigg[\varphi_1^{ⓢ}\left(\frac{-\Omega_1[\|\mathfrak{X}_1\|^{r_1} + 2\|\mathfrak{X}_3\|^{r_1} + \|\mathfrak{X}_5\|^{r_1}]}{(8-2\mathfrak{P}_1)\mathcal{T}}\right) \\
& \qquad \bigotimes_{\mathrm{TN}} \varphi_1^{ⓢ}\left(\frac{-\Omega_1[\|\mathfrak{X}_1\|^{r_1} + \|\mathfrak{X}_3\|^{r_1} + 2^{r_1}\|\mathfrak{X}_5\|^{r_1}]}{(4-\mathfrak{P}_1)\mathcal{T}}\right) \\
& \qquad\qquad \bigotimes_{\mathrm{TN}} \varphi_1^{ⓢ}\left(\frac{-\Omega_1[\|\mathfrak{X}_1\|^{r_1} + \|\mathfrak{X}_3\|^{r_1} + 2\|\mathfrak{X}_5\|^{r_1}]}{\frac{(4-\mathfrak{P}_1)}{2}\mathcal{T}}\right), \ldots, \\
& \qquad \varphi_n^{ⓢ}\left(\frac{-\Omega_n[\|\mathfrak{X}_1\|^{r_n} + 2\|\mathfrak{X}_3\|^{r_n} + \|\mathfrak{X}_5\|^{r_n}]}{(8-2\mathfrak{P}_n)\mathcal{T}}\right) \\
& \qquad \bigotimes_{\mathrm{TN}} \varphi_n^{ⓢ}\left(\frac{-\Omega_n[\|\mathfrak{X}_1\|^{r_n} + \|\mathfrak{X}_3\|^{r_n} + 2^{r_n}\|\mathfrak{X}_5\|^{r_n}]}{(4-\mathfrak{P}_n)\mathcal{T}}\right) \\
& \qquad\qquad \bigotimes_{\mathrm{TN}} \varphi_n^{ⓢ}\left(\frac{-\Omega_n[\|\mathfrak{X}_1\|^{r_n} + \|\mathfrak{X}_3\|^{r_n} + 2\|\mathfrak{X}_5\|^{r_n}]}{\frac{(4-\mathfrak{P}_n)}{2}\mathcal{T}}\right)\Bigg],
\end{aligned}
\tag{16.42}
$$

for every $\underbrace{\mathfrak{P}_i}_{i=1,\ldots,n} = 2^{\overbrace{r_i}^{i=1,\ldots,n}+1}$, $\mathfrak{X}_1, \mathfrak{X}_3, \mathfrak{X}_5 \in \mho_1$, $\curlywedge \in ⫛$ *and* $\mathcal{T} > 0$.

Proof Put $\underbrace{\mathfrak{P}_i}_{i=1,\ldots,n} = 2^{\overbrace{r_i}^{i=1,\ldots,n}+1}$, It follows immediately from the relation (16.40) and Theorem 16.2.

Corollary 16.3 *Consider the MVFB–algebra* $(\mathfrak{M}, \mathscr{N}, \bigotimes_{\mathrm{GTN}}, \bigoplus_{\mathrm{GTN}})$. *Let* $n \in \mathbb{N}$, *and* $\underbrace{r_i}_{i=1,\ldots,n} > 4$ *and* $\underbrace{\Omega_i}_{i=1,\ldots,n} > 0$ *be in* $\mathbb{R}$, *and* $\mathcal{Q} : ⫛ \times \mathfrak{M}^3 \to \mathfrak{M}$ *be a mapping satisfying* $\mathcal{Q}(\curlywedge, \mathfrak{X}_1, 0, \mathfrak{X}_5) = \mathcal{Q}(\curlywedge, 0, \mathfrak{X}_3, \mathfrak{X}_5) = \mathcal{Q}(\curlywedge, \mathfrak{X}_1, \mathfrak{X}_3, 0) = 0$ *and*

$$
\begin{aligned}
& \mathscr{N}\Big(\mathcal{Q}(\curlywedge, \Lambda_1(\mathfrak{X}_1 + \mathfrak{X}_2), \Lambda_2(\mathfrak{X}_3 - \mathfrak{X}_4), \Lambda_3(\mathfrak{X}_5 + \mathfrak{X}_6)) \\
& \qquad + \mathcal{Q}(\curlywedge, \Lambda_1(\mathfrak{X}_1 - \mathfrak{X}_2), \Lambda_2(\mathfrak{X}_3 + \mathfrak{X}_4), \Lambda_3(\mathfrak{X}_5 - \mathfrak{X}_6)) \\
& \qquad\qquad - \Lambda_1\Lambda_2\Lambda_3(2\mathcal{Q}(\curlywedge, \mathfrak{X}_1, \mathfrak{X}_3, \mathfrak{X}_5) - 2\mathcal{Q}(\curlywedge, \mathfrak{X}_1, \mathfrak{X}_4, \mathfrak{X}_6) \\
& \qquad\qquad\qquad + 2\mathcal{Q}(\curlywedge, \mathfrak{X}_2, \mathfrak{X}_3, \mathfrak{X}_6) - 2\mathcal{Q}(\curlywedge, \mathfrak{X}_2, \mathfrak{X}_4, \mathfrak{X}_5)), \mathcal{T}\Big) \\
& \geq \mathscr{N}\Big(\Upsilon[2\mathcal{Q}(\curlywedge, \frac{\mathfrak{X}_1 + \mathfrak{X}_2}{2}, \mathfrak{X}_3 - \mathfrak{X}_4, \mathfrak{X}_5 + \mathfrak{X}_6) + 2\mathcal{Q}(\curlywedge, \frac{\mathfrak{X}_1 - \mathfrak{X}_2}{2}, \mathfrak{X}_3 + \mathfrak{X}_4, \mathfrak{X}_5 - \mathfrak{X}_6] \\
& \qquad - 2\mathcal{Q}(\mathfrak{X}_1, \mathfrak{X}_3, \mathfrak{X}_5) + 2\mathcal{Q}(\curlywedge, \mathfrak{X}_1, \mathfrak{X}_4, \mathfrak{X}_6) \\
& \qquad\qquad - 2\mathcal{Q}(\mathfrak{X}_2, \mathfrak{X}_3, \mathfrak{X}_6) + 2\mathcal{Q}(\curlywedge, \mathfrak{X}_2, \mathfrak{X}_4, \mathfrak{X}_5)], \mathcal{T}\Big) \\
& \quad \bigotimes_{\mathrm{GTN}} diag\Big[\varphi_1^{ⓢ}\left(\frac{-\Omega_1[\|\mathfrak{X}_1\|^{r_1} + \|\mathfrak{X}_2\|^{r_1} + \|\mathfrak{X}_3\|^{r_1} + \|\mathfrak{X}_4\|^{r_1} + \|\mathfrak{X}_5\|^{r_1} + \|\mathfrak{X}_6\|^{r_1}]}{\mathcal{T}}\right)
\end{aligned}
\tag{16.43}
$$

$$\varphi_n^{\circledS}\left(\frac{-\Omega_n[\|\mathfrak{X}_1\|^{r_n}+\|\mathfrak{X}_2\|^{r_n}+\|\mathfrak{X}_3\|^{r_n}+\|\mathfrak{X}_4\|^{r_n}+\|\mathfrak{X}_5\|^{r_n}+\|\mathfrak{X}_6\|^{r_n}]}{\mathcal{T}}\right)\Bigg],$$

for every $\Lambda_1, \Lambda_2, \Lambda_3 \in \Delta^1$ *and every* $\mathfrak{X}_1, \mathfrak{X}_2, \mathfrak{X}_3, \mathfrak{X}_4, \mathfrak{X}_5, \mathfrak{X}_6 \in \mathfrak{M}$, $\curlywedge \in \pitchfork$ *and* $\mathcal{T} > 0$. *Then, we can find a unique* $\mathbb{C}$*–trilinear mapping* $\mathfrak{Z} : \pitchfork \times \mathfrak{M}^3 \to \mathfrak{M}$ *satisfying*

$$\mathcal{N}\left(\mathcal{Q}(\curlywedge, \mathfrak{X}_1, \mathfrak{X}_3, \mathfrak{X}_5) - \mathfrak{Z}(\curlywedge, \mathfrak{X}_1, \mathfrak{X}_3, \mathfrak{X}_5), \mathcal{T}\right) \succeq diag[\sigma_1, \ldots, \sigma_n],$$

in which $diag[\sigma_1, \ldots, \sigma_n]$ *satisfies* (16.39) *for every* $\mathfrak{X}_1, \mathfrak{X}_3, \mathfrak{X}_5 \in \mathfrak{M}$ *and* $\curlywedge \in \pitchfork$.

Also, if the mapping $\mathcal{Q} : \pitchfork \times \mathfrak{M}^3 \to \mathfrak{M}$ *satisfies* $\mathcal{Q}(\curlywedge, 2\mathfrak{X}_1, \mathfrak{X}_3, \mathfrak{X}_5) = 2\mathcal{Q}(\curlywedge, \mathfrak{X}_1, \mathfrak{X}_3, \mathfrak{X}_5)$ *and*

$$\begin{aligned}&\mathcal{N}\left(\mathcal{Q}(\curlywedge, \mathfrak{X}_1\mathfrak{X}_2, \mathfrak{X}_3, \mathfrak{X}_5) - \mathcal{Q}(\curlywedge, \mathfrak{X}_1, \mathfrak{X}_3, \mathfrak{X}_5)\mathfrak{X}_2 - \mathfrak{X}_1\mathcal{Q}(\curlywedge, \mathfrak{X}_2, \mathfrak{X}_3, \mathfrak{X}_5), \mathcal{T}\right) \\ &\succeq diag\Bigg[\varphi_1^{\circledS}\left(\frac{-\Omega_1[\|\mathfrak{X}_1\|^{r_1}+\|\mathfrak{X}_2\|^{r_1}+\|\mathfrak{X}_3\|^{r_1}+\|\mathfrak{X}_4\|^{r_1}+\|\mathfrak{X}_5\|^{r_1}+\|\mathfrak{X}_6\|^{r_1}]}{\mathcal{T}}\right), \ldots, \\ &\qquad \varphi_n^{\circledS}\left(\frac{-\Omega_n[\|\mathfrak{X}_1\|^{r_n}+\|\mathfrak{X}_2\|^{r_n}+\|\mathfrak{X}_3\|^{r_n}+\|\mathfrak{X}_4\|^{r_n}+\|\mathfrak{X}_5\|^{r_n}+\|\mathfrak{X}_6\|^{r_n}]}{\mathcal{T}}\right)\Bigg],\end{aligned} \tag{16.44}$$

and

$$\begin{aligned}&\mathcal{N}\left(\mathcal{Q}(\curlywedge, \mathfrak{X}'_{\delta(1)}, \mathfrak{X}'_{\delta(2)}, \mathfrak{X}'_{\delta(3)}) - \mathcal{Q}(\curlywedge, \mathfrak{X}'_1, \mathfrak{X}'_2, \mathfrak{X}'_3), \mathcal{T}\right) \\ &\succeq diag\Bigg[\varphi_1^{\circledS}\left(\frac{-\Omega_1[\|\mathfrak{X}_1\|^{r_1}+\|\mathfrak{X}_2\|^{r_1}+\|\mathfrak{X}_3\|^{r_1}+\|\mathfrak{X}_4\|^{r_1}+\|\mathfrak{X}_5\|^{r_1}+\|\mathfrak{X}_6\|^{r_1}]}{\mathcal{T}}\right), \ldots, \\ &\qquad \varphi_n^{\circledS}\left(\frac{-\Omega_n[\|\mathfrak{X}_1\|^{r_n}+\|\mathfrak{X}_2\|^{r_n}+\|\mathfrak{X}_3\|^{r_n}+\|\mathfrak{X}_4\|^{r_n}+\|\mathfrak{X}_5\|^{r_n}+\|\mathfrak{X}_6\|^{r_n}]}{\mathcal{T}}\right)\Bigg],\end{aligned} \tag{16.45}$$

for every $\mathfrak{X}_1, \mathfrak{X}_2, \mathfrak{X}_3, \mathfrak{X}'_1, \mathfrak{X}'_2, \mathfrak{X}'_3, \mathfrak{X}_5 \in \mathfrak{M}$, $\curlywedge \in \pitchfork$, $\mathcal{T} > 0$ *and for every permutation* $(\delta(1), \delta(2), \delta(3))$ *of* $(1, 2, 3)$, *then, the* $\mathbb{C}$*–trilinear mapping* $\mathfrak{Z} : \pitchfork \times \mathfrak{M}^3 \to \mathfrak{M}$ *is a permuting tri–derivation.*

Proof Let $\underbrace{\mathfrak{P}_i}_{i=1,\ldots,n} = 2^{\overbrace{r_i}^{i=1,\ldots,n}} {}^{-1}$. It is the coefficient that defines the control function $\underbrace{\varphi_i}_{i=1,\ldots,n}$ by (16.40). Then, we apply Theorem 16.3 to complete the proof.

Corollary 16.4 *Consider the MVFB–algebra* $(\mathfrak{M}, \mathcal{N}, \bigotimes_{\text{GTN}}, \bigoplus_{\text{GTN}})$. *Let* $n \in \mathbb{N}$, *and* $\underbrace{r_i}_{i=1,\ldots,n} < 3$ *and* $\underbrace{\Omega_i}_{i=1,\ldots,n} > 0$ *be in* $\mathbb{R}$, *and* $\mathcal{Q} : \pitchfork \times \mathfrak{M}^3 \to \mathfrak{M}$ *be a mapping satisfying* (16.43) *and* $\mathcal{Q}(\curlywedge, \mathfrak{X}_1, 0, \mathfrak{X}_5) = \mathcal{Q}(\curlywedge, 0, \mathfrak{X}_3, \mathfrak{X}_5) = \mathcal{Q}(\curlywedge, \mathfrak{X}_1, \mathfrak{X}_3, 0) = 0$, *for every* $\mathfrak{X}_1, \mathfrak{X}_3, \mathfrak{X}_5 \in \mathfrak{M}$ *and* $\curlywedge \in \pitchfork$. *Then, we can find a unique* $\mathbb{C}$*–trilinear mapping* $\mathfrak{Z} : \pitchfork \times \mathfrak{M}^3 \to \mathfrak{M}$ *satisfying*

$$\mathcal{N}\left(\mathcal{Q}(\curlywedge, \mathfrak{X}_1, \mathfrak{X}_3, \mathfrak{X}_5) - \mathfrak{Z}(\curlywedge, \mathfrak{X}_1, \mathfrak{X}_3, \mathfrak{X}_5), \mathcal{T}\right) \succeq diag[\sigma'_1, \ldots, \sigma'_n], \tag{16.46}$$

in which $diag[\sigma_1', \ldots, \sigma_n']$ *satisfies* (16.42), *for every* $\mathfrak{X}_1, \mathfrak{X}_3, \mathfrak{X}_5 \in \mathfrak{M}$, $\leftthreetimes \in \pitchfork$ *and* $\mathcal{T} > 0$.

Also, if the mapping $\mathcal{Q} : \mathfrak{M}^3 \to \mathfrak{M}$ *satisfies* (16.44), (16.45) *and* $\mathcal{Q}(\leftthreetimes, 2\mathfrak{X}_1, \mathfrak{X}_3, \mathfrak{X}_5) = 2\mathcal{Q}(\leftthreetimes, \mathfrak{X}_1, \mathfrak{X}_3, \mathfrak{X}_5)$, *for every* $\mathfrak{X}_1, \mathfrak{X}_3, \mathfrak{X}_5 \in \mathfrak{M}$ *and* $\leftthreetimes \in \pitchfork$, *then, the* $\mathbb{C}$*–trilinear mapping* $\mathfrak{Z} : \pitchfork \times \mathfrak{M}^3 \to \mathfrak{M}$ *is a permuting tri–derivation.*

Proof Let $\underbrace{\mathfrak{P}_i}_{i=1,\ldots,n} = 2^{\overbrace{r_i}^{i=1,\ldots,n}+1}$. It is the coefficient that defines the control function $\underbrace{\varphi_i}_{i=1,\ldots,n}$ by (16.40). Then, we apply Theorem 16.4 to complete the proof.

Corollary 16.5 *Consider the unital MVFC–*$\diamond$*–algebra* $(\mathfrak{M}, \mathcal{N}, \bigotimes_{\mathrm{GTN}}, \bigoplus_{\mathrm{GTN}})$. *Let* $n \in \mathbb{N}$, *and* $\underbrace{r_i}_{i=1,\ldots,n} > 4$ *and* $\underbrace{\Omega_i}_{i=1,\ldots,n} > 0$ *be in* $\mathbb{R}$, *and* $\mathcal{Q} : \pitchfork \times \mathfrak{M}^3 \to \mathfrak{M}$ *be a mapping satisfying* (16.43) *and* $\mathcal{Q}(\leftthreetimes, \mathfrak{X}_1, 0, \mathfrak{X}_5) = \mathcal{Q}(\leftthreetimes, 0, \mathfrak{X}_3, \mathfrak{X}_5) = \mathcal{Q}(\leftthreetimes, \mathfrak{X}_1, \mathfrak{X}_3, 0) = 0$, *for every* $\mathfrak{X}_1, \mathfrak{X}_3, \mathfrak{X}_5 \in \mathfrak{M}$ *and* $\leftthreetimes \in \pitchfork$. *Then, we can find a unique* $\mathbb{C}$*–trilinear mapping* $\mathfrak{Z} : \pitchfork \times \mathfrak{M}^3 \to \mathfrak{M}$ *satisfying* (16.44).

Also, if the mapping $\mathcal{Q} : \pitchfork \times \mathfrak{M}^3 \to \mathfrak{M}$ *satisfies* (16.45), *and* $\mathcal{Q}(\leftthreetimes, 2\beta, \mathfrak{X}_3, \mathfrak{X}_5) = 2\mathcal{Q}(\leftthreetimes, \beta, \mathfrak{X}_3, \mathfrak{X}_5)$ *and*

$$\begin{aligned} &\mathcal{N}\Big(\mathcal{Q}(\leftthreetimes, \beta\mathfrak{X}_2, \mathfrak{X}_3, \mathfrak{X}_5) - \mathcal{Q}(\leftthreetimes, \beta, \mathfrak{X}_3, \mathfrak{X}_5)\mathfrak{X}_2 - \beta\mathcal{Q}(\leftthreetimes, \mathfrak{X}_2, \mathfrak{X}_3, \mathfrak{X}_5), \mathcal{T}\Big) \\ &\succeq diag\Bigg[\varphi_1^{\circledS}\left(\frac{-\Omega_1[1 + \|\mathfrak{X}_2\|^{r_1} + 2\|\mathfrak{X}_3\|^{r_1} + 2\|\mathfrak{X}_5\|^{r_1}]}{\mathcal{T}}\right), \ldots, \\ &\qquad \varphi_n^{\circledS}\left(\frac{-\Omega_n[1 + \|\mathfrak{X}_2\|^{r_n} + 2\|\mathfrak{X}_3\|^{r_n} + 2\|\mathfrak{X}_5\|^{r_n}]}{\mathcal{T}}\right)\Bigg], \end{aligned} \tag{16.47}$$

for every $\beta \in U(\mathfrak{M})$ *and every* $\mathfrak{X}_2, \mathfrak{X}_3, \mathfrak{X}_5 \in \mathfrak{M}$, $\leftthreetimes \in \pitchfork$, *and* $\mathcal{T} > 0$, *then, the* $\mathbb{C}$*–trilinear mapping* $\mathfrak{Z} : \pitchfork \times \mathfrak{M}^3 \to \mathfrak{M}$ *is a permuting tri–derivation.*

Proof It is the coefficient that defines the control function $\underbrace{\varphi_i}_{i=1,\ldots,n}$ by (16.40). Then, we apply Theorem 16.5 to complete the proof.

Corollary 16.6 *Consider the unital MVFC–*$\diamond$*–algebra* $(\mathfrak{M}, \mathcal{N}, \bigotimes_{\mathrm{GTN}}, \bigoplus_{\mathrm{GTN}})$. *Let* $n \in \mathbb{N}$, *and* $\underbrace{r_i}_{i=1,\ldots,n} < 3$ *and* $\underbrace{\Omega_i}_{i=1,\ldots,n} > 0$ *be in* $\mathbb{R}$, *and* $\mathcal{Q} : \pitchfork \times \mathfrak{M}^3 \to \mathfrak{M}$ *be a mapping satisfying* (16.43) *and* $\mathcal{Q}(\leftthreetimes, \mathfrak{X}_1, 0, \mathfrak{X}_5) = \mathcal{Q}(\leftthreetimes, 0, \mathfrak{X}_3, \mathfrak{X}_5) = \mathcal{Q}(\leftthreetimes, \mathfrak{X}_1, \mathfrak{X}_3, 0) = 0$, *for every* $\mathfrak{X}_1, \mathfrak{X}_3, \mathfrak{X}_5 \in \mathfrak{M}$ *and* $\leftthreetimes \in \pitchfork$. *Then, we can find a unique* $\mathbb{C}$*–trilinear mapping* $\mathfrak{Z} : \pitchfork \times \mathfrak{M}^3 \to \mathfrak{M}$ *satisfying* (16.46).

Also, if the mapping $\mathcal{Q} : \pitchfork \times \mathfrak{M}^3 \to \mathfrak{M}$ *satisfies* (16.45), *(16.47) and* $\mathcal{Q}(\leftthreetimes, 2\beta, \mathfrak{X}_3, \mathfrak{X}_5) = 2\mathcal{Q}(\leftthreetimes, \beta, \mathfrak{X}_3, \mathfrak{X}_5)$, *for every* $\beta \in U(\mathfrak{M})$ *and every* $\mathfrak{X}_3, \mathfrak{X}_5 \in \mathfrak{M}$ *and* $\leftthreetimes \in \pitchfork$, *then, the* $\mathbb{C}$*–trilinear mapping* $\mathfrak{Z} : \pitchfork \times \mathfrak{M}^3 \to \mathfrak{M}$ *is a permuting tri–derivation.*

Proof It is the coefficient that defines the control function $\underbrace{\varphi_i}_{i=1,\ldots,n}$ by (16.40). Then, we apply Theorem 16.6 to complete the proof.

Corollary 16.7 *Consider the MVFB–algebras* $(\mathfrak{M}, \mathcal{N}, \bigotimes_{\mathrm{GTN}}, \bigoplus_{\mathrm{GTN}})$ *and* $(\mathfrak{N}, \mathcal{N}, \bigotimes_{\mathrm{GTN}}, \bigoplus_{\mathrm{GTN}})$. *Let* $n \in \mathbb{N}$, *and* $\underbrace{r_i}_{i=1,\ldots,n} > 6$ *and* $\underbrace{\Omega_i}_{i=1,\ldots,n} > 0$ *be in* $\mathbb{R}$, *and* $\mathcal{Q} : \pitchfork \times \mathfrak{M}^3 \to \mathfrak{N}$ *be a mapping satisfying* (16.43) *and* $\mathcal{Q}(\leftthreetimes, \mathfrak{X}_1, 0, \mathfrak{X}_5) = \mathcal{Q}(\leftthreetimes, 0, \mathfrak{X}_3, \mathfrak{X}_5) = \mathcal{Q}(\leftthreetimes, \mathfrak{X}_1, \mathfrak{X}_3, 0) = 0$, *for every* $\mathfrak{X}_1, \mathfrak{X}_2, \mathfrak{X}_3, \mathfrak{X}_4, \mathfrak{X}_5, \mathfrak{X}_6 \in \mathfrak{M}$ *and* $\leftthreetimes \in \pitchfork$. *Then, we can find a unique* $\mathbb{C}$*–trilinear mapping* $\mathcal{E} : \pitchfork \times \mathfrak{M}^3 \to \mathfrak{N}$ *satisfying*

$$\mathcal{N}\left(\mathcal{Q}(\leftthreetimes, \mathfrak{X}_1, \mathfrak{X}_3, \mathfrak{X}_5) - \mathcal{E}(\leftthreetimes, \mathfrak{X}_1, \mathfrak{X}_3, \mathfrak{X}_4), \mathcal{T}\right) \succeq diag[\sigma_1, \ldots, \sigma_n] \quad (16.48)$$

in which $diag[\sigma_1, \ldots, \sigma_n]$ *satisfies* (16.39), *for every* $\mathfrak{X}_1, \mathfrak{X}_3, \mathfrak{X}_5 \in \mathfrak{M}$, $\leftthreetimes \in \pitchfork$ *and* $\mathcal{T} > 0$.

Also, if the mapping $\mathcal{Q} : \pitchfork \times \mathfrak{M}^3 \to \mathfrak{N}$ *satisfies* (16.45) *and*

$$\begin{aligned} &\mathcal{N}\left(\mathcal{Q}(\leftthreetimes, \mathfrak{X}_1\mathfrak{X}_2, \mathfrak{X}_3\mathfrak{X}_4, \mathfrak{X}_5\mathfrak{X}_6) - \mathcal{Q}(\leftthreetimes, \mathfrak{X}_1, \mathfrak{X}_3, \mathfrak{X}_5)\mathcal{Q}(\leftthreetimes, \mathfrak{X}_2, \mathfrak{X}_4, \mathfrak{X}_6), \mathcal{T}\right) \quad (16.49)\\ &\succeq diag\Bigg[\varphi_1^{\circledS}\left(\frac{-\Omega_1[\|\mathfrak{X}_1\|^{r_1} + \|\mathfrak{X}_2\|^{r_1} + \|\mathfrak{X}_3\|^{r_1} + \|\mathfrak{X}_4\|^{r_1} + \|\mathfrak{X}_5\|^{r_1} + \|\mathfrak{X}_6\|^{r_1}]}{\mathcal{T}}\right), \ldots,\\ &\qquad \varphi_n^{\circledS}\left(\frac{-\Omega_n[\|\mathfrak{X}_1\|^{r_n} + \|\mathfrak{X}_2\|^{r_n} + \|\mathfrak{X}_3\|^{r_n} + \|\mathfrak{X}_4\|^{r_n} + \|\mathfrak{X}_5\|^{r_n} + \|\mathfrak{X}_6\|^{r_n}]}{\mathcal{T}}\right)\Bigg], \end{aligned}$$

for every $\mathfrak{X}_1, \mathfrak{X}_2, \mathfrak{X}_3, \mathfrak{X}_4, \mathfrak{X}_5, \mathfrak{X}_6 \in \mathfrak{M}$, $\leftthreetimes \in \pitchfork$ *and* $\mathcal{T} > 0$, *then, the* $\mathbb{C}$*–trilinear mapping* $\mathcal{E} : \pitchfork \times \mathfrak{M}^3 \to \mathfrak{N}$ *is a permuting tri–homomorphism.*

Proof It is the coefficient that defines the control function $\underbrace{\varphi_i}_{i=1,\ldots,n}$ by (16.40). Then, we apply Theorem 16.7 to complete the proof.

Corollary 16.8 *Consider the MVFB–algebras* $(\mathfrak{M}, \mathcal{N}, \bigotimes_{\mathrm{GTN}}, \bigoplus_{\mathrm{GTN}})$ *and* $(\mathfrak{N}, \mathcal{N}, \bigotimes_{\mathrm{GTN}}, \bigoplus_{\mathrm{GTN}})$. *Let* $n \in \mathbb{N}$, *and* $\underbrace{r_i}_{i=1,\ldots,n} < 3$ *and* $\underbrace{\Omega_i}_{i=1,\ldots,n} > 0$ *be in* $\mathbb{R}$, *and* $\mathcal{Q} : \pitchfork \times \mathfrak{M}^3 \to \mathfrak{N}$ *be a mapping satisfying* (16.43) *and* $\mathcal{Q}(\leftthreetimes, \mathfrak{X}_1, 0, \mathfrak{X}_5) = \mathcal{Q}(\leftthreetimes, 0, \mathfrak{X}_3, \mathfrak{X}_5) = \mathcal{Q}(\leftthreetimes, \mathfrak{X}_1, \mathfrak{X}_3, 0) = 0$, *for every* $\mathfrak{X}_1, \mathfrak{X}_2, \mathfrak{X}_3, \mathfrak{X}_4, \mathfrak{X}_5, \mathfrak{X}_6 \in \mathfrak{M}$ *and* $\leftthreetimes \in \pitchfork$. *Then, we can find a unique* $\mathbb{C}$*–trilinear mapping* $\mathcal{E} : \pitchfork \times \mathfrak{M}^3 \to \mathfrak{N}$ *satisfying*

$$\begin{aligned} &\mathcal{N}\left(\mathcal{Q}(\leftthreetimes, \mathfrak{X}_1, \mathfrak{X}_3, \mathfrak{X}_4) - \mathcal{E}(\leftthreetimes, \mathfrak{X}_1, \mathfrak{X}_3, \mathfrak{X}_4), \mathcal{T}\right) \quad (16.50)\\ &\succeq diag[\sigma'_1, \ldots, \sigma'_1] \end{aligned}$$

in which $diag[\sigma'_1, \ldots, \sigma'_1]$ *satisfies* (16.42), *for every* $\mathfrak{X}_1, \mathfrak{X}_3, \mathfrak{X}_5 \in \mathfrak{M}$, $\leftthreetimes \in \pitchfork$ *and* $\mathcal{T} > 0$.

Also, if the mapping $\mathcal{Q} : \pitchfork \times \mathfrak{M}^3 \to \mathfrak{N}$ *satisfies* (16.45) *and* (16.49), *then, the* $\mathbb{C}$*–trilinear mapping* $\mathcal{E} : \pitchfork \times \mathfrak{M}^3 \to \mathfrak{N}$ *is a permuting tri–homomorphism.*

Proof It is the coefficient that defines the control function $\underbrace{\varphi_i}_{i=1,\ldots,n}$ by (16.40). Then, we apply Theorem 16.8 to complete the proof.

Corollary 16.9 *Consider the unital MVFC–⋄–algebras* $(\mathfrak{M}, \mathscr{N}, \bigotimes_{\mathrm{GTN}}, \bigoplus_{\mathrm{GTN}})$ *and* $(\mathfrak{N}, \mathscr{N}, \bigotimes_{\mathrm{GTN}}, \bigoplus_{\mathrm{GTN}})$*. Let* $n \in \mathbb{N}$*, and* $\underbrace{r_i}_{i=1,\ldots,n} > 6$ *and* $\underbrace{\Omega_i}_{i=1,\ldots,n} > 0$ *be in* $\mathbb{R}$*, and* $\mathcal{Q} : ⫛ \times \mathfrak{M}^3 \to \mathfrak{N}$ *be a mapping satisfying* (16.43) *and* $\mathcal{Q}(\leftthreetimes, \mathfrak{X}_1, 0, \mathfrak{X}_5) = \mathcal{Q}(\leftthreetimes, 0, \mathfrak{X}_3, \mathfrak{X}_5) = \mathcal{Q}(\leftthreetimes, \mathfrak{X}_1, \mathfrak{X}_3, 0) = 0$*, for every* $\mathfrak{X}_1, \mathfrak{X}_2, \mathfrak{X}_3, \mathfrak{X}_4, \mathfrak{X}_5, \mathfrak{X}_6 \in \mathfrak{M}$ *and* $\leftthreetimes \in ⫛$*. Then, we can find a unique* $\mathbb{C}$*–trilinear mapping* $\mathcal{E} : ⫛ \times \mathfrak{M}^3 \to \mathfrak{N}$ *satisfying* (16.48).

Also, if the mapping $\mathcal{Q} : ⫛ \times \mathfrak{M}^3 \to \mathfrak{N}$ *satisfies* (16.45) *and*

$$\mathscr{N}\left(\mathcal{Q}(\leftthreetimes, \beta_1\alpha_1, \beta_2\alpha_2, \beta_3\alpha_3) - \mathcal{Q}(\leftthreetimes, \beta_1, \beta_2, \beta_3)\mathcal{Q}(\leftthreetimes, \alpha_1, \alpha_2, \alpha_3), \mathcal{T}\right)$$
$$\succeq diag\,\underbrace{[6\mathcal{T}, \ldots, 6\mathcal{T}]}_{n}, \tag{16.51}$$

for every $\leftthreetimes \in ⫛$*,* $\mathcal{T} > 0$*, and* $\beta_i, \alpha_i \in U(\mathfrak{M})$*,* $i = 1, 2, 3$*, then, the* $\mathbb{C}$*–trilinear mapping* $\mathcal{E} : ⫛ \times \mathfrak{M}^3 \to \mathfrak{N}$ *is a permuting tri–homomorphism.*

Proof It is the coefficient that defines the control function $\underbrace{\varphi_i}_{i=1,\ldots,n}$ by (16.40). Then, we apply Theorem 16.9 to complete the proof.

Corollary 16.10 *Consider the unital MVFC–⋄–algebras* $(\mathfrak{M}, \mathscr{N}, \bigotimes_{\mathrm{GTN}}, \bigoplus_{\mathrm{GTN}})$ *and* $(\mathfrak{N}, \mathscr{N}, \bigotimes_{\mathrm{GTN}}, \bigoplus_{\mathrm{GTN}})$*. Let* $n \in \mathbb{N}$*, and* $\underbrace{r_i}_{i=1,\ldots,n} < 3$ *and* $\underbrace{\Omega_i}_{i=1,\ldots,n} > 0$ *be in* $\mathbb{R}$*, and* $\mathcal{Q} : ⫛ \times \mathfrak{M}^3 \to \mathfrak{N}$ *be a mapping satisfying* (16.43) *and* $\mathcal{Q}(\leftthreetimes, \mathfrak{X}_1, 0, \mathfrak{X}_5) = \mathcal{Q}(\leftthreetimes, 0, \mathfrak{X}_3, \mathfrak{X}_5) = \mathcal{Q}(\leftthreetimes, \mathfrak{X}_1, \mathfrak{X}_3, 0) = 0$*, for every* $\mathfrak{X}_1, \mathfrak{X}_2, \mathfrak{X}_3, \mathfrak{X}_4, \mathfrak{X}_5, \mathfrak{X}_6 \in \mathfrak{M}$ *and* $\leftthreetimes \in ⫛$*. Then, we can find a unique* $\mathbb{C}$*–trilinear mapping* $\mathcal{E} : ⫛ \times \mathfrak{M}^3 \to \mathfrak{N}$ *satisfying* (16.50).

Also, if the mapping $\mathcal{Q} : ⫛ \times \mathfrak{M}^3 \to \mathfrak{N}$ *satisfies* (16.45) *and* (16.51)*, then, the* $\mathbb{C}$*–trilinear mapping* $\mathcal{E} : ⫛ \times \mathfrak{M}^3 \to \mathfrak{N}$ *is a permuting tri–homomorphism.*

Proof It is the coefficient that defines the control function $\underbrace{\varphi_i}_{i=1,\ldots,n}$ by (16.40). Then, we apply Theorem 16.10 to complete the proof.

We refer the interested readers to [4–14].

References

1. Aderyani SR, Saadati R, Mesiar R (2022) Estimation of permuting tri-homomorphisms and permuting tri-derivations associated with the tri-additive Υ-random operator inequality in matrix MB-algebra. Int J Gen Syst 51(6):547–569
2. Bae J, Park W (2010) Approximate bi-homomorphisms and bi-derivations in C^*-ternary algebras. Bull Korean Math Soc 47:195–209
3. Park C, Jin Y, Shin D, Zhang X, Govindan V (2020) Permuting triderivations and permuting trihomomorphisms in Banach algebras, Rocky mountain. J Math 50(5):1793–1806; MR4170688
4. Rezaei Aderyani S, Saadati R, Rassias TM (2022) Park C (2022) Best approximation of $(\mathcal{G}_1, \mathcal{G}_2)$-random operator inequality in matrix Menger Banach algebras with application of stochastic Mittag-Leffler and H $\mathbb{H}$-Fox control functions. J Inequal Appl 1:1–17
5. Aderyani SR, Saadati R (2021) Best approximations of the ϕ-Hadamard fractional Volterra integro-differential equation by matrix valued fuzzy control functions. Adv Differ Equ 154
6. Aderyani SR, Saadati R, Yang XJ (2021) Radu-Miheţ method for UHML stability for a class of ξ-Hilfer fractional differential equations in matrix valued fuzzy Banach spaces. Math Methods Appl Sci 44(18):14619–14631
7. Aderyani SR, Saadati R, Allahviranloo T (2022) Existence, uniqueness and matrix-valued fuzzy Mittag-Leffler-Hypergeometric-Wright stability for P-Hilfer fractional differential equations in matrix-valued fuzzy Banach space. Comput Appl Math 41(6):234
8. Aderyani SR, Saadati R (2022) Approximation of derivation-homomorphism fuzzy functional inequalities in matrix valued FC-$\diamond$-algebras. In: 2022 9th Iranian joint congress on fuzzy and intelligent systems (CFIS). IEEE, pp 1–6
9. Aderyani SR, Saadati R, Allahviranloo T, Abbasbandy S, Catak M (2023) Fuzzy approximation of a fractional Lorenz system and a fractional financial crisis. Iran J Fuzzy Syst 20(7):27–36
10. Aderyani SR, Saadati R, Abdeljawad T, Mlaiki N (2022) Multi-stability of non homogenous vector-valued fractional differential equations in matrix-valued Menger spaces. Alexandria Eng J 61(12):10913–10923
11. Aderyani SR, Saadati R (2023) Stability and controllability results by n-ary aggregation functions in matrix valued fuzzy n-normed spaces. Inf Sci 119265
12. O'Regan D, Aderyani SR, Saadati R, Li C (2023) Stability results and parametric delayed Mittag-Leffler matrices in symmetric fuzzy-random spaces with application. Symmetry 15(10):1880
13. Aderyani SR, Saadati R, O'Regan D, Li C (2023) On a new approach for stability and controllability analysis of functional equations. Mathematics 11(16):3458
14. Aderyani SR, Saadati R, O'Regan D, Alshammari FS (2023) Fuzzy approximate solutions of matrix-valued fractional differential equations by fuzzy control functions. Mathematics 11(6):1386

Chapter 17
Multi Stability of Additive–Additive Fuzzy Functional Inequalities in MVFC–⋄–Ternary Algebras

In this chapter, we study the multi stability of the following additive–additive (α, β)–fuzzy functional inequality [1]

$$
\begin{aligned}
&\mathcal{N}(\mu(\lambda, a+b) - \mu(\lambda, a) - \mu(\lambda, b), \mathcal{T}) \\
&\bigotimes_{\text{GTN}} \mathcal{N}(2\eta(\lambda, \frac{a+b}{2}) - \eta(\lambda, a) - \eta(\lambda, b), \mathcal{T}) \\
&\quad \succeq \mathcal{N}(\alpha[2\mu(\lambda, \frac{a+b}{2}) - \mu(\lambda, a) - \mu(\lambda, b)], \mathcal{T}) \\
&\quad \bigotimes_{\text{GTN}} \mathcal{N}(\beta[\eta(\lambda, a+b) - \eta(\lambda, a) - \eta(\lambda, b)], \mathcal{T}),
\end{aligned} \tag{17.1}
$$

where $0 \neq \alpha, \beta \in \mathbb{C}$ and $\max\{|\alpha|, |\beta|\} < 1$. In addition, we study the multi stability of homomorphisms and derivations in MVFB–algebras and unital MVFC–⋄–algebras associated with the additive–additive (α, β)–fuzzy functional inequality (17.1).

Furthermore, we investigate the multi stability of the following additive–additive (α, β)–fuzzy functional inequality [1]

$$
\begin{aligned}
&\mathcal{N}(\mu(\lambda, a+b+c) - \mu(\lambda, a) - \mu(\lambda, b) - \mu(\lambda, c), \mathcal{T}) \\
&\bigotimes_{\text{GTN}} \mathcal{N}(3\eta(\lambda, \frac{a+b+c}{3}) \\
&\quad + \eta(\lambda, a-2b+c) + \eta(\lambda, a+b-2c) - 3\eta(\lambda, a), \mathcal{T}) \\
&\succeq \mathcal{N}(\alpha[3\mu(\lambda, \frac{a+b+c}{3}) - \mu(\lambda, a) - \mu(\lambda, b) - \mu(\lambda, c)], \mathcal{T}) \\
&\quad \bigotimes_{\text{GTN}} \mathcal{N}(\beta[\eta(\lambda, a+b+c) + \eta(\lambda, a-2b+c) \\
&\qquad + \eta(\lambda, a+b-2c) - 3\eta(\lambda, a)], \mathcal{T}),
\end{aligned} \tag{17.2}
$$

S. Rezaei Aderyani et al., *Towards Ulam Type Multi Stability Analysis*, Studies in Systems, Decision and Control 523, https://doi.org/10.1007/978-3-031-55564-0_17

where $0 \neq \alpha, \beta \in \mathbb{C}$ and $\max\{|\alpha|, |\beta|\} < 1$. Also, we study the multi stability of FC–$\diamond$–ternary derivations and FC–$\diamond$–ternary homomorphisms in MVFC–$\diamond$–ternary algebras, associated with the additive–additive (α, β)–fuzzy functional inequality (17.2).

Here, we consider the unital MVFC–$\diamond$–algebra $(\mathcal{W}, \mathcal{N}, \bigotimes_{\text{GTN}}, \bigoplus_{\text{GTN}})$ with unite e and unitary group $U(e) = \{\Bbbk \in \mathcal{W} : \Bbbk^\diamond \Bbbk = \Bbbk\Bbbk^\diamond = e\}$.

Definition 17.1 ([1]) Consider the MVFB–algebra $(\mathcal{W}, \mathcal{N}, \bigotimes_{\text{GTN}}, \bigoplus_{\text{GTN}})$. A $\mathbb{C}$–linear fuzzy operator $\mu : \pitchfork \times \mathcal{W} \to \mathcal{W}$ is a derivation if,

$$\mu(\leftthreetimes, ab) = \mu(\leftthreetimes, a)b + a\mu(\leftthreetimes, b),$$

for every $a, b \in \mathcal{W}$, and $\leftthreetimes \in \pitchfork$. And a $\mathbb{C}$–linear fuzzy operator $\eta : \pitchfork \times \mathcal{W} \to \mathcal{W}$ is a homomorphism if,

$$\eta(\leftthreetimes, ab) = \eta(\leftthreetimes, a)\eta(\leftthreetimes, b),$$

for every $a, b \in \mathcal{W}$ and $\leftthreetimes \in \pitchfork$.

Definition 17.2 ([1]) Let $(\mathcal{W}, \mathcal{N}, \bigotimes_{\text{GTN}}, \bigoplus_{\text{GTN}})$ be an MVFC–$\diamond$–ternary algebra. A $\mathbb{C}$–linear fuzzy operator $\mu : \pitchfork \times \mathcal{W} \to \mathcal{W}$ is a FC–$\diamond$–ternary derivation if

$$\mu(\leftthreetimes, [a, b, c]) = [\mu(\leftthreetimes, a), b, c] + [a, \mu(\leftthreetimes, b), c] + [a, b, \mu(\leftthreetimes, c)],$$

for every $a, b, c \in \mathcal{W}$ and $\leftthreetimes \in \pitchfork$. And a $\mathbb{C}$–linear fuzzy operator $\eta : \pitchfork \times \mathcal{W} \to \mathcal{W}$ is a FC–$\diamond$–ternary homomorphism if

$$\eta(\leftthreetimes, [a, b, c]) = [\eta(\leftthreetimes, a), \eta(\leftthreetimes, b), \eta(\leftthreetimes, c)],$$

for every $a, b, c \in \mathcal{W}$ and $\leftthreetimes \in \pitchfork$.

17.1 Multi Stability of Derivation–Homomorphism Fuzzy Functional Inequalities in MVFB–Algebras

In this part, we study the multy stability of derivations and homomorphisms associated with the additive–additive (α, β)–fuzzy functional inequality (17.1) in MVFB–algebras by the fixed point method.

Throughout this section, let $(\mathcal{W}, \mathcal{N}, \bigotimes_{\text{GTN}}, \bigotimes_{\text{GTN}})$ be an MVFB–algebra.

Lemma 17.1 *If fuzzy operators $\mu, \eta : \pitchfork \times \mathcal{W} \to \mathcal{W}$ satisfy the fuzzy functional inequality* (17.1) *and $\mu(\leftthreetimes, 0) = \eta(\leftthreetimes, 0) = 0$, then, the fuzzy operators μ, η are additive.*

Proof Letting $a = b$ in functional inequality (17.1), we get

$$\mathcal{N}(\mu(\leftthreetimes, 2a) - 2\mu(\leftthreetimes, a), \mathcal{T}) \succeq \mathcal{N}(\beta[\eta(\leftthreetimes, 2a) - 2\eta(\leftthreetimes, a)], \mathcal{T}).$$

for every $a \in \mathcal{W}$, $\leftthreetimes \in \pitchfork$ and $\mathcal{T} > 0$. Now, let $b = 0$ in (17.1),

$$\mathscr{N}(2\eta(\leftthreetimes, \frac{a}{2}) - \eta(\leftthreetimes, a), \mathcal{T}) \succeq \mathscr{N}(\alpha[2\mu(\leftthreetimes, \frac{a}{2}) - \mu(\leftthreetimes, a)], \mathcal{T}),$$

and so

$$\mathscr{N}(2\eta(\leftthreetimes, a) - \eta(\leftthreetimes, 2a), \mathcal{T}) \succeq \mathscr{N}(\alpha[2\mu(\leftthreetimes, a) - \mu(\leftthreetimes, 2a)], \mathcal{T}),$$

for every $a \in \mathcal{W}$, $\leftthreetimes \in \pitchfork$ and $\mathcal{T} > 0$. Thus, we have that

$$\mathscr{N}(\mu(\leftthreetimes, 2a) - 2\mu(\leftthreetimes, a), \mathcal{T}) \succeq \mathscr{N}(\beta\alpha[\mu(\leftthreetimes, 2a) - 2\mu(\leftthreetimes, a)], \mathcal{T}),$$

and

$$\mathscr{N}(2\eta(\leftthreetimes, a) - \eta(\leftthreetimes, 2a), \mathcal{T}) \succeq \mathscr{N}(\alpha\beta[2\eta(\leftthreetimes, a) - \eta(\leftthreetimes, 2a)], \mathcal{T}),$$

for every $a \in \mathcal{W}$, $\leftthreetimes \in \pitchfork$ and $\mathcal{T} > 0$.

Since $\max\{|\alpha|, |\beta|\} < 1$, thus, $\eta(\leftthreetimes, 2a) = 2\eta(\leftthreetimes, a)$ and $\mu(\leftthreetimes, 2a) = 2\mu(\leftthreetimes, a)$, for every $a \in \mathcal{W}$ and $\leftthreetimes \in \pitchfork$. Now, according to (17.1), we have that

$$\begin{aligned}
&\mathscr{N}(\mu(\leftthreetimes, a+b) - \mu(\leftthreetimes, a) - \mu(\leftthreetimes, b), \mathcal{T}) \\
&\bigotimes_{\mathrm{GTN}} \mathscr{N}(\eta(\leftthreetimes, a+b) - \eta(\leftthreetimes, a) - \eta(\leftthreetimes, b), \mathcal{T}) \\
&\quad \succeq \mathscr{N}(\alpha[\mu(\leftthreetimes, a+b) - \mu(\leftthreetimes, a) - \mu(\leftthreetimes, b)], \mathcal{T}) \\
&\quad \bigotimes_{\mathrm{GTN}} \mathscr{N}(\beta[\eta(\leftthreetimes, a+b) - \eta(\leftthreetimes, a) - \eta(\leftthreetimes, b)], \mathcal{T}),
\end{aligned}$$

for every $a, b \in \mathcal{W}$, $\leftthreetimes \in \pitchfork$ and $\mathcal{T} > 0$. Thus, $\mu(\leftthreetimes, a+b) - \mu(\leftthreetimes, a) - \mu(\leftthreetimes, b) = 0$ and $\eta(\leftthreetimes, a+b) - \eta(\leftthreetimes, a) - \eta(\leftthreetimes, b) = 0$.

Lemma 17.2 *([1]) Let $\gamma : \pitchfork \times \mathcal{W} \to \mathcal{W}$ be a fuzzy operator, such that*

$$\gamma(\leftthreetimes, \lambda(a+b)) = \lambda\gamma(\leftthreetimes, a) + \lambda\gamma(\leftthreetimes, b),$$

for every $\lambda \in \Xi := \{\rho \in \mathbb{C} : |\rho| = 1\}$. Then, γ is $\mathbb{C}$–linear.

Theorem 17.1 *Let $n \in \mathbb{N}$, and $\underbrace{\varphi_i}_{i=1,\dots,n} : \mathcal{W}^2 \times (0, \infty) \to [0, 1]$, be fuzzy controllers, such that*

$$\lim_{k\to\infty} \underbrace{\varphi_i}_{i=1,\dots,n} (\frac{a}{2^k}, \frac{b}{2^k}, \frac{\mathcal{T}}{2^k}) = 1. \tag{17.3}$$

Let $\mu, \eta : \pitchfork \times \mathcal{W} \to \mathcal{W}$ be fuzzy operators satisfying $\mu(\leftthreetimes, 0) = \eta(\leftthreetimes, 0) = 0$ and the inequality below,

$$\begin{aligned}&\mathcal{N}(\mu(\curlywedge, \lambda(a+b)) - \lambda\mu(\curlywedge, a) - \lambda\mu(\curlywedge, b), \mathcal{T})\\ &\bigotimes_{\text{GTN}} \mathcal{N}(2\eta(\curlywedge, \lambda\frac{a+b}{2}) - \lambda\eta(\curlywedge, a) - \lambda\eta(\curlywedge, b), \mathcal{T})\\ &\quad\succeq \mathcal{N}(\alpha[2\mu(\curlywedge, \lambda\frac{a+b}{2}) - \lambda\mu(\curlywedge, a) - \lambda\mu(\curlywedge, b)], \mathcal{T})\\ &\quad\bigotimes_{\text{GTN}} \mathcal{N}(\beta[\eta(\curlywedge, \lambda(a+b)) - \lambda\eta(\curlywedge, a) - \lambda\eta(\curlywedge, b)], \mathcal{T}),\end{aligned} \tag{17.4}$$

for every $\lambda \in \Xi$, $a, b \in \mathcal{W}$, $\curlywedge \in \pitchfork$ *and* $\mathcal{T} \in (0, \infty)$. *If* μ, η *satisfy*

$$\begin{aligned}&\mathcal{N}(\mu(\curlywedge, ab) - \mu(\curlywedge, a)b - a\mu(\curlywedge, b), \mathcal{T})\\ &\bigotimes_{\text{GTN}} \mathcal{N}(\eta(\curlywedge, ab) - \eta(\curlywedge, a)\eta(\curlywedge, b), \mathcal{T})\\ &\succeq diag\Big[\varphi_1(a, b, \mathcal{T}), \ldots, \varphi_n(a, b, \mathcal{T})\Big],\end{aligned} \tag{17.5}$$

for every $\mathcal{T} \in (0, \infty)$, *then, the fuzzy operator* μ *is a derivation and the fuzzy operator* η *is a homomorphism.*

Proof Let $\lambda = 1$, in (17.4). In view of Lemma 17.1, η and μ are additive. It follows from (17.4) that

$$\begin{aligned}&\mathcal{N}(\mu(\curlywedge, \lambda(a+b)) - \lambda\mu(\curlywedge, a) - \lambda\mu(\curlywedge, b), \mathcal{T})\\ &\bigotimes_{\text{GTN}} \mathcal{N}(\eta(\curlywedge, \lambda(a+b)) - \lambda\eta(\curlywedge, a) - \lambda\eta(\curlywedge, b), \mathcal{T})\\ &\quad\succeq \mathcal{N}(\alpha[\mu(\curlywedge, \lambda(a+b)) - \lambda\mu(\curlywedge, a) - \lambda\mu(\curlywedge, b)], \mathcal{T})\\ &\quad\bigotimes_{\text{GTN}} \mathcal{N}(\beta[\eta(\curlywedge, \lambda(a+b)) - \lambda\eta(\curlywedge, a) - \lambda\eta(\curlywedge, b)], \mathcal{T}),\end{aligned} \tag{17.6}$$

for every $\mathcal{T} \in (0, \infty)$, $\lambda \in \Xi$, $a, b \in \mathcal{W}$ and $\curlywedge \in \pitchfork$. Since $\max\{|\alpha|, |\beta|\} < 1$, thus, we get

$$\mu(\curlywedge, \lambda(a+b)) - \lambda\mu(\curlywedge, a) - \lambda\mu(\curlywedge, b) = 0,$$

and

$$\eta(\curlywedge, \lambda(a+b)) - \lambda\eta(\curlywedge, a) - \lambda\eta(\curlywedge, b) = 0,$$

for every $\mathcal{T} \in (0, \infty)$, $\lambda \in \Xi$, $a, b \in \mathcal{W}$ and $\curlywedge \in \pitchfork$. Using Lemma 17.2, η and μ are $\mathbb{C}-$linear.

According to (17.5), and the additivity of μ and η, we have that

$$\begin{aligned}&\mathcal{N}(\mu(\curlywedge, ab) - \mu(\curlywedge, a)b - a\mu(\curlywedge, b), \mathcal{T})\\ &\bigotimes_{\text{GTN}} \mathcal{N}(\eta(\curlywedge, ab) - \eta(\curlywedge, a)\eta(\curlywedge, b), \mathcal{T})\end{aligned}$$

$$= \mathcal{N}(2^{2k}[\mu(\leftthreetimes, \frac{ab}{2^{2k}}) - \mu(\leftthreetimes, \frac{a}{2^k})\frac{b}{2^k} - \frac{a}{2^k}\mu(\leftthreetimes, \frac{a}{2^k})], \mathcal{T})$$
$$\bigotimes_{\text{GTN}} \mathcal{N}(2^{2k}[\eta(\leftthreetimes, \frac{ab}{2^{2k}}) - \eta(\leftthreetimes, \frac{a}{2^k})\eta(\leftthreetimes, \frac{b}{2^k})], \mathcal{T})$$
$$\succeq \operatorname{diag}\left[\varphi_1(\frac{a}{2^k}, \frac{b}{2^k}, \frac{\mathcal{T}}{2^k}), \ldots, \varphi_n(\frac{a}{2^k}, \frac{b}{2^k}, \frac{\mathcal{T}}{2^k})\right],$$

for every $\mathcal{T} \in (0, \infty)$, $a, b \in \mathcal{W}$ and $\leftthreetimes \in ⫛$.

Since $\lim_{k\to\infty} \underbrace{\varphi_i}_{i=1,\ldots,n} (\frac{a}{2^k}, \frac{b}{2^k}, \frac{\mathcal{T}}{2^k}) = 1$, then, $\mu(\leftthreetimes, ab) - \mu(\leftthreetimes, a)b - a\mu(\leftthreetimes, b) = 0$ and $\eta(\leftthreetimes, ab) = \eta(\leftthreetimes, a)\eta(\leftthreetimes, b)$, for every $a, b \in \mathcal{W}$ and $\leftthreetimes \in ⫛$. Thus, η is a homomorphism and μ is a derivation.

Theorem 17.2 *Let $n \in \mathbb{N}$, and $\underbrace{\varphi_i}_{i=1,\ldots,n} : \mathcal{W}^2 \times (0, \infty) \to [0, 1]$, be fuzzy control functions, such that*

$$\lim_{k\to\infty} \underbrace{\varphi_i}_{i=1,\ldots,n} (2^k a, 2^k b, 2^k \mathcal{T}) = 1, \tag{17.7}$$

for every $a, b \in \mathcal{W}$, $\leftthreetimes \in ⫛$ and $\mathcal{T} > 0$. Let $\mu, \eta : ⫛ \times \mathcal{W} \to \mathcal{W}$ be fuzzy operators satisfying $\mu(\leftthreetimes, 0) = \eta(\leftthreetimes, 0) = 0$, (17.4) and (17.5), for every $\leftthreetimes \in ⫛$. then, μ is a derivation and η is a homomorphism.

Proof Via the same reasoning presented in the proof of Theorem 17.1, we can show that μ and η are $\mathbb{C}-$linear. Making use of (17.5) and the additivity of η and μ, we have that

$$\mathcal{N}(\mu(\leftthreetimes, ab) - \mu(\leftthreetimes, a)b - a\mu(\leftthreetimes, b), \mathcal{T})$$
$$\bigotimes_{\text{GTN}} \mathcal{N}(\eta(\leftthreetimes, ab) - \eta(\leftthreetimes, a)\eta(\leftthreetimes, b), \mathcal{T})$$
$$= \mathcal{N}(\frac{1}{2^{2k}}[\mu(\leftthreetimes, 2^{2k}ab) - \mu(\leftthreetimes, 2^k a)2^k b - 2^k a\mu(\leftthreetimes, 2^k b)], \mathcal{T})$$
$$\bigotimes_{\text{GTN}} \mathcal{N}(\frac{1}{2^{2k}}[\eta(\leftthreetimes, 2^{2k}ab) - \eta(\leftthreetimes, 2^k a)\eta(\leftthreetimes, 2^k b)], \mathcal{T})$$
$$\succeq \operatorname{diag}\left[\varphi_1(2^k a, 2^k b, 2^k \mathcal{T}), \ldots, \varphi_n(2^k a, 2^k b, 2^k \mathcal{T})\right],$$

for every $a, b \in \mathcal{W}$, $\leftthreetimes \in ⫛$ and $\mathcal{T} > 0$. Which tend **1**, as $n \longrightarrow \infty$. So $\mu(\leftthreetimes, ab) - \mu(\leftthreetimes, a)b - a\mu(\leftthreetimes, b) = 0$ and $\eta(\leftthreetimes, ab) = \eta(\leftthreetimes, a)\eta(\leftthreetimes, b)$. Thus, η is a homomorphism and μ is a derivation.

Theorem 17.3 *Let $n \in \mathbb{N}$, and $\underbrace{\varphi_i}_{i=1,\ldots,n} : \mathcal{W}^2 \times (0, \infty) \to [0, 1]$ be fuzzy control functions such that there exists a $\varpi < 1$, with*

$$\underbrace{\varphi_i}_{i=1,\ldots,n}\left(\frac{a}{2},\frac{b}{2},\mathcal{T}\right) \geq \underbrace{\varphi_i}_{i=1,\ldots,n}\left(a,b,\frac{2\mathcal{T}}{\varpi}\right), \tag{17.8}$$

for every $a, b \in \mathcal{W}$, *and* $\mathcal{T} > 0$. *Let* $\mu, \eta : \pitchfork \times \mathcal{W} \to \mathcal{W}$ *be fuzzy operators satisfying* $\mu(\leftthreetimes, 0) = \eta(\leftthreetimes, 0) = 0$, (17.4) *and* (17.5) *for every* $\leftthreetimes \in \pitchfork$, *then,* μ *is a derivation and* η *is a homomorphism.*

Proof It follows from (17.8),

$$\lim_{k\to\infty}\underbrace{\varphi_i}_{i=1,\ldots,n}\left(\frac{a}{2^k},\frac{b}{2^k},\frac{\mathcal{T}}{2^{2k}}\right) \geq \lim_{k\to\infty}\underbrace{\varphi_i}_{i=1,\ldots,n}\left(a,b,\frac{\mathcal{T}}{(2\varpi)^k}\right) = 1.$$

Theorem 17.4 *Let* $n \in \mathbb{N}$, *and* $\underbrace{\varphi_i}_{i=1,\ldots,n} : \mathcal{W}^2 \times (0,\infty) \to [0,1]$ *be fuzzy controllers such that there exists a* $\varpi < 1$, *with*

$$\underbrace{\varphi_i}_{i=1,\ldots,n}(a,b,\mathcal{T}) \geq \underbrace{\varphi_i}_{i=1,\ldots,n}\left(\frac{a}{2}\frac{b}{2},\frac{\mathcal{T}}{2\varpi}\right), \tag{17.9}$$

for every $a, b \in \mathcal{W}$, $\leftthreetimes \in \pitchfork$ *and* $\mathcal{T} > 0$. *Let* $\mu, \eta : \pitchfork \times \mathcal{W} \to \mathcal{W}$ *be fuzzy operators satisfying* $\mu(\leftthreetimes, 0) = \eta(\leftthreetimes, 0) = 0$, (17.4) *and* (17.5) *for every* $\leftthreetimes \in \pitchfork$, *then,* μ *is a derivation and* η *is a homomorphism.*

Proof It follows from (17.9),

$$\lim_{k\to\infty}\underbrace{\varphi_i}_{i=1,\ldots,n}(2^k a, 2^k b, 2^{2k}\mathcal{T}) \geq \lim_{k\to\infty}\underbrace{\varphi_i}_{i=1,\ldots,n}\left(a,b,\frac{2^{2k}\mathcal{T}}{(2\varpi)^k}\right) = 1.$$

Remark 17.1 In (17.5), the pair (η, μ) of a homomorphism η and a derivation μ can be replaced by the pair of a homomorphism and a homomorphism or the pair of a derivation and a derivation.

17.2 Multi Stability of Derivation–Homomorphism Fuzzy Functional Inequalities in MVFC–⋄–Algebras

In this part, we study the multy stability of derivations and homomorphisms associated with the additive–additive (α, β)–fuzzy functional inequality (17.1) in MVFC–⋄–algebras by the fixed point method.

Throughout this section, let $(\mathcal{W}, \mathcal{N}, \bigotimes_{\text{GTN}}, \bigotimes_{\text{GTN}})$ be a unital MVFC–⋄–algebra with unitary group $U(\mathcal{W})$.

Theorem 17.5 *Let* $n \in \mathbb{N}$, *and* $\underbrace{\varphi_i}_{i=1,\ldots,n} : \mathcal{W}^2 \times (0,\infty) \to [0,1]$, *be fuzzy control functions satisfying* (17.3). *Let* $\mu, \eta : \pitchfork \times \mathcal{W} \to \mathcal{W}$ *be fuzzy operators satisfying* $\mu(\curlywedge, 0) = \eta(\curlywedge, 0) = 0$ *and* (17.4), *for every* $\curlywedge \in \pitchfork$. *If* μ, η *satisfy*

$$\begin{aligned} &\mathcal{N}(\mu(\curlywedge, \zeta\xi) - \mu(\curlywedge, \zeta)\xi - \zeta\mu(\curlywedge, \xi)), \mathcal{T}) \\ &\bigotimes_{\mathrm{GTN}} \mathcal{N}(\eta(\curlywedge, \zeta\xi) - \eta(\curlywedge, \zeta)\eta(\curlywedge, \xi), \mathcal{T}) \\ &\succeq diag\left[\varphi_1(\zeta, \xi, \mathcal{T}), \ldots, \varphi_n(\zeta, \xi, \mathcal{T})\right], \end{aligned} \tag{17.10}$$

for every $\zeta, \xi \in U(\mathcal{W})$, *then,* η *and* μ *are a homomorphism and a derivation, respectively.*

Proof Set $\lambda = 1$, in (17.4). In view of Lemma 17.1, η and μ are additive. It follows from (17.4),

$$\begin{aligned} &\mathcal{N}(\mu(\curlywedge, \lambda(a+b)) - \lambda\mu(\curlywedge, a) - \lambda\mu(\curlywedge, b), \mathcal{T}) \\ &\bigotimes_{\mathrm{GTN}} \mathcal{N}(\eta(\curlywedge, a+b) - \lambda\eta(\curlywedge, a) - \lambda\eta(\curlywedge, b), \mathcal{T}) \\ &\quad \succeq \mathcal{N}(\alpha[\mu(\curlywedge, a+b) - \lambda\mu(\curlywedge, a) - \lambda\mu(\curlywedge, b)], \mathcal{T}) \\ &\quad \bigotimes_{\mathrm{GTN}} \mathcal{N}(\beta[\eta(\curlywedge, \lambda(a+b)) - \lambda\eta(\curlywedge, a) - \lambda\eta(\curlywedge, b)], \mathcal{T}), \end{aligned} \tag{17.11}$$

for every $\lambda \in \Xi$, $a, b \in \mathcal{W}$, $\curlywedge \in \pitchfork$ and $\mathcal{T} \in (0,\infty)$. Since $\max\{|\alpha|, |\beta|\} < 1$, we have that

$$\mu(\curlywedge, \lambda(a+b)) - \lambda\mu(\curlywedge, a) - \lambda\mu(\curlywedge, b) = 0,$$

and

$$\eta(\curlywedge, \lambda(a+b)) - \lambda\eta(\curlywedge, a) - \lambda\eta(\curlywedge, b) = 0,$$

for every $\lambda \in \Xi$, $a, b \in \mathcal{W}$, $\curlywedge \in \pitchfork$ and $\mathcal{T} \in (0,\infty)$. Hence, by Lemma 17.2, η and μ are $\mathbb{C}-$linear. By a similar method to the proof of Theorem 3.1 in [2], the proof will be completed.

Theorem 17.6 *Let* $n \in \mathbb{N}$, *and* $\underbrace{\varphi_i}_{i=1,\ldots,n} : \mathcal{W}^2 \times (0,\infty) \to [0,1]$, *be fuzzy control functions satisfying* (17.7) *and* $\mu, \eta : \pitchfork \times \mathcal{W} \to \mathcal{W}$ *be fuzzy operators satisfying* $\mu(\curlywedge, 0) = \eta(\curlywedge, 0) = 0$, *and* (17.4) *and* (17.10), *for every* $\curlywedge \in \pitchfork$. *Then,* η *and* μ *are a homomorphism and a derivation, respectively.*

Proof The proof is similar to the proofs of Theorems 17.2 and 17.5.

Remark 17.2 Using the fixed point method given in Theorems 17.3 and 17.4, we can get the same results in Theorems 17.5 and 17.6.

17.3 Multi Stability of FC–⋄–Ternary Derivation–Homomorphism Fuzzy Functional Inesqualities in MVFC–⋄–Ternary Algebras

In this part, we study the multy stability of FC–⋄–ternary derivations and homomorphisms associated with the additive–additive (α, β)–fuzzy functional inequality (17.2) in MVFC–⋄–ternary algebras by the fixed point method.

Here, let $(\mathcal{W}, \mathcal{N}, \bigotimes_{\text{GTN}}, \bigotimes_{\text{GTN}})$ be an MVFC–⋄–ternary algebra.

Lemma 17.3 *If fuzzy operators* $\mu, \eta : \pitchfork \times \mathcal{W} \to \mathcal{W}$ *satisfy* (17.2) *and* $\mu(\leftthreetimes, 0) = \eta(\leftthreetimes, 0) = 0$, *for every* $\leftthreetimes \in \pitchfork$, *then,* μ *and* η *are additive.*

Proof Let $a = b = c$ in (17.2). Thus, we have

$$\mathcal{N}(\mu(\leftthreetimes, 3a) - 3\mu(\leftthreetimes, a), \mathcal{T}) \succeq \mathcal{N}(\beta[\eta(\leftthreetimes, 3a) - 3\eta(\leftthreetimes, a)], \mathcal{T}),$$

for every $\mathcal{T} \in (0, \infty)$, $\leftthreetimes \in \pitchfork$, and $a \in \mathcal{W}$. Now, let $b = c = 0$ in (17.1), then, we have that

$$\mathcal{N}(3\eta(\leftthreetimes, \frac{a}{3}) - 3\eta(\leftthreetimes, a), \mathcal{T}) \succeq \mathcal{N}(\alpha[3\mu(\leftthreetimes, \frac{a}{3}) - \mu(\leftthreetimes, a)], \mathcal{T}),$$

and so

$$\mathcal{N}(3\eta(\leftthreetimes, a) - \eta(\leftthreetimes, 3a), \mathcal{T}) \succeq \mathcal{N}(\alpha[3\mu(\leftthreetimes, a) - \mu(\leftthreetimes, 3a)], \mathcal{T}),$$

for every $\mathcal{T} \in (0, \infty)$, $\leftthreetimes \in \pitchfork$, and $a \in \mathcal{W}$. Thus, we get

$$\mathcal{N}(\mu(\leftthreetimes, 3a) - 3\mu(\leftthreetimes, a), \mathcal{T}) \succeq \mathcal{N}(\alpha\beta[\mu(\leftthreetimes, 3a) - 3\mu(\leftthreetimes, a)], \mathcal{T}),$$

and

$$\mathcal{N}(3\eta(\leftthreetimes, a) - \eta(\leftthreetimes, 3a), \mathcal{T}) \succeq \mathcal{N}(\alpha\beta[3\eta(\leftthreetimes, a) - \eta(\leftthreetimes, 3a)], \mathcal{T}),$$

for every $\mathcal{T} \in (0, \infty)$, $\leftthreetimes \in \pitchfork$, and $a \in \mathcal{W}$. Since $\max\{|a|, |b|\} < 1$, then, $\eta(\leftthreetimes, 3a) = 3\eta(\leftthreetimes, a)$ and $\mu(\leftthreetimes, 3a) = 3\mu(\leftthreetimes, a)$ for every $a \in \mathcal{W}$ and $\leftthreetimes \in \pitchfork$. Based on (17.1), we have

$$\begin{aligned}
&\mathcal{N}(\mu(\leftthreetimes, a+b+c) - \mu(\leftthreetimes, a) - \mu(\leftthreetimes, b) - \mu(\leftthreetimes, c), \mathcal{T}) \\
&\bigotimes_{\text{GTN}} \mathcal{N}(\eta(\leftthreetimes, a+b+c) + \eta(\leftthreetimes, a-2b+c) \\
&\qquad +\eta(\leftthreetimes, a+b-2c) - 3\eta(\leftthreetimes, a), \mathcal{T}) \\
&\succeq \mathcal{N}(\alpha[\mu(\leftthreetimes, a+b+c) - \mu(\leftthreetimes, a) - \mu(\leftthreetimes, b) - \mu(\leftthreetimes, c)], \mathcal{T}) \\
&\quad \bigotimes_{\text{GTN}} \mathcal{N}(\beta[\eta(\leftthreetimes, a+b+c) + \eta(\leftthreetimes, a-2b+c)
\end{aligned}$$

$$+\eta(\leftthreetimes, a+b-2c) - 3\eta(\leftthreetimes, a)], \mathcal{T}),$$

for every $\mathcal{T} \in (0, \infty)$, $\leftthreetimes \in \pitchfork$, and $a, b, c \in \mathcal{W}$. Since $\max\{|a|, |b|\} < 1$, then,

$$\mu(\leftthreetimes, a+b+c) - \mu(\leftthreetimes, a) - \mu(\leftthreetimes, b) - \mu(\leftthreetimes, c) = 0,$$

and

$$\eta(\leftthreetimes, a+b+c) + \eta(\leftthreetimes, a-2b+c) + \eta(\leftthreetimes, a+b-2c) - 3\eta(\leftthreetimes, a) = 0,$$

for every $\mathcal{T} \in (0, \infty)$, $\leftthreetimes \in \pitchfork$, and $a, b, c \in \mathcal{W}$. So η and μ are additive.

Theorem 17.7 *Let $n \in \mathbb{N}$, and $\underbrace{\varphi_i}_{i=1,\ldots,n} : \mathcal{W}^3 \times (0, \infty) \to [0, 1]$, be control fuzzy functions such that*

$$\lim_{k\to\infty} \underbrace{\varphi_i}_{i=1,\ldots,n} \left(\frac{a}{2^k}, \frac{b}{2^k}, \frac{c}{2^k}, \frac{\mathcal{T}}{2^k}\right) = 1, \tag{17.12}$$

for every $a, b, c \in \mathcal{W}$, and $\mathcal{T} > 0$. Let $\mu, \eta : \pitchfork \times \mathcal{W} \to \mathcal{W}$ be fuzzy operators satisfying $\mu(\leftthreetimes, 0) = \eta(\leftthreetimes, 0) = 0$, and

$$\begin{aligned}
&\mathcal{N}(\mu(\leftthreetimes, \lambda(a+b+c)) - \lambda\mu(\leftthreetimes, a) - \lambda\mu(\leftthreetimes, b) - \lambda\mu(\leftthreetimes, c), \mathcal{T}) \\
&\bigotimes_{\mathrm{GTN}} \mathcal{N}(3\eta(\leftthreetimes, \lambda\frac{a+b+c}{3}) \\
&\qquad +\lambda\eta(\leftthreetimes, a-2b+c) + \lambda\eta(\leftthreetimes, a+b-2c) - 3\lambda\eta(\leftthreetimes, a), \mathcal{T}) \\
&\succeq \mathcal{N}(\alpha[3\mu(\leftthreetimes, \lambda\frac{a+b+c}{3}) - \lambda\mu(\leftthreetimes, a) - \lambda\mu(\leftthreetimes, b) - \lambda\mu(\leftthreetimes, c)], \mathcal{T}) \\
&\quad \bigotimes_{\mathrm{GTN}} \mathcal{N}(\beta[\eta(\leftthreetimes, \lambda(a+b+c)) + \lambda\eta(\leftthreetimes, a-2b+c) \\
&\qquad +\lambda\eta(\leftthreetimes, a+b-2c) - 3\lambda\eta(\leftthreetimes, a)], \mathcal{T}),
\end{aligned} \tag{17.13}$$

for every $\lambda \in \Xi$, $\mathcal{T} \in (0, \infty)$, $a, b, c \in \mathcal{W}$ and $\leftthreetimes \in \pitchfork$. If μ and η satisfy

$$\begin{aligned}
&\mathcal{N}\Big(\mu(\leftthreetimes, [a, b, c]) - [\mu(\leftthreetimes, a), b, c] - [a, \mu(\leftthreetimes, b), c] - [a, b, \mu(\leftthreetimes, c)], \mathcal{T}\Big) \\
&\bigotimes_{\mathrm{GTN}} \mathcal{N}\Big(\eta(\leftthreetimes, [a, b, c]) - [\eta(\leftthreetimes, a), \eta(\leftthreetimes, b), \eta(\leftthreetimes, c)], \mathcal{T}\Big) \\
&\succeq diag\Big[\varphi_1(a, b, c, \mathcal{T}), \ldots, \varphi_n(a, b, c, \mathcal{T})\Big],
\end{aligned} \tag{17.14}$$

for every $\mathcal{T} \in (0, \infty)$, then, η is a FC–⋄–ternary homomorphism and μ is a FC–⋄–ternary derivation.

Proof Let $\lambda = 1$, in (17.13). In view of Lemma 17.3, η and μ are additive. Now, according to (17.13), we have that

$$\mathcal{N}\Big(\mu(\gimel, \lambda(a+b+c)) - \lambda\mu(\gimel, a) - \lambda\mu(\gimel, b) - \lambda\mu(\gimel, c), \mathcal{T}\Big)$$
$$\bigotimes_{\text{GTN}} \mathcal{N}\Big(\eta(\gimel, \lambda(a+b+c)) + \lambda\eta(\gimel, a-2b+c) + \lambda\eta(\gimel, a+b-2c) - 3\lambda\eta(\gimel, a), \mathcal{T}\Big)$$
$$\succeq \mathcal{N}\Big(\alpha[\mu(\gimel, \lambda(a+b+c)) - \lambda\mu(\gimel, a) - \lambda\mu(\gimel, b) - \lambda\mu(\gimel, c)], \mathcal{T}\Big)$$
$$\bigotimes_{\text{GTN}} \mathcal{N}\Big(\beta[\eta(\gimel, \lambda(a+b+c)) + \lambda\eta(\gimel, a-2b+c) + \lambda\eta(\gimel, a+b-2c) - 3\lambda\eta(\gimel, a)], \mathcal{T}\Big),$$

for every $\lambda \in \Xi$, $\mathcal{T} \in (0, \infty)$, $\gimel \in \pitchfork$ and $a, b, c \in \mathcal{W}$. Since $\max\{|a|, |b|\} < 1$, then, $\eta(\gimel, \lambda(a+b+c)) + \lambda\eta(\gimel, a-2b+c) + \lambda\eta(\gimel, a+b-2c) - 3\lambda\eta(\gimel, a) = 0$ and $\mu(\gimel, \lambda(a+b+c)) - \lambda\mu(\gimel, a) - \lambda\mu(\gimel, b) - \lambda\mu(\gimel, c) = 0$ for any $\lambda \in \Xi$ and $\mathcal{T} \in (0, \infty)$. Thus, by Lemma 17.2, η and μ are $\mathbb{C}-$linear. Makinguse of (17.14) and the additive property of μ and η, we get

$$\mathcal{N}\Big(\mu(\gimel, [a, b, c]) - [\mu(\gimel, a), b, c] - [a, \mu(\gimel, b), c] - [a, b, \mu(\gimel, c)], \mathcal{T}\Big)$$
$$\bigotimes_{\text{GTN}} \mathcal{N}\Big(\eta(\gimel, [a, b, c]) - [\eta(\gimel, a), \eta(\gimel, b), \eta(\gimel, c)], \mathcal{T}\Big)$$
$$= \mathcal{N}\Big(2^{3k}(\mu(\gimel, \frac{[a, b, c]}{2^{3k}}) - [\mu(\gimel, \frac{a}{2^k}), \frac{b}{2^k}, \frac{c}{2^k}] - [\frac{a}{2^k}, \mu(\gimel, \frac{b}{2^k}), \frac{c}{2^k}] - [\frac{a}{2^k}, \frac{b}{2^k}, \mu(\gimel, \frac{c}{2^k})]), \mathcal{T}\Big)$$
$$\bigotimes_{\text{GTN}} \mathcal{N}\Big(2^{3k}(\eta(\gimel, \frac{[a, b, c]}{2^{3k}}) - [\eta(\gimel, \frac{a}{2^k}), \eta(\gimel, \frac{b}{2^k}), \eta(\gimel, \frac{c}{2^k})]), \mathcal{T}\Big)$$
$$\succeq \operatorname{diag}\Big[\varphi_1(\frac{a}{2^k}, \frac{b}{2^k}, \frac{c}{2^k}, \frac{\mathcal{T}}{2^k}), \ldots, \varphi_n(\frac{a}{2^k}, \frac{b}{2^k}, \frac{c}{2^k}, \frac{\mathcal{T}}{2^k})\Big],$$

which tend to $\mathbf{1}$ as $n \to \infty$, for every $\mathcal{T} > 0$. Hence,

$$\mu(\gimel, [a, b, c]) - [\mu(\gimel, a), b, c] - [a, \mu(\gimel, b), c] - [a, b, \mu(\gimel, c)] = 0,$$

and

$$\eta(\gimel, [a, b, c]) - [\eta(\gimel, a), \eta(\gimel, b), \eta(\gimel, c)] = 0,$$

for every $a, b, c \in \mathcal{W}$ and $\leftthreetimes \in \pitchfork$. Then, η is a FC–⋄–ternary homomorphism and μ is a FC–⋄–ternary derivation.

Theorem 17.8 *Let $n \in \mathbb{N}$, and $\underbrace{\varphi_i}_{i=1,\dots,n} : \mathcal{W}^3 \times (0,\infty) \to [0,1]$ and $\mu, \eta : \mathcal{W} \to \mathcal{W}$ be fuzzy operators satisfying* (17.13), (17.14), *and $\mu(\leftthreetimes, 0) = \eta(\leftthreetimes, 0) = 0$, and*

$$\lim_{k\to\infty} \underbrace{\varphi_i}_{i=1,\dots,n} (2^k a, 2^k b, 2^k c, 2^k \mathcal{T}) = 1, \tag{17.15}$$

for every $\mathcal{T} \in (0,\infty)$. Then, η is a FC–⋄–ternary homomorphism and μ is a FC–⋄–ternary derivation.

Proof By the same reasoning as in the proof of Theorem 17.7, μ and η are $\mathbb{C}-$linear. According to (17.14) and the additive property of μ and η, we have that

$$\begin{aligned}
&\mathcal{N}\Big(\mu(\leftthreetimes, [a,b,c]) - [\mu(\leftthreetimes, a), b, c] - [a, \mu(\leftthreetimes, b), c] - [a, b, \mu(\leftthreetimes, c)], \mathcal{T}\Big) \\
&\bigotimes_{\text{GTN}} \mathcal{N}\Big(\eta(\leftthreetimes, [a,b,c]) - [\eta(\leftthreetimes, a), \eta(\leftthreetimes, b), \eta(\leftthreetimes, c)], \mathcal{T}\Big) \\
&= \mathcal{N}\Big(\frac{1}{2^{3k}}(\mu(\leftthreetimes, 2^{3k}[a,b,c]) - [\mu(\leftthreetimes, 2^k a), 2^k b, 2^k c] \\
&\qquad -[2^k a, \mu(\leftthreetimes, 2^k b), 2^k c] - [2^k a, 2^k b, \mu(\leftthreetimes, 2^k c)]), \mathcal{T}\Big) \\
&\bigotimes_{\text{GTN}} \mathcal{N}\Big(\frac{1}{2^{3k}}(\eta(\leftthreetimes, 2^{3k}[a,b,c]) - [\eta(\leftthreetimes, 2^k a), \eta(\leftthreetimes, 2^k b), \eta(\leftthreetimes, 2^k c)]), \mathcal{T}\Big) \\
&\succeq \operatorname{diag}\Big[\varphi_1(2^k a, 2^k b, 2^k c, 2^k \mathcal{T}), \dots, \varphi_n(2^k a, 2^k b, 2^k c, 2^k \mathcal{T})\Big],
\end{aligned}$$

which tends to **1**, as $n \to \infty$. So

$$\mu(\leftthreetimes, [a,b,c]) - [\mu(\leftthreetimes, a), b, c] - [a, \mu(\leftthreetimes, b), c] - [a, b, \mu(\leftthreetimes, c)] = 0,$$

and

$$\eta(\leftthreetimes, [a,b,c]) - [\eta(\leftthreetimes, a), \eta(\leftthreetimes, b), \eta(\leftthreetimes, c)] = 0,$$

for every $a, b, c \in \mho$ and $\leftthreetimes \in \pitchfork$. Then, η is a FC–⋄–ternary homomorphism and μ is a FC–⋄–ternary derivation.

Theorem 17.9 *Let $n \in \mathbb{N}$, and $\underbrace{\varphi_i}_{i=1,\dots,n} : \mathcal{W}^3 \times (0,\infty) \to [0,1]$ be fuzzy control functions such that there exists a $\varpi < 1$ with*

$$\underbrace{\varphi_i}_{i=1,\ldots,n}(\frac{a}{2},\frac{b}{2},\frac{c}{2},\mathcal{T}) \geq \underbrace{\varphi_i}_{i=1,\ldots,n}(a,b,c,\frac{2\mathcal{T}}{\varpi}), \tag{17.16}$$

for every $\mathcal{T} \in (0,\infty)$. *Let* $\mu,\eta : \mathcal{W} \to \mathcal{W}$ *be fuzzy operators satisfying* (17.13), (17.14), *and* $\mu(\curlywedge, 0) = \eta(\curlywedge, 0) = 0$, *for every* $\curlywedge \in \pitchfork$. *Then,* η *is a FC–⋄–ternary homomorphism and* μ *is a FC–⋄–ternary derivation.*

Proof

$$\lim_{k\to\infty}\underbrace{\varphi_i}_{i=1,\ldots,n}(\frac{a}{2^k},\frac{b}{2^k},\frac{c}{2^k},\frac{\mathcal{T}}{2^{3k}}) \geq \lim_{k\to\infty}\underbrace{\varphi_i}_{i=1,\ldots,n}(a,b,c,\frac{\mathcal{T}}{(4\varpi)^k}) = 1.$$

Theorem 17.10 *Let* $n \in \mathbb{N}$, *and* $\underbrace{\varphi_i}_{i=1,\ldots,n} : \mathcal{W}^3 \times (0,\infty) \to [0,1]$ *be fuzzy control functions such that there exists a* $\varpi < 1$ *with*

$$\underbrace{\varphi_i}_{i=1,\ldots,n}(a,b,c,\mathcal{T}) \geq \underbrace{\varphi_i}_{i=1,\ldots,n}(\frac{a}{2},\frac{b}{2},\frac{c}{2},\frac{\mathcal{T}}{2\varpi}), \tag{17.17}$$

for every $\mathcal{T} \in (0,\infty)$. *Let* $\mu,\eta : \mathcal{W} \to \mathcal{W}$ *be fuzzy operators satisfying* (17.13), (17.14), *and* $\mu(\curlywedge, 0) = \eta(\curlywedge, 0) = 0$, *for every* $\curlywedge \in \pitchfork$. *Then,* η *is a FC–⋄–ternary homomorphism and* μ *is a FC–⋄–ternary derivation.*

Proof

$$\lim_{k\to\infty}\underbrace{\varphi_i}_{i=1,\ldots,n}(2^k a, 2^k b, 2^k c, 2^{3k}\mathcal{T}) \geq \lim_{k\to\infty}\underbrace{\varphi_i}_{i=1,\ldots,n}(a,b,c,\frac{\mathcal{T}}{(\frac{4}{\varpi})^k}) = 1.$$

Remark 17.3 In (17.14), the pair (η, μ) of a FC–⋄–ternary homomorphism η and a FC–⋄–ternary derivation μ can be replaced by the pair of a FC–⋄–ternary homomorphism and a FC–⋄–ternary homomorphism or the pair of a a FC–⋄–ternary derivation and a FC–⋄–ternary derivation.

17.4 Applications

Consider the special controllers $\underbrace{\varphi_i^{\circledS}}_{i=1,\ldots,n\in\mathbb{N}}$ defined in Sects. 9.8.1 and 9.8.2.

Let

$$\underbrace{\varphi_i}_{i=1,\ldots,n}(a,b,\mathcal{T}) := \underbrace{\varphi_i^{\circledS}}_{i=1,\ldots,n}\Big(-\frac{\mathsf{W}(\|a\|^{\mathsf{M}} + \|b\|^{\mathsf{M}})}{\mathcal{T}}\Big), \tag{17.18}$$

$$\underbrace{\varphi_i}_{i=1,\dots,n}(a,b,c,\mathcal{T}) := \underbrace{\varphi_i^{\circledS}}_{i=1,\dots,n}\left(-\frac{\mathsf{W}(\|a\|^{\mathsf{M}}+\|b\|^{\mathsf{M}}+\|c\|^{\mathsf{M}})}{\mathcal{T}}\right), \quad (17.19)$$

for every $\mathcal{T} \in (0,\infty)$ and $n \in \mathbb{N}$.

Notice that if,

- $\varpi < \frac{1}{2}$, then,

$$\underbrace{\varphi_i}_{i=1,\dots,n}(a,b,\mathcal{T}) \succeq \underbrace{\varphi_i}_{i=1,\dots,n}(2a,2b,\frac{2\mathcal{T}}{\varpi}),$$

$$\underbrace{\varphi_i}_{i=1,\dots,n}(a,b,c,\mathcal{T}) \succeq \underbrace{\varphi_i}_{i=1,\dots,n}(2a,2b,2c,\frac{2\mathcal{T}}{\varpi}),$$

  for every $\mathsf{W} > 0$ and $\mathsf{M} > 1$.

- $\varpi > 2$, then,

$$\underbrace{\varphi_i}_{i=1,\dots,n}(a,b,\mathcal{T}) \succeq \underbrace{\varphi_i}_{i=1,\dots,n}(\frac{a}{2},\frac{b}{2},\frac{\mathcal{T}}{2\varpi}),$$

$$\underbrace{\varphi_i}_{i=1,\dots,n}(a,b,c,\mathcal{T}) \succeq \underbrace{\varphi_i}_{i=1,\dots,n}(\frac{a}{2},\frac{b}{2},\frac{c}{2},\frac{\mathcal{T}}{2\varpi}).$$

  for every $\mathsf{W} > 0$ and $\mathsf{M} < 1$.

Making use of (17.18) and (17.19), we have the multi stability results via special functions for Theorems 17.1 and 17.2 and Theorems 17.7 and 17.8, respectively.

We refer the interested readers to [3–13].

References

1. Aderyani SR, Saadati R (2022) Approximation of derivation–homomorphism fuzzy functional inequalities in matrix valued FC $\diamond$-algebras. In: 2022 9th Iranian joint congress on fuzzy and intelligent systems (CFIS). IEEE, pp 1–6
2. Choonkil P, Wu XY (2020) Homomorphism-derivation functional inequalities in C*-algebras. AIMS Math 5:4482–4493
3. Rezaei Aderyani S, Saadati R, Rassias TM (2022) Park C (2022) Best approximation of $(\mathcal{G}_1, \mathcal{G}_2)$-random operator inequality in matrix Menger Banach algebras with application of stochastic Mittag-Leffler and H $\mathbb{H}$-Fox control functions. J Inequal Appl 1:1–17
4. Aderyani SR, Saadati R (2021) Best approximations of the ϕ-Hadamard fractional Volterra integro-differential equation by matrix valued fuzzy control functions. Adv Differ Equ 154

5. Aderyani SR, Saadati R, Yang XJ (2021) Radu-Miheţ method for UHML stability for a class of ξ-Hilfer fractional differential equations in matrix valued fuzzy Banach spaces. Math Methods Appl Sci 44(18):14619–14631
6. Aderyani SR, Saadati R, Allahviranloo T (2022) Existence, uniqueness and matrix-valued fuzzy Mittag-Leffler-Hypergeometric-Wright stability for P-Hilfer fractional differential equations in matrix-valued fuzzy Banach space. Comput Appl Math 41(6):234
7. Aderyani SR, Saadati R, Mesiar R (2022) Estimation of permuting tri-homomorphisms and permuting tri-derivations associated with the tri-additive Υ-random operator inequality in matrix MB-algebra. Int J Gen Syst 51(6):547–569
8. Aderyani SR, Saadati R, Allahviranloo T, Abbasbandy S, Catak M (2023) Fuzzy approximation of a fractional Lorenz system and a fractional financial crisis. Iran J Fuzzy Syst 20(7):27–36
9. Aderyani SR, Saadati R, Abdeljawad T, Mlaiki N (2022) Multi-stability of non homogenous vector-valued fractional differential equations in matrix-valued Menger spaces. Alexandria Eng J 61(12):10913–10923
10. Aderyani SR, Saadati R (2023) Stability and controllability results by n-ary aggregation functions in matrix valued fuzzy n-normed spaces. Inf Sci 119265
11. O'Regan D, Aderyani SR, Saadati R, Li C (2023) Stability results and parametric delayed mittag-leffler matrices in symmetric fuzzy-random spaces with application. Symmetry 15(10):1880
12. Aderyani SR, Saadati R, O'Regan D, Li C (2023) On a new approach for stability and controllability analysis of functional equations. Mathematics 11(16):3458
13. Aderyani SR, Saadati R, O'Regan D, Alshammari FS (2023) Fuzzy approximate solutions of matrix-valued fractional differential equations by fuzzy control functions. Mathematics 11(6):1386

Chapter 18
Multi Stability of Pseudo Stochastic Additive Fuzzy Operator Inequalities in MVFB–Algebras

In this chapter, we investigate the multi stability results for the following pseudo stochastic additive $(\mathcal{G}_1, \mathcal{G}_2)$–fuzzy operator inequalities [1] in MVFB–algebras, by means of both the direct method and the fixed point method.

$$\mathcal{N}\Big(\mathcal{Q}(\leftthreetimes, \mathsf{S}+\mathsf{R}+\mathsf{A}) - \mathcal{Q}(\leftthreetimes, \mathsf{S}) - \mathcal{Q}(\leftthreetimes, \mathsf{R}) - \mathcal{Q}(\leftthreetimes, \mathsf{A}), (\mathcal{T}_1, \ldots, \mathcal{T}_2)\Big) \tag{18.1}$$
$$\succeq \mathcal{N}\Big(\mathcal{G}_1[\mathcal{Q}(\leftthreetimes, \mathsf{S}+\mathsf{A}) - \mathcal{Q}(\leftthreetimes, \mathsf{S}) - \mathcal{Q}(\leftthreetimes, \mathsf{A})], (\mathcal{T}_1, \ldots, \mathcal{T}_2)\Big)$$
$$\bigotimes_{\text{GTN}} \mathcal{N}\Big(\mathcal{G}_2[\mathcal{Q}(\leftthreetimes, \mathsf{R}+\mathsf{A}) - \mathcal{Q}(\leftthreetimes, \mathsf{R}) - \mathcal{Q}(\leftthreetimes, \mathsf{A})], (\mathcal{T}_1, \ldots, \mathcal{T}_2)\Big),$$

where $0 \neq \mathcal{G}_1, \mathcal{G}_2 \in \mathbb{C}$ are fixed and $\max\{|\mathcal{G}_1|, |\mathcal{G}_2|\} < 2$.

18.1 Multi Stability of Additive Functional Inequalities In MVFB–Algebras by Direct Method

In this part, we apply a direct method to get the multi stability results for the additive $(\mathcal{G}_1, \mathcal{G}_2)$–fuzzy operator inequality (18.1) in MVFB–algebras $(\mho_1, \mathcal{N}, \bigotimes_{\text{GTN}}, \bigoplus_{\text{GTN}})$ and $(\mho_2, \mathcal{N}, \bigotimes_{\text{GTN}}, \bigoplus_{\text{GTN}})$.

Lemma 18.1 *Let* $\mathcal{Q} : \pitchfork \times \mho_1 \to \mho_2$ *be a fuzzy operator satisfying* (18.1), *for every* $\mathsf{S}, \mathsf{R}, \mathsf{A} \in \mho_1$ *and* $\leftthreetimes \in \pitchfork$. *Then,* $\mathcal{Q} : \pitchfork \times \mho_1 \to \mho_2$ *is additive.*

Proof Putting $\mathsf{S} = \mathsf{R} = \mathsf{A} = 0$ in (18.1), we have

$$\mathcal{N}\Big(2\mathcal{Q}(\leftthreetimes, 0), (\mathcal{T}_1, \ldots, \mathcal{T}_n)\Big) \tag{18.2}$$
$$\succeq \mathcal{N}\Big(\mathcal{G}_1\mathcal{Q}(\leftthreetimes, 0), (\mathcal{T}_1, \ldots, \mathcal{T}_n)\Big) \bigotimes_{\text{GTN}} \mathcal{N}\Big(\mathcal{G}_2\mathcal{Q}(\leftthreetimes, 0), (\mathcal{T}_1, \ldots, \mathcal{T}_n)\Big),$$

S. Rezaei Aderyani et al., *Towards Ulam Type Multi Stability Analysis*, Studies in Systems, Decision and Control 523, https://doi.org/10.1007/978-3-031-55564-0_18

and so

$$\mathcal{N}\left(\mathcal{Q}(\leftthreetimes, 0), \left(\frac{\mathcal{T}_1}{2}, \dots, \frac{\mathcal{T}_n}{2}\right)\right) \tag{18.3}$$
$$\succeq \mathcal{N}\left(\mathcal{Q}(\leftthreetimes, 0), \left(\frac{\mathcal{T}_1}{\max\{|\mathcal{G}_1|, |\mathcal{G}_2|\}}, \dots, \frac{\mathcal{T}_n}{\max\{|\mathcal{G}_1|, |\mathcal{G}_2|\}}\right), \right),$$

since $\max\{|\mathcal{G}_1|, |\mathcal{G}_2|\} < 2$, $\mathcal{Q}(\leftthreetimes, 0) = 0$, for every $\leftthreetimes \in ⋔$.

Now, putting $\mathsf{R} = 0$ in (18.1), we have that

$$\mathcal{N}\Big(\mathcal{Q}(\leftthreetimes, \mathsf{S}+\mathsf{A}) - \mathcal{Q}(\leftthreetimes, \mathsf{S}) - \mathcal{Q}(\leftthreetimes, \mathsf{A}), (\mathcal{T}_1, \dots, \mathcal{T}_n)\Big) \tag{18.4}$$
$$\succeq \mathcal{N}\Big(\mathcal{G}_1[\mathcal{Q}(\leftthreetimes, \mathsf{S}+\mathsf{A}) - \mathcal{Q}(\leftthreetimes, \mathsf{S}) - \mathcal{Q}(\leftthreetimes, \mathsf{A})], (\mathcal{T}_1, \dots, \mathcal{T}_n)\Big),$$

and

$$\mathcal{Q}(\leftthreetimes, \mathsf{S}+\mathsf{A}) - \mathcal{Q}(\leftthreetimes, \mathsf{S}) - \mathcal{Q}(\leftthreetimes, \mathsf{A}) = 0,$$

or

$$\mathcal{Q}(\leftthreetimes, \mathsf{S}+\mathsf{A}) = \mathcal{Q}(\leftthreetimes, \mathsf{S}) + \mathcal{Q}(\leftthreetimes, \mathsf{A}),$$

for every $\mathsf{S}, \mathsf{A} \in \mho_1$ and $\leftthreetimes \in ⋔$, since $|\mathcal{G}_1| < 2$. So $\mathcal{Q}$ is additive.

Theorem 18.1 *Consider the following assumptions:*

- $(\mho_1, \mathcal{N}, \bigotimes_{\mathrm{GTN}}, \bigoplus_{\mathrm{GTN}})$ *and* $(\mho_2, \mathcal{N}, \bigotimes_{\mathrm{GTN}}, \bigoplus_{\mathrm{GTN}})$ *be MVFB–algebras,*
- $\underbrace{\varphi_i}_{i=1,\dots,n} : \mho_1^3 \times (0, +\infty) \to [0, 1]$ *be fuzzy sets, fore every* $n \in \mathbb{N}$,
- *there exist* $\underbrace{\mathfrak{P}_i}_{i=1,\dots,n} < 1$, *such that* $\underbrace{\varphi_i}_{i=1,\dots,n}(\frac{\mathsf{S}}{2}, \frac{\mathsf{R}}{2}, \frac{\mathsf{A}}{2}, \underbrace{\mathcal{T}_i}_{i=1,\dots,n}) \geq \underbrace{\varphi_i}_{i=1,\dots,n}(\mathsf{S}, \mathsf{R}, \mathsf{A}, \frac{2\overbrace{\mathcal{T}_i}^{i=1,\dots,n}}{\underbrace{\mathfrak{P}_i}_{i=1,\dots,n}})$, *for every* $\mathsf{S}, \mathsf{R}, \mathsf{A} \in \mho_1$ *and* $\underbrace{\mathcal{T}_i}_{i=1,\dots,n} > 0$, *and* $n \in \mathbb{N}$,
- *for every* $\mathsf{S}, \mathsf{R}, \mathsf{A} \in \mho_1$, $\underbrace{\mathcal{T}_i}_{i=1,\dots,n} > 0$, *and* $n \in \mathbb{N}$;

$$\lim_{k\to\infty} \underbrace{\varphi_i}_{i=1,\dots,n}\left(\frac{\mathsf{S}}{2^k}, \frac{\mathsf{R}}{2^k}, \frac{\mathsf{A}}{2^k}, \frac{\overbrace{\mathcal{T}_i}^{i=1,\dots,n}}{2^k}\right) = 1, \tag{18.5}$$

- *the fuzzy operator* $\mathcal{Q} : ⋔ \times \mho_1 \to \mho_2$ *satisfies* $\mathcal{Q}(\leftthreetimes, 0) = 0$ *and*

$$\mathcal{N}\Big(\mathcal{Q}(\lambda, \mathsf{S}+\mathsf{R}+\mathsf{A}) - \mathcal{Q}(\lambda, \mathsf{S}) - \mathcal{Q}(\lambda, \mathsf{R}) - \mathcal{Q}(\lambda, \mathsf{A}), (\mathcal{T}_1, \ldots, \mathcal{T}_n)\Big)$$
$$\succeq \mathcal{N}\Big(\mathcal{G}_1[\mathcal{Q}(\lambda, \mathsf{S}+\mathsf{A}) - \mathcal{Q}(\lambda, \mathsf{S}) - \mathcal{Q}(\lambda, \mathsf{A})], (\mathcal{T}_1, \ldots, \mathcal{T}_n)\Big)$$
$$\bigotimes_{\mathrm{GTN}} \mathcal{N}\Big(\mathcal{G}_2[\mathcal{Q}(\lambda, \mathsf{R}+\mathsf{A}) - \mathcal{Q}(\lambda, \mathsf{R}) - \mathcal{Q}(\lambda, \mathsf{A})], (\mathcal{T}_1, \ldots, \mathcal{T}_n)\Big)$$
$$\bigotimes_{\mathrm{GTN}} diag\Big[\varphi_1(\mathsf{S}, \mathsf{R}, \mathsf{A}, \mathcal{T}_1), \ldots, \varphi_n(\mathsf{S}, \mathsf{R}, \mathsf{A}, \mathcal{T}_n)\Big], \tag{18.6}$$

for every $\mathsf{S}, \mathsf{R}, \mathsf{A} \in \mho_1$, $\lambda \in \pitchfork$ *and* $\underbrace{\mathcal{T}_i}_{i=1,\ldots,n} > 0$.

Then, we can find a unique additive fuzzy operator $\mathcal{V} : \pitchfork \times \mho_1 \to \mho_2$ *such that*

$$\mathcal{N}\Big(\mathcal{Q}(\lambda, \mathsf{S}) - \mathcal{V}(\lambda, \mathsf{S}), (\mathcal{T}_1, \ldots, \mathcal{T}_n)\Big) \tag{18.7}$$
$$\succeq diag\Big[\varphi_1\Big(\mathsf{S}, \mathsf{S}, 0, \frac{2(1-\mathfrak{P}_1)\mathcal{T}_1}{\mathfrak{P}_1}\Big), \ldots, \varphi_n\Big(\mathsf{S}, \mathsf{S}, 0, \frac{2(1-\mathfrak{P}_n)\mathcal{T}_n}{\mathfrak{P}_n}\Big)\Big],$$

for every $\mathsf{S}, \mathsf{R}, \mathsf{A} \in \mho_1$, $\lambda \in \pitchfork$ *and* $\underbrace{\mathcal{T}_i}_{i=1,\ldots,n} > 0$.

Proof Putting $\mathsf{A} = 0$ and $\mathsf{S} = \mathsf{R}$ in (18.6), we get

$$\mathcal{N}\Big(\mathcal{Q}(\lambda, 2\mathsf{S}) - 2\mathcal{Q}(\lambda, \mathsf{S}), (\mathcal{T}_1, \ldots, \mathcal{T}_n)\Big) \tag{18.8}$$
$$\succeq \mathrm{diag}\Big[\varphi_1\Big(\mathsf{S}, \mathsf{S}, 0, \mathcal{T}_1\Big), \ldots, \varphi_n\Big(\mathsf{S}, \mathsf{S}, 0, \mathcal{T}_n\Big)\Big],$$

for every $\mathsf{S} \in \mho_1$, $\lambda \in \pitchfork$ and $\underbrace{\mathcal{T}_i}_{i=1,\ldots,n} > 0$. Thus, we have that

$$\mathcal{N}\Big(\mathcal{Q}(\lambda, \mathsf{S}) - 2\mathcal{Q}(\lambda, \frac{\mathsf{S}}{2}), (\mathcal{T}_1, \ldots, \mathcal{T}_n)\Big) \tag{18.9}$$
$$\succeq \mathrm{diag}\Big[\varphi_1\Big(\frac{\mathsf{S}}{2}, \frac{\mathsf{S}}{2}, 0, \mathcal{T}_1\Big), \ldots, \varphi_n\Big(\frac{\mathsf{S}}{2}, \frac{\mathsf{S}}{2}, 0, \mathcal{T}_n\Big)\Big]$$
$$\succeq \mathrm{diag}\Big[\varphi_1\Big(\mathsf{S}, \mathsf{S}, 0, \frac{2\mathcal{T}_1}{\mathfrak{P}_1}\Big), \ldots, \varphi_n\Big(\mathsf{S}, \mathsf{S}, 0, \frac{2\mathcal{T}_n}{\mathfrak{P}_n}\Big)\Big],$$

for every $\mathsf{S} \in \mho_1$, $\lambda \in \pitchfork$ and $\underbrace{\mathcal{T}_i}_{i=1,\ldots,n} > 0$. Replacing S by $\frac{\mathsf{S}}{2^k}$ in (18.9), we get

$$\begin{aligned}
&\mathcal{N}\Big(2^k\mathcal{Q}(\curlywedge,\frac{\mathsf{S}}{2^k})-2^{k+1}\mathcal{Q}\Big(\curlywedge,\frac{\mathsf{S}}{2^{k+1}}\Big),(\mathcal{T}_1,\ldots,\mathcal{T}_n)\Big) \qquad (18.10)\\
&\succeq \mathrm{diag}\Big[\varphi_1\Big(\frac{\mathsf{S}}{2^k},\frac{\mathsf{S}}{2^k},0,\frac{\mathcal{T}_1}{2^{k-1}\mathfrak{P}_1}\Big),\ldots,\varphi_n\Big(\frac{\mathsf{S}}{2^k},\frac{\mathsf{S}}{2^k},0,\frac{\mathcal{T}_n}{2^{k-1}\mathfrak{P}_n}\Big)\Big]\\
&\succeq \mathrm{diag}\Big[\varphi_1\Big(\frac{\mathsf{S}}{2^{k-1}},\frac{\mathsf{S}}{2^{k-1}},0,\frac{2}{\mathfrak{P}_1}(\frac{\mathcal{T}_1}{2^{k-1}\mathfrak{P}_1})\Big),\ldots,\varphi_n\Big(\frac{\mathsf{S}}{2^{k-1}},\frac{\mathsf{S}}{2^{k-1}},0,\frac{2}{\mathfrak{P}_n}(\frac{\mathcal{T}_n}{2^{k-1}\mathfrak{P}_n})\Big)\Big]\\
&\succeq \ldots\\
&\succeq \mathrm{diag}\Big[\varphi_1\Big(\mathsf{S},\mathsf{S},0,\frac{2}{\mathfrak{P}_1^{k+1}}\mathcal{T}_1\Big),\ldots,\varphi_n\Big(\mathsf{S},\mathsf{S},0,\frac{2}{\mathfrak{P}_n^{k+1}}\mathcal{T}_n\Big)\Big].
\end{aligned}$$

It follows from

$$2^k\mathcal{Q}\Big(\curlywedge,\frac{\mathsf{S}}{2^k}\Big)-\mathcal{Q}(\curlywedge,\mathsf{S})=\sum_{j=1}^{k}\Big(2^j\mathcal{Q}\Big(\curlywedge,\frac{\mathsf{S}}{2^j}\Big)-2^{j-1}\mathcal{Q}\Big(\curlywedge,\frac{\mathsf{S}}{2^{j-1}}\Big)\Big),$$

and (18.10) that

$$\begin{aligned}
&\mathcal{N}\Big(2^k\mathcal{Q}\Big(\curlywedge,\frac{\mathsf{S}}{2^k}\Big)-\mathcal{Q}(\curlywedge,\mathsf{S}),\Big(\sum_{j=1}^{k}\frac{\mathfrak{P}_1^j}{2}\mathcal{T}_1,\ldots,\sum_{j=1}^{k}\frac{\mathfrak{P}_n^j}{2}\mathcal{T}_n\Big)\Big) \qquad (18.11)\\
&\succeq \mathrm{diag}\Big[\varphi_1\Big(\mathsf{S},\mathsf{S},0,\mathcal{T}_1\Big),\ldots,\varphi_n\Big(\mathsf{S},\mathsf{S},0,\mathcal{T}_n\Big)\Big]\bigotimes_{\mathrm{GTN}}\\
&\qquad\cdots\bigotimes_{\mathrm{GTN}}\mathrm{diag}\Big[\varphi_1\Big(\mathsf{S},\mathsf{S},0,\mathcal{T}_1\Big),\ldots,\varphi_n\Big(\mathsf{S},\mathsf{S},0,\mathcal{T}_n\Big)\Big]\\
&= \mathrm{diag}\Big[\varphi_1\Big(\mathsf{S},\mathsf{S},0,\mathcal{T}_1\Big),\ldots,\varphi_n\Big(\mathsf{S},\mathsf{S},0,\mathcal{T}_n\Big)\Big],
\end{aligned}$$

for every $\mathsf{S}\in\mho_1$, $\curlywedge\in\pitchfork$, $\underbrace{\mathcal{T}_i}_{i=1,\ldots,n}>0$. That is,

$$\begin{aligned}
&\mathcal{N}\Big(2^k\mathcal{Q}\Big(\curlywedge,\frac{\mathsf{S}}{2^k}\Big)-\mathcal{Q}(\curlywedge,\mathsf{S}),(\mathcal{T}_1,\ldots,\mathcal{T}_n)\Big) \qquad (18.12)\\
&\succeq \mathrm{diag}\Big[\varphi_1\Big(\mathsf{S},\mathsf{S},0,\frac{\mathcal{T}_1}{\sum_{j=1}^{k}\frac{\mathfrak{P}_1^j}{2}}\Big),\ldots,\varphi_n\Big(\mathsf{S},\mathsf{S},0,\frac{\mathcal{T}_n}{\sum_{j=1}^{k}\frac{\mathfrak{P}_n^j}{2}}\Big)\Big].
\end{aligned}$$

Replacing S with $\frac{\mathsf{S}}{2^m}$ in (18.12), we get

$$\begin{aligned}
&\mathcal{N}\Big(2^{k+m}\mathcal{Q}(\curlywedge,\frac{\mathsf{S}}{2^{k+m}})-2^m\mathcal{Q}\Big(\curlywedge,\frac{\mathsf{S}}{2^m}\Big),(\mathcal{T}_1,\ldots,\mathcal{T}_n)\Big) \qquad (18.13)\\
&\succeq \mathrm{diag}\Big[\varphi_1\Big(\mathsf{S},\mathsf{S},0,\frac{\mathcal{T}_1}{\sum_{j=m+1}^{k+m}\frac{\mathfrak{P}_1^j}{2}}\Big),\ldots,\varphi_n\Big(\mathsf{S},\mathsf{S},0,\frac{\mathcal{T}_n}{\sum_{j=m+1}^{k+m}\frac{\mathfrak{P}_n^j}{2}}\Big)\Big].
\end{aligned}$$

Since $\underbrace{\varphi_i}_{i=1,\ldots,n}\left(\mathsf{S},\mathsf{S},0,\frac{\overbrace{\mathcal{T}_i}^{i=1,\ldots,n}}{\sum_{j=m+1}^{k+m}\frac{\overbrace{\mathfrak{P}_i^j}^{i=1,\ldots,n}}{2}}\right)$ tends to 1, as $k,m\to\infty$, it concludes that the sequence $\{2^k\mathcal{Q}(\curlywedge,\frac{\mathsf{S}}{2^k})\}$ is Cauchy, for every $\mathsf{S}\in\mho_1$, $\curlywedge\in\pitchfork$. Since $\mho_2$ is complete, the sequence $\{2^k\mathcal{Q}(\curlywedge,\frac{\mathsf{S}}{2^k})\}$ is convergent. Consider the fuzzy operator $\mathcal{V}:\pitchfork\times\mho_1\to\mho_2$ defined by

$$\mathcal{V}(\curlywedge,\mathsf{S}):=\lim_{k\to\infty}2^k\mathcal{Q}\left(\curlywedge,\frac{\mathsf{S}}{2^k}\right),$$

for every $\mathsf{S}\in\mho_1$, $\curlywedge\in\pitchfork$. Putting $m=0$ and letting $k\to\infty$ in (18.13), we obtain

$$\mathcal{N}\Big(\mathcal{Q}(\curlywedge,\mathsf{S})-\mathcal{V}(\curlywedge,\mathsf{S}),(\mathcal{T}_1,\ldots,\mathcal{T}_n)\Big) \tag{18.14}$$
$$\succeq \operatorname{diag}\left[\varphi_1\left(\mathsf{S},\mathsf{S},0,\frac{2(1-\mathfrak{P}_1)\mathcal{T}_1}{\mathfrak{P}_1}\right),\ldots,\varphi_n\left(\mathsf{S},\mathsf{S},0,\frac{2(1-\mathfrak{P}_n)\mathcal{T}_n}{\mathfrak{P}_n}\right)\right],$$

for every $\mathsf{S}\in\mho_1$, $\curlywedge\in\pitchfork$ and $\underbrace{\mathcal{T}_i}_{i=1,\ldots,n}>0$.

Now, (18.6) implies that

$$\mathcal{N}\Big(\mathcal{V}(\curlywedge,\mathsf{S}+\mathsf{R}+\mathsf{A})-\mathcal{V}(\curlywedge,\mathsf{S})-\mathcal{V}(\curlywedge,\mathsf{R})+\mathcal{V}(\curlywedge,\mathsf{A}),(\mathcal{T}_1,\ldots,\mathcal{T}_n)\Big)$$
$$=\lim_{k\to\infty}\mathcal{N}\Big(2^k(\mathcal{Q}(\curlywedge,\frac{\mathsf{S}+\mathsf{R}+\mathsf{A}}{2^k})-\mathcal{Q}(\curlywedge,\frac{\mathsf{S}}{2^k})-\mathcal{Q}(\curlywedge,\frac{\mathsf{R}}{2^k})$$
$$-\mathcal{Q}(\curlywedge,\frac{\mathsf{A}}{2^k})),(\mathcal{T}_1,\ldots,\mathcal{T}_n)\Big)$$
$$\succeq\lim_{k\to\infty}\mathcal{N}\Big(2^k\mathcal{G}_1[\mathcal{Q}(\curlywedge,\frac{\mathsf{S}+\mathsf{A}}{2^k})-\mathcal{Q}(\curlywedge,\frac{\mathsf{S}}{2^k})-\mathcal{Q}(\curlywedge,\frac{\mathsf{A}}{2^k})],(\mathcal{T}_1,\ldots,\mathcal{T}_n)\Big)$$
$$\bigotimes_{\mathrm{GTN}}\mathcal{N}\Big(2^k\mathcal{G}_2[\mathcal{Q}(\curlywedge,\frac{\mathsf{R}+\mathsf{A}}{2^k})-\mathcal{Q}(\curlywedge,\frac{\mathsf{R}}{2^k})-\mathcal{Q}(\curlywedge,\frac{\mathsf{A}}{2^k})],(\mathcal{T}_1,\ldots,\mathcal{T}_n)\Big)$$
$$\bigotimes_{\mathrm{GTN}}\lim_{k\to\infty}\operatorname{diag}\left[\varphi_1\left(\frac{\mathsf{S}}{2^k},\frac{\mathsf{S}}{2^k},0,\frac{\mathcal{T}_1}{2^k}\right),\ldots,\varphi_n\left(\frac{\mathsf{S}}{2^k},\frac{\mathsf{S}}{2^k},0,\frac{\mathcal{T}_n}{2^k}\right)\right]$$
$$\succeq\mathcal{N}\Big(\mathcal{G}_1[\mathcal{V}(\curlywedge,\mathsf{S}+\mathsf{A})-\mathcal{V}(\curlywedge,\mathsf{S})-\mathcal{V}(\curlywedge,\mathsf{A})],(\mathcal{T}_1,\ldots,\mathcal{T}_n)\Big)$$
$$\bigotimes_{\mathrm{GTN}}\mathcal{N}\Big(\mathcal{G}_2[\mathcal{V}(\curlywedge,\mathsf{R}+\mathsf{A})-\mathcal{V}(\curlywedge,\mathsf{R})-\mathcal{V}(\curlywedge,\mathsf{A})],(\mathcal{T}_1,\ldots,\mathcal{T}_n)\Big),$$

for every $\mathsf{S}, \mathsf{R}, \mathsf{A} \in \mho_1, \leftthreetimes \in \pitchfork$, and $\underbrace{\mathcal{T}_i}_{i=1,\ldots,n} > 0$, since $\underbrace{\varphi_i}_{i=1,\ldots,n}\left(\frac{x}{2^k}, \frac{\mathsf{S}}{2^k}, 0, \frac{\overbrace{\mathcal{T}_i}^{i=1,\ldots,n}}{2^k}\right)$ tends to 1 as $k \to \infty$. Thus, we have that

$$\begin{aligned}
&\mathcal{N}\Big(\mathcal{V}(\leftthreetimes, \mathsf{S}+\mathsf{R}+\mathsf{A}) - \mathcal{V}(\leftthreetimes, \mathsf{S}) - \mathcal{V}(\leftthreetimes, \mathsf{R}) - \mathcal{V}(\leftthreetimes, \mathsf{A}), (\mathcal{T}_1, \ldots, \mathcal{T}_n)\Big) \\
&\succeq \mathcal{N}\Big(\mathcal{G}_1[\mathcal{V}(\leftthreetimes, \mathsf{S}+\mathsf{A}) - \mathcal{V}(\leftthreetimes, \mathsf{S}) - \mathcal{V}(\leftthreetimes, \mathsf{A})], (\mathcal{T}_1, \ldots, \mathcal{T}_n)\Big) \\
&\bigotimes_{\text{GTN}} \mathcal{N}\Big(\mathcal{G}_2[\mathcal{V}(\leftthreetimes, \mathsf{R}+\mathsf{A}) - \mathcal{V}(\leftthreetimes, \mathsf{R}) - \mathcal{V}(\leftthreetimes, \mathsf{A})], (\mathcal{T}_1, \ldots, \mathcal{T}_n)\Big),
\end{aligned}$$

for every $\mathsf{S}, \mathsf{R}, \mathsf{A} \in \mho_1, \leftthreetimes \in \pitchfork$, and $\underbrace{\mathcal{T}_i}_{i=1,\ldots,n} > 0$. Lemma 18.1 implies that the fuzzy operator $\mathcal{V} : \pitchfork \times \mho_1 \to \mho_2$ is stochastic additive.

Now, to prove the uniqueness of the fuzzy operator $\mathcal{V}$, suppose that there exists a stochastic additive fuzzy operator $\mathcal{V}' : \pitchfork \times \mho_1 \to \mho_2$ which satisfies (18.7). Then, we get

$$\begin{aligned}
&\mathcal{N}\Big(\mathcal{V}(\leftthreetimes, \mathsf{S}) - \mathcal{V}'(\leftthreetimes, \mathsf{S}), (\mathcal{T}_1, \ldots, \mathcal{T}_n)\Big) \\
&= \lim_{k\to\infty} \mathcal{N}\left(2^k\mathcal{V}\left(\leftthreetimes, \frac{\mathsf{S}}{2^k}\right) - 2^k\mathcal{V}'\left(\leftthreetimes, \frac{\mathsf{S}}{2^k}\right), (\mathcal{T}_1, \ldots, \mathcal{T}_n)\right),
\end{aligned}$$

and

$$\begin{aligned}
&\mathcal{N}\left(2^k\mathcal{V}\left(\leftthreetimes, \frac{\mathsf{S}}{2^k}\right) - 2^k\mathcal{V}'\left(\leftthreetimes, \frac{\mathsf{S}}{2^k}\right), (\mathcal{T}_1, \ldots, \mathcal{T}_n)\right), \\
&\succeq \mathcal{N}\left(2^k\mathcal{V}\left(\leftthreetimes, \frac{\mathsf{S}}{2^k}\right) - 2^k\mathcal{Q}\left(\leftthreetimes, \frac{\mathsf{S}}{2^k}\right), (\frac{\mathcal{T}_1}{2}, \ldots, \frac{\mathcal{T}_n}{2})\right) \\
&\bigotimes_{\text{GTN}} \mathcal{N}\left(2^k\mathcal{V}'\left(\leftthreetimes, \frac{\mathsf{S}}{2^k}\right) - 2^k\mathcal{Q}\left(\leftthreetimes, \frac{\mathsf{S}}{2^k}\right), (\frac{\mathcal{T}_1}{2}, \ldots, \frac{\mathcal{T}_n}{2})\right), \\
&\succeq \operatorname{diag}\left[\varphi_1\left(\frac{\mathsf{S}}{2^k}, \frac{\mathsf{S}}{2^k}, 0, \frac{2(1-\mathfrak{P}_1)\mathcal{T}_1}{2^k\mathfrak{P}_1}\right), \ldots, \varphi_n\left(\frac{\mathsf{S}}{2^k}, \frac{\mathsf{S}}{2^k}, 0, \frac{2(1-\mathfrak{P}_n)\mathcal{T}_n}{2^k\mathfrak{P}_n}\right)\right] \\
&\succeq \operatorname{diag}\left[\varphi_1\left(\mathsf{S}, \mathsf{S}, 0, \frac{2(1-\mathfrak{P}_1)\mathcal{T}_1}{\mathfrak{P}_1^{k+1}}\right), \ldots, \varphi_n\left(\mathsf{S}, \mathsf{S}, 0, \frac{2(1-\mathfrak{P}_n)\mathcal{T}_n}{\mathfrak{P}_n^{k+1}}\right)\right],
\end{aligned}$$

for every $\mathsf{S}, \mathsf{R}, \mathsf{A} \in \mho_1$, $\leftthreetimes \in ⫛$, and $\underbrace{\mathcal{T}_i}_{i=1,\dots,n} > 0$. Since $\lim_{k\to\infty} \dfrac{2(1-\overbrace{\mathfrak{P}_i}^{i=1,\dots,n})}{\underbrace{\mathfrak{P}_i^{k+1}}_{i=1,\dots,n}} = \infty$,

we get $\underbrace{\varphi_i}_{i=1,\dots,n}\left(\mathsf{S}, \mathsf{S}, 0, \dfrac{2(1-\overbrace{\mathfrak{P}_i}^{i=1,\dots,n})\overbrace{\mathcal{T}_i}^{i=1,\dots,n}}{\underbrace{\mathfrak{P}_i^{k+1}}_{i=1,\dots,n}}\right)$ tends to 1, as $k \to \infty$.

Thus, it concludes that

$$\mathcal{N}\left(2^k\mathcal{V}\left(\leftthreetimes, \frac{\mathsf{S}}{2^k}\right) - 2^k\mathcal{V}'\left(\leftthreetimes, \frac{\mathsf{S}}{2^k}\right), (\mathcal{T}_1, \dots, \mathcal{T}_n)\right) = \mathbf{1},$$

for every $\mathsf{S} \in \mho_1, \leftthreetimes \in ⫛$, and $(\mathcal{T}_1, \dots, \mathcal{T}_n) > \underbrace{(0, \dots, 0)}_{n}$. So we obtain that $\mathcal{V}(\leftthreetimes, \mathsf{S}) = \mathcal{V}'(\leftthreetimes, \mathsf{S})$, for every $\mathsf{S} \in \mho_1$ and $\leftthreetimes \in ⫛$.

Theorem 18.2 *Let $n \in \mathbb{N}$, and $(\underbrace{\mho_j}_{j=1,2}, \mathcal{N}, \bigotimes_{\mathrm{GTN}}, \bigoplus_{\mathrm{GTN}})$ be MVFB–algebras and that $\underbrace{\varphi_i}_{i=1,\dots,n} : \mho_1^3 \times (0, +\infty) \to [0, 1]$ be fuzzy sets, such that there exist $\underbrace{\mathfrak{P}_i}_{i=1,\dots,n} < 1$ with*

$$\underbrace{\varphi_i}_{i=1,\dots,n}\left(\mathsf{S}, \mathsf{R}, 0, \underbrace{\mathcal{T}_i}_{i=1,\dots,n}\right) \geq \underbrace{\varphi_i}_{i=1,\dots,n}\left(\frac{\mathsf{S}}{2}, \frac{\mathsf{R}}{2}, 0, \frac{\mathcal{T}_i}{\underbrace{2\mathfrak{P}_i}_{i=1,\dots,n}}\right),$$

for every $\mathsf{S}, \mathsf{R} \in \mho_1$, and $\underbrace{\mathcal{T}_i}_{i=1,\dots,n} > 0$, and

$$\lim_{k\to\infty} \underbrace{\varphi_i}_{i=1,\dots,n}\left(2^k\mathsf{S}, 2^k\mathsf{R}, 0, 2^k \underbrace{\mathcal{T}_i}_{i=1,\dots,n}\right) = 1,$$

for every $\mathsf{S}, \mathsf{R} \in \mho_1$ and $\underbrace{\mathcal{T}_i}_{i=1,\dots,n} > 0$. Assume that fuzzy operator $\mathcal{Q} : ⫛ \times \mho_1 \to \mho_2$ satisfies (18.6) *and $\mathcal{Q}(\leftthreetimes, \mathsf{S}) = 0$, for every $\mathsf{S}, \mathsf{R} \in \mho_1$ and $\leftthreetimes \in ⫛$. Then, there exists a unique additive fuzzy operator $\mathcal{V} : ⫛ \times \mho_1 \to \mho_2$ such that*

$$\mathcal{N}\Big(\mathcal{Q}(\leftthreetimes, \mathsf{S}) - \mathcal{V}(\leftthreetimes, \mathsf{S}), (\mathcal{T}_1, \dots, \mathcal{T}_n)\Big) \tag{18.15}$$
$$\succeq diag\Big[\varphi_1\Big(\mathsf{S}, \mathsf{S}, 0, 2(1-\mathfrak{P}_1)\mathcal{T}_1\Big), \dots, \varphi_n\Big(\mathsf{S}, \mathsf{S}, 0, 2(1-\mathfrak{P}_n)\mathcal{T}_n\Big)\Big],$$

for every $\mathsf{S} \in \mho_1, \leftthreetimes \in \pitchfork$ *and* $\underbrace{\mathcal{T}_i}_{i=1,\ldots,n} > 0.$

Proof Putting $\mathsf{A} = 0$ and $\mathsf{S} = \mathsf{R}$ in (18.6), we have that

$$\mathcal{N}\left(\frac{1}{2}\mathcal{Q}(\leftthreetimes, 2\mathsf{S}) - \mathcal{Q}(\leftthreetimes, \mathsf{S}), (\mathcal{T}_1, \ldots, \mathcal{T}_n)\right) \tag{18.16}$$
$$\succeq \operatorname{diag}\left[\varphi_1\left(\mathsf{S}, \mathsf{S}, 0, 2\mathcal{T}_1\right), \ldots, \varphi_n\left(\mathsf{S}, \mathsf{S}, 0, 2\mathcal{T}_n\right)\right],$$

for every $\mathsf{S} \in \mho_1, \leftthreetimes \in \pitchfork$ and $\underbrace{\mathcal{T}_i}_{i=1,\ldots,n} > 0$. Thus, we get

$$\mathcal{N}\left(\frac{1}{2}\mathcal{Q}(\leftthreetimes, \mathsf{S}) - \mathcal{Q}\left(\leftthreetimes, 2\mathsf{S}\right), (\mathcal{T}_1, \ldots, \mathcal{T}_n)\right) \tag{18.17}$$
$$\succeq \operatorname{diag}\left[\varphi_1\left(2\mathsf{S}, 2\mathsf{S}, 0, \mathcal{T}_1\right), \ldots, \varphi_n\left(2\mathsf{S}, 2\mathsf{S}, 0, \mathcal{T}_n\right)\right]$$
$$\succeq \operatorname{diag}\left[\varphi_1\left(\mathsf{S}, \mathsf{S}, 0, \frac{\mathcal{T}_1}{2\mathfrak{P}_1}\right), \ldots, \varphi_n\left(\mathsf{S}, \mathsf{S}, 0, \frac{\mathcal{T}_n}{2\mathfrak{P}_n}\right)\right],$$

for every $\mathsf{S} \in \mho_1, \leftthreetimes \in \pitchfork$ and $\underbrace{\mathcal{T}_i}_{i=1,\ldots,n} > 0$. Replacing S by $2^n\mathsf{S}$ in (18.17), we have

$$\mathcal{N}\left(\frac{1}{2^k}\mathcal{Q}(\leftthreetimes, 2^k\mathsf{S}) - \frac{1}{2^{k+1}}\mathcal{Q}\left(\leftthreetimes, 2^{k+1}\mathsf{S}\right), (\mathcal{T}_1, \ldots, \mathcal{T}_n)\right) \tag{18.18}$$
$$\succeq \operatorname{diag}\left[\varphi_1\left(2^k\mathsf{S}, 2^k\mathsf{S}, 0, 2\times 2^k\mathcal{T}_1\right), \ldots, \varphi_n\left(2^k\mathsf{S}, 2^k\mathsf{S}, 0, 2\times 2^k\mathcal{T}_n\right)\right]$$
$$\succeq \operatorname{diag}\left[\varphi_1\left(\mathsf{S}, \mathsf{S}, 0, \frac{2\times 2^k}{(2\mathfrak{P}_1)^k}\mathcal{T}_1\right), \ldots, \varphi_n\left(\mathsf{S}, \mathsf{S}, 0, \frac{2\times 2^k}{(2\mathfrak{P}_n)^k}\mathcal{T}_n\right)\right].$$

From

$$\frac{1}{2^k}\mathcal{Q}\left(\leftthreetimes, 2^k\mathsf{S}\right) - \mathcal{Q}\left(\leftthreetimes, \mathsf{S}\right) = \sum_{j=0}^{k-1}\left(\frac{1}{2^{j+1}}\mathcal{Q}\left(\leftthreetimes, 2^{j+1}\mathsf{S}\right) - \frac{1}{2^j}\mathcal{Q}\left(\leftthreetimes, 2^j\mathsf{S}\right)\right),$$

and (18.18), we get

$$\mathcal{N}\left(\frac{1}{2^k}\mathcal{Q}\left(\leftthreetimes, 2^k\mathsf{S}\right) - \mathcal{Q}\left(\leftthreetimes, \mathsf{S}\right), \left(\sum_{j=0}^{k-1}\frac{(2\mathfrak{P}_1)^j}{2\times 2^j}\mathcal{T}_1, \ldots, \sum_{j=0}^{k-1}\frac{(2\mathfrak{P}_n)^j}{2\times 2^j}\mathcal{T}_n\right)\right)$$
$$\succeq \operatorname{diag}\left[\varphi_1\left(\mathsf{S}, \mathsf{S}, 0, \mathcal{T}_1\right), \ldots, \varphi_n\left(\mathsf{S}, \mathsf{S}, 0, \mathcal{T}_n\right)\right],$$

for every $\mathsf{S} \in \mho_1$, $\leftthreetimes \in \pitchfork$ and $\underbrace{\mathcal{T}_i}_{i=1,\ldots,n} > 0$. That is,

$$\mathcal{N}\left(\frac{1}{2^k}\mathcal{Q}\left(\leftthreetimes, 2^k\mathsf{S}\right) - \mathcal{Q}\left(\leftthreetimes, \mathsf{S}\right), (\mathcal{T}_1, \ldots, \mathcal{T}_n)\right) \tag{18.19}$$
$$\succeq \operatorname{diag}\left[\varphi_1\left(\mathsf{S}, \mathsf{S}, 0, \frac{\mathcal{T}_1}{\sum_{j=0}^{k-1}\frac{(2\mathfrak{P}_1)^j}{2\times 2^j}}\right), \ldots, \varphi_n\left(\mathsf{S}, \mathsf{S}, 0, \frac{\mathcal{T}_n}{\sum_{j=0}^{k-1}\frac{(2\mathfrak{P}_n)^j}{2\times 2^j}}\right)\right].$$

Replacing S by $2^m\mathsf{S}$ in (18.19), we get

$$\mathcal{N}\left(\frac{1}{2^{k+m}}\mathcal{Q}(\leftthreetimes, 2^{k+m}\mathsf{S}) - \frac{1}{2^m}\mathcal{Q}\left(\leftthreetimes, 2^m\mathsf{S}\right), (\mathcal{T}_1, \ldots, \mathcal{T}_n)\right) \tag{18.20}$$
$$\succeq \operatorname{diag}\left[\varphi_1\left(\mathsf{S}, \mathsf{S}, 0, \frac{\mathcal{T}_1}{\sum_{j=m}^{k+m}\frac{(2\mathfrak{P}_1)^j}{2\times 2^j}}\right), \ldots, \varphi_n\left(\mathsf{S}, \mathsf{S}, 0, \frac{\mathcal{T}_n}{\sum_{j=m}^{k+m}\frac{(2\mathfrak{P}_n)^j}{2\times 2^j}}\right)\right].$$

As $m, k \to \infty$, $\underbrace{\varphi_i}_{i=1,\ldots,n}\left(\mathsf{S}, \mathsf{S}, 0, \overbrace{\frac{\mathcal{T}_i}{\sum_{j=m}^{k+m}\frac{(2\mathfrak{P}_i)^j}{2\times 2^j}}}^{i=1,\ldots,n}\right)$ tend to 1. It implies that the sequence $\{\frac{1}{2^k}\mathcal{Q}(\leftthreetimes, 2^k\mathsf{S})\}$ is Cauchy, for every $\mathsf{S} \in \mho_1$ and $\leftthreetimes \in \pitchfork$. Since $\mho_2$ is a complete, the sequence $\{\frac{1}{2^k}\mathcal{Q}(\leftthreetimes, 2^k\mathsf{S})\}$ converges.

Now, we define the fuzzy operator $\mathcal{V} : \pitchfork \times \mho_1 \to \mho_2$ as follows:

$$\mathcal{V}(\leftthreetimes, \mathsf{S}) := \lim_{k\to\infty}\frac{1}{2^k}\mathcal{Q}\left(\leftthreetimes, 2^k\mathsf{S}\right),$$

for every $\mathsf{S} \in \mho_1$ and $\leftthreetimes \in \pitchfork$. Putting $m = 0$ and letting $k \to \infty$ in (18.20), we get

$$\mathcal{N}\left(\mathcal{Q}(\leftthreetimes, \mathsf{S}) - \mathcal{V}(\leftthreetimes, \mathsf{S}), (\mathcal{T}_1, \ldots, \mathcal{T}_n)\right) \tag{18.21}$$
$$\succeq \operatorname{diag}\left[\varphi_1\left(\mathsf{S}, \mathsf{S}, 0, 2(1-\mathfrak{P}_1)\mathcal{T}_1\right), \ldots, \varphi_n\left(\mathsf{S}, \mathsf{S}, 0, 2(1-\mathfrak{P}_n)\mathcal{T}_n\right)\right],$$

for every $\mathsf{S} \in \mho_1$, $\leftthreetimes \in \pitchfork$ and $\underbrace{\mathcal{T}_i}_{i=1,\ldots,n} > 0$. Using Theorem 18.1, completes the proof.

18.2 Multi Stability of Additive Functional Inequalities in MVFB–Algebras by an Alternative Fixed Point Method

We apply the fixed point method to get the multi stability results for the additive $(\mathcal{G}_1, \mathcal{G}_2)$–fuzzy operator inequality (18.1) in the MVFB–algebras $(\mho_1, \mathcal{N}, \bigotimes_{\text{GTN}}, \bigoplus_{\text{GTN}})$ and $(\mho_2, \mathcal{N}, \bigotimes_{\text{GTN}}, \bigoplus_{\text{GTN}})$.

Theorem 18.3 *Consider the following assumptions:*

- $(\mho_1, \mathcal{N}, \bigotimes_{\text{GTN}}, \bigoplus_{\text{GTN}})$ *and* $(\mho_2, \mathcal{N}, \bigotimes_{\text{GTN}}, \bigoplus_{\text{GTN}})$ *be MVFB–algebras,*
- $\underbrace{\varphi_i}_{i=1,\ldots,n} : \mho_1^3 \times (0, +\infty) \to [0, 1]$ *be fuzzy sets satisfy* (18.5), *for every* $\mathsf{S}, \mathsf{R}, \mathsf{A} \in \mho_1$, $n \in \mathbb{N}$ *and* $\underbrace{\mathcal{T}_i}_{i=1,\ldots,n} > 0$,
- *there exist* $\underbrace{\mathfrak{P}_i}_{i=1,\ldots,n} < 1$, *such that* $\underbrace{\varphi_i}_{i=1,\ldots,n}\left(\frac{\mathsf{S}}{2}, \frac{\mathsf{R}}{2}, \frac{\mathsf{A}}{2}, \overbrace{\mathcal{T}_i}^{i=1,\ldots,n}\right) \geq \underbrace{\varphi_i}_{i=1,\ldots,n}\left(\mathsf{S}, \mathsf{R}, \mathsf{A}, \overbrace{\frac{2\mathcal{T}_i}{\mathfrak{P}_i}}^{i=1,\ldots,n}\right)$, *for every* $\mathsf{S}, \mathsf{R}, \mathsf{A} \in \mho_1$, $n \in \mathbb{N}$ *and* $\underbrace{\mathcal{T}_i}_{i=1,\ldots,n} > 0$,
- *the fuzzy operator* $\mathcal{Q} : ⫛ \times \mho_1 \to \mho_2$ *satisfies* $\mathcal{Q}(\leftthreetimes, 0) = 0$ *and* (18.6) *for every* $\mathsf{S}, \mathsf{R}, \mathsf{A} \in \mho_1$, $\leftthreetimes \in ⫛$, $n \in \mathbb{N}$ *and* $\underbrace{\mathcal{T}_i}_{i=1,\ldots,n} > 0$,

Then, we can find a unique additive fuzzy operator $\mathcal{V} : ⫛ \times \mho_1 \to \mho_2$ *satisfying* (18.7), *for every* $\mathsf{S}, \mathsf{R}, \mathsf{A} \in \mho_1$, $\leftthreetimes \in ⫛$, $n \in \mathbb{N}$ *and* $\underbrace{\mathcal{T}_i}_{i=1,\ldots,n} > 0$.

Proof Putting $\mathsf{S} = \mathsf{R}$ and $\mathsf{A} = 0$ in (18.6), we have that

$$\begin{aligned} &\mathcal{N}\left(2\mathcal{Q}\left(\leftthreetimes, \frac{\mathsf{S}}{2}\right) - \mathcal{Q}(\leftthreetimes, \mathsf{S}), (\mathcal{T}_1, \ldots, \mathcal{T}_n)\right) \\ &\succeq \operatorname{diag}\left[\varphi_1\left(\frac{\mathsf{S}}{2}, \frac{\mathsf{S}}{2}, 0, \mathcal{T}_1\right), \ldots, \varphi_n\left(\frac{\mathsf{S}}{2}, \frac{\mathsf{S}}{2}, 0, \mathcal{T}_n\right)\right], \end{aligned} \tag{18.22}$$

almost everywhere, for every $\mathsf{S} \in \mho_1$ and $\underbrace{\mathcal{T}_i}_{i=1,\ldots,n} > 0$.

Consider the set below

$$\xi := \{\mathcal{Z} : ⫛ \times \mho_1 \to \mho_2, \ \ \mathcal{Z}(\leftthreetimes, 0) = 0\},$$

and the generalized metric defined as follows:

$$\zeta(\mathcal{P}, \mathcal{Z}) = \inf\Bigg\{(\hbar_1, \ldots, \hbar_n) \in \mathbb{R}_+^n : \mathcal{N}\Big(\mathcal{P}(\curlywedge, \mathsf{S}) - \mathcal{Z}(\curlywedge, \mathsf{S}), (\hbar_1 \mathcal{T}_1, \ldots, \hbar_n \mathcal{T}_n)\Big) \succeq \operatorname{diag}\Big[\varphi_1(\mathsf{S}, \mathsf{S}, 0, \mathcal{T}_1), \ldots, \varphi_n(\mathsf{S}, \mathsf{S}, 0, \mathcal{T}_n)\Big],\ \mathsf{S} \in \mho_1,\ \underbrace{\mathcal{T}_i}_{i=1,\ldots,n} > 0\Bigg\}.$$

In [2], Miheţ and Radu showed that (ξ, ζ) is complete.

We define the linear function $\mathfrak{L} : \xi \to \xi$ as

$$\mathfrak{L}\mathcal{H}(\curlywedge, \mathsf{S}) := 2\mathcal{H}\left(\curlywedge, \frac{\mathsf{S}}{2}\right),$$

almost everywhere, for every $\mathsf{S} \in \mho_1$ and $\curlywedge \in \pitchfork$. Consider $\mathcal{H}, \mathcal{K} \in \xi$ such that $\zeta(\mathcal{H}, \mathcal{K}) = (\varepsilon_1, \ldots, \varepsilon_n)$. Then, we get

$$\begin{aligned}
&\mathcal{N}\Big(\mathcal{H}(\curlywedge, \mathsf{S}) - \mathcal{K}(\curlywedge, \mathsf{S}), (\varepsilon_1 \mathcal{T}_1, \ldots, \varepsilon_n \mathcal{T}_n)\Big)\\
&\succeq \operatorname{diag}\Big[\varphi_1\Big(\mathsf{S}, \mathsf{S}, 0, \mathcal{T}_1\Big), \ldots, \varphi_n\Big(\mathsf{S}, \mathsf{S}, 0, \mathcal{T}_n\Big)\Big],
\end{aligned}$$

almost everywhere, for every $\mathsf{S} \in \mho_1$, $\curlywedge \in \pitchfork$ and $\underbrace{\mathcal{T}_i}_{i=1,\ldots,n} > 0$, and also

$$\begin{aligned}
&\mathcal{N}\Big(\mathfrak{L}\mathcal{H}(\curlywedge, \mathsf{S}) - \mathfrak{L}\mathcal{K}(\curlywedge, \mathsf{S}), (\mathfrak{P}_1 \varepsilon_1 \mathcal{T}_1, \ldots, \mathfrak{P}_n \varepsilon_n \mathcal{T}_n)\Big)\\
&= \mathcal{N}\left(\mathcal{H}\left(\curlywedge, \frac{\mathsf{S}}{2}\right) - \mathcal{K}\left(\curlywedge, \frac{\mathsf{S}}{2}\right), \left(\frac{\mathfrak{P}_1 \varepsilon_1 \mathcal{T}_1}{2}, \ldots, \frac{\mathfrak{P}_n \varepsilon_n \mathcal{T}_n}{2}\right)\right),\\
&\succeq \operatorname{diag}\left[\varphi_1\left(\frac{\mathsf{S}}{2}, \frac{\mathsf{S}}{2}, 0, \frac{\mathfrak{P}_1 \mathcal{T}_1}{2}\right), \ldots, \varphi_n\left(\frac{\mathsf{S}}{2}, \frac{\mathsf{S}}{2}, 0, \frac{\mathfrak{P}_n \mathcal{T}_n}{2}\right)\right]\\
&\succeq \operatorname{diag}\Big[\varphi_1\Big(\mathsf{S}, \mathsf{S}, 0, \mathcal{T}_1\Big), \ldots, \varphi_n\Big(\mathsf{S}, \mathsf{S}, 0, \mathcal{T}_n\Big)\Big],
\end{aligned}$$

almost everywhere, for every $\mathsf{S} \in \mho_1$ and $\underbrace{\mathcal{T}_i}_{i=1,\ldots,n} > 0$. Thus, from $\zeta(\mathcal{H}, \mathcal{K}) = (\varepsilon_1, \ldots, \varepsilon_n)$, we conclude that $\zeta(\mathfrak{L}\mathcal{H}, \mathfrak{L}\mathcal{K}) \preceq (\mathfrak{P}_1 \varepsilon_1, \ldots, \mathfrak{P}_n \varepsilon_n)$, and so

$$\zeta(\mathfrak{L}\mathcal{H}, \mathfrak{L}\mathcal{K}) \preceq (\mathfrak{P}_1, \ldots, \mathfrak{P}_n)\zeta(\mathcal{H}, \mathcal{K}),$$

for every $\mathcal{H}, \mathcal{K} \in \xi$.

By (18.22) we have that

$$\mathscr{N}\left(2\mathcal{Q}\left(\curlywedge, \frac{\mathsf{S}}{2}\right) - \mathcal{Q}(\curlywedge, \mathsf{S}), \left(\frac{\mathfrak{P}_1 \mathcal{T}_1}{2}, \dots, \frac{\mathfrak{P}_n \mathcal{T}_n}{2}\right)\right)$$
$$\succeq \operatorname{diag}\left[\varphi_1\left(\mathsf{S}, \mathsf{S}, 0, \mathcal{T}_1\right), \cdots, \varphi_n\left(\mathsf{S}, \mathsf{S}, 0, \mathcal{T}_n\right)\right],$$

almost everywhere, for every $\mathsf{S} \in \mho_1$, $\curlywedge \in \pitchfork$ and $\underbrace{\mathcal{T}_i}_{i=1,\dots,n} > 0$, which implies that $\zeta(\mathcal{Q}, \mathfrak{L}\mathcal{Q}) \preceq (\frac{\mathfrak{P}_1}{2}, \dots, \frac{\mathfrak{P}_n}{2})$.

Theorem 1.8 implies that there exists a fuzzy operator $\mathcal{V} : \pitchfork \times \mho_1 \to \mho_2$ such that:
(1) A fixed point for the function $\mathfrak{L}$, is $\mathcal{V}$,

$$\mathcal{V}(\curlywedge, \mathsf{S}) = 2\mathcal{V}\left(\curlywedge, \frac{\mathsf{S}}{2}\right), \tag{18.23}$$

almost everywhere for every $\mathsf{S} \in \mho_1$ and $\curlywedge \in \pitchfork$, which is unique in the set

$$\mathsf{D} = \left\{\mathcal{H} \in \xi : \zeta(\mathcal{H}, \mathcal{K}) \prec \underbrace{(\infty, \dots, \infty)}_{n}\right\};$$

(2) $\zeta(\mathfrak{L}^p \mathcal{Q}, \mathcal{V}) \to \underbrace{(0, \dots, 0)}_{n}$, as $p \to \infty$, which implies that

$$\lim_{p\to\infty} 2^p \mathcal{Q}\left(\curlywedge, \frac{\mathsf{S}}{2^p}\right) = \mathcal{V}(\curlywedge, \mathsf{S}), \tag{18.24}$$

almost everywhere for every $\mathsf{S} \in \mho_1$ and $\curlywedge \in \pitchfork$;
(3) $\zeta(\mathcal{Q}, \mathcal{V}) \preceq (\frac{1}{1-\mathfrak{P}_1}, \dots, \frac{1}{1-\mathfrak{P}_n})\zeta(\mathcal{Q}, \mathfrak{L}\mathcal{Q})$, which implies that

$$\mathscr{N}\left(\mathcal{Q}(\curlywedge, \mathsf{S}) - \mathcal{V}(\curlywedge, \mathsf{S}), (\mathcal{T}_1, \dots, \mathcal{T}_n)\right)$$
$$\succeq \operatorname{diag}\left[\varphi_1\left(\mathsf{S}, \mathsf{S}, 0, \frac{2(1-\mathfrak{P}_1)}{\mathfrak{P}_1}\mathcal{T}_1\right), \dots, \varphi_n\left(\mathsf{S}, \mathsf{S}, 0, \frac{2(1-\mathfrak{P}_n)}{\mathfrak{P}_n}\mathcal{T}_n\right)\right],$$

almost everywhere for every $\mathsf{S} \in \mho_1$ and $\underbrace{\mathcal{T}_i}_{i=1,\dots,n} > 0$.

Using (18.6) and (18.24), we have

$$\begin{aligned}
&\mathscr{N}\Big(\mathcal{V}(\leftthreetimes, \mathsf{S}+\mathsf{R}+\mathsf{A}) - \mathcal{V}(\leftthreetimes, \mathsf{S}) - \mathcal{V}(\leftthreetimes, \mathsf{R}) + \mathcal{V}(\leftthreetimes, \mathsf{A}), (\mathcal{T}_1, \ldots, \mathcal{T}_n)\Big), \\
&= \lim_{k\to\infty} \mathscr{N}\Big(2^k(\mathcal{Q}(\leftthreetimes, \frac{\mathsf{S}+\mathsf{R}+\mathsf{A}}{2^k}) - \mathcal{Q}(\leftthreetimes, \frac{\mathsf{S}}{2^k}) \\
&\qquad\qquad -\mathcal{Q}(\leftthreetimes, \frac{\mathsf{R}}{2^k}) - \mathcal{Q}(\leftthreetimes, \frac{\mathsf{A}}{2^k})), (\mathcal{T}_1, \ldots, \mathcal{T}_n)\Big), \\
&\succeq \lim_{k\to\infty} \mathscr{N}\Big(2^k\mathcal{G}_1[\mathcal{Q}(\leftthreetimes, \frac{\mathsf{S}+\mathsf{A}}{2^k}) - \mathcal{Q}(\leftthreetimes, \frac{\mathsf{S}}{2^k}) - \mathcal{Q}(\leftthreetimes, \frac{\mathsf{A}}{2^k})], (\mathcal{T}_1, \ldots, \mathcal{T}_n)\Big) \\
&\quad \bigotimes_{\mathrm{GTN}} \mathscr{N}\Big(2^k\mathcal{G}_2[\mathcal{Q}(\leftthreetimes, \frac{\mathsf{R}+\mathsf{A}}{2^k}) - \mathcal{Q}(\leftthreetimes, \frac{\mathsf{R}}{2^k}) - \mathcal{Q}(\leftthreetimes, \frac{\mathsf{A}}{2^k})], (\mathcal{T}_1, \ldots, \mathcal{T}_n)\Big) \\
&\quad \bigotimes_{\mathrm{GTN}} \lim_{k\to\infty} \operatorname{diag}\Big[\varphi_1\Big(\frac{\mathsf{S}}{2^k}, \frac{\mathsf{S}}{2^k}, 0, \frac{\mathcal{T}_1}{2^k}\Big), \ldots, \varphi_n\Big(\frac{\mathsf{S}}{2^k}, \frac{\mathsf{S}}{2^k}, 0, \frac{\mathcal{T}_n}{2^k}\Big)\Big] \\
&\succeq \mathscr{N}\Big(\mathcal{G}_1[\mathcal{V}(\leftthreetimes, \mathsf{S}+\mathsf{A}) - \mathcal{V}(\leftthreetimes, \mathsf{S}) - \mathcal{V}(\leftthreetimes, \mathsf{A})], (\mathcal{T}_1, \ldots, \mathcal{T}_n)\Big) \\
&\quad \mathscr{N}\Big(\mathcal{G}_2[\mathcal{V}(\leftthreetimes, \mathsf{R}+\mathsf{A}) - \mathcal{V}(\leftthreetimes, \mathsf{R}) - \mathcal{V}(\leftthreetimes, \mathsf{A})], (\mathcal{T}_1, \ldots, \mathcal{T}_n)\Big),
\end{aligned}$$

almost everywhere for every $\mathsf{S}, \mathsf{R}, \mathsf{A} \in \mho_1$ and $\underbrace{\mathcal{T}_i}_{i=1,\ldots,n} > 0$. Thus, we get

$$\begin{aligned}
&\mathscr{N}\Big(\mathcal{V}(\leftthreetimes, \mathsf{S}+\mathsf{R}+\mathsf{A}) - \mathcal{V}(\leftthreetimes, \mathsf{S}) - \mathcal{V}(\leftthreetimes, \mathsf{R}) - \mathcal{V}(\leftthreetimes, \mathsf{A}), (\mathcal{T}_1, \ldots, \mathcal{T}_n)\Big), \\
&\succeq \mathscr{N}\Big(\mathcal{G}_1[\mathcal{V}(\leftthreetimes, \mathsf{S}+\mathsf{A}) - \mathcal{V}(\leftthreetimes, \mathsf{S}) - \mathcal{V}(\leftthreetimes, \mathsf{A})], (\mathcal{T}_1, \ldots, \mathcal{T}_n)\Big) \\
&\quad \bigotimes_{\mathrm{GTN}} \mathscr{N}\Big(\mathcal{G}_2[\mathcal{V}(\leftthreetimes, \mathsf{R}+\mathsf{A}) - \mathcal{V}(\leftthreetimes, \mathsf{R}) - \mathcal{V}(\leftthreetimes, \mathsf{A})], (\mathcal{T}_1, \ldots, \mathcal{T}_n)\Big),
\end{aligned}$$

almost everywhere for every $\mathsf{S}, \mathsf{R}, \mathsf{A} \in \mho_1, \leftthreetimes \in \pitchfork$ and $\underbrace{\mathcal{T}_i}_{i=1,\ldots,n} > 0$. Now, Lemma 18.1 implies that $\mathcal{V}$ is an additive fuzzy operator.

Theorem 18.4 *Let $n \in \mathbb{N}$, and $(\underbrace{\mho_j}_{j=1,\ldots,2}, \mathscr{N}, \bigotimes_{\mathrm{GTN}}, \bigoplus_{\mathrm{GTN}})$ be MVFB–algebras and that $\underbrace{\varphi_i}_{i=1,\ldots,n} : \mho_1^3 \times (0, +\infty) \to [0, 1]$ be fuzzy sets such that there exist $\underbrace{\mathfrak{P}_i}_{i=1,\ldots,n} < 1$ with*

$$\underbrace{\varphi_i}_{i=1,\ldots,n}\Big(\mathsf{S}, \mathsf{R}, 0, \overbrace{\mathcal{T}_i}^{i=1,\ldots,n}\Big) \geq \underbrace{\varphi_i}_{i=1,\ldots,n}\Big(\frac{\mathsf{S}}{2}, \frac{\mathsf{R}}{2}, 0, \overbrace{\frac{\mathcal{T}_i}{2\mathfrak{P}_i}}^{i=1,\ldots,n}\Big),$$

for every $\mathsf{S}, \mathsf{R} \in \mho_1$ *and* $\underbrace{\mathcal{T}_i}_{i=1,\ldots,n}$, *and*

$$\lim_{k\to\infty} \underbrace{\varphi_i}_{i=1,\ldots,n} \left(2^k\mathsf{S}, 2^k\mathsf{R}, 0, 2^k \underbrace{\mathcal{T}_i}_{i=1,\ldots,n}\right) = 1,$$

for every $\mathsf{S}, \mathsf{R} \in \mho_1$ *and* $\underbrace{\mathcal{T}_i}_{i=1,\ldots,n} > 0$. *Assume that the fuzzy operator* $\mathcal{Q} : \pitchfork \times \mho_1 \to \mho_2$ *satisfies* (18.6) *and* $\mathcal{Q}(\leftthreetimes, \mathsf{S}) = 0$, *for every* $\mathsf{S}, \mathsf{R} \in \mho_1$ *and* $\leftthreetimes \in \pitchfork$. *Then, there exists a unique additive fuzzy operator* $\mathcal{V} : \pitchfork \times \mho_1 \to \mho_2$ *satisfying* (18.15), *for every* $\mathsf{S} \in \mho_1, \leftthreetimes \in \pitchfork$ *and* $\underbrace{\mathcal{T}_i}_{i=1,\ldots,n} > 0$.

Proof Let (ξ, ζ) be the same as in the proof of Theorem 18.1. We define the linear function $\mathfrak{L} : \xi \to \xi$ as

$$\mathfrak{L}\mathcal{H}(\leftthreetimes, \mathsf{S}) := \frac{1}{2}\mathcal{H}(\leftthreetimes, 2\mathsf{S}),$$

almost everywhere for every $\mathsf{S} \in \mho_1$ and $\leftthreetimes \in \pitchfork$. Using (18.7), we obtain

$$\begin{aligned}
&\mathscr{N}\left(\frac{\mathcal{Q}(\leftthreetimes, 2\mathsf{S})}{2} - \mathcal{Q}(\leftthreetimes, \mathsf{S}), (\mathcal{T}_1, \ldots, \mathcal{T}_n)\right)\\
&\succeq \operatorname{diag}\left[\varphi_1\left(\frac{\mathsf{S}}{2}, \frac{\mathsf{S}}{2}, 0, \frac{\mathcal{T}_1}{\mathfrak{P}_1}\right), \ldots, \varphi_n\left(\frac{\mathsf{S}}{2}, \frac{\mathsf{S}}{2}, 0, \frac{\mathcal{T}_n}{\mathfrak{P}_n}\right)\right],
\end{aligned}$$

almost everywhere for every $\mathsf{S} \in \mho_1, \leftthreetimes \in \pitchfork$ and $\underbrace{\mathcal{T}_i}_{i=1,\ldots,n} > 0$.

By a similar method to the proof of Theorem 18.3, the proof will be completed.

18.3 Application

Consider the special controllers $\underbrace{\varphi_i^{\circledS}}_{1\le i\le n\in\mathbb{N}}$ defined in Sects. 9.8.1 and 9.8.2.

Corollary 18.1 *Let* $n \in \mathbb{N}$, *and* $(\mho_1, \mathscr{N}, \bigotimes_{\mathrm{GTN}}, \bigotimes_{\mathrm{GTN}})$ *and* $(\mho_2, \mathscr{N}, \bigotimes_{\mathrm{GTN}}, \bigotimes_{\mathrm{GTN}})$ *be MVFB–algebras. Let* $\underbrace{\mathsf{M}_i}_{i=1,\ldots,n} > 1$ *and* $\underbrace{\mathsf{W}_i}_{i=1,\ldots,n}$ *be nonnegative real numbers and* $\mathcal{Q} : \pitchfork \times \mho_1 \to \mho_2$ *be a fuzzy operator satisfying* $\mathcal{Q}(\leftthreetimes, 0) = 0$ *and*

$$\mathscr{N}\Big(\mathcal{Q}(\curlywedge, \mathsf{S}+\mathsf{R}+\mathsf{A}) - \mathcal{Q}(\curlywedge, \mathsf{S}) - \mathcal{Q}(\curlywedge, \mathsf{R}) - \mathcal{Q}(\curlywedge, \mathsf{A}), (\mathcal{T}_1, \dots, \mathcal{T}_n)\Big) \tag{18.25}$$
$$\succeq \mathscr{N}\Big(\mathcal{G}_1[\mathcal{Q}(\curlywedge, \mathsf{S}+\mathsf{A}) - \mathcal{Q}(\curlywedge, \mathsf{S}) - \mathcal{Q}(\curlywedge, \mathsf{A})], (\mathcal{T}_1, \dots, \mathcal{T}_n)\Big)$$
$$\bigotimes_{\mathrm{GTN}} \mathscr{N}\Big(\mathcal{G}_2[\mathcal{Q}(\curlywedge, \mathsf{R}+\mathsf{A}) - \mathcal{Q}(\curlywedge, \mathsf{R}) - \mathcal{Q}(\curlywedge, \mathsf{A})], (\mathcal{T}_1, \dots, \mathcal{T}_n)\Big),$$
$$\bigotimes_{\mathrm{GTN}} \mathrm{diag}\Big[\varphi_1^{\circledS}\Big(-\frac{\mathsf{W}_1(\|\mathsf{S}\|^{\mathsf{M}_1} + \|\mathsf{R}\|^{\mathsf{M}_1} + \|\mathsf{A}\|^{\mathsf{M}_1})}{\mathcal{T}_1}\Big), \dots,$$
$$\varphi_n^{\circledS}\Big(-\frac{\mathsf{W}_n(\|\mathsf{S}\|^{\mathsf{M}_n} + \|\mathsf{R}\|^{\mathsf{M}_n} + \|\mathsf{A}\|^{\mathsf{M}_n})}{\mathcal{T}_n}\Big)\Big],$$

for every $\mathsf{S}, \mathsf{R}, \mathsf{A} \in \mho_1$, $\curlywedge \in \pitchfork$ *and* $\underbrace{\mathcal{T}_i}_{i=1,\dots,n} > 0$. *Then, there exists a unique additive fuzzy operator* $\mathcal{V} : \pitchfork \times \mho_1 \to \mho_2$ *such that*

$$\mathscr{N}\Big(\mathcal{Q}(\curlywedge, \mathsf{S}) - \mathcal{V}(\curlywedge, \mathsf{S}), (\mathcal{T}_1, \dots, \mathcal{T}_n)\Big) \tag{18.26}$$
$$\succeq \mathrm{diag}\Big[\varphi_1^{\circledS}\Big(-\frac{2^{\mathsf{M}_1+2}\mathsf{W}_1\|\mathsf{S}\|^{\mathsf{M}_1}}{2(2^{\mathsf{M}_1}-2)\mathcal{T}_1}\Big), \dots, \varphi_n^{\circledS}\Big(-\frac{2^{\mathsf{M}_n+2}\mathsf{W}_n\|\mathsf{S}\|^{\mathsf{M}_n}}{2(2^{\mathsf{M}_n}-2)\mathcal{T}_n}\Big)\Big],$$

for every $\mathsf{S} \in \mho_1$, $\curlywedge \in \pitchfork$ *and* $\underbrace{\mathcal{T}_i}_{i=1,\dots,n} > 0$.

Proof It follows from Theorem 18.1 by putting

$$\mathrm{diag}\Big[\varphi_1\Big(\mathsf{S}, \mathsf{R}, \mathsf{A}, \mathcal{T}_1\Big), \dots, \varphi_n\Big(\mathsf{S}, \mathsf{R}, \mathsf{A}, \mathcal{T}_n\Big)\Big]$$
$$:= \mathrm{diag}\Big[\varphi_1^{\circledS}\Big(-\frac{\mathsf{W}_1(\|\mathsf{S}\|^{\mathsf{M}_1} + \|\mathsf{R}\|^{\mathsf{M}_1} + \|\mathsf{A}\|^{\mathsf{M}_1})}{\mathcal{T}_1}\Big), \dots,$$
$$\varphi_n^{\circledS}\Big(-\frac{\mathsf{W}_n(\|\mathsf{S}\|^{\mathsf{M}_n} + \|\mathsf{R}\|^{\mathsf{M}_n} + \|\mathsf{A}\|^{\mathsf{M}_n})}{\mathcal{T}_n}\Big)\Big],$$

for every $\mathsf{S}, \mathsf{R}, \mathsf{A} \in \mho_1$, $\underbrace{\mathcal{T}_i}_{i=1,\dots,n} > 0$ and $\underbrace{\mathfrak{P}_i}_{i=1,\dots,n} = 2^{1-\overbrace{\mathsf{M}_i}^{i=1,\dots,n}}$.

Corollary 18.2 *Let* $n \in \mathbb{N}$, *and* $(\mho_1, \mathscr{N}, \bigotimes_{\mathrm{GTN}}, \bigoplus_{\mathrm{GTN}})$ *and* $(\mho_2, \mathscr{N}, \bigotimes_{\mathrm{GTN}}, \bigoplus_{\mathrm{GTN}})$ *be MVFB–algebras. Let* $\underbrace{\mathsf{M}_i}_{i=1,\dots,n} < 1$, $\underbrace{\mathsf{W}_i}_{i=1,\dots,n} \geq 0$, *and* $\mathcal{Q} : \pitchfork \times \mho_1 \to \mho_2$ *be a fuzzy operator satisfying* (18.25) *and* $\mathcal{Q}(\curlywedge, \mathsf{S}) = 0$, *for every* $\mathsf{S} \in \mho_1$ *and* $\curlywedge \in \pitchfork$. *Then, there exists a unique additive fuzzy operator* $\mathcal{V} : \pitchfork \times \mho_1 \to \mho_2$ *such that*

$$\mathcal{N}\Big(\mathcal{Q}(\lambda, \mathsf{S}) - \mathcal{V}(\lambda, \mathsf{S}), (\mathcal{T}_1, \ldots, \mathcal{T}_n)\Big) \tag{18.27}$$
$$\succeq \operatorname{diag}\left[\varphi_1^{\circledS}\left(-\frac{2\mathsf{W}_1\|\mathsf{S}\|^{\mathsf{M}_1}}{(2-2^{\mathsf{M}_1})\mathcal{T}_1}\right), \cdots, \varphi_{\mathrm{n}}^{\circledS}\left(-\frac{2\mathsf{W}_{\mathrm{n}}\|\mathsf{S}\|^{\mathsf{M}_{\mathrm{n}}}}{(2-2^{\mathsf{M}_{\mathrm{n}}})\mathcal{T}_{\mathrm{n}}}\right)\right],$$

for every $\mathsf{S} \in \mho_1, \lambda \in \pitchfork$, *and* $\underbrace{\mathcal{T}_i}_{i=1,\ldots,n} > 0.$

Proof It follows from Theorem 18.2 by putting

$$\operatorname{diag}\Big[\varphi_1\Big(\mathsf{S}, \mathsf{R}, \mathsf{A}, \mathcal{T}_1\Big), \ldots, \varphi_n\Big(\mathsf{S}, \mathsf{R}, \mathsf{A}, \mathcal{T}_n\Big)\Big]$$
$$:= \operatorname{diag}\left[\varphi_1^{\circledS}\left(-\frac{\mathsf{W}_1(\|\mathsf{S}\|^{\mathsf{M}_1} + \|\mathsf{R}\|^{\mathsf{M}_1} + \|\mathsf{A}\|^{\mathsf{M}_1})}{\mathcal{T}_1}\right), \ldots,\right.$$
$$\left.\varphi_n^{\circledS}\left(-\frac{\mathsf{W}_n(\|\mathsf{S}\|^{\mathsf{M}_n} + \|\mathsf{R}\|^{\mathsf{M}_n} + \|\mathsf{A}\|^{\mathsf{M}_n})}{\mathcal{T}_n}\right)\right],$$

for every $\mathsf{S} \in \mho_1, \lambda \in \pitchfork$, and $\underbrace{\mathcal{T}_i}_{i=1,\ldots,n} > 0$, and $\underbrace{\mathfrak{P}_i}_{i=1,\ldots,n} = 2^{\overbrace{\mathsf{M}_i}^{i=1,\ldots,n} -1}$.

We refer the interested readers to [3–13].

References

1. Rezaei Aderyani S, Saadati R, Rassias TM, Park C (2022) Best approximation of $(\mathcal{G}_1, \mathcal{G}_2)$-random operator inequality in matrix Menger Banach algebras with application of stochastic Mittag-Leffler and H $\mathbb{H}$-Fox control functions. J Inequal Appl 2022(1):1–17
2. Mihet D, Radu V (2009) On the stability of the additive Cauchy functional equation in random normed spaces. J Math Anal Appl 343:567–572
3. Aderyani SR, Saadati R, Mesiar R (2022) Estimation of permuting tri-homomorphisms and permuting tri-derivations associated with the tri-additive Υ-random operator inequality in matrix MB-algebra. Int J Gen Syst 51(6):547–569
4. Aderyani SR, Saadati R (2021) Best approximations of the ϕ-Hadamard fractional Volterra integro-differential equation by matrix valued fuzzy control functions. Adv Differ Equ 154
5. Aderyani SR, Saadati R, Yang XJ (2021) Radu-Miheţ method for UHML stability for a class of ξ-Hilfer fractional differential equations in matrix valued fuzzy Banach spaces. Math Methods Appl Sci 44(18):14619–14631
6. Aderyani SR, Saadati R, Allahviranloo T (2022) Existence, uniqueness and matrix-valued fuzzy Mittag-Leffler-Hypergeometric-Wright stability for P-Hilfer fractional differential equations in matrix-valued fuzzy Banach space. Comput Appl Math 41(6):234
7. Aderyani SR, Saadati R (2022) Approximation of derivation–homomorphism fuzzy functional inequalities in matrix valued FC-$\diamond$-algebras. In: 2022 9th Iranian joint congress on fuzzy and intelligent systems (CFIS). IEEE, pp 1–6
8. Aderyani SR, Saadati R, Allahviranloo T, Abbasbandy S, Catak M (2023) Fuzzy approximation of a fractional Lorenz system and a fractional financial crisis. Iran J Fuzzy Syst 20(7):27–36

9. Aderyani SR, Saadati R, Abdeljawad T, Mlaiki N (2022) Multi-stability of non homogenous vector-valued fractional differential equations in matrix-valued Menger spaces. Alexandria Eng J 61(12):10913–10923
10. Aderyani SR, Saadati R (2023) Stability and controllability results by n-ary aggregation functions in matrix valued fuzzy n-normed spaces. Inf Sci 119265
11. O'Regan D, Aderyani SR, Saadati R, Li C (2023) Stability results and parametric delayed Mittag-Leffler matrices in symmetric fuzzy-random spaces with application. Symmetry 15(10):1880
12. Aderyani SR, Saadati R, O'Regan D, Li C (2023) On a new approach for stability and controllability analysis of functional equations. Mathematics 11(16):3458
13. Aderyani SR, Saadati R, O'Regan D, Alshammari FS (2023) Fuzzy approximate solutions of matrix-valued fractional differential equations by fuzzy control functions. Mathematics 11(6):1386

Chapter 19
Multi Stability of Fractional–Order Volterra Equations in MVFB–Spaces

In this chapter, we investigate the multi stabilty results for nonlinear ϕ–Hadamard fractional Volterra integro–differential equations in MVFB–spaces.

19.1 Multi Stability of Hadamard Fractional Volterra Integro–Differential Equations in MVFB–Spaces

In this section, we investigate the multi stability of fractional Volterra integro–differential equations in Hadamard sense by the fixed point method in MVFB–spaces $(\mathcal{W}, \mathscr{N}, \bigotimes_{\mathrm{GTN}})$.

Consider the following ϕ–Hadamard fractional Volterra integro–differential equation [1] defined by

$$ {}^{HC}_{0^+}D^{\kappa,\vartheta}_{\phi(\varsigma)}\eta(\varsigma) = \mu(\varsigma, \eta(\varsigma)) + \int_0^{\varsigma} \mathcal{K}(\varsigma, \iota, \eta(\iota))d\iota, \tag{19.1} $$

where $\kappa \in (0, 1)$, $\vartheta \in [0, 1]$ and $\mu : [0, \mathsf{T}] \times \mathcal{W} \to \mathcal{W}$, $\mathcal{K} : [0, \mathsf{T}] \times [0, \mathsf{T}] \times \mathcal{W} \to \mathcal{W}$ and $\mathsf{T} > 0$.

Definition 19.1 The fractional Volterra equation (19.1) is said to be the multi stable, if $\eta(\varsigma)$ is a given differentiable function, satisfying the inequality below

$$ \mathscr{N}\left({}^{HC}_{0^+}D^{\kappa,\vartheta}_{\phi(\varsigma)}\eta(\varsigma) - \mu(\varsigma, \eta(\varsigma)) - \int_0^{\varsigma} \mathcal{K}(\varsigma, \iota, \eta(\iota))d\iota, (\mathcal{T}_1, \dots, \mathcal{T}_n)\right) $$
$$ \succeq \mathrm{diag}\Big[\varphi_1(\varsigma, \mathcal{T}_1), \dots, \varphi_n(\varsigma, \mathcal{T}_n)\Big], $$

S. Rezaei Aderyani et al., *Towards Ulam Type Multi Stability Analysis*, Studies in Systems, Decision and Control 523, https://doi.org/10.1007/978-3-031-55564-0_19

for evry $\varsigma \in [0,1]$, $n \in \mathbb{N}$, $\underbrace{\mathcal{T}_i}_{i=1,\dots,n} > 0$, and all fuzzy sets $\underbrace{\varphi_i}_{i=1,\dots,n} : \mathcal{W} \times (0,+\infty) \to (0,1]$, and we can find a solution $\eta'(\varsigma)$ of equation (19.1) such that for some $\partial > 0$,

$$\mathcal{N}\Big(\eta(\varsigma) - \eta'(\varsigma), (\mathcal{T}_1, \dots, \mathcal{T}_n)\Big) \succeq \mathrm{diag}\left[\varphi_1\left(\varsigma, \frac{\mathcal{T}_1}{\partial}\right), \dots, \varphi_n\left(\varsigma, \frac{\mathcal{T}_n}{\partial}\right)\right].$$

Consider the MVFB–space $(\mathcal{W}, \mathcal{N}, \bigotimes_{\mathrm{GTN}})$ and the fuzzy sets $\underbrace{\varphi_i}_{i=1,\dots,n} : \mathcal{W} \times (0,+\infty) \to (0,1]$. We set

$$\Phi := \{\eta : [0, \mathrm{T}] \to \mathcal{W}, \eta \text{ is differentiable}\},$$

and define a mapping δ from $\Phi \times \Phi$ to $[0,+\infty]^n$ by

$$\delta(\eta,\eta') = \inf\Bigg\{(\lambda_1,\dots,\lambda_n) \in \mathbb{R}_+^n : \mathcal{N}\Big({}_{0^+}^{HC}D_{\phi(\varsigma)}^{\kappa,\vartheta}\eta(\varsigma) - {}_{0^+}^{HC}D_{\phi(\varsigma)}^{\kappa,\vartheta}\eta'(\varsigma), (\mathcal{T}_1,\dots,\mathcal{T}_n)\Big) \bigotimes_{\mathrm{GTN}} \mathcal{N}\Big(\eta(\varsigma) - \eta'(\varsigma), (\mathcal{T}_1,\dots,\mathcal{T}_n)\Big) \succeq \mathrm{diag}\left[\varphi_1\left(\varsigma, \frac{\mathcal{T}_1}{\lambda_1}\right), \dots, \varphi_n\left(\varsigma, \frac{\mathcal{T}_n}{\lambda_n}\right)\right], \eta, \eta' \in \Phi,\ \varsigma \in [0,\mathrm{T}],\ (\mathcal{T}_1,\dots,\mathcal{T}_n) \in (0,+\infty)^n\Bigg\}.$$

Theorem 19.1 (Φ, δ) *is a complete* $[0,+\infty]^n$*–valued metric space.*

Proof First, we show that (Φ, δ) is a $[0,+\infty]^n$–valued metric space.

We show that $\delta(\eta,\eta') = \underbrace{(0,\dots,0)}_{n}$ if and only if $\eta = \eta'$. Let $\delta(\eta,\eta') = \underbrace{(0,\dots,0)}_{n}$. Then, we have

$$\inf\Bigg\{(\lambda_1,\dots,\lambda_n) \in (0,+\infty)^n : \mathcal{N}\Big({}_{0^+}^{HC}D_{\phi(\varsigma)}^{\kappa,\vartheta}\eta(\varsigma) - {}_{0^+}^{HC}D_{\phi(\varsigma)}^{\kappa,\vartheta}\eta'(\varsigma), (\mathcal{T}_1,\dots,\mathcal{T}_n)\Big) \bigotimes_{\mathrm{GTN}} \mathcal{N}\Big(\eta(\varsigma) - \eta'(\varsigma), (\mathcal{T}_1,\dots,\mathcal{T}_n)\Big) \succeq \mathrm{diag}\left[\varphi_1\left(\varsigma, \frac{\mathcal{T}_1}{\lambda_1}\right), \dots, \varphi_n\left(\varsigma, \frac{\mathcal{T}_n}{\lambda_n}\right)\right],$$

$$\forall\, \eta, \eta' \in \Phi,\ \varsigma \in [0, \mathsf{T}],\ (\mathcal{T}_1, \ldots, \mathcal{T}_n) \in (0, +\infty)^n \Big\} = \underbrace{(0, \ldots, 0)}_{n},$$

and so

$$\mathscr{N}\left({}^{HC}_{0^+}D^{\kappa,\vartheta}_{\phi(\varsigma)}\eta(\varsigma) - {}^{HC}_{0^+}D^{\kappa,\vartheta}_{\phi(\varsigma)}\eta'(\varsigma), (\mathcal{T}_1, \ldots, \mathcal{T}_n)\right) \bigotimes_{\text{GTN}} \mathscr{N}\left(\eta(\varsigma) - \eta'(\varsigma), (\mathcal{T}_1, \ldots, \mathcal{T}_n)\right) \succeq \operatorname{diag}\left[\varphi_1\left(\varsigma, \frac{\mathcal{T}_1}{\lambda_1}\right), \ldots, \varphi_n\left(\varsigma, \frac{\mathcal{T}_n}{\lambda_1}\right)\right],$$

for every $\underbrace{\lambda_i}_{i=1,\ldots,n} \in (0, +\infty)$ and $\underbrace{\mathcal{T}_i}_{i=1,\ldots,n} > 0$. Tending $\underbrace{\lambda_i}_{i=1,\ldots,n}$ to zero in the above inequality, we get

$$\mathscr{N}\left({}^{HC}_{0^+}D^{\kappa,\vartheta}_{\phi(\varsigma)}\eta(\varsigma) - {}^{HC}_{0^+}D^{\kappa,\vartheta}_{\phi(\varsigma)}\eta'(\varsigma), (\mathcal{T}_1, \ldots, \mathcal{T}_n)\right) \bigotimes_{\text{GTN}} \mathscr{N}\left(\eta(\varsigma) - \eta'(\varsigma), (\mathcal{T}_1, \ldots, \mathcal{T}_n)\right) = \mathbf{1}$$

and so

$$\mathscr{N}\left(\eta(\varsigma) - \eta'(\varsigma), (\mathcal{T}_1, \ldots, \mathcal{T}_n)\right) = \mathbf{1},$$

thus, $\eta(\varsigma) = \eta'(\varsigma)$, for every $\varsigma \in [0, \mathsf{T}]$, and vice versa. It is straight forward to show $\delta(\eta, \eta') = \delta(\eta', \eta)$, for every $\eta, \eta' \in \Phi$. Now, let $\delta(\eta, \rho) = (\ell_1, \ldots, \ell_n) \in (0, +\infty)^n$ and $\delta(\rho, \eta') = (\jmath_1, \ldots, \jmath_n) \in (0, +\infty)^n$. Then, we have

$$\mathscr{N}\left({}^{HC}_{0^+}D^{\kappa,\vartheta}_{\phi(\varsigma)}\eta(\varsigma) - {}^{HC}_{0^+}D^{\kappa,\vartheta}_{\phi(\varsigma)}\rho(\varsigma), (\mathcal{T}_1, \ldots, \mathcal{T}_n)\right) \bigotimes_{\text{GTN}} \mathscr{N}\left(\eta(\varsigma) - \rho(\varsigma), (\mathcal{T}_1, \ldots, \mathcal{T}_n)\right) \succeq \operatorname{diag}\left[\varphi_1\left(\varsigma, \frac{\mathcal{T}_1}{\ell_1}\right), \ldots, \varphi_n\left(\varsigma, \frac{\mathcal{T}_n}{\ell_n}\right)\right],$$

and

$$\mathscr{N}\left({}^{HC}_{0^+}D^{\kappa,\vartheta}_{\phi(\varsigma)}\rho(\varsigma) - {}^{HC}_{0^+}D^{\kappa,\vartheta}_{\phi(\varsigma)}\eta'(\varsigma), (\mathcal{T}_1, \ldots, \mathcal{T}_n)\right) \bigotimes_{\text{GTN}} \mathscr{N}\left(\rho(\varsigma) - \eta'(\varsigma), (\mathcal{T}_1, \ldots, \mathcal{T}_n)\right) \succeq \operatorname{diag}\left[\varphi_1\left(\varsigma, \frac{\mathcal{T}_1}{\jmath_1}\right), \ldots, \varphi_n\left(\varsigma, \frac{\mathcal{T}_n}{\jmath_n}\right)\right].$$

Then, we have that

$$\begin{aligned}
&\mathcal{N}\left({}^{HC}_{0^+}D^{\kappa,\vartheta}_{\phi(\varsigma)}\eta(\varsigma) - {}^{HC}_{0^+}D^{\kappa,\vartheta}_{\phi(\varsigma)}\eta'(\varsigma), \left((\ell_1 + \jmath_1)\mathcal{T}_1, \dots, (\ell_n + \jmath_n)\mathcal{T}_n\right)\right)\\
&\quad \bigotimes_{\text{GTN}} \mathcal{N}\left(\eta(\varsigma) - \eta'(\varsigma), \left((\ell_1 + \jmath_1)\mathcal{T}_1, \dots, (\ell_n + \jmath_n)\mathcal{T}_n\right)\right)\\
&\succeq \Big[\mathcal{N}\left({}^{HC}_{0^+}D^{\kappa,\vartheta}_{\phi(\varsigma)}\eta(\varsigma) - {}^{HC}_{0^+}D^{\kappa,\vartheta}_{\phi(\varsigma)}\rho(\varsigma), (\ell_1\mathcal{T}_1, \dots, \ell_n\mathcal{T}_n)\right)\\
&\quad \bigotimes_{\text{GTN}} \mathcal{N}\left({}^{HC}_{0^+}D^{\kappa,\vartheta}_{\phi(\varsigma)}\rho(\varsigma) - {}^{HC}_{0^+}D^{\kappa,\vartheta}_{\phi(\varsigma)}\eta'(\varsigma), (\jmath_1\mathcal{T}_1, \dots, \jmath_n\mathcal{T}_n)\right)\Big]\\
&\bigotimes_{\text{GTN}} \Big[\mathcal{N}\left(\eta(\varsigma) - \rho(\varsigma), (\ell_1\mathcal{T}_1, \dots, \ell_n\mathcal{T}_n)\right) \bigotimes_{\text{GTN}} \mathcal{N}\left(\rho(\varsigma) - \eta'(\varsigma), (\jmath_1\mathcal{T}_1, \dots, \jmath_n\mathcal{T}_n)\right)\Big]\\
&\succeq \text{diag}\Big[\varphi_1(\varsigma, \mathcal{T}_1), \dots, \varphi_n(\varsigma, \mathcal{T}_n)\Big] \bigotimes_{\text{GTN}} \text{diag}\Big[\varphi_1(\varsigma, \mathcal{T}_1), \dots, \varphi_n(\varsigma, \mathcal{T}_n)\Big]\\
&= \text{diag}\Big[\varphi_1(\varsigma, \mathcal{T}_1), \dots, \varphi_n(\varsigma, \mathcal{T}_n)\Big],
\end{aligned}$$

and so $\delta(\eta, \eta') \leq (\ell_1 + \jmath_1, \dots, \ell_n + \jmath_n)$. Thus, $\delta(\eta, \eta') \leq \delta(\eta, \rho) + \delta(\rho, \eta')$. Now, we are ready to prove (Φ, δ) is complete. Let $\{\eta_k\}_k$ be a Cauchy sequence in (Φ, δ). Let $\varsigma \in [0, \mathsf{T}]$ be fixed. Assume that $(\sigma_1, \dots, \sigma_n) \in (0, +\infty)^n$ and $\underbrace{\Omega_i}_{i=1,\dots,n} \in (0, 1)$ are arbitrary and consider $(\mathcal{T}_1, \dots, \mathcal{T}_n) \in (0, +\infty)^n$, such that $\underbrace{\varphi_i}_{i=1,\dots,n}(\varsigma, \underbrace{\mathcal{T}_i}_{i=1,\dots,n}) \succ 1 - \underbrace{\Omega_i}_{i=1,\dots,n}$. For $\underbrace{\varepsilon_i\mathcal{T}_i}_{i=1,\dots,n} < \underbrace{\sigma_i}_{i=1,\dots,n}$ choose $k'' \in \mathbb{N}$ such that

$$\delta(\eta_k, \eta_{k'}) < (\varepsilon_1, \dots, \varepsilon_n), \quad \forall\, k, k' \geq k''.$$

Then, we gat

$$\begin{aligned}
&\mathcal{N}\left({}^{HC}_{0^+}D^{\kappa,\vartheta}_{\phi(\varsigma)}\eta_k(\varsigma) - {}^{HC}_{0^+}D^{\kappa,\vartheta}_{\phi(\varsigma)}\eta_{k'}(\varsigma), (\sigma_1, \dots, \sigma_n)\right)\\
&\quad \bigotimes_{\text{GTN}} \mathcal{N}\left(\eta(\varsigma) - \eta'(\varsigma), (\sigma_1, \dots, \sigma_n)\right)\\
&\succeq \mathcal{N}\left({}^{HC}_{0^+}D^{\kappa,\vartheta}_{\phi(\varsigma)}\eta_k(\varsigma) - {}^{HC}_{0^+}D^{\kappa,\vartheta}_{\phi(\varsigma)}\eta_{k'}(\varsigma), (\varepsilon_1\mathcal{T}_1, \dots, \varepsilon_n\mathcal{T}_n)\right)\\
&\quad \bigotimes_{\text{GTN}} \mathcal{N}\left(\eta(\varsigma) - \eta'(\varsigma), (\varepsilon_1\mathcal{T}_1, \dots, \varepsilon_n\mathcal{T}_n)\right)\\
&\succeq \text{diag}\Big[\varphi_1(\varsigma, \mathcal{T}_1), \dots, \varphi_n(\varsigma, \mathcal{T}_n)\Big]\\
&\succ \text{diag}[1 - \Omega_1, \dots, 1 - \Omega_n].
\end{aligned}$$

Then, we have that

$$\mathcal{N}\left({}^{HC}_{0^+}D^{\kappa,\vartheta}_{\phi(\varsigma)}\eta_k(\varsigma) - {}^{HC}_{0^+}D^{\kappa,\vartheta}_{\phi(\varsigma)}\eta_{k'}(\varsigma), (\sigma_1, \ldots, \sigma_n)\right)$$
$$\succ \operatorname{diag}[1 - \Omega_1, \ldots, 1 - \Omega_n],$$

and

$$\mathcal{N}\left(\eta_k(\varsigma) - \eta_{k'}(\varsigma), (\sigma_1, \ldots, \sigma_n)\right)$$
$$\succ \operatorname{diag}[1 - \Omega_1, \ldots, 1 - \Omega_n],$$

i.e., the sequences both $\{\eta_k(\varsigma)\}_k$ and $\{{}^{HC}_{0^+}D^{\kappa,\vartheta}_{\phi(\varsigma)}\eta_k(\varsigma)\}_k$ are Cauchy in complete space $(\mathcal{W}, \mathcal{N}, \bigotimes_{\text{GTN}})$ on compact set $[0, \mathsf{T}]$, so they are uniformly convergent to the mapping $\eta : [0, \mathsf{T}] \to \mathcal{W}$ and ${}^{HC}_{0^+}D^{\kappa,\vartheta}_{\phi(\varsigma)}\eta$ respectively. The uniform convergence leads us to the fact that η is differentiable, i.e., an element of Φ and then (Φ, δ) is complete.

Now, we are ready to study the multi stability of the ϕ–Hadamard fractional Volterra integro–differential equation (19.1).

Theorem 19.2 *Let* $(\mathcal{W}, \mathcal{N}, \bigotimes_{\text{GTN}})$ *be an MVFB–space and* $\Theta_1, \Theta_2, \Theta_3, \Theta_4$ *and* T *be positive constants such that* $\max[\Theta_1, \Theta_2\Theta_3, \Theta_1\Theta_4, \Theta_2\Theta_3\Theta_4] < 0.5$. *Assume that the continuous mappings* $\mu : [0, \mathsf{T}] \times \mathcal{W} \to \mathcal{W}$, $\mathcal{K} : [0, \mathsf{T}] \times [0, \mathsf{T}] \times \mathcal{W} \to \mathcal{W}$ *with fuzzy sets* $\underbrace{\varphi_i}_{i=1,\ldots,n} : \mathcal{W} \times (0, +\infty) \to (0, 1]$ *satisfy*

$$\mathcal{N}\left(\mu(\varsigma, \eta(\varsigma)) - \mu(\varsigma, \eta'(\varsigma)), (\mathcal{T}_1, \ldots, \mathcal{T}_n)\right) \tag{19.2}$$
$$\succeq \mathcal{N}\left(\eta(\varsigma) - \eta'(\varsigma), \left(\frac{\mathcal{T}_1}{\Theta_1}, \ldots, \frac{\mathcal{T}_n}{\Theta_1}\right)\right),$$

$$\mathcal{N}\left(\mathcal{K}(\varsigma, \iota, \eta(\iota)) - \mathcal{K}(\varsigma, \iota, \eta'(\iota)), (\mathcal{T}_1, \ldots, \mathcal{T}_n)\right) \tag{19.3}$$
$$\succeq \mathcal{N}\left(\eta(\iota) - \eta'(\iota), \left(\frac{\mathcal{T}_1}{\Theta_2}, \ldots, \frac{\mathcal{T}_n}{\Theta_2}\right)\right), \quad \iota \le \varsigma,$$

$$\inf_{\mathfrak{F} \in [0, \mathsf{T}]} \underbrace{\varphi_i}_{i=1,\ldots,n} (\mathfrak{F}, \underbrace{\mathcal{T}_i}_{i=1,\ldots,n}) \ge \underbrace{\varphi_i}_{i=1,\ldots,n} \left(\varsigma, \frac{\mathsf{T}\ \overbrace{\mathcal{T}_i}^{i=1,\ldots,n}}{\Theta_3}\right), \tag{19.4}$$

and

$$\mathcal{N}(\eta(\varsigma), (\mathcal{T}_1, \ldots, \mathcal{T}_n)) \succeq diag[\varphi_1(\varsigma, \mathcal{T}_1), \ldots, \varphi_n(\varsigma, \mathcal{T}_n)], \tag{19.5}$$
$$\text{implies that } \mathcal{N}\left({}^{H}_{0^+}I^{\kappa,\vartheta}_{\phi(\varsigma)}\eta(\iota), (\mathcal{T}_1, \ldots, \mathcal{T}_n)\right) \succeq \mathcal{N}\left(\varsigma, \left(\frac{\mathcal{T}_1}{\Theta_4}, \ldots, \frac{\mathcal{T}_n}{\Theta_4}\right)\right),$$

for every $\varsigma \in [0, \mathrm{T}]$, $\eta, \eta' : [0, \mathrm{T}] \to \mathcal{W}$ $\underbrace{\mathcal{T}_i}_{i=1,\ldots,n} \in (0, +\infty)$, *and* $n \in \mathbb{N}$. *Let* $\gamma : [0, \mathrm{T}] \to \mathcal{W}$ *be a differentiable function satisfying*

$$\mathcal{N}\left({}^{HC}_{0^+}D^{\kappa,\vartheta}_{\phi(\varsigma)}\gamma(\varsigma) - \mu(\varsigma, \gamma(\varsigma)) - \int_0^{\varsigma} \mathcal{K}(\varsigma, \iota, \gamma(\iota))d\iota, (\mathcal{T}_1, \ldots, \mathcal{T}_n)\right) \tag{19.6}$$
$$\succeq diag[\varphi_1(\varsigma, \mathcal{T}_1), \ldots, \varphi_n(\varsigma, \mathcal{T}_n)],$$

for every $\varsigma \in [0, \mathrm{T}]$ *and* $\underbrace{\mathcal{T}_i}_{i=1,\ldots,n} \in (0, +\infty)$. *Then, we have a unique differentiable function* $\gamma_0 : [0, \mathrm{T}] \to \mathcal{W}$ *such that*

$${}^{HC}_{0^+}D^{\kappa,\vartheta}_{\phi(\varsigma)}\gamma_0(\varsigma) = \mu(\varsigma, \gamma_0(\varsigma)) + \int_0^{\varsigma} \mathcal{K}(\varsigma, \iota, \gamma(\iota))d\iota, \tag{19.7}$$

and

$$\mathcal{N}\left({}^{HC}_{0^+}D^{\kappa,\vartheta}_{\phi(\varsigma)}\gamma(\varsigma) - {}^{HC}_{0^+}D^{\kappa,\vartheta}_{\phi(\varsigma)}\gamma_0(\varsigma), (\mathcal{T}_1, \ldots, \mathcal{T}_n)\right) \tag{19.8}$$
$$\bigotimes_{\mathrm{GTN}} \mathcal{N}\left(\gamma(\varsigma) - \gamma_0(\varsigma), (\mathcal{T}_1, \ldots, \mathcal{T}_n)\right)$$
$$\succeq diag\left[\varphi_1\left(\varsigma, \frac{(1 - 2\max[\Theta_1, \Theta_2\Theta_3, \Theta_1\Theta_4, \Theta_2\Theta_3\Theta_4])\mathcal{T}_1}{\max[1, \Theta_4]}\right), \ldots, \right.$$
$$\left. \varphi_n\left(\varsigma, \frac{(1 - 2\max[\Theta_1, \Theta_2\Theta_3, \Theta_1\Theta_4, \Theta_2\Theta_3\Theta_4])\mathcal{T}_n}{\max[1, \Theta_4]}\right)\right],$$

for every $\varsigma \in [0, \mathrm{T}]$ *and* $\underbrace{\mathcal{T}_i}_{i=1,\ldots,n} \in (0, +\infty)$.

Proof We set

$$\Phi := \{\eta : [0, \mathrm{T}] \to \mathcal{W}, \eta \text{ is differentiable}\},$$

and introduce the $[0, +\infty]^n$–valued metric on Φ as,

$$\delta(\eta,\eta')=\inf\Bigg\{(\lambda_1,\ldots,\lambda_n)\in(0,+\infty)^n: \mathcal{N}\left({}_{0^+}^{HC}D_{\phi(\varsigma)}^{\kappa,\vartheta}\eta(\varsigma)-{}_{0^+}^{HC}D_{\phi(\varsigma)}^{\kappa,\vartheta}\eta'(\varsigma),(\mathcal{T}_1,\ldots,\mathcal{T}_n)\right) \bigotimes_{\mathrm{GTN}}\mathcal{N}\left(\eta(\varsigma)-\eta'(\varsigma),(\mathcal{T}_1,\ldots,\mathcal{T}_n)\right) \succeq \operatorname{diag}\left[\varphi_1\left(\varsigma,\frac{\mathcal{T}_1}{\lambda_1}\right),\ldots,\varphi_n\left(\varsigma,\frac{\mathcal{T}_n}{\lambda_n}\right)\right], \eta,\eta'\in\Phi,\varsigma\in[0,\mathsf{T}],\underbrace{\mathcal{T}_i}_{i=1,\ldots,n}\in(0,+\infty),n\in\mathbb{N}\Bigg\}.$$

By Theorem 19.1, we have (Φ,δ) is a complete $[0,+\infty]^n$–valued metric space.

Now, we define the mapping $\propto$ from Φ to Φ by

$$\propto(\eta(\varsigma))={}_{0^+}^{H}I_{\phi(\varsigma)}^{\kappa,\vartheta}(\mu(\iota,\eta(\iota)))+{}_{0^+}^{H}I_{\phi(\varsigma)}^{\kappa,\vartheta}\left(\int_0^{\varsigma}\mathcal{K}(\iota,\varpi,\eta(\varpi))d\varpi\right), \tag{19.9}$$

where $\kappa\in(0,1)$, $\vartheta\in[0,1]$, $\mu:[0,\mathsf{T}]\times\mathcal{W}\to\mathcal{W}$, $\mathcal{K}:[0,\mathsf{T}]\times[0,\mathsf{T}]\times\mathcal{W}\to\mathcal{W}$. We prove $\propto$ is a strictly contractive mapping. Let $\eta,\eta'\in\Phi$, $\underbrace{\epsilon_i}_{i=1,\ldots,n}\in(0,+\infty)$ and $\delta(\eta,\eta')<(\epsilon_1,\ldots,\epsilon_n)$, then, we have

$$\mathcal{N}\left({}_{0^+}^{HC}D_{\phi(\varsigma)}^{\kappa,\vartheta}\eta(\varsigma)-{}_{0^+}^{HC}D_{\phi(\varsigma)}^{\kappa,\vartheta}\eta'(\varsigma),(\epsilon_1\mathcal{T}_1,\ldots,\epsilon_n\mathcal{T}_n)\right) \bigotimes_{\mathrm{GTN}}\mathcal{N}\left(\eta(\varsigma)-\eta'(\varsigma),(\epsilon_1\mathcal{T}_1,\ldots,\epsilon_n\mathcal{T}_n)\right) \succeq \operatorname{diag}\left[\varphi_1\left(\varsigma,\mathcal{T}_1\right),\ldots,\varphi_n\left(\varsigma,\mathcal{T}_n\right)\right],\quad \eta,\eta'\in\Phi,\ \varsigma\in[0,\mathsf{T}],\ \underbrace{\mathcal{T}_i}_{i=1,\ldots,n}\in(0,+\infty).$$

Using properties (D-2) and (D-3) of Definition 15.5 and (19.9), we have

$$\begin{aligned}&\mathcal{N}\left(\left({}_{0^+}^{HC}D_{\phi(\varsigma)}^{\kappa,\vartheta}\propto(\eta(\varsigma))\right)-\left({}_{0^+}^{HC}D_{\phi(\varsigma)}^{\kappa,\vartheta}\propto(\eta'(\varsigma))\right),(2\epsilon_1\mathcal{T}_1,\ldots,2\epsilon_n\mathcal{T}_n)\right)\\ &\quad\bigotimes_{\mathrm{GTN}}\mathcal{N}\left(\propto(\eta(\varsigma))-\propto(\eta'(\varsigma)),(2\epsilon_1\mathcal{T}_1,\ldots,2\epsilon_n\mathcal{T}_n)\right)\\ &=\mathcal{N}\Bigg(\left[\mu(\varsigma,\eta(\varsigma))-\mu(\varsigma,\eta'(\varsigma))\right]+\\ &\qquad\int_0^{\varsigma}\left[\mathcal{K}(\varsigma,\iota,\eta(\iota))-\mathcal{K}(\varsigma,\iota,\eta'(\iota))\right]d\iota,(2\epsilon_1\mathcal{T}_1,\ldots,2\epsilon_n\mathcal{T}_n)\Bigg)\\ &\quad\bigotimes_{\mathrm{GTN}}\mathcal{N}\Bigg({}_{0^+}^{H}I_{\phi(\varsigma)}^{\kappa,\vartheta}\left(\mu(\varsigma,\eta(\varsigma))-\mu(\varsigma,\eta'(\varsigma))\right)\\ &\qquad+{}_{0^+}^{H}I_{\phi(\varsigma)}^{\kappa,\vartheta}\left(\int_0^{s}\left[\mathcal{K}(\varsigma,\iota,\eta(\iota))-\mathcal{K}(\varsigma,\iota,\eta'(\iota))\right]d\iota\right),(2\epsilon_1\mathcal{T}_1,\ldots,2\epsilon_n\mathcal{T}_n)\Bigg)\end{aligned} \tag{19.10}$$

$$
\begin{aligned}
&\succeq \mathscr{N}\Big(\mu(\varsigma,\eta(\varsigma)) - \mu(\varsigma,\eta'(\varsigma)), (\epsilon_1\mathcal{T}_1,\ldots,\epsilon_n\mathcal{T}_n)\Big)\\
&\quad\bigotimes_{\mathrm{GTN}} \mathscr{N}\Big(\int_0^{\varsigma}\Big[\mathcal{K}(\varsigma,\iota,\eta(\iota)) - \mathcal{K}(\varsigma,\iota,\eta'(\iota))\Big]d\iota, (\epsilon_1\mathcal{T}_1,\ldots,\epsilon_n\mathcal{T}_n)\Big)\\
&\quad\bigotimes_{\mathrm{GTN}} \mathscr{N}\Big({}^{H}_{0+}I^{\kappa,\vartheta}_{\phi(\varsigma)}\Big(\mu(\iota,\eta(\iota)) - \mu(\iota,\eta'(\iota))\Big), (\epsilon_1\mathcal{T}_1,\ldots,\epsilon_n\mathcal{T}_n)\Big)\\
&\quad\bigotimes_{\mathrm{GTN}} \mathscr{N}\Big({}^{H}_{0+}I^{\kappa,\vartheta}_{\phi(\varsigma)}\Big(\int_0^{\iota}\Big[\mathcal{K}(\iota,\varsigma,\eta(\varsigma)) - \mathcal{K}(\iota,\varsigma,\eta'(\varsigma))\Big]d\varsigma\Big), (\epsilon_1\mathcal{T}_1,\ldots,\epsilon_n\mathcal{T}_n)\Big).
\end{aligned}
$$

In the last part of (19.10) there are four formulas, in the next steps we work on them to get new formulas derived from the control functions $\underbrace{\varphi_i}_{i=1,\ldots,n}$. Let $0 = \Xi_1 < \Xi_2 < \ldots < \Xi_{\Bbbk} = \varsigma$, $\Delta\Xi_j = \Xi_j - \Xi_{j-1} = \frac{\varsigma}{\Bbbk}$, $j = 1, 2, \ldots, \Bbbk$ and $\|\Delta\Xi\| = \max_{1\le j\le \Bbbk}\left(\Delta\Xi_j\right)$.

Step one. From (19.2) we have

$$
\begin{aligned}
&\mathscr{N}\Big(\mu(\varsigma,\eta(\varsigma)) - \mu(\varsigma,\eta'(\varsigma)), (\epsilon_1\mathcal{T}_1,\ldots,\epsilon_n\mathcal{T}_n)\Big)\\
&\succeq \mathscr{N}\left(\eta(\varsigma) - \eta'(\varsigma), \left(\frac{\epsilon_1\mathcal{T}_1}{\Theta_1},\ldots,\frac{\epsilon_n\mathcal{T}_n}{\Theta_1}\right)\right)\\
&\succeq \operatorname{diag}\left[\varphi_1\left(\varsigma,\frac{\mathcal{T}_1}{\Theta_1}\right),\ldots,\varphi_n\left(\varsigma,\frac{\mathcal{T}_n}{\Theta_1}\right)\right].
\end{aligned}
\tag{19.11}
$$

Step two. Using (D-2) and (D-3) of Definition 15.5, continuity property of matrix fuzzy set $\mathscr{N}$, (19.3) and (19.4), we get

$$
\begin{aligned}
&\mathscr{N}\left(\int_0^{\varsigma}\Big[\mathcal{K}(\varsigma,\iota,\eta(\iota)) - \mathcal{K}(\varsigma,\iota,\eta'(\iota))\Big]d\iota, (\epsilon_1\mathcal{T}_1,\ldots,\epsilon_n\mathcal{T}_n)\right) \qquad (19.12)\\
&= \mathscr{N}\left(\lim_{\|\Delta\Xi\|\to 0}\sum_{j=1}^{\Bbbk}\Big[\mathcal{K}(\varsigma,\Xi_j,\eta(\Xi_j)) - \mathcal{K}(\varsigma,\Xi_j,\eta'(\Xi_j))\Big]\Delta\Xi_i, (\epsilon_1\mathcal{T}_1,\ldots,\epsilon_n\mathcal{T}_n)\right)\\
&= \lim_{\|\Delta\Xi\|\to 0}\mathscr{N}\left(\sum_{j=1}^{\Bbbk}\Big[\mathcal{K}(\varsigma,\Xi_j,\eta(\Xi_j)) - \mathcal{K}(\varsigma,\Xi_j,\upsilon(\Xi_j))\Big]\Delta\Xi_i, (\epsilon_1\mathcal{T}_1,\ldots,\epsilon_n\mathcal{T}_n)\right)\\
&\succeq \lim_{\|\Delta\Xi\|\to 0}\bigwedge_{j=1}^{\Bbbk}\mathscr{N}\left(\Big[\mathcal{K}(\varsigma,\Xi_j,\eta(\Xi_j)) - \mathcal{K}(\varsigma,\Xi_j,\eta'(\Xi_j))\Big]\Delta\Xi_i, \left(\frac{\epsilon_1\mathcal{T}_1}{\Bbbk},\ldots,\frac{\epsilon_n\mathcal{T}_n}{\Bbbk}\right)\right)\\
&\succeq \inf_{\eth\in[0,\mathsf{T}]}\mathscr{N}\left(\mathcal{K}(\varsigma,\eth,\eta(\eth)) - \mathcal{K}(\varsigma,\eth,\eta'(\eth)), \left(\frac{\Bbbk\epsilon_1\mathcal{T}_1}{\Bbbk\mathsf{T}},\ldots,\frac{\Bbbk\epsilon_n\mathcal{T}_n}{\Bbbk\mathsf{T}}\right)\right)\\
&\succeq \inf_{\eth\in[0,\mathsf{T}]}\mathscr{N}\left(\eta(\eth) - \eta'(\eth), \left(\frac{\epsilon_1\mathcal{T}_1}{\Theta_2\mathsf{T}},\ldots,\frac{\epsilon_n\mathcal{T}_n}{\Theta_2\mathsf{T}}\right)\right)
\end{aligned}
$$

$$\succeq \inf_{\eth\in[0,\mathrm{T}]} \operatorname{diag}\left[\varphi_1\left(\eth, \frac{\mathcal{T}_1}{\Theta_2\mathrm{T}}\right), \ldots, \varphi_n\left(\eth, \frac{\mathcal{T}_n}{\Theta_2\mathrm{T}}\right)\right]$$
$$\succeq \operatorname{diag}\left[\varphi_1\left(\varsigma, \frac{\mathcal{T}_1}{\Theta_2\Theta_3}\right), \ldots, \varphi_n\left(\varsigma, \frac{\mathcal{T}_n}{\Theta_2\Theta_3}\right)\right].$$

Step three. Using (19.5) and (19.11), we get

$$\mathscr{N}\left({}^{H}_{0^+}I^{\kappa,\vartheta}_{\phi(\varsigma)}\Big(\mu(\iota,\eta(\iota)) - \mu(\iota,\eta'(\iota))\Big), (\epsilon_1\mathcal{T}_1, \ldots, \epsilon_n\mathcal{T}_n)\right)$$
$$\succeq \operatorname{diag}\left[\varphi_1(\varsigma, \frac{\mathcal{T}_1}{\Theta_1\Theta_4}), \ldots, \varphi_n(\varsigma, \frac{\mathcal{T}_n}{\Theta_1\Theta_4})\right]. \tag{19.13}$$

Step four. Using (19.5) and (19.12), we get

$$\mathscr{N}\left({}^{H}_{0^+}I^{\kappa,\vartheta}_{\varsigma}\left(\int_0^{\iota}\Big[\mathcal{K}(\iota,\varpi,\eta(\varpi)) - \mathcal{K}(\iota,\varpi,\eta'(\varpi))\Big]d\varpi\right), (\epsilon_1\mathcal{T}_1, \ldots, \epsilon_n\mathcal{T}_n)\right)$$
$$\succeq \operatorname{diag}\left[\varphi_1\left(\varsigma, \frac{\mathcal{T}_1}{\Theta_2\Theta_3\Theta_4}\right), \ldots, \varphi_n\left(\varsigma, \frac{\mathcal{T}_n}{\Theta_2\Theta_3\Theta_4}\right)\right]. \tag{19.14}$$

Form (19.10), (19.11), (19.12), (19.13) and (19.14), we have

$$\mathscr{N}\left({}^{HC}_{0^+}D^{\kappa,\vartheta}_{\phi(\varsigma)} \propto (\eta(\varsigma)) - {}^{HC}_{0^+}D^{\kappa,\vartheta}_{\phi(\varsigma)} \propto (\eta'(\varsigma)), (2\epsilon_1\mathcal{T}_1, \ldots, 2\epsilon_n\mathcal{T}_n)\right)$$
$$\bigotimes_{\mathrm{GTN}} \mathscr{N}\left(\propto (\eta(\varsigma)) - \propto (\eta'(\varsigma)), (2\epsilon_1\mathcal{T}_1, \ldots, 2\epsilon_n\mathcal{T}_n)\right)$$
$$\succeq \operatorname{diag}\left[\varphi_1\left(\varsigma, \frac{\mathcal{T}_1}{\Theta_1}\right), \ldots, \varphi_n\left(\varsigma, \frac{\mathcal{T}_n}{\Theta_1}\right)\right]$$
$$\bigotimes_{\mathrm{GTN}} \operatorname{diag}\left[\varphi_1\left(\varsigma, \frac{\mathcal{T}_1}{\Theta_2\Theta_3}\right), \ldots, \varphi_n\left(\varsigma, \frac{\mathcal{T}_n}{\Theta_2\Theta_3}\right)\right]$$
$$\bigotimes_{\mathrm{GTN}} \operatorname{diag}\left[\varphi_1\left(\varsigma, \frac{\mathcal{T}_1}{\Theta_1\Theta_4}\right), \ldots, \varphi_n\left(\varsigma, \frac{\mathcal{T}_n}{\Theta_1\Theta_4}\right)\right]$$
$$\bigotimes_{\mathrm{GTN}} \operatorname{diag}\left[\varphi_1\left(\varsigma, \frac{\mathcal{T}_1}{\Theta_2\Theta_3\Theta_4}\right), \ldots, \varphi_n\left(\varsigma, \frac{\mathcal{T}_n}{\Theta_2\Theta_3\Theta_4}\right)\right]$$
$$\succeq \operatorname{diag}\left[\varphi_1\left(\varsigma, \frac{\mathcal{T}_1}{\max[\Theta_1, \Theta_2\Theta_3, \Theta_1\Theta_4, \Theta_2\Theta_3\Theta_4]}\right), \ldots, \varphi_n\left(\varsigma, \frac{\mathcal{T}_n}{\max[\Theta_1, \Theta_2\Theta_3, \Theta_1\Theta_4, \Theta_2\Theta_3\Theta_4]}\right)\right]. \tag{19.15}$$

and so

$$\begin{aligned}
&\mathcal{N}\left({}^{HC}_{0^+}D^{\kappa,\vartheta}_{\phi(\varsigma)} \propto (\eta(\varsigma)) - {}^{HC}_{0^+}D^{\kappa,\vartheta}_{\phi(\varsigma)} \propto (\eta'(\varsigma)), (\epsilon_1\mathcal{T}_1, \ldots, \epsilon_n\mathcal{T}_n)\right) \\
&\quad \bigotimes_{\text{GTN}} \mathcal{N}\left(\propto (\eta(\varsigma)) - \propto (\eta'(\varsigma)), (\epsilon_1\mathcal{T}_1, \ldots, \epsilon_n\mathcal{T}_n)\right) \qquad (19.16)\\
&\succeq \operatorname{diag}\left[\varphi_1\left(\varsigma, \frac{\mathcal{T}_1}{2\max[\Theta_1, \Theta_2\Theta_3, \Theta_1\Theta_4, \Theta_2\Theta_3\Theta_4]}\right), \ldots, \right.\\
&\qquad\qquad \left.\varphi_n\left(\varsigma, \frac{\mathcal{T}_n}{2\max[\Theta_1, \Theta_2\Theta_3, \Theta_1\Theta_4, \Theta_2\Theta_3\Theta_4]}\right)\right],
\end{aligned}$$

which implies that

$$\begin{aligned}
&\delta(\propto (\eta), \propto (\eta')) \\
&\preceq \left(2\max[\Theta_1, \Theta_2\Theta_3, \Theta_1\Theta_4, \Theta_2\Theta_3\Theta_4]\epsilon_1, \ldots, 2\max[\Theta_1, \Theta_2\Theta_3, \Theta_1\Theta_4, \Theta_2\Theta_3\Theta_4]\epsilon_n\right),
\end{aligned} \tag{19.17}$$

and so

$$\begin{aligned}
&\delta(\propto (\eta), \propto (\eta')) \\
&\preceq \underbrace{\left(2\max[\Theta_1, \Theta_2\Theta_3, \Theta_1\Theta_4, \Theta_2\Theta_3\Theta_4], \ldots, 2\max[\Theta_1, \Theta_2\Theta_3, \Theta_1\Theta_4, \Theta_2\Theta_3\Theta_4]\right)}_{n} \delta(\eta, \eta').
\end{aligned} \tag{19.18}$$

Then, $\propto$ is a strictly contractive mapping with the Lipschitz constant $2\max[\Theta_1, \Theta_2\Theta_3, \Theta_1\Theta_4, \Theta_2\Theta_3\Theta_4]$.

Let $\gamma \in \Phi$, we show that $\delta(\propto (\gamma), \gamma) < \underbrace{(\infty, \ldots, \infty)}_{n}$. Using (19.5) and (19.6) we get

$$\begin{aligned}
&\mathcal{N}\left({}^{HC}_{0^+}D^{\kappa,\vartheta}_{\phi(\varsigma)}\left[\propto (\gamma(\varsigma)) - \gamma(\varsigma)\right], (\mathcal{T}_1, \ldots, \mathcal{T}_n)\right) \\
&\quad \bigotimes_{\text{GTN}} \mathcal{N}\left(\propto (\gamma(\varsigma)) - \gamma(\varsigma), (\mathcal{T}_1, \ldots, \mathcal{T}_n)\right) \\
&= \mathcal{N}\left(\mu(\varsigma, \gamma(\varsigma)) + \int_0^{\varsigma} \mathcal{K}(\varsigma, \iota, \gamma(\iota))d\iota - {}^{HC}_{0^+}D^{\kappa,\vartheta}_{\phi(\varsigma)}\gamma(\varsigma), (\mathcal{T}_1, \ldots, \mathcal{T}_n)\right) \\
&\quad \bigotimes_{\text{GTN}} \mathcal{N}\left({}^{H}_{0^+}I^{\kappa,\vartheta}_{\phi(\varsigma)}(\mu(\iota, \gamma(\iota))) + {}^{H}_{0^+}I^{\kappa,\vartheta}_{\phi(\varsigma)}\left(\int_0^{\iota} \mathcal{K}(\iota, \varpi, \gamma(\varpi))d\varpi\right)\right. \\
&\qquad\qquad \left. - {}^{H}_{0^+}I^{\kappa,\vartheta}_{\phi(\varsigma)}{}^{HC}_{0^+}D^{\kappa,\vartheta}_{\phi(\iota)}\gamma(\iota), (\mathcal{T}_1, \ldots, \mathcal{T}_n)\right)
\end{aligned}$$

$$= \mathcal{N}\left(\mu(\varsigma,\gamma(\varsigma)) + \int_0^{\varsigma} \mathcal{K}(\varsigma,\iota,\gamma(\iota))d\iota - {}_{0^+}^{HC}D_{\phi(\varsigma)}^{\kappa,\vartheta}\gamma(\varsigma), (\mathcal{T}_1,\ldots,\mathcal{T}_n)\right)$$

$$\bigotimes_{\mathrm{GTN}} \mathcal{N}\left({}_{0^+}^{H}I_{\phi(\varsigma)}^{\kappa,\vartheta}\left[\mu(\iota,\gamma(\iota)) + \int_0^{\iota} \mathcal{K}(\iota,\varpi,\gamma(\varpi))d\varpi - {}_{0^+}^{HC}D_{\phi(\varsigma)}^{\kappa,\vartheta}\gamma(\iota)\right], (\mathcal{T}_1,\ldots,\mathcal{T}_n)\right)$$

$$\succeq \mathrm{diag}\left[\varphi_1(\varsigma,\mathcal{T}_1),\ldots,\varphi_n(\varsigma,\mathcal{T}_n)\right] \bigotimes_{\mathrm{GTN}} \mathrm{diag}\left[\varphi_1\left(\varsigma,\frac{\mathcal{T}_1}{\Theta_4}\right),\ldots,\varphi_n\left(\varsigma,\frac{\mathcal{T}_n}{\Theta_4}\right)\right],$$

$$\succeq \mathrm{diag}\left[\varphi_1\left(\varsigma,\frac{\mathcal{T}_1}{\max[1,\Theta_4]}\right),\ldots,\varphi_n\left(\varsigma,\frac{\mathcal{T}_n}{\max[1,\Theta_4]}\right)\right], \tag{19.19}$$

for every $\underbrace{\mathcal{T}_i}_{i=1,\ldots,n} \in (0,+\infty)$. Then we have $\delta(\propto(\gamma),\gamma) < \underbrace{(\max[1,\Theta_4],\ldots,\max[1,\Theta_4])}_{n} < \underbrace{(\infty,\ldots,\infty)}_{n}$.

Now, Theorem 1.8 enables us to find an element γ_0 in Φ satisfying the following:

(1) γ_0 is a fixed point of $\propto$, i.e.,

$$\begin{aligned}\gamma_0(\varsigma) &= \propto(\gamma_0(\varsigma))\\ &= {}_{0^+}^{H}I_{\phi(\varsigma)}^{\kappa,\vartheta}(\mu(\iota,\gamma_0(\iota))) + {}_{0^+}^{H}I_{\phi(\varsigma)}^{\kappa,\vartheta}\left(\int_0^{\iota} \mathcal{K}(\iota,\varpi,\gamma_0(\varpi))d\varpi\right),\end{aligned} \tag{19.20}$$

which is unique in the set

$$\Phi^* = \{\eta \in \Phi : \delta(\propto(\gamma),\eta) < \underbrace{(\infty,\ldots,\infty)}_{n}\}.$$

Taking ${}_{0^+}^{HC}D_{\phi(\varsigma)}^{\kappa,\vartheta}$ from (19.20), we get

$${}_{0^+}^{HC}D_{\phi(\varsigma)}^{\kappa,\vartheta}\gamma_0(\varsigma) = \mu(\varsigma,\gamma_0(\varsigma)) + \int_0^{\varsigma} \mathcal{K}(\varsigma,\iota,\gamma_0(\iota))d\iota, \tag{19.21}$$

where $\kappa \in (0,1)$, $\vartheta \in [0,1]$, $\mu : [0,\mathrm{T}] \times \mathcal{W} \to \mathcal{W}$, $\mathcal{K} : [0,\mathrm{T}] \times [0,\mathrm{T}] \times \mathcal{W} \to \mathcal{W}$.

(2) $\delta(\propto^{\tau}(\gamma),\gamma_0) \to \underbrace{(0,\ldots,0)}_{n}$, as $\tau \to \infty$;

(3) we have

$$\begin{aligned}&\delta(\gamma,\gamma_0)\\ &\leq \Bigg(\frac{1}{1-2\max[\Theta_1,\Theta_2\Theta_3,\Theta_1\Theta_4,\Theta_2\Theta_3\Theta_4]},\ldots,\\ &\qquad \frac{1}{1-2\max[\Theta_1,\Theta_2\Theta_3,\Theta_1\Theta_4,\Theta_2\Theta_3\Theta_4]}\Bigg)\delta(\propto(\gamma),\gamma)\end{aligned}$$

$$\leq \underbrace{\left(\frac{\max[1,\Theta_4]}{1-2\max[\Theta_1,\Theta_2\Theta_3,\Theta_1\Theta_4,\Theta_2\Theta_3\Theta_4]},\ldots,\frac{\max[1,\Theta_4]}{1-2\max[\Theta_1,\Theta_2\Theta_3,\Theta_1\Theta_4,\Theta_2\Theta_3\Theta_4]}\right)}_{n},$$

which implies that

$$\begin{aligned}
&\mathcal{N}\left({}^{HC}_{0^+}D^{\kappa,\vartheta}_{\phi(\varsigma)}\gamma(\varsigma) - {}^{HC}_{0^+}D^{\kappa,\vartheta}_{\phi(\varsigma)}\gamma_0(\varsigma),(\mathcal{T}_1,\ldots,\mathcal{T}_n)\right)\\
&\bigotimes_{\text{GTN}}\mathcal{N}\left(\gamma(\varsigma)-\gamma_0(\varsigma),(\mathcal{T}_1,\ldots,\mathcal{T}_n)\right)\\
&\succeq \operatorname{diag}\Bigg[\varphi_1\left(\varsigma,\frac{\left(1-2\max\left[\Theta_1,\Theta_2\Theta_3,\Theta_1\Theta_4,\Theta_2\Theta_3\Theta_4\right]\right)\mathcal{T}_1}{\max[1,\Theta_4]}\right),\ldots,\\
&\qquad\varphi_n\left(\varsigma,\frac{\left(1-2\max\left[\Theta_1,\Theta_2\Theta_3,\Theta_1\Theta_4,\Theta_2\Theta_3\Theta_4\right]\right)\mathcal{T}_n}{\max[1,\Theta_4]}\right)\Bigg],
\end{aligned} \tag{19.22}$$

for every $\varsigma \in [0,\mathtt{T}]$, $\underbrace{\mathcal{T}_i}_{i=1,\ldots,n} \in (0,+\infty)$ and $n \in \mathbb{N}$.

Now, we show that the fixed point in Φ^* is unique. Let γ_0' be an element of Φ satisfying (19.7) and (19.8), we prove that $\gamma_0' = \gamma_0$ and $\gamma_0' \in \Phi^*$. From (19.7) we get

$${}^{HC}_{0^+}D^{\kappa,\vartheta}_{\phi(\varsigma)}\gamma_0'(\varsigma) = \mu(\varsigma,\gamma_0'(\varsigma)) + \int_0^{\varsigma}\mathcal{K}(\varsigma,\iota,\gamma_0'(\iota))d\iota, \tag{19.23}$$

and so

$$\begin{aligned}
\gamma_0'(\varsigma) &= {}^{H}_{0^+}I^{\kappa,\vartheta}_{\phi(\varsigma)}\mu(\iota,\gamma_0'(\iota)) + {}^{H}_{0^+}I^{\kappa,\vartheta}_{\phi(\varsigma)}\int_0^{\iota}\mathcal{K}(\iota,\varpi,\gamma_0'(\varpi))d\varpi\\
&=\propto(\gamma_0'(\varsigma)),
\end{aligned} \tag{19.24}$$

where $\kappa \in (0,1)$, $\vartheta \in [0,1]$, $\mu : [0,\mathtt{T}] \times \mathcal{W} \to \mathcal{W}$, $\mathcal{K} : [0,\mathtt{T}] \times [0,\mathtt{T}] \times \mathcal{W} \to \mathcal{W}$.

Now, we show that

$$\gamma_0' \in \{\eta \in \Phi : \delta(\propto(\gamma),\eta) < \underbrace{(\infty,\ldots,\infty)}_{n}\},$$

i.e., $\delta(\propto(\gamma),\gamma_0') < \underbrace{(\infty,\ldots,\infty)}_{n}$. We set $\theta = \frac{1-2\max[\Theta_1,\Theta_2\Theta_3,\Theta_1\Theta_4,\Theta_2\Theta_3\Theta_4]}{\max[1,\Theta_4]}$, from (19.8), we get

$$\mathscr{N}\left({}^{HC}_{0^+}D^{\kappa,\vartheta}_{\phi(\varsigma)}\gamma(\varsigma) - {}^{HC}_{0^+}D^{\kappa,\vartheta}_{\phi(\varsigma)}\gamma_0'(\varsigma), (\mathcal{T}_1, \ldots, \mathcal{T}_n)\right)$$
$$\bigotimes_{\text{GTN}} \mathscr{N}\Big(\gamma(\varsigma) - \gamma_0'(\varsigma), (\mathcal{T}_1, \ldots, \mathcal{T}_n)\Big)$$
$$\succeq \operatorname{diag}\Big[\varphi_1(\varsigma, \theta\mathcal{T}_1), \ldots, \varphi_n(\varsigma, \theta\mathcal{T}_n)\Big], \tag{19.25}$$

for every $\varsigma \in [0, \mathrm{T}]$ and $\underbrace{\mathcal{T}_i}_{i=1,\ldots,n} \in (0, +\infty)$.

Form (19.2) and (19.25) we get

$$\mathscr{N}\Big(\mu(\varsigma, \gamma(\varsigma)) - \mu(\varsigma, \gamma_0'(\varsigma)), (\mathcal{T}_1, \ldots, \mathcal{T}_n)\Big)$$
$$\succeq \mathscr{N}\left(\gamma(\varsigma) - \gamma_0'(\varsigma), \left(\frac{\mathcal{T}_1}{\Theta_1}, \ldots, \frac{\mathcal{T}_n}{\Theta_1}\right)\right)$$
$$\succeq \operatorname{diag}\left[\varphi_1\left(\varsigma, \theta\frac{\mathcal{T}_1}{\Theta_1}\right), \ldots, \varphi_n\left(\varsigma, \theta\frac{\mathcal{T}_n}{\Theta_1}\right)\right], \tag{19.26}$$

also, from (19.3) and (19.25) we get

$$\mathscr{N}\Big(\mathcal{K}(\varsigma, \iota, \gamma(\iota)) - \mathcal{K}(\varsigma, \iota, \gamma_0'(\iota)), (\mathcal{T}_1, \ldots, \mathcal{T}_n)\Big)$$
$$\succeq \mathscr{N}\left(\gamma(\iota) - \gamma_0'(\iota), \left(\frac{\mathcal{T}_1}{\Theta_2}, \ldots, \frac{\mathcal{T}_n}{\Theta_2}\right)\right)$$
$$\succeq \operatorname{diag}\left[\varphi_1\left(\varsigma, \theta\frac{\mathcal{T}_1}{\Theta_2}\right), \ldots, \varphi_n\left(\varsigma, \theta\frac{\mathcal{T}_n}{\Theta_2}\right)\right], \tag{19.27}$$

for every $\varsigma \in [0, \mathrm{T}]$, $\iota \leq \varsigma$ and $\underbrace{\mathcal{T}_i}_{i=1,\ldots,n} \in (0, +\infty)$. Now, using step two and (19.27) we get

$$\mathscr{N}\left(\int_0^{\varsigma}\Big[\mathcal{K}(\varsigma, \iota, \gamma(\iota)) - \mathcal{K}(\varsigma, \iota, \gamma_0'(\iota))\Big]d\iota, (\mathcal{T}_1, \ldots, \mathcal{T}_n)\right)$$
$$\succeq \operatorname{diag}\left[\varphi_1\left(\varsigma, \frac{\mathcal{T}_1}{\Theta_2\Theta_3}\right), \ldots, \varphi_n\left(\varsigma, \frac{\mathcal{T}_n}{\Theta_2\Theta_3}\right)\right]$$
$$\succeq \operatorname{diag}\left[\varphi_1\left(\varsigma, \theta\frac{\mathcal{T}_1}{\Theta_2\Theta_3}\right), \ldots, \varphi_n\left(\varsigma, \theta\frac{\mathcal{T}_n}{\Theta_2\Theta_3}\right)\right]. \tag{19.28}$$

Using triangular inequality (D-3), (19.26) and (19.28) we get

$$\mathcal{N}\left(\mu(\varsigma,\gamma(\varsigma))-\mu(\varsigma,\gamma_0'(\varsigma))+\int_0^{\varsigma}\left[\mathcal{K}(\varsigma,\iota,\gamma(\iota))-\mathcal{K}(\varsigma,\iota,\gamma_0'(\iota))\right]d\iota,(2\mathcal{T}_1,\ldots,2\mathcal{T}_n)\right)$$
$$\succeq \mathcal{N}\Big(\mu(\varsigma,\gamma(\varsigma))-\mu(\varsigma,\gamma_0'(\varsigma)),(\mathcal{T}_1,\ldots,\mathcal{T}_n)\Big)$$
$$\bigotimes_{\mathrm{GTN}}\mathcal{N}\left(\int_0^{\varsigma}\left[\mathcal{K}(\varsigma,\iota,\gamma(\iota))-\mathcal{K}(\varsigma,\iota,\gamma_0'(\iota))\right]d\iota,(\mathcal{T}_1,\ldots,\mathcal{T}_n)\right)$$
$$\succeq \operatorname{diag}\left[\varphi_1\left(\varsigma,\theta\frac{\mathcal{T}_1}{\Theta_1}\right),\ldots,\varphi_n\left(\varsigma,\theta\frac{\mathcal{T}_n}{\Theta_1}\right)\right]$$
$$\bigotimes_{\mathrm{GTN}}\operatorname{diag}\left[\varphi_1\left(\varsigma,\theta\frac{\mathcal{T}_1}{\Theta_2\Theta_3}\right),\ldots,\varphi_n\left(\varsigma,\theta\frac{\mathcal{T}_n}{\Theta_2\Theta_3}\right)\right]$$
$$\succeq \operatorname{diag}\left[\varphi_1\left(\varsigma,\theta\frac{\mathcal{T}_1}{\max[\Theta_1,\Theta_2\Theta_3]}\right),\ldots,\varphi_n\left(\varsigma,\theta\frac{\mathcal{T}_n}{\max[\Theta_1,\Theta_2\Theta_3]}\right)\right], \tag{19.29}$$

and so

$$\mathcal{N}\left(\mu(\varsigma,\gamma(\varsigma))-\mu(\varsigma,\gamma_0'(\varsigma))+\int_0^{\varsigma}\left[\mathcal{K}(\varsigma,\iota,\gamma(\iota))-\mathcal{K}(\varsigma,\iota,\gamma_0'(\iota))\right]d\iota,(\mathcal{T}_1,\ldots,\mathcal{T}_n)\right)$$
$$\succeq \operatorname{diag}\left[\varphi_1\left(\varsigma,\theta\frac{\mathcal{T}_1}{2\max[\Theta_1,\Theta_2\Theta_3]}\right),\ldots,\varphi_n\left(\varsigma,\theta\frac{\mathcal{T}_n}{2\max[\Theta_1,\Theta_2\Theta_3]}\right)\right]. \tag{19.30}$$

We apply (19.5) to get

$$\mathcal{N}\Bigg({}^{H}_{0^+}I^{\kappa,\vartheta}_{\phi(\varsigma)}\Big[\mu(\iota,\gamma(\iota))-\mu(\iota,\gamma_0'(\iota))\Big]$$
$$+\,{}^{H}_{0^+}I^{\kappa,\vartheta}_{\phi(\varsigma)}\left(\int_0^{\iota}\left[\mathcal{K}(\iota,\varpi,\gamma(\varpi))-\mathcal{K}(\iota,\varpi,\gamma_0'(\varpi))\right]d\varpi\right),(\mathcal{T}_1,\ldots,\mathcal{T}_n)\Bigg)$$
$$\succeq \operatorname{diag}\left[\varphi_1\left(\varsigma,\theta\frac{\mathcal{T}_1}{2\Theta_4\max[\Theta_1,\Theta_2\Theta_3]}\right),\ldots,\varphi_n\left(\varsigma,\theta\frac{\mathcal{T}_n}{2\Theta_4\max[\Theta_1,\Theta_2\Theta_3]}\right)\right], \tag{19.31}$$

for every $\varsigma\in[0,\mathrm{T}]$, $\iota\leq\varsigma$ and $\underbrace{\mathcal{T}_i}_{i=1,\ldots,n}\in(0,+\infty)$.

Using (19.30) and (19.31) we get

$$\mathcal{N}\left({}^{HC}_{0^+}D^{\kappa,\vartheta}_{\phi(\varsigma)}\Big[\propto(\gamma(\varsigma))-\gamma_0'(\varsigma)\Big],(\mathcal{T}_1,\ldots,\mathcal{T}_n)\right)$$
$$\bigotimes_{\mathrm{GTN}}\mathcal{N}\Big(\propto(\gamma(\varsigma))-\gamma_0'(\varsigma),(\mathcal{T}_1,\ldots,\mathcal{T}_n)\Big)$$

$$= \mathscr{N}\left(\mu(\varsigma,\gamma(\varsigma)) - \mu(\varsigma,\gamma_0'(\varsigma)) + \int_0^\varsigma \left[\mathcal{K}(\varsigma,\iota,\gamma(\iota)) - \mathcal{K}(\varsigma,\iota,\gamma_0'(\iota))\right]d\iota, (\mathcal{T}_1,\ldots,\mathcal{T}_n)\right)$$
$$\bigotimes_{\text{GTN}} \mathscr{N}\left({}^H_{0^+}I^{\kappa,\vartheta}_{\phi(\varsigma)}\left[\mu(\iota,\gamma(\iota)) - \mu(\iota,\gamma_0'(\iota))\right]\right.$$
$$\left. + {}^H_{0^+}I^{\kappa,\vartheta}_{\phi(\varsigma)}\left(\int_0^\iota \left[\mathcal{K}(\iota,\varpi,\gamma(\varpi)) - \mathcal{K}(\iota,\varpi,\gamma_0'(\varpi))\right]d\varpi\right), (\mathcal{T}_1,\ldots,\mathcal{T}_n)\right)$$
$$\succeq \operatorname{diag}\left[\varphi_1\left(\varsigma, \theta\frac{\mathcal{T}_1}{2\max[\Theta_1,\Theta_2\Theta_3]}\right),\ldots,\varphi_n\left(\varsigma,\theta\frac{\mathcal{T}_n}{2\max[\Theta_1,\Theta_2\Theta_3]}\right)\right]$$
$$\bigotimes_{\text{GTN}} \operatorname{diag}\left[\varphi_1\left(\varsigma,\theta\frac{\mathcal{T}_1}{2\Theta_4\max[\Theta_1,\Theta_2\Theta_3]}\right),\ldots,\varphi_n\left(\varsigma,\theta\frac{\mathcal{T}_n}{2\Theta_4\max[\Theta_1,\Theta_2\Theta_3]}\right)\right]$$
$$\succeq \operatorname{diag}\left[\varphi_1\left(\varsigma,\theta\frac{\mathcal{T}_1}{2\max[\Theta_1,\Theta_2\Theta_3](1+\Theta_4)}\right),\ldots,\varphi_n\left(\varsigma,\theta\frac{\mathcal{T}_n}{2\max[\Theta_1,\Theta_2\Theta_3](1+\Theta_4)}\right)\right],$$

which implies that $\delta(\alpha(\gamma),\gamma_0') \preceq \underbrace{\left(\frac{2\max[\Theta_1,\Theta_2\Theta_3](1+\Theta_4)}{\theta},\ldots,\frac{2\max[\Theta_1,\Theta_2\Theta_3](1+\Theta_4)}{\theta}\right)}_{n}$ $\prec \underbrace{(\infty,\ldots,\infty)}_{n}$, then $\gamma_0' \in \Phi^*$.

19.2 Multi Stability of Hadamard Fractional Volterra Integral Equations in MVFB–Spaces

In this section, we investigate the multi stability of fractional Volterra integral equations in Hadamard sense by the fixed point method in MVFB–spaces $(\mathcal{W}, \mathscr{N}, \bigotimes_{\text{GTN}})$.

Now, we are ready to study the multi stability of the following ϕ–Hadamard fractional Volterra integral equation [1],

$$\eta(\varsigma) = \mu(\varsigma,\eta(\varsigma)) + {}^H_{0^+}I^{\kappa,\vartheta}_{\phi(\varsigma)}\mathcal{K}(\varsigma,\iota,\eta(\iota)), \tag{19.32}$$

where $\kappa \in (0,1)$, $\vartheta \in [0,1]$, $\mu : [0,\mathsf{T}] \times W \to W$, $\mathcal{K} : [0,\mathsf{T}] \times [0,\mathsf{T}] \times W \to W$.

Theorem 19.3 *Let* $(\mathcal{W}, \mathscr{N}, \bigotimes_{\text{GTN}})$ *be an MVFB–space and* $\Theta_1, \Theta_2, \Theta_3, \Theta_4$ *and* T *be positive constant such that* $\max[\Theta_1, \Theta_2\Theta_3, \Theta_1\Theta_4, \Theta_2\Theta_3\Theta_4] < 0.5$. *Assume that the continuous mappings* $\mu : [0,\mathsf{T}] \times \mathcal{W} \to \mathcal{W}$, $\mathcal{K} : [0,\mathsf{T}] \times [0,\mathsf{T}] \times \mathcal{W} \to \mathcal{W}$ *with fuzzy sets* $\underbrace{\varphi_i}_{i=1,\ldots,n} : \mathcal{W} \times (0,+\infty) \to (0,1]$ *satisfy* (19.2), (19.3), (19.4) *and* (19.5), *for every* $n \in \mathbb{N}$.

Let $\gamma : [0,\mathsf{T}] \to \mathcal{W}$ *be a differentiable function satisfying*

$$\mathscr{N}\left(\gamma(\varsigma) - \mu(\varsigma,\gamma(\varsigma)) - {}^H_{0^+}I^{\kappa,\vartheta}_{\phi(\varsigma)}\mathcal{K}(\varsigma,\iota,w(\iota)), (\mathcal{T}_1,\ldots,\mathcal{T}_n)\right)$$
$$\succeq \mathit{diag}\left[\varphi_1(\varsigma,\mathcal{T}_1),\ldots,\varphi_n(\varsigma,\mathcal{T}_n)\right], \tag{19.33}$$

for every $\varsigma \in [0, \mathrm{T}]$ *and* $\underbrace{\mathcal{T}_i}_{i=1,\ldots,n} \in (0, +\infty)$. *Then, we have a unique differentiable function* $\gamma_0 : [0, \mathrm{T}] \to W$ *such that*

$$\gamma_0(\varsigma) = \mu(\varsigma, \gamma_0(\varsigma)) + {}^{H}_{0^+}I^{\kappa,\vartheta}_{\phi(\varsigma)}\mathcal{K}(\varsigma, \iota, \gamma_0(\iota)), \tag{19.34}$$

and

$$\begin{aligned} &\mathcal{N}\left(\gamma(\varsigma) - \gamma_0(\varsigma), (\mathcal{T}_1, \ldots, \mathcal{T}_n)\right) \\ &\succeq diag\left[\varphi_1\left(\varsigma, \frac{\left(1 - 2\max\left[\Theta_1, \Theta_2\Theta_3, \Theta_1\Theta_4, \Theta_2\Theta_3\Theta_4\right]\right)\mathcal{T}_1}{\max[1, \Theta_4]}\right), \ldots, \right. \\ &\qquad \left. \varphi_n\left(\varsigma, \frac{\left(1 - 2\max\left[\Theta_1, \Theta_2\Theta_3, \Theta_1\Theta_4, \Theta_2\Theta_3\Theta_4\right]\right)\mathcal{T}_n}{\max[1, \Theta_4]}\right)\right], \end{aligned} \tag{19.35}$$

for every $\varsigma \in [0, \mathrm{T}]$ *and* $\underbrace{\mathcal{T}_i}_{i=1,\ldots,n} \in (0, +\infty)$.

Proof We set

$$\Phi := \{\eta : [0, \mathrm{T}] \to \mathcal{W}, \eta \text{ is differentiable}\}$$

and introduce the $[0, +\infty]^n$–valued metric on Φ as,

$$\begin{aligned} \delta(\eta, \eta') = \inf &\left\{(\lambda_1, \ldots, \lambda_n) \in (0, +\infty)^n : \mathcal{N}\left(\eta(\varsigma) - \eta'(\varsigma), (\mathcal{T}_1, \ldots, \mathcal{T}_n)\right)\right. \\ &\left. \succeq \mathrm{diag}\left[\varphi_1\left(\varsigma, \frac{\mathcal{T}_1}{\lambda_1}\right), \ldots, \varphi_n\left(\varsigma, \frac{\mathcal{T}_n}{\lambda_n}\right)\right], \eta, \eta' \in \Phi, \varsigma \in [0, \mathrm{T}], \underbrace{\mathcal{T}_i}_{i=1,\ldots,n} \in (0, +\infty), n \in \mathbb{N}\right\}. \end{aligned}$$

By Theorem 19.1, we have (Φ, δ) is a complete $[0, +\infty]^n$–valued metric space.

Now, we define the mapping $\propto$ from Φ to Φ by

$$\propto (\eta(\varsigma)) = \mu(\varsigma, \eta(\varsigma)) + {}^{H}_{0^+}I^{\kappa,\vartheta}_{\phi(\varsigma)}(\mathcal{K}(\varsigma, \iota, \eta(\iota))), \tag{19.36}$$

where $\kappa \in (0, 1)$, $\vartheta \in [0, 1]$, $\mu : [0, \mathrm{T}] \times \mathcal{W} \to \mathcal{W}$, $\mathcal{K} : [0, \mathrm{T}] \times [0, \mathrm{T}] \times \mathcal{W} \to \mathcal{W}$. We prove $\propto$ is a strictly contractive mapping. Let $\eta, \eta' \in \Phi$, $\underbrace{\epsilon_i}_{i=1,\ldots,n} \in (0, +\infty)$ and $\delta(\eta, \eta') < \underbrace{(\epsilon_1, \ldots, \epsilon_n)}_{n}$, then, we have

$$\mathscr{N}\Big(\eta(\varsigma)-\eta'(\varsigma),(\epsilon_1\mathcal{T}_1,\ldots,\epsilon_n\mathcal{T}_n)\Big)$$
$$\succeq \operatorname{diag}\Big[\varphi_1(\varsigma,\mathcal{T}_1),\ldots,\varphi_n(\varsigma,\mathcal{T}_n)\Big],\quad \eta,\eta'\in\Phi,\ \varsigma\in[0,\intercal],\ \underbrace{\mathcal{T}_i}_{i=1,\ldots,n}\in(0,+\infty).$$

Using properties (D-2) and (D-3) of Definition 15.5, (19.2), (19.3), (19.4), (19.5) and (19.36), we have

$$\begin{aligned}
&\mathscr{N}\Big(\propto(\eta(\varsigma))-\propto(\eta'(\varsigma)),(2\epsilon_1\mathcal{T}_1,\ldots,2\epsilon_n\mathcal{T}_n)\Big)\\
&=\mathscr{N}\Big(\Big[\mu(\varsigma,\eta(\varsigma))-\mu(\varsigma,\eta'(\varsigma))\Big]\\
&\qquad+{}^{H}_{0^+}I^{\kappa,\vartheta}_{\phi(\varsigma)}\Big[\mathcal{K}(\varsigma,\iota,\eta(\iota))-\mathcal{K}(\varsigma,\iota,\eta'(\iota))\Big],(2\epsilon_1\mathcal{T}_1,\ldots,2\epsilon_n\mathcal{T}_n)\Big)\\
&\succeq\mathscr{N}\Big(\mu(\varsigma,\eta(\varsigma))-\mu(\varsigma,\eta'(\varsigma)),(\epsilon_1\mathcal{T}_1,\ldots,\epsilon_n\mathcal{T}_n)\Big)\\
&\qquad\bigotimes_{\mathrm{GTN}}\mathscr{N}\Big(\mathcal{K}(\varsigma,\iota,\eta(\iota))-\mathcal{K}(\varsigma,\iota,\eta'(\iota)),(\epsilon_1\mathcal{T}_1,\ldots,\epsilon_n\mathcal{T}_n)\Big)\\
&\succeq\operatorname{diag}\Big[\varphi_1\Big(\varsigma,\frac{\mathcal{T}_1}{\Theta_1}\Big),\ldots,\varphi_n\Big(\varsigma,\frac{\mathcal{T}_n}{\Theta_1}\Big)\Big]\\
&\qquad\bigotimes_{\mathrm{GTN}}\operatorname{diag}\Big[\varphi_1\Big(\varsigma,\frac{\mathcal{T}_1}{\Theta_2\Theta_4}\Big),\ldots,\varphi_n\Big(\varsigma,\frac{\mathcal{T}_n}{\Theta_2\Theta_4}\Big)\Big]\\
&\succeq\operatorname{diag}\Big[\varphi_1\Big(\varsigma,\frac{\mathcal{T}_1}{\max[\Theta_1,\Theta_2\Theta_4]}\Big),\ldots,\varphi_n\Big(\varsigma,\frac{\mathcal{T}_n}{\max[\Theta_1,\Theta_2\Theta_4]}\Big)\Big],
\end{aligned}$$

and so

$$\mathscr{N}\Big(\propto(\eta(\varsigma))-\propto(\eta'(\varsigma)),(\epsilon_1\mathcal{T}_1,\ldots,\epsilon_n\mathcal{T}_n)\Big)$$
$$\succeq\operatorname{diag}\Big[\varphi_1\Big(\varsigma,\frac{\mathcal{T}_1}{2\max[\Theta_1,\Theta_2\Theta_4]}\Big),\ldots,\varphi_n\Big(\varsigma,\frac{\mathcal{T}_n}{2\max[\Theta_1,\Theta_2\Theta_4]}\Big)\Big],$$

for every $\varsigma\in[0,\intercal]$ and $\underbrace{\mathcal{T}_i}_{i=1,\ldots,n}\in(0,+\infty)$. Then,

$$\delta(\propto(\eta),\propto(\eta'))\preceq\Big(2\max[\Theta_1,\Theta_2\Theta_4]\epsilon_1,\ldots,2\max[\Theta_1,\Theta_2\Theta_4]\epsilon_n\Big),$$

and so

$$\delta(\propto(\eta), \propto(\eta')) \leq \underbrace{\Big(2\max[\Theta_1, \Theta_2\Theta_4], \ldots, 2\max[\Theta_1, \Theta_2\Theta_4]\Big)}_{n} \delta(\eta, \eta').$$

Then, $\propto$ is a strictly contractive function with the Lipschitz constant $2\max[\Theta_1, \Theta_2\Theta_4]$.

Let $\gamma \in \Phi$. Then, we show that $\delta(\propto(\gamma), \gamma) < \underbrace{(\infty, \ldots, \infty)}_{n}$. Using (19.33) we get

$$\begin{aligned}
&\mathscr{N}\Big(\propto(\gamma(\varsigma)) - \gamma(\varsigma), (\mathcal{T}_1, \ldots, \mathcal{T}_n)\Big) \\
&= \mathscr{N}\Big(\mu(\varsigma, \gamma(\varsigma)) + {}^{H}_{0^+}I^{\kappa,\vartheta}_{\phi(\varsigma)}\mathcal{K}(\varsigma, \iota, \gamma(\iota)) - \gamma(\varsigma), (\mathcal{T}_1, \ldots, \mathcal{T}_n)\Big) \\
&\succeq \operatorname{diag}\Big[\varphi_1(\varsigma, \mathcal{T}_1), \ldots, \varphi_n(\varsigma, \mathcal{T}_n)\Big],
\end{aligned} \tag{19.37}$$

for every $\underbrace{\mathcal{T}_i}_{i=1,\ldots,n} \in (0, +\infty)$. Then we have $\delta(\propto(\gamma), \gamma) < \underbrace{(1, \ldots, 1)}_{n}$.

Now, Theorem 2.20 enables us to find an element γ_0 in Φ satisfying the following:

(1) γ_0 is a fixed point of $\propto$, i.e.,

$$\begin{aligned}
\gamma_0(\varsigma) &= \propto(\gamma_0(\varsigma)) \\
&= \mu(\varsigma, \gamma_0(\varsigma)) + {}^{H}_{0^+}I^{\kappa,\vartheta}_{\phi(\varsigma)}(\mathcal{K}(\varsigma, \iota, \gamma_0(\iota))),
\end{aligned}$$

which is unique in the set

$$\Phi^* = \Big\{\eta \in \Phi : \delta(\propto(\gamma), \eta) < \underbrace{(\infty, \ldots, \infty)}_{n}\Big\}.$$

(2) $\delta(\propto^{\tau}(\gamma), \gamma_0) \to \underbrace{(0, \ldots, 0)}_{n}$ as $\tau \to \infty$;

(3) We have

$$\begin{aligned}
&\delta(\gamma, \gamma_0) \\
&\leq \underbrace{\left(\frac{1}{1 - 2\max[\Theta_1, \Theta_2\Theta_4]}, \ldots, \frac{1}{1 - 2\max[\Theta_1, \Theta_2\Theta_4]}\right)}_{n} \delta(\propto(\gamma), \gamma) \\
&\leq \underbrace{\left[\frac{1}{1 - 2\max[\Theta_1, \Theta_2\Theta_4]}, \ldots, \frac{1}{1 - 2\max[\Theta_1, \Theta_2\Theta_4]}\right]}_{n},
\end{aligned}$$

which implies that

$$\begin{aligned}&\mathcal{N}\Big(\gamma(\varsigma)-\gamma_0(\varsigma),(\mathcal{T}_1,\ldots,\mathcal{T}_n)\Big)\\&\succeq \operatorname{diag}\Big[\varphi_1\Big(\varsigma,\Big(1-2\max[\Theta_1,\Theta_2\Theta_4]\Big)\mathcal{T}_1\Big),\ldots,\varphi_n\Big(\varsigma,\Big(1-2\max[\Theta_1,\Theta_2\Theta_4]\Big)\mathcal{T}_n\Big)\Big],\end{aligned}$$

for every $\varsigma \in [0, \mathsf{T}]$, $\underbrace{\mathcal{T}_i}_{i=1,\ldots,n} \in (0, +\infty)$, and $n \in \mathbb{N}$.

19.3 Application

In this part, we apply the main results to solve some examples.

Consider the special controllers $\underbrace{\varphi_i^{\circledS}}_{1\le i\le n\in\mathbb{N}}$ defined in Sects. 9.8.1 and 9.8.2.

Example 19.1 Let $(\mathbb{R}, \mathcal{N}, \bigotimes_{\mathrm{GTN}})$ be an MVFB–space. Consider $\eta, \eta' : [0, \mathsf{T}] \to \mathbb{R}$ and define $\mu(\varsigma, \eta(\varsigma)) = \ln\sqrt{|\eta(\varsigma)|}$. Define $\mathcal{K} : [0, \mathsf{T}] \times [0, \mathsf{T}] \times \mathbb{R} \to \mathbb{R}$ as $\mathcal{K}(\varsigma, \iota, \eta(\iota)) = \exp(\varsigma - \iota)\eta(\iota)$ for every $\varsigma \in [0, \mathsf{T}]$ and $\iota \le \varsigma$.

Then, we have

$$\begin{aligned}&\mathcal{N}\Big(\mathcal{K}(\varsigma,\iota,\eta(\iota))-\mathcal{K}(\varsigma,\iota,\eta'(\iota)),(\mathcal{T}_1,\ldots,\mathcal{T}_n)\Big)\\&=\mathcal{N}\Big(\exp(\varsigma-\iota)[\eta(\iota)-\eta'(\iota)],(\mathcal{T}_1,\ldots,\mathcal{T}_n)\Big)\\&\succeq\mathcal{N}\left(\eta(\iota)-\eta'(\iota),\left(\frac{\mathcal{T}_1}{|\exp(\varsigma-\iota)|},\ldots,\frac{\mathcal{T}_n}{|\exp(\varsigma-\iota)|}\right)\right)\\&\succeq\mathcal{N}\left(\eta(\iota)-\eta'(\iota),\left(\frac{\mathcal{T}_1}{\Lambda},\ldots,\frac{\mathcal{T}_n}{\Lambda}\right)\right),\end{aligned}$$

$$\begin{aligned}&\mathcal{N}\Big(\mu(\varsigma,\eta(\varsigma))-\mu(\varsigma,\eta'(\varsigma)),(\mathcal{T}_1,\ldots,\mathcal{T}_n)\Big)\\&=\mathcal{N}\Big(\ln\sqrt{|\eta(\varsigma)|}-\ln\sqrt{|\eta'(\varsigma)|},(\mathcal{T}_1,\ldots,\mathcal{T}_n)\Big)\\&=\mathcal{N}\left(\ln\frac{\sqrt{|\eta(\varsigma)|}}{\sqrt{|\eta'(\varsigma)|}},(\mathcal{T}_1,\ldots,\mathcal{T}_n)\right)\\&\succeq\mathcal{N}\left(\ln\left(1+\left(\frac{\sqrt{|\eta(\varsigma)|}}{\sqrt{|\eta'(\varsigma)|}}-1\right)\right),(\mathcal{T}_1,\ldots,\mathcal{T}_n)\right)\\&\succeq\mathcal{N}\left(\frac{\sqrt{|\eta(\varsigma)|}}{\sqrt{|\eta'(\varsigma)|}}-1,(\mathcal{T}_1,\ldots,\mathcal{T}_n)\right)\end{aligned}$$

$$
\begin{aligned}
&\succeq \mathcal{N}\Bigg(\frac{\sqrt{|\eta(\varsigma)|}-\sqrt{|\eta'(\varsigma)|}}{\sqrt{\min(|\eta'(\varsigma)|)}},(\mathcal{T}_1,\ldots,\mathcal{T}_n)\Bigg)\\
&\succeq \mathcal{N}\Bigg(\sqrt{|\eta(\varsigma)|}-\sqrt{|\eta'(\varsigma)|},(\mathcal{T}_1,\ldots,\mathcal{T}_n)\Bigg)\\
&\succeq \mathcal{N}\Bigg(\frac{|\eta(\varsigma)|-|\eta'(\varsigma)|}{\sqrt{|\eta(\varsigma)|}+\sqrt{|\eta'(\varsigma)|}},(\mathcal{T}_1,\ldots,\mathcal{T}_n)\Bigg)\\
&\succeq \mathcal{N}\Bigg(\frac{|\eta(\varsigma)-\eta'(\varsigma)|}{\sqrt{\max(|\eta(\varsigma)|,|\eta'(\varsigma)|)}},(\mathcal{T}_1,\ldots,\mathcal{T}_n)\Bigg)\\
&\succeq \mathcal{N}\Bigg(\mathtt{C}\Big(\eta(\varsigma)-\eta'(\varsigma)\Big),(\mathcal{T}_1,\ldots,\mathcal{T}_n)\Bigg)\\
&= \mathcal{N}\Bigg(\eta(\varsigma)-\eta'(\varsigma),(\frac{\mathcal{T}_1}{\mathtt{C}},\ldots,\frac{\mathcal{T}_n}{\mathtt{C}})\Bigg),
\end{aligned}
$$

for some $\Lambda \in (0,+\infty)$ and $\mathtt{C} \in (0,1]$. Consider fuzzy sets $\underbrace{\varphi_i}_{i=1,\ldots,n} : [0,\mathtt{T}] \times (0,+\infty) \to (0,1]$ that is defined as follows:

$$
\begin{aligned}
&\operatorname{diag}\Big[\varphi_1(\varsigma,\mathcal{T}_1),\ldots,\varphi_n(\varsigma,\mathcal{T}_n)\Big]\\
&:= \operatorname{diag}\Bigg[\varphi_1^{\circledS}\Bigg(-\frac{|(\phi(\varsigma)-\phi(0))^{\kappa}|}{\ni_1\,\mathcal{T}_1}\Bigg),\ldots,\varphi_n^{\circledS}\Bigg(-\frac{|(\phi(\varsigma)-\phi(0))^{\kappa}|}{\ni_n\,\mathcal{T}_n}\Bigg)\Bigg],
\end{aligned}
\tag{19.38}
$$

for every $\varsigma \in [0,\mathtt{T}]$, $\underbrace{\mathcal{T}_i}_{i=1,\ldots,n} \in (0,+\infty)$, $\kappa \in (0,1)$, $\underbrace{\ni_i}_{i=1,\ldots,n} > 0$ and $n \in \mathbb{N}$.

Let $\gamma : [0,\mathtt{T}] \to \mathbb{R}$ be a differentiable function satisfying

$$
\begin{aligned}
&\mathcal{N}\Bigg({}^{HC}_{0^+}D^{\kappa,\vartheta}_{\phi(\varsigma)}\gamma(\varsigma)-\ln\sqrt{|\gamma(\varsigma)|}-\int_0^{\varsigma}\exp(\varsigma-\iota)\gamma(\iota)d\iota,(\mathcal{T}_1,\ldots,\mathcal{T}_n)\Bigg)\\
&\succeq \operatorname{diag}\Bigg[\varphi_1^{\circledS}\Bigg(-\frac{|(\phi(\varsigma)-\phi(0))^{\kappa}|}{\ni_1\,\mathcal{T}_1}\Bigg),\ldots,\varphi_n^{\circledS}\Bigg(-\frac{|(\phi(\varsigma)-\phi(0))^{\kappa}|}{\ni_n\,\mathcal{T}_n}\Bigg)\Bigg],
\end{aligned}
$$

for every $\varsigma \in [0,\mathtt{T}]$ and $\underbrace{\mathcal{T}_i}_{i=1,\ldots,n} \in (0,+\infty)$. Now, Theorem 19.2 implies that if,

$$
\max[\mathtt{C},\Lambda\Theta_3,\mathtt{C}\Theta_4,\Lambda\Theta_3\Theta_4] < 0.5,
$$

we can find a unique differentiable function $\gamma_0 : [0, \mathrm{T}] \to \mathbb{R}$ such that

$$ {}^{HC}_{0^+}D^{\kappa,\vartheta}_{\phi(\varsigma)}\gamma_0(\varsigma) = \ln\sqrt{|\gamma_0(\varsigma)|} + \int_0^{\varsigma} \exp(\varsigma - \iota)\gamma_0(\iota)d\iota, $$

and

$$ \begin{aligned} &\mathcal{N}\Big({}^{HC}_{0^+}D^{\kappa,\vartheta}_{\phi(\varsigma)}\gamma(\varsigma) - {}^{HC}_{0^+}D^{\kappa,\vartheta}_{\phi(\varsigma)}\gamma_0(\varsigma), (\mathcal{T}_1, \dots, \mathcal{T}_n)\Big) \\ &\quad \bigotimes_{\mathrm{GTN}} \mathcal{N}\Big(\gamma(\varsigma) - \gamma_0(\varsigma), (\mathcal{T}_1, \dots, \mathcal{T}_n)\Big) \\ &\succeq \mathrm{diag}\Bigg[\varphi_1^{\circledS}\Bigg(\frac{-|(\phi(\varsigma) - \phi(0))^{\kappa}|}{\frac{\ni_1\left(1-2\max\left[\mathrm{C}, \Lambda\Theta_3, \mathrm{C}\Theta_4, \Lambda\Theta_3\Theta_4\right]\right)\mathcal{T}_1}{\max[1,\Theta_4]}}\Bigg), \dots, \\ &\qquad \varphi_n^{\circledS}\Bigg(\frac{-|(\phi(\varsigma) - \phi(0))^{\kappa}|}{\frac{\ni_n\left(1-2\max\left[\mathrm{C}, \Lambda\Theta_3, \mathrm{C}\Theta_4, \Lambda\Theta_3\Theta_4\right]\right)\mathcal{T}_n}{\max[1,\Theta_4]}}\Bigg)\Bigg], \end{aligned} $$

for every $\varsigma \in [0, \mathrm{T}]$ and $\underbrace{\mathcal{T}_i}_{i=1,\dots,n} \in (0, +\infty)$.

Example 19.2 Let $(\mathbb{R}, \mathcal{N}, \bigotimes_{\mathrm{GTN}})$ be an MVFB–space. Consider $\eta, \eta' : [0, \mathrm{T}] \to \mathbb{R}$ and define $\mu(\varsigma, \eta(\varsigma)) = \ln\sqrt{|\eta(\varsigma)|}$. Define $\mathcal{K} : [0, \mathrm{T}] \times [0, \mathrm{T}] \times \mathbb{R} \to \mathbb{R}$ as $\mathcal{K}(\varsigma, \iota, \eta(\iota)) = \exp(\varsigma - \iota)\eta(\iota)$ for every $\varsigma \in [0, \mathrm{T}]$ and $\iota \le \varsigma$.

Let $\underbrace{\varphi_i}_{i=1,\dots,n} : [0, \mathrm{T}] \times (0, +\infty) \to (0, 1]$ be defined as (19.38).

Let $\gamma : [0, \mathrm{T}] \to \mathbb{R}$ be a differentiable function satisfying

$$ \begin{aligned} &\mathcal{N}\Big(\gamma(\varsigma) - \ln\sqrt{|\eta(\varsigma)|} - {}^{H}_{0^+}I^{\kappa,\vartheta}_{\phi(\varsigma)}\exp(\varsigma - \iota)\gamma(\iota), (\mathcal{T}_1, \dots, \mathcal{T}_n)\Big) \\ &\succeq \mathrm{diag}\Bigg[\varphi_1^{\circledS}\Bigg(-\frac{|(\phi(\varsigma) - \phi(0))^{\kappa}|}{\ni_1\,\mathcal{T}_1}\Bigg), \dots, \varphi_n^{\circledS}\Bigg(-\frac{|(\phi(\varsigma) - \phi(0))^{\kappa}|}{\ni_n\,\mathcal{T}_n}\Bigg)\Bigg], \end{aligned} $$

for every $\varsigma \in [0, \mathrm{T}]$ and $\underbrace{\mathcal{T}_i, \ni_i}_{i=1,\dots,n} \in (0, +\infty)$. Now, Theorem 19.3 implies that, if $\max[\mathrm{C}, \Lambda\Theta_4] < 0.5$, we can find a unique differentiable function $\gamma_0 : [0, \mathrm{T}] \to \mathbb{R}$ such that

$$ \gamma_0(\varsigma) = \ln\sqrt{|\eta(\varsigma)|} + {}^{H}_{0^+}I^{\kappa,\vartheta}_{\phi([0,\mathrm{T}])}\exp(\varsigma - \iota)\gamma(\iota), $$

and

$$\mathscr{N}\left(\gamma(\varsigma)-\gamma_0(\varsigma),(\mathcal{T}_1,\ldots,\mathcal{T}_n)\right)$$
$$\succeq \operatorname{diag}\left[\varphi_1^{ⓢ}\left(\frac{-|(\phi(\varsigma)-\phi(0))^{\kappa}|}{\frac{\ni_1\left(1-2\max\left[\mathtt{C},\Lambda\Theta_4\right]\right)\mathcal{T}_1}{\max[1,\Theta_4]}}\right),\ldots,\varphi_n^{ⓢ}\left(\frac{-|(\phi(\varsigma)-\phi(0))^{\kappa}|}{\frac{\ni_n\left(1-2\max\left[\mathtt{C},\Lambda\Theta_4\right]\right)\mathcal{T}_n}{\max[1,\Theta_4]}}\right)\right],$$

for every $\varsigma \in [0, \mathsf{T}]$ and $\underbrace{\mathcal{T}_i}_{i=1,\ldots,n} \in (0,+\infty)$.

We refer the interested readers to [2–12].

References

1. Aderyani SR, Saadati R (2021) Best approximations of the ϕ-Hadamard fractional Volterra integro-differential equation by matrix valued fuzzy control functions. Adv Diff Equ 154
2. Rezaei Aderyani S, Saadati R, Rassias TM (2022) Park C (2022) Best approximation of $(\mathcal{G}_1, \mathcal{G}_2)$-random operator inequality in matrix Menger Banach algebras with application of stochastic Mittag-Leffler and H $\mathbb{H}$-Fox control functions. J Inequalities Appl 1:1–17
3. Aderyani SR, Saadati R, Mesiar R (2022) Estimation of permuting tri-homomorphisms and permuting tri-derivations associated with the tri-additive Υ-random operator inequality in matrix MB-algebra. Int J Gen Syst 51(6):547–569
4. Aderyani SR, Saadati R, Yang XJ (2021) Radu-Miheţ method for UHML stability for a class of ξ-Hilfer fractional differential equations in matrix valued fuzzy Banach spaces. Math Methods Appl Sci 44(18):14619–14631
5. Aderyani SR, Saadati R, Allahviranloo T (2022) Existence, uniqueness and matrix-valued fuzzy mittag-leffler-hypergeometric-wright stability for P-Hilfer fractional differential equations in matrix-valued fuzzy Banach space. Comput Appl Math 41(6):234
6. Aderyani SR, Saadati R (2022) Approximation of derivation–homomorphism fuzzy functional inequalities in matrix valued FC-diamond⋄-algebras. In: 2022 9th Iranian joint congress on fuzzy and intelligent systems (CFIS). IEEE, pp 1–6
7. Aderyani SR, Saadati R, Allahviranloo T, Abbasbandy S, Catak M (2023) Fuzzy approximation of a fractional Lorenz system and a fractional financial crisis. Iran J Fuzzy Syst 20(7):27–36
8. Aderyani SR, Saadati R, Abdeljawad T, Mlaiki N (2022) Multi-stability of non homogenous vector-valued fractional differential equations in matrix-valued Menger spaces. Alex Eng J 61(12):10913–10923
9. Aderyani SR, Saadati R (2023) Stability and controllabilit results by n-ary aggregation functions in matrix valued fuzzy n-normed spaces. Inf Sci 119265
10. O'Regan D, Aderyani SR, Saadati R, Li C (2023) Stability results and parametric delayed mittag-leffler matrices in symmetric fuzzy-random spaces with application. Symmetry 15(10):1880
11. Aderyani SR, Saadati R, O'Regan D, Li C (2023) On a new approach for stability and controllability analysis of functional equations. Mathematics 11(16):3458
12. Aderyani SR, Saadati R, O'Regan D, Alshammari FS (2023) Fuzzy approximate solutions of matrix-valued fractional differential equations by fuzzy control functions. Mathematics 11(6):1386

Chapter 20
Multi Stability of Fractional–Order System I in MVFB–Spaces

Applying the fixed point theory, we study the existence, uniqueness and the multi stability of ξ–Hilfer fractional differential systems in MVFB–spaces $(\mathcal{W}, \mathcal{N}, \bigotimes_{\mathrm{GTN}})$.

Now, consider the fractional–order systems I [1], as follows:

$$\begin{cases} {}^{\mathscr{H}}D_{0^+}^{\mathcal{A},\mathcal{B}:\xi}\Phi(\zeta) = \mu(\zeta, \Phi(\zeta), \Phi(\eta(\zeta))), & \zeta \in (0, \mathrm{T}], \\ I_{0^+}^{1-\mathcal{C}:\xi}\Phi(0^+) = f_0, & f_0 \in \mathbb{R}, \end{cases} \tag{20.1}$$

where ${}^{\mathscr{H}}D_{0^+}^{\mathcal{A},\mathcal{B}:\xi}(.)$ is the ξ–Hilfer fractional derivative of order $0 < \mathcal{A} \leq 1$ and type $0 \leq \mathcal{B} \leq 1$, $I_{0^+}^{1-\mathcal{C}:\xi}(.)$ is the Riemann–Liouville fractional integral of order $1 - \mathcal{C}$, $\mathcal{C} = \mathcal{A} + \mathcal{B}(1 - \mathcal{A})$ with respect to the function ξ, and $\mu : (0, \mathrm{T}] \times \mathbb{R}^2 \longrightarrow \mathbb{R}$ is a given function, for every $\mathrm{T} > 0$.

20.1 Multi Stability of Hilfer Fractional System I on Compact Domains

In this part, we prove the existence, uniqueness and the multi stability results of the ξ–Hilfer fractional systems (20.1) defined on compact domains by the fixed point theorem.

For $\mu \in C((0, \mathrm{T}] \times \mathbb{R}^2, \mathbb{R})$ and $\theta > 0$, consider the following fractional equations

$${}^{\mathscr{H}}D_{0^+}^{\mathcal{A},\mathcal{B}:\xi}\Phi(\zeta) = \mu(\zeta, \Phi(\zeta), \Phi(\eta(\zeta))), \qquad \zeta \in (0, \mathrm{T}], \tag{20.2}$$

$$I_{0^+}^{1-\mathcal{C}:\xi}\Phi(0^+) = f_0, \qquad f_0 \in \mathbb{R}, \tag{20.3}$$

and the inequality below

S. Rezaei Aderyani et al., *Towards Ulam Type Multi Stability Analysis*, Studies in Systems, Decision and Control 523, https://doi.org/10.1007/978-3-031-55564-0_20

$$\mathcal{N}\left({}^{\mathscr{H}}D_{0^+}^{\mathcal{A},\mathcal{B};\xi}\Phi(\zeta)-\mu(\zeta,\Phi(\zeta),\Phi(\eta(\zeta))),(\mathcal{T}_1,\ldots,\mathcal{T}_n)\right) \tag{20.4}$$
$$\succeq \operatorname{diag}\left[\varphi_1^{\circledS}\left(\frac{-|(\xi(\zeta)-\xi(0))^{\mathcal{A}}|}{\theta\mathcal{T}_1}\right),\ldots,\varphi_n^{\circledS}\left(\frac{-|(\xi(\zeta)-\xi(0))^{\mathcal{A}}|}{\theta\mathcal{T}_n}\right)\right],$$

where $\zeta\in(0,\mathsf{T}]$, $\underbrace{\mathcal{T}_i}_{i=1,\ldots,n}\in(0,+\infty)$, $n\in\mathbb{N}$, and $\underbrace{\varphi_i^{\circledS}}_{i=1,\ldots,n}$ are fuzzy controllers defined in Sects. 9.8.1 and 9.8.2.

Definition 20.1 Fractional equations (20.2)–(20.3) have the multi stability with respect to $\operatorname{diag}[\varphi_1^{\circledS}((\xi(\zeta)-\xi(0))^{\mathcal{A}}),\ldots,\varphi_n^{\circledS}((\xi(\zeta)-\xi(0))^{\mathcal{A}})]$ if there exists $\eth>0$ such that, for every $\theta>0$, and every solution $\Psi\in C((0,\mathsf{T}],\mathbb{R})$ to (20.4) and $I_{0^+}^{1-\mathcal{C};\xi}\Psi(0^+)=f_0\in\mathbb{R}$, there exists a solution $\Phi\in C((0,\mathsf{T}],\mathbb{R})$ to (20.2)–(20.3) with

$$\mathcal{N}\left(\Psi(\zeta)-\Phi(\zeta),(\mathcal{T}_1,\ldots,\mathcal{T}_n)\right)$$
$$\succeq \operatorname{diag}\left[\varphi_1^{\circledS}\left(\frac{-|(\xi(\zeta)-\xi(0))^{\mathcal{A}}|}{\eth\theta\mathcal{T}_1}\right),\ldots,\varphi_n^{\circledS}\left(\frac{-|(\xi(\zeta)-\xi(0))^{\mathcal{A}}|}{\eth\theta\mathcal{T}_n}\right)\right],$$

for every $\zeta\in(0,\mathsf{T}]$, $\underbrace{\mathcal{T}_i}_{i=1,\ldots,n}\in(0,+\infty)$, and $n\in\mathbb{N}$.

Lemma 20.1 ([2]) *Let* $\mu:(0,\mathsf{T}]\times\mathbb{R}\times\mathbb{R}\longrightarrow\mathbb{R}$ *be a continuous function. Then, fractional equations* (20.2)–(20.3) *are equivalent to*

$$\Phi(\zeta)=\frac{(\xi(\zeta)-\xi(0))^{\mathcal{C}-1}}{\Gamma(\mathcal{C})}f_0+\frac{1}{\Gamma(\mathcal{A})}\int_0^{\zeta}\xi'(\zeta_\circ)(\xi(\zeta)-\xi(\zeta_\circ))^{\mathcal{A}-1}\mu(\zeta_\circ,\Phi(\zeta_\circ),\Phi(\eta(\zeta_\circ)))d\zeta_\circ.$$

Remark 20.1 ([1]) Let $\Psi\in C((0,\mathsf{T}],\mathbb{R})$ be a solution of the inequality (20.4), and $I_{0^+}^{1-\mathcal{C};\xi}\Psi(0^+)=f_0\in\mathbb{R}$. Then, Ψ is a solution of the following integral inequality

$$\mathcal{N}\left(\Psi(\zeta)-\frac{(\xi(\zeta)-\xi(0))^{\mathcal{C}-1}}{\Gamma(\mathcal{C})}f_0\right.$$
$$\left.-\frac{1}{\Gamma(\mathcal{A})}\int_0^{\zeta}\xi'(\zeta_\circ)(\xi(\zeta)-\xi(\zeta_\circ))^{\mathcal{A}-1}\mu(\zeta_\circ,\Psi(\zeta_\circ),\Psi(\eta(\zeta_\circ)))d\zeta_\circ,(\mathcal{T}_1,\ldots,\mathcal{T}_n)\right)$$
$$\succeq \operatorname{diag}\left[\varphi_1^{\circledS}\left(\frac{-|(\xi(\zeta)-\xi(0))^{\mathcal{A}}|}{\theta\mathcal{T}_1}\right),\ldots,\varphi_n^{\circledS}\left(\frac{-|(\xi(\zeta)-\xi(0))^{\mathcal{A}}|}{\theta\mathcal{T}_n}\right)\right],$$

where $\zeta\in(0,\mathsf{T}]$, $\underbrace{\mathcal{T}_i}_{i=1,\ldots,n}\in(0,+\infty)$ and $n\in\mathbb{N}$.

Proof We only prove the following special cases:

$$
\begin{aligned}
&\mathcal{N}\Bigg(\Psi(\zeta)-\frac{(\xi(\zeta)-\xi(0))^{\mathcal{C}-1}}{\Gamma(\mathcal{C})}f_0 \\
&\qquad -\frac{1}{\Gamma(\mathcal{A})}\int_0^{\zeta}\xi'(\zeta_\circ)(\xi(\zeta)-\xi(\zeta_\circ))^{\mathcal{A}-1}\mu(\zeta_\circ,\Psi(\zeta_\circ),\Psi(\eta(\zeta_\circ)))d\zeta_\circ,(\mathcal{T}_1,\ldots,\mathcal{T}_n)\Bigg)\\
&\succeq \mathcal{N}\Bigg(\int_0^{\zeta}\xi'(\zeta_\circ)(\xi(\zeta)-\xi(\zeta_\circ))^{\mathcal{A}-1}\mathbb{E}_{\mathcal{A}}((\xi(\zeta_\circ)-\xi(0))^{\mathcal{A}})d\zeta_\circ,(\mathcal{T}_1,\ldots,\mathcal{T}_n)\Bigg)\\
&\succeq \mathcal{N}\Bigg(\frac{\theta}{\Gamma(\mathcal{A})}\int_0^{\zeta}\xi'(\zeta_\circ)(\xi(\zeta)-\xi(\zeta_\circ))^{\mathcal{A}-1}\sum_{k=0}^{\infty}\frac{(\xi(\zeta_\circ)-\xi(0))^{k\mathcal{A}}}{\Gamma(k\mathcal{A}+1)}d\zeta_\circ,(\mathcal{T}_1,\ldots,\mathcal{T}_n)\Bigg)\\
&\succeq \mathcal{N}\Bigg(\frac{\theta}{\Gamma(\mathcal{A})}\sum_{k=0}^{\infty}\frac{1}{\Gamma(k\mathcal{A}+1)}\\
&\qquad\times\int_0^{\zeta}(\xi(\zeta)-\xi(\zeta_\circ))^{\mathcal{A}-1}(\xi(\zeta_\circ)-\xi(0))^{k\mathcal{A}}d\xi(\zeta_\circ),(\mathcal{T}_1,\ldots,\mathcal{T}_n)\Bigg)\\
&\succeq \mathcal{N}\Bigg(\frac{\theta}{\Gamma(\mathcal{A})}\sum_{k=0}^{\infty}\frac{1}{\Gamma(k\mathcal{A}+1)}\int_0^{\xi(\zeta)-\xi(0)}(\xi(\zeta)-\xi(0)-\mathcal{E})^{\mathcal{A}-1}\mathcal{E}^{k\mathcal{A}}d\mathcal{E},(\mathcal{T}_1,\ldots,\mathcal{T}_n)\Bigg)\\
&\quad(\text{let }\ \mathcal{E}=\xi(\zeta_\circ)-\xi(0))\\
&\succeq \mathcal{N}\Bigg(\frac{\theta}{\Gamma(\mathcal{A})}\sum_{k=0}^{\infty}\frac{1}{\Gamma(k\mathcal{A}+1)}(\xi(\zeta)-\xi(0))^{\mathcal{A}-1}\\
&\qquad\times\int_0^{\xi(\zeta)-\xi(0)}\left(1-\frac{\mathcal{E}}{\xi(\zeta)-\xi(0)}\right)^{\mathcal{A}-1}\mathcal{E}^{k\mathcal{A}}d\mathcal{E},(\mathcal{T}_1,\ldots,\mathcal{T}_n)\Bigg)\\
&\succeq \mathcal{N}\Bigg(\frac{\theta}{\Gamma(\mathcal{A})}\sum_{k=0}^{\infty}\frac{1}{\Gamma(k\mathcal{A}+1)}(\xi(\zeta)-\xi(0))^{(k+1)\mathcal{A}}\int_0^{1}(1-\mathcal{S})^{\mathcal{A}-1}\mathcal{S}^{k\mathcal{A}}d\mathcal{S},(\mathcal{T}_1,\ldots,\mathcal{T}_n)\Bigg)\\
&\quad\left(\text{let }\ \mathcal{S}=\frac{\mathcal{E}}{\xi(\zeta)-\xi(0)}\right)\\
&\succeq \mathcal{N}\Bigg(\frac{\theta}{\Gamma(\mathcal{A})}\sum_{k=0}^{\infty}\frac{1}{\Gamma(k\mathcal{A}+1)}(\xi(\zeta)-\xi(0))^{(k+1)\mathcal{A}}\frac{\Gamma(k\mathcal{A}+1)\Gamma(\mathcal{A})}{\Gamma((k+1)\mathcal{A}+1)},(\mathcal{T}_1,\ldots,\mathcal{T}_n)\Bigg)\\
&\succeq \mathcal{N}\Bigg(\theta\sum_{n=0}^{\infty}\frac{(\xi(\zeta)-\xi(0))^{n\mathcal{A}}}{\Gamma(n\mathcal{A}+1)},(\mathcal{T}_1,\ldots,\mathcal{T}_n)\Bigg)\\
&\succeq \operatorname{diag}\left[\mathbb{E}_{\mathcal{A}}\left(\frac{-|(\xi(\zeta)-\xi(0))^{\mathcal{A}}|}{\theta\mathcal{T}_1}\right),\ldots,\mathbb{E}_{\mathcal{A}}\left(\frac{-|(\xi(\zeta)-\xi(0))^{\mathcal{A}}|}{\theta\mathcal{T}_n}\right)\right].
\end{aligned}
$$

$$
\mathcal{N}\Bigg(\Psi(\zeta)-\frac{(\xi(\zeta)-\xi(0))^{\mathcal{C}-1}}{\Gamma(\mathcal{C})}f_0
$$

$$
\begin{aligned}
&- \frac{1}{\Gamma(\mathcal{A})}\int_0^{\zeta} \xi'(\zeta_\circ)(\xi(\zeta)-\xi(\zeta_\circ))^{\mathcal{A}-1}\mu(\zeta_\circ,\Psi(\zeta_\circ),\Psi(\eta(\zeta_\circ)))d\zeta_\circ,(\mathcal{T}_1,\ldots,\mathcal{T}_n)\Big) \\
&\succeq \mathcal{N}\Big(\frac{\theta}{\Gamma(\mathcal{A})}\int_0^{\zeta} \xi'(\zeta_\circ)(\xi(\zeta)-\xi(\zeta_\circ))^{\mathcal{A}-1}{}_2\mathbb{F}_1\Big(\mathsf{A},\mathsf{B};\mathsf{C};(\xi(\zeta_\circ)-\xi(0))^{\mathcal{A}}\Big)d\zeta_\circ,(\mathcal{T}_1,\ldots,\mathcal{T}_n)\Big) \\
&\succeq \mathcal{N}\Big(\frac{\theta}{\Gamma(\mathcal{A})}\int_0^{\zeta} \xi'(\zeta_\circ)(\xi(\zeta)-\xi(\zeta_\circ))^{\mathcal{A}-1}\frac{\Gamma(\mathsf{C})}{\Gamma(\mathsf{A})\Gamma(\mathsf{B})} \\
&\qquad \times\sum_{k=0}^{\infty}\frac{\Gamma(\mathsf{A}+k)\Gamma(\mathsf{B}+k)}{\Gamma(\mathsf{C}+k)}\frac{(\xi(\zeta_\circ)-\xi(0))^{k\mathcal{A}}}{k!}d\zeta_\circ,(\mathcal{T}_1,\ldots,\mathcal{T}_n)\Big) \\
&= \mathcal{N}\Big(\frac{\theta}{\Gamma(\mathcal{A})}\frac{\Gamma(\mathsf{C})}{\Gamma(\mathsf{A})\Gamma(\mathsf{B})}\sum_{k=0}^{\infty}\frac{\Gamma(\mathsf{A}+k)\Gamma(\mathsf{B}+k)}{\Gamma(\mathsf{C}+k)}\frac{1}{k!} \\
&\qquad \times\int_0^{\zeta}(\xi(\zeta)-\xi(\zeta_\circ))^{\mathcal{A}-1}(\xi(\zeta_\circ)-\xi(0))^{k\mathcal{A}}d\xi(\zeta_\circ),(\mathcal{T}_1,\ldots,\mathcal{T}_n)\Big) \\
&= \mathcal{N}\Big(\frac{\theta}{\Gamma(\mathcal{A})}\frac{\Gamma(\mathsf{C})}{\Gamma(\mathsf{A})\Gamma(\mathsf{B})}\sum_{k=0}^{\infty}\frac{\Gamma(\mathsf{A}+k)\Gamma(\mathsf{B}+k)}{\Gamma(\mathsf{C}+k)}\frac{1}{k!} \\
&\qquad \times\int_0^{\xi(\zeta)-\xi(0)}(\xi(\zeta)-\xi(0)-u)^{\mathcal{A}-1}(u)^{k\mathcal{A}}du,(\mathcal{T}_1,\ldots,\mathcal{T}_n)\Big) \\
&(u=\xi(\zeta_\circ)-\xi(0)) \\
&\succeq \mathcal{N}\Big(\frac{\theta}{\Gamma(\mathcal{A})}\frac{\Gamma(\mathsf{C})}{\Gamma(\mathsf{A})\Gamma(\mathsf{B})}\sum_{k=0}^{\infty}\frac{\Gamma(\mathsf{A}+k)\Gamma(\mathsf{B}+k)}{\Gamma(\mathsf{C}+k)}\frac{1}{k!}(\xi(\zeta)-\xi(0))^{\mathcal{A}-1} \\
&\qquad \int_0^{\xi(\zeta)-\xi(0)}(1-\frac{u}{\xi(\zeta)-\xi(0)})^{\mathcal{A}-1}u^{k\mathcal{A}}du,(\mathcal{T}_1,\ldots,\mathcal{T}_n)\Big) \\
&= \mathcal{N}\Big(\frac{\theta}{\Gamma(\mathcal{A})}\frac{\Gamma(\mathsf{C})}{\Gamma(\mathsf{A})\Gamma(\mathsf{B})}\sum_{k=0}^{\infty}\frac{\Gamma(\mathsf{A}+k)\Gamma(\mathsf{B}+k)}{\Gamma(\mathsf{C}+k)}\frac{1}{k!}(\xi(\zeta)-\xi(0))^{(k+1)\mathcal{A}} \\
&\qquad \int_0^{1}(1-v)^{\mathcal{A}-1}v^{k\mathcal{A}}dv,(\mathcal{T}_1,\ldots,\mathcal{T}_n)\Big) \\
&\Big(v=\frac{u}{\xi(\zeta)-\xi(0)}\Big) \\
&= \mathcal{N}\Big(\frac{\theta}{\Gamma(\mathcal{A})}\frac{\Gamma(\mathsf{C})}{\Gamma(\mathsf{A})\Gamma(\mathsf{B})}\sum_{k=0}^{\infty}\frac{\Gamma(\mathsf{A}+k)\Gamma(\mathsf{B}+k)}{\Gamma(\mathsf{C}+k)}\frac{1}{k!} \\
&\qquad \times(\xi(\zeta)-\xi(0))^{(k+1)\mathcal{A}}\frac{\Gamma(k\mathcal{A}+1)\Gamma(\mathcal{A})}{\Gamma((k+1)\mathcal{A}+1)},(\mathcal{T}_1,\ldots,\mathcal{T}_n)\Big) \\
&\succeq \mathcal{N}\Big(\theta\frac{\Gamma(\mathsf{C})}{\Gamma(\mathsf{A})\Gamma(\mathsf{B})}\sum_{k=0}^{\infty}\frac{\Gamma(\mathsf{A}+k)\Gamma(\mathsf{B}+k)}{\Gamma(\mathsf{C}+k)}\frac{(\xi(\zeta)-\xi(0))^{k\mathcal{A}}}{k!},(\mathcal{T}_1,\ldots,\mathcal{T}_n)\Big)
\end{aligned}
$$

$$= \text{diag}\left[{}_2\mathbb{F}_1\left(\mathsf{A}, \mathsf{B}, \mathsf{C}; \frac{-|(\xi(\zeta) - \xi(0)|^{\mathcal{A}}}{\theta\mathcal{T}_1}\right), \ldots, {}_2\mathbb{F}_1\left(\mathsf{A}, \mathsf{B}; \mathsf{C}; \frac{-|(\xi(\zeta) - \xi(0)|^{\mathcal{A}}}{\theta\mathcal{T}_n}\right)\right],$$

and

$$\mathcal{N}\left(\Psi(\zeta) - \frac{(\xi(\zeta) - \xi(0))^{\mathcal{C}-1}}{\Gamma(\mathcal{C})} f_0 - \frac{1}{\Gamma(\mathcal{A})}\int_0^{\zeta} \xi'(\zeta_\circ)(\xi(\zeta) - \xi(\zeta_\circ))^{\mathcal{A}-1}\mu(\zeta_\circ, \Psi(\zeta_\circ), \Psi(\eta(\zeta_\circ)))d\zeta_\circ, (\mathcal{T}_1, \ldots, \mathcal{T}_n)\right)$$

$$\succeq \mathcal{N}\left(\frac{\theta}{\Gamma(\mathcal{A})}\int_0^{\zeta} \xi'(\zeta_\circ)(\xi(\zeta) - \xi(\zeta_\circ))^{\mathcal{A}-1}\mathbb{K}\left(\vartheta, b, (\xi(\zeta_\circ) - \xi(0))^{\mathcal{A}}\right)d\zeta_\circ, (\mathcal{T}_1, \ldots, \mathcal{T}_n)\right)$$

$$\succeq \mathcal{N}\left(\frac{\theta}{\Gamma(\mathcal{A})}\int_0^{\zeta} \xi'(\zeta_\circ)(\xi(\zeta) - \xi(\zeta_\circ))^{\mathcal{A}-1}\sum_{k=0}^{\infty}\frac{(\xi(\zeta_\circ) - \xi(0))^{k\mathcal{A}}}{k!\Gamma(\vartheta k + b)}d\zeta_\circ, (\mathcal{T}_1, \ldots, \mathcal{T}_n)\right)$$

$$= \mathcal{N}\left(\frac{\theta}{\Gamma(\mathcal{A})}\sum_{k=0}^{\infty}\frac{1}{k!\Gamma(\vartheta k + b)}\int_0^{\zeta}(\xi(\zeta) - \xi(\zeta_\circ))^{\mathcal{A}-1}(\xi(\zeta_\circ) - \xi(0))^{k\mathcal{A}}d\xi(\zeta_\circ), (\mathcal{T}_1, \ldots, \mathcal{T}_n)\right)$$

$$= \mathcal{N}\left(\frac{\theta}{\Gamma(\mathcal{A})}\sum_{k=0}^{\infty}\frac{1}{k!\Gamma(\vartheta k + b)}\int_0^{\xi(\zeta)-\xi(0)}(\xi(\zeta) - \xi(0) - u)^{\mathcal{A}-1}(u)^{k\mathcal{A}}du, (\mathcal{T}_1, \ldots, \mathcal{T}_n)\right)$$

$(u = \xi(\zeta_\circ) - \xi(0))$

$$\succeq \mathcal{N}\left(\frac{\theta}{\Gamma(\mathcal{A})}\sum_{k=0}^{\infty}\frac{1}{k!\Gamma(\vartheta k + b)}(\xi(\zeta) - \xi(0))^{\mathcal{A}-1}\int_0^{\xi(\zeta)-\xi(0)}(1 - \frac{u}{\xi(\zeta) - \xi(0)})^{\mathcal{A}-1}u^{k\mathcal{A}}du, (\mathcal{T}_1, \ldots, \mathcal{T}_n)\right)$$

$$= \mathcal{N}\left(\frac{\theta}{\Gamma(\mathcal{A})}\sum_{k=0}^{\infty}\frac{1}{k!\Gamma(\vartheta k + b)}(\xi(\zeta) - \xi(0))^{(k+1)\mathcal{A}}\int_0^1(1 - v)^{\mathcal{A}-1}v^{k\mathcal{A}}dv, (\mathcal{T}_1, \ldots, \mathcal{T}_n)\right)$$

$$\left(v = \frac{u}{\xi(\zeta) - \xi(0)}\right)$$

$$= \mathcal{N}\left(\frac{\theta}{\Gamma(\mathcal{A})}\sum_{k=0}^{\infty}\frac{1}{k!\Gamma(\vartheta k + b)}(\xi(\zeta) - \xi(0))^{(k+1)\mathcal{A}}\frac{\Gamma(k\mathcal{A} + 1)\Gamma(\mathcal{A})}{\Gamma((k+1)\mathcal{A} + 1)}, (\mathcal{T}_1, \ldots, \mathcal{T}_n)\right)$$

$$\succeq \mathcal{N}\left(\theta\sum_{k=0}^{\infty}\frac{(\xi(\zeta) - \xi(0))^{k\mathcal{A}}}{k!\Gamma(\vartheta k + b)}, (\mathcal{T}_1, \ldots, \mathcal{T}_n)\right)$$

$$= \text{diag}\left[\mathbb{K}\left(\vartheta, b, \frac{-|(\xi(\zeta) - \xi(0)|^{\mathcal{A}}}{\theta\mathcal{T}_1}\right), \ldots, \mathbb{K}\left(\vartheta, b, \frac{-|(\xi(\zeta) - \xi(0)|^{\mathcal{A}}}{\theta\mathcal{T}_n}\right)\right],$$

where $\mathbb{E}_{\mathcal{A}}(.)$, ${}_2\mathbb{F}_1(\mathsf{A}, \mathsf{B}; \mathsf{C}; .)$ and $\mathbb{K}(\vartheta, b, .)$ are the one–parameter mittag–Leffler function, the Hypergeometric function and the Wright functions, respectively, with

$\underbrace{\mathcal{T}_i}_{i=1,\ldots,n} \in (0, +\infty)$, $n \in \mathbb{N}$, $\theta > 0$, $\mathcal{A} \in (0, 1]$, $\vartheta, b, \mathsf{A}, \mathsf{B}, \mathsf{C} \in \mathbb{C}$, and $\Re(\vartheta)$, $\Re(b)$, $\Re(\mathsf{A})$, $\Re(\mathsf{B})$, $\Re(\mathsf{C}) > 0$.

Let us suppose the following axioms are satisfied:

($\mathcal{F}_1$) $\mu \in C((0, \mathsf{T}] \times \mathbb{R}^2, \mathbb{R})$, $\eta \in C([0, \mathsf{T}], [0, \mathsf{T}])$ and $\Psi, \Phi \in C([0, \mathsf{T}], \mathbb{R})$, such that

$$\begin{aligned}
&\mathscr{N}\Big(\Phi(\eta(\zeta)) - \Psi(\eta(\zeta)), (\mathcal{T}_1, \ldots, \mathcal{T}_n)\Big)\\
&\succeq \mathscr{N}\Big(\Phi(\zeta) - \Psi(\zeta), (\mathcal{T}_1, \ldots, \mathcal{T}_n)\Big),
\end{aligned}$$

for every $\underbrace{\mathcal{T}_i}_{i=1,\ldots,n} \in (0, +\infty)$ and $n \in \mathbb{N}$.

($\mathcal{F}_2$) There exists $\Delta > 0$ such that

$$\begin{aligned}
&\mathscr{N}\Big(\mu(\zeta, \vartheta_1, \vartheta_2) - \mu(\zeta, \kappa_1, \kappa_2), (\mathcal{T}_1, \ldots, \mathcal{T}_n)\Big)\\
&\succeq \mathscr{N}\Big(\Delta \sum_{j=1}^{2} (\vartheta_j - \kappa_j), (\mathcal{T}_1, \ldots, \mathcal{T}_n)\Big),
\end{aligned}$$

for every $\underbrace{\mathcal{T}_i}_{i=1,\ldots,n} \in (0, +\infty)$, $n \in \mathbb{N}$, $\zeta \in (0, \mathsf{T}]$, $\vartheta_j, \kappa_j \in \mathbb{R}$, and $j = 1, 2$.

($\mathcal{F}_3$) There exist $0 < \underbrace{\mathcal{M}_i}_{i=1,\ldots,n} < 1$, such that

$$\begin{aligned}
&I_{0^+}^{\mathcal{A};\xi} \underbrace{\varphi_i^{\circledS}}_{i=1,\ldots,n} \left(\frac{-|(\xi(\zeta) - \xi(0))^{\mathcal{A}}|}{\theta \underbrace{\mathcal{T}_i}_{i=1,\ldots,n}}\right)\\
&= \frac{1}{\Gamma(\mathcal{A})} \int_{0^+}^{\zeta} \xi'(\zeta_\circ)(\xi(\zeta) - \xi(\zeta_\circ))^{\mathcal{A}-1} \underbrace{\varphi_i^{\circledS}}_{i=1,\ldots,n} \left(\frac{-|(\xi(\zeta) - \xi(0))^{\mathcal{A}}|}{\theta \underbrace{\mathcal{T}_i}_{i=1,\ldots,n}}\right)\\
&\succeq \underbrace{\varphi_i^{\circledS}}_{i=1,\ldots,n} \left(\frac{-|(\xi(\zeta) - \xi(0))^{\mathcal{A}}|}{\theta \underbrace{\mathcal{M}_i \mathcal{T}_i}_{i=1,\ldots,n}}\right)
\end{aligned}$$

where $\zeta \in (0, \mathsf{T}]$, $\theta > 0$, $n \in \mathbb{N}$ and $\underbrace{\mathcal{T}_i}_{i=1,\ldots,n} \in (0, +\infty)$.

($\mathcal{F}_4$) we have $(2\Delta\mathcal{M}_1, \ldots, 2\Delta\mathcal{M}_n) < \underbrace{(1, \ldots, 1)}_{n}$, for every $n \in \mathbb{N}$.

Proposition 20.1 *Let $\lambda, \gamma \in \mathbb{R}$ be two integrable functions. If for every $\underbrace{\mathcal{T}_i}_{i=1,\ldots,n} \in (0, +\infty)$, we have*

$$\mathcal{N}\Big(\lambda, (\mathcal{T}_1, \ldots, \mathcal{T}_n)\Big) \succeq \mathcal{N}\Big(\gamma, (\mathcal{T}_1, \ldots, \mathcal{T}_n)\Big),$$

then, for every $\underbrace{\mathcal{T}_i}_{i=1,\ldots,n} \in (0, +\infty)$, we get

$$\mathcal{N}\Big(I_{0^+}^{\mathcal{A},\xi}\lambda, (\mathcal{T}_1, \ldots, \mathcal{T}_n)\Big) \succeq \mathcal{N}\Big(I_{0^+}^{\mathcal{A},\xi}\gamma, (\mathcal{T}_1, \ldots, \mathcal{T}_n)\Big).$$

Proof We have $\lambda \leq \gamma$, thus for every $\underbrace{\mathcal{T}_i}_{i=1,\ldots,n} \in (0, +\infty)$,

$$\begin{aligned}
\mathcal{N}\Big(I_{0^+}^{\mathcal{A},\xi}\lambda, (\mathcal{T}_1, \ldots, \mathcal{T}_n)\Big) &= \mathcal{N}\left(1, \frac{(\mathcal{T}_1, \ldots, \mathcal{T}_n)}{I_{0^+}^{\mathcal{A},\xi}\lambda}\right) \\
&\succeq \mathcal{N}\left(1, \frac{(\mathcal{T}_1, \ldots, \mathcal{T}_n)}{I_{0^+}^{\mathcal{A},\xi}\gamma}\right) \\
&= \mathcal{N}\Big(I_{0^+}^{\mathcal{A},\xi}\gamma, (\mathcal{T}_1, \ldots, \mathcal{T}_n)\Big).
\end{aligned}$$

Theorem 20.1 *Assume that $(\mathcal{F}_1)$, $(\mathcal{F}_2)$, $(\mathcal{F}_3)$ and $(\mathcal{F}_4)$ are satisfied, and suppose $\Phi \in C[0, \mathsf{T}]$ satisfies (20.4) and $I_{0^+}^{1-\mathcal{C};\xi}\Phi(0^+) = f_0 \in \mathbb{R}$. Then, (20.2)–(20.3) have a unique solution $\Upsilon \in C[0, \mathsf{T}]$, such that*

$$\begin{aligned}
\Upsilon(\zeta) = {} & \frac{(\xi(\zeta) - \xi(0))^{\mathcal{C}-1}}{\Gamma(\mathcal{C})} f_0 \\
& + \frac{1}{\Gamma(\mathcal{A})} \int_0^{\zeta} \xi'(\zeta_\circ)(\xi(\zeta) - \xi(\zeta_\circ))^{\mathcal{A}-1} \mu(\zeta_\circ, \Upsilon(\zeta_\circ), \Upsilon(\eta(\zeta_\circ)))d\zeta_\circ, \quad \zeta \in (0, \mathsf{T}],
\end{aligned} \tag{20.5}$$

in which $I_{0^+}^{1-\mathcal{C};\xi}\Upsilon(0^+) = f_0 \in \mathbb{R}$, and for every $\zeta \in (0, \mathsf{T}]$, $n \in \mathbb{N}, \theta > 0$, and $\underbrace{\mathcal{T}_i}_{i=1,\ldots,n} \in (0, +\infty)$, we have that

$$\begin{aligned}
&\mathcal{N}\Big(\Phi(\zeta) - \Upsilon(\zeta), (\mathcal{T}_1, \ldots, \mathcal{T}_n)\Big) \\
&\succeq diag\left[\varphi_1^{\circledS}\left(\frac{-|(\xi(\zeta) - \xi(0))^{\mathcal{A}}|}{\mathcal{D}_1\, \theta \mathcal{T}_1}\right), \ldots, \varphi_n^{\circledS}\left(\frac{-|(\xi(\zeta) - \xi(0))^{\mathcal{A}}|}{\mathcal{D}_n\, \theta \mathcal{T}_n}\right)\right],
\end{aligned} \tag{20.6}$$

where

$$\underbrace{\mathcal{D}_i}_{i=1,\dots,n} := \frac{1}{1-2\Delta \underbrace{\mathcal{M}_i}_{i=1,\dots,n}}.$$

Proof Set $\mho = \{\varpi \in C(0, \mathsf{T}], \varpi \text{ is continuous}\}$ and define a mapping $\Omega : \mho \times \mho \to [0, +\infty]^n$ by

$$\Omega(\varpi, \varpi') = \inf\Bigg\{(\mathtt{C}_1, \dots, \mathtt{C}_n) \in (0, +\infty)^n : \mathscr{N}\Big(\varpi(\zeta) - \varpi'(\zeta), (\mathcal{T}_1, \dots, \mathcal{T}_n)\Big) \\ \succeq \operatorname{diag}\Bigg[\varphi_1^{\circledS}\Bigg(\frac{-|(\xi(\zeta)-\xi(0))^{\mathcal{A}}|}{\theta\frac{\mathcal{T}_1}{\mathtt{C}_1}}\Bigg), \dots, \varphi_n^{\circledS}\Bigg(\frac{-|(\xi(\zeta)-\xi(0))^{\mathcal{A}}|}{\theta\frac{\mathcal{T}_n}{\mathtt{C}_n}}\Bigg)\Bigg], \\ \varpi, \varpi' \in \mho,\ \zeta \in (0, \mathsf{T}],\ \underbrace{\mathcal{T}_i}_{i=1,\dots,n} \in (0, +\infty), n \in \mathbb{N}\Bigg\}. \tag{20.7}$$

First, we show that $(\mho, \Omega)$ is a $[0, +\infty]^n$–valued metric space.

We show that $\Omega(\varpi, \varpi') = \underbrace{(0, \dots, 0)}_{n}$ if and only if $\varpi = \varpi'$. Let $\Omega(\varpi, \varpi') = \underbrace{(0, \dots, 0)}_{n}$, we have

$$\inf\Bigg\{(\mathtt{C}_1, \dots, \mathtt{C}_n) \in (0, +\infty)^n : \mathscr{N}\Big(\varpi(\zeta) - \varpi'(\zeta), (\mathcal{T}_1, \dots, \mathcal{T}_n)\Big) \\ \succeq \operatorname{diag}\Bigg[\varphi_1^{\circledS}\Bigg(\frac{-|(\xi(\zeta)-\xi(0))^{\mathcal{A}}|}{\theta\frac{\mathcal{T}_1}{\mathtt{C}_1}}\Bigg), \dots, \varphi_n^{\circledS}\Bigg(\frac{-|(\xi(\zeta)-\xi(0))^{\mathcal{A}}|}{\theta\frac{\mathcal{T}_n}{\mathtt{C}_n}}\Bigg)\Bigg], \\ \varpi, \varpi' \in \mho,\ \zeta \in (0, \mathsf{T}],\ \underbrace{\mathcal{T}_i}_{i=1,\dots,n} \in (0, +\infty), n \in \mathbb{N}\Bigg\} = 0.$$

and so

$$\mathscr{N}\Big(\varpi(\zeta) - \varpi'(\zeta), (\mathcal{T}_1, \dots, \mathcal{T}_n)\Big) \\ \succeq \operatorname{diag}\Bigg[\varphi_1^{\circledS}\Bigg(\frac{-|(\xi(\zeta)-\xi(0))^{\mathcal{A}}|}{\theta\frac{\mathcal{T}_1}{\mathtt{C}_1}}\Bigg), \dots, \varphi_n^{\circledS}\Bigg(\frac{-|(\xi(\zeta)-\xi(0))^{\mathcal{A}}|}{\theta\frac{\mathcal{T}_n}{\mathtt{C}_n}}\Bigg)\Bigg],$$

for every $\underbrace{\mathtt{C}_i}_{i=1,\dots,n} \in (0, +\infty)$. Tend $\underbrace{\mathtt{C}_i}_{i=1,\dots,n}$ to zero in the above inequality, we get

$$\mathscr{N}\Big(\varpi(\zeta) - \varpi'(\zeta), (\mathcal{T}_1, \dots, \mathcal{T}_n)\Big) = \mathbf{1}$$

thus, $\varpi(\zeta) = \varpi'(\zeta)$, for every $\zeta \in [0, \mathrm{T}]$, and vice versa. It is straight forward to show $\Omega(\varpi, \varpi') = \Omega(\varpi', \varpi)$, for every $\varpi, \varpi' \in \mho$. Now, let $\Omega(\varpi, \rho) = (\ell_1, \ldots, \ell_n) \in (0, +\infty)^n$ and $\Omega(\rho, \varpi') = (\jmath_1, \ldots, \jmath_n) \in (0, +\infty)^n$. Then, we have

$$\mathscr{N}\Big(\varpi(\zeta) - \rho(\zeta), (\mathcal{T}_1, \ldots, \mathcal{T}_n)\Big) \succeq \operatorname{diag}\left[\varphi_1^{\circledS}\left(\frac{-|(\xi(\zeta) - \xi(0))^{\mathcal{A}}|}{\theta\frac{\mathcal{T}_1}{\ell_1}}\right), \ldots, \varphi_n^{\circledS}\left(\frac{-|(\xi(\zeta) - \xi(0))^{\mathcal{A}}|}{\theta\frac{\mathcal{T}_n}{\ell_n}}\right)\right],$$

and

$$\mathscr{N}\Big(\rho(\zeta) - \varpi'(\zeta), (\mathcal{T}_1, \ldots, \mathcal{T}_n)\Big) \succeq \operatorname{diag}\left[\varphi_1^{\circledS}\left(\frac{-|(\xi(\zeta) - \xi(0))^{\mathcal{A}}|}{\theta\frac{\mathcal{T}_1}{\jmath_1}}\right), \ldots, \varphi_n^{\circledS}\left(\frac{-|(\xi(\zeta) - \xi(0))^{\mathcal{A}}|}{\theta\frac{\mathcal{T}_n}{\jmath_n}}\right)\right],$$

for every $\underbrace{\mathcal{T}_i}_{i=1,\ldots,n} \in (0, +\infty)$, then, we have that

$$\begin{aligned}
&\mathscr{N}\Big(\varpi(\zeta) - \varpi'(\zeta), \Big((\ell_1 + \jmath_1)\mathcal{T}_1, \ldots, (\ell_n + \jmath_n)\mathcal{T}_n\Big)\Big) \\
&\succeq \mathscr{N}\Big(\varpi(\zeta) - \rho(\zeta), (\ell_1\mathcal{T}_1, \ldots, \ell_n\mathcal{T}_n)\Big) \bigotimes_{\mathrm{GTN}} \mathscr{N}\Big(\rho(\zeta) - \varpi'(\zeta), (\jmath_1\mathcal{T}_1, \ldots, \jmath_n\mathcal{T}_n)\Big) \\
&\succeq \operatorname{diag}\left[\varphi_1^{\circledS}\left(\frac{-|(\xi(\zeta) - \xi(0))^{\mathcal{A}}|}{\theta\mathcal{T}_1}\right), \ldots, \varphi_n^{\circledS}\left(\frac{-|(\xi(\zeta) - \xi(0))^{\mathcal{A}}|}{\theta\mathcal{T}_n}\right)\right] \\
&\qquad \bigotimes_{\mathrm{GTN}} \operatorname{diag}\left[\varphi_1^{\circledS}\left(\frac{-|(\xi(\zeta) - \xi(0))^{\mathcal{A}}|}{\theta\mathcal{T}_1}\right), \ldots, \varphi_n^{\circledS}\left(\frac{-|(\xi(\zeta) - \xi(0))^{\mathcal{A}}|}{\theta\mathcal{T}_n}\right)\right] \\
&\succeq \operatorname{diag}\left[\varphi_1^{\circledS}\left(\frac{-|(\xi(\zeta) - \xi(0))^{\mathcal{A}}|}{\theta\mathcal{T}_1}\right), \ldots, \varphi_n^{\circledS}\left(\frac{-|(\xi(\zeta) - \xi(0))^{\mathcal{A}}|}{\theta\mathcal{T}_n}\right)\right],
\end{aligned}$$

and so $\Omega(\varpi, \varpi') \leq (\ell_1 + \jmath_1, \ldots, \ell_n + \jmath_n)$. Thus, $\Omega(\varpi, \varpi') \leq \Omega(\varpi, \rho) + \Omega(\rho, \varpi')$.

Now, we are ready to prove $(\mho, \Omega)$ is complete. Let $\{\varpi_k\}_k$ be a Cauchy sequence in $(\mho, \Omega)$. Let $\zeta \in [0, \mathrm{T}]$ be fixed, and $\underbrace{\sigma_i}_{i=1,\ldots,n} \in (0, +\infty)$ and $\underbrace{\ni_i}_{i=1,\ldots,n} \in (0, 1)$, be arbitrary and consider $\underbrace{\mathcal{T}_i}_{i=1,\ldots,n} \in (0, +\infty)$, such that

$$\underbrace{\varphi_i^{\circledS}}_{i=1,\ldots,n}\left(\frac{-|(\xi(\zeta)-\xi(0))^{\mathcal{A}}|}{\theta\underbrace{\mathcal{T}_i}_{i=1,\ldots,n}}\right)\succ 1-\underbrace{\ni_i}_{i=1,\ldots,n},$$

for every $n\in\mathbb{N}$.

For $\underbrace{\varepsilon_i\mathcal{T}_i}_{i=1,\ldots,n}<\underbrace{\sigma_i}_{i=1,\ldots,n}$ choose $k''\in\mathbb{N}$, such that

$$\Omega(\varpi_k,\varpi_{k'})<(\varepsilon_1,\ldots,\varepsilon_n),\quad \forall\, k,k'\geq k''.$$

Then, we get

$$\begin{aligned}
&\mathscr{N}\Bigg(\varpi_k(\zeta)-\varpi'_{k'}(\zeta),(\sigma_1,\ldots,\sigma_n)\Bigg)\\
&\succeq\mathscr{N}\Bigg(\varpi_k(\zeta)-\varpi'_{k'}(\zeta),(\varepsilon_1\mathcal{T}_1,\ldots,\varepsilon_n\mathcal{T}_n)\Bigg)\\
&\succeq\operatorname{diag}\left[\varphi_1^{\circledS}\left(\frac{-|(\xi(\zeta)-\xi(0))^{\mathcal{A}}|}{\theta\mathcal{T}_1}\right),\ldots,\varphi_n^{\circledS}\left(\frac{-|(\xi(\zeta)-\xi(0))^{\mathcal{A}}|}{\theta\mathcal{T}_n}\right)\right]\\
&\succ\operatorname{diag}\underbrace{[1-\ni_1,\ldots,1-\ni_n]}_{n},
\end{aligned}$$

Thus,

$$\mathscr{N}\Bigg(\varpi_k(\zeta)-\varpi_{k'}(\zeta),(\sigma_1,\ldots,\sigma_n)\Bigg)\succ\operatorname{diag}\underbrace{[1-\ni_1,\ldots,1-\ni_n]}_{n}.$$

and i.e., the sequence $\{\varpi_k(\zeta)\}_k$ is Cauchy in complete space $(\mathcal{W},\mathscr{N},\bigotimes_{\mathrm{GTN}})$ on compact set $[0,\mathsf{T}]$, so they are uniformly convergent to the mapping $\varpi\in C(0,\mathsf{T}]$. The uniform convergence leads us to the fact that ϖ is continuous, i.e., an element of $\mho$ and then $(\mho,\Omega)$ is complete.

In view of Lemma 20.1, we have that (20.2)–(20.3) are equivalent to the following system

$$\begin{aligned}
\Psi(\zeta)=&\frac{(\xi(\zeta)-\xi(0))^{\mathcal{C}-1}}{\Gamma(\mathcal{C})}f_0 \qquad (20.8)\\
&+\frac{1}{\Gamma(\mathcal{A})}\int_0^{\zeta}\xi'(\zeta_\circ)(\xi(\zeta)-\xi(\zeta_\circ))^{\mathcal{A}-1}\mu(\zeta_\circ,\Psi(\zeta_\circ),\Psi(\eta(\zeta_\circ)))d\zeta_\circ,\ \ \zeta\in(0,\mathsf{T}].
\end{aligned}$$

In fact, applying the fractional integral operator $I_{0^+}^{\mathcal{A};\xi}(.)$ on both sides of the fractional equation (20.1) and using Theorem 2.5 we can earn (20.8).

On the other hand, if Ψ satisfies (20.8), then Ψ satisfies (20.1). However, applying the fractional derivative ${}^{\mathscr{H}}D_{0^+}^{\mathcal{A},\mathcal{B};\xi}(.)$ on both sides of (20.8), we have

$$ {}^{\mathscr{H}}D_{0^+}^{\mathcal{A},\mathcal{B};\xi}\Psi(\zeta) = {}^{\mathscr{H}}D_{0^+}^{\mathcal{A},\mathcal{B};\xi}\left[\frac{(\xi(\zeta)-\xi(0))^{\mathcal{C}-1}}{\Gamma(\mathcal{C})}f_0\right] + {}^{\mathscr{H}}D_{0^+}^{\mathcal{A},\mathcal{B};\xi}I_{0^+}^{\mathcal{A};\xi}\mu(\xi,\Psi(\xi),\Psi(\eta(\xi))). $$

Using Theorem 2.4 and the expression

$$ {}^{\mathscr{H}}D_{0^+}^{\mathcal{A},\mathcal{B};\xi}(\xi(\zeta)-\xi(0))^{\mathcal{C}-1} = 0, \quad 0 < \mathcal{C} < 1, $$

we conclude that, $\Psi(\zeta)$ satisfies initial problem equation (20.1) if, and only if, $\Psi(\zeta)$ satisfies the integral equation (20.8).

Consider $\mathbf{H} := \mho \longrightarrow \mho$, such that

$$ \mathbf{H}(\Phi(\zeta)) = s\frac{(\xi(\zeta)-\xi(0))^{\mathcal{C}-1}}{\Gamma(\mathcal{C})}f_0 + \frac{1}{\Gamma(\mathcal{A})}\int_0^\zeta \xi'(\zeta_\circ)(\xi(\zeta)-\xi(\zeta_\circ))^{\mathcal{A}-1}\mu(\zeta_\circ,\Psi(\zeta_\circ),\Psi(\eta(\zeta_\circ)))d\zeta_\circ, \quad \zeta \in (0,\mathrm{T}]. \tag{20.9} $$

Notice that if Φ be a continuous function, then $\mathbf{H}\Phi$ is also continuous. Actually,

$$ \begin{aligned} &\mathscr{N}\Big(\mathbf{H}(\Phi(\zeta)) - \mathbf{H}(\Phi(\zeta_0)), (\mathcal{T}_1,\dots,\mathcal{T}_n)\Big) \\ &= \mathscr{N}\Bigg(\frac{(\xi(\zeta)-\xi(0))^{\mathcal{C}-1}}{\Gamma(\mathcal{C})}f_0 \\ &\quad + \frac{1}{\Gamma(\mathcal{A})}\int_0^\zeta \Psi'(\zeta_\circ)(\Psi(\zeta)-\Psi(\zeta_\circ))^{\mathcal{A}-1}\mu(\zeta_\circ,\Psi(\zeta_\circ),\Psi(\eta(\zeta_\circ)))d\zeta_\circ \\ &\quad - \frac{(\xi(\zeta_0)-\xi(0))^{\mathcal{C}-1}}{\Gamma(\mathcal{C})}f_0 \\ &\quad - \frac{1}{\Gamma(\mathcal{A})}\int_0^\zeta \Psi'(\zeta_\circ)(\Psi(\zeta_0)-\Psi(\zeta_\circ))^{\mathcal{A}-1}\mu(\zeta_\circ,\Psi(\zeta_\circ),\Psi(\eta(\zeta_\circ)))d\zeta_\circ, (\mathcal{T}_1,\dots,\mathcal{T}_n)\Bigg) \\ &\longrightarrow \mathbf{1}, \end{aligned} $$

as $\zeta \longrightarrow \zeta_0$. Next, we show that the self-mapping $\mathbf{H}$ is contractive on $\mho$. Consider $\mathbf{H} : \mho \longrightarrow \mho$ defined in (20.9). Suppose $\Phi, \Psi \in C[0,\mathrm{T}]$, $\underbrace{\Bbbk_i}_{i=1,\dots,n} \in [0,+\infty]$, and $\Omega(\Phi(\zeta),\Psi(\zeta)) \preceq (\Bbbk_1,\dots,\Bbbk_n)$. So for every $\zeta \in [0,\mathrm{T}]$,

$$ \begin{aligned} &\mathscr{N}\Big(\Phi(\zeta)-\Psi(\zeta),(\mathcal{T}_1,\dots,\mathcal{T}_n)\Big) \\ &\succeq \operatorname{diag}\left[\varphi_1^{\circledS}\left(\frac{-|(\xi(\zeta)-\xi(0))^{\mathcal{A}}|}{\theta\frac{\mathcal{T}_1}{\Bbbk_1}}\right),\dots,\varphi_n^{\circledS}\left(\frac{-|(\xi(\zeta)-\xi(0))^{\mathcal{A}}|}{\theta\frac{\mathcal{T}_n}{\Bbbk_n}}\right)\right]. \end{aligned} $$

For every $\zeta \in (0,\mathrm{T}]$ and $\underbrace{\mathcal{T}_i}_{i=1,\dots,n} \in (0,+\infty)$ with $n \in \mathbb{N}$, we have

$$
\begin{aligned}
&\mathscr{N}\Big(\mathbf{H}(\Phi(\zeta)) - \mathbf{H}(\Psi(\zeta)), (\mathcal{T}_1, \ldots, \mathcal{T}_n)\Big)\\
\succeq\ &\mathscr{N}\Big(\frac{1}{\Gamma(\mathcal{A})}\int_0^{\zeta} \xi'(\zeta_\circ)(\xi(\zeta) - \xi(\zeta_\circ))^{\mathcal{A}-1}\Big(\mu(\zeta_\circ, \Phi(\zeta_\circ), \Phi(\eta(\zeta_\circ)))\\
&\qquad -\mu(\zeta_\circ, \Psi(\zeta_\circ), \Psi(\eta(\zeta_\circ)))\Big)d\zeta_\circ, (\mathcal{T}_1, \ldots, \mathcal{T}_n)\Big)\\
\succeq\ &\mathscr{N}\Big(\frac{\Delta}{\Gamma(\mathcal{A})}\int_0^{\zeta} \xi'(\zeta_\circ)(\xi(\zeta) - \xi(\zeta_\circ))^{\mathcal{A}-1}\\
&\qquad \Big[\Big(\Phi(\zeta_\circ) - \Psi(\zeta_\circ)\Big) + \Big(\Phi(\eta(\zeta_\circ)) - \Psi(\eta(\zeta_\circ))\Big)\Big]d\zeta_\circ, (\mathcal{T}_1, \ldots, \mathcal{T}_n)\Big)\\
\succeq\ &\mathscr{N}\Big(\frac{\Delta}{\Gamma(\mathcal{A})}\int_0^{\zeta} \xi'(\zeta_\circ)(\xi(\zeta) - \xi(\zeta_\circ))^{\mathcal{A}-1}\Big(\Phi(\zeta_\circ) - \Psi(\zeta_\circ)\Big)d\zeta_\circ, \frac{(\mathcal{T}_1, \ldots, \mathcal{T}_n)}{2}\Big)\\
\bigotimes_{\mathrm{GTN}}\ &\mathscr{N}\Big(\frac{\Delta}{\Gamma(\mathcal{A})}\int_0^{\zeta} \xi'(\zeta_\circ)(\xi(\zeta) - \xi(\zeta_\circ))^{\mathcal{A}-1}\Big(\Phi(\eta(\zeta_\circ)) - \Psi(\eta(\zeta_\circ))\Big)d\zeta_\circ, \frac{(\mathcal{T}_1, \ldots, \mathcal{T}_n)}{2}\Big)\\
\succeq\ &\mathscr{N}\Big(\frac{\Delta}{\Gamma(\mathcal{A})}\int_0^{\zeta} \xi'(\zeta_\circ)(\xi(\zeta) - \xi(\zeta_\circ))^{\mathcal{A}-1}\Big(\Phi(\zeta_\circ) - \Psi(\zeta_\circ)\Big)d\zeta_\circ, \frac{(\mathcal{T}_1, \ldots, \mathcal{T}_n)}{2}\Big)\\
\bigotimes_{\mathrm{GTN}}\ &\mathscr{N}\Big(\frac{\Delta}{\Gamma(\mathcal{A})}\int_0^{\zeta} \xi'(\zeta_\circ)(\xi(\zeta) - \xi(\zeta_\circ))^{\mathcal{A}-1}\Big(\Phi(\zeta_\circ) - \Psi(\zeta_\circ)\Big)d\zeta_\circ, \frac{(\mathcal{T}_1, \ldots, \mathcal{T}_n)}{2}\Big)\\
\succeq\ &\mathscr{N}\Big(\frac{\Delta}{\Gamma(\mathcal{A})}\int_0^{\zeta} \xi'(\zeta_\circ)(\xi(\zeta) - \xi(\zeta_\circ))^{\mathcal{A}-1}\Big(\Phi(\zeta_\circ) - \Psi(\zeta_\circ)\Big)d\zeta_\circ, \frac{(\mathcal{T}_1, \ldots, \mathcal{T}_n)}{2}\Big)\\
\succeq\ &\frac{1}{\Gamma(\mathcal{A})}\int_0^{\zeta} \xi'(\zeta_\circ)(\xi(\zeta) - \xi(\zeta_\circ))^{\mathcal{A}-1}\\
&\qquad \times\mathrm{diag}\Big[\varphi_1^{\circledS}\Big(\frac{-|(\xi(\zeta) - \xi(0))^{\mathcal{A}}|}{2\Delta\theta\frac{\mathcal{T}_1}{\Bbbk_1}}\Big), \ldots, \varphi_n^{\circledS}\Big(\frac{-|(\xi(\zeta) - \xi(0))^{\mathcal{A}}|}{2\Delta\theta\frac{\mathcal{T}_n}{\Bbbk_n}}\Big)\Big]d\zeta_\circ\\
\succeq\ &\mathrm{diag}\Big[\varphi_1^{\circledS}\Big(\frac{-|(\xi(\zeta) - \xi(0))^{\mathcal{A}}|}{2\Delta\mathcal{M}_1\theta\frac{\mathcal{T}_1}{\Bbbk_1}}\Big), \ldots, \varphi_n^{\circledS}\Big(\frac{-|(\xi(\zeta) - \xi(0))^{\mathcal{A}}|}{2\Delta\mathcal{M}_n\theta\frac{\mathcal{T}_n}{\Bbbk_n}}\Big)\Big].
\end{aligned}
$$

So, we get

$$
\begin{aligned}
&\Omega\Big(\mathbf{H}(\Phi(\zeta)) - \mathbf{H}(\Psi(\zeta)), (\mathcal{T}_1, \ldots, \mathcal{T}_n)\Big)\\
&\preceq (2\Delta\mathcal{M}_1, \ldots, 2\Delta\mathcal{M}_n)\, \Omega\Big(\Phi(\zeta) - \Psi(\zeta), (\mathcal{T}_1, \ldots, \mathcal{T}_n)\Big),
\end{aligned}
$$

which implies the contractively property of $\mathbf{H}$, because $2\Delta \underbrace{\mathcal{M}_i}_{i=1,\dots,n} < 1$.

Let $\psi \in \mho$, we show that $\Omega(\mathbf{H}\psi, \psi) < \underbrace{(\infty, \dots, \infty)}_{n}$. Making use of (20.4) and Remark 20.1, we get

$$\begin{aligned}&\mathcal{N}\Big(\psi(\zeta) - \frac{(\xi(\zeta) - \xi(0))^{\mathcal{C}-1}}{\Gamma(\mathcal{C})} f_0\\&\qquad - \frac{1}{\Gamma(\mathcal{A})}\int_0^\zeta \xi'(\zeta_\circ)(\xi(\zeta) - \xi(\zeta_\circ))^{\mathcal{A}-1}\mu(\zeta_\circ, \psi(\zeta_\circ), \psi(\eta(\zeta_\circ)))d\zeta_\circ, (\mathcal{T}_1, \dots, \mathcal{T}_n)\Big)\\&\succeq \operatorname{diag}\Big[\varphi_1^{\circledS}\Big(\frac{-|(\xi(\zeta) - \xi(0))^{\mathcal{A}}|}{\theta\mathcal{T}_1}\Big), \dots, \varphi_n^{\circledS}\Big(\frac{-|(\xi(\zeta) - \xi(0))^{\mathcal{A}}|}{\theta\mathcal{T}_n}\Big)\Big],\end{aligned}$$

for every $\underbrace{\mathcal{T}_i}_{i=1,\dots,n} \in (0, +\infty)$ and $\zeta \in (0, \mathsf{T}]$, where $I_{0^+}^{1-\mathcal{C};\xi}\psi(0^+) = f_0 \in \mathbb{R}$. Then, we have $\Omega(\mathbf{H}\psi, \psi) < \underbrace{(1, \dots, 1)}_{n}$, for every $n \in \mathbb{N}$.

Now, Theorem 2.20 enables us to find an element Υ in $\mho$ satisfying the following:

(1) Υ is a fixed point of $\mathbf{H}$, i.e.,

$$\begin{aligned}\mathbf{H}(\Upsilon(\zeta)) = {}& \frac{(\xi(\zeta) - \xi(0))^{\mathcal{C}-1}}{\Gamma(\mathcal{C})} f_0\\&+ \frac{1}{\Gamma(\mathcal{A})}\int_0^\zeta \xi'(\zeta_\circ)(\xi(\zeta) - \xi(\zeta_\circ))^{\mathcal{A}-1}\mu(\zeta_\circ, \Upsilon(\zeta_\circ), \Upsilon(\eta(\zeta_\circ)))d\zeta_\circ, \quad \zeta \in (0, \mathsf{T}],\end{aligned}$$

which is unique in the set

$$\mho^* = \{\phi \in \mho : \Omega(\mathbf{H}\psi, \phi) < \underbrace{(\infty, \dots, \infty)}_{n}\},$$

where $I_{0^+}^{1-\mathcal{C};\xi}\Upsilon(0^+) = f_0 \in \mathbb{R}$.

(2) $\Omega(\mathbf{H}^n(\psi), \Upsilon) \to \underbrace{(0, \dots, 0)}_{n}$ as $n \to \infty$;

(3) $$\Omega(\psi, \Upsilon) \le \underbrace{(\frac{1}{1 - 2\Delta\mathcal{M}_1}, \dots, \frac{1}{1 - 2\Delta\mathcal{M}_n})}_{n}\Omega(\mathbf{H}\psi, \psi) \le$$

$\underbrace{(\frac{1}{1 - 2\Delta\mathcal{M}_1}, \dots, \frac{1}{1 - 2\Delta\mathcal{M}_n})}_{n}$, which implies that

$$\mathcal{N}\Big(\psi(\zeta)-\Upsilon(\zeta),(\mathcal{T}_1,\ldots,\mathcal{T}_n)\Big)$$
$$\succeq \operatorname{diag}\Bigg[\varphi_1\Bigg(\frac{-|(\xi(\zeta)-\xi(0))^{\mathcal{A}}|}{\mathcal{D}_1\,\theta\mathcal{T}_1}\Bigg),\ldots,\varphi_n\Bigg(\frac{-|(\xi(\zeta)-\xi(0))^{\mathcal{A}}|}{\mathcal{D}_n\,\theta\mathcal{T}_n}\Bigg)\Bigg],$$

where

$$\underbrace{\mathcal{D}_i}_{i=1,\ldots,n} := \frac{1}{1-2\Delta\underbrace{\mathcal{M}_i}_{i=1,\ldots,n}},$$

for every $\zeta \in (0,\mathsf{T}]$, $n \in \mathbb{N}$, and $\underbrace{\mathcal{T}_i}_{i=1,\ldots,n} \in (0,+\infty)$.

Now, we show that the fixed point in $\mho^*$ is unique. Let Λ be an element of $\mho$ satisfying (20.5) and (20.6), we prove that $\Lambda=\Upsilon$ and $\Lambda \in \mho^*$. From (20.5) we get

$$\Lambda(\zeta)=\frac{(\xi(\zeta)-\xi(0))^{\mathcal{C}-1}}{\Gamma(\mathcal{C})}f_0 \tag{20.10}$$
$$+\frac{1}{\Gamma(\mathcal{A})}\int_0^{\zeta}\xi'(\zeta_\circ)(\xi(\zeta)-\xi(\zeta_\circ))^{\mathcal{A}-1}\mu(\zeta_\circ,\Lambda(\zeta_\circ),\Lambda(\eta(\zeta_\circ)))d\zeta_\circ,\quad \zeta\in(0,\mathsf{T}],$$

and so

$$\mathbf{H}(\Lambda(\zeta))=\frac{(\xi(\zeta)-\xi(0))^{\mathcal{C}-1}}{\Gamma(\mathcal{C})}f_0 \tag{20.11}$$
$$+\frac{1}{\Gamma(\mathcal{A})}\int_0^{\zeta}\xi'(\zeta_\circ)(\xi(\zeta)-\xi(\zeta_\circ))^{\mathcal{A}-1}\mu(\zeta_\circ,\Lambda(\zeta_\circ),\Lambda(\eta(\zeta_\circ)))d\zeta_\circ,\quad \zeta\in(0,\mathsf{T}],$$

where $\mu \in C((0,\mathsf{T}]\times\mathbb{R}^2,\mathbb{R})$, $\eta \in C([0,\mathsf{T}],[0,\mathsf{T}])$, and $I_{0^+}^{1-\mathcal{C};\xi}\Lambda(0^+)=f_0\in\mathbb{R}$.

Now, we show that

$$\Lambda \in \{\phi\in\mho : \Omega(\mathbf{H}(\psi),\phi) < \underbrace{(\infty,\ldots,\infty)}_{n}\},$$

i.e., $\Omega(\mathbf{H}(\psi),\Lambda) < \underbrace{(\infty,\ldots,\infty)}_{n}$. From (20.6) we get

$$\mathcal{N}\Big(\psi(\zeta)-\Lambda(\zeta),(\mathcal{T}_1,\ldots,\mathcal{T}_n)\Big) \tag{20.12}$$
$$\succeq \operatorname{diag}\Bigg[\varphi_1^{\circledS}\Bigg(\frac{-|(\xi(\zeta)-\xi(0))^{\mathcal{A}}|}{\mathcal{D}_1\,\theta\mathcal{T}_1}\Bigg),\ldots,\varphi_n^{\circledS}\Bigg(\frac{-|(\xi(\zeta)-\xi(0))^{\mathcal{A}}|}{\mathcal{D}_n\,\theta\mathcal{T}_n}\Bigg)\Bigg],$$

where

$$\underbrace{\mathcal{D}_i}_{i=1,\ldots,n} := \frac{1}{1-2\Delta \underbrace{\mathcal{M}_i}_{i=1,\ldots,n}}.$$

for every $\zeta \in (0, \mathtt{T}]$, $n \in \mathbb{N}$ and $\underbrace{\mathcal{T}_i}_{i=1,\ldots,n} \in (0, +\infty)$.

Using triangular inequality, (20.11), (20.12), (20.4) and Remark 20.1, we have

$$\begin{aligned}
&\mathscr{N}\Big(\mathbf{H}(\psi(\zeta)) - \Lambda(\zeta), (\mathcal{T}_1, \ldots, \mathcal{T}_n)\Big)\\
&= \mathscr{N}\Bigg(\frac{(\xi(\zeta)-\xi(0))^{\mathcal{C}-1}}{\Gamma(\mathcal{C})} f_0\\
&\qquad + \frac{1}{\Gamma(\varsigma)} \int_0^{\zeta} \xi'(\zeta_\circ)(\xi(\zeta)-\xi(\zeta_\circ))^{\mathcal{A}-1}\mu(\zeta_\circ, \psi(\zeta_\circ), \psi(\eta(\zeta_\circ)))d\zeta_\circ\\
&\qquad\qquad + \psi(\zeta) - \psi(\zeta) - \Lambda(\zeta), (\mathcal{T}_1, \ldots, \mathcal{T}_n)\Bigg)\\
&\succeq \mathscr{N}\Bigg(\frac{(\xi(\zeta)-\xi(0))^{\mathcal{C}-1}}{\Gamma(\mathcal{C})} f_0\\
&\qquad + \frac{1}{\Gamma(\mathcal{C})} \int_0^{\zeta} \xi'(\zeta_\circ)(\xi(\zeta)-\xi(\zeta_\circ))^{\mathcal{A}-1}\mu(\zeta_\circ, \psi(\zeta_\circ), \psi(\eta(\zeta_\circ)))d\zeta_\circ - \psi(\zeta), \frac{(\mathcal{T}_1, \ldots, \mathcal{T}_n)}{2}\Bigg)\\
&\quad \bigotimes_{\mathrm{GTN}} \mathscr{N}\left(\psi(\zeta) - \Lambda(\zeta), \frac{(\mathcal{T}_1, \ldots, \mathcal{T}_n)}{2}\right)\\
&\succeq \operatorname{diag}\left[\varphi_1^{\circledS}\left(\frac{-|(\xi(\zeta)-\xi(0))^{\mathcal{A}}|}{2\theta\mathcal{T}_1}\right), \ldots, \varphi_n^{\circledS}\left(\frac{-|(\xi(\zeta)-\xi(0))^{\mathcal{A}}|}{2\theta\mathcal{T}_n}\right)\right]\\
&\quad \bigotimes_{\mathrm{GTN}} \operatorname{diag}\left[\varphi_1^{\circledS}\left(\frac{-|(\xi(\zeta)-\xi(0))^{\mathcal{A}}|}{2\mathcal{D}_1\,\theta\mathcal{T}_1}\right), \ldots, \varphi_n^{\circledS}\left(\frac{-|(\xi(\zeta)-\xi(0))^{\mathcal{A}}|}{2\mathcal{D}_n\,\theta\mathcal{T}_n}\right)\right]\\
&\succeq \operatorname{diag}\left[\varphi_1^{\circledS}\left(\frac{-|(\xi(\zeta)-\xi(0))^{\mathcal{A}}|}{2\max\{1, \mathcal{D}_1\}\theta\mathcal{T}_1}\right), \ldots, \varphi_n^{\circledS}\left(\frac{-|(\xi(\zeta)-\xi(0))^{\mathcal{A}}|}{2\max\{1, \mathcal{D}_n\}\theta\mathcal{T}_n}\right)\right].
\end{aligned}$$

For every $\zeta \in (0, \mathtt{T}]$, $\theta > 0$, $n \in \mathbb{N}$ and $\underbrace{\mathcal{T}_i}_{i=1,\ldots,n} \in (0, +\infty)$.

Which implies that $\Omega(\mathbf{H}\psi, \Lambda) \preceq \underbrace{(2\max\{1, \mathcal{D}_1\}, \ldots, 2\max\{1, \mathcal{D}_n\})}_{n} < \underbrace{(\infty, \ldots, \infty)}_{n}$, then $\Lambda \in \mho^*$.

20.2 Application

In this part, we apply the main results to solve an example.

Example 20.1 Let $(\mathbb{R}, \mathscr{N}, \bigotimes_{\mathrm{GTN}})$ be an MVFB–space. Consider the following fractional system

$$\begin{cases} {}^{\mathscr{H}}D_{0^+}^{\mathcal{A},\mathcal{B};\xi}\Phi(\zeta) = \sin^2(\Phi(\zeta)) + \cos(\Phi(\frac{\zeta}{2})) + 1, & \zeta \in (0, \mathrm{T}], \\ I_{0^+}^{1-\mathcal{C};\xi}\Phi(0^+) = f_0, & f_0 \in \mathbb{R}, \end{cases}$$

where ${}^{\mathscr{H}}D_{0^+}^{\mathcal{A},\mathcal{B};\xi}(.)$ is the ξ–Hilfer fractional derivative of order $0 < \mathcal{A} \leq 1$ and type $0 \leq \mathcal{B} \leq 1$, $I_{0^+}^{1-\mathcal{C};\xi}(.)$ is the Riemann–Liouville fractional integral of order $1 - \mathcal{C}$, $\mathcal{C} = \mathcal{A} + \mathcal{B}(1 - \mathcal{A})$ with respect to the function ξ, and $\Phi, \Psi \in C[0, \mathrm{T}]$ are given functions, such that for every $\zeta \in [0, \mathrm{T}]$ and $\underbrace{\mathcal{T}_i}_{i=1,\dots,n} \in (0, +\infty)$,

$$\mathscr{N}\left(\Phi(\frac{1}{2}\zeta) - \Psi(\frac{1}{2}\zeta), (\mathcal{T}_1, \dots, \mathcal{T}_n)\right) \succeq \mathscr{N}\left(\Phi(\zeta) - \Psi(\zeta), (\mathcal{T}_1, \dots, \mathcal{T}_n)\right),$$

and

$$\mathscr{N}\left(\Phi(\zeta) - \Psi(\zeta), (\mathcal{T}_1, \dots, \mathcal{T}_n)\right) \succeq \operatorname{diag}\left[\varphi_1^{\circledS}\left(\frac{-|(\xi(\zeta) - \xi(0))^{\mathcal{A}}|}{\theta\mathcal{T}_1}\right), \dots, \varphi_n^{\circledS}\left(\frac{-|(\xi(\zeta) - \xi(0))^{\mathcal{A}}|}{\theta\mathcal{T}_n}\right)\right],$$

where $\underbrace{\varphi_i^{\circledS}}_{i=1,\dots,n}$ with $n \in \mathbb{N}$, are fuzzy controllers defined in Sects. 9.8.1 and 9.8.2.

For every $\zeta \in [0, \mathrm{T}]$ and $\underbrace{\mathcal{T}_i}_{i=1,\dots,n} \in (0, +\infty)$ we have,

$$\begin{aligned} &\mathscr{N}\left(\sin^2(\Phi(\zeta)) - \sin^2(\Psi(\zeta)), (\mathcal{T}_1, \dots, \mathcal{T}_n)\right) \\ &= \mathscr{N}\left([\sin(\Phi(\zeta)) - \sin(\Psi(\zeta))][\sin(\Phi(\zeta)) + \sin(\Psi(\zeta))], (\mathcal{T}_1, \dots, \mathcal{T}_n)\right) \\ &\succeq \mathscr{N}\left((\Phi(\zeta) - \Psi(\zeta))\left[2\max\left(\sin(\Phi(\zeta)), \sin(\Psi(\zeta))\right)\right], (\mathcal{T}_1, \dots, \mathcal{T}_n)\right) \\ &\succeq \mathscr{N}\left(\Delta_1(\Phi(\zeta) - \Psi(\zeta)), (\mathcal{T}_1, \dots, \mathcal{T}_n)\right), \end{aligned}$$

and

$$
\begin{aligned}
&\mathcal{N}\left(\cos(\Phi(\frac{1}{2}\zeta)) - \cos(\Psi(\frac{1}{2}\zeta)), (\mathcal{T}_1, \ldots, \mathcal{T}_n)\right)\\
&= \mathcal{N}\left(2\sin\left(\frac{\Phi(\frac{1}{2}\zeta) + \Psi(\frac{1}{2}\zeta)}{2}\right)\sin\left(\frac{\Phi(\frac{1}{2}\zeta) - \Psi(\frac{1}{2}\zeta)}{2}\right), (\mathcal{T}_1, \ldots, \mathcal{T}_n)\right)\\
&\succeq \mathcal{N}\left(2\Delta_2\sin\left(\frac{\Phi(\frac{1}{2}\zeta) - \Psi(\frac{1}{2}\zeta)}{2}\right), (\mathcal{T}_1, \ldots, \mathcal{T}_n)\right)\\
&\succeq \mathcal{N}\left(2\Delta_2\frac{(\Phi(\frac{1}{2}\zeta) - \Psi(\frac{1}{2}\zeta))}{2}, (\mathcal{T}_1, \ldots, \mathcal{T}_n)\right)\\
&\succeq \mathcal{N}\left(\Delta_2(\Phi(\frac{1}{2}\zeta) - \Psi(\frac{1}{2}\zeta)), (\mathcal{T}_1, \ldots, \mathcal{T}_n)\right)\\
&\succeq \mathcal{N}\left(\Delta_2(\Phi(\zeta) - \Psi(\zeta)), (\mathcal{T}_1, \ldots, \mathcal{T}_n)\right),
\end{aligned}
$$

where $\Delta_1, \Delta_2 \in (0, +\infty)$.

Thus, for every $\zeta \in [0, \mathsf{T}]$ and $\underbrace{\mathcal{T}_i}_{i=1,\ldots,n} \in (0, +\infty)$ we have that

$$
\begin{aligned}
&\mathcal{N}\left(\left[\sin^2(\Phi(\zeta)) + \cos(\Phi(\frac{1}{2}\zeta)) + 1\right]\right.\\
&\qquad\left. -\left[\sin^2(\Psi(\zeta)) + \cos(\Psi(\frac{1}{2}\zeta)) + 1\right], (\mathcal{T}_1, \ldots, \mathcal{T}_n)\right)\\
&\succeq \mathcal{N}\left(\Delta_1(\Phi(\zeta) - \Psi(\zeta)), \frac{(\mathcal{T}_1, \ldots, \mathcal{T}_n)}{2}\right)\\
&\qquad \bigotimes_{\mathrm{GTN}} \mathcal{N}\left(\Delta_2(\Phi(\frac{1}{2}\zeta) - \Psi(\frac{1}{2}\zeta)), \frac{(\mathcal{T}_1, \ldots, \mathcal{T}_n)}{2}\right)\\
&\succeq \mathcal{N}\left(\Delta_0(\Phi(\zeta) - \Psi(\zeta)), (\mathcal{T}_1, \ldots, \mathcal{T}_n)\right),
\end{aligned}
$$

where $\Delta_0 \in (0, +\infty)$.

Let $\Upsilon \in C[0, \mathsf{T}]$ satisfies

$$
\begin{aligned}
&\mathcal{N}\left({}^{\mathscr{H}}D_{0^+}^{\mathcal{A},\mathcal{B};\xi}\Upsilon(\zeta) - \sin^2(\Upsilon(\zeta)) - \cos(\Upsilon(\frac{1}{2}\zeta)) - 1, (\mathcal{T}_1, \ldots, \mathcal{T}_n)\right)\\
&\succeq \mathrm{diag}\left[\varphi_1^{\circledS}\left(\frac{-|(\xi(\zeta) - \xi(0))^{\mathcal{A}}|}{\theta\mathcal{T}_1}\right), \ldots, \varphi_n^{\circledS}\left(\frac{-|(\xi(\zeta) - \xi(0))^{\mathcal{A}}|}{\theta\mathcal{T}_n}\right)\right],
\end{aligned}
$$

for every $\zeta \in [0, \mathsf{T}]$, $n \in \mathbb{N}$ and $\underbrace{\mathcal{T}_i}_{i=1,\ldots 0,n} \in (0, +\infty)$. Now, Theorem 2.20 implies that, if $2\Delta \underbrace{\mathcal{M}_i}_{i=1,\ldots,n} < 1$, we can find a unique differentiable function $\Lambda \in C[0, \mathsf{T}]$, such that

$$
{}^{\mathscr{H}}D_{0^+}^{\mathcal{A},\mathcal{B};\xi}\Lambda(\zeta) = \sin^2(\Lambda(\zeta)) + \cos(\Lambda(\frac{1}{2}\zeta)) + 1, \qquad \zeta \in [0, \mathsf{T}],
$$

and

$$
\mathscr{N}\Big(\Lambda(\zeta) - \Upsilon(\zeta), (\mathcal{T}_1, \ldots, \mathcal{T}_n)\Big)
$$
$$
\succeq \operatorname{diag}\left[\varphi_1^{\circledS}\left(\frac{-|(\xi(\zeta) - \xi(0))^{\mathcal{A}}|}{\mathcal{D}_1\,\theta\mathcal{T}_1}\right), \ldots, \varphi_n^{\circledS}\left(\frac{-|(\xi(\zeta) - \xi(0))^{\mathcal{A}}|}{\mathcal{D}_n\,\theta\mathcal{T}_n}\right)\right],
$$

where

$$
\underbrace{\mathcal{D}_i}_{i=1,\ldots,n} := \frac{1}{1 - 2\Delta \underbrace{\mathcal{M}_i}_{i=1,\ldots,n}},
$$

for every $\zeta \in [0, \mathsf{T}]$, $n \in \mathbb{N}$, $\theta > 0$ and $\underbrace{\mathcal{T}_i}_{i=1,\ldots,n} \in (0, +\infty)$.

We refer the interested readers to [3–13].

References

1. Aderyani SR, Saadati R, Yang XJ (2021) Radu-Mihet method for UHML stability for a class of ξ-Hilfer fractional differential equations in matrix valued fuzzy Banach spaces. Math Methods Appl Sci 44(18):14619–14631
2. Capelas de Olivera E, Vanterler de C, Sousa J (2018) Ulam-Hyers-Rassias stability for a class of fractional integro-differential equations. Results Math 73:111
3. Rezaei Aderyani S, Saadati R, Rassias TM, Park C (2022) Best approximation of $(\mathcal{G}_1, \mathcal{G}_2)$-random operator inequality in matrix Menger Banach algebras with application of stochastic Mittag-Leffler and H $\mathbb{H}$-Fox control functions. J Inequal Appl 2022(1):1–17
4. Aderyani SR, Saadati R (2021) Best approximations of the ϕ-Hadamard fractional Volterra integro-differential equation by matrix valued fuzzy control functions. Adv Differ Equ 154
5. Aderyani SR, Saadati R, Mesiar R (2022) Estimation of permuting tri-homomorphisms and permuting tri-derivations associated with the tri-additive Υ-random operator inequality in matrix MB-algebra. Int J Gen Syst 51(6):547–569
6. Aderyani SR, Saadati R, Allahviranloo T (2022) Existence, uniqueness and matrix-valued fuzzy Mittag-Leffler-Hypergeometric-Wright stability for P-Hilfer fractional differential equations in matrix-valued fuzzy Banach space. Comput Appl Math 41(6):234
7. Aderyani SR, Saadati R (2022) Approximation of derivation–homomorphism fuzzy functional inequalities in matrix valued FC-$\diamond$-algebras. In: 2022 9th Iranian joint congress on fuzzy and intelligent systems (CFIS). IEEE, pp 1–6

8. Aderyani SR, Saadati R, Allahviranloo T, Abbasbandy S, Catak M (2023) Fuzzy approximation of a fractional Lorenz system and a fractional financial crisis. Iran J Fuzzy Syst 20(7):27–36
9. Aderyani SR, Saadati R, Abdeljawad T, Mlaiki N (2022) Multi-stability of non homogenous vector-valued fractional differential equations in matrix-valued Menger spaces. Alexandria Eng J 61(12):10913–10923
10. Aderyani SR, Saadati R (2023) Stability and controllability results by n-ary aggregation functions in matrix valued fuzzy n-normed spaces. Inf Sci 119265
11. O'Regan D, Aderyani SR, Saadati R, Li C (2023) Stability results and parametric delayed Mittag-Leffler matrices in symmetric fuzzy-random spaces with application. Symmetry 15(10):1880
12. Aderyani SR, Saadati R, O'Regan D, Li C (2023) On a new approach for stability and controllability analysis of functional equations. Mathematics 11(16):3458
13. Aderyani SR, Saadati R, O'Regan D, Alshammari FS (2023) Fuzzy approximate solutions of matrix-valued fractional differential equations by fuzzy control functions. Mathematics 11(6):1386

Chapter 21
Multi Stability of Fractional–Order System II in MVFB–Spaces

In this chapter, we apply the fixed point theorem to investigate existence, uniqueness and the multi stability of a class of ξ–Hilfer fractional differential systems on compact domains and in MVFB–spaces $(\mathcal{W}, \mathscr{N}, \bigotimes_{\mathrm{GTN}})$. Next, we show the main results for unbounded domains. An example is given to illustrate the multi stability for a fractional system.

Now, consider the fractional–order systems II [1], as follows:

$$\begin{cases} {}^{\mathscr{H}}D_{\lambda^+}^{\mathcal{A},\mathcal{B};\xi}\Phi(\zeta) = F(\zeta, \Phi(\zeta), {}^{\mathscr{H}}D_{\lambda^+}^{\mathcal{A},\mathcal{B};\xi}\Phi(\zeta)), \\ I_{\lambda^+}^{1-\mathcal{C};\xi}\Phi(\lambda) = \mathfrak{z}_\lambda, \end{cases} \tag{21.1}$$

where ${}^{\mathscr{H}}D_{\lambda^+}^{\mathcal{A},\mathcal{B};\xi}(.)$ is the ξ–Hilfer fractional derivative of order $0 < \mathcal{A} \leq 1$ and type $0 \leq \mathcal{B} \leq 1$, $I_{\lambda^+}^{1-\mathcal{C};\xi}(.)$ is the Riemann–Liouville fractional integral of order $1-\mathcal{C}$, $\mathcal{C} = \mathcal{A} + \mathcal{B}(1-\mathcal{A})$, with respect to the function ξ, and $F : \Delta \times \mathbb{R}^2 \longrightarrow \mathbb{R}$ is a given function. Let $\zeta \in \Delta = [\lambda, D]$ with $D > \lambda$ and $\mathfrak{z}_\lambda \in \mathbb{R}$.

21.1 Multi Stability of Hilfer Fractional System II on Compact Domains

In this part, we prove the existence, uniqueness and the multi stability results of the ξ–Hilfer fractional system (21.1) defined on compact domains by the fixed point theorem.

For $F : \Delta \times \mathbb{R}^2 \to \mathbb{R}$ and $\theta > 0$, consider the following fractional equations defined by

$${}^{\mathscr{H}}D_{\lambda^+}^{\mathcal{A},\mathcal{B};\xi}\Phi(\zeta) = F(\zeta, \Phi(\zeta), {}^{\mathscr{H}}D_{\lambda^+}^{\mathcal{A},\mathcal{B};\xi}\Phi(\zeta)), \tag{21.2}$$

$$I_{\lambda^+}^{\mathcal{C};\xi}\Phi(\lambda) = \mathfrak{z}_\lambda, \quad \mathfrak{z}_\lambda \in \mathbb{R}, \tag{21.3}$$

S. Rezaei Aderyani et al., *Towards Ulam Type Multi Stability Analysis*, Studies in Systems, Decision and Control 523, https://doi.org/10.1007/978-3-031-55564-0_21

and consider the inequality below

$$\mathcal{N}\left({}^{\mathcal{H}}D_{\lambda^+}^{\mathcal{A},\mathcal{B};\xi}\Psi(\zeta) - F(\zeta, \Psi(\zeta), {}^{\mathcal{H}}D_{\lambda^+}^{\mathcal{A},\mathcal{B};\xi}\Phi(\zeta)), (\mathcal{T}_1, \ldots, \mathcal{T}_n)\right) \tag{21.4}$$
$$\succeq \operatorname{diag}\left[\varphi_1^{\circledS}(\frac{-|\xi(\zeta) - \xi(0)|^{\mathcal{A}}}{\theta\mathcal{T}_1}), \ldots, \varphi_n^{\circledS}(\frac{-|\xi(\zeta) - \xi(0)|^{\mathcal{A}}}{\theta\mathcal{T}_n})\right],$$

where $\zeta \in \Delta, n \in \mathbb{N}$, $\underbrace{\mathcal{T}_i}_{i=1,\ldots,n} \in (0, +\infty)$ and $\underbrace{\varphi_i^{\circledS}}_{i=1,\ldots,n}$ are fuzzy contrllers given in Sects. 9.8.1 and 9.8.2.

Definition 21.1 Equations (21.2)–(21.3) have the multi stability with respect to the matrix valued fuzzy set below

$$\operatorname{diag}\left[\varphi_1^{\circledS}\left((\xi(\zeta) - \xi(0))^{\mathcal{A}}\right), \ldots, \varphi_n^{\circledS}\left((\xi(\zeta) - \xi(0))^{\mathcal{A}}\right)\right], \quad \zeta \in \Delta,$$

in which $\underbrace{\varphi_i^{\circledS}}_{i=1,\ldots,n}$ are fuzzy controllers with $n \in \mathbb{N}$, and $\mathcal{A} \in (0, 1]$, if there exists $c > 0$, such that, for every $\theta > 0$ and every solution $\Psi \in C(\Delta, \mathbb{R})$ to (21.4) and $I_{\lambda^+}^{1-\mathcal{C};\xi}\Phi(\lambda) = \mathfrak{Z}_\lambda$, there exists a solution $\Phi \in C(\Delta, \mathbb{R})$ to (21.2)–(21.3) with

$$\mathcal{N}\left(\Psi(\zeta) - \Phi(\zeta), (\mathcal{T}_1, \ldots, \mathcal{T}_n)\right)$$
$$\succeq \operatorname{diag}\left[\varphi_1^{\circledS}\left(\frac{-|(\xi(\zeta) - \xi(0)|^{\mathcal{A}}}{c\theta\mathcal{T}_1}\right), \ldots, \varphi_n^{\circledS}\left(\frac{-|(\xi(\zeta) - \xi(0)|^{\mathcal{A}}}{c\theta\mathcal{T}_n}\right)\right],$$

for every $\zeta \in \Delta, n \in \mathbb{N}$, and $\underbrace{\mathcal{T}_i}_{i=1,\ldots,n} \in (0, +\infty)$.

Lemma 21.1 ([2]) *Consider a continuous function $F(\zeta, \kappa, \upsilon) : \Delta \times \mathbb{R}^2 \to \mathbb{R}$. Then, Eq.* (21.1) *is equivalent to:*

$$\Phi(\zeta) = \frac{(\xi(\zeta) - \xi(\lambda))^{\mathcal{C}-1}}{\Gamma(\mathcal{C})}\mathfrak{Z}_\lambda + I_{\lambda^+}^{\mathcal{A};\xi}G(\zeta), \tag{21.5}$$

where $G \in C(\Delta, \mathbb{R})$ satisfies the functional equation below

$$G(\zeta) = F\left(\zeta, \frac{(\xi(\zeta) - \xi(\lambda))^{\mathcal{C}-1}}{\Gamma(\mathcal{C})}\mathfrak{Z}_\lambda + I_{\lambda^+}^{\mathcal{A};\xi}G(\zeta), G(\zeta)\right).$$

Remark 21.1 Let $\Psi \in C(\Delta, \mathbb{R})$ be a solution of inequality (21.4) and $I_{\lambda^+}^{1-\mathcal{C};\xi}\Phi(\lambda) = \mathfrak{Z}_\lambda$. Then, Ψ is a solution of the following inequality:

$$\mathscr{N}\left(\Psi(\zeta) - \frac{(\xi(\zeta)-\xi(0))^{\mathcal{C}-1}}{\Gamma(\mathcal{C})}\mathfrak{Z}_{\lambda} - \frac{1}{\Gamma(\mathcal{A})}\int_{0}^{\zeta}\xi'(s)(\xi(\zeta)-\xi(s))^{\mathcal{A}-1}G_{\Psi}(s)ds, (\mathcal{T}_1,\ldots,\mathcal{T}_n)\right)$$
$$\succeq \operatorname{diag}\left[\varphi_1^{\circledS}\left(\frac{-|\xi(\zeta)-\xi(0)|^{\mathcal{A}}}{\theta\mathcal{T}_1}\right),\ldots,\varphi_n^{\circledS}\left(\frac{-|\xi(\zeta)-\xi(0)|^{\mathcal{A}}}{\theta\mathcal{T}_n}\right)\right],$$

where $G_{\Psi} \in C(\Delta, \mathbb{R})$ satisfies the functional equation:

$$G_{\Psi}(\zeta) = F\left(\zeta, \frac{(\xi(\zeta)-\xi(0))^{\mathcal{C}-1}}{\Gamma(\mathcal{C})}\xi_{\lambda} + I_{\lambda^+}^{\mathcal{A}:\xi}G_{\Psi}(\zeta), G_{\Psi}(\zeta)\right), \tag{21.6}$$

for every $\underbrace{\mathcal{T}_i}_{i=1,\ldots,n} \in (0, +\infty)$.

Proof See [1] and Remark 20.1.

Let us suppose the following axioms are satisfied:

($\mathcal{E}_1$) $F : \Delta \times \mathbb{R}^2 \to \mathbb{R}$ is continuous.

($\mathcal{E}_2$) There exist $0 < \underbrace{\eth_i}_{i=1,\ldots,n} < 1$, such that

$$I_{\lambda^+}^{\mathcal{A}:\xi}\underbrace{\varphi_i^{\circledS}}_{i=1,\ldots,n}\left(\frac{-|(\xi(\zeta)-\xi(0)|^{\mathcal{A}}}{\theta\underbrace{\mathcal{T}_i}_{i=1,\ldots,n}}\right)$$
$$= \frac{1}{\Gamma(\mathcal{A})}\int_{\lambda^+}^{\zeta}\xi'(\zeta_\circ)(\xi(\zeta)-\xi(\zeta_\circ))^{\mathcal{A}-1}\underbrace{\varphi_i^{\circledS}}_{i=1,\ldots,n}\left(\frac{-|(\xi(\zeta)-\xi(0)|^{\mathcal{A}}}{\theta\underbrace{\mathcal{T}_i}_{i=1,\ldots,n}}\right)$$
$$\geq \underbrace{\varphi_i^{\circledS}}_{i=1,\ldots,n}\left(\frac{-|(\xi(\zeta)-\xi(0)|^{\mathcal{A}}}{\frac{\theta}{\underbrace{\eth_i}_{i=1,\ldots,n}}\underbrace{\mathcal{T}_i}_{i=1,\ldots,n}}\right)$$

where $\zeta \in \Delta$, $\underbrace{\mathcal{T}_i}_{i=1,\ldots,n} \in (0, +\infty)$, and $n \in \mathbb{N}$.

($\mathcal{E}_3$) There exist $0 < \Theta_1, \Theta_2$, such that $\dfrac{\overbrace{\eth_i}^{i=1,\ldots,n}}{(\Theta_1+\Theta_2)} < 1$ and

$$\mathcal{N}\left(F(\zeta,\kappa,\upsilon)-F(\overline{\zeta},\overline{\kappa},\overline{\upsilon}),(\mathcal{T}_1,\ldots,\mathcal{T}_n)\right)$$
$$\succeq \mathcal{N}\left(\Theta_1(\kappa-\overline{\kappa}),(\mathcal{T}_1,\ldots,\mathcal{T}_n)\right)\bigotimes_{\mathrm{GTN}}\mathcal{N}\left(\Theta_2(\upsilon-\overline{\upsilon}),(\mathcal{T}_1,\ldots,\mathcal{T}_n)\right)$$

for every $\kappa,\upsilon,\overline{\kappa},\overline{\upsilon}\in\mathbb{R}$, $\zeta\in\Delta$, $\underbrace{\mathcal{T}_i}_{i=1,\ldots,n}\in(0,+\infty)$ and $n\in\mathbb{N}$.

Theorem 21.1 *Assume that assumptions* $(\mathcal{E}_1)$, $(\mathcal{E}_2)$ *and* $(\mathcal{E}_3)$ *are satisfied. Then,* (21.2)–(21.3) *have a unique solution* Υ *in* $C(\Delta,\mathbb{R})$, *such that*

$$\Upsilon(\zeta)=\frac{(\xi(\zeta)-\xi(\lambda))^{\mathcal{C}-1}}{\Gamma(\mathcal{C})}\mathfrak{Z}_\lambda+I_{\lambda^+}^{\mathcal{A};\xi}G_\Upsilon(\zeta), \tag{21.7}$$

where $G_\Upsilon\in C(\Delta,\mathbb{R})$ *satisfies the functional equation:*

$$G_\Upsilon(\zeta)=F\left(\zeta,\frac{(\xi(\zeta)-\xi(\lambda))^{\mathcal{C}-1}}{\Gamma(\mathcal{C})}\mathfrak{Z}_\lambda+I_{\lambda^+}^{\mathcal{A};\xi}G_\Upsilon(\zeta),G_\Upsilon(\zeta)\right).$$

and for every $\zeta\in\Delta$ *and* $\underbrace{\mathcal{T}_i}_{i=1,\ldots,n}\in(0,+\infty)$, *with* $n\in\mathbb{N}$,

$$\mathcal{N}\left(\Phi(\zeta)-\Upsilon(\zeta),(\mathcal{T}_1,\ldots,\mathcal{T}_n)\right) \tag{21.8}$$
$$\succeq diag\left[\varphi_1^{\circledS}\left(\frac{-|(\xi(\zeta)-\xi(0)|^{\mathcal{A}}}{\mathcal{D}_1\theta\mathcal{T}_1}\right),\ldots,\varphi_n^{\circledS}\left(\frac{-|(\xi(\zeta)-\xi(0)|^{\mathcal{A}}}{\mathcal{D}_n\theta\mathcal{T}_n}\right)\right],$$

where

$$\underbrace{\mathcal{D}_i}_{i=1,\ldots,n}:=\frac{1}{1-\dfrac{\overbrace{\eth_i}^{i=1,\ldots,n}}{(\Theta_1+\Theta_2)}}.$$

Proof Set $\nabla=C(\Delta,\mathbb{R})$ and define a mapping $\partial:\nabla\times\nabla\to[0,+\infty]^n$ by

$$\partial(\mathcal{J},\mathcal{J}')=\inf\Bigg\{(\mathtt{C}_1,\ldots,\mathtt{C}_n)\in(0,+\infty)^n:\mathcal{N}\left(\mathcal{J}(\zeta)-\mathcal{J}'(\zeta),(\mathcal{T}_1,\ldots,\mathcal{T}_n)\right) \tag{21.9}$$
$$\succeq \mathrm{diag}\left[\varphi_1^{\circledS}\left(\frac{-|(\xi(\zeta)-\xi(0)|^{\mathcal{A}}}{\frac{\theta}{\mathtt{C}_1}\mathcal{T}_1}\right),\ldots,\varphi_n^{\circledS}\left(\frac{-|(\xi(\zeta)-\xi(0)|^{\mathcal{A}}}{\frac{\theta}{\mathtt{C}_n}\mathcal{T}_n}\right)\right],$$
$$\mathcal{J},\mathcal{J}'\in\nabla,\ \zeta\in\Delta,\ \underbrace{\mathcal{T}_i}_{i=1,\ldots,n}\in(0,+\infty),\ n\in\mathbb{N}\Bigg\}.$$

We prove that (∇,∂) is a metric space.

First we show that $\partial(\mathscr{J}, \mathscr{J}') = 0$ if and only if $\mathscr{J} = \mathscr{J}'$. Let $\partial(\mathscr{J}, \mathscr{J}') = \underbrace{(0, \ldots, 0)}_{n}$, we have that

$$\inf\Bigg\{(\mathtt{C}_1, \ldots, \mathtt{C}_n) \in (0, +\infty)^n : \mathscr{N}\Big(\mathscr{J}(\zeta) - \mathscr{J}'(\zeta), (\mathcal{T}_1, \ldots, \mathcal{T}_n)\Big) \tag{21.10}$$

$$\succeq \operatorname{diag}\left[\varphi_1^{Ⓢ}\left(\frac{-|(\xi(\zeta) - \xi(0)|^{\mathcal{A}}}{\frac{\theta}{\mathtt{C}_1}\mathcal{T}_1}\right), \ldots, \varphi_n^{Ⓢ}\left(\frac{-|(\xi(\zeta) - \xi(0)|^{\mathcal{A}}}{\frac{\theta}{\mathtt{C}_n}\mathcal{T}_n}\right)\right],$$

$$\mathscr{J}, \mathscr{J}' \in \nabla, \zeta \in \Delta, \underbrace{\mathcal{T}_i}_{i=1,\ldots,n} \in (0, +\infty), n \in \mathbb{N}\Bigg\} = \underbrace{(0, \ldots, 0)}_{n}.$$

and so

$$\mathscr{N}\Big(\mathscr{J}(\zeta) - \mathscr{J}'(\zeta), (\mathcal{T}_1, \ldots, \mathcal{T}_n)\Big)$$
$$\succeq \operatorname{diag}\left[\varphi_1^{Ⓢ}\left(\frac{-|(\xi(\zeta) - \xi(0)|^{\mathcal{A}}}{\frac{\theta}{\mathtt{C}_1}\mathcal{T}_1}\right), \ldots, \varphi_n^{Ⓢ}\left(\frac{-|(\xi(\zeta) - \xi(0)|^{\mathcal{A}}}{\frac{\theta}{\mathtt{C}_n}\mathcal{T}_n}\right)\right],$$

for every $\underbrace{\mathtt{C}_i}_{i=1,\ldots,n} \in (0, +\infty)$. In the above inequality, when $\underbrace{\mathtt{C}_i}_{i=1,\ldots,n} \to 0$, we have

$$\mathscr{N}\Big(\mathscr{J}(\zeta) - \mathscr{J}'(\zeta), (\mathcal{T}_1, \ldots, \mathcal{T}_n)\Big) = \mathbf{1},$$

thus, $\mathscr{J}(\zeta) = \mathscr{J}'(\zeta)$, for every $\zeta \in \Delta$, and vice versa. It is straight forward to show $\partial(\mathscr{J}, \mathscr{J}') = \partial(\mathscr{J}', \mathscr{J})$, for every $\mathscr{J}, \mathscr{J}' \in \nabla$. Now, let $\partial(\mathscr{J}, \mathscr{E}) = (\ell_1, \ldots, \ell_n) \in (0, +\infty)^n$ and $\partial(\mathscr{E}, \mathscr{J}') = (\jmath_1, \ldots, \jmath_n) \in (0, +\infty)^n$. Then, we have

$$\mathscr{N}\Big(\mathscr{J}(\zeta) - \mathscr{E}(\zeta), (\mathcal{T}_1, \ldots, \mathcal{T}_n)\Big)$$
$$\succeq \operatorname{diag}\left[\varphi_1^{Ⓢ}\left(\frac{-|(\xi(\zeta) - \xi(0)|^{\mathcal{A}}}{\frac{\theta}{\ell_1}\mathcal{T}_1}\right), \ldots, \varphi_n^{Ⓢ}\left(\frac{-|(\xi(\zeta) - \xi(0)|^{\mathcal{A}}}{\frac{\theta}{\ell_n}\mathcal{T}_n}\right)\right],$$

and

$$\mathscr{N}\Big(\mathscr{E}(\zeta) - \mathscr{J}'(\zeta), (\mathcal{T}_1, \ldots, \mathcal{T}_n)\Big)$$
$$\succeq \operatorname{diag}\left[\varphi_1^{Ⓢ}\left(\frac{-|(\xi(\zeta) - \xi(0)|^{\mathcal{A}}}{\frac{\theta}{\jmath_1}\mathcal{T}_1}\right), \ldots, \varphi_n^{Ⓢ}\left(\frac{-|(\xi(\zeta) - \xi(0)|^{\mathcal{A}}}{\frac{\theta}{\jmath_n}\mathcal{T}_n}\right)\right],$$

for every $\underbrace{\mathcal{T}_i}_{i=1,\dots,n} \in (0, +\infty)$, then, we have that

$$\begin{aligned}
&\mathcal{N}\Big(\mathcal{J}(\zeta) - \mathcal{J}'(\zeta), \Big((\ell_1 + \jmath_1)\mathcal{T}_1, \dots, (\ell_n + \jmath_n)\mathcal{T}_n\Big)\Big) \\
&\succeq \mathcal{N}\Big(\mathcal{J}(\zeta) - \mathcal{E}(\zeta), (\ell_1\mathcal{T}_1, \dots, \ell_n\mathcal{T}_n)\Big) \\
&\qquad \bigotimes_{\text{GTN}} \mathcal{N}\Big(\mathcal{E}(\zeta) - \mathcal{J}'(\zeta), (\jmath_1\mathcal{T}_1, \dots, \jmath_n\mathcal{T}_n)\Big) \\
&\succeq \operatorname{diag}\left[\varphi_1^{\circledS}\left(\frac{-|(\xi(\zeta) - \xi(0)|^{\mathcal{A}}}{\theta\mathcal{T}_1}\right), \dots, \varphi_n^{\circledS}\left(\frac{-|(\xi(\zeta) - \xi(0)|^{\mathcal{A}}}{\theta\mathcal{T}_n}\right)\right] \\
&\qquad \bigotimes_{\text{GTN}} \operatorname{diag}\left[\varphi_1^{\circledS}\left(\frac{-|(\xi(\zeta) - \xi(0)|^{\mathcal{A}}}{\theta\mathcal{T}_1}\right), \dots, \varphi_n^{\circledS}\left(\frac{-|(\xi(\zeta) - \xi(0)|^{\mathcal{A}}}{\theta\mathcal{T}_n}\right)\right] \\
&\succeq \operatorname{diag}\left[\varphi_1^{\circledS}\left(\frac{-|(\xi(\zeta) - \xi(0)|^{\mathcal{A}}}{\theta\mathcal{T}_1}\right), \dots, \varphi_n^{\circledS}\left(\frac{-|(\xi(\zeta) - \xi(0)|^{\mathcal{A}}}{\theta\mathcal{T}_n}\right)\right],
\end{aligned}$$

and so $\partial(\mathcal{J}, \mathcal{J}') \leq (\ell_1 + \jmath_1, \dots, \ell_n + \jmath_n)$. Thus, $\partial(\mathcal{J}, \mathcal{J}') \leq \partial(\mathcal{J}, \mathcal{E}) + \partial(\mathcal{E}, \mathcal{J}')$.

Now, we show that (∇, ∂) is complete. Assume $\{\mathcal{J}_k\}_k$ is a Cauchy sequence in (∇, ∂). Assume $\zeta \in [0, \mathsf{T}]$ is fixed and $\underbrace{\mathcal{T}_i}_{i=1,\dots,n} \in (0, +\infty)$. Consider arbitrary $\underbrace{\sigma_i}_{i=1,\dots,n} \in (0, +\infty)$ and $\underbrace{\ni_i}_{i=1,\dots,n} \in (0, 1)$, such that

$$\underbrace{\varphi_i^{\circledS}}_{i=1,\dots,n}\left(\frac{-|(\xi(\zeta) - \xi(0)|^{\mathcal{A}}}{\theta \underbrace{\mathcal{T}_i}_{i=1,\dots,n}}\right) \succ 1 - \underbrace{\ni_i}_{i=1,\dots,n}.$$

For $\underbrace{\varepsilon_i\mathcal{T}_i}_{i=1,\dots,n} < \underbrace{\sigma_i}_{i=1,\dots,n}$ choose $k'' \in \mathbb{N}$, such that

$$\partial(\mathcal{J}_k, \mathcal{J}_{k'}) < (\varepsilon_1, \dots, \varepsilon_n), \quad \text{for ever } k', k \geq k''.$$

Then, we get

$$\begin{aligned}
&\mathcal{N}\Big(\mathcal{J}_k(\zeta) - \mathcal{J}'_{k'}(\zeta), (\sigma_1, \dots, \sigma_n)\Big) \\
&\succeq \mathcal{N}\Big(\mathcal{J}_k(\zeta) - \mathcal{J}'_{k'}(\zeta), (\varepsilon_1\mathcal{T}_1, \dots, \varepsilon_n\mathcal{T}_n)\Big)
\end{aligned}$$

$$\succeq \operatorname{diag}\left[\varphi_1^{\circledS}\left(\frac{-|(\xi(\zeta)-\xi(0)|^{\mathcal{A}}}{\theta\mathcal{T}_1}\right),\ldots,\varphi_n^{\circledS}\left(\frac{-|(\xi(\zeta)-\xi(0)|^{\mathcal{A}}}{\theta\mathcal{T}_n}\right)\right]$$
$$\succ \operatorname{diag}\underbrace{[1-\ni_1,\ldots,1-\ni_n]}_{n}.$$

Thus, we have that

$$\mathscr{N}\left(\mathscr{J}_k(\zeta)-\mathscr{J}_{k'}(\zeta),(\sigma_1,\ldots,\sigma_n)\right)\succ \operatorname{diag}\underbrace{[1-\ni_1,\ldots,1-\ni_n]}_{n},$$

and i.e., the sequence $\{\mathscr{J}_k(\zeta)\}_k$ is Cauchy in complete space $(\mathcal{W},\mathscr{N},\bigotimes_{\text{GTN}})$ on compact set $[0,\mathsf{T}]$, then, they are uniformly convergent to the mapping $\mathscr{J}\in C(\Delta,\mathbb{R})$. It implies that $\mathscr{J}$ is continuous, and then (∇,∂) is complete.

Making use of Lemma 21.1, we have that (21.2)–(21.3) are equivalent to the following system

$$\Phi(\zeta)=\frac{(\xi(\zeta)-\xi(\lambda))^{\mathcal{C}-1}}{\Gamma(\mathcal{C})}\mathfrak{Z}_\lambda+I_{\lambda^+}^{\mathcal{A}:\xi}G(\zeta), \tag{21.11}$$

where $G\in C(\Delta,\mathbb{R})$ satisfies the functional equation:

$$G(\zeta)=F\left(\zeta,\frac{(\xi(\zeta)-\xi(\lambda))^{\mathcal{C}-1}}{\Gamma(\mathcal{C})}\mathfrak{Z}_\lambda+I_{\lambda^+}^{\mathcal{A}:\xi}G(\zeta),G(\zeta)\right).$$

for every $\zeta\in\Delta$. To see this note applying $I_{\lambda^+}^{\mathcal{A}:\xi}(.)$ on both sides of the fractional equation (21.1) and using Theorem 2.5, we have

$$\Phi(\zeta)-\frac{(\xi(\zeta)-\xi(\lambda))^{\mathcal{C}-1}}{\Gamma(\mathcal{C})}I_{\lambda^+}^{(1-\mathcal{A})(1-\mathcal{B}):\xi}\Phi(\lambda)=I_{\lambda^+}^{\mathcal{A}:\xi}G(\zeta).$$

Thus,

$$\Phi(\zeta)=\frac{(\xi(\zeta)-\xi(\lambda))^{\mathcal{C}-1}}{\Gamma(\mathcal{C})}\mathfrak{Z}_\lambda+I_{\lambda^+}^{\mathcal{A}:\xi}G(\zeta). \tag{21.12}$$

On the other hand, if Φ satisfies (21.12), then it satisfies equation (21.1). However applying ${}^{\mathscr{H}}D_{\lambda^+}^{\mathcal{A},\mathcal{B}:\xi}(.)$ on both sides of equation (21.12) and using Theorem 2.4, we get

$${}^{\mathscr{H}}D_{\lambda^+}^{\mathcal{A},\mathcal{B}:\xi}\Phi(\zeta)={}^{\mathscr{H}}D_{\lambda^+}^{\mathcal{A},\mathcal{B}:\xi}\frac{(\xi(\zeta)-\xi(\lambda))^{\mathcal{C}-1}}{\Gamma(\mathcal{C})}\mathfrak{Z}_\lambda+{}^{\mathscr{H}}D_{\lambda^+}^{\mathcal{A},\mathcal{B}:\xi}I_{\lambda^+}^{\mathcal{A}:\xi}G(\zeta)=G(\zeta),$$

where for $\mathcal{C}\in(0,1)$, we have ${}^{\mathscr{H}}D_{\lambda^+}^{\mathcal{A},\mathcal{B}:\xi}(\xi(\zeta)-\xi(\lambda))^{\mathcal{C}-1}=0$,

which implies that, $\Phi(\zeta)$ satisfies the initial value problem equation (21.1) if and only if $\Phi(\zeta)$ satisfies the integral equation (21.11).

Consider $\mathscr{T} := \nabla \longrightarrow \nabla$, such that for $\rho \in \nabla$,

$$\begin{aligned}&\mathscr{T}(\rho(\zeta)) \\ &= \frac{(\xi(\zeta)-\xi(0))^{\mathcal{C}-1}}{\Gamma(\mathcal{C})}\mathfrak{Z}_\lambda + \frac{1}{\Gamma(\mathcal{A})}\int_0^\zeta \xi'(\mathcal{A})(\xi(\zeta)-\xi(\xi))^{\mathcal{A}-1}G_\rho(\xi)d\xi\end{aligned} \tag{21.13}$$

in which $G_\rho \in C(\varDelta, \mathbb{R})$ satisfies the functional equation:

$$G_\rho(\zeta) = F\Big(\zeta, \rho(\zeta), G_\rho(\zeta)\Big). \tag{21.14}$$

Notice that if ρ be a continuous function, then $\mathscr{T}\rho$ is also continuous. Actually, for $\rho \in \nabla$ and $\underbrace{T_i}_{i=1,\dots,n} \in (0, +\infty)$ we have

$$\begin{aligned}&\mathscr{N}\Big(\mathscr{T}(\rho(\zeta)) - \mathscr{T}(\rho(\zeta_0)), (T_1, \dots, T_n)\Big) \\ &= \mathscr{N}\Big(\frac{(\xi(\zeta)-\xi(0))^{\mathcal{C}-1}}{\Gamma(\mathcal{C})}\mathfrak{Z}_\lambda + \frac{1}{\Gamma(\mathcal{A})}\int_0^\zeta \xi'(\mathcal{A})(\xi(\zeta)-\xi(\xi))^{\mathcal{A}-1}G_\rho(\xi)d\xi \\ &\quad -\frac{(\xi(\zeta_0)-\xi(0))^{\mathcal{C}-1}}{\Gamma(\mathcal{C})}\mathfrak{Z}_\lambda + \frac{1}{\Gamma(\mathcal{A})}\int_0^\zeta \xi'(\mathcal{A})(\xi(\zeta_0)-\xi(\xi))^{\mathcal{A}-1}G_\rho(\xi)d\xi, (T_1, \dots, T_n)\Big) \\ &\longrightarrow \mathbf{1},\end{aligned}$$

as $\zeta \longrightarrow \zeta_0$. Then, $\mathscr{T}(\rho(\zeta)) = \mathscr{T}(\rho(\zeta_0))$, which implies the continuity property of $\mathscr{T}(\rho)$ and so $\mathscr{T}(\rho) \in \nabla$.

Next, we show that the self–mapping $\mathscr{T}$ is contractive on ∇. Consider $\mathscr{T} : \nabla \longrightarrow \nabla$ defined in (21.13). Suppose $\rho, \varrho \in C(\varDelta, \mathbb{R})$, $\underbrace{\Bbbk_i}_{i=1,\dots,n} \in [0, +\infty]$, and $\partial(\rho(\zeta), \varrho(\zeta)) \preceq (\Bbbk_1, \dots, \Bbbk_n)$. So for every $\zeta \in \varDelta$,

$$\begin{aligned}&\mathscr{N}\Big(\rho(\zeta) - \varrho(\zeta), (T_1, \dots, T_n)\Big) \\ &\succeq \operatorname{diag}\Bigg[\varphi_1^{\circledS}\Bigg(\frac{-|(\xi(\zeta)-\xi(0)|^{\mathcal{A}}}{\frac{\theta}{\Bbbk_1}T_1}\Bigg), \dots, \varphi_n^{\circledS}\Bigg(\frac{-|(\xi(\zeta)-\xi(0)|^{\mathcal{A}}}{\frac{\theta}{\Bbbk_n}T_n}\Bigg)\Bigg].\end{aligned}$$

For every $\zeta \in \varDelta$ and $\underbrace{T_i}_{i=1,\dots,n} \in (0, +\infty)$, we have

$$\mathscr{N}\left(\mathscr{T}(\rho(\zeta)) - \mathscr{T}(\varrho(\zeta)), (\mathcal{T}_1, \ldots, \mathcal{T}_n)\right) \tag{21.15}$$
$$\succeq \mathscr{N}\left(\frac{1}{\Gamma(\mathcal{A})}\int_0^{\zeta} \xi'(\mathcal{A})(\xi(\zeta) - \xi(\xi))^{\mathcal{A}-1}(G_\rho(\xi) - G_\varrho(\xi))d\xi, (\mathcal{T}_1, \ldots, \mathcal{T}_n)\right),$$

where

$$G_\rho(\zeta) = F(\zeta, \rho(\zeta), G_\rho(\zeta)),$$

and

$$G_\varrho(\zeta) = F(\zeta, \varrho(\zeta), G_\varrho(\zeta)).$$

Using hypothesis $\mathcal{E}_3$ and two inequalities above for $G_\rho(.)$ and $G_\varrho(.)$, we have, for any $\zeta \in \Delta$ and any $\underbrace{\mathcal{T}_i}_{i=1,\ldots,n} \in (0, +\infty)$,

$$\mathscr{N}\left(G_\rho(\zeta) - G_\varrho(\zeta), (\mathcal{T}_1, \ldots, \mathcal{T}_n)\right)$$
$$\succeq \mathscr{N}\left(\Theta_1(\rho(\zeta) - \varrho(\zeta)), (\mathcal{T}_1, \ldots, \mathcal{T}_n)\right)$$
$$\bigotimes_{\text{GTN}} \mathscr{N}\left(\Theta_2(G_\rho(\zeta) - G_\varrho(\zeta)), (\mathcal{T}_1, \ldots, \mathcal{T}_n)\right),$$

which can be written as

$$\mathscr{N}\left(G_\rho(\zeta) - G_\varrho(\zeta), (\mathcal{T}_1, \ldots, \mathcal{T}_n)\right) \tag{21.16}$$
$$\succeq \mathscr{N}\left((\Theta_1 + \Theta_2)(\rho(\zeta) - \varrho(\zeta)), (\mathcal{T}_1, \ldots, \mathcal{T}_n)\right).$$

Now using Remark 21.1, Eqs. (21.15) and (21.16), we have that

$$\mathscr{N}\left(\mathscr{T}(\rho(\zeta)) - \mathscr{T}(\varrho(\zeta)), (\mathcal{T}_1, \ldots, \mathcal{T}_n)\right)$$
$$\succeq \mathscr{N}\left(\frac{\Theta_1 + \Theta_2}{\Gamma(\mathcal{A})}\int_0^{\zeta} \xi'(\mathcal{A})(\xi(\zeta) - \xi(\xi))^{\mathcal{A}-1}(\rho(\xi) - \varrho(\xi))d\xi, (\mathcal{T}_1, \ldots, \mathcal{T}_n)\right)$$
$$\succeq \frac{1}{\Gamma(\mathcal{A})}\int_0^{\zeta} \xi'(\mathcal{A})(\xi(\zeta) - \xi(\xi))^{\mathcal{A}-1}$$

$$\times \text{diag}\left[\varphi_1^{\circledS}\left(\frac{-(\xi(\xi)-\xi(0))^{\mathcal{A}}}{\frac{(\Theta_1+\Theta_2)\theta}{\Bbbk_1}\mathcal{T}_1}\right),\ldots,\varphi_n^{\circledS}\left(\frac{-(\xi(\xi)-\xi(0))^{\mathcal{A}}}{\frac{(\Theta_1+\Theta_2)\theta}{\Bbbk_n}\mathcal{T}_n}\right)\right]d\xi$$

$$\succeq \text{diag}\left[\varphi_1^{\circledS}\left(\frac{-(\xi(\zeta)-\xi(0))^{\mathcal{A}}}{\frac{(\Theta_1+\Theta_2)\theta}{\eth_1\Bbbk_1}\mathcal{T}_1}\right),\ldots,\varphi_n^{\circledS}\left(\frac{-(\xi(\zeta)-\xi(0))^{\mathcal{A}}}{\frac{(\Theta_1+\Theta_2)\theta}{\eth_n\Bbbk_n}\mathcal{T}_n}\right)\right].$$

which implies that

$$\partial(\mathscr{T}(\rho(\zeta))-\mathscr{T}(\varrho(\zeta)))\geq\left(\frac{\Bbbk_1\eth_1}{(\Theta_1+\Theta_2)},\ldots,\frac{\Bbbk_n\eth_n}{(\Theta_1+\Theta_2)}\right),$$

and so

$$\partial(\mathscr{T}(\rho(\zeta))-\mathscr{T}(\varrho(\zeta)))\geq\left(\frac{\eth_1}{(\Theta_1+\Theta_2)},\ldots,\frac{\eth_n}{(\Theta_1+\Theta_2)}\right)\partial(\rho(\zeta)-\varrho(\zeta)),$$

which implies that the contractively property of $\mathscr{T}$, because $\frac{\overbrace{\eth_i}^{i=1,\ldots,n}}{(\Theta_1+\Theta_2)}<1$, for every $n\in\mathbb{N}$.

Let $\Psi\in\nabla$, we show that $\partial(\mathscr{T}\Psi,\Psi)<\underbrace{(\infty,\ldots,\infty)}_{n}$. Using (21.4) and Remark 21.1, we get

$$\mathscr{N}\left(\mathscr{T}(\Psi(\zeta))-\Psi(\zeta),(\mathcal{T}_1,\ldots,\mathcal{T}_n)\right)$$
$$\succeq\mathscr{N}\left(\Psi(\zeta)-\frac{(\xi(\zeta)-\xi(0))^{\mathcal{C}-1}}{\Gamma(\mathcal{C})}\mathfrak{Z}_\lambda-\frac{1}{\Gamma(\mathcal{A})}\int_0^\zeta\xi'(\mathcal{A})(\xi(\zeta)-\xi(\xi))^{\mathcal{A}-1}G_\Psi(\xi)d\xi,(\mathcal{T}_1,\ldots,\mathcal{T}_n)\right)$$
$$\succeq\text{diag}\left[\varphi_1^{\circledS}\left(\frac{-|(\xi(\zeta)-\xi(0)|^{\mathcal{A}}}{\theta\mathcal{T}_1}\right),\ldots,\varphi_n^{\circledS}\left(\frac{-|(\xi(\zeta)-\xi(0)|^{\mathcal{A}}}{\theta\mathcal{T}_n}\right)\right].$$

for every $\underbrace{\mathcal{T}_i}_{i=1,\ldots,n}\in(0,+\infty)$ and $\zeta\in\Delta$. Note that $\partial(\mathscr{T}\Psi,\Psi)=\underbrace{(1,\ldots,1)}_{n}$, for every $\underbrace{\mathcal{T}_i}_{i=1,\ldots,n}\in(0,+\infty)$, with $n\in\mathbb{N}$, and $\zeta\in\Delta$. Then, we have $\partial(\mathscr{T}\Psi,\Psi)<\underbrace{(1,\ldots,1)}_{n}$.

Now, Theorem 2.20 helps us to find $\Upsilon\in\nabla$, such that:

(1) Υ is a fixed point of $\mathscr{T}$, i.e.,

$$\mathscr{T}(\Upsilon(\zeta)) \tag{21.17}$$
$$=\frac{(\xi(\zeta)-\xi(0))^{\mathcal{C}-1}}{\Gamma(\mathcal{C})}\mathfrak{Z}_\lambda+\frac{1}{\Gamma(\mathcal{A})}\int_0^\zeta\xi'(\mathcal{A})(\xi(\zeta)-\xi(\xi))^{\mathcal{A}-1}G_\Upsilon(\xi)d\xi,$$

in which $G_{\Upsilon} \in C(\Delta, \mathbb{R})$ satisfies the functional equation:

$$G_{\Upsilon}(\zeta) = F\Bigg(\zeta, \frac{(\xi(\zeta) - \xi(0))^{\mathcal{C}-1}}{\Gamma(\mathcal{C})}\mathfrak{Z}_{\lambda} + I_{\lambda^+}^{\mathcal{A};\xi} G_{\Upsilon}(\zeta), G_{\Upsilon}(\zeta)\Bigg), \tag{21.18}$$

which is unique in the set

$$\nabla^* = \{\phi \in \nabla : \partial(\mathscr{T}\Psi, \phi) < \underbrace{(\infty, \ldots, \infty)}_{n}\}.$$

(2) $\partial(\mathscr{T}^n(\Psi), \Upsilon) \to \underbrace{(0, \ldots, 0)}_{n}$ as $n \to \infty$;

(3) we have

$$\begin{aligned}
\partial(\Psi, \Upsilon) &\leq \underbrace{(\frac{1}{1 - \eth_1(\Theta_1 + \Theta_2)}, \ldots, \frac{1}{1 - \eth_n(\Theta_1 + \Theta_2)})}_{n} \partial(\mathscr{T}\Psi, \Psi) \\
&\leq \underbrace{(\frac{1}{1 - \eth_1(\Theta_1 + \Theta_2)}, \ldots, \frac{1}{1 - \eth_n(\Theta_1 + \Theta_2)})}_{n},
\end{aligned}$$

which implies that

$$\begin{aligned}
&\mathscr{N}\Bigg(\Psi(\zeta) - \Upsilon(\zeta), (\mathcal{T}_1, \ldots, \mathcal{T}_n)\Bigg) \\
&\succeq \operatorname{diag}\Bigg[\varphi_1^{\circledS}\Bigg(\frac{-|(\xi(\zeta) - \xi(0)|^{\mathcal{A}}}{\mathcal{D}_1 \theta \mathcal{T}_1}\Bigg), \ldots, \varphi_n^{\circledS}\Bigg(\frac{-|(\xi(\zeta) - \xi(0)|^{\mathcal{A}}}{\mathcal{D}_n \theta \mathcal{T}_n}\Bigg)\Bigg],
\end{aligned} \tag{21.19}$$

where

$$\underbrace{\mathcal{D}_i}_{i=1,\ldots,n} := \frac{1}{1 - \dfrac{\overbrace{\eth_i}^{i=1,\ldots,n}}{(\Theta_1 + \Theta_2)}}, \tag{21.20}$$

for every $\zeta \in \Delta$ and $\underbrace{\mathcal{T}_i}_{i=1,\ldots,n} \in (0, +\infty)$, with $n \in \mathbb{N}$.

Here, we prove that the fixed point in ∇^* is unique. Assume $\Lambda \in \nabla$ satisfying (21.7) and (21.8), we show that $\Lambda = \Upsilon$ and $\Lambda \in \nabla^*$. Using (21.7) we get

$$\Lambda(\zeta) = \frac{(\xi(\zeta) - \xi(0))^{\mathcal{C}-1}}{\Gamma(\mathcal{C})}\mathfrak{Z}_{\lambda} + \frac{1}{\Gamma(\mathcal{A})}\int_0^{\zeta} \xi'(\mathcal{A})(\xi(\zeta) - \xi(\xi))^{\mathcal{A}-1} G_{\Lambda}(\xi) d\xi \tag{21.21}$$

in which $G_{\Lambda} \in C(\Delta, \mathbb{R})$ satisfies the functional equation:

$$G_{\Lambda}(\zeta) = F\left(\zeta, \frac{(\xi(\zeta) - \xi(0))^{\mathcal{C}-1}}{\Gamma(\mathcal{C})}\mathfrak{Z}_{\lambda} + I_{\lambda^+}^{\mathcal{A};\xi} G_{\Lambda}(\zeta), G_{\Lambda}(\zeta)\right). \tag{21.22}$$

and so

$$\mathscr{T}(\Lambda(\zeta)) = \frac{(\xi(\zeta) - \xi(0))^{\mathcal{C}-1}}{\Gamma(\mathcal{C})}\mathfrak{Z}_{\lambda} + \frac{1}{\Gamma(\mathcal{A})}\int_0^{\zeta} \xi'(\mathcal{A})(\xi(\zeta) - \xi(\xi))^{\mathcal{A}-1} G_{\Lambda}(\xi) d\xi. \tag{21.23}$$

Now, we show that

$$\Lambda \in \left\{\phi \in \nabla : \partial(\mathscr{T}(\Psi), \phi) < \underbrace{(\infty, \ldots, \infty)}_{n}\right\},$$

i.e., $\partial(\mathscr{T}(\Psi), \Lambda) < \underbrace{(\infty, \ldots, \infty)}_{n}$. From (21.8), we get

$$\begin{aligned} &\mathscr{N}\left(\Psi(\zeta) - \Lambda(\zeta), (\mathcal{T}_1, \ldots, \mathcal{T}_n)\right) \\ &\succeq \operatorname{diag}\left[\varphi_1^{\circledS}\left(\frac{-|(\xi(\zeta) - \xi(0)|^{\mathcal{A}}}{\mathcal{D}_1 \theta \mathcal{T}_1}\right), \ldots, \varphi_n^{\circledS}\left(\frac{-|(\xi(\zeta) - \xi(0)|^{\mathcal{A}}}{\mathcal{D}_n \theta \mathcal{T}_n}\right)\right], \end{aligned} \tag{21.24}$$

where $\underbrace{\mathcal{D}_i}_{i=1,\ldots,n}$ is defined in (21.20), for every $\zeta \in \Delta$ and $\underbrace{\mathcal{T}_i}_{i=1,\ldots,n} \in (0, +\infty)$.

Using triangular inequality, (21.23), (21.24), (21.4) and Remark 21.1, we have

$$\begin{aligned} &\mathscr{N}\left(\mathscr{T}(\Psi(\zeta)) - \Lambda(\zeta), (\mathcal{T}_1, \ldots, \mathcal{T}_n)\right) \\ &= \mathscr{N}\left(\frac{(\xi(\zeta) - \xi(0))^{\mathcal{C}-1}}{\Gamma(\mathcal{C})}\mathfrak{Z}_{\lambda}\right. \\ &\qquad \left. + \frac{1}{\Gamma(\mathcal{A})}\int_0^{\zeta} \xi'(\mathcal{A})(\xi(\zeta) - \xi(\xi))^{\mathcal{A}-1} G_{\Psi}(\xi) d\xi + \Psi(\zeta) - \Psi(\zeta) - \Lambda(\zeta), (\mathcal{T}_1, \ldots, \mathcal{T}_n)\right) \\ &\succeq \mathscr{N}\left(\frac{(\xi(\zeta) - \xi(0))^{\mathcal{C}-1}}{\Gamma(\mathcal{C})}\mathfrak{Z}_{\lambda} + \frac{1}{\Gamma(\mathcal{A})}\int_0^{\zeta} \xi'(\mathcal{A})(\xi(\zeta) - \xi(\xi))^{\mathcal{A}-1} G_{\Psi}(\xi) d\xi - \Psi(\zeta), \frac{(\mathcal{T}_1, \ldots, \mathcal{T}_n)}{2}\right) \\ &\qquad \bigotimes_{\mathrm{GTN}} \mathscr{N}\left(\Psi(\zeta) - \Lambda(\zeta), \frac{(\mathcal{T}_1, \ldots, \mathcal{T}_n)}{2}\right) \\ &\succeq \operatorname{diag}\left[\varphi_1^{\circledS}\left(\frac{-|(\xi(\zeta) - \xi(0)|^{\mathcal{A}}}{2\theta \mathcal{T}_1}\right), \ldots, \varphi_n^{\circledS}\left(\frac{-|(\xi(\zeta) - \xi(0)|^{\mathcal{A}}}{2\theta \mathcal{T}_n}\right)\right] \\ &\qquad \bigotimes_{\mathrm{GTN}} \operatorname{diag}\left[\varphi_1^{\circledS}\left(\frac{-|(\xi(\zeta) - \xi(0)|^{\mathcal{A}}}{2\mathcal{D}_1 \theta \mathcal{T}_1}\right), \ldots, \varphi_n^{\circledS}\left(\frac{-|(\xi(\zeta) - \xi(0)|^{\mathcal{A}}}{2\mathcal{D}_n \theta \mathcal{T}_n}\right)\right] \\ &\succeq \operatorname{diag}\left[\varphi_1^{\circledS}\left(-\frac{|(\xi(\zeta) - \xi(0)|^{\mathcal{A}}}{2\max\{1, \mathcal{D}_1\}\theta \mathcal{T}_1}\right), \ldots, \varphi_n^{\circledS}\left(-\frac{|(\xi(\zeta) - \xi(0)|^{\mathcal{A}}}{2\max\{1, \mathcal{D}_n\}\theta \mathcal{T}_n}\right)\right], \end{aligned}$$

for enery $\zeta \in \Delta$, $n \in \mathbb{N}$ and $\underbrace{\mathcal{T}_i}_{i=1,\dots,n} \in (0, +\infty)$. Which implies that

$$\partial(\mathscr{T}\Psi, \Lambda) \leq (2\max\{1, \mathcal{D}_1\}, \dots, 2\max\{1, \mathcal{D}_n\}) < \infty,$$

then, $\Lambda \in \nabla^*$.

21.2 Multi Stability of Hilfer Fractional System II on Unbounded Domains

In the former theorem, we have studied the multi stability of the ξ–Hilfer fractional differential equation (21.1) defined on compact domains. Now, we want to prove the last theorem for the case of unbounded domains. More precisely, Theorem 21.1 is also true if Δ is replaced by an unbounded interval $(-\infty, 0]$, $\mathbb{R}$ or $[0, +\infty)$, as we see in the following theorem. Let $\mathsf{S} = C(\mathbb{R})$.

Theorem 21.2 *Assume that* $(\mathcal{E}_1)$, $(\mathcal{E}_2)$ *and* $(\mathcal{E}_3)$ *are satisfied. If* Ψ *in* S *satisfies* (21.4) *and* $\mathcal{I}_{\lambda^+}^{1-\mathcal{C};\xi}\Phi_n(\lambda^+) = \mathfrak{Z}_\lambda$, *then, there exists a unique function* Φ *satisfying in* (21.2)–(21.3), *such that satisfies* (21.8), *for every* $\zeta \in \mathbb{R}$.

Proof We want to prove the theorem for all real numbers. We also can prove our theorem for $(-\infty, 0]$ or $[0, +\infty)$ by the same way.

For every $m \in \mathbb{N}$, and $\mathfrak{I} > 0$, we consider $\mathsf{P}_m = [\mathfrak{I} - m, \mathfrak{I} + m]$. As stated by theorem 21.1, there is a unique function $\Phi_m \in C(\mathsf{P}_m, \mathbb{R})$, such that

$$^{\mathscr{H}}D_{\lambda^+}^{\mathcal{A},\mathcal{B};\xi}\Phi_m(\zeta) = \mathcal{C}(\zeta, \Phi_m(\zeta), \Phi_m(\eta(\zeta))), \qquad \zeta \in \mathsf{P}_m, \tag{21.25}$$

$$\mathcal{I}_{\lambda^+}^{1-\mathcal{C};\xi}\Phi_m(\lambda^+) = \mathfrak{Z}_\lambda, \qquad \mathfrak{Z}_\lambda \in \mathbb{R}, \tag{21.26}$$

and

$$\mathscr{N}\Big(\Phi_m(\zeta) - \Psi(\zeta), (\mathcal{T}_1, \dots, \mathcal{T}_n)\Big) \tag{21.27}$$
$$\succeq \operatorname{diag}\left[\varphi_1^{\circledS}\left(\frac{-|(\xi(\zeta) - \xi(0)|^{\mathcal{A}}}{\mathcal{D}_1\theta\mathcal{T}_1}\right), \dots, \varphi_n^{\circledS}\left(\frac{-|(\xi(\zeta) - \xi(0)|^{\mathcal{A}}}{\mathcal{D}_n\theta\mathcal{T}_n}\right)\right],$$

for every $\underbrace{\mathcal{T}_i}_{i=1,\dots,n} \in (0, +\infty)$, and $n \in \mathbb{N}$, where

$$\underbrace{\mathcal{D}_i}_{i=1,\dots,n} := \frac{1}{1 - \dfrac{\overbrace{\eth_i}^{i=1,\dots,n}}{(\Theta_1 + \Theta_2)}},$$

for every $\zeta \in \mathsf{P}_m$. The uniqueness of Φ_m implies that if $\zeta \in \mathsf{P}_m$, then,

$$\Phi_m(\zeta) = \Phi_{m+1}(\zeta) = \Phi_{m+2}(\zeta) = \dots, \tag{21.28}$$

for every $\zeta \in \mathbb{R}$, let us define $m(\zeta) \in \mathbb{N}$ as

$$m(\zeta) = \min\{m \in \mathbb{N} \mid \zeta \in \mathsf{P}_m\}.$$

Moreover, we define a function Φ, by

$$\Phi(\zeta) = \Phi_{m(\zeta)}(\zeta), \quad \zeta \in \mathbb{R},$$

and we claim that $\Phi \in \mathsf{S}$. For a real number ζ_1, we consider the integer $m_1 = m(\zeta_1)$. Thus, ζ_1 belongs to the interior of P_{m_1+1} and there exists an $\epsilon > 0$ that $\Phi(\zeta) = \Phi_{m_1+1}(\zeta)$, for every ζ, with $\zeta_1 - \epsilon < \zeta < \zeta_1 + \epsilon$. Next we want to show that Φ satisfies (21.2), (21.3) and (21.8) for each $\zeta \in \mathbb{R}$. For any $\zeta \in \mathbb{R}$, we consider the integer $m(\zeta)$. Thus, it infers that $\zeta \in \mathsf{P}_{m(\zeta)}$ and it concludes from (21.25) and (21.26) that for $G_\Phi, G_{\Phi_{m(\zeta)}} \in \mathsf{S}$ which is defined in the functional equation (21.18), we have

$$\begin{aligned}\Phi(\zeta) = &\ \Phi_{m(\zeta)}(\zeta) \\ = &\ \frac{(\xi(\zeta)-\xi(0))^{\mathcal{C}-1}}{\Gamma(\mathcal{C})}\mathfrak{Z}_\lambda + \frac{1}{\Gamma(\mathcal{A})}\int_0^\zeta \xi'(\mathcal{A})(\xi(\zeta)-\xi(\measuredangle))^{\mathcal{A}-1} G_{\Phi_{m(\zeta)}}(\measuredangle)d\measuredangle \\ = &\ \frac{(\xi(\zeta)-\xi(0))^{\mathcal{C}-1}}{\Gamma(\mathcal{C})}\mathfrak{Z}_\lambda + \frac{1}{\Gamma(\mathcal{A})}\int_0^\zeta \xi'(\mathcal{A})(\xi(\zeta)-\xi(\measuredangle))^{\mathcal{A}-1} G_{\Phi}(\measuredangle)d\measuredangle,\end{aligned} \tag{21.29}$$

where the equality above remains true because $m(\measuredangle) \le m(\zeta)$, for every $\measuredangle \in \mathsf{P}_{m(\zeta)}$ and it concludes from (21.28) that

$$\Phi(\measuredangle) = \Phi_{m(\measuredangle)}(\measuredangle) = \Phi_{m(\zeta)}(\measuredangle). \tag{21.30}$$

Since $\Phi(\zeta) = \Phi_{m(\zeta)}(\zeta)$ and $\zeta \in \mathsf{P}_{m(\zeta)}$, for every $\zeta \in \mathbb{R}, n \in \mathbb{N}$ and $\underbrace{\mathcal{T}_i}_{i=1,\dots,n} \in (0, +\infty)$, the inequality (21.27) implies that

$$\begin{aligned}&\mathscr{N}\Big(\Phi(\zeta) - \Psi(\zeta), (\mathcal{T}_1, \dots, \mathcal{T}_n)\Big) \\ = &\ \mathscr{N}\Big(\Phi_{m(\zeta)}(\zeta) - \Psi(\zeta), (\mathcal{T}_1, \dots, \mathcal{T}_n)\Big) \\ \succeq &\ \operatorname{diag}\left[\varphi_1^{\circledS}\left(\frac{-|(\xi(\zeta)-\xi(0)|^{\mathcal{A}}}{\mathcal{D}_1\theta\mathcal{T}_1}\right), \dots, \varphi_n^{\circledS}\left(\frac{-|(\xi(\zeta)-\xi(0)|^{\mathcal{A}}}{\mathcal{D}_n\theta\mathcal{T}_n}\right)\right].\end{aligned} \tag{21.31}$$

Finally, we claim that Φ is unique. Let that $\Phi' \in \mathsf{S}$ be another function, which satisfies (21.2), (21.3) and (21.8), for all $\zeta \in \mathbb{R}$. Let $\zeta \in \mathbb{R}$. Since $\Phi|_{\mathsf{P}_{m(\zeta)}} (= \Phi_{m(\zeta)})$ and $\Phi'|_{\mathsf{P}_{m(\zeta)}}$ both satisfy (21.2), (21.3) and (21.8), for all $\zeta \in \mathsf{P}_{m(\zeta)}$, the uniqueness of $\Phi_{m(\zeta)} = \Phi|_{\mathsf{P}_{m(\zeta)}}$ implies that

$$\Phi(\zeta) = \Phi|_{\mathsf{P}_{m(\zeta)}}(\zeta) = \Phi'|_{\mathsf{P}_{m(\zeta)}}(\zeta) = \Phi'(\zeta),$$

as required.

21.3 Application

Here, as an application, we apply our main results to prove an example.

Example 21.1 Consider the nonlinear implicate fractional differential equations of the form:

$$\begin{cases} {}^{\mathscr{H}}D_{\lambda^+}^{\mathcal{A},\mathcal{B};\xi}\Phi(\zeta) = F(\zeta, \Phi(\zeta), {}^{\mathscr{H}}D_{\lambda^+}^{\mathcal{A},\mathcal{B};\xi}\Phi(\zeta)), \\ I_{\lambda^+}^{1-\mathcal{C};\xi}\Phi(\lambda) = 1, \end{cases} \tag{21.32}$$

where ${}^{\mathscr{H}}D_{\lambda^+}^{\mathcal{A},\mathcal{B};\xi}(.)$ is the ξ–Hilfer fractional derivative of order $0 < \mathcal{A} \leq 1$ and type $0 \leq \mathcal{B} \leq 1$, $I_{\lambda^+}^{1-\mathcal{C};\xi}(.)$ is the Riemann–Liouville fractional integral of order $1-\mathcal{C}$, $\mathcal{C} = \mathcal{A} + \mathcal{B}(1-\mathcal{A})$ with respect to the function ξ, and $F : \Delta \times \mathbb{R}^2 \longrightarrow \mathbb{R}$ is a given function. Let $\zeta \in \Delta = [\lambda, D]$ with $D > \lambda$.

For $\xi(\zeta) = \zeta$, $\lambda = 1$, $D = 1$, $\mathcal{A} = 0.5$, taking the limit $\mathcal{B} \to 0.5$, $\Phi : [0, 1] \to \mathbb{R}$ and $F : [0, 1] \times \mathbb{R}^2 \to \mathbb{R}$, a nonlinear function defined by

$$F\left(\zeta, \Phi(\zeta), {}^{\mathscr{H}}D_{0^+}^{0.5,0.5;\zeta}\Phi(\zeta)\right) = \frac{\chi}{20}\mathbb{E}_{0.5}(\chi^{0.5}\zeta^{0.5})\Phi(\zeta) + \frac{\chi}{10}{}^{\mathscr{H}}D_{0^+}^{0.5,0.5;\zeta}\Phi(\zeta) \tag{21.33}$$

for every $\zeta \in [0, 1]$ and with $0 < \chi < \dfrac{13}{10}$. Note that $\mathbb{E}$ is the one–parameter Mittag–Leffler function defined in Chap. 4. For every $\kappa, \upsilon, \overline{\kappa}, \overline{\upsilon} \in \mathbb{R}$, $\underbrace{\mathcal{T}_i}_{i=1,\ldots,n} \in (0, +\infty)$, $n \in \mathbb{N}$ and $\zeta \in [0, 1]$, we get

$$\begin{aligned} &\mathscr{N}\left(F(\zeta, \kappa, \upsilon) - F(\zeta, \overline{\kappa}, \overline{\upsilon}), (\mathcal{T}_1, \ldots, \mathcal{T}_n)\right) \qquad (21.34) \\ &= \mathscr{N}\left(\frac{\chi}{20}\mathbb{E}_{0.5}(\chi^{0.5}\zeta^{0.5})\kappa + \frac{\chi}{10}\upsilon - \frac{\chi}{20}\mathbb{E}_{0.5}(\chi^{0.5}\zeta^{0.5})\overline{\kappa} - \frac{\chi}{10}\overline{\upsilon}, (\mathcal{T}_1, \ldots, \mathcal{T}_n)\right) \\ &\succeq \mathscr{N}\left(\frac{\chi}{20}\mathbb{E}_{0.5}(\chi^{0.5}\zeta^{0.5})(\kappa - \overline{\kappa}) + \frac{\chi}{10}(\upsilon - \overline{\upsilon}), (\mathcal{T}_1, \ldots, \mathcal{T}_n)\right) \\ &\succeq \mathscr{N}\left(\frac{\chi}{20}\mathbb{E}_{0.5}(\chi^{0.5}\zeta^{0.5})(\kappa - \overline{\kappa}), (\mathcal{T}_1, \ldots, \mathcal{T}_n)\right) \bigotimes_{\mathrm{GTN}} \mathscr{N}\left(\frac{\chi}{10}(\upsilon - \overline{\upsilon}), (\mathcal{T}_1, \ldots, \mathcal{T}_n)\right). \end{aligned}$$

Let $\Upsilon \in C(\Delta, \mathbb{R})$ satisfies

$$\mathcal{N}\Big({}^{\mathcal{H}}D_{\lambda+}^{\mathcal{A},\mathcal{B};\xi}\Upsilon(\zeta) - \frac{\chi}{20}\mathbb{E}_{0.5}(\chi^{0.5}\zeta^{0.5})\Upsilon(\zeta) - \frac{\chi}{10}{}^{\mathcal{H}}D_{0+}^{0.5,0.5;\zeta}\Upsilon(\zeta), (\mathcal{T}_1, \ldots, \mathcal{T}_n)\Big)$$
$$\succeq \operatorname{daig}\Big[\varphi_1^{\circledS}\Big(\frac{-|\xi(\zeta) - \xi(0)|^{\mathcal{A}}}{\theta \mathcal{T}_1}\Big), \ldots, \varphi_n^{\circledS}\Big(\frac{-|\xi(\zeta) - \xi(0)|^{\mathcal{A}}}{\theta \mathcal{T}_n}\Big)\Big], \tag{21.35}$$

for every $\zeta \in \Delta$ and $\underbrace{\mathcal{T}_i}_{i=1,\ldots,n} \in (0, +\infty)$, in which $\underbrace{\varphi_i^{\circledS}}_{1\le i\le n}$ with $n \in \mathbb{N}$, are fuzzy controllers defined in Sects. 9.8.1 and 9.8.2.

Now, Theorem 21.1 implies that, if $\dfrac{\overbrace{\eth_i}^{i=1,\ldots,n}}{(\Theta_1 + \Theta_2)} < 1$, we can find a unique differentiable function $\Lambda \in C(\Delta, \mathbb{R})$, such that

$${}^{\mathcal{H}}D_{\lambda+}^{\mathcal{A},\mathcal{B};\xi}\Lambda(\zeta) - \frac{\chi}{20}\mathbb{E}_{0.5}(\chi^{0.5}\zeta^{0.5})\Lambda(\zeta) - \frac{\chi}{10}{}^{\mathcal{H}}D_{0+}^{0.5,0.5;\zeta}\Lambda(\zeta) \tag{21.36}$$

and

$$\mathcal{N}\Big(\Lambda(\zeta) - \Upsilon(\zeta), (\mathcal{T}_1, \ldots, \mathcal{T}_n)\Big)$$
$$\succeq \operatorname{diag}\Big[\varphi_1^{\circledS}\Big(\frac{-|(\xi(\zeta) - \xi(0)|^{\mathcal{A}}}{\mathcal{D}_1 \theta \mathcal{T}_1}\Big), \ldots, \varphi_n^{\circledS}\Big(\frac{-|(\xi(\zeta) - \xi(0)|^{\mathcal{A}}}{\mathcal{D}_n \theta \mathcal{T}_1}\Big)\Big],$$

where

$$\underbrace{\mathcal{D}_i}_{i=1,\ldots,n} := \frac{1}{1 - \dfrac{\overbrace{\eth_i}^{i=1,\ldots,n}}{(\Theta_1 + \Theta_2)}},$$

for every $n \in \mathbb{N}$.

We refer the interested readers to [3–13].

References

1. Aderyani SR, Saadati R, Allahviranloo T (2022) Existence, uniqueness and matrix-valued fuzzy Mittag-Leffler-Hypergeometric-Wright stability for P-Hilfer fractional differential equations in matrix-valued fuzzy Banach space. Comput. Appl. Math. 41(6):234
2. Vanterler J, Sousa DC, Capelas de Olivera E (2018) On the Ulam-Hyers-Russias stability for nonlinear fractional differential equations using the ψ-Hilfer operator. J Fixed Point Theory Appl

3. Rezaei Aderyani S, Saadati R, Rassias TM, Park C (2022) Best approximation of $(\mathcal{G}_1, \mathcal{G}_2)$-random operator inequality in matrix Menger Banach algebras with application of stochastic Mittag-Leffler and H $\mathbb{H}$-Fox control functions. J Inequalities Appl 2022(1):1–17
4. Aderyani SR, Saadati R (2021) Best approximations of the ϕ-Hadamard fractional Volterra integro-differential equation by matrix valued fuzzy control functions. Adv Differ Equ 154
5. Aderyani SR, Saadati R, Yang XJ (2021) Radu-Mihe method for UHML stability for a class of ξ-Hilfer fractional differential equations in matrix valued fuzzy Banach spaces. Math Methods Appl Sci 44(18):14619–14631
6. Aderyani SR, Saadati R, Mesiar R (2022) Estimation of permuting tri-homomorphisms and permuting tri-derivations associated with the tri-additive Υ-random operator inequality in matrix MB-algebra. Int J Gen Syst 51(6):547–569
7. Aderyani SR, Saadati R (2022) Approximation of derivation–homomorphism fuzzy functional inequalities in matrix valued FC-$\diamond$-algebras. In: 2022 9th Iranian joint congress on fuzzy and intelligent systems (CFIS). IEEE, pp 1–6
8. Aderyani SR, Saadati R, Allahviranloo T, Abbasbandy S, Catak M (2023) Fuzzy approximation of a fractional Lorenz system and a fractional financial crisis. Iran J Fuzzy Syst 20(7):27–36
9. Aderyani SR, Saadati R, Abdeljawad T, Mlaiki N (2022) Multi-stability of non homogenous vector-valued fractional differential equations in matrix-valued Menger spaces. Alex Eng J 61(12):10913–10923
10. Aderyani SR, Saadati R (2023) Stability and controllability results by n-ary aggregation functions in matrix valued fuzzy n-normed spaces. Inf Sci 119265
11. O'Regan D, Aderyani SR, Saadati R, Li C (2023) Stability results and parametric delayed Mittag-Leffler matrices in symmetric fuzzy-random spaces with application. Symmetry 15(10):1880
12. Aderyani SR, Saadati R, O'Regan D, Li C (2023) On a new approach for stability and controllability analysis of functional equations. Mathematics 11(16):3458
13. Aderyani SR, Saadati R, O'Regan D, Alshammari FS (2023) Fuzzy approximate solutions of matrix-valued fractional differential equations by fuzzy control functions. Mathematics 11(6):1386

Chapter 22
The Matrix Mittag–Leffler Function and Nonlinear Fractional Integro–Differential Equations

Fractional differential equations with initial conditions or boundary value problems have recently attracted many researchers and emerged as an important field of research due to their applications in various areas of science and engineering such as control theory, wave propagation, mechanics, and biology. There is very little known in the current literature on boundary value problems of fractional integro-differential equations with integral boundary conditions and variable coefficients. Clearly, one important topic in the area is to study uniqueness and existence of solutions to nonlinear fractional differential or integro-differential equations, including partial differential equations. This chapter is comprised of seven independent and self-contained sections which deal with nonlinear equations with various boundary conditions or initial value problems, based on the matrix Mittag-Leffler function given in Sect. 4.8, fixed point theory, as well as Babenko's approach.

22.1 Nonlocal Boundary Condition

22.1.1 Preliminaries

Let $T > 0$. The Riemann-Liouville fractional integral I^α of order $\alpha \in R^+$ is defined for function $u(x)$ as (see [1, 2])

$$(I^\alpha u)(x) = \frac{1}{\Gamma(\alpha)} \int_0^x (x-t)^{\alpha-1} u(t)dt, \quad x \in [0, T].$$

In particular,

$$(I^0 u)(x) = u(x),$$

from [3].

Let $l \in N = \{1, 2, 3, \cdots\}$. The Liouville-Caputo derivative of fractional order $\alpha \in R^+$ of function $u(x)$ is defined as

S. Rezaei Aderyani et al., *Towards Ulam Type Multi Stability Analysis*, Studies in Systems, Decision and Control 523, https://doi.org/10.1007/978-3-031-55564-0_22

$$({}_C D^\alpha u)(x) = I^{l-\alpha} \frac{d^l}{dx^l} u(x) = \frac{1}{\Gamma(l-\alpha)} \int_0^x (x-t)^{l-\alpha-1} u^{(l)}(t)dt,$$

where $l - 1 < \alpha \leq l$.

We define the Banach space $C[0, T]$ (the set of all continuous functions over $[0, T]$) with the norm

$$\|u\| = \max_{x \in [0,T]} |u(x)| < +\infty.$$

Let $a(x) \in C[0, T]$, $g : [0, T] \times R \to R$ and $f : C[0, T] \to R$. We shall study the uniqueness of solutions for the following nonlinear integro-differential equation with nonlocal boundary condition and variable coefficients for $l < \alpha \leq l + 1$:

$$\begin{cases} {}_C D^\alpha u(x) + a(x)\, I^\beta u(x) = g(x, u(x)), \quad x \in [0, T], \\ u(0) = -f(u), \quad u''(0) = \cdots = u^{(l)}(0) = 0, \\ \int_0^T u(x)dx = \lambda, \end{cases} \tag{22.1}$$

where λ is a constant. In particular for $l = 1$, Eq. (22.1) turns out to be

$$\begin{cases} {}_C D^\alpha u(x) + a(x)\, I^\beta u(x) = g(x, u(x)), \quad x \in [0, T], \\ u(0) = -f(u), \quad \int_0^T u(x)dx = \lambda. \end{cases} \tag{22.2}$$

The two-parameter Mittag-Leffler function [2] is defined by

$$E_{\alpha,\beta}(z) = \sum_{k=0}^{\infty} \frac{z^k}{\Gamma(\alpha k + \beta)}, \quad z \in C, \ \alpha, \beta > 0.$$

Babenko's approach [4] is a powerful tool in solving differential and integro-differential equations with initial conditions by treating bounded integral operators as normal variables. The method itself is similar to the Laplace transform while working on differential and integral equations with constant coefficients, but it can be applied to equations with continuous and bounded variable coefficients. As an example to demonstrate the technique, we are going to show the following lemma, which plays an essential role in deriving our main theorem.

Lemma 22.1 *Let $a, h : [0, T] \to R$ be continuous functions and $f : C[0, T] \to R$ be a continuous functional. Assume that*

$$\frac{2MT^{\alpha+\beta}}{\Gamma(\alpha+\beta+2)} E_{\alpha+\beta,1}\left(MT^{\alpha+\beta}\right) < 1,$$

where M is a constant satisfying

$$\|a\| = \max_{x\in[0,T]} |a(x)| \le M.$$

Then $u(x)$ is a solution to the following nonlinear integro-differential equation

$$\begin{cases} {}_C D^{\alpha} u(x) + a(x)\, I^{\beta} u(x) = h(x), \quad l < \alpha \le l+1, \quad x \in [0,T], \\ u(0) = -f(u), \quad u''(0) = \cdots = u^{(l)}(0) = 0, \\ \displaystyle\int_0^T u(x)dx = \lambda, \end{cases} \tag{22.3}$$

if and only if $u(x)$ satisfies the integral equation

$$\begin{aligned} u(x) = {} & \sum_{j=0}^{\infty}(-1)^j \left(I^{\alpha} a(x)\, I^{\beta}\right)^j I^{\alpha} h(x) - f(u) \sum_{j=0}^{\infty}(-1)^j \left(I^{\alpha} a(x)\, I^{\beta}\right)^j \cdot 1 \\ & - \frac{2}{\Gamma(\alpha+1)T^2} \int_0^T (T-t)^{\alpha} h(t)dt \sum_{j=0}^{\infty}(-1)^j \left(I^{\alpha} a(x)\, I^{\beta}\right)^j x \\ & + \frac{2}{\Gamma(\alpha+1)\Gamma(\beta)T^2} \int_0^T (T-x_1)^{\alpha} a(x_1)dx_1 \int_0^{x_1} (x_1-t)^{\beta-1} u(t)dt \cdot \\ & \sum_{j=0}^{\infty}(-1)^j \left(I^{\alpha} a(x)\, I^{\beta}\right)^j x + \frac{2\lambda}{T^2} \sum_{j=0}^{\infty}(-1)^j \left(I^{\alpha} a(x)\, I^{\beta}\right)^j x \\ & + \frac{2f(u)}{T} \sum_{j=0}^{\infty}(-1)^j \left(I^{\alpha} a(x)\, I^{\beta}\right)^j x, \end{aligned}$$

in the space $C[0,T]$.

Proof Clearly,

$$I^{\alpha}({}_C D^{\alpha})u(x) = u(x) + f(u) + c_1 x,$$

using

$$u(0) = -f(u), \quad u''(0) = \cdots = u^{(l)}(0) = 0.$$

Hence, applying the operator I^{α} to both sides of equation

$${}_C D^{\alpha} u(x) + a(x)\, I^{\beta} u(x) = h(x),$$

we get

$$u(x) + f(u) + c_1 x + I^{\alpha} a(x)\, I^{\beta} u(x) = I^{\alpha} h(x).$$

Obviously,

$$\int_0^T I^\alpha h(x)dx = \frac{1}{\Gamma(\alpha)}\int_0^T dx \int_0^x (x-t)^{\alpha-1}h(t)dt = \frac{1}{\Gamma(\alpha)}\int_0^T h(t)dt \int_t^T (x-t)^{\alpha-1}dx$$
$$= \frac{1}{\Gamma(\alpha+1)}\int_0^T (T-t)^\alpha h(t)dt.$$

Similarly,

$$\int_0^T I^\alpha a(x)\, I^\beta u(x)dx = \frac{1}{\Gamma(\alpha+1)\Gamma(\beta)}\int_0^T (T-x_1)^\alpha a(x_1)dx_1 \int_0^{x_1}(x_1-t)^{\beta-1}u(t)dt.$$

Thus,

$$\int_0^T u(x)dx + \int_0^T f(u)dx + c_1\int_0^T xdx + \int_0^T I^\alpha a(x)\, I^\beta u(x)dx = \int_0^T I^\alpha h(x)dx,$$

which implies that

$$\lambda + f(u)T + \frac{T^2}{2}c_1 = \int_0^T I^\alpha h(x)dx - \int_0^T I^\alpha a(x)\, I^\beta u(x)dx,$$

by noting that $f(u) \in R$. So,

$$\begin{aligned} c_1 = & \frac{2}{\Gamma(\alpha+1)T^2}\int_0^T (T-t)^\alpha h(t)dt \\ & -\frac{2}{\Gamma(\alpha+1)\Gamma(\beta)T^2}\int_0^T (T-x_1)^\alpha a(x_1)dx_1 \int_0^{x_1}(x_1-t)^{\beta-1}u(t)dt \\ & -\frac{2\lambda}{T^2} - \frac{2f(u)}{T}, \end{aligned}$$

and

$$\left(1 + I^\alpha a(x)\, I^\beta\right)u(x) = I^\alpha h(x) - f(u) - c_1 x.$$

Treating the factor $\left(1 + I^\alpha a(x)\, I^\beta\right)$ as a variable, we deduce that, by Babenko's approach

$$\begin{aligned} u(x) = & \left(1 + I^\alpha a(x)\, I^\beta\right)^{-1}\left[I^\alpha h(x) - f(u) - c_1 x\right] \\ = & \sum_{j=0}^\infty (-1)^j \left(I^\alpha a(x)\, I^\beta\right)^j I^\alpha h(x) - f(u)\sum_{j=0}^\infty (-1)^j \left(I^\alpha a(x)\, I^\beta\right)^j \cdot 1 \\ & -c_1\sum_{j=0}^\infty (-1)^j \left(I^\alpha a(x)\, I^\beta\right)^j x \\ = & \sum_{j=0}^\infty (-1)^j \left(I^\alpha a(x)\, I^\beta\right)^j I^\alpha h(x) - f(u)\sum_{j=0}^\infty (-1)^j \left(I^\alpha a(x)\, I^\beta\right)^j \cdot 1 \end{aligned}$$

$$
\begin{aligned}
&-\frac{2}{\Gamma(\alpha+1)T^2}\int_0^T (T-t)^{\alpha}h(t)\,dt\cdot\sum_{j=0}^{\infty}(-1)^j\left(I^{\alpha}a(x)\,I^{\beta}\right)^j x\\
&+\frac{2}{\Gamma(\alpha+1)\Gamma(\beta)T^2}\int_0^T (T-x_1)^{\alpha}a(x_1)dx_1\int_0^{x_1}(x_1-t)^{\beta-1}u(t)dt\,\cdot\\
&\sum_{j=0}^{\infty}(-1)^j\left(I^{\alpha}a(x)\,I^{\beta}\right)^j x+\frac{2\lambda}{T^2}\sum_{j=0}^{\infty}(-1)^j\left(I^{\alpha}a(x)\,I^{\beta}\right)^j x\\
&+\frac{2f(u)}{T}\sum_{j=0}^{\infty}(-1)^j\left(I^{\alpha}a(x)\,I^{\beta}\right)^j x.
\end{aligned}
$$

It remains to be shown that $u\in C[0,T]$. Clearly,

$$
\|I^{\alpha}h\|=\frac{1}{\Gamma(\alpha)}\max_{x\in[0,T]}\left|\int_0^x (x-t)^{\alpha-1}h(t)dt\right|\le\frac{T^{\alpha}}{\Gamma(\alpha+1)}\|h\|.
$$

Hence, we infer that

$$
\begin{aligned}
\|u\|\le\ &\sum_{j=0}^{\infty}\left\|\left(I^{\alpha}a(x)\,I^{\beta}\right)^j I^{\alpha}\right\|\|h\|+|f(u)|\sum_{j=0}^{\infty}\left\|\left(I^{\alpha}a(x)\,I^{\beta}\right)^j\right\|\\
&+\frac{2\|h\|}{\Gamma(\alpha+2)T^2}T^{\alpha+1}\sum_{j=0}^{\infty}\left\|\left(I^{\alpha}a(x)\,I^{\beta}\right)^j\right\|\|x\|\\
&+\frac{2M\|u\|\,T^{\alpha+\beta-1}}{\Gamma(\alpha+\beta+2)}\sum_{j=0}^{\infty}\left\|\left(I^{\alpha}a(x)\,I^{\beta}\right)^j\right\|\|x\|\\
&+\frac{2|\lambda|}{T^2}\sum_{j=0}^{\infty}\left\|\left(I^{\alpha}a(x)\,I^{\beta}\right)^j\right\|\|x\|+\frac{2|f(u)|}{T}\sum_{j=0}^{\infty}\left\|\left(I^{\alpha}a(x)\,I^{\beta}\right)^j\right\|\|x\|\\
\le\ &\|h\|\sum_{j=0}^{\infty}\frac{M^j\,T^{(\alpha+\beta)j+\alpha}}{\Gamma((\alpha+\beta)j+\alpha+1)}+|f(u)|\sum_{j=0}^{\infty}\frac{M^j\,T^{(\alpha+\beta)j}}{\Gamma((\alpha+\beta)j+1)}\\
&+\frac{2\|h\|\,T^{\alpha}}{\Gamma(\alpha+2)}\sum_{j=0}^{\infty}\frac{M^j\,T^{(\alpha+\beta)j}}{\Gamma((\alpha+\beta)j+1)}+\frac{2M\|u\|\,T^{\alpha+\beta}}{\Gamma(\alpha+\beta+2)}\sum_{j=0}^{\infty}\frac{M^j\,T^{(\alpha+\beta)j}}{\Gamma((\alpha+\beta)j+1)}\\
&+\left[\frac{2|\lambda|}{T}+2|f(u)|\right]\sum_{j=0}^{\infty}\frac{M^j\,T^{(\alpha+\beta)j}}{\Gamma((\alpha+\beta)j+1)}.
\end{aligned}
\tag{22.4}
$$

This implies that

$$
\begin{aligned}
&\left(1-\frac{2MT^{\alpha+\beta}}{\Gamma(\alpha+\beta+2)}E_{\alpha+\beta,1}\left(MT^{\alpha+\beta}\right)\right)\|u\|=\\
&\|h\|\,T^{\alpha}E_{\alpha+\beta,\alpha+1}\left(MT^{\alpha+\beta}\right)+\left[3|f(u)|+\frac{2\|h\|\,T^{\alpha}}{\Gamma(\alpha+2)}+\frac{2|\lambda|}{T}\right]E_{\alpha+\beta,1}\left(MT^{\alpha+\beta}\right)<+\infty.
\end{aligned}
$$

By our assumption,

$$1-\frac{2MT^{\alpha+\beta}}{\Gamma(\alpha+\beta+2)}E_{\alpha+\beta,1}\left(MT^{\alpha+\beta}\right)>0,$$

which infers that $\|u\|$ is bounded. Furthermore, $u(x)$ is clearly continuous over the interval $[0,T]$. This completes the proof of Lemma 22.1.

22.1.2 Main Results

We are now ready to present our key results in this subsection regarding the uniqueness of solutions to Eq. (22.1) using Banach's contractive principle.

Theorem 22.1 *Assume $a(x)\in C[0,T]$, $g:[0,T]\times R\to R$ is continuous and $f: C[0,T]\to R$ is a continuous functional, and there exist nonnegative constants L and L_1, such that*

$$|g(x,y_1)-g(x,y_2)|\le L|y_1-y_2|,\quad y_1,\,y_2\in R,$$
$$|f(u)-f(v)|\le L_1\|u-v\|,\quad u,\,v\in C[0,T].$$

Furthermore,

$$q=LT^{\alpha}E_{\alpha+\beta,\alpha+1}\left(MT^{\alpha+\beta}\right)+\left[3L_1+\frac{2LT^{\alpha}}{\Gamma(\alpha+2)}+\frac{2MT^{\alpha+\beta}}{\Gamma(\alpha+\beta+2)}\right]E_{\alpha+\beta,1}\left(MT^{\alpha+\beta}\right)$$
$$<1.$$

Then Eq. (22.1) has a unique solution in the space $C[0,T]$.

Proof From Lemma 22.1, we define a nonlinear mapping T over the space $C[0,T]$ by

$$\begin{aligned}(Tu)(x)=&\sum_{j=0}^{\infty}(-1)^j\left(I^{\alpha}a(x)\,I^{\beta}\right)^j I^{\alpha}g(x,u)-f(u)\sum_{j=0}^{\infty}(-1)^j\left(I^{\alpha}a(x)\,I^{\beta}\right)^j\cdot 1\\&-\frac{2}{\Gamma(\alpha+1)T^2}\int_0^T(T-t)^{\alpha}g(t,u(t))\,dt\cdot\sum_{j=0}^{\infty}(-1)^j\left(I^{\alpha}a(x)\,I^{\beta}\right)^j x\\&+\frac{2}{\Gamma(\alpha+1)\Gamma(\beta)T^2}\int_0^T(T-x_1)^{\alpha}a(x_1)dx_1\int_0^{x_1}(x_1-t)^{\beta-1}u(t)dt\cdot\\&\sum_{j=0}^{\infty}(-1)^j\left(I^{\alpha}a(x)\,I^{\beta}\right)^j x+\frac{2\lambda}{T^2}\sum_{j=0}^{\infty}(-1)^j\left(I^{\alpha}a(x)\,I^{\beta}\right)^j x\end{aligned}$$

$$+\frac{2f(u)}{T}\sum_{j=0}^{\infty}(-1)^j\left(I^{\alpha}a(x)\,I^{\beta}\right)^j x.$$

Clearly,

$$|g(x,u)| = |g(x,u)-g(x,0)+g(x,0)| \le L|u| + |g(x,0)|,$$

which implies that

$$\max_{x\in[0,T]}|g(x,u)| \le L\,\|u\| + \max_{x\in[0,T]}|g(x,0)| < +\infty,$$

since $g(x,0)\in C[0,T]$. It follows from inequality (22.4) that $(Tu)(x)\in C[0,T]$ by noting that

$$1-\frac{2MT^{\alpha+\beta}}{\Gamma(\alpha+\beta+2)}E_{\alpha+\beta,1}\left(MT^{\alpha+\beta}\right) > 0$$

in Lemma 22.1, as $q<1$. Further, we need to prove that T is contractive. Indeed,

$$(Tu)(x)-(Tv)(x) = \sum_{j=0}^{\infty}(-1)^j\left(I^{\alpha}a(x)\,I^{\beta}\right)^j I^{\alpha}(g(x,u)-g(x,v))$$

$$-(f(u)-f(v))\sum_{j=0}^{\infty}(-1)^j\left(I^{\alpha}a(x)\,I^{\beta}\right)^j\cdot 1$$

$$-\frac{2}{\Gamma(\alpha+1)T^2}\int_0^T(T-t)^{\alpha}(g(t,u(t))-g(t,v(t)))\,dt\cdot\sum_{j=0}^{\infty}(-1)^j\left(I^{\alpha}a(x)\,I^{\beta}\right)^j x$$

$$+\frac{2}{\Gamma(\alpha+1)\Gamma(\beta)T^2}\int_0^T(T-x_1)^{\alpha}a(x_1)dx_1\int_0^{x_1}(x_1-t)^{\beta-1}(u(t)-v(t))dt\,\cdot$$

$$\sum_{j=0}^{\infty}(-1)^j\left(I^{\alpha}a(x)\,I^{\beta}\right)^j x+\frac{2(f(u)-f(v))}{T}\sum_{j=0}^{\infty}(-1)^j\left(I^{\alpha}a(x)\,I^{\beta}\right)^j x.$$

Thus,

$$|(Tu)(x)-(Tv)(x)| \le LT^{\alpha}\,\|u-v\|\sum_{j=0}^{\infty}\frac{M^j\,T^{(\alpha+\beta)j}}{\Gamma((\alpha+\beta)j+\alpha+1)}$$

$$+3L_1\,\|u-v\|\sum_{j=0}^{\infty}\frac{M^j\,T^{(\alpha+\beta)j}}{\Gamma((\alpha+\beta)j+1)}+\frac{2L\,\|u-v\|\,T^{\alpha}}{\Gamma(\alpha+2)}\sum_{j=0}^{\infty}\frac{M^j\,T^{(\alpha+\beta)j}}{\Gamma((\alpha+\beta)j+1)}$$

$$\frac{2M\,\|u-v\|\,T^{\alpha+\beta}}{\Gamma(\alpha+\beta+2)}\sum_{j=0}^{\infty}\frac{M^j\,T^{(\alpha+\beta)j}}{\Gamma((\alpha+\beta)j+1)} = q\,\|u-v\|.$$

Since $q < 1$, Eq. (22.1) has a unique solution in the space $C[0, T]$ by Banach's fixed point theorem. This completes the proof of Theorem 22.1.

Example 22.1 The following nonlinear integro-differential equation with nonlocal boundary condition and variable coefficients:

$$\begin{cases} {}_C D^{3.5} u(x) + \dfrac{2x^2}{x^2+1} I^{1.1} u(x) = \dfrac{1}{2}\cos(x^2 u(x)) + x^3, \quad x \in [0, 1], \\ u(0) = \dfrac{1}{9}\sin u(1/2), \quad u''(0) = u'''(0) = 0, \\ \displaystyle\int_0^1 u(x)dx = \sqrt{2}, \end{cases}$$

has a unique solution in $C[0, 1]$.

Proof Clearly,

$$g(x, u) = \frac{1}{2}\cos(x^2 u(x)) + x^3,$$

and

$$|g(x, u) - g(x, v)| \leq \frac{1}{2}|x^2 u - x^2 v| \leq \frac{1}{2}|u - v|.$$

Therefore, $L = 1/2$ as $x \in [0, 1]$. On the other hand,

$$f(u) = \frac{1}{9}\sin u(1/2),$$

and

$$|f(u) - f(v)| \leq \frac{1}{9}|\sin u(1/2) - \sin v(1/2)| \leq \frac{1}{9}|u(1/2) - v(1/2)| \leq \frac{1}{9}\|u - v\|,$$

which indicates that $L_1 = 1/9$. It follows from Theorem 22.1 that

$$q = \frac{1}{2}E_{4.6, 4.5}(2) + \left[\frac{3}{9} + \frac{1}{\Gamma(5.5)} + \frac{4}{\Gamma(6.6)}\right] E_{4.6, 1}(2).$$

Using online calculators from the site https://www.wolframalpha.com/ (accessed on 09 October 2022), we get

$$E_{4.6, 4.5}(2) = \sum_{j=0}^{\infty} \frac{2^j}{\Gamma(4.6j + 4.5)} \approx 0.0860118,$$

$$E_{4.6, 1}(2) = \sum_{j=0}^{\infty} \frac{2^j}{\Gamma(4.6j + 1)} \approx 1.0325,$$

$$\frac{1}{\Gamma(5.5)} \approx 0.0191048, \quad \frac{4}{\Gamma(6.6)} \approx 0.0116042.$$

These clearly demonstrate that $q < 1$. By Theorem 22.1, the equation has a unique solution in $C[0, 1]$.

22.1.3 Summary

We have investigated the uniqueness of solutions to the nonlinear fractional integro-differential equation (22.1) with nonlocal boundary condition and variable coefficients using the Mittag-Leffler function, Babenko's approach, and Banach's contractive principle, and presented an applicable example. The technique used clearly opens up new directions for studying other types of boundary conditions or with different fractional derivatives, such as boundary value problems of the nonlinear fractional partial integro-differential equations with variable coefficients as well as nonlinear integro-differential equations with the Hilfer fractional derivatives.

22.2 Application of a Matrix Mittag–Leffler Function to the Nonlinear Equation

22.2.1 Preliminaries

Let $y \in \Omega = [0, 1]^n \subset \mathbb{R}^n$ $(n \geq 1)$ and $x \in [0, 1]$. It follows from [7] for $\alpha_1 \geq 0, \cdots, \alpha_n \geq 0$ that

$$I_1^{\alpha_1} \cdots I_n^{\alpha_n} Z(x, y) = \frac{1}{\Gamma(\alpha_1) \cdots \Gamma(\alpha_n)} \int_0^{y_1} \cdots \int_0^{y_n} \prod_{k=1}^{n} (y_k - s_k)^{\alpha_k - 1} Z(x, s_1, \cdots, s_n) ds_n \cdots ds_1.$$

In particular for $\alpha_2 = \cdots = \alpha_n = 0$, we have

$$I_1^{\alpha_1} Z(x, y) = \frac{1}{\Gamma(\alpha_1)} \int_0^{y_1} (y_1 - s_1)^{\alpha_1 - 1} Z(x, s_1, y_2, \cdots, y_n) ds_1. \tag{22.5}$$

Let $1 < \alpha \leq 2$. Then from [8]

$$I_x^{\alpha} \, {}^c\frac{\partial^\alpha}{\partial x^\alpha} Z(x, y) = Z(x, y) - Z(0, y) - Z_x'(0, y) x, \tag{22.6}$$

where the operator I_x^α is the partial Riemann-Liouville fractional integral of order $\alpha > 0$ with respect to x with initial point zero [1],

$$(I_x^\alpha Z)(x, y) = \frac{1}{\Gamma(\alpha)} \int_0^x (x - \zeta)^{\alpha - 1} Z(\zeta, y) d\zeta, \tag{22.7}$$

and $\frac{{}_c\partial^\alpha}{\partial x^\alpha}$ is the partial Liouville-Caputo fractional derivative of order α with respect to x

$$\left(\frac{{}_c\partial^\alpha}{\partial x^\alpha} Z\right)(x, y) = \frac{1}{\Gamma(2-\alpha)} \int_0^x (x-s)^{-\alpha+1} W_x''(s, y) ds, \quad 1 < \alpha \le 2. \quad (22.8)$$

$\mathcal{S}([0, 1] \times \Omega)$ (the set of all continuous functions over $[0, 1] \times \Omega$) is defined as the Banach space with the norm given as

$$\|Z\|_{\mathcal{S}} = \sup_{x \in [0,1],\ y \in \Omega} |Z(x, y)| \quad \text{for} \quad Z \in \mathcal{S}([0, 1] \times \Omega).$$

In this section, we study the uniqueness of solutions to the following equation with the boundary condition for $1 < \alpha \le 2$ and $m \in \mathbb{N}$ in the space $\mathcal{S}([0, 1] \times \Omega)$ for constants a_k:

$$\begin{cases} \frac{{}_c\partial^\alpha}{\partial x^\alpha} Z(x, y) + \sum_{k=1}^m a_k I_1^{\alpha_{1k}} \cdots I_n^{\alpha_{nk}} Z(x, y) = f(x, y, Z(x, y)), \\ Z(0, y) = Z(1, y) = 0, \quad (x, y) \in [0, 1] \times \Omega, \end{cases} \quad (22.9)$$

where $\alpha_{ik} \ge 0$ for all $i = 1, \cdots, n$, $k = 1, \cdots, m$, and $f : [0, 1] \times \Omega \times \mathbb{R} \to \mathbb{R}$ is a function satisfying certain conditions to be given.

A function Z is a solution of the problem (22.9) if it satisfies the equation over $\mathcal{S}([0, 1] \times \Omega)$ and its boundary condition $Z(0, y) = Z(1, y) = 0$.

Let $C[0, 1]$ be the set of all continuous functions over $[0, 1]$ with the norm

$$\|x\| = \max_{t \in [0,1]} |x(t)|.$$

Then $C[0, 1]$ is a Banach space.

Babenko's approach [4] is an efficient tool for dealing with integral and differential equations with initial conditions or boundary value problems [7, 9]. To show this method in detail, we will derive the solution to the following equation with the initial conditions in the space $C[0, 1]$ for constant λ:

$$\begin{cases} {}_cD_0^\alpha x(t) + \lambda\, {}_cD_0^\beta x(t) = t^2, \quad 0 < \beta < \alpha, \quad 1 < \alpha \le 2, \\ x(0) = x'(0) = 0, \end{cases} \quad (22.10)$$

where

$${}_cD_0^\alpha x(t) = \frac{1}{\Gamma(2-\alpha)} \int_0^t (t-s)^{-\alpha+1} x''(s) ds, \quad 1 < \alpha \le 2,$$

and

$${}_cD_0^\beta x(t) = \frac{1}{\Gamma(1-\beta)} \int_0^t (t-s)^{-\beta} x'(s) ds, \quad \text{if } 0 < \beta \le 1.$$

From (22.6) one has

$$I_0^{\alpha}\left({}_cD_0^{\alpha}x(t)\right) = x(t) - x(0) - x'(0)t = x(t). \tag{22.11}$$

Applying I_0^{α} to the first equation in (22.10), we have

$$\left(1 + \lambda I_0^{\alpha-\beta}\right)x(t) = I_0^{\alpha}t^2 = \frac{2}{\Gamma(\alpha+3)}t^{\alpha+2}.$$

Treating the operator $\left(1 + \lambda I_0^{\alpha-\beta}\right)$ as a variable, we informally derive by Babenko's approach

$$x(t) = \frac{2}{\Gamma(\alpha+3)}\left(1 + \lambda I_0^{\alpha-\beta}\right)^{-1} t^{\alpha+2} = \frac{2}{\Gamma(\alpha+3)}\sum_{l=0}^{\infty}(-1)^l\lambda^l I_0^{l(\alpha-\beta)}\, t^{\alpha+2}$$

$$= 2t^{\alpha+2}\sum_{l=0}^{\infty}\frac{\left(-\lambda t^{\alpha-\beta}\right)^l}{\Gamma(l(\alpha-\beta)+\alpha+3)} = 2t^{\alpha+2}E_{\alpha-\beta,\,\alpha+3}\left(-\lambda t^{\alpha-\beta}\right),$$

using

$$I_0^{l(\alpha-\beta)}\, t^{\alpha+2} = \frac{\Gamma(\alpha+3)}{\Gamma(l(\alpha-\beta)+\alpha+3)}t^{l(\alpha-\beta)+\alpha+2}.$$

This infers that

$$\|x\| \leq 2\sum_{l=0}^{\infty}\frac{|\lambda|^l}{\Gamma(l(\alpha-\beta)+\alpha+3)} = 2E_{\alpha-\beta,\,\alpha+3}(|\lambda|) < +\infty,$$

which claims that the series solution

$$x(t) = 2t^{\alpha+2}\sum_{l=0}^{\infty}\frac{\left(-\lambda t^{\alpha-\beta}\right)^l}{\Gamma(l(\alpha-\beta)+\alpha+3)} = 2t^{\alpha+2}E_{\alpha-\beta,\,\alpha+3}\left(-\lambda t^{\alpha-\beta}\right),$$

is in $C[0, 1]$.

22.2.2 Main Results

Theorem 22.2 *Let* $1 < \alpha \leq 2$, $\alpha_{ij} \geq 0$ *for* $i = 1, \cdots, n$, $j = 1, \cdots, m$, *and*

$$M_k = \begin{bmatrix} \alpha \cdots & \alpha & 2 \\ \alpha_{11} \cdots & \alpha_{1m} & \alpha_{1k}+1 \\ & \cdots & \\ \alpha_{n1} \cdots & \alpha_{nm} & \alpha_{nk}+1 \end{bmatrix}, \tag{22.12}$$

for $k = 1, \cdots, m$. We assume that

$$A = 1 - \frac{1}{\Gamma(\alpha+1)} \sum_{k=1}^{m} |a_k| E_{M_k}(|a_1|, \cdots, |a_m|) > 0.$$

Then $Z(x, y)$ is a solution of the problem (22.9) if and only if it is bounded and satisfies the following implicit integral equation in the space $\mathcal{S}([0, 1] \times \Omega)$:

$$\begin{aligned} Z(x, y) =& \sum_{l=0}^{\infty} (-1)^l \sum_{\substack{l_1+\cdots+l_m=l \\ l_1 \geq 0, \cdots, l_m \geq 0}} \binom{l}{l_1, \cdots, l_m} a_1^{l_1} \cdots a_m^{l_m} I_x^{l_1\alpha+\cdots+l_m\alpha+\alpha} \\ &\cdot I_1^{\alpha_{11}l_1+\cdots+\alpha_{1m}l_m} \cdots I_n^{\alpha_{n1}l_1+\cdots+\alpha_{nm}l_m} f(x, y, Z(x, y)) \\ &- \sum_{l=0}^{\infty} (-1)^l \sum_{\substack{l_1+\cdots+l_m=l \\ l_1 \geq 0, \cdots, l_m \geq 0}} \binom{l}{l_1, \cdots, l_m} a_1^{l_1} \cdots a_m^{l_m} \frac{x^{\alpha l_1+\cdots+\alpha l_m+1}}{\Gamma(\alpha l_1+\cdots+\alpha l_m+2)} \\ &\cdot I_1^{\alpha_{11}l_1+\cdots+\alpha_{1m}l_m} \cdots I_n^{\alpha_{n1}l_1+\cdots+\alpha_{nm}l_m} I_{x=1}^{\alpha} f(x, y, Z(x, y)) + \sum_{k=1}^{m} a_k \sum_{l=0}^{\infty} (-1)^l \\ &\cdot \sum_{\substack{l_1+\cdots+l_m=l \\ l_1 \geq 0, \cdots, l_m \geq 0}} \binom{l}{l_1, \cdots, l_m} a_1^{l_1} \cdots a_m^{l_m} \frac{x^{\alpha l_1+\cdots+\alpha l_m+1}}{\Gamma(\alpha l_1+\cdots+\alpha l_m+2)} \\ &\cdot I_1^{\alpha_{11}l_1+\cdots+\alpha_{1m}l_m+\alpha_{1k}} \cdots I_n^{\alpha_{n1}l_1+\cdots+\alpha_{nm}l_m+\alpha_{nk}} I_{x=1}^{\alpha} Z(x, y). \end{aligned} \tag{22.13}$$

Furthermore,

$$\|Z\| \leq \frac{1}{q} \left(E_{Q_\alpha}(|a_1|, \cdots, |a_m|) + \frac{1}{\Gamma(\alpha+1)} E_{Q_1}(|a_1|, \cdots, |a_m|) \right) \cdot \sup_{(x,y)\in[0,1]\times\Omega,\, w\in R} |f(x, y, w)| < +\infty,$$

where

$$Q_\alpha = \begin{bmatrix} \alpha \cdots & \alpha & \alpha+1 \\ \alpha_{11} \cdots & \alpha_{1m} & 1 \\ & \cdots & \\ \alpha_{n1} \cdots & \alpha_{nm} & 1 \end{bmatrix},$$

and we define

$$Q_1 = \begin{bmatrix} \alpha \cdots & \alpha & 2 \\ \alpha_{11} \cdots & \alpha_{1m} & 1 \\ & \cdots & \\ \alpha_{n1} \cdots & \alpha_{nm} & 1 \end{bmatrix}.$$

Proof Applying the operator I_x^α to both sides of the first equation in (22.9), we get

$$I_x^\alpha \, {}^c\frac{\partial^\alpha}{\partial x^\alpha} Z(x, y) + \sum_{k=1}^{m} a_k I_x^\alpha I_1^{\alpha_{1k}} \cdots I_n^{\alpha_{nk}} Z(x, y) = I_x^\alpha f(x, y, Z(x, y)).$$

According to (22.6), we get

$$Z(x, y) - Z(0, y) - Z_x'(0, y)x = I_x^\alpha f(x, y, Z(x, y)) - \sum_{k=1}^{m} a_k I_x^\alpha I_1^{\alpha_{1k}} \cdots I_n^{\alpha_{nk}} Z(x, y),$$

which implies

$$-Z_x'(0, y) = I_{x=1}^\alpha f(x, y, Z(x, y)) - \sum_{k=1}^{m} a_k I_1^{\alpha_{1k}} \cdots I_n^{\alpha_{nk}} I_{x=1}^\alpha Z(x, y),$$

using the boundary condition

$$Z(0, y) = Z(1, y) = 0.$$

Hence,

$$Z_x'(0, y) = -I_{x=1}^\alpha f(x, y, Z(x, y)) + \sum_{k=1}^{m} a_k I_1^{\alpha_{1k}} \cdots I_n^{\alpha_{nk}} I_{x=1}^\alpha Z(x, y),$$

which infers that

$$Z(x, y) + \sum_{k=1}^{m} a_k I_x^\alpha I_1^{\alpha_{1k}} \cdots I_n^{\alpha_{nk}} Z(x, y) =$$

$$I_x^\alpha f(x, y, Z(x, y)) - x I_{x=1}^\alpha f(x, y, Z(x, y)) + \sum_{k=1}^{m} a_k I_1^{\alpha_{1k}} \cdots I_n^{\alpha_{nk}} x I_{x=1}^\alpha Z(x, y).$$

Therefore,

$$\left(1 + \sum_{k=1}^{m} a_k I_x^\alpha I_1^{\alpha_{1k}} \cdots I_n^{\alpha_{nk}}\right) Z(x, y) =$$

$$I_x^\alpha f(x, y, Z(x, y)) - x I_{x=1}^\alpha f(x, y, Z(x, y)) + \sum_{k=1}^{m} a_k I_1^{\alpha_{1k}} \cdots I_n^{\alpha_{nk}} x I_{x=1}^\alpha Z(x, y).$$

Using Babenko's method and the multinomial theorem, we have

$$Z(x, y) = \left(1 + \sum_{k=1}^{m} a_k I_x^\alpha I_1^{\alpha_{1k}} \cdots I_n^{\alpha_{nk}}\right)^{-1}$$

$$\cdot\left(I_x^\alpha f(x,y,Z(x,y)) - xI_{x=1}^\alpha f(x,y,Z(x,y)) + \sum_{k=1}^m a_k I_1^{\alpha_{1k}}\cdots I_n^{\alpha_{nk}} x I_{x=1}^\alpha Z(x,y)\right)$$

$$=\sum_{l=0}^\infty (-1)^l \left(\sum_{k=1}^m a_k I_x^\alpha I_1^{\alpha_{1k}}\cdots I_n^{\alpha_{nk}}\right)^l$$

$$\cdot\left(I_x^\alpha f(x,y,Z(x,y)) - xI_{x=1}^\alpha f(x,y,Z(x,y)) + \sum_{k=1}^m a_k I_1^{\alpha_{1k}}\cdots I_n^{\alpha_{nk}} x I_{x=1}^\alpha Z(x,y)\right)$$

$$=\sum_{l=0}^\infty (-1)^l \sum_{\substack{l_1+\cdots+l_m=l\\ l_1\ge 0,\cdots,l_m\ge 0}} \binom{l}{l_1,\cdots,l_m} a_1^{l_1}\cdots a_m^{l_m}\left(I_x^\alpha I_1^{\alpha_{11}}\cdots I_n^{\alpha_{n1}}\right)^{l_1}\cdots\left(I_x^\alpha I_1^{\alpha_{1m}}\cdots I_n^{\alpha_{nm}}\right)^{l_m}$$

$$\cdot\left(I_x^\alpha f(x,y,Z(x,y)) - xI_{x=1}^\alpha f(x,y,Z(x,y)) + \sum_{k=1}^m a_k I_1^{\alpha_{1k}}\cdots I_n^{\alpha_{nk}} x I_{x=1}^\alpha Z(x,y)\right)$$

$$=\sum_{l=0}^\infty (-1)^l \sum_{\substack{l_1+\cdots+l_m=l\\ l_1\ge 0,\cdots,l_m\ge 0}} \binom{l}{l_1,\cdots,l_m} a_1^{l_1}\cdots a_m^{l_m} I_x^{l_1\alpha+\cdots+l_m\alpha+\alpha}$$

$$\cdot I_1^{\alpha_{11}l_1+\cdots+\alpha_{1m}l_m}\cdots I_n^{\alpha_{n1}l_1+\cdots+\alpha_{nm}l_m} f(x,y,Z(x,y))$$

$$-\sum_{l=0}^\infty (-1)^l \sum_{\substack{l_1+\cdots+l_m=l\\ l_1\ge 0,\cdots,l_m\ge 0}} \binom{l}{l_1,\cdots,l_m} a_1^{l_1}\cdots a_m^{l_m} \frac{x^{\alpha l_1+\cdots+\alpha l_m+1}}{\Gamma(\alpha l_1+\cdots+\alpha l_m+2)}$$

$$\cdot I_1^{\alpha_{11}l_1+\cdots+\alpha_{1m}l_m}\cdots I_n^{\alpha_{n1}l_1+\cdots+\alpha_{nm}l_m} I_{x=1}^\alpha f(x,y,Z(x,y)) + \sum_{k=1}^m a_k \sum_{l=0}^\infty (-1)^l$$

$$\cdot \sum_{\substack{l_1+\cdots+l_m=l\\ l_1\ge 0,\cdots,l_m\ge 0}} \binom{l}{l_1,\cdots,l_m} a_1^{l_1}\cdots a_m^{l_m} \frac{x^{\alpha l_1+\cdots+\alpha l_m+1}}{\Gamma(\alpha l_1+\cdots+\alpha l_m+2)}$$

$$\cdot I_1^{\alpha_{11}l_1+\cdots+\alpha_{1m}l_m+\alpha_{1k}}\cdots I_n^{\alpha_{n1}l_1+\cdots+\alpha_{nm}l_m+\alpha_{nk}} I_{x=1}^\alpha Z(x,y).$$

Clearly, all above steps are reversible. It remains to be shown that $Z \in \mathcal{S}([0,1]\times\Omega)$. In fact,

$$\|Z\| \le \sum_{l=0}^\infty \sum_{\substack{l_1+\cdots+l_m=l\\ l_1\ge 0,\cdots,l_m\ge 0}} \binom{l}{l_1,\cdots,l_m} \frac{|a_1|^{l_1}\cdots|a_m|^{l_m}}{\Gamma(l_1\alpha+\cdots+l_m\alpha+\alpha+1)}$$

$$\cdot\frac{1}{\Gamma(\alpha_{11}l_1+\cdots+\alpha_{1m}l_m+1)}\cdots\frac{1}{\Gamma(\alpha_{n1}l_1+\cdots+\alpha_{nm}l_m+1)}\sup_{(x,y)\in[0,1]\times\Omega,\, w\in R}|f(x,y,w)|$$

$$+\frac{1}{\Gamma(\alpha+1)}\sum_{l=0}^\infty \sum_{\substack{l_1+\cdots+l_m=l\\ l_1\ge 0,\cdots,l_m\ge 0}} \binom{l}{l_1,\cdots,l_m} \frac{|a_1|^{l_1}\cdots|a_m|^{l_m}}{\Gamma(l_1\alpha+\cdots+l_m\alpha+2)}$$

$$\cdot\frac{1}{\Gamma(\alpha_{11}l_1+\cdots+\alpha_{1m}l_m+1)}\cdots\frac{1}{\Gamma(\alpha_{n1}l_1+\cdots+\alpha_{nm}l_m+1)}\sup_{(x,y)\in[0,1]\times\Omega,\, w\in R}|f(x,y,w)|$$

$$+\frac{1}{\Gamma(\alpha+1)}\sum_{k=1}^m |a_k| \sum_{l=0}^\infty \sum_{\substack{l_1+\cdots+l_m=l\\ l_1\ge 0,\cdots,l_m\ge 0}} \binom{l}{l_1,\cdots,l_m} \frac{|a_1|^{l_1}\cdots|a_m|^{l_m}}{\Gamma(l_1\alpha+\cdots+l_m\alpha+2)}$$

$$\cdot\frac{1}{\Gamma(\alpha_{11}l_1+\cdots+\alpha_{1m}l_m+\alpha_{1k}+1)}\cdots\frac{1}{\Gamma(\alpha_{n1}l_1+\cdots+\alpha_{nm}l_m+\alpha_{nk}+1)}\|Z\|$$

$$=\left(E_{Q_\alpha}(|a_1|,\cdots,|a_m|)+\frac{1}{\Gamma(\alpha+1)}E_{Q_1}(|a_1|,\cdots,|a_m|)\right)\sup_{(x,y)\in[0,1]\times\Omega,\, w\in R}|f(x,y,w)|$$

$$+\frac{1}{\Gamma(\alpha+1)}\sum_{k=1}^{m}|a_k|E_{M_k}(|a_1|,\cdots,|a_m|)\,\|Z\|\,.$$

Since,

$$A=1-\frac{1}{\Gamma(\alpha+1)}\sum_{k=1}^{m}|a_k|E_{M_k}(|a_1|,\cdots,|a_m|)>0,$$

we come to

$$\|Z\|\le\frac{1}{A}\left(E_{Q_\alpha}(|a_1|,\cdots,|a_m|)+\frac{1}{\Gamma(\alpha+1)}E_{Q_1}(|a_1|,\cdots,|a_m|)\right)\cdot\sup_{(x,y)\in[0,1]\times\Omega,\,w\in R}|f(x,y,Z)|<+\infty,$$

since f is bounded. Hence $Z\in\mathcal{S}([0,1]\times\Omega)$. This completes the proof of Theorem 22.2.

We are ready to present the key theorem in this section.

Theorem 22.3 *Let $f:[0,1]\times\Omega\times\mathbb{R}\to\mathbb{R}$ be a continuous and bounded function satisfying the following Lipschitz condition for a positive constant $\mathcal{B}$:*

$$|f(x,y,w_1)-f(x,y,w_2)|\le\mathcal{B}|w_1-w_2|,\quad w_1,w_2\in\mathbb{R}.$$

In addition, we assume that $1<\alpha\le 2$ and

$$D=\mathcal{B}\left(E_{Q_\alpha}(|a_1|,\cdots,|a_m|)+\frac{1}{\Gamma(\alpha+1)}E_{Q_1}(|a_1|,\cdots,|a_m|)\right)+\frac{1}{\Gamma(\alpha+1)}\sum_{k=1}^{m}|a_k|E_{M_k}(|a_1|,\cdots,|a_m|)<1.$$

Then the problem (22.9) has a unique solution in the space $\mathcal{S}([0,1]\times\Omega)$.

Proof The nonlinear mapping $\mathcal{T}$ is defined over the space $\mathcal{S}([0,1]\times\Omega)$ by

$$\begin{aligned}(\mathcal{T}Z)(x,y)&=\sum_{l=0}^{\infty}(-1)^l\sum_{\substack{l_1+\cdots+l_m=l\\ l_1\ge 0,\cdots,l_m\ge 0}}\binom{l}{l_1,\cdots,l_m}a_1^{l_1}\cdots a_m^{l_m}I_x^{l_1\alpha+\cdots+l_m\alpha+\alpha}\\&\cdot I_1^{\alpha_{11}l_1+\cdots+\alpha_{1m}l_m}\cdots I_n^{\alpha_{n1}l_1+\cdots+\alpha_{nm}l_m}f(x,y,Z(x,y))\\&-\sum_{l=0}^{\infty}(-1)^l\sum_{\substack{l_1+\cdots+l_m=l\\ l_1\ge 0,\cdots,l_m\ge 0}}\binom{l}{l_1,\cdots,l_m}a_1^{l_1}\cdots a_m^{l_m}\frac{x^{\alpha l_1+\cdots+\alpha l_m+1}}{\Gamma(\alpha l_1+\cdots+\alpha l_m+2)}\\&\cdot I_1^{\alpha_{11}l_1+\cdots+\alpha_{1m}l_m}\cdots I_n^{\alpha_{n1}l_1+\cdots+\alpha_{nm}l_m}I_{x=1}^{\alpha}f(x,y,Z(x,y))+\sum_{k=1}^{m}a_k\sum_{l=0}^{\infty}(-1)^l\end{aligned}$$

$$\cdot \sum_{\substack{l_1+\cdots+l_m=l \\ l_1\geq 0,\cdots,l_m\geq 0}} \binom{l}{l_1,\cdots,l_m} a_1^{l_1}\cdots a_m^{l_m} \frac{x^{\alpha l_1+\cdots+\alpha l_m+1}}{\Gamma(\alpha l_1+\cdots+\alpha l_m+2)}$$
$$\cdot I_1^{\alpha_{11}l_1+\cdots+\alpha_{1m}l_m+\alpha_{1k}} \cdots I_n^{\alpha_{n1}l_1+\cdots+\alpha_{nm}l_m+\alpha_{nk}} I_{x=1}^{\alpha} Z(x,y).$$

It follows from Theorem 22.2 that $\mathcal{T}Z \in \mathcal{S}([0,1]\times\Omega)$. We shall show that $\mathcal{T}$ is contractive. Indeed, for $Z_1, Z_2 \in \mathcal{C}([0,1]\times\Omega)$, we get

$$\begin{aligned}\|\mathcal{T}Z_1-\mathcal{T}Z_2\| \leq\ & \mathcal{B}E_{Q_\alpha}(|a_1|,\cdots,|a_m|)\,\|Z_1-Z_2\| \\ &+\frac{\mathcal{B}}{\Gamma(\alpha+1)}E_{Q_1}(|a_1|,\cdots,|a_m|)\,\|Z_1-Z_2\| \\ &+\frac{1}{\Gamma(\alpha+1)}\sum_{k=1}^{m}|a_k|E_{M_k}(|a_1|,\cdots,|a_m|)\,\|Z_1-Z_2\| \\ =\ & D\,\|Z_1-Z_2\|,\end{aligned}$$

from the proof of Theorem 22.2 and noting that

$$|f(x,y,Z_1)-f(x,y,Z_2)| \leq \mathcal{B}|Z_1-Z_2|.$$

Since $D<1$, the problem (22.9) has a unique solution in the space $\mathcal{S}([0,1]\times\Omega)$ by Banach's contractive principle. This completes the proof of Theorem 22.3.

Example 22.2 Consider the following equation:

$$\begin{cases}\dfrac{{}^c\partial^{1.5}}{\partial x^{1.5}}Z(x,y)+\displaystyle\sum_{k=1}^{3}\frac{1}{2^k}I_1^{\alpha_{1k}}\cdots I_4^{\alpha_{4k}}Z(x,y)=\frac{19}{50}\cos(x^2y_2Z(z,y))+x^2+|y|\sin x,\\ Z(0,y)=Z(1,y)=0,\quad (x,y)\in[0,1]\times[0,1]^4,\end{cases} \tag{22.14}$$

where

$$(\alpha_{ij})_{1\leq i\leq 3,\,1\leq j\leq 4}=\begin{bmatrix}1.1 & 1.2 & 1.3 & 1.4\\ 1.2 & 1.3 & 1.4 & 1.5\\ 1.3 & 1.4 & 1.5 & 1.6\end{bmatrix}.$$

Then the problem (22.14) has a unique solution in the space $\mathcal{S}([0,1]\times[0,1]^4)$.

Proof Clearly,

$$f(x,y,Z)=\frac{19}{50}\cos(x^2y_2Z(x,y))+x^2+|y|\sin x,$$

and

$$|f(x,y,z_1)-f(x,y,z_2)| \leq \frac{19}{50}|\cos(x^2y_2z_1)-\cos(x^2y_2z_2)| \leq \frac{19}{50}|z_1-z_2|,$$

for all $z_1, z_2 \in \mathbb{R}$, by noting that $(x, y) \in [0, 1] \times [0, 1]^4$. Therefore $\mathcal{B} = 19/50$. We evaluate the following D given in Theorem 22.3 by our Python code to get

$$D = \frac{19}{50}\left(E_{Q_{1.5}}(1/2, 1/4, 1/8) + \frac{1}{\Gamma(2.5)}E_{Q_1}(1/2, 1/4, 1/8)\right) \\ + \frac{1}{\Gamma(2.5)}\sum_{k=1}^{m}\frac{1}{2^k}E_{M_k}(1/2, 1/4, 1/8) \approx 0.9896 < 1.$$

By Theorem 22.3, the problem (22.14) has a unique solution in the space $\mathcal{C}([0, 1] \times [0, 1]^4)$.

22.2.3 Summary

We have derived the sufficient condition for the uniqueness of solutions to the new boundary value problem of the fractional nonlinear partial integro-differential equation (22.9) in $\mathbb{R}^n$ using the matrix Mittag-Leffler function, Babenko's approach as well as Banach's contractive principle. Finally, we presented one example showing the applications of the key results derived by a Python code computing the approximate value of the matrix Mittag-Leffler function.

22.3 Nonlocal Initial Value Conditions

22.3.1 Preliminaries

Let $b > 0$ and $c_0, c_1, c_2 \in C[0, b]$. In this section, we consider the uniqueness of solutions for the following for $2 < \varsigma \leq 3, 0 < \varepsilon \leq 1$ and $\upsilon > 0$:

$$\begin{cases} \frac{{}^c\partial^\varsigma}{\partial\theta^\varsigma}\Phi(\theta, \sigma) + c_0(\sigma)\frac{{}^c\partial^\varepsilon}{\partial\theta^\varepsilon}\Phi(\theta, \sigma) + c_1(\sigma)\Phi(\theta, \sigma) + c_2(\sigma)I_\sigma^\upsilon\Phi(\theta, \sigma) = F(\theta, \sigma, \Phi(\theta, \sigma)), \\ \Phi(0, \sigma) = \eta\int_0^1 \Phi(\theta, \sigma)d\theta, \quad \frac{\partial}{\partial\theta}\Phi(0, \sigma) = \int_0^1 \psi(\theta)\Phi(\theta, \sigma)d\theta, \quad \Phi''_\theta(0, \sigma) = 0, \end{cases} \tag{22.15}$$

where $(\theta, \sigma) \in [0, 1] \times [0, b]$, η is a constant, $\psi \in C[0, 1]$ and $F : [0, 1] \times [0, b] \times R \to R$ satisfies certain conditions to be given later. Assume that $\upsilon \in R^+$ and $\sigma \in [0, b]$, the operator I_σ^υ is the partial Riemann-Liouville fractional integral of order υ w.r.t σ with initial point zero (see [1]),

$$(I_\sigma^\upsilon\Phi)(\theta, \sigma) = \frac{1}{\Gamma(\upsilon)}\int_0^\sigma (\sigma - \zeta)^{\upsilon-1}\Phi(\theta, \zeta)d\zeta,$$

and $\frac{{}^c\partial^\varsigma}{\partial\theta^\varsigma}$ is the partial Liouville-Caputo fractional derivative of order ς with respect to $\theta \in [0, 1]$

$$\left(\frac{{}^c\partial^\varsigma}{\partial\theta^\varsigma}\Phi\right)(\theta,\sigma) = \frac{1}{\Gamma(3-\varsigma)}\int_0^\theta (\theta-s)^{2-\varsigma}\Phi_s^{(3)}(s,\sigma)ds, \ \ 2<\varsigma\le 3,$$

and similarly,

$$\left(\frac{{}^c\partial^\varepsilon}{\partial\theta^\varepsilon}\Phi\right)(\theta,\sigma) = \frac{1}{\Gamma(1-\varepsilon)}\int_0^\theta (\theta-s)^{-\varepsilon}\Phi_s'(s,\sigma)ds, \ \ 0<\varepsilon\le 1.$$

A function Φ is said to be a solution of Eq. (22.15) if it satisfies the equation and its three initial conditions.

If $\psi(\theta) = \eta$ for all $\theta \in [0, 1]$, then Eq. (22.15) is

$$\begin{cases} \frac{{}^c\partial^\varsigma}{\partial\theta^\varsigma}\Phi(\theta,\sigma) + c_0(\sigma)\frac{{}^c\partial^\varepsilon}{\partial\theta^\varepsilon}\Phi(\theta,\sigma) + c_1(\sigma)\Phi(\theta,\sigma) + c_2(\sigma)I_\sigma^\upsilon\Phi(\theta,\sigma) = F(\theta,\sigma,\Phi(\theta,\sigma)), \\ \Phi(0,\sigma) = \frac{\partial}{\partial\theta}\Phi(0,\sigma) = \eta\int_0^1 \Phi(\theta,\sigma)d\theta, \ \ \Phi_t''(0,\sigma) = 0. \end{cases}$$

It follows from [3] that

$$(I_\sigma^0\Phi)(\theta,\sigma) = \Phi(\theta,\sigma).$$

22.3.2 *Main Results*

Let $C([0, 1] \times [0, b])$ be the set of all continuous functions over $[0, 1] \times [0, b]$ with the norm

$$\|\Phi\| = \max_{(\theta,\sigma)\in[0,1]\times[0,b]} |\Phi(\theta,\sigma)|.$$

Then it is a Banach space.

Theorem 22.4 *Let $b > 0$, $2 < \varsigma \le 3$, $0 < \varepsilon \le 1$, $\upsilon > 0$, η be a constant and $c_0, c_1, c_2, \psi \in C[0, b]$. We assume $F : [0, 1] \times [0, b] \times R \to R$ is a bounded and continuous function, and*

$$2\left(|\eta| + m_0|\eta| + M\right) E_{(\varsigma-\varepsilon,\,\varsigma,\,\varsigma),\,1}(m_0, m_1, m_2 b^\upsilon) < 1,$$

where

$$m_i = \max_{\sigma\in[0,b]} |c_i(\sigma)|, \ \ \text{and} \ \ M = \max_{\sigma\in[0,b]} |\psi(\sigma)|,$$

for $i = 0, 1, 2$. Then Φ is a solution of Eq. (22.15) if and only if it satisfies the following implicit integral equation in the space $C([0, 1] \times [0, b])$:

$$\begin{aligned}\Phi(\theta,\sigma)=&\sum_{h=0}^{\infty}(-1)^h\sum_{\substack{h_1+h_2+h_3=h\\ h_1\geq 0,\,h_2\geq 0,\,h_3\geq 0}}\binom{h}{h_1,h_2,h_3}c_0^{h_1}(\sigma)c_1^{h_2}(\sigma)\\
&\cdot I_t^{(\varsigma-\varepsilon)h_1+\varsigma h_2+\varsigma h_3+\varsigma}(c_2(\sigma)I_\sigma^{\upsilon})^{h_3}\digamma(\theta,\sigma,\Phi(\theta,\sigma))\\
&+\sum_{h=0}^{\infty}(-1)^h\sum_{\substack{h_1+h_2+h_3=h\\ h_1\geq 0,\,h_2\geq 0,\,h_3\geq 0}}\binom{h}{h_1,h_2,h_3}c_0^{h_1}(\sigma)c_1^{h_2}(\sigma)\\
&\cdot\frac{\theta^{(\varsigma-\varepsilon)h_1+\varsigma h_2+\varsigma h_3}}{\Gamma((\varsigma-\varepsilon)h_1+\varsigma h_2+\varsigma h_3+1)}\eta\int_0^1(c_2(\sigma)I_\sigma^{\upsilon})^{h_3}\Phi(\theta,\sigma)d\theta\\
&+\sum_{h=0}^{\infty}(-1)^h\sum_{\substack{h_1+h_2+h_3=h\\ h_1\geq 0,\,h_2\geq 0,\,h_3\geq 0}}\binom{h}{h_1,h_2,h_3}c_0^{h_1}(\sigma)c_1^{h_2}(\sigma)\\
&\cdot\frac{\theta^{(\varsigma-\varepsilon)h_1+\varsigma h_2+\varsigma h_3+\varsigma-\varepsilon}}{\Gamma((\varsigma-\beta)h_1+\varsigma h_2+\varsigma h_3+\varsigma-\beta+1)}\eta\int_0^1(c_2(\sigma)I_\sigma^{\upsilon})^{h_3}c_0(\sigma)\Phi(\theta,\sigma)d\theta\\
&+\sum_{h=0}^{\infty}(-1)^h\sum_{\substack{h_1+h_2+h_3=h\\ h_1\geq 0,\,h_2\geq 0,\,h_3\geq 0}}\binom{h}{h_1,h_2,h_3}c_0^{h_1}(\sigma)c_1^{h_2}(\sigma)\\
&\cdot\frac{\theta^{(\varsigma-\varepsilon)h_1+\varsigma h_2+\varsigma h_3+1}}{\Gamma((\varsigma-\varepsilon)h_1+\varsigma h_2+\varsigma h_3+2)}\int_0^1\psi(\theta)(c_2(\sigma)I_\sigma^{\upsilon})^{h_3}\Phi(\theta,\sigma)d\theta. \qquad (22.16)\end{aligned}$$

Proof It can be clearly seen that

$$\begin{aligned}&I_\theta^{\varsigma}\,\frac{{}^c\partial^{\varsigma}}{\partial\theta^{\varsigma}}\Phi(\theta,\sigma)=\Phi(\theta,\sigma)-\Phi(0,\sigma)-\Phi_t'(0,\sigma)\theta-\frac{\Phi_t''(0,\sigma)}{2!}\theta^2\\
&=\Phi(\theta,\sigma)-\eta\int_0^1\Phi(\theta,\sigma)d\theta-\int_0^1\psi(\theta)\Phi(\theta,\sigma)d\theta\;\theta,\end{aligned}$$

and, since $0<\varepsilon\leq 1$,

$$\begin{aligned}&I_\theta^{\varsigma}\;c_0(\sigma)\,\frac{{}^c\partial^{\varepsilon}}{\partial\theta^{\varepsilon}}\Phi(\theta,\sigma)=c_0(\sigma)\,I_\theta^{\varsigma-\beta}\,I_\theta^{\varepsilon}\,\frac{{}^c\partial^{\varepsilon}}{\partial\theta^{\beta}}\Phi(\theta,\sigma)\\
&=c_0(\sigma)\,I_\theta^{\varsigma-\varepsilon}\;(\Phi(\theta,\sigma)-\Phi(0,\sigma))=c_0(\sigma)I_\theta^{\varsigma-\varepsilon}\Phi(\theta,\sigma)-\frac{c_0(\sigma)\,\theta^{\varsigma-\beta}\eta}{\Gamma(\varsigma-\beta+1)}\int_0^1\Phi(\theta,\sigma)d\theta.\end{aligned}$$

Applying the operator I_θ^{ς} to the equation

$$\frac{{}^c\partial^{\varsigma}}{\partial\theta^{\varsigma}}\Phi(\theta,\sigma)+c_0(\sigma)\frac{{}^c\partial^{\varepsilon}}{\partial\theta^{\varepsilon}}\Phi(\theta,\sigma)+c_1(\sigma)\Phi(\theta,\sigma)+c_2(\sigma)I_\sigma^{\upsilon}\Phi(\theta,\sigma)=\digamma(\theta,\sigma,\Phi(\theta,\sigma)),$$

we get

$$\begin{aligned}&\left(1+c_0(\sigma)I_\theta^{\varsigma-\varepsilon}+c_1(\sigma)I_\theta^{\varsigma}+c_2(\sigma)I_\theta^{\varsigma}I_\sigma^{\upsilon}\right)\Phi(\theta,\sigma)\\
&=I_\theta^{\varsigma}\digamma(\theta,\sigma,\Phi(\theta,\sigma))+\left(1+\frac{c_0(\sigma)\theta^{\varsigma-\varepsilon}}{\Gamma(\varsigma-\varepsilon+1)}\right)\eta\int_0^1\Phi(\theta,\sigma)d\theta+\int_0^1\psi(\theta)\Phi(\theta,\sigma)d\theta\;\theta.\end{aligned}$$

Using Babenko's approach, we obtain

$$\begin{aligned}\Phi(\theta,\sigma) = &\left(1 + c_0(\sigma) I_\theta^{\varsigma-\varepsilon} + c_1(\sigma) I_\theta^{\varsigma} + c_2(\sigma) I_\theta^{\varsigma} I_\sigma^{\upsilon}\right)^{-1} \\ &\cdot\left[I_\theta^{\varsigma} F(\theta,\sigma,\Phi(\theta,\sigma)) + \left(1 + \frac{c_0(\sigma)\theta^{\varsigma-\varepsilon}}{\Gamma(\varsigma-\varepsilon+1)}\right)\eta\int_0^1 \Phi(\theta,\sigma)d\theta + \int_0^1 \psi(\theta)\Phi(\theta,\sigma)d\theta\ \theta\right] \\ = &\sum_{h=0}^{\infty}(-1)^h \left(c_0(\sigma) I_\theta^{\varsigma-\varepsilon} + c_1(\sigma) I_\theta^{\varsigma} + c_2(\sigma) I_\theta^{\varsigma} I_\sigma^{\upsilon}\right)^h \\ &\cdot\left[I_\theta^{\varsigma} F(\theta,\sigma,\Phi(\theta,\sigma)) + \left(1 + \frac{c_0(\sigma)\theta^{\varsigma-\varepsilon}}{\Gamma(\varsigma-\varepsilon+1)}\right)\eta\int_0^1 \Phi(\theta,\sigma)d\theta + \int_0^1 \psi(\theta)\Phi(\theta,\sigma)d\theta\ \theta\right] \\ = &\sum_{h=0}^{\infty}(-1)^h \sum_{\substack{h_1+h_2+h_3=h \\ h_1\ge 0,\, h_2\ge 0,\, h_3\ge 0}} \binom{h}{h_1,\, h_2,\, h_3} \left(c_0(\sigma) I_\theta^{\varsigma-\varepsilon}\right)^{h_1} \left(c_1(\sigma) I_\theta^{\varsigma}\right)^{h_2} \left(c_2(\sigma) I_\theta^{\varsigma} I_\sigma^{\upsilon}\right)^{h_3} \\ &\cdot\left[I_\theta^{\varsigma} F(\theta,\sigma,\Phi(\theta,\sigma)) + \left(1 + \frac{c_0(\sigma)\theta^{\varsigma-\varepsilon}}{\Gamma(\varsigma-\varepsilon+1)}\right)\eta\int_0^1 \Phi(\theta,\sigma)d\theta + \int_0^1 \psi(\theta)\Phi(\theta,\sigma)d\theta\ \theta\right].\end{aligned}$$

Clearly,

$$\left(c_0(\sigma) I_\theta^{\varsigma-\varepsilon}\right)^{h_1} = c_0(\sigma) I_\theta^{\varsigma-\varepsilon} c_0(\sigma) I_\theta^{\varsigma-\varepsilon} \cdots c_0(\sigma) I_\theta^{\varsigma-\varepsilon} = c_0^{h_1}(\sigma) I_\theta^{(\varsigma-\varepsilon)h_1}$$

and $\left(c_1(\sigma) I_\theta^{\varsigma}\right)^{h_2} = c_1^{h_2}(\sigma) I_\theta^{\varsigma h_2}$. As for the term $\left(c_2(\sigma) I_\theta^{\varsigma} I_\sigma^{\upsilon}\right)^{h_3}$, we have

$$\left(c_2(\sigma) I_\theta^{\varsigma} I_\sigma^{\upsilon}\right)^{h_3} = \left(c_2(\sigma) I_\theta^{\varsigma} I_\sigma^{\upsilon}\right)\left(c_2(\sigma) I_\theta^{\varsigma} I_\sigma^{\upsilon}\right)\cdots\left(c_2(\sigma) I_\theta^{\varsigma} I_\sigma^{\upsilon}\right) = I_\theta^{\varsigma h_3}(c_2(\sigma) I_\sigma^{\upsilon})^{h_3}.$$

Therefore,

$$\begin{aligned}\Phi(\theta,\sigma) = &\sum_{h=0}^{\infty}(-1)^h \sum_{\substack{h_1+h_2+h_3=h \\ h_1\ge 0,\, h_2\ge 0,\, h_3\ge 0}} \binom{h}{h_1,\, h_2,\, h_3} c_0^{h_1}(\sigma) c_1^{h_2}(\sigma) I_\theta^{(\varsigma-\varepsilon)h_1+\varsigma h_2+\varsigma h_3} (c_2(\sigma) I_\sigma^{\upsilon})^{h_3} \\ &\cdot\left[I_\theta^{\varsigma} F(\theta,\sigma,\Phi(\theta,\sigma)) + \left(1 + \frac{c_0(\sigma)\theta^{\varsigma-\varepsilon}}{\Gamma(\varsigma-\varepsilon+1)}\right)\eta\int_0^1 \Phi(\theta,\sigma)d\theta + \int_0^1 \psi(\theta)\Phi(\theta,\sigma)d\theta\ \theta\right] \\ = &\sum_{h=0}^{\infty}(-1)^h \sum_{\substack{h_1+h_2+h_3=h \\ h_1\ge 0,\, h_2\ge 0,\, h_3\ge 0}} \binom{h}{h_1,\, h_2,\, h_3} c_0^{h_1}(\sigma) c_1^{h_2}(\sigma) \\ &\cdot I_\theta^{(\varsigma-\varepsilon)h_1+\varsigma h_2+\varsigma h_3+\varsigma} (c_2(\sigma) I_\sigma^{\upsilon})^{h_3} F(\theta,\sigma,\Phi(\theta,\sigma)) \\ &+ \sum_{h=0}^{\infty}(-1)^h \sum_{\substack{h_1+h_2+h_3=h \\ h_1\ge 0,\, h_2\ge 0,\, h_3\ge 0}} \binom{h}{h_1,\, h_2,\, h_3} c_0^{h_1}(\sigma) c_1^{h_2}(\sigma) \\ &\cdot \frac{\theta^{(\varsigma-\varepsilon)h_1+\varsigma h_2+\varsigma h_3}}{\Gamma((\varsigma-\varepsilon)h_1+\varsigma h_2+\varsigma h_3+1)}\eta\int_0^1 (c_2(\sigma) I_\sigma^{\upsilon})^{h_3}\Phi(\theta,\sigma)d\theta \\ &+ \sum_{h=0}^{\infty}(-1)^h \sum_{\substack{h_1+h_2+h_3=h \\ h_1\ge 0,\, h_2\ge 0,\, h_3\ge 0}} \binom{h}{h_1,\, h_2,\, h_3} c_0^{h_1}(\sigma) c_1^{h_2}(\sigma) \\ &\cdot \frac{I_\theta^{(\varsigma-\varepsilon)h_1+\varsigma h_2+\varsigma h_3}\theta^{\varsigma-\varepsilon}}{\Gamma(\varsigma-\beta+1)}\eta\int_0^1 (c_2(\sigma) I_\sigma^{\upsilon})^{h_3} c_0(\sigma)\Phi(\theta,\sigma)d\theta\end{aligned}$$

$$+\sum_{h=0}^{\infty}(-1)^h \sum_{\substack{h_1+h_2+h_3=h\\ h_1\geq 0,\, h_2\geq 0,\, h_3\geq 0}} \binom{h}{h_1, h_2, h_3} c_0^{h_1}(\sigma) c_1^{h_2}(\sigma)$$

$$\cdot I_\theta^{(\varsigma-\varepsilon)h_1+\varsigma h_2+\varsigma h_3}\, \theta \int_0^1 \psi(\theta)(c_2(\sigma) I_\sigma^{\upsilon})^{h_3} \Phi(\theta,\sigma) d\theta.$$

Using

$$I_\theta^{(\varsigma-\varepsilon)h_1+\varsigma h_2+\varsigma h_3}\, \theta^{\varsigma-\varepsilon}$$
$$= \frac{\Gamma(\varsigma-\varepsilon+1)}{\Gamma((\varsigma-\varepsilon)h_1+\varsigma h_2+\varsigma h_3+\varsigma-\beta+1)} \theta^{(\varsigma-\beta)h_1+\varsigma h_2+\varsigma h_3+\varsigma-\beta},$$
$$I_\theta^{(\varsigma-\varepsilon)h_1+\varsigma h_2+\varsigma h_3}\, \theta = \frac{\theta^{(\varsigma-\varepsilon)h_1+\varsigma h_2+\varsigma h_3+1}}{\Gamma((\varsigma-\beta)h_1+\varsigma h_2+\varsigma h_3+2)},$$

we have

$$\Phi(\theta,\sigma) = \sum_{h=0}^{\infty}(-1)^h \sum_{\substack{h_1+h_2+h_3=h\\ h_1\geq 0,\, h_2\geq 0,\, h_3\geq 0}} \binom{h}{h_1, h_2, h_3} c_0^{h_1}(\sigma) c_1^{h_2}(\sigma)$$
$$\cdot I_\theta^{(\varsigma-\varepsilon)h_1+\varsigma h_2+\varsigma h_3+\varsigma} (c_2(\sigma) I_\sigma^{\upsilon})^{h_3} F(\theta,\sigma,\Phi(\theta,\sigma))$$
$$+\sum_{h=0}^{\infty}(-1)^h \sum_{\substack{h_1+h_2+h_3=h\\ h_1\geq 0,\, h_2\geq 0,\, h_3\geq 0}} \binom{h}{h_1, h_2, h_3} c_0^{h_1}(\sigma) c_1^{h_2}(\sigma)$$
$$\cdot \frac{\theta^{(\varsigma-\varepsilon)h_1+\varsigma h_2+\varsigma h_3}}{\Gamma((\varsigma-\varepsilon)h_1+\varsigma h_2+\varsigma h_3+1)} \eta \int_0^1 (c_2(\sigma) I_\sigma^{\upsilon})^{h_3} \Phi(\theta,\sigma) d\theta$$
$$+\sum_{h=0}^{\infty}(-1)^h \sum_{\substack{h_1+h_2+h_3=h\\ h_1\geq 0,\, h_2\geq 0,\, h_3\geq 0}} \binom{h}{h_1, h_2, h_3} c_0^{h_1}(\sigma) c_1^{h_2}(\sigma)$$
$$\cdot \frac{\theta^{(\varsigma-\varepsilon)h_1+\varsigma h_2+\varsigma h_3+\varsigma-\varepsilon}}{\Gamma((\varsigma-\beta)h_1+\varsigma h_2+\varsigma h_3+\varsigma-\beta+1)} \eta \int_0^1 (c_2(\sigma) I_\sigma^{\upsilon})^{h_3} c_0(\sigma) \Phi(\theta,\sigma) d\theta$$
$$+\sum_{h=0}^{\infty}(-1)^h \sum_{\substack{h_1+h_2+h_3=h\\ h_1\geq 0,\, h_2\geq 0,\, h_3\geq 0}} \binom{h}{h_1, h_2, h_3} c_0^{h_1}(\sigma) c_1^{h_2}(\sigma)$$
$$\cdot \frac{\theta^{(\varsigma-\varepsilon)h_1+\varsigma h_2+\varsigma h_3+1}}{\Gamma((\varsigma-\varepsilon)h_1+\varsigma h_2+\varsigma h_3+2)} \int_0^1 \psi(\theta)(c_2(\sigma) I_\sigma^{\upsilon})^{h_3} \Phi(\theta,\sigma) d\theta.$$

The function $\Phi(\theta,\sigma)$ satisfies the initial conditions: First $\Phi(0,\sigma) = \eta \int_0^1 \Phi(\theta,\sigma) d\sigma$, which is the term

$$(-1)^0 \sum_{\substack{h_1+h_2+h_3=0\\ h_1\geq 0,\, h_2\geq 0,\, h_3\geq 0}} \binom{0}{h_1, h_2, h_3} c_0^{h_1}(\sigma) c_1^{h_2}(\sigma)$$

$$\cdot \frac{\theta^{(\varsigma-\varepsilon)h_1+\varsigma h_2+\varsigma h_3}}{\Gamma((\varsigma-\varepsilon)h_1+\varsigma h_2+\varsigma h_3+1)}\eta\int_0^1 (c_2(\sigma)I_\sigma^\upsilon)^{h_3}\Phi(\theta,\sigma)d\theta,$$

and the rest is zero by setting $\theta = 0$. Similarly, $\frac{\partial}{\partial\theta}\Phi(0,\sigma) = \int_0^1 \psi(\theta)\Phi(\theta,\sigma)d\theta$, $\Phi''_\theta(0,\sigma) = 0$, by noting that $\varsigma - \varepsilon > 1$. Hence Φ is a solution of Eq. (22.15) if and only if it satisfies Eq. (22.16). Next, in $C([0,1]\times[0,b])$, we show the convergence of the right side of Eq. (22.16), which is a series. Since $c_0,\ c_1,\ c_2 \in C[0,b]$, we let $m_i = \max_{\sigma\in[0,b]}|c_i(\sigma)|$, for $i = 0, 1, 2$. Then,

$$\begin{aligned}
\|\Phi\| \le & \sum_{h=0}^{\infty}\sum_{\substack{h_1+h_2+h_3=h\\ h_1\ge 0,\, h_2\ge 0,\, h_3\ge 0}}\binom{h}{h_1,h_2,h_3} m_0^{h_1}m_1^{h_2}m_2^{h_3}\\
&\cdot\frac{b^{\upsilon h_3}}{\Gamma((\varsigma-\varepsilon)h_1+\varsigma h_2+\varsigma h_3+\varsigma+1)\,\Gamma(\upsilon h_3+1)}\sup_{(\theta,\sigma)\in[0,1]\times[0,b]}|F(\theta,\sigma,\Phi(\theta,\sigma))|\\
&+\sum_{h=0}^{\infty}\sum_{\substack{h_1+h_2+h_3=h\\ h_1\ge 0,\, h_2\ge 0,\, h_3\ge 0}}\binom{h}{h_1,h_2,h_3} m_0^{h_1}m_1^{h_2}m_2^{h_3}\\
&\cdot\frac{b^{\upsilon h_3}|\eta|}{\Gamma((\varsigma-\varepsilon)h_1+\varsigma h_2+\varsigma h_3+1)\,\Gamma(\upsilon h_3+1)}\|\Phi\|\\
&+\sum_{h=0}^{\infty}\sum_{\substack{h_1+h_2+h_3=h\\ h_1\ge 0,\, h_2\ge 0,\, h_3\ge 0}}\binom{h}{h_1,h_2,h_3} m_0^{h_1+1}m_1^{h_2}m_2^{h_3}\\
&\cdot\frac{b^{\upsilon h_3}|\eta|}{\Gamma((\varsigma-\varepsilon)h_1+\varsigma h_2+\varsigma h_3+\varsigma-\varepsilon+1)\,\Gamma(\upsilon h_3+1)}\|\Phi\|\\
&+\sum_{h=0}^{\infty}\sum_{\substack{h_1+h_2+h_3=h\\ h_1\ge 0,\, h_2\ge 0,\, h_3\ge 0}}\binom{h}{h_1,h_2,h_3} m_0^{h_1}m_1^{h_2}m_2^{h_3}\\
&\cdot\frac{M\, b^{\upsilon h_3}}{\Gamma((\varsigma-\varepsilon)h_1+\varsigma h_2+\varsigma h_3+2)\,\Gamma(\upsilon h_3+1)}\|\Phi\|,
\end{aligned}$$

where $M = \max_{\sigma\in[0,b]}|\psi(\sigma)|$. Applying the property $\Gamma(\upsilon h_3+1)\ge 1/2$, we deduce

$$\begin{aligned}
\|\Phi\| \le\ & 2E_{(\varsigma-\varepsilon,\,\varsigma,\,\varsigma),\,\varsigma+1}(m_0,m_1,m_2b^\upsilon)\sup_{(\theta,\sigma)\in[0,1]\times[0,b]}|F(\theta,\sigma,\Phi(\theta,\sigma))|\\
&+2|\eta|E_{(\varsigma-\varepsilon,\,\varsigma,\,\varsigma),\,1}(m_0,m_1,m_2b^\upsilon)\,\|\Phi\|\\
&+2\,m_0|\eta|\,E_{(\varsigma-\varepsilon,\,\varsigma,\,\varsigma),\,\varsigma-\varepsilon+1}(m_0,m_1,m_2b^\upsilon)\,\|\Phi\|\\
&+2\,M\,E_{(\varsigma-\varepsilon,\,\varsigma,\,\varsigma),\,2}(m_0,m_1,m_2b^\upsilon)\,\|\Phi\|\\
\le\ & 2E_{(\varsigma-\varepsilon,\,\varsigma,\,\varsigma),\,\varsigma+1}(m_0,m_1,m_2b^\upsilon)\sup_{(\theta,\sigma)\in[0,1]\times[0,b]}|F(\theta,\sigma,\Phi(\theta,\sigma))|\\
&+2\,(|\eta|+m_0|\eta|+M)\,E_{(\varsigma-\varepsilon,\,\varsigma,\,\varsigma),\,1}(m_0,m_1,m_2b^\upsilon)\,\|\Phi\|.
\end{aligned}$$

From $q = 1-2\,(|\eta|+m_0|\eta|+M)\,E_{(\varsigma-\varepsilon,\,\varsigma,\,\varsigma),\,1}(m_0,m_1,m_2b^\upsilon) > 0$, we have

$$\|\Phi\| \leq \frac{2}{q} E_{(\varsigma-\varepsilon,\,\varsigma,\,\varsigma),\,\varsigma+1}(m_0, m_1, m_2 b^{\upsilon}) \sup_{(\theta,\sigma)\in[0,1]\times[0,b]} |F(\theta, \sigma, \Phi(\theta, \sigma))| < +\infty,$$

since F is a bounded function. This implies that the series is well defined in the space $C([0, 1] \times [0, b])$. This completes the proof.

In the following, we present and prove the uniqueness of the solutions of Eq. (22.15) in the form of a theorem.

Theorem 22.5 *Let $b > 0$, $2 < \varsigma \leq 3$, $0 < \varepsilon \leq 1$, $\upsilon > 0$, η be a constant and c_0, c_1, c_2, $f \in C[0, b]$. We assume the continuous and bounded function $F : [0, 1] \times [0, b] \times R \to R$ satisfies the following Lipschitz condition with a constant $L \geq 0$:*

$$|F(\theta, \sigma, y_1) - F(\theta, \sigma, y_2)| \leq L|y_1 - y_2|, \quad (\theta, \sigma) \in [0, 1] \times [0, b], \; y_1, y_2 \in R.$$

In addition, we suppose

$$2\,(L + |\eta| + m_0|\eta| + M)\, E_{(\varsigma-\varepsilon,\,\varsigma,\,\varsigma),\,1}(m_0, m_1, m_2 b^{\upsilon}) < 1,$$

where

$$m_i = \max_{\sigma\in[0,b]} |c_i(\sigma)|, \quad \text{and } M = \max_{\sigma\in[0,b]} |\psi(\sigma)|,$$

for $i = 0, 1, 2$. Then Eq. (22.15) has a unique solution in the space $C([0, 1] \times [0, b])$.

Proof We consider the nonlinear mapping T over the space $C([0, 1] \times [0, b])$ and define it as follows:

$$\begin{aligned}
(T\Phi)(\theta, \sigma) = {} & \sum_{h=0}^{\infty}(-1)^h \sum_{\substack{h_1+h_2+h_3=h\\ h_1\geq 0,\, h_2\geq 0,\, h_3\geq 0}} \binom{h}{h_1, h_2, h_3} c_0^{h_1}(\sigma) c_1^{h_2}(\sigma) \\
& \cdot I_t^{(\varsigma-\varepsilon)h_1+\varsigma h_2+\varsigma h_3+\varsigma} (c_2(\sigma) I_\sigma^{\upsilon})^{h_3} F(\theta, \sigma, \Phi(\theta, \sigma)) \\
& + \sum_{h=0}^{\infty}(-1)^h \sum_{\substack{h_1+h_2+h_3=h\\ h_1\geq 0,\, h_2\geq 0,\, h_3\geq 0}} \binom{h}{h_1, h_2, h_3} c_0^{h_1}(\sigma) c_1^{h_2}(\sigma) \\
& \cdot \frac{\theta^{(\varsigma-\varepsilon)h_1+\varsigma h_2+\varsigma h_3}}{\Gamma((\varsigma-\varepsilon)h_1+\varsigma h_2+\varsigma h_3+1)} \eta \int_0^1 (c_2(\sigma) I_\sigma^{\upsilon})^{h_3} \Phi(\theta, \sigma) d\theta \\
& + \sum_{h=0}^{\infty}(-1)^h \sum_{\substack{h_1+h_2+h_3=h\\ h_1\geq 0,\, h_2\geq 0,\, h_3\geq 0}} \binom{h}{h_1, h_2, h_3} c_0^{h_1}(\sigma) c_1^{h_2}(\sigma) \\
& \cdot \frac{\theta^{(\varsigma-\varepsilon)h_1+\varsigma h_2+\varsigma h_3+\varsigma-\varepsilon}}{\Gamma((\varsigma-\beta)h_1+\varsigma h_2+\varsigma h_3+\varsigma-\beta+1)} \eta \int_0^1 (c_2(\sigma) I_\sigma^{\upsilon})^{h_3} c_0(\sigma) \Phi(\theta, \sigma) d\theta \\
& + \sum_{h=0}^{\infty}(-1)^h \sum_{\substack{h_1+h_2+h_3=h\\ h_1\geq 0,\, h_2\geq 0,\, h_3\geq 0}} \binom{h}{h_1, h_2, h_3} c_0^{h_1}(\sigma) c_1^{h_2}(\sigma)
\end{aligned}$$

$$\cdot \frac{\theta^{(\varsigma-\varepsilon)h_1+\varsigma h_2+\varsigma h_3+1}}{\Gamma((\varsigma-\varepsilon)h_1+\varsigma h_2+\varsigma h_3+2)} \int_0^1 \psi(\theta)(c_2(\sigma)I_\sigma^\upsilon)^{h_3}\Phi(\theta,\sigma)d\theta.$$

It follows from the proof of Theorem 22.4 that $T\Phi \in C([0,1]\times[0,b])$. We now show that T is contractive. Indeed, similar to that in the proof of Theorem 22.4,

$$\begin{aligned}
\|T\Phi - T\vartheta\| \le\ & 2E_{(\varsigma-\varepsilon,\varsigma,\varsigma),\varsigma+1}(m_0,m_1,m_2b^\upsilon) \sup_{(\theta,\sigma)\in[0,1]\times[0,b]} |F(\theta,\sigma,\Phi(\theta,\sigma)) - F(\theta,\sigma,\vartheta(\theta,\sigma))| \\
&+ 2|\eta| E_{(\varsigma-\varepsilon,\varsigma,\varsigma),1}(m_0,m_1,m_2b^\upsilon)\,\|\Phi-\vartheta\| \\
&+ 2m_0|\eta|\, E_{(\varsigma-\varepsilon,\varsigma,\varsigma),\varsigma-\varepsilon+1}(m_0,m_1,m_2b^\upsilon)\,\|\Phi-\vartheta\| \\
&+ 2M\, E_{(\varsigma-\varepsilon,\varsigma,\varsigma),2}(m_0,m_1,m_2b^\upsilon)\,\|\Phi-\vartheta\| \\
\le\ & 2LE_{(\varsigma-\varepsilon,\varsigma,\varsigma),\varsigma+1}(m_0,m_1,m_2b^\upsilon)\,\|\Phi-\vartheta\| \\
&+ 2(|\eta| + m_0|\eta| + M)\, E_{(\varsigma-\varepsilon,\varsigma,\varsigma),1}(m_0,m_1,m_2b^\upsilon)\,\|\Phi-\vartheta\| \\
\le\ & 2(L + |\eta| + m_0|\eta| + M)\, E_{(\varsigma-\varepsilon,\varsigma,\varsigma),1}(m_0,m_1,m_2b^\upsilon)\,\|\Phi-\vartheta\|.
\end{aligned}$$

Since $2(L+|\eta|+m_0|\eta|+M)\,E_{(\varsigma-\varepsilon,\varsigma,\varsigma),1}(m_0,m_1,m_2b^\upsilon) < 1$, Eq. (22.15) has a unique solution in the space $C([0,1]\times[0,b])$ by Banach's contractive principle.

Example 22.3 We consider the following equation

$$\begin{cases}
{}^c\dfrac{\partial^{2.5}}{\partial\theta^{2.5}}\Phi(\theta,\sigma) + \sin\sigma\ {}^c\dfrac{\partial^{0.5}}{\partial\theta^{0.5}}\Phi(\theta,\sigma) + \dfrac{\sigma^2}{1+\sigma^2}\Phi(\theta,\sigma) + \cos\sigma\ I_\sigma\Phi(\theta,\sigma) = \dfrac{1}{21}\sin|\Phi(\theta,\sigma)|, \\
\Phi(0,\sigma) = \dfrac{1}{40}\displaystyle\int_0^1 \Phi(\theta,\sigma)d\theta,\quad \dfrac{\partial}{\partial\theta}\Phi(0,\sigma) = \dfrac{1}{22}\displaystyle\int_0^1 \theta\Phi(\theta,\sigma)d\theta,\quad \Phi''_\theta(0,\sigma) = 0.
\end{cases}$$

Then, it has a unique solution in the space $C([0,1]\times[0,1])$.

Proof Clearly, $m_0 = m_1 = m_2 = 1$, and $F(\theta,\sigma,y) = \frac{1}{21}\sin|y|$, is a continuous and bounded function satisfying

$$|F(\theta,\sigma,y_1) - F(\theta,\sigma,y_2)| \le \frac{1}{21}|y_1 - y_2|.$$

Thus $L = 1/21$. Moreover, $M = 1/22$ and $\eta = 1/40$. Finally, we note

$$\begin{aligned}
&2(L+|\eta|+m_0|\eta|+M)\,E_{(\varsigma-\varepsilon,\varsigma,\varsigma),1}(m_0,m_1,m_2b^\upsilon) \\
&= 2(1/21 + 1/40 + 1/40 + 1/22)\,E_{(2,2.5,2.5),1}(1,1,1) \\
&\le \frac{3}{10}E_{(2,2.5,2.5),1}(1,1,1),
\end{aligned}$$

and

$$\begin{aligned}
E_{(2,2.5,2.5),1}(1,1,1) &= \sum_{h=0}^{\infty} \sum_{\substack{h_1+h_2+h_3=h \\ h_1\ge0,\,h_2\ge0,\,h_3\ge0}} \binom{h}{h_1,h_2,h_3} \frac{1}{\Gamma(2h_1+2.5h_2+2.5h_3+1)} \\
&\le \sum_{h=0}^{\infty} \frac{3^h}{(2h)!} = \cosh(\sqrt{3}) \approx 2.9146,
\end{aligned}$$

since

$$\sum_{\substack{h_1+h_2+h_3=h \\ h_1\geq 0,\, h_2\geq 0,\, h_3\geq 0}} \binom{h}{h_1,\, h_2,\, h_3} = 3^h,$$

$$\frac{1}{\Gamma(2h_1 + 2.5h_2 + 2.5h_3 + 1)} \leq \frac{1}{\Gamma(2h+1)} = \frac{1}{(2h)!}.$$

Hence,

$$2\,(L + |\eta| + m_0|\eta| + M)\, E_{(\varsigma-\varepsilon,\,\varsigma,\,\varsigma),\,1}(m_0, m_1, m_2 b^{\upsilon}) < 1.$$

By Theorem 22.5, the equation has a unique solution in the space $C([0, 1] \times [0, 1])$.

22.3.3 Summary

In this section, we have considered the nonlinear partial integro-differential equation with nonlocal initial value conditions. Our main work was to investigate the uniqueness of solutions using an implicit integral equation, Babenko's approach, the multivariate Mittag-Leffler function, and Banach's contractive principle. In the final part, we have presented the numerical results of our key theorem by giving an illustrative example.

22.4 The Existence Problem

22.4.1 Preliminaries

Let $\mathcal{M}$ be a positive constant. We consider the following nonlinear system in this section

$$\begin{cases} \mathrm{v}^{(\ell)}(x) + \sum_{\imath=0}^{\varsigma} \mathrm{p}_\imath(x)\, {}_C D^{\beta_\imath} \mathrm{v}(x) + \sum_{\jmath=0}^{m} \mathrm{q}_\jmath(x) I^{\alpha_\jmath} \mathrm{v}(x) = f(x, \mathrm{v}'(x), \cdots, \mathrm{v}^{(\ell)}(x)) \\ \qquad\qquad\qquad\qquad + g(x, \mathrm{v}(x)), \quad x \in [0, \mathcal{M}], \\ \mathrm{v}(0) = \mathrm{v}'(0) = \cdots = \mathrm{v}^{(\ell-1)}(0) = 0, \end{cases} \tag{22.17}$$

where ς and m are nonnegative integers, $0 \leq \beta_0 < \beta_1 < \cdots < \beta_\varsigma < \ell$, $0 < \alpha_0 < \alpha_1 < \cdots < \alpha_m$, and f and g are mappings from $[0, \mathcal{M}] \times \mathbb{R}^\ell$ to $\mathbb{R}$ and from $[0, \mathcal{M}] \times \mathbb{R}$ to $\mathbb{R}$ respectively, with certain conditions. Our main goal is to investigate the existence of solutions for (22.17) in a Banach space with initial conditions and continuous coefficients $\mathrm{p}_\imath(x)$ and $\mathrm{q}_\jmath(x)$.

Define the space $C^{\ell}[0, \mathcal{M}]$ as

$$C^{\ell}[0, \mathcal{M}] = \left\{ \mathrm{v} : [0, \mathcal{M}] \to \mathbb{R} : \ \mathrm{v}^{(\ell)}(x) \text{ is continuous on } [0, \mathcal{M}] \text{ and } \|\mathrm{v}\|_{\ell} < +\infty \right\},$$

for $\mathcal{M} > 0$, $\ell \in \mathbb{N} = \{1, 2, \cdots\}$ and

$$\|\mathrm{v}\|_{\ell} = \max\left\{ \max_{x \in [0, \mathcal{M}]} |\mathrm{v}(x)|, \cdots, \max_{x \in [0, \mathcal{M}]} |\mathrm{v}^{(\ell)}(x)| \right\}.$$

Clearly, $C^{\ell}[0, \mathcal{M}]$ is a Banach space based on the following theorem in reference [10].

Theorem 22.6 *Consider a sequence of differentiable functions as* $\{\mathrm{v}_k\}$ *such that for some* $x_0 \in [\mathrm{p}, \mathrm{q}]$, $\lim_{k \to \infty} \mathrm{v}_k(x_0)$ *exists and is finite. Also, if the sequence* $\{\mathrm{v}'_k\}$ *converges uniformly on* $[\mathrm{p}, \mathrm{q}]$, *then* v_k *converges uniformly to a function* v *on* $[\mathrm{p}, \mathrm{q}]$, *and for* $x \in [\mathrm{p}, \mathrm{q}]$, $\mathrm{v}'(x) = \lim_{k \to \infty} \mathrm{v}'_k(x)$.

Definition 22.1 The fractional (or the Riemann-Liouville) integral I^{α} of fractional order $\alpha \in \mathbb{R}^+$ ($\mathbb{R}^+ = (0, +\infty)$) of function u is defined in [1, 2] as

$$(I^{\alpha}\mathrm{v})(x) = \frac{1}{\Gamma(\alpha)} \int_0^x (x - t)^{\alpha - 1} \mathrm{v}(t) dt.$$

In particular for $\alpha = 0$, using the Dirac delta function $\delta(x) = \frac{x_+^{-1}}{\Gamma(0)}$ (see [3] for more details), we have

$$(I^0\mathrm{v})(x) = \delta(x) * \mathrm{v}(x) = \mathrm{v}(x).$$

Definition 22.2 Let $m \in \mathbb{N}$. The derivative of the fractional order $\alpha \in \mathbb{R}^+$ of the Riemann-Liouville-Caputo type of function v is defined as

$$({}_C D^{\alpha}\mathrm{v})(x) = I^{m-\alpha} \frac{d^m}{dx^m} \mathrm{v}(x) = \frac{1}{\Gamma(m - \alpha)} \int_0^x (x - t)^{m - \alpha - 1} \mathrm{v}^{(m)}(t) dt,$$

for function v and $\alpha \in (m - 1, m)$. Assume that $\mathrm{v}^{(j)}(0) = 0$, for $j = 0, 1, 2, \cdots, \ell - 1$, then

$$I^{\ell} \frac{d^{\ell}}{dx^{\ell}} \mathrm{v}(x) = \mathrm{v}(x).$$

In the following, we state Krasnoselskii's Fixed Point Theorem [11], which is to be used soon.

Theorem 22.7 *Let* X *be a Banach space and* $X_1 \subseteq X$ *be a closed, convex, and non-empty set. Also, let* $\mathcal{N}_1, \mathcal{N}_2$ *be mappings such that*

- $\mathcal{N}_1\mathrm{v} + \mathcal{N}_2\mathrm{z} \in X_1$ *whenever* $\mathrm{v}, \mathrm{z} \in X_1$,

- *operator* $\mathcal{N}_1$ *is continuous and compact,*
- *mapping* $\mathcal{N}_2$ *is a contraction.*

Then there exists a $w \in X_1$ *satisfying* $w = \mathcal{N}_1 w + \mathcal{N}_2 w$.

22.4.2 Main Results

Theorem 22.8 *Let* ς *and* m *be nonnegative integers,* $0 \leq \beta_0 < \beta_1 < \cdots < \beta_\varsigma < \ell$, $0 < \alpha_0 < \alpha_1 < \cdots < \alpha_m$, *and* $\mathrm{p}_\imath(x)$ *and* $\mathrm{q}_\jmath(x)$ *be continuous functions over* $[0, \mathcal{M}]$ *for* $\imath = 0, 1, \cdots, \varsigma$ *and* $\jmath = 0, 1, \cdots, m$. *In addition, suppose that* $f : [0, \mathcal{M}] \times \mathbb{R}^\ell \to \mathbb{R}$ *is a continuous mapping, satisfying the following condition for nonnegative constants* $\mathcal{H}_0, \mathcal{H}_1, \cdots, \mathcal{H}_\ell$ *and* $(y_1, \cdots, y_\ell) \in \mathbb{R}^\ell$

$$|f(x, y_1, \cdots, y_\ell)| \leq \mathcal{H}_0 + \mathcal{H}_1|y_1| + \cdots + \mathcal{H}_\ell|y_\ell|,$$

and g *is a continuous mapping from* $[0, \mathcal{M}] \times \mathbb{R}$ *to* $\mathbb{R}$, *with the condition*

$$|g(x, z_1) - g(x, z_2)| \leq C_1|z_1 - z_2|,$$

for all $z_1, z_2 \in \mathbb{R}$ *and nonnegative* C_1. *Furthermore, assume that*

$$\begin{aligned} \mu = \; & C_1 \max\{\mathcal{M}^\ell E_{(\ell-\beta_0,\cdots,\ell+\alpha_m),\ell+1}\left(\mathcal{M}^{\ell-\beta_0}\mathrm{P}_0, \cdots, \mathcal{M}^{\ell+\alpha_m}\mathrm{Q}_m\right), \cdots, \\ & E_{(\ell-\beta_0,\cdots,\ell+\alpha_m),1}\left(\mathcal{M}^{\ell-\beta_0}\mathrm{P}_0, \cdots, \mathcal{M}^{\ell+\alpha_m}\mathrm{Q}_m\right)\} < 1, \end{aligned}$$

where

$$\max_{x\in[0,\mathcal{M}]} |\mathrm{p}_\imath(x)| \leq \mathrm{P}_\imath, \quad \max_{x\in[0,\mathcal{M}]} |\mathrm{q}_\jmath(x)| \leq \mathrm{Q}_\jmath.$$

Then Eq. (22.17) has a solution in the Banach space $C^\ell[0, \mathcal{M}]$.

Proof Equation (22.17) with its initial conditions is equivalent to

$$\begin{aligned} & \mathrm{v}(x) + \sum_{\imath=0}^{\varsigma} I^\ell \, \mathrm{p}_\imath(x)\, {}_C D^{\beta_\imath} \mathrm{v}(x) + \sum_{\jmath=0}^{m} I^\ell \, \mathrm{q}_\jmath(x) I^{\alpha_\jmath} \mathrm{v}(x) \\ & = I^\ell \, f(x, \mathrm{v}'(x), \cdots, \mathrm{v}^{(\ell)}(x)) + \; I^\ell \, g(x, \mathrm{v}(x)). \end{aligned}$$

By Babenko's approach, we come to

$$\begin{aligned} \mathrm{v}(x) = & \left(1 + \sum_{\imath=0}^{\varsigma} I^\ell \, \mathrm{p}_\imath(x)\, {}_C D^{\beta_\imath} + \sum_{\jmath=0}^{m} I^\ell \, \mathrm{q}_\jmath(x) I^{\alpha_\jmath}\right)^{-1} \\ & \cdot \left(I^\ell \, f(x, \mathrm{v}'(x), \cdots, \mathrm{v}^{(\ell)}(x)) + \; I^\ell \, g(x, \mathrm{v}(x))\right) \end{aligned}$$

$$= \sum_{\lambda=0}^{\infty}(-1)^{\lambda}\left(\sum_{\imath=0}^{\varsigma} I^{\ell}\, \mathrm{p}_{\imath}(x)\, {}_{C}D^{\beta_{\imath}} + \sum_{\jmath=0}^{m} I^{\ell}\, \mathrm{q}_{\jmath}(x) I^{\alpha_{\jmath}}\right)^{\lambda}$$
$$\cdot\left(I^{\ell} f(x, \mathrm{v}'(x), \cdots, \mathrm{v}^{(\ell)}(x)) + I^{\ell} g(x, \mathrm{v}(x))\right)$$
$$= \sum_{\lambda=0}^{\infty}(-1)^{\lambda} \sum_{\lambda_0+\cdots+\lambda_{m+\varsigma+1}=\lambda} \binom{\lambda}{\lambda_0, \lambda_1, \cdots, \lambda_{m+\varsigma+1}} (I^{\ell}\, \mathrm{p}_0(x)\, {}_{C}D^{\beta_0})^{\lambda_0} \cdots$$
$$\cdot(I^{\ell}\, \mathrm{q}_m(x)\, I^{\alpha_m})^{\lambda_{m+\varsigma+1}} \cdot \left(I^{\ell} f(x, \mathrm{v}'(x), \cdots, \mathrm{v}^{(\ell)}(x)) + I^{\ell} g(x, \mathrm{v}(x))\right)$$
$$= \sum_{\lambda=0}^{\infty}(-1)^{\lambda} \sum_{\lambda_0+\cdots+\lambda_{m+\varsigma+1}=\lambda} I^{\ell}\binom{\lambda}{\lambda_0, \lambda_1, \cdots, \lambda_{m+\varsigma+1}} (\mathrm{p}_0(x)\, I^{\ell-\beta_0})^{\lambda_0} \cdots$$
$$\cdot(\mathrm{q}_m(x)\, I^{\ell+\alpha_m})^{\lambda_{m+\varsigma+1}} \cdot \left(f(x, \mathrm{v}'(x), \cdots, \mathrm{v}^{(\ell)}(x)) + g(x, \mathrm{v}(x))\right),$$

using

$$(I^{\ell}\, \mathrm{p}_0(x)\, {}_{C}D^{\beta_0})^{\lambda_0} \cdots \cdot \left(I^{\ell}\, \mathrm{q}_m(x)\, I^{\alpha_m}\right)^{\lambda_{m+\varsigma+1}} I^{\ell}$$
$$= I^{\ell}(\mathrm{p}_0(x)\, I^{\ell-\beta_0})^{\lambda_0} \cdots (\mathrm{q}_m(x)\, I^{\ell+\alpha_m})^{\lambda_{m+\varsigma+1}}.$$

We define the (nonlinear) operator $\mathcal{N}_1$ over $C^{\ell}[0, \mathcal{M}]$ as

$$(\mathcal{N}_1\mathrm{v})(x) = \sum_{\lambda=0}^{\infty}(-1)^{\lambda} \sum_{\lambda_0+\cdots+\lambda_{m+\varsigma+1}=\lambda} I^{\ell}\binom{\lambda}{\lambda_0, \lambda_1, \cdots, \lambda_{m+\varsigma+1}} (\mathrm{p}_0(x)\, I^{\ell-\beta_0})^{\lambda_0} \cdots$$
$$\cdot(\mathrm{q}_m(x)\, I^{\ell+\alpha_m})^{\lambda_{m+\varsigma+1}} \cdot f(x, \mathrm{v}'(x), \cdots, \mathrm{v}^{(\ell)}(x)),$$

and the (nonlinear) operator $\mathcal{N}_2$ over $C^{\ell}[0, \mathcal{M}]$ as

$$(\mathcal{N}_2\mathrm{v})(x) = \sum_{\lambda=0}^{\infty}(-1)^{\lambda} \sum_{\lambda_0+\cdots+\lambda_{m+\varsigma+1}=\lambda} I^{\ell}\binom{\lambda}{\lambda_0, \lambda_1, \cdots, \lambda_{m+\varsigma+1}} (\mathrm{p}_0(x)\, I^{\ell-\beta_0})^{\lambda_0} \cdots$$
$$\cdot(\mathrm{q}_m(x)\, I^{\ell+\alpha_m})^{\lambda_{m+\varsigma+1}} \cdot g(x, \mathrm{v}(x)).$$

We first show that $\mathcal{N}_2$ is a contraction. Indeed,

$$\|\mathcal{N}_2\mathrm{v}\|_C \leq \sum_{\lambda=0}^{\infty} \sum_{\lambda_0+\cdots+\lambda_{m+\varsigma+1}=\lambda} \binom{\lambda}{\lambda_0, \lambda_1, \cdots, \lambda_{m+\varsigma+1}} \mathrm{P}_0^{\lambda_0} \cdots \mathrm{Q}_m^{\lambda_{m+\varsigma+1}}$$
$$\cdot \left\| I^{\ell+(\ell-\beta_0)\lambda_0+\cdots+(\ell+\alpha_m)\lambda_{m+\varsigma+1}} \right\|_C \|g(x, \mathrm{v}(x))\|_C\,.$$

Clearly,

$$\begin{aligned}\|g(x, v(x))\|_C &= \max_{x\in[0,\mathcal{M}]} |g(x, v(x))| = \max_{x\in[0,\mathcal{M}]} |g(x, v(x)) - g(x,0) + g(x,0)| \\ &\le \max_{x\in[0,\mathcal{M}]} |g(x,0)| + \max_{x\in[0,\mathcal{M}]} |g(x, v(x)) - g(x,0)| \\ &\le \max_{x\in[0,\mathcal{M}]} |g(x,0)| + C_1 \max_{x\in[0,\mathcal{M}]} |v(x)| = \max_{x\in[0,\mathcal{M}]} |g(x,0)| + C_1 \|v\|_C < +\infty,\end{aligned}$$

since $g(x,0)$ is continuous on $[0, \mathcal{M}]$. Moreover,

$$\left\| I^{\ell+(\ell-\beta_0)\lambda_0+\cdots+(\ell+\alpha_m)\lambda_{m+\varsigma+1}} \right\|_C \le \frac{\mathcal{M}^{\ell+(\ell-\beta_0)\lambda_0+\cdots+(\ell+\alpha_m)\lambda_{m+\varsigma+1}}}{\Gamma(\ell+1+(\ell-\beta_0)\lambda_0+\cdots+(\ell+\alpha_m)\lambda_{m+\varsigma+1})}.$$

Hence,

$$\|\mathcal{N}_2 v\|_C \le \mathcal{M}^{\ell} E_{(\ell-\beta_0,\cdots,\ell+\alpha_m),\ell+1}\left(\mathcal{M}^{\ell-\beta_0}\mathrm{P}_0,\cdots,\mathcal{M}^{\ell+\alpha_m}\mathrm{Q}_m\right) \|g(x, v(x))\|_C < +\infty.$$

Similarly,

$$\begin{aligned}\left\| \frac{d}{dx}\mathcal{N}_2 v \right\|_C &\le \mathcal{M}^{\ell-1} E_{(\ell-\beta_0,\cdots,\ell+\alpha_m),\ell}\left(\mathcal{M}^{\ell-\beta_0}\mathrm{P}_0,\cdots,\mathcal{M}^{\ell+\alpha_m}\mathrm{Q}_m\right) \\ &\quad \cdot \|g(x, v(x))\|_C < +\infty, \\ &\cdots, \\ \left\| \frac{d^{\ell}}{dx^{\ell}}\mathcal{N}_2 v \right\|_C &\le E_{(\ell-\beta_0,\cdots,\ell+\alpha_m),1}\left(\mathcal{M}^{\ell-\beta_0}\mathrm{P}_0,\cdots,\mathcal{M}^{\ell+\alpha_m}\mathrm{Q}_m\right) \\ &\quad \cdot \|g(x, v(x))\|_C < +\infty.\end{aligned}$$

These inequalities claim that $\mathcal{N}_2$ is a mapping from $C^{\ell}[0, \mathcal{M}]$ to $C^{\ell}[0, \mathcal{M}]$ and is continuous since g is continuous. To prove that it is contractive, we note that for $v, z \in C^{\ell}[0, \mathcal{M}]$,

$$\begin{aligned}&\|\mathcal{N}_2 v - \mathcal{N}_2 z\|_C \le C_1 \mathcal{M}^{\ell} E_{(\ell-\beta_0,\cdots,\ell+\alpha_m),\ell+1}\left(\mathcal{M}^{\ell-\beta_0}\mathrm{P}_0,\cdots,\mathcal{M}^{\ell+\alpha_m}\mathrm{Q}_m\right) \|v - z\|_C \\ &\le C_1 \mathcal{M}^{\ell} E_{(\ell-\beta_0,\cdots,\ell+\alpha_m),\ell+1}\left(\mathcal{M}^{\ell-\beta_0}\mathrm{P}_0,\cdots,\mathcal{M}^{\ell+\alpha_m}\mathrm{Q}_m\right) \|v - z\|_{\ell}, \\ &\cdots, \\ &\left\| \frac{d^{\ell}}{dx^{\ell}}(\mathcal{N}_2 v - \mathcal{N}_2 z) \right\|_C \le C_1 E_{(\ell-\beta_0,\cdots,\ell+\alpha_m),1}\left(\mathcal{M}^{\ell-\beta_0}\mathrm{P}_0,\cdots,\mathcal{M}^{\ell+\alpha_m}\mathrm{Q}_m\right) \|v - z\|_C \\ &\le C_1 E_{(\ell-\beta_0,\cdots,\ell+\alpha_m),1}\left(\mathcal{M}^{\ell-\beta_0}\mathrm{P}_0,\cdots,\mathcal{M}^{\ell+\alpha_m}\mathrm{Q}_m\right) \|v - z\|_{\ell}.\end{aligned}$$

Therefore,

$$\begin{aligned}\|\mathcal{N}_2 v - \mathcal{N}_2 z\|_{\ell} \le\ & C_1 \max\{\mathcal{M}^{\ell} E_{(\ell-\beta_0,\cdots,\ell+\alpha_m),\ell+1}\left(\mathcal{M}^{\ell-\beta_0}\mathrm{P}_0,\cdots,\mathcal{M}^{\ell+\alpha_m}\mathrm{Q}_m\right), \\ & \cdots, E_{(\ell-\beta_0,\cdots,\ell+\alpha_m),1}\left(\mathcal{M}^{\ell-\beta_0}\mathrm{P}_0,\cdots,\mathcal{M}^{\ell+\alpha_m}\mathrm{Q}_m\right)\} \|v - z\|_{\ell} \\ =\ & \mu \|v - z\|_{\ell},\end{aligned}$$

which indicates that $\mathcal{N}_2$ is a contraction. To show that $\mathcal{N}_1$ is continuous and compact, we first find that

$$\begin{aligned}
\|\mathcal{N}_1 \mathrm{v}\|_C &\le \mathcal{M}^\ell E_{(\ell-\beta_0,\cdots,\ell+\alpha_m),\ell+1}\left(\mathcal{M}^{\ell-\beta_0}\mathrm{P}_0,\cdots,\mathcal{M}^{\ell+\alpha_m}\mathrm{Q}_m\right) \\
&\qquad \cdot \left\| f(x,\mathrm{v}'(x),\cdots,\mathrm{v}^{(\ell)}(x))\right\|_C \\
&\le \mathcal{M}^\ell E_{(\ell-\beta_0,\cdots,\ell+\alpha_m),\ell+1}\left(\mathcal{M}^{\ell-\beta_0}\mathrm{P}_0,\cdots,\mathcal{M}^{\ell+\alpha_m}\mathrm{Q}_m\right) \\
&\qquad \cdot \left(\mathcal{H}_0+\mathcal{H}_1\left\|\mathrm{v}'\right\|_C+\cdots+\mathcal{H}_\ell\left\|\mathrm{v}^{(\ell)}\right\|_C\right)<+\infty, \\
&\cdots, \\
\left\|\frac{d^\ell}{dx^\ell}\mathcal{N}_1\mathrm{v}\right\|_C &\le E_{(\ell-\beta_0,\cdots,\ell+\alpha_m),1}\left(\mathcal{M}^{\ell-\beta_0}\mathrm{P}_0,\cdots,\mathcal{M}^{\ell+\alpha_m}\mathrm{Q}_m\right) \\
&\qquad \cdot \left(\mathcal{H}_0+\mathcal{H}_1\left\|\mathrm{v}'\right\|_C+\cdots+\mathcal{H}_\ell\left\|\mathrm{v}^{(\ell)}\right\|_C\right)<+\infty.
\end{aligned}$$

So $\mathcal{N}_1$ is a mapping from $C^\ell[0,\mathcal{M}]$ to itself and is continuous since f is continuous and the series

$$\sum_{\lambda=0}^{\infty}(-1)^\lambda \sum_{\lambda_0+\cdots+\lambda_{m+\varsigma+1}=\lambda} I^\ell \binom{\lambda}{\lambda_0,\lambda_1,\cdots,\lambda_{m+\varsigma+1}} \left(\mathrm{p}_0(x)\, I^{\ell-\beta_0}\right)^{\lambda_0}\cdots$$
$$\cdot\left(\mathrm{q}_m(x)\, I^{\ell+\alpha_m}\right)^{\lambda_{m+\varsigma+1}}\cdot f(x,\mathrm{v}'(x),\cdots,\mathrm{v}^{(\ell)}(x)),$$

is uniformly convergent in $C[0,\mathcal{M}]$ from above. Hence, $\mathcal{N}_1\mathrm{v}+\mathcal{N}_2\mathrm{z}\in C^\ell[0,\mathcal{M}]$ whenever $\mathrm{v},\mathrm{z}\in C^\ell[0,\mathcal{M}]$. Suppose S is a bounded subset of $C^\ell[0,\mathcal{M}]$, there exists a constant $W>0$ such that

$$\left\| f(x,\mathrm{v}',\cdots,\mathrm{v}^{(\ell)})\right\|_C \le W$$

for all $\mathrm{v}\in S$, which implies that

$$\begin{aligned}
\|\mathcal{N}_1\mathrm{v}\|_C &\le W\mathcal{M}^\ell E_{(\ell-\beta_0,\cdots,\ell+\alpha_m),\ell+1}\left(\mathcal{M}^{\ell-\beta_0}\mathrm{P}_0,\cdots,\mathcal{M}^{\ell+\alpha_m}\mathrm{Q}_m\right)\le\frac{\mu W}{C_1}, \\
&\cdots, \\
\left\|\frac{d^\ell}{dx^\ell}\mathcal{N}_1\mathrm{v}\right\|_C &\le W E_{(\ell-\beta_0,\cdots,\ell+\alpha_m),1}\left(\mathcal{M}^{\ell-\beta_0}\mathrm{P}_0,\cdots,\mathcal{M}^{\ell+\alpha_m}\mathrm{Q}_m\right)\le\frac{\mu W}{C_1}.
\end{aligned}$$

These infer that $\mathcal{N}_1 S$ is uniformly bounded in the space $C^\ell[0,\mathcal{M}]$. It remains to be shown that $\mathcal{N}_1 u$ is equicontinuous over the interval $[0,\mathcal{M}]$. Let $0\le\tau_1<\tau_2\le\mathcal{M}$, we get

$$\begin{aligned}
&|(\mathcal{N}_1\mathrm{v})(\tau_2)-(\mathcal{N}_1\mathrm{v})(\tau_1)| \\
&\quad\le\sum_{\lambda=0}^{\infty}\sum_{\lambda_0+\cdots+\lambda_{m+\varsigma+1}=\lambda}\binom{\lambda}{\lambda_0,\lambda_1,\cdots,\lambda_{m+\varsigma+1}}\mathrm{P}_0^{\lambda_0}\cdots\mathrm{Q}_m^{\lambda_{m+\varsigma+1}}
\end{aligned}$$

$$\cdot \frac{1}{\Gamma(s)} \left\{ \int_0^{\tau_1} \left((\tau_2 - t)^{s-1} - (\tau_1 - t)^{s-1} \right) |f(t, \mathrm{v}'(t), \cdots, \mathrm{v}^{(\ell)}(t))| dt \right.$$
$$\left. + \int_{\tau_1}^{\tau_2} (\tau_2 - t)^{s-1} |f(t, \mathrm{v}'(t), \cdots, \mathrm{v}^{(\ell)}(t))| dt \right\},$$

where

$$s = \ell + (\ell - \beta_0)\lambda_0 + \cdots + (\ell + \alpha_m)\lambda_{m+\varsigma+1} \geq 1.$$

For $\mathrm{v} \in S$, we derive that

$$\int_0^{\tau_1} \left((\tau_2 - t)^{s-1} - (\tau_1 - t)^{s-1} \right) |f(t, \mathrm{v}'(t), \cdots, \mathrm{v}^{(\ell)}(t))| dt$$
$$\leq W \int_0^{\tau_1} \left((\tau_2 - t)^{s-1} - (\tau_1 - t)^{s-1} \right) dt$$
$$= \frac{W}{s}(\tau_2^s - (\tau_2 - \tau_1)^s - \tau_1^s) < \frac{W}{s}(\tau_2^s - \tau_1^s).$$

By the mean value theorem,

$$\frac{\tau_2^s - \tau_1^s}{\tau_2 - \tau_1} = s\,\theta^{s-1} < s\,\mathcal{M}^{s-1},$$

where $\theta \in (\tau_1, \tau_2)$. Hence,

$$\int_0^{\tau_1} \left((\tau_2 - t)^{s-1} - (\tau_1 - t)^{s-1} \right) |f(t, \mathrm{v}'(t), \cdots, \mathrm{v}^{(\ell)}(t))| dt < W(\tau_2 - \tau_1)\mathcal{M}^{s-1}.$$

Furthermore,

$$\int_{\tau_1}^{\tau_2} (\tau_2 - t)^{s-1} |f(t, \mathrm{v}'(t), \cdots, \mathrm{v}^{(\ell)}(t))| dt \leq W(\tau_2 - \tau_1)\mathcal{M}^{s-1}.$$

In summary,

$$|(\mathcal{N}_1 \mathrm{v})(\tau_2) - (\mathcal{N}_1 \mathrm{v})(\tau_1)|$$
$$\leq W(\tau_2 - \tau_1) \sum_{\lambda=0}^{\infty} \sum_{\lambda_0 + \cdots + \lambda_{m+\varsigma+1} = \lambda} \binom{\lambda}{\lambda_0, \lambda_1, \cdots, \lambda_{m+\varsigma+1}} \mathrm{P}_0^{\lambda_0} \cdots \mathrm{Q}_m^{\lambda_{m+\varsigma+1}}$$
$$\cdot \frac{\mathcal{M}^{\ell-1+(\ell-\beta_0)\lambda_0+\cdots+(\ell+\alpha_m)\lambda_{\ell+\varsigma+1}}}{\Gamma(\ell + (\ell - \beta_0)\lambda_0 + \cdots + (\ell + \alpha_m)\lambda_{n+\varsigma+1})}$$
$$= W\mathcal{M}^{\ell-1}(\tau_2 - \tau_1) E_{(\ell-\beta_0, \cdots, \ell+\alpha_m), \ell} \left(\mathcal{M}^{\ell-\beta_0} \mathrm{P}_0, \cdots, \mathcal{M}^{\ell+\alpha_m} \mathrm{Q}_m \right),$$

which proves that $\mathcal{N}_1 \mathrm{v}$ is equicontinuous in the space $C[0, \mathcal{M}]$. Following the above step, we can show that $\mathcal{N}_1 \mathrm{v}$ is equicontinuous in the space $C^\ell[0, \mathcal{M}]$ by noting the integral operator I^{ℓ} in the right-hand side series. Using the Arzela–Ascoli theorem, $\mathcal{N}_1$ is a compact operator. Hence, Eq. (22.17) has a solution in the Banach space $C^{\ell}[0, \mathcal{M}]$ by Krasnoselskii's fixed point theorem.

22.4.3 Existence of Positive Solutions

Now, we investigate the existence of a positive solution to a variant of equation (22.17). We consider the following equation

$$\begin{cases} \mathrm{v}^{(\ell)}(x) - \sum_{\imath=0}^{\varsigma} \mathrm{p}_\imath(x)\, {}_C D^{\beta_\imath} \mathrm{v}(x) - \sum_{\jmath=0}^{m} \mathrm{q}_\jmath(x) I^{\alpha_\jmath} \mathrm{v}(x) = f(x, \mathrm{v}'(x), \cdots, \mathrm{v}^{(\ell)}(x)) \\ \qquad\qquad + g(x, \mathrm{v}(x)), \\ \mathrm{v}(0) = \mathrm{v}'(0) = \cdots = \mathrm{v}^{(\ell-1)}(0) = 0. \end{cases} \tag{22.18}$$

Let $\mathbb{R}_0^+ = [0, +\infty)$ and $\ell \in \mathbb{N} = \{1, 2, \cdots\}$. Define the space $C_+^{\ell}[0, \mathcal{M}]$ as

$$C_+^{\ell}[0, \mathcal{M}] = \left\{ \mathrm{v}(x) : [0, \mathcal{M}] \to \mathbb{R}_0^+ : \ \mathrm{v}(x) \in C^{\ell}[0, \mathcal{M}] \text{ and } \|\mathrm{v}\|_{\ell} < +\infty \right\},$$

where

$$\|\mathrm{v}\|_{\ell} = \max \left\{ \max_{x \in [0, \mathcal{M}]} \mathrm{v}(x), \max_{x \in [0, \mathcal{M}]} |\mathrm{v}'(x)|, \cdots, \max_{x \in [0, \mathcal{M}]} |\mathrm{v}^{(\ell)}(x)| \right\}.$$

Clearly, $C_+^{\ell}[0, \mathcal{M}]$ is a Banach space.

We have the following theorem dealing with the existence of positive solutions to Eq. (22.18).

Theorem 22.9 *Let ς and m be nonnegative integers, $0 \le \beta_0 < \beta_1 < \cdots < \beta_\varsigma < \ell$, $0 < \alpha_0 < \alpha_1 < \cdots < \alpha_m$, and $\mathrm{p}_\imath(x)$ and $\mathrm{q}_\jmath(x)$ be nonnegative and continuous functions over $[0, \mathcal{M}]$ for $\imath = 0, 1, \cdots, \varsigma$ and $\jmath = 0, 1, \cdots, m$. In addition, suppose that $f : [0, \mathcal{M}] \times \mathbb{R}^{\ell} \to \mathbb{R}$ is a continuous mapping, satisfying the following condition for nonnegative constants $\mathcal{H}_0, \mathcal{H}_1, \cdots, \mathcal{H}_\ell$ and $(y_1, \cdots, y_\ell) \in \mathbb{R}^{\ell}$*

$$0 \le f(x, y_1, \cdots, y_\ell) \le \mathcal{H}_0 + \mathcal{H}_1 |y_1| + \cdots + \mathcal{H}_\ell |y_\ell|,$$

and g is a nonnegative and continuous mapping from $[0, \mathcal{M}] \times \mathbb{R}$ to $\mathbb{R}$, with the condition

$$|g(x, z_1) - g(x, z_2)| \le C_1 |z_1 - z_2|$$

for all $z_1, z_2 \in \mathbb{R}$ and nonnegative C_1. Furthermore, assume that

$$\mu = C_1 \max\{\mathcal{M}^\ell E_{(\ell-\beta_0,\cdots,\ell+\alpha_m),\ell+1}\left(\mathcal{M}^{\ell-\beta_0}\mathrm{P}_0,\cdots,\mathcal{M}^{\ell+\alpha_m}\mathrm{Q}_m\right),\cdots,$$
$$E_{(\ell-\beta_0,\cdots,\ell+\alpha_m),1}\left(\mathcal{M}^{\ell-\beta_0}\mathrm{P}_0,\cdots,\mathcal{M}^\ell+\alpha_m\mathrm{Q}_m\right)\} < 1,$$

where

$$\max_{x\in[0,\mathcal{M}]} \mathrm{p}_\iota(x) \leq \mathrm{P}_\iota, \quad \max_{x\in[0,\mathcal{M}]} \mathrm{q}_\jmath(x) \leq \mathrm{Q}_\jmath.$$

Then Eq. (22.18) has a nonnegative solution in the Banach space $C_+^\ell[0,\mathcal{M}]$.

Proof Clearly, Eq. (22.18) with its initial conditions is equivalent to

$$\mathrm{v}(x) = \sum_{\lambda=0}^{\infty} \sum_{\lambda_0+\cdots+\lambda_{m+\varsigma+1}=\lambda} I^\ell \binom{\lambda}{\lambda_0,\lambda_1,\cdots,\lambda_{m+\varsigma+1}} (\,\mathrm{p}_0(x)\, I^{\ell-\beta_0})^{\lambda_0}\cdots$$
$$\cdot(\mathrm{q}_m(x)\, I^{\ell+\alpha_m})^{\lambda_{m+\varsigma+1}} \cdot \left(f(x,\mathrm{v}'(x),\cdots,\mathrm{v}^{(\ell)}(x)) + g(x,\mathrm{v}(x))\right),$$

which is nonnegative as f, g, $\mathrm{p}_\iota(x)$, $\mathrm{q}_\jmath(x)$ are all nonnegative for $\iota = 0, 1, \cdots, \varsigma$ and $\jmath = 0, 1, \cdots, m$. Then the result follows immediately from the proof of Theorem 22.8.

Example 22.4 Consider the following nonlinear and implicit equation

$$\begin{cases} \mathrm{v}''(x) + x\ {}_C D^{0.9}\mathrm{v}(x) + x^2\ {}_C D^{0.8}\mathrm{v}(x) + \cos x\ I^{0.1}\mathrm{v}(x) = \\ \qquad \dfrac{1}{(\mathrm{v}')^2(x) + (\mathrm{v}'')^2(x) + 1} + \dfrac{1}{21}\sin \mathrm{v}(x), \\ \mathrm{v}(0) = \mathrm{v}'(0) = 0. \end{cases} \tag{22.19}$$

Then (22.19) has a solution in the space $C^2[0,1]$.

Proof It is easy to show

$$f(x,\mathrm{v}',\mathrm{v}'') = \frac{1}{(\mathrm{v}')^2(x) + (\mathrm{v}'')^2(x) + 1}$$

is bounded by $M_0 = 1$, and

$$g(x,\mathrm{v}) = \frac{1}{21}\sin \mathrm{v}(x),$$

satisfying the condition

$$|g(x,\mathrm{v}) - g(x,\mathrm{z})| \leq \frac{1}{21}|\mathrm{v}-\mathrm{z}|,$$

with $C_1 = \dfrac{1}{21}$. In addition,

$$\max_{x\in[0,1]}|x|\le 1,\quad \max_{x\in[0,1]}|x^2|\le 1,\quad \max_{x\in[0,1]}|\cos x|\le 1,$$

and

$$\begin{aligned}\mu &= \frac{1}{21}\max\left\{E_{(1.1,1.2,2.1),3}(1,1,1),\, E_{(1.1,1.2,2.1),2}(1,1,1),\, E_{(1.1,1.2,2.1),1}(1,1,1)\right\}\\ &= \frac{1}{21}E_{(1.1,1.2,2.1),1}(1,1,1).\end{aligned}$$

Evidently,

$$E_{(1.1,1.2,2.1),1}(1,1,1) = \sum_{\lambda=0}^{\infty}\sum_{\lambda_1+\lambda_2+\lambda_3=\lambda}\binom{\lambda}{\lambda_1,\lambda_2,\lambda_3}\frac{1}{\Gamma(1.1\lambda_1+1.2\lambda_2+2.1\lambda_3+1)},$$

$$\sum_{\lambda_1+\lambda_2+\lambda_3=\lambda}\binom{\lambda}{\lambda_1,\lambda_2,\lambda_3} = 3^{\lambda},$$

$$\frac{1}{\Gamma(1.1\lambda_1+1.2\lambda_2+2.1\lambda_3+1)} \le \frac{1}{\Gamma(\lambda+1)}.$$

So

$$E_{(1.1,1.2,2.1),1}(1,1,1) \le \sum_{\lambda=0}^{\infty}\frac{3^{\lambda}}{\lambda!} < 20.4125,$$

see the bottom of page 11 in [12] for more details. This implies that $\mu < 1$. By Theorem 22.8, the equation has a solution in the space $C^2[0,1]$.

Example 22.5 Consider the following nonlinear and implicit equation

$$\begin{cases} \mathrm{v}''(x) - x^3\, {}_C D^{0.9}\mathrm{v}(x) - \dfrac{1}{x^2+1}\, {}_C D^{0.8}\mathrm{v}(x) - \sin x\, I^{0.1}\mathrm{v}(x) = \\ \qquad\qquad \dfrac{1}{(\mathrm{v}')^2(x)+1} + \dfrac{1}{21}|\cos \mathrm{v}(x)|, \\ \mathrm{v}(0) = \mathrm{v}'(0) = 0. \end{cases} \tag{22.20}$$

Then (22.20) has a nonnegative solution in the space $C_+^2[0,1]$.

Proof Clearly, all the functions

$$\frac{1}{(\mathrm{v}')^2(x)+1},\quad x^3,\quad \frac{1}{x^2+1},\quad \text{and}\quad \sin x$$

are nonnegative over the interval [0, 1] and bounded by 1. Furthermore,

$$g(x,\mathrm{v}) = \frac{1}{21}|\cos \mathrm{v}(x)| \ge 0$$

satisfying the condition

$$|g(x, \mathrm{v}) - g(x, \mathrm{z})| \leq \frac{1}{21}|\mathrm{v} - \mathrm{z}|,$$

with $C_1 = \dfrac{1}{21}$. It follows from Example 22.4 that $\mu < 1$. By Theorem 22.9, the equation has a nonnegative solution in the space $C_+^2[0, 1]$.

22.4.4 Summary

In this section, we have investigated an implicit and nonlinear system with initial conditions and variable coefficients, and presented the sufficient condition for the existence of solutions, as well as the results regarding the existence of positive solutions to the variant equation, using Babenko's approach, Krasnoselskii's fixed point theorem, and the Multivariate Mittag-Leffler function. We have also presented examples showing the applications of the main theorems based on the Mittag-Leffler function of three variables.

22.5 The Existence of Solutions for a System of Nonlinear Hadamard–Type

22.5.1 Preliminaries

Let $0 < a < b < \infty$ and $C[a, b]$ be the space given by

$$C[a, b] = \left\{u : [a, b] \to \mathbb{R} : \ u \text{ is continuous on } [a, b] \text{ and } \|u\|_C = \max_{x \in [a,b]} |u(x)| < +\infty\right\}.$$

Clearly, $C[a, b]$ is a Banach space. Furthermore, the product space $C[a, b] \times C[a, b]$ (also a Banach space) is defined as

$$C[a, b] \times C[a, b] = \{(u, v) : \ u, v \in C[a, b]\},$$

with the norm

$$\|(u, v)\| = \|u\|_C + \|v\|_C .$$

The Hadamard-type fractional integral of order $\alpha > 0$ for a function u is defined in [1, 2, 13, 14] (see also the recent developments which are reported in [15]) as follows:

$$\left(\mathcal{J}_{a+,\mu}^{\alpha}u\right)(x) = \frac{1}{\Gamma(\alpha)}\int_a^x \left(\frac{t}{x}\right)^{\mu}\left(\log\frac{x}{t}\right)^{\alpha-1} u(t)\frac{dt}{t}, \quad a < x < b,$$

where $\log(\cdot) = \log_e(\cdot)$ and $\mu \in \mathbb{R}$. In particular, we let

$$\left(\mathcal{J}_{a+}^{\alpha}u\right)(x) = \left(\mathcal{J}_{a+,0}^{\alpha}u\right)(x) = \frac{1}{\Gamma(\alpha)}\int_a^x \left(\log\frac{x}{t}\right)^{\alpha-1} u(t)\frac{dt}{t}.$$

Then for $u \in C[a,b]$,

$$\left\|\mathcal{J}_{a+}^{\alpha}u\right\|_C \leq \frac{1}{\Gamma(\alpha+1)}\left(\log\frac{b}{a}\right)^{\alpha}\|u\|_C.$$

Indeed, we have

$$\begin{aligned}\left\|\mathcal{J}_{a+}^{\alpha}u\right\|_C &= \frac{1}{\Gamma(\alpha)}\max_{x\in[a,b]}\int_a^x \left(\log\frac{x}{t}\right)^{\alpha-1} u(t)\frac{dt}{t} \\ &\leq \frac{1}{\Gamma(\alpha)}\max_{x\in[a,b]}\int_a^x \left(\log\frac{x}{t}\right)^{\alpha-1}\frac{dt}{t}\|u\|_C \leq \frac{1}{\Gamma(\alpha+1)}\left(\log\frac{b}{a}\right)^{\alpha}\|u\|_C.\end{aligned}$$

Let $X_\mu(a,b)$ be the space of those Lebesgue measurable functions u on $[a,b]$ for which $x^{\mu-1}u(x)$ is absolutely integrable [14]:

$$X_\mu(a,b) = \left\{ u : [a,b] \to C : \quad \|u\|_{X_\mu} = \int_a^b x^{\mu-1}|u(x)|dx < \infty \right\}.$$

Obviously, $C[a,b] \subset X_\mu(a,b)$. Then it follows from Lemma 2.2 in [14] that the following semigroup property holds

$$\mathcal{J}_{a+}^{\alpha}\mathcal{J}_{a+}^{\beta}u = \mathcal{J}_{a+}^{\alpha+\beta}u,$$

for all $\alpha,\ \beta > 0$ and $u \in C[a,b]$.

The goal of this section is to study the existence of solutions for the following nonlinear integro-differential system involving the fractional Hadamard-type operators by using Leray-Schauder's alternative and the multivariate Mittag-Leffler function in the product space $C[a,b] \times C[a,b]$:

$$\begin{cases} u(x) + a_n(\mathcal{J}_{a+}^{\alpha_n}u)(x) + \cdots + a_1(\mathcal{J}_{a+}^{\alpha_1}u)(x) = f_1(x, u(x), v(x)), \\ v(x) + b_n(\mathcal{J}_{a+}^{\beta_n}v)(x) + \cdots + b_1(\mathcal{J}_{a+}^{\beta_1}v)(x) = f_2(x, u(x), v(x)), \end{cases} \tag{22.21}$$

where $\alpha_n > \alpha_{n-1} > \cdots > \alpha_1 > 1$, $\beta_n > \beta_{n-1} > \cdots > \beta_1 > 1$ and the functions f_1 and f_2 are mappings from $[a,b] \times \mathbb{R}^2$ to $\mathbb{R}$ satisfying certain conditions.

Theorem 22.10 (Leray-Schauder's alternative [16]) *Consider the continuous and compact mapping T of a Banach space S into itself. The boundedness of*

$$\{x \in S : x = \lambda T x \text{ for some } 0 \leqq \lambda \leqq 1\}$$

implies that T has a fixed point.

Leray-Schauder's alternative is a useful theorem for showing existence of solutions to nonlinear fractional differential equations [17–24].

22.5.2 Main Results

We shall present our main theorem dealing with the existence of solutions to the nonlinear system (22.21) by Babenko's approach, Leray-Schauder's alternative and the multivariate Mittag-Leffler function.

Theorem 22.11 *Assume that $\alpha_n > \alpha_{n-1} > \cdots > \alpha_1 > 1$, $\beta_n > \beta_{n-1} > \cdots > \beta_1 > 1$, and the functions $f_1(x, z_1, z_2)$ and $f_2(x, z_1, z_2)$ are continuous mappings from $[a, b] \times \mathbb{R}^2$ to $\mathbb{R}$ satisfying the following conditions for nonnegative constants C_0, C_1, and C_2 :*

$$|f_1(x, y_1, y_2)| \leqq C_0 + C_1|y_1| + C_2|y_2|$$

and

$$|f_2(x, y_1, y_2)| \leqq C_0 + C_1|y_1| + C_2|y_2|.$$

In addition, suppose that f'_{1x} and f'_{2x} are bounded and

$$\max\{C_1, C_2\}(E_{(\alpha_1,\cdots,\alpha_n,\, 1)}\left(|a_1|\left(\log\frac{b}{a}\right)^{\alpha_1}, \cdots, |a_n|\left(\log\frac{b}{a}\right)^{\alpha_n}\right)$$
$$+\ E_{(\beta_1,\cdots,\beta_n,\, 1)}\left(|b_1|\left(\log\frac{b}{a}\right)^{\beta_1}, \cdots, |b_n|\left(\log\frac{b}{a}\right)^{\beta_n}\right)) < 1.$$

Then there exists a solution to the system (22.21) in the space $C[a, b] \times C[a, b]$.

Proof Let $f \in C[a, b]$ with $0 < a < b < \infty$. Then the following equation

$$u(x) + a_n(\mathcal{J}_{a+}^{\alpha_n}u)(x) + \cdots + a_1(\mathcal{J}_{a+}^{\alpha_1}u)(x) = f(x),$$

has a unique and global solution in the space $C[a, b]$ by Babenko's approach and the semigroup property

$$u(x) = \sum_{k=0}^{\infty}(-1)^k \sum_{k_1+\cdots+k_n=k} \binom{k}{k_1, k_2, \cdots, k_n} a_n^{k_n} \cdots a_1^{k_1} (\mathcal{J}_{a+}^{k_n\alpha_n+\cdots+k_1\alpha_1} f)(x),$$

where we define

$$\mathcal{J}_{a+}^{0} f(x) = f(x).$$

Indeed,

$$\begin{aligned}
\|u\|_C &\leqq \sum_{k=0}^{\infty} \sum_{k_1+\cdots+k_n=k} \binom{k}{k_1, k_2, \cdots, k_n} |a_n|^{k_n} \cdots |a_1|^{k_1} \left\| \mathcal{J}_{a+}^{k_n\alpha_n+\cdots+k_1\alpha_1} f \right\|_C \\
&\leqq \sum_{k=0}^{\infty} \sum_{k_1+\cdots+k_n=k} \binom{k}{k_1, k_2, \cdots, k_n} |a_n|^{k_n} \cdots |a_1|^{k_1} \\
&\quad \cdot \frac{1}{\Gamma(k_n\alpha_n + \cdots + k_1\alpha_1 + 1)} \left(\log \frac{b}{a}\right)^{k_n\alpha_n+\cdots+k_1\alpha_1} \|f\|_C \\
&= E_{(\alpha_1,\cdots,\alpha_n,\,1)} \left(|a_1| \left(\log \frac{b}{a}\right)^{\alpha_1}, \cdots, |a_n| \left(\log \frac{b}{a}\right)^{\alpha_n} \right) \|f\|_C < +\infty.
\end{aligned} \tag{22.22}$$

This claims that the series is uniformly convergent on $[a, b]$ and hence $u(x)$ is continuous.

Let $(u, v) \in C[a,] \times C[a, b]$. Define a mapping T on the space $C[a, b] \times C[a, b]$ as

$$T(u, v) = (T_1(u, v),\ T_2(u, v)),$$

where

$$\begin{aligned}
T_1(u, v) = \sum_{k=0}^{\infty} (-1)^k \sum_{k_1+\cdots+k_n=k} \binom{k}{k_1, k_2, \cdots, k_n} a_n^{k_n} \cdots a_1^{k_1} \\
\cdot (\mathcal{J}_{a+}^{k_n\alpha_n+\cdots+k_1\alpha_1} f_1)(x, u(x), v(x)),
\end{aligned}$$

and

$$\begin{aligned}
T_2(u, v) = \sum_{k=0}^{\infty} (-1)^k \sum_{k_1+\cdots+k_n=k} \binom{k}{k_1, k_2, \cdots, k_n} b_n^{k_n} \cdots b_1^{k_1} \\
\cdot (\mathcal{J}_{a+}^{k_n\beta_n+\cdots+k_1\beta_1} f_2)(x, u(x), v(x)).
\end{aligned}$$

It follows from the inequality (22.22) that

$$\begin{aligned}
\|T_1(u, v)\|_C &\leqq E_{(\alpha_1,\cdots,\alpha_n,\,1)} \left(|a_1| \left(\log \frac{b}{a}\right)^{\alpha_1}, \cdots, |a_n| \left(\log \frac{b}{a}\right)^{\alpha_n} \right) \|f_1(x, u, v)\|_C \\
&\leqq E_{(\alpha_1,\cdots,\alpha_n,\,1)} \left(|a_1| \left(\log \frac{b}{a}\right)^{\alpha_1}, \cdots, |a_n| \left(\log \frac{b}{a}\right)^{\alpha_n} \right) \\
&\quad \cdot (C_0 + C_1 \|u\|_C + C_2 \|v\|_C),
\end{aligned}$$

$$\|T_2(u,v)\|_C \leqq E_{(\beta_1,\cdots,\beta_n,1)}\left(|b_1|\left(\log\frac{b}{a}\right)^{\beta_1},\cdots,|b_n|\left(\log\frac{b}{a}\right)^{\beta_n}\right)\|f_2(x,u,v)\|_C$$
$$\leqq E_{(\beta_1,\cdots,\beta_n,1)}\left(|b_1|\left(\log\frac{b}{a}\right)^{\beta_1},\cdots,|b_n|\left(\log\frac{b}{a}\right)^{\beta_n}\right)$$
$$\cdot\left(C_0+C_1\|u\|_C+C_2\|v\|_C\right).$$

Therefore, T is a continuous mapping from the space $C[a,b]\times C[a,b]$ to itself, since f_1 and f_2 are continuous.

Suppose that B is a proper bounded subset of $C[a,b]\times C[a,b]$, then we can find constants $W_1,\ W_2>0$ such that

$$\|f_1(x,u,v)\|_C\leqq W_1,\quad \|f_2(x,u,v)\|_C\leqq W_2,$$

for all $(u,v)\in B$, which deduces that

$$\|T_1(u,v)\|_C\leqq W_1E_{(\alpha_1,\cdots,\alpha_n,1)}\left(|a_1|\left(\log\frac{b}{a}\right)^{\alpha_1},\cdots,|a_n|\left(\log\frac{b}{a}\right)^{\alpha_n}\right),\quad\text{and}$$
$$\|T_2(u,v)\|_Y\leqq W_2E_{(\beta_1,\cdots,\beta_n,1)}\left(|b_1|\left(\log\frac{b}{a}\right)^{\beta_1},\cdots,|b_n|\left(\log\frac{b}{a}\right)^{\beta_n}\right).$$

Thus, TB is uniformly bounded in the space $C[a,b]\times C[a,b]$. We need to show that T is equicontinuous on $C[a,b]\times C[a,b]$. Let $\tau_1,\tau_2\in[a,b]$ with $\tau_1<\tau_2$, we come to

$$|T_1(u,v)(\tau_2)-T_1(u,v)(\tau_1)|\leqq|f_1(\tau_2,u(\tau_2),v(\tau_2))-f_1(\tau_1,u(\tau_1),v(\tau_1))|$$
$$+\sum_{k=1}^{\infty}\sum_{k_1+\cdots+k_n=k}\binom{k}{k_1,\cdots,k_n}|a_1|^{k_1}\cdots|a_n|^{k_n}\frac{1}{\Gamma(\lambda_1)}$$
$$\cdot\left|\int_a^{\tau_2}\left(\log\frac{\tau_2}{t}\right)^{\lambda_1-1}f_1(t,u(t),v(t))\frac{dt}{t}-\int_a^{\tau_1}\left(\log\frac{\tau_1}{t}\right)^{\lambda_1-1}f_1(t,u(t),v(t))\frac{dt}{t}\right|,$$

where

$$\lambda_1=k_1\alpha_1+\cdots+k_n\alpha_n\geqq\alpha_1>1,$$

for $k=k_1+\cdots+k_n\geqq 1$.

Since f'_{1x} is bounded, there is a constant $M_1>0$ such that

$$|f_1(\tau_2,u(\tau_2),v(\tau_2))-f_1(\tau_1,u(\tau_1),v(\tau_1))|\leqq M_1(\tau_2-\tau_1),$$

by the mean value theorem.

Furthermore,

$$\int_a^{\tau_2}\left(\log\frac{\tau_2}{t}\right)^{\lambda_1-1} f_1(t,u(t),v(t))\frac{dt}{t} = \int_a^{\tau_1}\left(\log\frac{\tau_2}{t}\right)^{\lambda_1-1} f_1(t,u(t),v(t))\frac{dt}{t}$$
$$+\int_{\tau_1}^{\tau_2}\left(\log\frac{\tau_2}{t}\right)^{\lambda_1-1} f_1(t,u(t),v(t))\frac{dt}{t}.$$

Thus, we have

$$\left|\int_a^{\tau_2}\left(\log\frac{\tau_2}{t}\right)^{\lambda_1-1} f_1(t,u(t),v(t))\frac{dt}{t} - \int_a^{\tau_1}\left(\log\frac{\tau_1}{t}\right)^{\lambda_1-1} f_1(t,u(t),v(t))\frac{dt}{t}\right|$$
$$\leqq \left|\int_a^{\tau_1}\left(\left(\log\frac{\tau_2}{t}\right)^{\lambda_1-1} - \left(\log\frac{\tau_1}{t}\right)^{\lambda_1-1}\right) f_1(t,u(t),v(t))\frac{dt}{t}\right|$$
$$+\int_{\tau_1}^{\tau_2}\left(\log\frac{\tau_2}{t}\right)^{\lambda_1-1} |f_1(t,u(t),v(t))|\frac{dt}{t}.$$

Obviously,

$$\int_{\tau_1}^{\tau_2}\left(\log\frac{\tau_2}{t}\right)^{\lambda_1-1} |f_1(t,u(t),v(t))|\frac{dt}{t} \leqq W_1\frac{\left(\log\frac{\tau_2}{\tau_1}\right)^{\lambda_1}}{\lambda_1},$$

and

$$\left|\int_a^{\tau_1}\left(\left(\log\frac{\tau_2}{t}\right)^{\lambda_1-1} - \left(\log\frac{\tau_1}{t}\right)^{\lambda_1-1}\right) f_1(t,u(t),v(t))\frac{dt}{t}\right|$$
$$\leqq W_1\int_a^{\tau_1}\left(\left(\log\frac{\tau_2}{t}\right)^{\lambda_1-1} - \left(\log\frac{\tau_1}{t}\right)^{\lambda_1-1}\right)\frac{dt}{t}$$
$$\leqq W_1\left(\frac{\left(\log\frac{\tau_2}{\tau_1}\right)^{\lambda_1}}{\lambda_1} + \frac{\left(\log\frac{\tau_2}{a}\right)^{\lambda_1}}{\lambda_1} - \frac{\left(\log\frac{\tau_1}{a}\right)^{\lambda_1}}{\lambda_1}\right).$$

Again, by the mean value theorem, we deduce that

$$0 < \frac{\left(\log\frac{\tau_2}{a}\right)^{\lambda_1}}{\lambda_1} - \frac{\left(\log\frac{\tau_1}{a}\right)^{\lambda_1}}{\lambda_1} \leqq (\tau_2-\tau_1)\left(\log\frac{b}{a}\right)^{\lambda_1-1}.$$

Hence, we have

$$\left|\int_a^{\tau_1}\left(\left(\log\frac{\tau_2}{t}\right)^{\lambda_1-1}-\left(\log\frac{\tau_1}{t}\right)^{\lambda_1-1}\right)f_1(t,u(t),v(t))\frac{dt}{t}\right|$$
$$\leqq W_1\frac{\left(\log\dfrac{\tau_2}{\tau_1}\right)^{\lambda_1}}{\lambda_1}+W_1(\tau_2-\tau_1)\left(\log\frac{b}{a}\right)^{\lambda_1-1}.$$

In summary, therefore, we find that

$$|T_1(u,v)(\tau_2)-T_1(u,v)(\tau_1)|\leqq M_1(\tau_2-\tau_1)$$
$$+\,2W_1\sum_{k=1}^{\infty}\sum_{k_1+\cdots+k_n=k}\binom{k}{k_1,\cdots,k_n}|a_1|^{k_1}\cdots|a_n|^{k_n}\frac{\left(\log\dfrac{\tau_2}{\tau_1}\right)^{k_1\alpha_1+\cdots+k_n\alpha_n}}{\Gamma(k_1\alpha_1+\cdots+k_n\alpha_n+1)}$$
$$+\,W_1\,(\tau_2-\tau_1)\sum_{k=1}^{\infty}\sum_{k_1+\cdots+k_n=k}\binom{k}{k_1,\cdots,k_n}|a_1|^{k_1}\cdots|a_n|^{k_n}$$
$$\cdot\frac{\left(\log\dfrac{b}{a}\right)^{k_1\alpha_1+\cdots+k_n\alpha_n}}{\Gamma(k_1\alpha_1+\cdots+k_n\alpha_n)}.$$

Noting that

$$2W_1\sum_{k=1}^{\infty}\sum_{k_1+\cdots+k_n=k}\binom{k}{k_1,\cdots,k_n}|a_1|^{k_1}\cdots|a_n|^{k_n}\frac{\left(\log\dfrac{\tau_2}{\tau_1}\right)^{k_1\alpha_1+\cdots+k_n\alpha_n}}{\Gamma(k_1\alpha_1+\cdots+k_n\alpha_n+1)}$$
$$\leqq 2W_1\sum_{k=1}^{\infty}\sum_{k_1+\cdots+k_n=k}\binom{k}{k_1,\cdots,k_n}|a_1|^{k_1}\cdots|a_n|^{k_n}\frac{\left(\log\dfrac{b}{a}\right)^{k_1\alpha_1+\cdots+k_n\alpha_n}}{\Gamma(k_1\alpha_1+\cdots+k_n\alpha_n+1)},$$

which implies that the series on the left-hand side is uniformly convergent on $[a,b]$, and every term in the series has the factor $\left(\log\dfrac{\tau_2}{\tau_1}\right)^{\alpha_1}$. Therefore, T_1 is equicontinuous on $C[a,b]$.

Regarding T_2, we let M_2 be a constant such that

$$|f'_{2x}|\leqq M_2.$$

Then it follows from a similar step that

$$|T_2(u,v)(\tau_2)-T_2(u,v)(\tau_1)|\leqq M_2(\tau_2-\tau_1)$$

$$+ 2W_2 \sum_{k=1}^{\infty} \sum_{k_1+\cdots+k_n=k} \binom{k}{k_1, \cdots, k_n} |b_1|^{k_1} \cdots |b_n|^{k_n} \frac{\left(\log \dfrac{\tau_2}{\tau_1}\right)^{k_1\beta_1+\cdots+k_n\beta_n}}{\Gamma(k_1\beta_1 + \cdots + k_n\beta_n + 1)}$$

$$+ W_2 (\tau_2 - \tau_1) \sum_{k=1}^{\infty} \sum_{k_1+\cdots+k_n=k} \binom{k}{k_1, \cdots, k_n} |b_1|^{k_1} \cdots |b_n|^{k_n}$$

$$\cdot \frac{\left(\log \dfrac{b}{a}\right)^{k_1\beta_1+\cdots+k_n\beta_n}}{\Gamma(k_1\beta_1 + \cdots + k_n\beta_n)}.$$

So, clearly, T_2 is also equicontinuous on $C[a, b]$. This further infers that T is a compact mapping by the Arzela–Ascoli theorem. It remains to be proven that the set

$$W = \{(u, v) \in C[a, b] \times C[a, b] : (u, v) = \lambda T(u, v) \text{ for some } 0 \leqq \lambda \leqq 1\}$$

is bounded.

For any $x \in [a, b]$,

$$u(x) = \lambda T_1(u, v)(x), \quad v(x) = \lambda T_2(u, v)(x).$$

From inequality (22.22), we have

$$\|u\|_C \leqq \|T_1(u, v)\|_C \leqq E_{(\alpha_1, \cdots, \alpha_n, 1)} \left(|a_1| \left(\log \frac{b}{a}\right)^{\alpha_1}, \cdots, |a_n| \left(\log \frac{b}{a}\right)^{\alpha_n} \right)$$
$$\cdot \left(C_0 + C_1 \|u\|_C + C_2 \|v\|_C\right), \quad \text{and}$$
$$\|v\|_C \leqq \|T_2(u, v)\|_C \leqq E_{(\beta_1, \cdots, \beta_n, 1)} \left(|b_1| \left(\log \frac{b}{a}\right)^{\beta_1}, \cdots, |b_n| \left(\log \frac{b}{a}\right)^{\beta_n} \right)$$
$$\cdot \left(C_0 + C_1 \|u\|_C + C_2 \|v\|_C\right).$$

Therefore,

$$A_1 \|u\|_C + A_2 \|v\|_C \leqq$$
$$\leqq C_0 E_{(\alpha_1, \cdots, \alpha_n, 1)} \left(|a_1| \left(\log \frac{b}{a}\right)^{\alpha_1}, \cdots, |a_n| \left(\log \frac{b}{a}\right)^{\alpha_n} \right)$$
$$+ C_0 E_{(\beta_1, \cdots, \beta_n, 1)} \left(|b_1| \left(\log \frac{b}{a}\right)^{\beta_1}, \cdots, |b_n| \left(\log \frac{b}{a}\right)^{\beta_n} \right),$$

where

$$A_1 = 1 - C_1 (E_{(\alpha_1, \cdots, \alpha_n, 1)} \left(|a_1| \left(\log \frac{b}{a}\right)^{\alpha_1}, \cdots, |a_n| \left(\log \frac{b}{a}\right)^{\alpha_n} \right)$$

$$+ \; E_{(\beta_1,\cdots,\beta_n,\,1)}\left(|b_1|\left(\log\frac{b}{a}\right)^{\beta_1},\cdots,|b_n|\left(\log\frac{b}{a}\right)^{\beta_n}\right)) > 0, \quad \text{and}$$

$$A_2 = 1 - C_2(E_{(\alpha_1,\cdots,\alpha_n,\,1)}\left(|a_1|\left(\log\frac{b}{a}\right)^{\alpha_1},\cdots,|a_n|\left(\log\frac{b}{a}\right)^{\alpha_n}\right)$$

$$+ \; E_{(\beta_1,\cdots,\beta_n,\,1)}\left(|b_1|\left(\log\frac{b}{a}\right)^{\beta_1},\cdots,|b_n|\left(\log\frac{b}{a}\right)^{\beta_n}\right)) > 0,$$

by our hypothesis.

Let

$$A_0 = \min\{A_1, A_2\} > 0,$$

then,

$$A_0(\|u\|_C + \|v\|_C) \leqq A_1\,\|u\|_C + A_2\,\|v\|_C\,,$$

and

$$\begin{aligned}\|u\|_C &+ \|v\|_C \\ &\leqq \frac{C_0}{A_0}E_{(\alpha_1,\cdots,\alpha_n,\,1)}\left(|a_1|\left(\log\frac{b}{a}\right)^{\alpha_1},\cdots,|a_n|\left(\log\frac{b}{a}\right)^{\alpha_n}\right) \\ &+ \frac{C_0}{A_0}E_{(\beta_1,\cdots,\beta_n,\,1)}\left(|b_1|\left(\log\frac{b}{a}\right)^{\beta_1},\cdots,|b_n|\left(\log\frac{b}{a}\right)^{\beta_n}\right).\end{aligned}$$

Hence, W is bounded for all $\lambda \in [0, 1]$. Using Leray-Schauder's alternative, we deduce that system (22.21) has a solution in the space $C[a, b] \times C[a, b]$.

Remark 22.1 From Theorem 22.11, we can derive that if f_1, f'_{1x}, f_2, and f'_{2x} are continuous and bounded (that is, $C_1 = C_2 = 0$), then the system (22.21) has a solution in the space $C[a, b] \times C[a, b]$.

Example 22.6 As an illustrative example, the following nonlinear Hadamard-type integro-differential system

$$\begin{cases} u(x) + 7.9\,\mathcal{J}_{a+}^{2.7}u(x) - 7\,\mathcal{J}_{a+}^{2.1}u(x) = \cos(x + u(x) + v(x)), \\ v(x) - 1.6\,\mathcal{J}_{a+}^{2.1}v(x) + \mathcal{J}_{a+}^{1.1}v(x) = \sin^2(x + u(x)), \end{cases}$$

has a solution in the space $C[a, b] \times C[a, b]$ $(0 < a < b < +\infty)$, since

$$\begin{aligned} f_1(x, y_1, y_2) &= \cos(z + y_1 + y_2) \;\text{ and} \\ f_2(x, y_1, y_2) &= \sin^2(x + y_1) \end{aligned}$$

are continuous and bounded with their partial derivatives with respect to x.

22.5.3 Summary

Using Babenko's approach, Leray-Schauder's alternative, and the multivariate Mittag-Leffler function, we have studied the existence of solutions to the nonlinear Hadamard-type integro-differential system in the Banach space $C[a, b] \times C[a, b]$. The results obtained are new and interesting. We have also included an example showing the application of the main theorem.

22.6 On the Uniqueness of the Bounded Solution with Approximations

22.6.1 Preliminaries

The operator I_θ^β is the partial Riemann-Liouville fractional integral of order $\beta > 0$ with respect to θ with initial point zero [1],

$$(I_\theta^\beta W)(\theta, \sigma) = \frac{1}{\Gamma(\beta)} \int_0^\theta (\theta - \zeta)^{\beta-1} W(\zeta, \sigma) d\zeta, \ \theta > 0,$$

and $\frac{{}_c\partial^\alpha}{\partial \theta^\alpha}$ is the partial Caputo fractional derivative of order α with respect to θ

$$\left(\frac{{}_c\partial^\alpha}{\partial \theta^\alpha} W \right)(\theta, \sigma) = \frac{1}{\Gamma(1-\alpha)} \int_0^\theta (\theta - s)^{-\alpha} W_s'(s, \sigma) ds, \quad 0 < \alpha \leq 1.$$

It follows from [8] that for $0 < \alpha \leq 1$,

$$I_\theta^\alpha \left(\frac{{}_c\partial^\alpha}{\partial \theta^\alpha} W \right)(\theta, \sigma) = W(\theta, \sigma) - W(0, \sigma).$$

We define the Banach space $C([0, 1] \times [0, b])$ with the norm for $b > 0$ as

$$\|W\| = \sup_{\theta \in [0,1], \ \sigma \in [0,b]} |W(\theta, \sigma)| \ \text{ for } \ W \in C([0, 1] \times [0, b]).$$

In this section, we consider the uniqueness of the bounded solution to the following fractional nonlinear partial integro-differential equation with the initial condition in the Caputo sense for $0 < \alpha_2 < \alpha_1 \leq 1$ in the space $C([0, 1] \times [0, b])$:

$$\begin{cases} \frac{{}_c\partial^{\alpha_1}}{\partial\theta^{\alpha_1}} W(\theta,\sigma) + C\frac{{}_c\partial^{\alpha_2}}{\partial\theta^{\alpha_2}} W(\theta,\sigma) + \sum_{i=1}^{m} \lambda_i(\sigma) I_\theta^{\gamma_i} I_\sigma^{\beta_i} W(\theta,\sigma) \\ \qquad = f(\theta,\sigma) + g(W(\theta,\sigma)), \\ W(0,\sigma) = \phi(\sigma), \quad (\theta,\sigma) \in [0,1]\times[0,b], \end{cases} \tag{22.23}$$

where $\lambda_i(\sigma)$, $f(\theta,\sigma)$, g, $\phi(\sigma)$ are functions satisfying certain conditions and C is a constant. In addition, we avoid computational difficulties in the series and construct the algorithm for finding analytic and approximate solutions to (22.23) by Adomian's decomposition method. We should note that any existing integral transform seems hard to deal with this equation due to variable coefficients and double integral operators involved.

22.6.2 Uniqueness of Bounded Solution

Theorem 22.12 *Let $\phi(\sigma)$, $\lambda_i(\sigma) \in C[0,b]$ for all $i = 1, 2, \cdots, m$, $f(\theta,\sigma) \in C([0,1]\times[0,b])$ and g be a continuous and bounded function on R (the set of all real numbers). In addition, we assume that $0 < \alpha_2 < \alpha_1 \le 1$, γ_i, $\beta_i \ge 0$ for all $i = 1, 2, \cdots, m$ and C is a constant in R. Then $W(\theta,\sigma)$ is a solution of Eq. (22.23) if and only if it is bounded and satisfies the following implicit integral equation in the space $C([0,1]\times[0,b])$:*

$$\begin{aligned} & W(\theta,\sigma) \\ &= \sum_{l=0}^{\infty}(-1)^l \sum_{\substack{l_1+\cdots+l_{m+1}=l \\ l_1\ge 0,\cdots,l_{m+1}\ge 0}} \binom{l}{l_1,\cdots,l_{m+1}} C^{l_{m+1}} I_\theta^{l_1(\alpha_1+\gamma_1)+\cdots+l_m(\alpha_1+\gamma_m)+l_{m+1}(\alpha_1-\alpha_2)+\alpha_1} \\ &\quad \cdot \left(\lambda_1(\sigma) I_\sigma^{\beta_1}\right)^{l_1} \cdots \left(\lambda_m(\sigma) I_\sigma^{\beta_m}\right)^{l_m} f(\theta,\sigma) \\ &\quad + \sum_{l=0}^{\infty}(-1)^l \sum_{\substack{l_1+\cdots+l_{m+1}=l \\ l_1\ge 0,\cdots,l_{m+1}\ge 0}} \binom{l}{l_1,\cdots,l_{m+1}} C^{l_{m+1}} I_\theta^{l_1(\alpha_1+\gamma_1)+\cdots+l_{m+1}(\alpha_1-\alpha_2)+\alpha_1} \\ &\quad \cdot \left(\lambda_1(\sigma) I_\sigma^{\beta_1}\right)^{l_1} \cdots \left(\lambda_m(\sigma) I_\sigma^{\beta_m}\right)^{l_m} g(W(\theta,\sigma)) + \sum_{l=0}^{\infty}(-1)^l \\ &\quad \cdot \sum_{\substack{l_1+\cdots+l_{m+1}=l \\ l_1\ge 0,\cdots,l_{m+1}\ge 0}} \binom{l}{l_1,\cdots,l_{m+1}} \frac{C^{l_{m+1}}\, \theta^{l_1(\alpha_1+\gamma_1)+\cdots+l_{m+1}(\alpha_1-\alpha_2)}}{\Gamma(l_1(\alpha_1+\gamma_1)+\cdots+l_{m+1}(\alpha_1-\alpha_2)+1)} \\ &\quad \cdot \left(\lambda_1(\sigma) I_\sigma^{\beta_1}\right)^{l_1} \cdots \left(\lambda_m(\sigma) I_\sigma^{\beta_m}\right)^{l_m} \phi(\sigma) + C\sum_{l=0}^{\infty}(-1)^l \\ &\quad \cdot \sum_{\substack{l_1+\cdots+l_{m+1}=l \\ l_1\ge 0,\cdots,l_{m+1}\ge 0}} \binom{l}{l_1,\cdots,l_{m+1}} \frac{C^{l_{m+1}}\, \theta^{l_1(\alpha_1+\gamma_1)+\cdots+l_{m+1}(\alpha_1-\alpha_2)+\alpha_1-\alpha_2}}{\Gamma(l_1(\alpha_1+\gamma_1)+\cdots+l_{m+1}(\alpha_1-\alpha_2)+\alpha_1-\alpha_2+1)} \\ &\quad \cdot \left(\lambda_1(\sigma) I_\sigma^{\beta_1}\right)^{l_1} \cdots \left(\lambda_m(\sigma) I_\sigma^{\beta_m}\right)^{l_m}. \end{aligned} \tag{22.24}$$

Furthermore,

$$\begin{aligned}\|W\| \leq & 2E_{(\alpha_1+\gamma_1,\cdots,\alpha_1+\gamma_m,\alpha_1-\alpha_2),\,\alpha_1+1}\left(Bb^{\beta_1},\cdots,Bb^{\beta_m},|C|\right)\\ & \cdot\left(\max_{(\theta,\sigma)\in[0,1]\times[0,b]}|f(\theta,\sigma)|+\sup_{x\in R}|g(x)|\right)\\ & +2E_{(\alpha_1+\gamma_1,\cdots,\alpha_1+\gamma_m,\alpha_1-\alpha_2),\,1}\left(Bb^{\beta_1},\cdots,Bb^{\beta_m},|C|\right)\max_{\sigma\in[0,b]}|\phi(\sigma)|\\ & +2E_{(\alpha_1+\gamma_1,\cdots,\alpha_1+\gamma_m,\alpha_1-\alpha_2),\,\alpha_1-\alpha_2+1}\left(Bb^{\beta_1},\cdots,Bb^{\beta_m},|C|\right)<+\infty,\end{aligned}$$

where B is a positive constant satisfying

$$\max_{\sigma\in[0,b]}|\lambda_i(\sigma)|\leq B,\quad \textit{for } i=1,2,\cdots,m.$$

Proof Applying the operator $I_\theta^{\alpha_1}$ to both sides of Eq. (22.23), we get

$$\begin{aligned}& I_\theta^{\alpha_1}\,{}^c\frac{\partial^{\alpha_1}}{\partial\theta^{\alpha_1}}W(\theta,\sigma)+CI_\theta^{\alpha_1-\alpha_2}I_\theta^{\alpha_2}\,{}^c\frac{\partial^{\alpha_2}}{\partial\theta^{\alpha_2}}W(\theta,\sigma)+\sum_{i=1}^m\lambda_i(\sigma)I_\theta^{\alpha_1+\gamma_i}I_\sigma^{\beta_i}W(\theta,\sigma)\\ &\quad =I_\theta^{\alpha_1}f(\theta,\sigma)+I_\theta^{\alpha_1}g(W(\theta,\sigma)).\end{aligned}$$

This implies that

$$\begin{aligned}& W(\theta,\sigma)-\phi(\sigma)+CI_\theta^{\alpha_1-\alpha_2}(W(\theta,\sigma)-\phi(\sigma))+\sum_{i=1}^m\lambda_i(\sigma)I_\theta^{\alpha_1+\gamma_i}I_\sigma^{\beta_i}W(\theta,\sigma)\\ &\quad =I_\theta^{\alpha_1}f(\theta,\sigma)+I_\theta^{\alpha_1}g(W(\theta,\sigma)).\end{aligned}$$

Hence,

$$\begin{aligned}&\left(1+CI_\theta^{\alpha_1-\alpha_2}+\sum_{i=1}^m\lambda_i(\sigma)I_\theta^{\alpha_1+\gamma_i}I_\sigma^{\beta_i}\right)W(\theta,\sigma)\\ &\quad =I_\theta^{\alpha_1}f(\theta,\sigma)+I_\theta^{\alpha_1}g(W(\theta,\sigma))+\phi(\sigma)\left(1+C\frac{\theta^{\alpha_1-\alpha_2}}{\Gamma(\alpha_1-\alpha_2+1)}\right).\end{aligned}$$

By Babenko's approach and the multinomial theorem,

$$\begin{aligned}& W(\theta,\sigma)\\ &=\left(1+CI_\theta^{\alpha_1-\alpha_2}+\sum_{i=1}^m\lambda_i(\sigma)I_\theta^{\alpha_1+\gamma_i}I_\sigma^{\beta_i}\right)^{-1}\\ &\quad\cdot\left(I_\theta^{\alpha_1}f(\theta,\sigma)+I_\theta^{\alpha_1}g(W(\theta,\sigma))+\phi(\sigma)\left(1+C\frac{\theta^{\alpha_1-\alpha_2}}{\Gamma(\alpha_1-\alpha_2+1)}\right)\right)\end{aligned}$$

$$
\begin{aligned}
&= \sum_{l=0}^{\infty}(-1)^l \left(C I_\theta^{\alpha_1-\alpha_2} + \sum_{i=1}^{m} \lambda_i(\sigma) I_\theta^{\alpha_1+\gamma_i} I_\sigma^{\beta_i} \right)^l \\
&\quad \cdot \left(I_\theta^{\alpha_1} f(\theta,\sigma) + I_\theta^{\alpha_1} g(W(\theta,\sigma)) + \phi(\sigma)\left(1 + C\frac{\theta^{\alpha_1-\alpha_2}}{\Gamma(\alpha_1-\alpha_2+1)}\right)\right) \\
&= \sum_{l=0}^{\infty}(-1)^l \sum_{\substack{l_1+\cdots+l_{m+1}=l\\ l_1\ge 0,\cdots,l_{m+1}\ge 0}} \binom{l}{l_1,\cdots,l_{m+1}} \left(\lambda_1(\sigma) I_\theta^{\alpha_1+\gamma_1} I_\sigma^{\beta_1}\right)^{l_1} \cdots \\
&\quad \cdot \left(\lambda_m(\sigma) I_\theta^{\alpha_1+\gamma_m} I_\sigma^{\beta_m}\right)^{l_m} \left(C I_\theta^{\alpha_1-\alpha_2}\right)^{l_{m+1}} \\
&\quad \cdot \left(I_\theta^{\alpha_1} f(\theta,\sigma) + I_\theta^{\alpha_1} g(W(\theta,\sigma)) + \phi(\sigma)\left(1 + C\frac{\theta^{\alpha_1-\alpha_2}}{\Gamma(\alpha_1-\alpha_2+1)}\right)\right) \\
&= \sum_{l=0}^{\infty}(-1)^l \sum_{\substack{l_1+\cdots+l_{m+1}=l\\ l_1\ge 0,\cdots,l_{m+1}\ge 0}} \binom{l}{l_1,\cdots,l_{m+1}} C^{l_{m+1}} I_\theta^{l_1(\alpha_1+\gamma_1)+\cdots+l_m(\alpha_1+\gamma_m)+l_{m+1}(\alpha_1-\alpha_2)+\alpha_1} \\
&\quad \cdot \left(\lambda_1(\sigma) I_\sigma^{\beta_1}\right)^{l_1} \cdots \left(\lambda_m(\sigma) I_\sigma^{\beta_m}\right)^{l_m} f(\theta,\sigma) \\
&\quad + \sum_{l=0}^{\infty}(-1)^l \sum_{\substack{l_1+\cdots+l_{m+1}=l\\ l_1\ge 0,\cdots,l_{m+1}\ge 0}} \binom{l}{l_1,\cdots,l_{m+1}} C^{l_{m+1}} I_\theta^{l_1(\alpha_1+\gamma_1)+\cdots+l_{m+1}(\alpha_1-\alpha_2)+\alpha_1} \\
&\quad \cdot \left(\lambda_1(\sigma) I_\sigma^{\beta_1}\right)^{l_1} \cdots \left(\lambda_m(\sigma) I_\sigma^{\beta_m}\right)^{l_m} g(W(\theta,\sigma)) + \sum_{l=0}^{\infty}(-1)^l \\
&\quad \cdot \sum_{\substack{l_1+\cdots+l_{m+1}=l\\ l_1\ge 0,\cdots,l_{m+1}\ge 0}} \binom{l}{l_1,\cdots,l_{m+1}} \frac{C^{l_{m+1}}\, \theta^{l_1(\alpha_1+\gamma_1)+\cdots+l_{m+1}(\alpha_1-\alpha_2)}}{\Gamma(l_1(\alpha_1+\gamma_1)+\cdots+l_{m+1}(\alpha_1-\alpha_2)+1)} \\
&\quad \cdot \left(\lambda_1(\sigma) I_\sigma^{\beta_1}\right)^{l_1} \cdots \left(\lambda_m(\sigma) I_\sigma^{\beta_m}\right)^{l_m} \phi(\sigma) + C\sum_{l=0}^{\infty}(-1)^l \\
&\quad \cdot \sum_{\substack{l_1+\cdots+l_{m+1}=l\\ l_1\ge 0,\cdots,l_{m+1}\ge 0}} \binom{l}{l_1,\cdots,l_{m+1}} \frac{C^{l_{m+1}}\, \theta^{l_1(\alpha_1+\gamma_1)+\cdots+l_{m+1}(\alpha_1-\alpha_2)+\alpha_1-\alpha_2}}{\Gamma(l_1(\alpha_1+\gamma_1)+\cdots+l_{m+1}(\alpha_1-\alpha_2)+\alpha_1-\alpha_2+1)} \\
&\quad \cdot \left(\lambda_1(\sigma) I_\sigma^{\beta_1}\right)^{l_1} \cdots \left(\lambda_m(\sigma) I_\sigma^{\beta_m}\right)^{l_m}.
\end{aligned}
$$

Clearly, all above steps are reversible. It remains to be shown that $W \in C([0,1] \times [0,b])$. Let B be a positive constant such that

$$
\max_{\sigma\in[0,b]} |\lambda_i(\sigma)| \le B, \quad \text{for } i = 1, 2, \cdots, m.
$$

Then,

$$
\begin{aligned}
&\|W\| \\
&\le \sum_{l=0}^{\infty} \sum_{\substack{l_1+\cdots+l_{m+1}=l\\ l_1\ge 0,\cdots,l_{m+1}\ge 0}} \binom{l}{l_1,\cdots,l_{m+1}} \frac{|C|^{l_{m+1}} B^{l_1}\cdots B^{l_m}}{\Gamma(l_1(\alpha_1+\gamma_1)+\cdots+l_{m+1}(\alpha_1-\alpha_2)+\alpha_1+1)}
\end{aligned}
$$

$$
\cdot \frac{b^{l_1\beta_1+\cdots+l_m\beta_m}}{\Gamma(l_1\beta_1+\cdots+l_m\beta_m+1)} \max_{(\theta,\sigma)\in[0,1]\times[0,b]} |f(\theta,\sigma)|
+\sum_{l=0}^{\infty} \sum_{\substack{l_1+\cdots+l_{m+1}=l \\ l_1\geq 0,\cdots,l_{m+1}\geq 0}} \binom{l}{l_1,\cdots,l_{m+1}} \frac{|C|^{l_{m+1}} B^{l_1}\ldots B^{l_m}}{\Gamma(l_1(\alpha_1+\gamma_1)+\cdots+l_{m+1}(\alpha_1-\alpha_2)+\alpha_1+1)}
$$

$$
\cdot \frac{b^{l_1\beta_1+\cdots+l_m\beta_m}}{\Gamma(l_1\beta_1+\cdots+l_m\beta_m+1)} \sup_{x\in R}|g(x)| + \sum_{l=0}^{\infty} \sum_{\substack{l_1+\cdots+l_{m+1}=l \\ l_1\geq 0,\cdots,l_{m+1}\geq 0}} \binom{l}{l_1,\cdots,l_{m+1}}
$$

$$
\frac{|C|^{l_{m+1}} B^{l_1}\ldots B^{l_m} b^{l_1\beta_1+\cdots+l_m\beta_m}}{\Gamma(l_1(\alpha_1+\gamma_1)+\cdots+l_{m+1}(\alpha_1-\alpha_2)+1)\,\Gamma(l_1\beta_1+\cdots+l_m\beta_m+1)} \max_{\sigma\in[0,b]}|\phi(\sigma)|
$$

$$
+|C|\sum_{l=0}^{\infty} \sum_{\substack{l_1+\cdots+l_{m+1}=l \\ l_1\geq 0,\cdots,l_{m+1}\geq 0}} \binom{l}{l_1,\cdots,l_{m+1}}
$$

$$
\cdot \frac{|C|^{l_{m+1}} (Bb^{\beta_1})^{l_1}\ldots (Bb^{\beta_m})^{l_m}}{\Gamma(l_1(\alpha_1+\gamma_1)+\cdots+l_{m+1}(\alpha_1-\alpha_2)+\alpha_1-\alpha_2+1)\,\Gamma(l_1\beta_1+\cdots+l_m\beta_m+1)}.
$$

Clearly,

$$
\Gamma(l_1\beta_1+\cdots+l_m\beta_m+1) \geq 1/2.
$$

Therefore,

$$
\begin{aligned}
\|W\| \leq & 2E_{(\alpha_1+\gamma_1,\cdots,\alpha_1+\gamma_m,\alpha_1-\alpha_2),\,\alpha_1+1}\left(Bb^{\beta_1},\cdots,Bb^{\beta_m},|C|\right) \\
& \left(\max_{(\theta,\sigma)\in[0,1]\times[0,b]}|f(\theta,\sigma)| + \sup_{x\in R}|g(x)|\right) \\
& + 2E_{(\alpha_1+\gamma_1,\cdots,\alpha_1+\gamma_m,\alpha_1-\alpha_2),\,1}\left(Bb^{\beta_1},\cdots,Bb^{\beta_m},|C|\right) \max_{\sigma\in[0,b]}|\phi(\sigma)| \\
& + 2E_{(\alpha_1+\gamma_1,\cdots,\alpha_1+\gamma_m,\alpha_1-\alpha_2),\,\alpha_1-\alpha_2+1}\left(Bb^{\beta_1},\cdots,Bb^{\beta_m},|C|\right) < +\infty.
\end{aligned}
$$

This completes the proof of Theorem 22.12.

Theorem 22.13 *Let* $\phi(\sigma),\ \lambda_i(\sigma) \in C[0,b]$ *for all* $i = 1, 2, \cdots, m,\ f(\theta,\sigma) \in C([0,1]\times[0,b])$ *and* g *be a continuous and bounded function on* R *satisfying the following Lipschitz condition for a positive constant* $\mathcal{L}$:

$$
|g(x)-g(y)| \leq \mathcal{L}|x-y|, \quad x,\ y \in R.
$$

In addition, we assume that $0 < \alpha_2 < \alpha_1 \leq 1,\ \gamma_i,\ \beta_i \geq 0$ *for all* $i = 1, 2, \cdots, m$, C *is a constant in* R *and*

$$
q = 2\mathcal{L}E_{(\alpha_1+\gamma_1,\cdots,\alpha_1+\gamma_m,\alpha_1-\alpha_2),\,\alpha_1+1}\left(Bb^{\beta_1},\cdots,Bb^{\beta_m},|C|\right) < 1,
$$

where B *is defined in Theorem 22.12. Then Eq. (22.23) has a unique solution in the space* $C([0,1]\times[0,b])$.

Proof We define a nonlinear mapping $\mathcal{T}$ over the space $C([0, 1] \times [0, b])$ by

$$
\begin{aligned}
&(\mathcal{T}W)(\theta, \sigma) \\
&= \sum_{l=0}^{\infty}(-1)^l \sum_{\substack{l_1+\cdots+l_{m+1}=l \\ l_1\geq 0,\cdots,l_{m+1}\geq 0}} \binom{l}{l_1, \cdots, l_{m+1}} C^{l_{m+1}} I_\theta^{l_1(\alpha_1+\gamma_1)+\cdots+l_m(\alpha_1+\gamma_m)+l_{m+1}(\alpha_1-\alpha_2)+\alpha_1} \\
&\quad \cdot \left(\lambda_1(\sigma) I_\sigma^{\beta_1}\right)^{l_1} \cdots \left(\lambda_m(\sigma) I_\sigma^{\beta_m}\right)^{l_m} f(\theta, \sigma) \\
&\quad + \sum_{l=0}^{\infty}(-1)^l \sum_{\substack{l_1+\cdots+l_{m+1}=l \\ l_1\geq 0,\cdots,l_{m+1}\geq 0}} \binom{l}{l_1, \cdots, l_{m+1}} C^{l_{m+1}} I_\theta^{l_1(\alpha_1+\gamma_1)+\cdots+l_{m+1}(\alpha_1-\alpha_2)+\alpha_1} \\
&\quad \cdot \left(\lambda_1(\sigma) I_\sigma^{\beta_1}\right)^{l_1} \cdots \left(\lambda_m(\sigma) I_\sigma^{\beta_m}\right)^{l_m} g(W(\theta, \sigma)) + \sum_{l=0}^{\infty}(-1)^l \\
&\quad \cdot \sum_{\substack{l_1+\cdots+l_{m+1}=l \\ l_1\geq 0,\cdots,l_{m+1}\geq 0}} \binom{l}{l_1, \cdots, l_{m+1}} \frac{C^{l_{m+1}}\, \theta^{l_1(\alpha_1+\gamma_1)+\cdots+l_{m+1}(\alpha_1-\alpha_2)}}{\Gamma(l_1(\alpha_1+\gamma_1)+\cdots+l_{m+1}(\alpha_1-\alpha_2)+1)} \\
&\quad \cdot \left(\lambda_1(\sigma) I_\sigma^{\beta_1}\right)^{l_1} \cdots \left(\lambda_m(\sigma) I_\sigma^{\beta_m}\right)^{l_m} \phi(\sigma) + C\sum_{l=0}^{\infty}(-1)^l \\
&\quad \cdot \sum_{\substack{l_1+\cdots+l_{m+1}=l \\ l_1\geq 0,\cdots,l_{m+1}\geq 0}} \binom{l}{l_1, \cdots, l_{m+1}} \frac{C^{l_{m+1}}\, \theta^{l_1(\alpha_1+\gamma_1)+\cdots+l_{m+1}(\alpha_1-\alpha_2)+\alpha_1-\alpha_2}}{\Gamma(l_1(\alpha_1+\gamma_1)+\cdots+l_{m+1}(\alpha_1-\alpha_2)+\alpha_1-\alpha_2+1)} \\
&\quad \cdot \left(\lambda_1(\sigma) I_\sigma^{\beta_1}\right)^{l_1} \cdots \left(\lambda_m(\sigma) I_\sigma^{\beta_m}\right)^{l_m}.
\end{aligned}
$$

It follows from Theorem 22.12 that $\mathcal{T}W \in C([0, 1] \times [0, b])$. We are going to show that $\mathcal{T}$ is contractive. Indeed, for $W_1, W_2 \in C([0, 1] \times [0, b])$, we come to

$$
\begin{aligned}
&\mathcal{T}W_1 - \mathcal{T}W_2 \\
&= \sum_{l=0}^{\infty}(-1)^l \sum_{\substack{l_1+\cdots+l_{m+1}=l \\ l_1\geq 0,\cdots,l_{m+1}\geq 0}} \binom{l}{l_1, \cdots, l_{m+1}} C^{l_{m+1}} I_\theta^{l_1(\alpha_1+\gamma_1)+\cdots+l_{m+1}(\alpha_1-\alpha_2)+\alpha_1} \\
&\quad \cdot \left(\lambda_1(\sigma) I_\sigma^{\beta_1}\right)^{l_1} \cdots \left(\lambda_m(\sigma) I_\sigma^{\beta_m}\right)^{l_m} g(W_1(\theta, \sigma)) \\
&\quad - \sum_{l=0}^{\infty}(-1)^l \sum_{\substack{l_1+\cdots+l_{m+1}=l \\ l_1\geq 0,\cdots,l_{m+1}\geq 0}} \binom{l}{l_1, \cdots, l_{m+1}} C^{l_{m+1}} I_\theta^{l_1(\alpha_1+\gamma_1)+\cdots+l_{m+1}(\alpha_1-\alpha_2)+\alpha_1} \\
&\quad \cdot \left(\lambda_1(\sigma) I_\sigma^{\beta_1}\right)^{l_1} \cdots \left(\lambda_m(\sigma) I_\sigma^{\beta_m}\right)^{l_m} g(W_2(\theta, \sigma)),
\end{aligned}
$$

and

$$
\begin{aligned}
&\|\mathcal{T}W_1 - \mathcal{T}W_2\| \\
&\leq 2\mathcal{L}E_{(\alpha_1+\gamma_1,\cdots,\alpha_1+\gamma_m,\alpha_1-\alpha_2),\,\alpha_1+1}\left(Bb^{\beta_1},\cdots,Bb^{\beta_m},|C|\right)\|W_1 - W_2\| \\
&= q\,\|W_1 - W_2\|,
\end{aligned}
$$

from the proof of Theorem 22.12 and noting that

$$|g(W_1) - g(W_2)| \leq \mathcal{L}|W_1 - W_2|.$$

Since $q < 1$, Eq. (22.23) has a unique solution in the space $C([0,1] \times [0,b])$ by Banach's contractive principle. This completes the proof of Theorem 22.13.

22.6.3 Analytic and Approximate Solutions

We begin deriving analytic and approximate solutions to Eq. (22.23) from the implicit integral equation (22.24) obtained previously using an initial value $W_0 = 0$ and the following recurrence (iterative method):

$$
\begin{aligned}
&W_{n+1}(\theta,\sigma) \\
&= \sum_{l=0}^{\infty}(-1)^l \sum_{\substack{l_1+\cdots+l_{m+1}=l \\ l_1\geq 0,\cdots,l_{m+1}\geq 0}} \binom{l}{l_1,\cdots,l_{m+1}} C^{l_{m+1}} I_\theta^{l_1(\alpha_1+\gamma_1)+\cdots+l_m(\alpha_1+\gamma_m)+l_{m+1}(\alpha_1-\alpha_2)+\alpha_1} \\
&\quad\cdot\left(\lambda_1(\sigma)I_\sigma^{\beta_1}\right)^{l_1}\cdots\left(\lambda_m(\sigma)I_\sigma^{\beta_m}\right)^{l_m} f(\theta,\sigma) \\
&\quad+\sum_{l=0}^{\infty}(-1)^l \sum_{\substack{l_1+\cdots+l_{m+1}=l \\ l_1\geq 0,\cdots,l_{m+1}\geq 0}} \binom{l}{l_1,\cdots,l_{m+1}} C^{l_{m+1}} I_\theta^{l_1(\alpha_1+\gamma_1)+\cdots+l_{m+1}(\alpha_1-\alpha_2)+\alpha_1} \\
&\quad\cdot\left(\lambda_1(\sigma)I_\sigma^{\beta_1}\right)^{l_1}\cdots\left(\lambda_m(\sigma)I_\sigma^{\beta_m}\right)^{l_m} g(W_n(\theta,\sigma)) + \sum_{l=0}^{\infty}(-1)^l \\
&\quad\cdot\sum_{\substack{l_1+\cdots+l_{m+1}=l \\ l_1\geq 0,\cdots,l_{m+1}\geq 0}} \binom{l}{l_1,\cdots,l_{m+1}} \frac{C^{l_{m+1}}\,\theta^{l_1(\alpha_1+\gamma_1)+\cdots+l_{m+1}(\alpha_1-\alpha_2)}}{\Gamma(l_1(\alpha_1+\gamma_1)+\cdots+l_{m+1}(\alpha_1-\alpha_2)+1)} \\
&\quad\cdot\left(\lambda_1(\sigma)I_\sigma^{\beta_1}\right)^{l_1}\cdots\left(\lambda_m(\sigma)I_\sigma^{\beta_m}\right)^{l_m} \phi(\sigma) + C\sum_{l=0}^{\infty}(-1)^l \\
&\quad\cdot\sum_{\substack{l_1+\cdots+l_{m+1}=l \\ l_1\geq 0,\cdots,l_{m+1}\geq 0}} \binom{l}{l_1,\cdots,l_{m+1}} \frac{C^{l_{m+1}}\,\theta^{l_1(\alpha_1+\gamma_1)+\cdots+l_{m+1}(\alpha_1-\alpha_2)+\alpha_1-\alpha_2}}{\Gamma(l_1(\alpha_1+\gamma_1)+\cdots+l_{m+1}(\alpha_1-\alpha_2)+\alpha_1-\alpha_2+1)} \\
&\quad\cdot\left(\lambda_1(\sigma)I_\sigma^{\beta_1}\right)^{l_1}\cdots\left(\lambda_m(\sigma)I_\sigma^{\beta_m}\right)^{l_m},
\end{aligned}
$$

for $n = 0, 1, \cdots$.

Let W_{exact} be an exact solution to Eq. (22.23). Then it follows from Theorem 22.13 that

$$\begin{aligned}
&\|W_{n+1}-W_{exact}\| \le q\,\|W_n-W_{exact}\| \le q^2\,\|W_{n-1}-W_{exact}\| \le \cdots \le q^n\,\|W_1-W_{exact}\| \\
&\le 2q^n E_{(\alpha_1+\gamma_1,\cdots,\alpha_1+\gamma_m,\alpha_1-\alpha_2),\alpha_1+1}\left(Bb^{\beta_1},\cdots,Bb^{\beta_m},|C|\right)\|g(W_0)-g(W_{exact})\| \\
&\le 4q^n E_{(\alpha_1+\gamma_1,\cdots,\alpha_1+\gamma_m,\alpha_1-\alpha_2),\alpha_1+1}\left(Bb^{\beta_1},\cdots,Bb^{\beta_m},|C|\right)\sup_{x\in R}|g(x)|,
\end{aligned}$$

which is the absolute error between W_{n+1} and W_{exact}.

We shall apply Adomian's decomposition method to construct the algorithm for finding analytic and approximate solutions to Eq. (22.23).

It follows from the proof of Theorem 22.12 that

$$\begin{aligned}
W(\theta,\sigma) = {}& \phi(\sigma)\left(1+\frac{C\theta^{\alpha_1-\alpha_2}}{\Gamma(\alpha_1-\alpha_2+1)}\right) - C I_\theta^{\alpha_1-\alpha_2} W(\theta,\sigma) - \sum_{i=1}^m \lambda_i(\sigma) I_\theta^{\alpha_1+\gamma_i} I_\sigma^{\beta_i} W(\theta,\sigma) \\
&+ I_\theta^{\alpha_1} f(\theta,\sigma) + I_\theta^{\alpha_1} g(W(\theta,\sigma)).
\end{aligned} \tag{22.25}$$

Let

$$W(\theta,\sigma) = W_1(\theta,\sigma) + W_2(\theta,\sigma) + W_3(\theta,\sigma) + \cdots, \tag{22.26}$$

by noting that $W_0 = 0$, and the nonlinear term

$$g(W(\theta,\sigma)) = A_1(W_1) + A_2(W_1, W_2) + \cdots. \tag{22.27}$$

Substituting (22.26) and (22.27) into (22.25), we have

$$\begin{aligned}
&W_1 + W_2 + W_3 + \cdots \\
&= \phi(\sigma)\left(1+\frac{C\theta^{\alpha_1-\alpha_2}}{\Gamma(\alpha_1-\alpha_2+1)}\right) + I_\theta^{\alpha_1} f(\theta,\sigma) - C I_\theta^{\alpha_1-\alpha_2}(W_1+W_2+W_3+\cdots) \\
&\quad - \sum_{i=1}^m \lambda_i(\sigma) I_\theta^{\alpha_1+\gamma_i} I_\sigma^{\beta_i}(W_1+W_2+W_3+\cdots) + I_\theta^{\alpha_1}(A_1+A_2+A_3+\cdots),
\end{aligned}$$

which implies by Adomian's decomposition method that

$$\begin{aligned}
W_1 &= \phi(\sigma)\left(1+\frac{C\theta^{\alpha_1-\alpha_2}}{\Gamma(\alpha_1-\alpha_2+1)}\right) + I_\theta^{\alpha_1} f(\theta,\sigma), \\
W_{k+1} &= -C I_\theta^{\alpha_1-\alpha_2} W_k - \sum_{i=1}^m \lambda_i(\sigma) I_\theta^{\alpha_1+\gamma_i} I_\sigma^{\beta_i} W_k + I_\theta^{\alpha_1} A_k,
\end{aligned} \tag{22.28}$$

for $k = 1, 2, \cdots$.

Here are a couple of examples to establish A_k from a given nonlinear function $g(W)$. First we assume that $g(W) = W\bar{W} = |W|^2$. Then,

$$(W_1 + W_2 + \cdots)(\bar{W_1} + \bar{W_2} + \cdots) = A_1 + A_2 + \cdots ,$$
$$= |W_1|^2 + 2W_1\bar{W_2} + \cdots + \sum_{i=0}^{n-1} W_{i+1}\bar{W}_{n-i} + \cdots$$

which derives that

$$A_1 = |W_1|^2,$$
$$A_2 = W_1\bar{W_2} + \bar{W_2}W_1 = 2W_1\bar{W_2},$$
$$\vdots$$
$$A_n = \sum_{i=0}^{n-1} W_{i+1}\bar{W}_{n-i},$$

for $n = 1, 2, \cdots$ from [25].

Secondly, we consider $g(W) = \sin W$. Using Taylor's series, we get

$$g(W) = \sin W \approx W - \frac{W^3}{3!} + \frac{W^5}{5!}.$$

Thus,

$$(W_1 + W_2 + \cdots) - \frac{(W_1 + W_2 + \cdots)^3}{3!} + \frac{(W_1 + W_2 + \cdots)^5}{5!}$$
$$= A_1 + A_2 + \cdots .$$

This infers that

$$A_1 = W_1 - \frac{W_1^3}{3!} + \frac{W_1^5}{5!},$$
$$A_2 = W_2 - \frac{W_1^2 W_2}{2!} + \frac{W_1^4 W_2}{4!},$$
$$\vdots$$

In summary, we are able to find analytic and approximate solutions to Eq. (22.23) as

$$W = W_1 + W_2 + W_3 + \cdots + W_n,$$

from (22.28) and recurrence of A_n for certain nonlinear functions $g(W)$.

22.6.4 Example

Example 22.7 Consider the following fractional nonlinear partial integro-differential equation with the initial condition:

$$\begin{cases} \frac{\partial}{\partial\theta}W(\theta,\sigma)+2\frac{{}^{c}\partial^{0.5}}{\partial\theta^{0.5}}W(\theta,\sigma)+(\sigma^2+1)I_{\theta}I_{\sigma}^{0.5}W(\theta,\sigma) \\ \qquad =\theta+\frac{1}{22216010}\cos(W(\theta,\sigma)), \\ W(0,\sigma)=\sigma, \quad (\theta,\sigma)\in[0,1]\times[0,1]. \end{cases} \tag{22.29}$$

Then Eq. (22.29) has a unique solution in the space $C([0,1]\times[0,1])$.

Proof Clearly, $g(W)=\frac{1}{22216010}\cos(W(\theta,\sigma))$ is a continuous and bounded function satisfying the following Lipschitz condition for $W\in C([0,1]\times[0,1])$:

$$|g(x)-g(y)|\leq\frac{1}{22216010}|x-y|, \quad x,y\in R.$$

Furthermore,

$$\max_{\sigma\in[0,1]}|\sigma^2+1|\leq B=2, \quad C=2,$$

and

$$\begin{aligned} q &= 2\mathcal{L}E_{(\alpha_1+\gamma_1,\cdots,\alpha_1+\gamma_m,\alpha_1-\alpha_2),\,\alpha_1+1}\left(Bb^{\beta_1},\cdots,Bb^{\beta_m},|C|\right) \\ &= \frac{2}{22216010}E_{(2,\,0.5),\,2}(2,2)=\frac{2}{22216010}\sum_{l=0}^{\infty}\sum_{l_1+l_2=l}\binom{l}{l_1,l_2}\frac{2^{l_1}2^{l_2}}{\Gamma(2l_1+0.5l_2+2)}. \end{aligned}$$

Using

$$\sum_{l_1+l_2=l}\binom{l}{l_1,l_2}=2^l,$$
$$\Gamma(2l_1+0.5l_2+2)=\Gamma(1.5l_1+0.5l+2)\geq\Gamma(0.5l+2),$$

we deduce that

$$q\leq\frac{2}{22216010}\sum_{l=0}^{\infty}\frac{4^l}{\Gamma(0.5l+2)}\approx\frac{2*1.1108*10^6}{22216010}\approx 0.1<1,$$

using online calculators from the site https://www.wolframalpha.com/ (accessed on 2 March 2023). By Theorem 22.13, Eq. (22.29) has a unique solution in the space $C([0,1]\times[0,1])$.

Finally, we are going to find an analytic and approximate solution to (22.29) using Recurrence (22.28). Let

$$g(W) = \frac{1}{22216010}\cos W \approx \frac{1}{22216010}\left(1 - \frac{W^2}{2!} + \frac{W^4}{4!}\right) = A_1 + A_2 + \cdots,$$

and

$$W = W_1 + W_2 + \cdots.$$

This infers that

$$\begin{aligned} A_1 &= \frac{1}{22216010}\left(1 - \frac{W_1^2}{2!} + \frac{W_1^4}{4!}\right), \\ A_2 &= \frac{1}{22216010}\left(-\frac{W_1 W_2}{1!} + \frac{W_1^3 W_2}{3!}\right), \\ &\vdots \end{aligned}$$

Moreover,

$$W_1 = \sigma\left(1 + \frac{2\,\theta^{0.5}}{\Gamma(0.5+1)}\right) + I_\theta \theta = \sigma\left(1 + \frac{4\,\theta^{0.5}}{\sqrt{\pi}}\right) + \frac{\theta^2}{2!},$$

and

$$W_2 = -2I_\theta^{0.5} W_1 - (\sigma^2+1) I_\theta^2 I_\sigma^{0.5} W_1 + I_\theta A_1.$$

By MATHEMATICA and manually simplifying, we finally get

$$\begin{aligned} W_{appro} \approx & W_1 + W_2 \\ = & \sigma + \frac{\theta(24 - 2132736960\sigma - 12\sigma^2 + \sigma^4)}{533184240} + \frac{2\theta^{3/2}\sigma^2(-6+\sigma^2)}{99972045\sqrt{\pi}} \\ & + \theta^2\left(\frac{1}{2} + \frac{\sigma^2(-2+\sigma^2)}{11108005\pi} - \frac{2\sigma^{3/2}(1+\sigma^2)}{3\sqrt{\pi}}\right) \\ & + \theta^{5/2}\left(-\frac{16}{15\sqrt{\pi}} + \frac{32\sigma^4}{166620075\pi^{3/2}} - \frac{64\sigma^{3/2}(1+\sigma^2)}{45\pi}\right) \\ & + \theta^3\left(\frac{16\sigma^4}{99972045\pi^2} + \frac{\sigma(-6+\sigma^2)}{799776360}\right) + \frac{\theta^{7/2}\sigma(-2+\sigma^2)}{77756035\sqrt{\pi}} \\ & + \theta^4\left(\frac{\sigma^3}{22216010\pi} - \frac{\sqrt{\sigma}(1+\sigma^2)}{12\sqrt{\pi}}\right) + \frac{16\theta^{9/2}\sigma^3}{299916135\pi^{3/2}} + \frac{\theta^5(-2+\sigma^2)}{1777280800} \\ & + \frac{\theta^{11/2}\sigma^2}{244376110\sqrt{\pi}} + \frac{\theta^6\sigma^2}{133296060\pi} + \frac{\theta^7\sigma}{7464579360} + \frac{\theta^{15/2}\sigma}{1999440900\sqrt{\pi}} + \frac{\theta^9}{76778530560}, \end{aligned}$$

which is an approximation in a separate form of θ and σ arranged in an increasing power of θ from 0 to 9. Clearly,

$$W_{appro}(0, \sigma) = \sigma,$$

which demonstrates that the initial condition is satisfied.

Clearly, the absolute error is

$$\left\| W_{appro} - W_{exact} \right\| \leq 4 * 0.1 * \frac{1.1168 * 10^6}{22216010} \approx 0.02. \tag{22.30}$$

Remark 22.2 Since the exact solution to Eq. (22.29) is unknown we are unable to set up a table showing absolute errors between the approximate and exact solutions at particular values. In addition, there is no existing research available to compare these with as the equation is new. However, Inequality (22.30) has plainly indicated the efficiency of our proposed method for the first two-term approximation only.

22.6.5 Summary

We have obtained the sufficient condition for the uniqueness of the bounded solution to the new Cauchy problem of the fractional nonlinear partial integro-differential equation (22.23) using the multivariate Mittag-Leffler function as well as Banach's contractive principle. In addition, we constructed the algorithm for finding analytic and approximate solutions based on Adomian's decomposition method and presented the approximation order analysis, with one example showing the applications of the main results derived. Clearly, the technique used in this work can be applied to other nonlinear differential equations with initial or boundary conditions as well as integral equations with variable coefficients.

22.7 Uniqueness of the Partial Integro–Differential Equations

22.7.1 Preliminaries

Let $0 < \omega_i < \infty$ for $i = 1, 2, \cdots, n$ and $x \in \Omega = [0, \omega_1] \times [0, \omega_2] \times \cdots \times [0, \omega_n] \subset R^n$. Let I_k^β be the partial Riemann-Liouville fractional integral of order $\beta \in R^+$ with respect to $x_k \in [0, \omega_k]$, with initial point zero [1],

$$(I_k^\beta u)(x) = \frac{1}{\Gamma(\beta)} \int_0^{x_k} (x_k - s)^{\beta - 1} u(x_1, \cdots, x_{k-1}, s, x_{k+1}, \cdots, x_n) ds,$$

for $k = 1, 2, \cdots, n$.

In particular for $\beta = 0$,

$$(I_k^0 u)(x) = u(x).$$

Clearly for $\beta_{ik} \geq 0$,

$$I_1^{\beta_{1k}} I_2^{\beta_{2k}} \cdots I_n^{\beta_{nk}} u(x)$$
$$= \frac{1}{\Gamma(\beta_{1k}) \cdots \Gamma(\beta_{nk})} \int_0^{x_1} \cdots \int_0^{x_n} (x_1 - s_1)^{\beta_{1k}-1} \cdots (x_n - s_n)^{\beta_{nk}-1}$$
$$u(s_1, \cdots, s_n) ds_n \cdots ds_1,$$

which is regarded as the partial Riemann-Liouville fractional integral with order $\beta_{1k} + \cdots + \beta_{nk}$ (β_{ik}th order in x_i-direction for $i = 1, 2, \cdots, n$) [26].

Let $t \in [0, t_0]$ with $0 < t_0 < \infty$. The space $L([0, t_0] \times \Omega)$ of Lebesgue integrable functions on $[0, t_0] \times \Omega$ is defined as

$$L([0, t_0] \times \Omega) = \left\{ u(t, x) : \ \|u\|_L = \int_{[0, t_0] \times \Omega} |u(t, x)| dt dx < \infty \right\}.$$

The space $C([0, t_0] \times \Omega)$ of continuous functions on $[0, t_0] \times \Omega$ is given by

$$C([0, t_0] \times \Omega) = \left\{ u(t, x) : \ \|u\|_C = \max_{t \in [0, t_0], \ x \in \Omega} |u(t, x)| < \infty \right\}.$$

Clearly,

$$C([0, t_0] \times \Omega) \subset L([0, t_0] \times \Omega).$$

We further define the space $S_m([0, t_0] \times \Omega)$, for $m \in N = \{1, 2, \cdots\}$, of those functions on $[0, t_0] \times \Omega$ with up to mth order continuous partial derivatives with respect to t by

$$S_m([0, t_0] \times \Omega) = \left\{ u(t, x) : \ \frac{\partial^m}{\partial t^m} u(t, x) \in C([0, t_0] \times \Omega) \ \text{ and } \ \|u\|_m < \infty \right\},$$

where

$$\|u\|_m = \max \left\{ \|u(t, x)\|_C, \left\| \frac{\partial}{\partial t} u(t, x) \right\|_C, \cdots, \left\| \frac{\partial^m}{\partial t^m} u(t, x) \right\|_C \right\}.$$

Obviously,

$$S_m([0, t_0] \times \Omega) \subset C([0, t_0] \times \Omega) \subset L([0, t_0] \times \Omega).$$

Furthermore, $S_m([0, t_0] \times \Omega)$ is a Banach space using Theorem 7.17 in [10] stated as:

Theorem If $\{u_n\}$ is a sequence of differentiable functions on $[a, b]$ such that $\lim_{n \to \infty} u_n(x_0)$ exists (and is finite) for some $x_0 \in [a, b]$ and the sequence $\{u_n'\}$ converges uniformly on $[a, b]$, then u_n converges uniformly to a function u on $[a, b]$, and $u'(x) = \lim_{n \to \infty} u_n'(x)$ for $x \in [a, b]$.

Let $\lambda_k(x)$ for $k = 1, 2, \cdots, l$ be continuous on Ω. In this paper, we shall consider the partial integro-differential equation in the space $S_m([0, t_0] \times \Omega)$ with a given function $g(t, x)$

$$\frac{\partial^m}{\partial t^m}u(t,x) - \sum_{k=1}^{l}\lambda_k(x)I_1^{\beta_{1k}}\cdots I_n^{\beta_{nk}}u(t,x) = g(t,x) \in C([0,t_0]\times\Omega), \quad (22.31)$$

with the initial conditions

$$u(0,x) = 0, \cdots, \frac{\partial^{m-1}}{\partial t^{m-1}}u(0,x) = 0. \quad (22.32)$$

In addition, we study the uniqueness of solutions for the partial integro-differential equation by Banach's contraction principle with condition (22.32)

$$\frac{\partial^m}{\partial t^m}u(t,x) - \sum_{k=1}^{l}\lambda_k(x)I_1^{\beta_{1k}}\cdots I_n^{\beta_{nk}}u(t,x) = f(t,x,u(t,x),\cdots,\frac{\partial^{m-1}}{\partial t^{m-1}}u(t,x)), \quad (22.33)$$

where $f(t, x, y_0, \cdots, y_{m-1})$ is continuous on $[0, t_0] \times \Omega \times R^m$ with a Lipchitz condition.

22.7.2 Main Results

We begin presenting our first theorem which shows the solution for Eq. (22.31) with condition (22.32) as a convergent series in the space $S_m([0, t_0] \times \Omega)$ by Babenko's approach.

Theorem 22.14 *Let $\beta_{ij} \geq 0$ for $i = 1, 2, \cdots, n$ and $j = 1, 2, \cdots, l$, and $\lambda_i(x)$ for $i = 1, 2, \cdots, l$ be continuous on Ω. Then Eq. (22.31) with condition (22.32) has a unique solution*

$$u(t,x) = I_t^m \sum_{j=0}^{\infty} I_t^{jm} \sum_{j_1+j_2+\cdots+j_l=j} \binom{j}{j_1, j_2, \cdots, j_l} \left(\lambda_1(x)I_1^{\beta_{11}}\cdots I_n^{\beta_{n1}}\right)^{j_1}\cdots \left(\lambda_l(x)I_1^{\beta_{1l}}\cdots I_n^{\beta_{nl}}\right)^{j_l} g(t,x) \quad (22.34)$$

in $S_m([0, t_0] \times \Omega)$.

Proof Clearly,

$$I_t^m\lambda_k(x)I_1^{\beta_{1k}}\cdots I_n^{\beta_{nk}} = \lambda_k(x)I_1^{\beta_{1k}}\cdots I_n^{\beta_{nk}}I_t^m,$$

and

$$u(t,x) - \sum_{k=1}^{l}\lambda_k(x)I_t^m I_1^{\beta_{1k}}\cdots I_n^{\beta_{nk}}u(t,x) = I_t^m g(t,x),$$

by applying the integral operator I_t^m to both sides of Eq. (22.31) and the initial conditions. By Babenko's approach,

$$
\begin{aligned}
u(t,x) &= \left(1-\sum_{k=1}^{l}\lambda_k(x)I_t^m I_1^{\beta_{1k}}\cdots I_n^{\beta_{nk}}u(t,x)\right)^{-1} I_t^m g(t,x)\\
&= \sum_{j=0}^{\infty}\left(\sum_{k=1}^{l}\lambda_k(x)I_t^m I_1^{\beta_{1k}}\cdots I_n^{\beta_{nk}}\right)^{j} I_t^m g(t,x)\\
&= \sum_{j=0}^{\infty}\sum_{j_1+j_2+\cdots+j_l=j}\binom{j}{j_1,j_2,\cdots,j_l}\left(\lambda_1(x)I_t^m I_1^{\beta_{11}}\cdots I_n^{\beta_{n1}}\right)^{j_1}\cdots\\
&\quad \left(\lambda_l(x)I_t^m I_1^{\beta_{1l}}\cdots I_n^{\beta_{nl}}\right)^{j_l} I_t^m g(t,x)\\
&= I_t^m\sum_{j=0}^{\infty}I_t^{jm}\sum_{j_1+j_2+\cdots+j_l=j}\binom{j}{j_1,j_2,\cdots,j_l}\left(\lambda_1(x)I_1^{\beta_{11}}\cdots I_n^{\beta_{n1}}\right)^{j_1}\cdots\\
&\quad \left(\lambda_l(x)I_1^{\beta_{1l}}\cdots I_n^{\beta_{nl}}\right)^{j_l} g(t,x).
\end{aligned}
$$

Obviously, $u(t,x)$ satisfies the initial conditions due to the integral operator I_t^m and the uniqueness immediately follows from the fact that the linear homogeneous integral equation

$$\frac{\partial^m}{\partial t^m}u(t,x)-\sum_{k=1}^{l}\lambda_k(x)I_1^{\beta_{1k}}\cdots I_n^{\beta_{nk}}u(t,x)=0,$$

with condition (22.32) has only the solution zero. It remains to be shown that $u(t,x)$ is in the space $S_m([0,t_0]\times\Omega)$ and is a solution of Eq. (22.31).

Since $\lambda_k(x)$, for $k=1,2,\cdots,l$, is continuous on Ω, it is bounded and there exists $M>0$ such that

$$\max_{x\in\Omega}|\lambda_k(x)|\le M.$$

Let $\lambda=\max\{t_0,\omega_1,\cdots,\omega_n\}>0$. Then the norm of the integral operator $I_i^{\beta_{ir}}$ for $i=1,2,\cdots,n$ and $r=1,2,\cdots,l$ on the space $C([0,t_0]\times\Omega)$ is

$$
\begin{aligned}
\left\|I_i^{\beta_{ir}}\right\|_C &= \max_{\|u\|_C\le 1}|I_i^{\beta_{ir}}u|\\
&= \max_{\|u\|_C\le 1}\left|\frac{1}{\Gamma(\beta_{ir})}\int_0^{x_i}(x_i-s)^{\beta_{ir}-1}u(t,x_1,\cdots,x_{i-1},s,x_{s+1},\cdots,x_n)ds\right|\\
&\le \frac{1}{\Gamma(\beta_{ir})}\int_0^{x_i}(x_i-s)^{\beta_{ir}-1}ds=\frac{x_i^{\beta_{ir}}}{\Gamma(\beta_{ir}+1)}\le\frac{\lambda^{\beta_{ir}}}{\Gamma(\beta_{ir}+1)}.
\end{aligned}
$$

Similarly for $j = 0, 1, \cdots,$

$$\left\| I_t^{jm+m} \right\|_C \leq \frac{\lambda^{jm+m}}{(jm+m)!}.$$

Therefore,

$$\begin{aligned}
&\left\| I_t^{mj+m} \sum_{j_1+j_2+\cdots+j_l=j} \binom{j}{j_1, j_2, \cdots, j_l} \left(\lambda_1(x) I_1^{\beta_{11}} \cdots I_n^{\beta_{n1}}\right)^{j_1} \cdots \left(\lambda_l(x) I_1^{\beta_{1l}} \cdots I_n^{\beta_{nl}}\right)^{j_l} \right\|_C \\
&\leq \frac{\lambda^{jm+m}}{(jm+m)!} M^j \sum_{j_1+\cdots+j_l=j} \binom{j}{j_1, j_2, \cdots, j_l} \left\| I_1^{\beta_{11} j_1+\cdots+\beta_{1l} j_l} \right\|_C \cdots \left\| I_n^{\beta_{n1} j_1+\cdots+\beta_{nl} j_l} \right\|_C.
\end{aligned}$$

Using

$$\begin{aligned}
&\sum_{j_1+\cdots+j_l=j} \binom{j}{j_1, j_2, \cdots, j_l} = l^j, \\
&\left\| I_1^{\beta_{11} j_1+\cdots+\beta_{1l} j_l} \right\|_C \leq \frac{\lambda^{\beta_{11} j_1+\cdots+\beta_{1l} j_l}}{\Gamma(\beta_{11} j_1 + \cdots + \beta_{1l} j_l + 1)} \leq \frac{5}{4} \lambda^{\beta_{11} j_1+\cdots+\beta_{1l} j_l}, \\
&\cdots, \\
&\left\| I_n^{\beta_{n1} j_1+\cdots+\beta_{nl} j_l} \right\|_C \leq \frac{\lambda^{\beta_{n1} j_1+\cdots+\beta_{nl} j_l}}{\Gamma(\beta_{n1} j_1 + \cdots + \beta_{nl} j_l + 1)} \leq \frac{5}{4} \lambda^{\beta_{n1} j_1+\cdots+\beta_{nl} j_l}.
\end{aligned}$$

By noting that for $y \geq 0$,

$$\Gamma(y+1) \geq \frac{4}{5}$$

proof the property of the Gamma function, we arrive at

$$\begin{aligned}
&\left\| I_t^{mj+m} \sum_{j_1+j_2+\cdots+j_l=j} \binom{j}{j_1, j_2, \cdots, j_l} \left(\lambda_1(x) I_1^{\beta_{11}} \cdots I_n^{\beta_{n1}}\right)^{j_1} \cdots \left(\lambda_l(x) I_1^{\beta_{1l}} \cdots I_n^{\beta_{nl}}\right)^{j_l} \right\|_C \\
&\leq \left(\frac{5}{4}\right)^n (S^n)^j l^j M^j \frac{\lambda^{jm+m}}{(jm+m)!},
\end{aligned}$$

where

$$S = \max\{\lambda^{\beta_{ir}}\} > 0.$$

Hence,

$$\|u(t,x)\|_C \leq \left(\frac{5}{4}\right)^n \lambda^m \|g(t,x)\|_C \sum_{j=0}^{\infty} \frac{(\lambda^m S^n l M)^j}{(jm+m)!}$$

$$= \left(\frac{5}{4}\right)^n \lambda^m \|g(t,x)\|_C \sum_{j=0}^{\infty} \frac{(\lambda^m S^n l M)^j}{\Gamma(jm+m+1)}$$
$$= \left(\frac{5}{4}\right)^n \lambda^m \|g(t,x)\|_C E_{m,\, m+1}(\lambda^m S^n l M) < \infty,$$

where

$$E_{\alpha,\,\beta}(z) = \sum_{j=0}^{\infty} \frac{z^n}{\Gamma(\alpha j+\beta)}, \quad \alpha,\, \beta > 0,\ z \in \mathbb{C},$$

is the Mittag-Leffler function.

Evidently,

$$\frac{\partial}{\partial t} u(t,x) = I_t^{m-1} \sum_{j=0}^{\infty} I_t^{jm} \sum_{j_1+j_2+\cdots+j_l=j} \binom{j}{j_1,\, j_2,\, \cdots,\, j_l} \left(\lambda_1(x) I_1^{\beta_{11}} \cdots I_n^{\beta_{n1}}\right)^{j_1} \cdots$$
$$\left(\lambda_l(x) I_1^{\beta_{1l}} \cdots I_n^{\beta_{nl}}\right)^{j_l} g(t,x),$$
$$\cdots,$$
$$\frac{\partial^m}{\partial t^m} u(t,x) = \sum_{j=0}^{\infty} I_t^{jm} \sum_{j_1+j_2+\cdots+j_l=j} \binom{j}{j_1,\, j_2,\, \cdots,\, j_l} \left(\lambda_1(x) I_1^{\beta_{11}} \cdots I_n^{\beta_{n1}}\right)^{j_1} \cdots$$
$$\left(\lambda_l(x) I_1^{\beta_{1l}} \cdots I_n^{\beta_{nl}}\right)^{j_l} g(t,x),$$

and

$$\left\| \frac{\partial}{\partial t} u(t,x) \right\|_C \le \left(\frac{5}{4}\right)^n \lambda^{m-1} \|g(t,x)\|_C E_{m,\, m}(\lambda^m S^n l M) < \infty,$$
$$\cdots,$$
$$\left\| \frac{\partial^m}{\partial t^m} u(t,x) \right\|_C \le \left(\frac{5}{4}\right)^n \|g(t,x)\|_C E_{m,\, 1}(\lambda^m S^n l M) < \infty.$$

Thus, $u(t,x) \in S_m([0,t_0] \times \Omega)$ and the series in Eq. (22.34) is absolutely and uniformly convergent. Finally, we substitute $u(t,x)$ in Eq. (22.34) into the left-hand side of Eq. (22.31),

$$\sum_{j=0}^{\infty} I_t^{jm} \sum_{j_1+j_2+\cdots+j_l=j} \binom{j}{j_1,\, j_2,\, \cdots,\, j_l} \left(\lambda_1(x) I_1^{\beta_{11}} \cdots I_n^{\beta_{n1}}\right)^{j_1} \cdots$$
$$\left(\lambda_l(x) I_1^{\beta_{1l}} \cdots I_n^{\beta_{nl}}\right)^{j_l} g(t,x) - \left(\sum_{k=1}^{l} \lambda_k(x) I_1^{\beta_{1k}} \cdots I_n^{\beta_{nk}}\right) \sum_{j=0}^{\infty} I_t^{jm+m}$$
$$\sum_{j_1+j_2+\cdots+j_l=j} \binom{j}{j_1,\, j_2,\, \cdots,\, j_l} \left(\lambda_1(x) I_1^{\beta_{11}} \cdots I_n^{\beta_{n1}}\right)^{j_1} \cdots$$

$$\left(\lambda_l(x)I_1^{\beta_{1l}}\cdots I_n^{\beta_{nl}}\right)^{j_l} g(t,x)$$
$$= g(t,x) + \sum_{j=1}^{\infty} I_t^{jm} \sum_{j_1+j_2+\cdots+j_l=j} \binom{j}{j_1, j_2, \cdots, j_l} \left(\lambda_1(x)I_1^{\beta_{11}}\cdots I_n^{\beta_{n1}}\right)^{j_1}\cdots$$
$$\left(\lambda_l(x)I_1^{\beta_{1l}}\cdots I_n^{\beta_{nl}}\right)^{j_l} g(t,x) - \left(\sum_{k=1}^{l}\lambda_k(x)I_1^{\beta_{1k}}\cdots I_n^{\beta_{nk}}\right)(I_t^m g(t,x) + \sum_{j=1}^{\infty} I_t^{mj+m}$$
$$\sum_{j_1+\cdots+j_l=j} \binom{j}{j_1, j_2, \cdots, j_l} \left(\lambda_1(x)I_1^{\beta_{11}}\cdots I_n^{\beta_{n1}}\right)^{j_1}\cdots\left(\lambda_l(x)I_1^{\beta_{1l}}\cdots I_n^{\beta_{nl}}\right)^{j_l} g(t,x)).$$

Clearly,

$$\sum_{j=1}^{\infty} I_t^{jm} \sum_{j_1+j_2+\cdots+j_l=j} \binom{j}{j_1, j_2, \cdots, j_l} \left(\lambda_1(x)I_1^{\beta_{11}}\cdots I_n^{\beta_{n1}}\right)^{j_1}\cdots$$
$$\left(\lambda_l(x)I_1^{\beta_{1l}}\cdots I_n^{\beta_{nl}}\right)^{j_l} g(t,x)$$
$$= \left(\sum_{k=1}^{l}\lambda_k(x)I_1^{\beta_{1k}}\cdots I_n^{\beta_{nk}}\right) I_t^m g(t,x)$$
$$+ \sum_{j=2}^{\infty} I_t^{mj} \sum_{j_1+j_2+\cdots+j_l=j} \binom{j}{j_1, j_2, \cdots, j_l} \left(\lambda_1(x)I_1^{\beta_{11}}\cdots I_n^{\beta_{n1}}\right)^{j_1}\cdots$$
$$\left(\lambda_l(x)I_1^{\beta_{1l}}\cdots I_n^{\beta_{nl}}\right)^{j_l} g(t,x).$$

Therefore,

$$\sum_{j=1}^{\infty} I_t^{jm} \sum_{j_1+j_2+\cdots+j_l=j} \binom{j}{j_1, j_2, \cdots, j_l} \left(\lambda_1(x)I_1^{\beta_{11}}\cdots I_n^{\beta_{n1}}\right)^{j_1}\cdots$$
$$\left(\lambda_l(x)I_1^{\beta_{1l}}\cdots I_n^{\beta_{nl}}\right)^{j_l} g(t,x) - \left(\sum_{k=1}^{l}\lambda_k(x)I_1^{\beta_{1k}}\cdots I_n^{\beta_{nk}}\right)(I_t^m g(t,x) + \sum_{j=1}^{\infty} I_t^{mj+m}$$
$$\sum_{j_1+\cdots+j_l=j} \binom{j}{j_1, j_2, \cdots, j_l} \left(\lambda_1(x)I_1^{\beta_{11}}\cdots I_n^{\beta_{n1}}\right)^{j_1}\cdots\left(\lambda_l(x)I_1^{\beta_{1l}}\cdots I_n^{\beta_{nl}}\right)^{j_l} g(t,x))$$
$$= \sum_{j=2}^{\infty} I_t^{mj} \sum_{j_1+j_2+\cdots+j_l=j} \binom{j}{j_1, j_2, \cdots, j_l} \left(\lambda_1(x)I_1^{\beta_{11}}\cdots I_n^{\beta_{n1}}\right)^{j_1}\cdots$$
$$\left(\lambda_l(x)I_1^{\beta_{1l}}\cdots I_n^{\beta_{nl}}\right)^{j_l} g(t,x)$$

$$- \left(\sum_{k=1}^{l} \lambda_k(x) I_1^{\beta_{1k}} \cdots I_n^{\beta_{nk}}\right) (\sum_{j=1}^{\infty} I_t^{mj+m} \sum_{j_1+\cdots+j_l=j} \binom{j}{j_1, j_2, \cdots, j_l}$$
$$\left(\lambda_1(x) I_1^{\beta_{11}} \cdots I_n^{\beta_{n1}}\right)^{j_1} \cdots \left(\lambda_l(x) I_1^{\beta_{1l}} \cdots I_n^{\beta_{nl}}\right)^{j_l} g(t, x)$$
$$= \sum_{j=3}^{\infty} I_t^{mj} \sum_{j_1+\cdots+j_l=j} \binom{j}{j_1, j_2, \cdots, j_l} \left(\lambda_1(x) I_1^{\beta_{11}} \cdots I_n^{\beta_{n1}}\right)^{j_1} \cdots$$
$$\left(\lambda_l(x) I_1^{\beta_{1l}} \cdots I_n^{\beta_{nl}}\right)^{j_l} g(t, x)$$
$$- \left(\sum_{k=1}^{l} \lambda_k(x) I_1^{\beta_{1k}} \cdots I_n^{\beta_{nk}}\right) (\sum_{j=2}^{\infty} I_t^{mj+m} \sum_{j_1+\cdots+j_l=j} \binom{j}{j_1, j_2, \cdots, j_l}$$
$$\left(\lambda_1(x) I_1^{\beta_{11}} \cdots I_n^{\beta_{n1}}\right)^{j_1} \cdots \left(\lambda_l(x) I_1^{\beta_{1l}} \cdots I_n^{\beta_{nl}}\right)^{j_l} g(t, x),$$

by noting that

$$I_t^{2m} \sum_{j_1+\cdots+j_l=2} \binom{2}{j_1, j_2, \cdots, j_l} \left(\lambda_1(x) I_1^{\beta_{11}} \cdots I_n^{\beta_{n1}}\right)^{j_1} \cdots$$
$$\left(\lambda_l(x) I_1^{\beta_{1l}} \cdots I_n^{\beta_{nl}}\right)^{j_l} g(t, x)$$
$$= \left(\sum_{k=1}^{l} \lambda_k(x) I_1^{\beta_{1k}} \cdots I_n^{\beta_{nk}}\right) (I_t^{2m} \sum_{j_1+\cdots+j_l=1} \binom{1}{j_1, j_2, \cdots, j_l}$$
$$\left(\lambda_1(x) I_1^{\beta_{11}} \cdots I_n^{\beta_{n1}}\right)^{j_1} \cdots \left(\lambda_l(x) I_1^{\beta_{1l}} \cdots I_n^{\beta_{nl}}\right)^{j_l} g(t, x).$$

Repeating this pattern, we show that $u(t, x)$ is a solution of Eq. (22.31). This completes the proof of Theorem 22.14.

The following is an example demonstrating the utility of Theorem 22.14.

Example 22.8 The partial integro-differential equation

$$\frac{\partial^2}{\partial t^2} t(t, x_1, x_2) - x_1 I_1^{0.5} t(t, x_1, x_2) - x_2^{1.5} I_2^{1.5} t(t, x_1, x_2) = \frac{t x_1^2 x_2^2}{4}$$

with the initial conditions

$$u(0, x_1, x_2) = \frac{\partial}{\partial t} u(0, x_1, x_2) = 0,$$

has a unique solution in the space $S_2([0, t_0] \times \Omega)$

$$u(t, x_1, x_2) = t^3 \sum_{j=0}^{\infty} \frac{t^{2j}}{(2j+3)!} \sum_{j_1+j_2=j} \binom{j}{j_1, j_2} A_{j_1} B_{j_2} \frac{x_1^{1.5j_1+2}}{\Gamma(1.5j_1+3)} \frac{x_2^{3j_2+2}}{(3j_2+2)!},$$

where

$$A_{j_1} = \begin{cases} 3.5 \cdot 5 \cdots (1.5j_1+2) & \text{if } j_1 \geq 1, \\ 1 & \text{if } j_1 = 0, \end{cases}$$

and

$$B_{j_2} = \begin{cases} \dfrac{5! \cdots (3j_2+2)!}{\Gamma(4.5) \cdots \Gamma(3j_2+1.5)} & \text{if } j_2 \geq 1, \\ 1 & \text{if } j_2 = 0. \end{cases}$$

Proof Indeed by Theorem 22.14,

$$\begin{aligned} u(t, x_1, x_2) &= \sum_{j=0}^{\infty} I_t^{2j+2} \sum_{j_1+j_2=j} \binom{j}{j_1, j_2} \left(x_1 I_1^{0.5}\right)^{j_1} \left(x_2^{1.5} I_2^{1.5}\right)^{j_2} \frac{1}{4} t x_1^2 x_2^2 \\ &= \sum_{j=0}^{\infty} I_t^{2j+2} t \sum_{j_1+j_2=j} \binom{j}{j_1, j_2} \left(x_1 I_1^{0.5}\right)^{j_1} \frac{x_1^2}{2} \left(x_2^{1.5} I_2^{1.5}\right)^{j_2} \frac{x_2^2}{2}. \end{aligned}$$

Let α and β be arbitrary complex numbers. Then it follows from [3] that

$$\Phi_\alpha * \Phi_\beta = \Phi_{\alpha+\beta}$$

where

$$\Phi_\alpha = \frac{x_+^{\alpha-1}}{\Gamma(\alpha)}.$$

So,

$$I_t^{2j+2} t = \Phi_{2j+2}(t) * \frac{t_+^{2-1}}{\Gamma(2)} = \Phi_{2j+2}(t) * \Phi_2(t) = \Phi_{2j+4}(t) = \frac{t_+^{2j+3}}{(2j+3)!} = \frac{t^{2j+3}}{(2j+3)!}$$

since $t \geq 0$.

Consider the term $\left(x_1 I_1^{0.5}\right)^{j_1} \frac{x_1^2}{2}$. Clearly,

$$\begin{aligned} x_1 I_1^{0.5} \frac{x_1^2}{2} &= x_1(\Phi_{0.5}(x_1) * \Phi_3(x_1)) = x_1 \Phi_{3.5}(x_1) = \frac{x_1^{3.5}}{\Gamma(3.5)}, \\ x_1 I_1^{0.5} \frac{x_1^{3.5}}{\Gamma(3.5)} &= \frac{\Gamma(4.5)}{\Gamma(3.5)} x_1(\Phi_{0.5} * \Phi_{4.5}) = \frac{\Gamma(4.5)}{\Gamma(3.5)} \frac{x_1^5}{\Gamma(5)}, \\ &\cdots, \\ \left(x_1 I_1^{0.5}\right)^{j_1} \frac{x_1^2}{2} &= \frac{\Gamma(4.5)\Gamma(6) \cdots \Gamma(1.5j_1+3)}{\Gamma(3.5)\Gamma(5) \cdots \Gamma(1.5j_1+2)} \frac{x_1^{1.5j_1+2}}{\Gamma(1.5j_1+3)} \end{aligned}$$

$$= 3.5 \cdot 5 \cdots (1.5j_1 + 2) \frac{x_1^{1.5j_1+2}}{\Gamma(1.5j_1 + 3)} = A_{j_1} \frac{x_1^{1.5j_1+2}}{\Gamma(1.5j_1 + 3)}.$$

Similarly,

$$\left(x_2^{1.5} I_2^{1.5}\right)^{j_2} \frac{x_2^2}{2} = \frac{\Gamma(6) \cdots \Gamma(3j_2 + 3)}{\Gamma(4.5) \cdots \Gamma(3j_2 + 1.5)} \frac{x_2^{3j_2+2}}{\Gamma(3j_2 + 3)}$$
$$= \frac{5! \cdots (3j_2 + 2)!}{\Gamma(4.5) \cdots \Gamma(3j_2 + 1.5)} \frac{x_2^{3j_2+2}}{(3j_2 + 2)!} = B_{j_2} \frac{x_2^{3j_2+2}}{(3j_2 + 2)!}.$$

This completes the proof of Example 22.8.

Remark 22.3 (i) Following the proof of Theorem 22.14, we can easily solve the following partial integro-differential equation with condition (22.32)

$$\frac{\partial^m}{\partial t^m} u(t, x) - \sum_{k=1}^{l} I_t^{\alpha_k} \lambda_k(x) I_1^{\beta_{1k}} \cdots I_n^{\beta_{nk}} u(t, x) = g(t, x) \in C([0, t_0] \times \Omega),$$

where $\alpha_k \geq 0$.

(ii) Similarly, we can work on the partial integral equation without any initial conditions

$$u(t, x) - \sum_{k=1}^{l} \lambda_k(x) I_1^{\beta_{1k}} \cdots I_n^{\beta_{nk}} u(t, x) = g(t, x) \in L([0, t_0] \times \Omega)$$

in the space $L([0, t_0] \times \Omega)$ using the same method.

We are ready to present the following theorem on the uniqueness of solutions for Eq. (22.33) by Banach's contraction principle.

Theorem 22.15 *Assume* $f(t, x, y_0, \cdots, y_{m-1})$ *is continuous on* $[0, t_0] \times \Omega \times R^m$ *and there exist nonnegative constants* $C_0, C_1, \cdots, C_{m-1}$ *such that*

$$|f(t, x, y_0, \cdots, y_{m-1}) - f(t, x, z_0, \cdots, z_{m-1})| \leq C_0|y_0 - z_0| + \cdots + C_{m-1}|y_{m-1} - z_{m-1}|.$$

Furthermore, suppose

$$q = m \max\{C_0, \cdots, C_{m-1}\} \left(\frac{5}{4}\right)^n \max\left\{\lambda^m E_{m,\, m+1}(\lambda^m S^n l M), \cdots, E_{m,\, 1}(\lambda^m S^n l M)\right\} < 1.$$

Then Eq. (22.33) with condition (22.32) has a unique solution in $S_m([0, t_0] \times \Omega)$.

Proof If $u(t, x) \in S_m([0, t_0] \times \Omega)$, then $f(t, x, u(t, x), \cdots, \frac{\partial^{m-1}}{\partial t^{m-1}} u(t, x))$ is continuous on $[0, t_0] \times \Omega$. Construct a mapping on the space $S_m([0, t_0] \times \Omega)$ by

$$T(u) = I_t^m \sum_{j=0}^{\infty} I_t^{jm} \sum_{j_1+j_2+\cdots+j_l=j} \binom{j}{j_1, j_2, \cdots, j_l} \left(\lambda_1(x) I_1^{\beta_{11}} \cdots I_n^{\beta_{n1}}\right)^{j_1} \cdots$$
$$\left(\lambda_l(x) I_1^{\beta_{1l}} \cdots I_n^{\beta_{nl}}\right)^{j_l} f(t, x, u(t, x), \cdots, \frac{\partial^{m-1}}{\partial t^{m-1}} u(t, x)).$$

We first show that T is a mapping from $S_m([0, t_0] \times \Omega)$ to itself. Indeed from the proof of Theorem 22.14,

$$\|T(u)\|_C \leq \left(\frac{5}{4}\right)^n \lambda^m E_{m,\, m+1}(\lambda^m S^n l M) \left\| f(t, x, u(t, x), \cdots, \frac{\partial^{m-1}}{\partial t^{m-1}} u(t, x)) \right\|_C < \infty,$$
$$\left\| \frac{\partial}{\partial t} T(u) \right\|_C \leq \left(\frac{5}{4}\right)^n \lambda^{m-1} E_{m,\, m}(\lambda^m S^n l M) \, \|f\|_C < \infty,$$
$$\cdots,$$
$$\left\| \frac{\partial^m}{\partial t^m} T(u) \right\|_C \leq \left(\frac{5}{4}\right)^n E_{m,\, 1}(\lambda^m S^n l M) \, \|f\|_C < \infty.$$

Thus,

$$\|T(u)\|_m = \max \left\{ \|T(u)\|_C, \left\| \frac{\partial}{\partial t} T(u) \right\|_C, \cdots, \left\| \frac{\partial^m}{\partial t^m} T(u) \right\|_C \right\} < \infty.$$

It remains to be shown that T is contractive by Banach's contraction principle. Then for $u, v \in S_m([0, t_0] \times \Omega)$,

$$\|T(u) - T(v)\|_m$$
$$= \max \left\{ \|T(u) - T(v)\|_C, \left\| \frac{\partial}{\partial t} (T(u) - T(v)) \right\|_C, \cdots, \left\| \frac{\partial^m}{\partial t^m} (T(u) - T(v)) \right\|_C \right\}.$$

Clearly,

$$\|T(u) - T(v)\|_C \leq \left(\frac{5}{4}\right)^n \lambda^m E_{m,\, m+1}(\lambda^m S^n l M)$$
$$\left\| f(t, x, u(t, x), \cdots, \frac{\partial^{m-1}}{\partial t^{m-1}} u(t, x)) - f(t, x, v(t, x), \cdots, \frac{\partial^{m-1}}{\partial t^{m-1}} v(t, x)) \right\|_C$$
$$= \left(\frac{5}{4}\right)^n \lambda^m E_{m,\, m+1}(\lambda^m S^n l M)$$
$$\max_{t \in [0, t_0],\, x \in \Omega} \left| f(t, x, u(t, x), \cdots, \frac{\partial^{m-1}}{\partial t^{m-1}} u(t, x)) - f(t, x, v(t, x), \cdots, \frac{\partial^{m-1}}{\partial t^{m-1}} v(t, x)) \right|$$
$$\leq \left(\frac{5}{4}\right)^n \lambda^m E_{m,\, m+1}(\lambda^m S^n l M)$$
$$\max_{t \in [0, t_0],\, x \in \Omega} \left\{ C_0 |u(t, x) - v(t, x)| + \cdots + C_{m-1} \left| \frac{\partial^{m-1}}{\partial t^{m-1}} (u(t, x) - v(t, x)) \right| \right\}$$

$$\leq m \max\{C_0, \cdots, C_{m-1}\} \left(\frac{5}{4}\right)^n \lambda^m E_{m,\, m+1}(\lambda^m S^n l M) \, \|u - v\|_m \, .$$

Similarly,

$$\begin{aligned}\left\| \frac{\partial}{\partial t}(T(u) - T(v)) \right\|_C &\leq m \max\{C_0, \cdots, C_{m-1}\} \\ &\quad \left(\frac{5}{4}\right)^n \lambda^{m-1} E_{m,\, m}(\lambda^m S^n l M) \, \|u - v\|_m \, , \\ &\quad \cdots, \\ \left\| \frac{\partial^m}{\partial t^m}(T(u) - T(v)) \right\|_C &\leq m \max\{C_0, \cdots, C_{m-1}\} \\ &\quad \left(\frac{5}{4}\right)^n E_{m,\, 1}(\lambda^m S^n l M) \, \|u - v\|_m \, .\end{aligned}$$

Hence,

$$\|T(u) - T(v)\|_m \leq q \, \|u - v\|_m \, ,$$

which implies that Eq. (22.33) has a unique solution in the space $S_m([0, t_0] \times \Omega)$. This completes the proof of Theorem 22.15.

Example 22.9 Let $(t, x_1, x_2) \in [0, 1/2] \times [0, 1/3] \times [0, 1]$. Then the partial integro-differential equation

$$\frac{\partial^2}{\partial t^2} u(t, x_1, x_2) - \frac{1}{2} \sin(x_1 x_2) I_1^{0.75} I_2^{1.33} u(t, x_1, x_2) - \frac{1}{x_1^2 + x_2^2 + 3} I_1^{0.5} I_2^{2.133} u(t, x_1, x_2)$$

$$= \cos(t(x_1 + x_2)) + \frac{1}{7} \sin u(t, x_1, x_2) + \frac{1}{5} \frac{\frac{\partial}{\partial t} u(t, x_1, x_2)}{1 + \left(\frac{\partial}{\partial t} u(t, x_1, x_2)\right)^2}$$

with the initial conditions

$$u(0, x_1, x_2) = \frac{\partial}{\partial t} u(0, x_1, x_2) = 0,$$

has a unique solution in the space $S_m([0, t_0] \times \Omega)$.

Proof Let

$$\lambda = \max\left\{\frac{1}{2}, \frac{1}{3}, 1\right\} = 1.$$

Hence,

$$S = \max\{\lambda^{\beta_{ir}}\} = 1$$

for $i = 1, 2$ and $r = 1, 2$. Obviously,

$$\left|\frac{1}{2}\sin(x_1x_2)\right| \le 1/2, \quad \left|\frac{1}{x_1^2+x_2^2+3}\right| \le 1/3$$

and $M = 1/2$. The function

$$f(t, x, y_0, y_1) = \cos(t(x_1 + x_2)) + \frac{1}{7}\sin y_0 + \frac{1}{5}\frac{y_1}{1+y_1^2}$$

is continuous and satisfies the condition

$$|f(t, x, y_0, y_1) - f(t, x, z_0, z_1)| \le \frac{1}{7}|y_0 - z_0| + \frac{1}{5}|y_1 - z_1|.$$

Thus,

$$\max\{C_0, C_1\} = 1/5$$

and

$$\begin{aligned}
&\max\left\{\lambda^2 E_{2,3}(\lambda^2 S^2 lM), \lambda E_{2,2}(\lambda^2 S^2 lM), E_{2,1}(\lambda^2 S^2 lM)\right\}\\
&= \sum_{j=0}^{\infty}\frac{1}{\Gamma(2j+1)} = \sum_{j=0}^{\infty}\frac{1}{(2j)!} = 1 + \frac{1}{2!} + \frac{1}{4!} + \frac{1}{6!} + \frac{1}{8!} + \cdots\\
&\le \left(\frac{1}{4}\right)^0 + \left[\left(\frac{1}{4}\right)^1 + \left(\frac{1}{4}\right)^1\right] + \left(\frac{1}{4}\right)^2 + \left(\frac{1}{4}\right)^3 + \left(\frac{1}{4}\right)^4 + \cdots\\
&= \left(\frac{1}{4}\right)^1 + \frac{1}{1-1/4} = \frac{19}{12}.
\end{aligned}$$

Therefore,

$$\begin{aligned}
&q = m\max\{C_0, \cdots, C_{m-1}\}\left(\frac{5}{4}\right)^n \max\left\{\lambda^m E_{m,m+1}(\lambda^m S^n lM), \cdots, E_{m,1}(\lambda^m S^n lM)\right\}\\
&\le 2\cdot\frac{1}{5}\left(\frac{5}{4}\right)^2\frac{19}{12} = \frac{950}{960} < 1.
\end{aligned}$$

By Theorem 22.15, it has a unique nonzero solution as zero is clearly not a solution. This completes the proof of Example 22.9.

22.7.3 Summary

Applying Babenko's approach and Banach's contraction principle, we have studied the uniqueness of solutions for the partial integro-differential equation (22.33) with the initial conditions (22.32) in the Banach space. Several examples were also presented for the illustration of main results by the Gamma function and convolutions.

References

1. Kilbas AA, Srivastava HM, Trujillo JJ (2006) Theory and applications of fractional differential equations. Elsevier, North-Holland
2. Samko SG, Kilbas AA, Marichev OI (1993) Fractional integrals and derivatives: theory and applications. Gordon and Breach (1993)
3. Li C (2015) Several results of fractional derivatives in $\mathcal{D}'(R_+)$. Fract Calc Appl Anal 18:192–207
4. Babenkos YI (1986) Heat and mass transfer. Khimiya, Leningrad (in Russian)
5. Li C (2023) Uniqueness of a nonlinear integro-differential equation with nonlocal boundary condition and variable coefficients. Bound Value Probl 2023:26
6. Cabada A, Hamdi Z (2014) Nonlinear fractional differential equations with integral boundary value conditions. Appl Math Comput 228:251–257
7. Li C (2021) Uniqueness of the partial integro-differential equations. J Integral Equ Appl 22:463–475
8. Li C, Saadati R, Srivastava R, Beaudin J (2020) On the boundary value problem of nonlinear fractional integro-differential equations. Mathematics 10:1971. https://doi.org/10.3390/math10121971
9. Li C, Saadati R, Allahviranloo T (2022) Conditions to guarantee the existence of solutions for a nonlinear and implicit integro-differential equation with variable coefficients. Math Methods Appl Sci 1–12
10. Rudin W (1976) Principle of mathematical analysis. McGraw Hill, New York
11. Burton TA, Furumochi T (2002) Krasnoselskii's fixed point theorem and stability. Nonlinear Anal Theory Methods Appl 49:445–454
12. Li C, Srivastava H (2021) Uniqueness of solutions of the generalized abel integral equations in Banach spaces. Fractal Fract 5:105. https://doi.org/10.3390/fractalfract5030105
13. Kilbas AA (2001) Hadamard-type fractional calculus. J Korean Math Soc 38:1191–1204
14. Kilbas AA (2003) Hadamard-type integral equations and fractional calculus operators. Oper Theory Adv Appl 142:175–188
15. Srivastava H (2020) Fractional-order derivatives and integrals: introductory overview and recent developments, Kyungpook Math J 60:73–116
16. Granas A, Dugundji J (2005) Fixed point theory. Springer, New York, USA
17. Yu C, Gao G (2005) Existence of fractional differential equations. J Math Anal Appl 310:26–29
18. Ahmad B, Ntouyas SK, Alsaedi A (2017) Existence of solutions for fractional differential equations with nonlocal and average type integral boundary conditions. J Appl Math Comput 53:129–145
19. Ntouyas SK, Al-Sulami HH (2020) A study of coupled systems of mixed order fractional differential equations and inclusions with coupled integral fractional boundary conditions. Adv Differ Equ 2020:73
20. Houas M (2018) Existence of solutions for fractional differential equations involving two Riemann-Liouville fractional orders. Anal Theory Appl 34:253–274
21. El-Shahed M, Nieto JJ (2010) Nontrivial solutions for a nonlinear multi-point boundary value problem of fractional order. Comput Math Appl 59:3438–3443

22. Zhou WX, Liu HZ (2014) Uniqueness and existence of solution for a system of fractional q-difference equations. Abstr Appl Anal 2014(Article ID 340159):11
23. Thongsalee N, Ntouyas SK, Tariboon J (2016) Nonlinear Riemann-Liouville fractional differential equations with nonlocal Erdélyi-Kober fractional integral conditions. Fract Calc Appl Anal 19:480–497
24. Li B, Sun S, Li Y, Zhao P (2014) Multi-point boundary value problems for a class of Riemann-Liouville fractional differential equations. Adv Differ Equ 2014:151
25. Li C, Nonlaopon K, Hrytsenko A, Beaudin J (2023) On the analytic and approximate solutions for the fractional nonlinear Schrodinger equations. J Nonlinear Sci Appl 16:51–59
26. Li CP, Zeng F (2015) Numerical methods for fractional calculus. Chapman and Hall/CRC, Boca Raton

The manufacturer's authorised representative in the EU is Springer Nature Customer Service Centre GmbH, Europaplatz 3, 69115 Heidelberg, Germany. If you have any concerns regarding our products, please contact ProductSafety@springernature.com

Printed and bound by CPI Group (UK) Ltd, Croydon, CR0 4YY
01/10/2026
02225897-0017